# The Optics Encyclopedia

*Th. G. Brown, K. Creath, H. Kogelnik,*
*M. A. Kriss, J. Schmit, M. J. Weber (Eds.)*

## Volume 2

# The Optics Encyclopedia

## Basic Foundations and Practical Applications

*Edited by*
*Thomas G. Brown, Katherine Creath, Herwig Kogelnik,*
*Michael A. Kriss, Joanna Schmit, Marvin J. Weber*

**Volume 2**

WILEY-VCH

WILEY-VCH Verlag GmbH & Co. KGaA

**Editors:**

*Thomas G. Brown*
University of Rochester, NY, USA
e-mail: brown@optics.rochester.edu

*Katherine Creath*
Creath Optineering Services, Tucson, AZ, USA
e-mail: kcreath@ieee.org

*Herwig Kogelnik*
Lucent Technologies, Bell Labs, Holmdel, NJ, USA
e-mail: herwig@lucent.com

*Michael A. Kriss*
Sharp Laboratories of America, Camas, WA, USA
e-mail: mkriss@sharplabs.com

*Joanna Schmit*
Veeco Instruments, Inc., Tucson, AZ, USA
e-mail: jschmit@veeco.com

*Marvin J. Weber*
Lawrence Berkeley National Laboratory,
Berkeley, CA, USA
e-mail: mjweber@lbl.gov



**Library of Congress Card No.: applied for**

**British Library Cataloging-in-Publication Data:**
A catalogue record for this book is available from the British Library

**Bibliographic information published by Die Deutsche Bibliothek**
Die Deutsche Bibliothek lists this publication in the Deutsche Nationalbibliografie; detailed bibliographic data is available in the Internet at <http://dnb.ddb.de>.



Printed in the Federal Republic of Germany
Printed on acid-free paper

**Composition** Laserwords Private Ltd, Chennai, India
**Printing** Druckhaus Darmstadt GmbH, Darmstadt
**Bookbinding** Buchbinderei Schaumann GmbH, Darmstadt
**ISBN** 3-527-40320-5

# Preface

This Encyclopedia is designed to serve as a general-purpose reference for the discipline of Optics. This unique compendium of 94 articles organized in 5 volumes and written by expert scientists and engineers covers all major areas of optics and offers an excellent guide for scientists, engineers, teachers, and undergraduate and graduate students working in the field.

From the basic theory of optics and optical methods to the most advanced research and applications of light, the Optics Encyclopedia offers readers both depth and breadth in the diverse field of Optics. In these volumes, the reader can find descriptions of both the first great works in optics and the latest cutting edge research in this broad field, including lasers and optical fibers. Basic concepts and fundamental properties of light are included. Some chapters provide overviews of optical methods in different disciplines such as optical metrology, microscopy and interferometry; these chapters offer a wealth of information on not only fundamental research but also industrial applications and solutions. Other articles guide the reader in a tutorial-like fashion through more focused topics like the theory of optical coatings or ellipsometry, while still others review different optical techniques that are used in fields such as biomedical or material science.

The scope of the Optics Encyclopedia extends from the application of light to determine properties of materials on a macroscale to a nanoscale level to the use of laser light to trap atoms, manipulate molecules and to transmit and store vast amounts of information. Often a major field in optics such as laser science and spectroscopy cannot be adequately covered in a single chapter. Thus, the wide range of work being done in certain areas has been parsed out into multiple articles. For example, in the field of spectroscopy two chapters on atomic and molecular spectroscopy provide a general overview of the topic while several other articles develop detailed descriptions of the emerging research and applications of spectrometers that use different ranges within the light spectrum.

Through their advanced and technical but also accessible and concise presentations, all the articles prepare the reader to delve further into the topic; to this goal, each chapter provides a glossary of terms, extensive bibliography, and suggested readings. Within each chapter of the Optics Encyclopedia frequent cross-referencing to other chapters allows the reader to follow his interests in a given topic without having to immediately access any different source. Thus, the reader with the Optics Encyclopedia alone can attain a decent level of expertise in any one field, and the extensive references and suggested readings can guide him to further levels of information.

From the research laboratory to industry and to the classroom, the broad scope of the Optics Encyclopedia offers readers a necessary introduction, compendium, and reference guide to optics.

For the editors: **Joanna Schmit**

**Tucson, October 2003**

# Contents

## Volume 2

# Geometric Optics

*Roland Shack*
*University of Arizona, Optical Sciences Center, Tucson, Arizona, USA*
*Phone: 520-621-1356; e-mail: roland.shack@optics.arizona.edu*

**Abstract**
This article reviews the most elementary theory of optics, namely geometric optics, and describes its connection to quantum optics and other high-level theories. The basics of light propagation in the form of rays and wave-fronts are described. The main emphasis is on optical image-forming systems and system aberrations. This article provides basic knowledge for any optical engineer.

**Keywords**
geometric optics; optical engineering; image formation; aberrations; imaging systems.

## 1 Introduction

Optics is blessed with several levels of theory. At the top are quantum optics and electromagnetic theory, which account respectively for the microscopic and the macroscopic interaction of light and matter. This is the domain of optical physics, and all optical phenomena are encompassed by them.

However, in applications such as electromagnetic diffraction theory, exact solutions to problems are extremely difficult. To deal practically with such problems there is a simpler approximate theory, scalar wave theory, which provides an adequately accurate result in most applications involving interference and diffraction. It does not, however, account for polarization nor does it predict the quantitative interactions with matter. This is the domain of physical optics.

At a still lower level of approximation, where not even interference and diffraction are accounted for, is geometrical optics. Geometrical optics is the lowest level of optical theory available today. It does not compete with physical optics in explaining optical phenomena, and can only provide a crude approximation to the detailed behavior of light. Why is it still in use?

Its overwhelming virtue is in its simplicity. It can provide useful, if somewhat inexact, answers to many practical problems, where the complexity of a more accurate theory would be prohibitive. It has been applied successfully to problems where one would not expect useful results, for example, in the analysis of waveguides.

Its primary area of application, however, is in optical design. It has proven itself to be indispensable for the efficient development of optical systems. No useful alternative has yet been developed.

Until the development of the computer and its application to optical design, virtually no consideration was given to the physical nature of light. However, in recent years the application of the wave theory of geometrical optics has made it possible to convert from geometrical optics to physical optics in the final image space of an optical system, and to calculate the structure of the image taking diffraction into account. The hard work of tracing the wave front through the optical system is done geometrically.

In this article, we deal with the basic concepts of both the corpuscular theory and the wave theory of geometrical optics. Applications as an approximation to physical optics and to radiometry are touched on, but the major part of this article deals with the geometrical concepts relevant to an understanding of image-forming optics.

## 2 Basic Concepts

Historically, there have been two different approaches to the theory of geometrical optics. The oldest and most prevalent one is the corpuscular theory in which it is assumed that light consists of particles that transport energy from a source to a receiver. The other is a wave theory, which goes as far back as Christian Huygens, and which is somewhat subtly different from physical wave optics.

### 2.1 Corpuscular Theory

#### 2.1.1 Rays as Trajectories of Energetic Particles

In the corpuscular theory, light is assumed to consist of energetic particles. Particles of different colors have different energies, the blue being more energetic than the red. In this respect, they can be thought of as simplified versions of physical photons.

These particles are emitted from a source and travel along trajectories to a receiver. The trajectories are identified as ray paths, or simply rays. In classical geometrical optics, the study of rays and their properties is paramount.

One important characteristic of geometrical particles of light is that they never collide or interfere with each other regardless of the intensity of the light beam. Intense beams can intersect each other with no interaction.

### 2.1.2 Refractive Index, Color, and Dispersion

In a vacuum, all light particles travel at the well-known speed of light. However, in transparent media they travel more slowly, and particles of different colors travel at different speeds. The ratio of the speed of light in a vacuum to the speed of light in a given medium is known as the refractive index of the medium. Blue light travels more slowly in the medium than red light, and so its refractive index is higher. The variation of refractive index with color is what is known as dispersion.

The designation of color by name results in a rather imprecise determination of dispersion, as indeed it was before Fraunhofer's discovery of spectrum lines. Today the wavelengths of the light producing the spectrum lines are used for a precise designation of color.

### 2.1.3 Fermat's Principle and Optical Paths

A fundamental property of ray paths is described by Fermat's principle. Given two points along a ray, the time it takes for light to go from one point to the other is given by

$$t_A - t_B = \int_{t_A}^{t_B} \mathrm{d}t = \frac{1}{c}\int_A^B n\,\mathrm{d}s, \qquad (1)$$

which is stationary, usually a minimum, along all possible neighboring paths. If the refractive index of the medium is constant along the path, the path will be a straight line. If the refractive index is not constant, the path will vary in such a way that Fermat's principle is satisfied.

The integral of the product of refractive index and geometrical path length is called the *optical path length*, and is the distance that light would travel in a vacuum in the same time interval. Because the speed of light in a vacuum is constant, Fermat's principle can be expressed in terms of optical path as well as transit time.

### 2.1.4 Snell's Law

Most optical systems consist of reasonably homogeneous media with constant refractive indices separated by abrupt discontinuities, for example, lenses in air. Given a boundary separating two dissimilar media, a ray will change its direction in going through the boundary unless it is normal to the boundary. This change in direction is governed by Snell's law,

$$n' \sin i' = n \sin i, \qquad (2)$$

where the primed quantities are in the emergent space, $i$ is the angle between the incident ray and the normal to the boundary at the intersection point, and $i'$ is the angle between the emergent ray and the normal.

Snell's law is the basic tool in designing and analyzing optical systems. It is simple and powerful, and can be derived from Fermat's principle.

## 2.2
## Wave Theory

### 2.2.1 The Point Source

The basic elementary source in geometrical optics is the point source. It is an infinitesimal region of space that emits light. Associated with every point source is a family of rays that trace the trajectories of the light particles emitted. Different point sources and their ray families behave independently of each other.

### 2.2.2 Geometrical Wave Fronts

Associated with each point source and its family of rays is a family of surfaces. These surfaces are surfaces of constant optical path from the source measured along the rays, and they are called *geometrical wave fronts* (see also WAVE OPTICS). Although there is no characteristic wavelength associated with these surfaces, they are called *wave fronts* because they are generally good approximations to physical wave fronts except in the neighborhood of geometrical shadow boundaries. Christian Huygens was the first to use the idea of geometrical wave fronts in discussing the properties of light.

### 2.2.3 Connection with Physical Optics

As stated above, the precise specification of a color in geometrical optics is in terms of the wavelength of spectral lines. This wavelength is of course a physical quantity that in itself has no other significance in geometrical optics. However, it is a unit of length, and geometrical optical path lengths can be measured in wavelength units. This combined with the geometrical wave front gives us a geometrical model of a physical wave field.

### 2.2.4 Significance in Imaging Systems

Traditionally, optical image-forming systems have been designed and evaluated using ray optics almost exclusively. Before the advent of computers, the stupendous amount of work required to do it any other way was out of the question. However, today it is possible to trace geometrical wave fronts through an optical system (admittedly using rays to do it) to the final image space where one can convert to physical optics in order to account for diffraction in the image-forming process.

# 3
# General Applications

There are a number of areas where geometrical optics is commonly applied. What they have in common is that the geometrical optical model is much easier to deal with than a more accurate higher-level model, and that, for these applications, the assumptions and approximations are adequate for the purpose. Two application areas are discussed briefly in this section, and the remainder of the article deals with the most important area, that of image-forming optics.

## 3.1
## Physical Optics

### 3.1.1 Wave-front Propagation

The principal general area of physical optics where geometrical optical tools are commonly employed is in the analysis and evaluation of instruments in which wave-front transmission is the principal phenomenon, such as spectrometers and interferometers. Geometrical optics has even been applied successfully to waveguide analysis.

#### 3.1.2 The Use of *Ad Hoc* Properties

A number of physical wave-field properties do not directly occur in the geometrical optical model. Only by assuming them as *ad hoc* properties can they be accounted for. One such property already mentioned is the use of wavelength as the unit for measuring optical paths. Other properties are wave amplitude, reflectance, transmittance, and polarization. When used with care, *ad hoc* properties are a significant enhancement to the geometrical optical model.

#### 3.1.3 Diffracted Rays

One interesting special application is in extending the geometrical model to include diffraction, at least by sharp-edged apertures. It is based on the Rabinowicz model of diffraction in which he decomposes a diffracted field into the coherent superposition of a standard geometrical field truncated by the aperture and an induced field that appears as that produced by a coherent line source on the edge of the aperture. In the geometrical model, additional "diffracted" rays are created on the boundary.

### 3.2 Radiometry

In most cases, a geometrical optical model is adequate for dealing with the radiometry of optical systems. Although point sources are certainly included in radiometric phenomena, extended sources generally play a greater role. In the geometrical model, an extended source is simply a dense array of independent point sources. The independence of the point sources means that in radiometry, extended sources are assumed to be noncoherent.

## 4 Imaging Systems: First Order

The principal application of geometrical optics is in the area of image-forming systems. Principles and concepts specific to imaging systems are fairly extensive, and the rest of this article deals with these.

No optical system forms images perfectly, and departures from ideal behavior are called *aberrations*. However, to deal with aberrations we must first define what constitutes ideal behavior. Although there are subtle differences between the terms, ideal behavior is variously described as first order, collinear, paraxial, and Gaussian. Each of these will be dealt with in further detail in the following conceptual development.

### 4.1 Objects and Images: Optical Spaces

In every optical image-forming system, there is at least one object space and at least one image space. In compound imaging systems, each element takes the image space of the preceding element as its object space and reimages that into its own image space, which in turn is the object space of the next element. At its most basic level, the elements of a compound system are the optical surfaces. A single lens is a compound of two surfaces and has a total of three spaces. A three-lens system has a total of seven spaces.

These spaces are called *optical spaces* and are of infinite extent. They are not bounded by the optical surfaces. That part of the optical space that lies between adjacent surfaces is called the *real part*. The rest of the optical space is its virtual part. In a conventional drawing of an optical system showing ray paths from surface to surface, only the real parts of each space are shown.

## 4.2
## Ideal Behavior: Collinear Mapping

An object consists of a three-dimensional array of point sources. Its image consists ideally of a corresponding array of point images. The ideal image is related to the object through a mapping process.

The most popular mapping to represent ideal behavior is a collinear mapping. The basic rule is that, in addition to there being a one-to-one correspondence between object and image points, there is also a one-to-one correspondence between straight lines in the two spaces. As a consequence, there is in addition a one-to-one correspondence between object and image planes. Corresponding elements are called *conjugate elements.*

## 4.3
## Axially Symmetric Systems

In the vast majority of optical systems, every surface and aperture in the system is intended to be rotationally symmetric about a single common mechanical axis. The basic concepts in the geometric theory of image-forming systems were developed in the context of such systems, and most of the following discussion assumes axial symmetry.

### 4.3.1 Focal Systems and Gaussian Imagery

Each space in an axially symmetric system has its own axis of symmetry. They all coincide with the mechanical axis of symmetry, but each space is distinct in all its properties.

Each point on the object axis has one and only one point on the image axis corresponding to it. If an object point at infinity on the axis has a conjugate at a finite location on the image axis, the conjugate image point is the rear focal point of the system. There is a corresponding front focal point that has as its conjugate a point at infinity on the image axis. Such a system is called a *focal system.*

**4.3.1.1 Cardinal Points** The two focal points of a focal system are two of the six cardinal points defined by Gauss. The other four are defined as follows.

A point off axis in the object space will have a conjugate image point off axis. The distance of such a point from the axis is called the *height of the point,* and the ratio of the image height to the object height is called *the magnification of the pair.*

An object plane perpendicular to the axis and containing the given point has as its conjugate an image plane, also perpendicular to the image axis, that contains the given image point. All conjugate pairs of points in the two planes have the same magnification. Every pair of conjugate planes has a unique magnification connecting them. No two pairs have the same magnification. The pair of planes having a unit positive magnification are called the *principal planes of the system.* Their axial points are called the *principal points.* These are also two of the cardinal points.

The distance from the principal point to the focal point in image space is called the *rear focal length* of the system, and the corresponding distance in object space is called the *front focal length.*

The last pair of cardinal points are the nodal points. These are conjugate axial points located so that a line passing through the front nodal point making some angle with the axis will have as its conjugate a line passing through the rear nodal point making the same angle with the axis. These make it possible to determine the image size in the rear focal plane of an object at infinity but with a finite angular extent.

The cardinal points of a system are used to establish the constants used in the collinear mapping of object space into image space. The term Gaussian imagery is used to describe this application of the cardinal points, but the underlying model is collinear.

4.3.1.2 **Gaussian Imagery** It is important in developing the mapping equations to be consistent in the assumed sign convention. We assume a standard Cartesian coordinate system in each space with the standard Cartesian sign conventions. We also identify the axis with the optical axis in each of the spaces.

The usual mapping equations take the origin of the coordinate system in each space to be located at the principal point. The distances to a pair of conjugate planes are given by

$$\frac{n'}{z'} = \frac{n}{z} + \phi, \qquad (3)$$

where

$$\phi = \frac{n'}{f'} = -\frac{n}{f} = \frac{1}{f_e} \qquad (4)$$

is the power of the system, $f$ and $f'$ are the front and rear focal lengths respectively, and $f_e$, the reciprocal of the power, is the effective focal length. The magnification is given by

$$m = \frac{z'/n'}{z/n}, \qquad (5)$$

and the transverse coordinates by

$$x' = mx, \quad y' = my. \qquad (6)$$

4.3.1.3 **Gaussian Reduction** It is a relatively simple thing to determine the Gaussian properties (the power and the location of the principal points) of a compound system if the Gaussian properties of the components are known. The reduction is carried out for two components at a time. The most elementary component is a single refracting surface, and the Gaussian properties of a single lens can be obtained by this method. Once one has the Gaussian properties of the individual lenses in a lens system, the properties of the system as a whole can be obtained by continuing the process.

If we have two components, where the power of the first is $\phi_1$ and the power of the second is $\phi_2$, there are three spaces involved. The object space for the first component is the object space for the system, and the image space of the second component is the image space for the system. Let the refractive index of the object space be $n_o$ and the image space $n'_o$.

The front principal point $P_1$ of the first component is in the object space, as is the front principal point $P$ of the system. The rear principal point $P'_2$ of the second component is in the image space, as is the rear principal point $P'$ of the system.

The space between the two components is common to both. It is the image space of the first component and the object space of the second component, and contains respectively the rear principal point $P'_1$ of the first component and the front principal point $P_2$ of the second component. Let the distance from $P'_1$ to $P_2$ be $t$ and the refractive index of the space be $n$. Then the power of the system is given by

$$\phi = \phi_1 + \phi_2 - \phi_1\phi_2 t/n, \qquad (7)$$

and the principal points of the system are located by

$$\frac{\overline{P_1P}}{n_o} = \left(\frac{\phi_2}{\phi}\right)\frac{t}{n}, \quad \frac{\overline{P'_2P'}}{n'_o} = -\left(\frac{\phi_1}{\phi}\right)\frac{t}{n}, \qquad (8)$$

where $\overline{P_1P}$ is the distance in object space from the front principal point of the first component to the front principal point of the system, and $\overline{P'_2P'}$ is the distance in image space from the rear principal point of the second component to the rear principal point of the system. The resultant system can in turn be a component in a larger system.

### 4.3.2 Afocal Systems

It is also possible for a system to have an image point at infinity corresponding to an object point at infinity. Such a system is called an *afocal system*.

Afocal systems have no cardinal points. Instead, they have a constant characteristic magnification; that is, all pairs of conjugate planes have the same magnification. An arbitrarily selected pair of conjugate planes can be used to establish the origins of the coordinate systems. The mapping equations are then

$$\frac{z'}{n} = \frac{m^2 z}{n}, \quad x' = mx, \quad y' = my. \quad (9)$$

### 4.3.3 Ray Tracing and Paraxial Optics

The general procedure for tracing a ray through an optical system is sequential. An incident ray at an optical surface is defined by its location on the surface and its direction of propagation. By applying Snell's law the ray is refracted, determining the direction of the emergent ray. The ray is transferred to the next surface, for which it is the incident ray, and the process is repeated. The operations are fairly complex, and very tedious if done by hand.

However, in a region sufficiently close to the axis that angles, sines, and tangents cannot be distinguished from each other, the ray tracing is considerably simplified. Both the refraction and the transfer operations reduce to simple linear equations. This paraxial behavior of the optical system is congruent with the paraxial behavior of the collinear model of the optical system, and because the ray-tracing equations are linear, they can be applied to the extended collinear model of the optical system without restriction to its paraxial region.

If the real ray-tracing equations had been represented by a power-series expansion, the first-order terms would have to be congruent with the paraxial behavior. This is where the idea of first-order optics comes from.

In the collinear model that represents the ideal optical system, every refracting surface is represented by its principal planes, which are coincident at the surface. In the ray tracing, refraction occurs at the principal planes and transfer occurs between adjacent principal planes. The rays should properly be called *collinear rays*, but they are commonly called *paraxial rays*, even though they are not restricted to the paraxial region.

In the refraction process, a ray incident on a surface is determined by its height at the surface and the angle that it makes with the optical axis. The emergent ray is determined by the refraction equation

$$n'u' = nu - y\phi, \quad (10)$$

where $\phi = (n'n)c$ is the power of the surface and $c$ is the curvature of the surface. The transfer to the next surface is governed by

$$y' = y + n'u'\left(\frac{t'}{n'}\right), \quad (11)$$

where $t'$ is the axial distance from the first surface to the second. This ray is now incident on the second surface, and the process can be repeated until the ray trace is complete.

**Fig. 1** Collinear (first-order) modeling. The top part represents a real system with real ray paths. The middle part shows the collinear model with collinear (paraxial) rays. The bottom part compares the real ray paths with the collinear ray paths

The cardinal points of the complete system can be determined by tracing just two rays without concern for the cardinal points of the components (see Fig. 1). In general, Gaussian reduction is more efficient if there are no more than three or four surfaces. For larger numbers, the ray trace is much more efficient.

### 4.3.4 Fields and Pupils

It is customary in describing an optical image-forming system to identify a unique object plane perpendicular to the axis as the object, and its conjugate as the image. The useful areas of these surfaces are also usually limited in extent, either by a physical aperture known as a field stop, or by a virtual limit imposed by the application. The resulting limited surfaces are called the *object and image fields,* and paraxial (collinear) ray tracing is used to find the location and size of the image field if the object field is given.

For an object point on axis, only a limited beam of light can actually get through the optical system to form the image point on axis. The aperture in the system that limits the size of this beam is called the *aperture stop* of the system, and it often is located internally. The object-space conjugate to the aperture stop is called the *entrance pupil* and the image-space conjugate the exit pupil. The beam in object space appears to be limited by the entrance pupil and in image space by the exit pupil.

### 4.3.5 Marginal and Chief Rays

Because of the axial symmetry assumed for the optical system, the axial beam of light is completely characterized by a single ray, one that goes from the axial object point to the edge of the aperture stop and on to the axial image point. This ray is called the *marginal ray.*

For an object point at the edge of the field, the beam connecting it with its image point is also limited by the aperture stop, and to a first approximation has the same cross section everywhere as the axial beam except for being displaced from the axis. The displacement is entirely characterized by a ray going from the point at the edge of the field through the center of the aperture stop and on to the conjugate image point. This ray is called the *chief ray.*

Only these two rays are necessary to specify the fields and pupils. Wherever the marginal ray crosses the axis we have an image, and its size is given by the chief ray height at that location. Wherever the chief ray crosses the axis we have a pupil, and its size is given by the marginal ray.

The chief ray properties are distinguished from the marginal ray properties by the variables being barred, whereas the marginal ray variables remain unbarred.

### 4.3.6 The Optical (Lagrange) Invariant

At any plane in the optical system, the marginal ray is characterized by its height on the plane and its angle with respect to the axis. The chief ray is similarly characterized. A very interesting and useful relationship between the two rays is obtained by forming the combination

$$L = n\bar{u}y - nu\bar{y}. \tag{12}$$

This quantity is constant throughout the system and is called the *optical*, or Lagrange, invariant. It is particularly useful at the field and pupil planes, having the same value at all of them.

One simple application involves the Lagrange invariant at the object and image planes. At the object and image planes respectively,

$$L = -nu\bar{y} = -n'y'\bar{y}', \tag{13}$$

and so we can express the magnification connecting the image with the object in terms of the marginal ray angles:

$$m = \frac{y'}{y} = \frac{nu}{n'u'}. \tag{14}$$

### 4.3.7 Numerical Aperture and the *f* Number

The magnitude of the quantity $nu$ is a paraxial approximation to a quantity known as the numerical aperture, which is strictly given by $n \sin u$ and is usually represented by the symbol NA. The numerical aperture is an important quantity for several reasons. In physical optics, it determines the resolution of the system, but even in geometrical optics the numerical aperture in the image space determines the irradiance on the image plane; the irradiance varies as the square of the numerical aperture.

In the early days of photography, it was recognized that some simple yet effective way of controlling the exposure of the photographic negative was needed. The amount of light getting to the film clearly was proportional to the area of the entrance pupil. Also, the irradiance on the film plane for a given pupil area was inversely proportional to the focal length of the lens. A long–focal-length lens with the same pupil area put less light on the film plane in the same exposure time. Thus, the ratio of the focal length of the lens to its entrance pupil diameter, or $f$ number, was established as a quantity that could be used to control the exposure. Lenses of the same $f$ number give the same exposure in the same exposure time, regardless of the focal length.

When the object is at infinity, the numerical aperture is equal to the reciprocal of twice the $f$ number.

Complications arose when the object was so close to the lens that the image distance was considerably larger than the focal length. The exposure of course went down, and because the definition of the $f$ number involved the focal length, not the image distance, some modification of the exposure factor involving the magnification was in order. Various fixes, some more ungainly than others, have persisted to this day. The fundamental problem is that the numerical aperture is the quantity that actually controls the exposure, and the $f$ number was an empirical creation, adequate for the conditions it was designed for. Today, the numerical aperture is as easy to measure as the $f$ number, and because it is completely general, one might, for the convenience of those who persist in using the $f$ number, define an effective $f$ number that is the reciprocal of twice the numerical aperture.

### 4.4
### Systems with No Axial Symmetry

Most of the above properties seem to have depended on the presence of an axis of rotational symmetry. If you do not have an axis, how can you have paraxial behavior?

The fact is, we do have a kind of axis. There is some point on the object defined as the center of the object field, and there is some element in the system that limits the beam of light coming from that point. The latter is the aperture stop, and its image in object space is the entrance pupil. A real ray traced from the central object point through the center of the entrance pupil will eventually pass through the center of the aperture stop and on through the system until it reaches the image surface at what is therefore defined as the center of the image field. This real ray is called the *optical-axis ray*, and is the nearest thing we have to an optical axis.

Unfortunately, rays that are paraxial to the optical-axis ray can in general have large angles of incidence at the refracting or reflecting surfaces, and the ray-tracing equations do not simplify the way they do for an axially symmetric system. On the other hand, the collinear mapping process does not require any imposed symmetry conditions. For a focal system, unique front and rear focal planes exist, and object planes parallel to the front focal plane have as conjugates planes parallel to the rear focal plane. Moreover, there is a line normal to the front focal plane that has as its conjugate a line normal to the rear focal plane. These lines can be identified as the axes in their respective spaces.

The mapping between any conjugate pair of the planes parallel to the focal planes is affine, that is, the image is at most anamorphic. This means that there are a pair of orthogonal directions in the image plane where the magnification is constant, although it may be different in the two directions. We can select the orientation of the $x$ and $y$ axes to correspond with these directions. The result is a set of mapping equations that are very nearly as simple as those for an axially symmetric system. The only added feature is the anamorphism between object and image planes.

The tough part of the problem is how to obtain the collinear mapping parameters from the system properties. A general solution to this problem has not yet been obtained.

## 5
## Imaging Systems: Aberrations

### 5.1
### Aberrations as Departures from Ideal Behavior

The collinear model represents ideal behavior in an optical image-forming system. Real systems depart from this behavior, and these departures are called *aberrations*. (For a more mathematical presentation, see OPTICAL ABERRATIONS.)

From a ray point of view, all real rays from any object point should pass through the ideal collinear image point. If a real ray misses the ideal image point, we say that the ray is aberrated. The transverse displacement of the ray intersection with image plane from the ideal image point is the usual measure of a ray aberration.

From a wave-front point of view, the wave front emerging from the exit pupil should be spherical and centered on the ideal image point. If it is not spherical, or if it is not centered at the ideal image point, we say that the wave front is aberrated. The ideal wave front is usually called the *reference sphere* from which the wave aberration is measured. At any given point on the reference sphere, the wave aberration is the optical path distance from the reference sphere to the aberrated wave front along the ray passing through that point. The wave aberration function is the wave aberration as a function of the position on the reference sphere. The latter is the proper surface of reference for the pupil.

## 5.2 Wave and Ray Aberrations

Rays are everywhere normal to geometrical wave fronts, and so ray aberrations and wave aberrations must be connected. As a function of pupil coordinates, the wave aberration function is a scalar function, whereas the transverse ray aberrations are vector quantities. If the pupil coordinates are chosen to be the transverse coordinates on the reference sphere, it can be shown that, to an excellent approximation, the transverse ray aberrations are proportional to the gradient of the wave aberration function.

Between the two types of aberration, the wave aberration is a simpler function of the pupil coordinates, and the ray aberrations are simply derived from it. The reverse is not true.

Another significant advantage of the wave aberration is that the geometrical wave front is a good approximation to a physical wave front, especially in the exit pupil where Fresnel diffraction effects are minimal. The wave aberration function is easily and simply converted into a phase variation over the reference sphere, the latter being the proper surface of integration for a diffraction integral.

## 5.3 Monochromatic Aberrations

In general, different wavelengths of light result in different aberration functions, and these variations with wavelength are identified as chromatic variations. Before dealing with them, we first must understand the aberrations at a given wavelength, the monochromatic aberrations.

### 5.3.1 Expansions and Classification

The wave aberration function is a function of pupil coordinates and is also parametrically a function of field coordinates. That is, for every point in the field there is a corresponding wave aberration function that may be different for different field points. Taking into account the axial symmetry of the optical system, it is possible to represent the wave aberration function by a two-dimensional power-series expansion in the pupil coordinates, the terms of which are the elementary wave aberrations.

The principal classification of the terms is by order determined dimensionally (see Fig. 2).

### 5.3.2 Elementary (Third-order) Aberrations

The lowest nontrivial order for monochromatic aberrations is called *third order*, although the terms are dimensionally fourth degree. The labeling was historically determined by a ray classification, and the ray aberrations are determined by the gradient of the wave aberration function. There are five significant third-order aberrations.

**5.3.2.1 Spherical Aberration** Spherical aberration is the only aberration that is

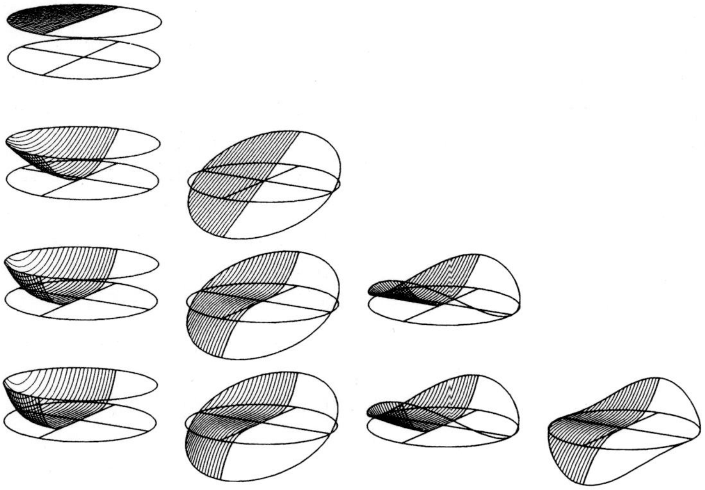

**Fig. 2** Wave aberration functions. Each row represents a different order of aberration. The third row down shows the third-order aberration functions for spherical aberration, coma, and astigmatism

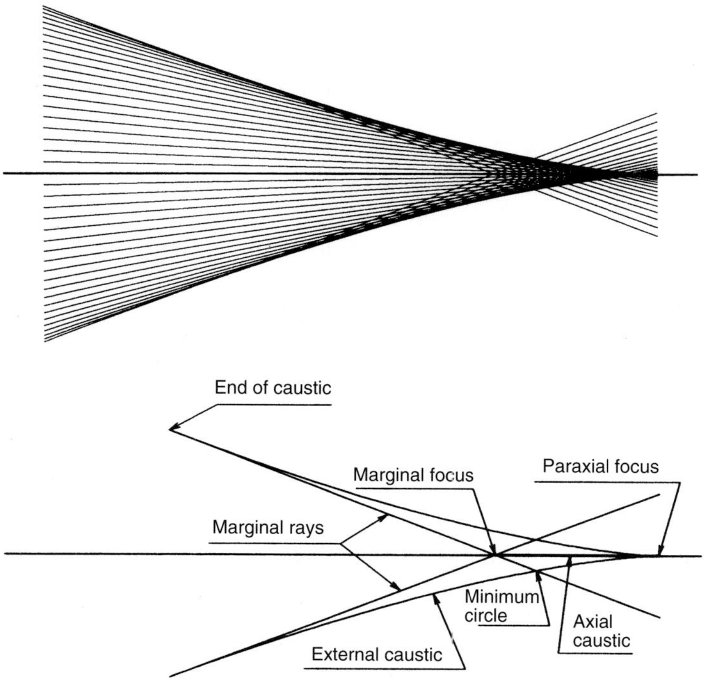


**Fig. 3** The caustic produced by spherical aberration. The top part shows the ray paths; the bottom part, at the same scale, shows the structure of the caustic

independent of the field. It is therefore the only monochromatic aberration to affect the axial image. Although its effect is felt over the entire field, it is simpler to describe as an axial aberration.

The wave aberration function is a rotationally symmetric, fourth-degree departure from the reference sphere. All the rays in a given pupil zone come to a common focus on the axis, but the axial location of these zonal foci varies with the radius of the zone. In addition to this zonal focal shift, the rays produce a trumpet-shaped external caustic, seen as the envelope of the rays, that is characteristic of spherical aberration (see Fig. 3).

The geometrically predicted image is represented by tracing a uniformly dense mesh of rays in the pupil and observing the array of intersection points with an observation plane in the neighborhood of the ideal image. This array of points is called a *spot diagram*. In the case of spherical aberration, if we move the observation plane through focus (see Fig. 4), we find that there is no symmetry with respect to focus.

5.3.2.2 **Coma** Coma is an aberration that varies linearly with field height and is therefore absent on axis. Its wave aberration function has a cubic cross section, and the ray aberrations are all on one side of the ideal image, either toward the center of the field or away from it, depending on the sign of the aberration coefficient. Moreover, at best focus the rays are contained within a 60° angle with the ideal image at its vertex.

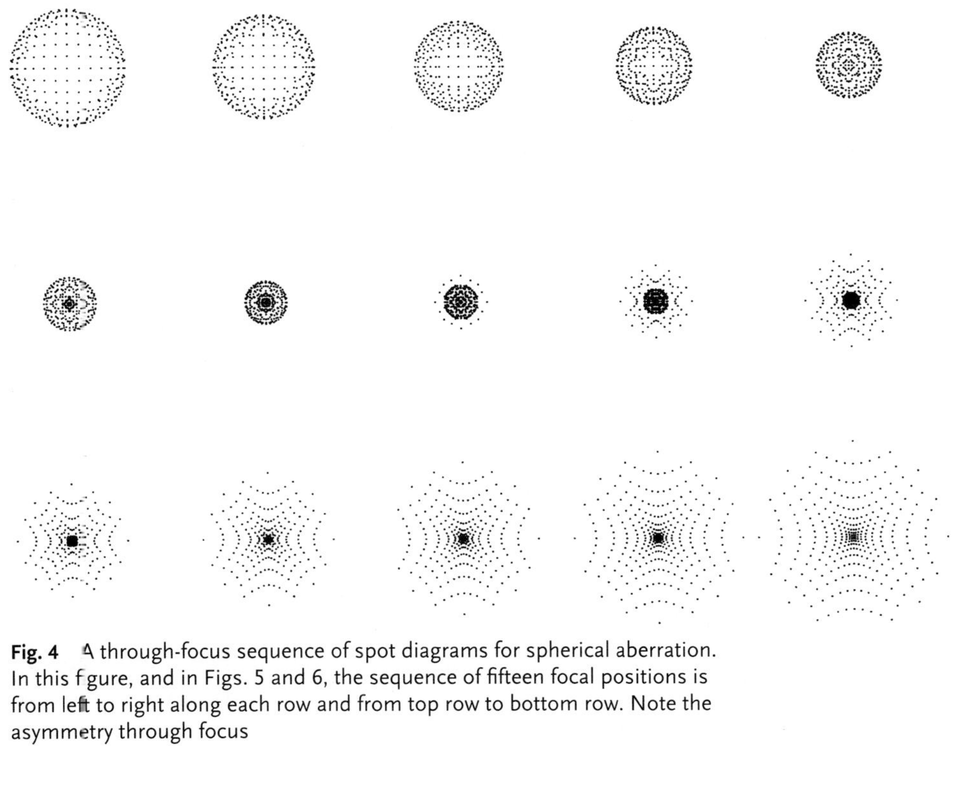

**Fig. 4** A through-focus sequence of spot diagrams for spherical aberration. In this figure, and in Figs. 5 and 6, the sequence of fifteen focal positions is from left to right along each row and from top row to bottom row. Note the asymmetry through focus

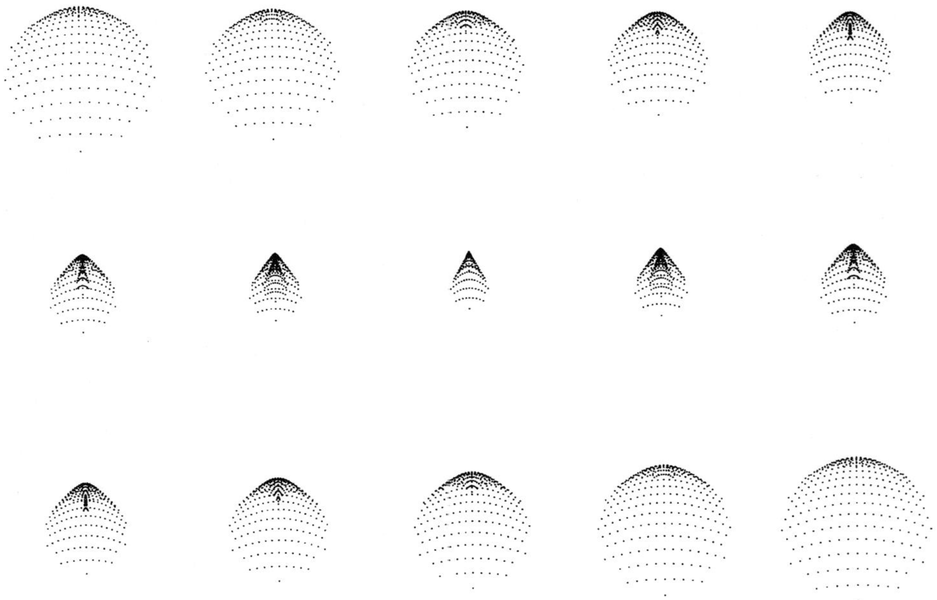

**Fig. 5** A through-focus sequence of spot diagrams for coma. The images are symmetrical through focus, but they are asymmetrical in the vertical direction

Coma is symmetrical through focus, as shown in Fig. 5.

5.3.2.3 **Astigmatism** Astigmatism is an aberration that varies as the square of the field height and is absent on axis. The wave aberration function is a quadratic cylinder, constant in a direction perpendicular to the meridional plane containing the ideal field point. (A meridional plane is a plane that contains the axis of rotational symmetry.)

The distinguishing feature of astigmatism is the presence of two linear foci, one in the meridional plane and the other perpendicular to it (see Fig. 6). The first is called the *sagittal focus* and the second the tangential focus. The two foci are separated along the chief ray by a distance that is proportional to the amount of astigmatism. In the absence of other aberrations the sagittal field is flat, but because the focus shift to the tangential image varies as the square of the field height, the surface containing the tangential images is curved.

5.3.2.4 **Field Curvature** Field curvature is an aberration in which the wave front is spherical and all the rays for a given image pass through a single point, but this point is shifted along the chief ray by an amount that increases quadratically with field height. As a consequence, in the absence of other aberrations, the images are locally very good, but they lie on a curved surface.

5.3.2.5 **Distortion** Distortion is also an aberration in which the wave front is spherical and all the rays pass through a single point, but in this case the points remain in the image plane. However, they are displaced toward or away from the axis by an amount that varies as the cube of the field height. The consequence of this is

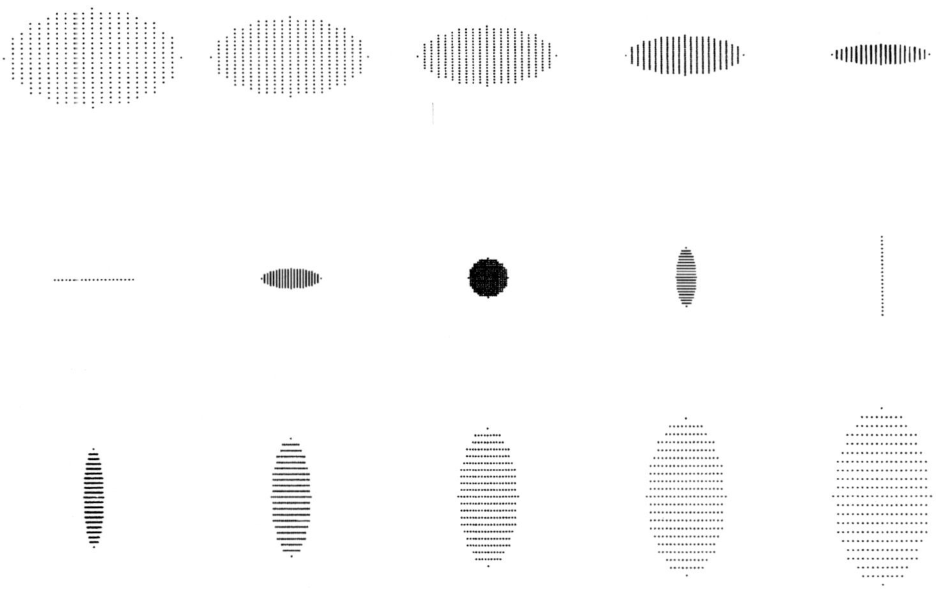

**Fig. 6** A through-focus sequence of spot diagrams for astigmatism. The images are symmetrical about the vertical and horizontal axes, but there is a twist symmetry through focus

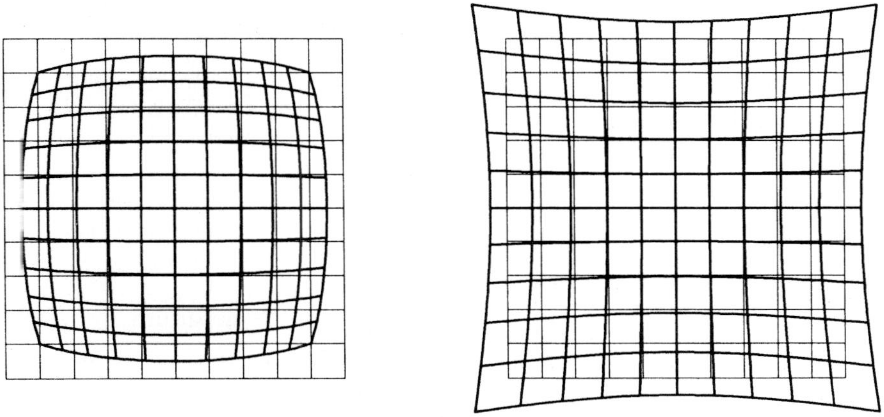

**Fig. 7** Distortion. The part on the left shows barrel distortion; the part on the right shows pincushion distortion

that images are radially distorted. A square centered on the axis is imaged with curved sides. If the sides are concave the result is called *pincushion distortion*, whereas if the sides are convex it is called *barrel distortion* (see Fig. 7).

### 5.4 Chromatic Aberrations

Chromatic aberrations arise because the optical properties of a lens system depend on the refractive indices of the components, and the refractive indices vary with wavelength.

### 5.4.1 Elementary (Primary) Chromatic Aberrations

Even the first-order properties of a system can change with wavelength. The resulting aberrations are the primary chromatic aberrations. There are only two, longitudinal chromatic aberration and transverse chromatic aberration. The first is a change in focal position with wavelength and the second is a change in image size, or magnification, with wavelength.

Mirror systems do not suffer from chromatic aberrations because they do not contain dispersive elements.

### 5.4.2 Chromatic Variation of Monochromatic Aberrations

In addition to the primary chromatic aberrations, the aberration types that have been identified as monochromatic can also vary with wavelength. In some cases, the chromatic variation of these aberrations is common enough for them to be given their own names; for example, the chromatic variation of spherical aberration is called *spherochromatism*.

## 5.5 The Control of Aberrations in Optical Design

Some of the aberrations of a single lens can be modified by bending the lens (changing the shape of the lens without changing its power) and by locating the aperture stop away from the lens. Others – for example, longitudinal chromatic aberration – are insensitive to these variables. At least two lenses of different glasses are required to correct chromatic aberrations.

All of the third-order aberrations and the primary chromatic aberrations are correctable with only three lenses, but the performance of this triplet, although fairly good, is limited by higher-order aberrations. The complexity of a high-speed camera lens is due entirely to the need for a high degree of aberration correction.

Even in a well-corrected lens, the state of correction is the result of balancing sometimes surprisingly large aberrations from one or more elements with correspondingly large aberrations of opposite sign from other elements. As a result, manufacturing tolerances are usually very tight.

In the case of mirror systems, there are no chromatic aberrations to be corrected. However, a mirror cannot be "bent" like a lens to modify its performance. Aberrations are primarily controlled by making the mirror surfaces aspheric.

Lens systems can also use aspheric surfaces, but until recently this has been avoided because of the expense involved in producing them. Aspheric mirrors are expensive too, but there is no other recourse.

## Glossary

**Aberration:** A departure from ideal behavior.

**Afocal Systems:** Systems that do not contain focal planes. An object plane at infinity has its conjugate image plane at infinity. Rays parallel to the axis in object space have conjugate rays parallel to the axis in image space.

**Aperture Stop:** That aperture that limits the size of a beam going from the axial object point to the axial image point.

**Cardinal Points:** The Focal Points, principal Points, and Nodal Points of a system taken as a set.

**Caustic:** The envelope of all rays emanating from a point and reflected or refracted by a curved surface.

**Chief Ray:** A ray that goes from the edge of the object field to the edge of the image field, passing through the center of the aperture stop on the way.

**Collinear Mapping:** A mapping of an object space into an image space in which there is a one-to-one correspondence between points, between straight lines, and between planes in the two spaces.

**Conjugate Elements:** Object and image elements (points, lines, or planes) that are in one-to-one correspondence with each other.

**Dispersion:** The variation in refractive index of a medium.

**Effective f Number:** The reciprocal of twice the numerical aperture. It is a property of the space (object or image) of the system.

**Entrance Pupil:** The image of the aperture stop in the object space of the system.

**Exit Pupil:** The image of the aperture stop in the image space of the system.

**f Number:** The ratio of the effective focal length of a system to its entrance pupil diameter. It is a property of the system.

**Focal Length (Effective):** The reciprocal of the power of a focal system.

**Focal Lengths (Front and Rear):** The distances from the principal points to the focal points.

**Focal Systems:** Systems that contain focal planes. An object plane at infinity has a conjugate plane at a finite distance from the optical system in the image space.

**Front Focal Plane:** Object plane that is conjugate to an image plane at infinity.

**Front Focal Point:** The axial point in the front focal plane.

**Height:** The distance from the axis of an object or image point.

**Ideal Behavior:** A Collinear Mapping.

**Image Space:** The optical space that contains the image of an optical element.

**Magnification:** The ratio of the image height to the object height for an off-axis object point.

**Marginal Ray:** A ray that goes from the axial object point to the axial image point, touching the edge of the aperture stop on its way.

**Meridional Plane:** A plane in an optical system that contains the optical axis.

**Nodal Points (Front and Rear):** Conjugate axial points where any object ray passing through the front nodal point and its conjugate ray passing through the rear nodal point make equal angles with the optical axis.

**Numerical Aperture:** The sine of the angle between the real marginal ray in the object or image space of a system multiplied by the refractive index of the space. It is a property of the space, and not of the system as a whole.

**Object Space:** The optical space that contains the object of an optical element.

**Optical Angles:** Ray angles multiplied by the refractive index of the space that contains them.

**Optical Path Length:** The product of the refractive index and the geometrical path length along a ray.

**Optical Space:** The object or image space of an optical element.

**Paraxial:** Close to the axis.

**Power:** The refractive strength of a system. It is equal to the reciprocal equivalent length of a focal system. It is zero for afocal systems.

**Principal Planes (Front and Rear):** The conjugate object and image planes that have a unit positive magnification connecting them.

**Principal Points (Front and Rear):** The axial points of the principal planes.

**Ray:** (1) The trajectory along which light particles travel from a source to a receiver. (2) The normal to a wave front.

**Ray Aberration (Transverse):** The departure of the ray intersection with the image plane from its ideal position.

**Real Image:** An image formed in the real part of an image space.

**Real Part of a Space:** That part of an optical space that is normally seen in a conventional drawing of an optical system.

**Rear Focal Plane:** The image plane that is conjugate to an object plane at infinity.

**Rear Focal Point:** The axial point in the rear focal plane.

**Reduced Distances:** Axial distances divided by the refractive index of the space in which they exist.

**Refractive Index:** The ratio of the speed of light in vacuum to the speed of light in a medium.

**Virtual Part of a Space:** That part of an optical space that is outside the real part.

**Wave Aberration:** The departure of an aberrated wave front in the exit pupil from the ideal wave front, measured along the ray.

**Wave Front:** A surface of constant optical path from the source.

## Further Reading

### Optical Design

Kidger, M. J. (2002), *Fundamental Optical Design*. Bellingham, WA: SPIE Press.

Kingslake, R. (1978), *Lens Design Fundamentals*. New York: Academic Press.

Kingslake, R. (1983), *Optical System Design*. New York: Academic Press.

Malacara, D., Thompson, B. J. (2001), *Handbook of Optical Engineering*. New York: Marcel-Dekker.

O'Shea, D. C. (1985), *Elements of Modern Optical Design*. New York: Wiley-Interscience.

Shannon, R. R. (1997), *The art and science of optical design*. Cambridge: University Press.

Smith, W. J. (1990), *Modern Optical Engineering*. New York: McGraw-Hill.

## Theoretical

Buchdahl, H. A. (1993), *An Introduction to Hamiltonian Optics*. New York: Dover Publications.

Hansen, R. C., Ed. (1981), *Geometric Theory of Diffraction*. New York: IEEE Press.

Luneburg, R. K. (1966), *Mathematical Theory of Optics* Berkeley, CA: University of California Press.

Mahajan, V. N. (1998), *Optical Imaging and Aberrations: Part I. Ray Geometrical Optics*. Bellingham, WA: SPIE Press.

Mahajan, V. N. (2001), *Optical Imaging and Aberrations: Part II. Wave Diffraction Optics*. Bellingham, WA: SPIE Press.

Sommerfeld, A. (1964), *Optics*. New York: Academic Press.

Stavroudis, O. N. (1972), *The Optics of Rays, Wavefronts, and Caustics*. New York: Academic Press.

Welford, W. T. (1986), *Aberrations of Optical Systems*. London: Adam Hilger.

# Holography

*Emmett N. Leith*
*Department of Electrical Engineering and Computer Science, The University of Michigan, Ann Arbor, Michigan, USA*
*Phone: (734) 764 9545; Fax: (734) 647 2718; e-mail: leith@engin.umich.edu*

**Abstract**
Holography, a method of photography that produces highly realistic images, fully three-dimensional, with the normal parallax relations of the actual object, is described. Basics of the theory are presented from both a mathematical and a physical viewpoint. Various areas of application are given, including three-dimensional displays, complex spatial filtering, hologram interferometry for nondestructive testing and metrology, product authentication and counterfeiting security, holographic optical elements, and imaging through inhomogeneities using holographic phase conjugation. The various forms of white light viewable holograms are described, including the Denisyuk or volume type, the Benton or rainbow type, and the integral type. The diffractive properties of holograms are analyzed, including Bragg diffraction, i.e., diffraction from a volume hologram. The construction of commercial display holograms is discussed. The present state of commercial and artistic holography is discussed, along with expected future developments.

**Keywords**
imaging; phase; interference; wavefront reconstruction; metrology; nondestructive testing; optical storage; displays; phase conjugation.

## 1 Introduction

Holography is a process of recording a complete wave field, both amplitude and phase. It differs from conventional photography in which only the amplitude or equivalently, the irradiance of the field is recorded. By recording the entirety of the wave, some rather remarkable effects are produced. In particular, the images formed with holography are startlingly realistic, being fully three dimensional and containing all of the parallax relations that are encountered in everyday seeing but which are lost when an image is recorded photographically.

Holography was invented by the Hungarian – British scientist Dennis Gabor in 1947. His aim was to improve the resolution of the electron microscope, which at that time suffered from noncorrectable spherical aberration [1–3]. The idea was to photographically record the complete wave field of the electron waves, both phase and amplitude, and then make reconstructions using visible light, that is, to produce light waves that were replicas of the electron waves, complete with aberrations. The aberrations could then be corrected by appropriately designed lenses. The idea, while bold and imaginative, was unsuccessful. However, since then holography has burgeoned and has found many applications, all quite remote from Gabor's original intent.

## 2 Holographic Principles

Photographic film is inherently insensitive to phase, as are indeed all detectors of radiation. Yet the holographic process requires the recording of the entire wave, phase as well as amplitude. This feat is accomplished by converting phase variations into amplitude variations, which the film can then record. The process is described with reference to Figs. 1 through 3. Figure 1 shows a plane wave of light reflecting off an object (equivalently, for a transparent object, it could be transmitted through the object). The reflected light is a complicated wave pattern; the waves are of irregular shape, and the amplitude (brightness) varies from point to point. The phase is essentially the shape of the wave front, and the irregularities of the wave front of Fig. 1 show a phase that changes rapidly from point to point. To record the phase means essentially to record the shape of the wave front. But the argument is just as before: photographic film is insensitive to the shape of the wave front.

Figure 2 shows how the introduction of a background wave can result in recording of the total field. For simplicity, we have sketched a rather simple object wave front, where the shape varies slowly across the recording aperture. This wave front is propagating toward the recording medium. Point A of the wave front is ahead of point B, and arrives at the recording plane first. But, as we stress, the recording plate is insensitive to the time of arrival of a wave front, being sensitive only to the amplitude; hence the photographic record has no information about the shape of the wave front. However, the introduction of a second wave called the *reference wave*, here a plane wave, effectively converts phase into amplitude through the process of interference. The interference of two beams of light produces a fringe pattern – a sequence of light and dark areas, where the two beams interfere constructively and destructively. The fringe spacing depends on the angle between the two beams. Hence, near the top of Fig. 2, where the angle between the waves is large, the fringe pattern is fine. Further down, where the two beams are nearly parallel, the fringe pattern is coarse. Finally, near the bottom, the angle has increased again, and the fringe pattern is again fine. Hence, one can, by observing the fringe pattern, gain an idea of the shape of the wave front. In short,

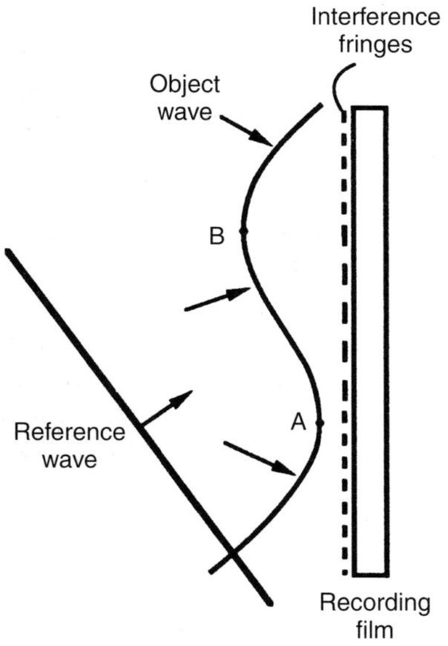


**Fig. 2** Recording of a wavefront by interference with a reference wave

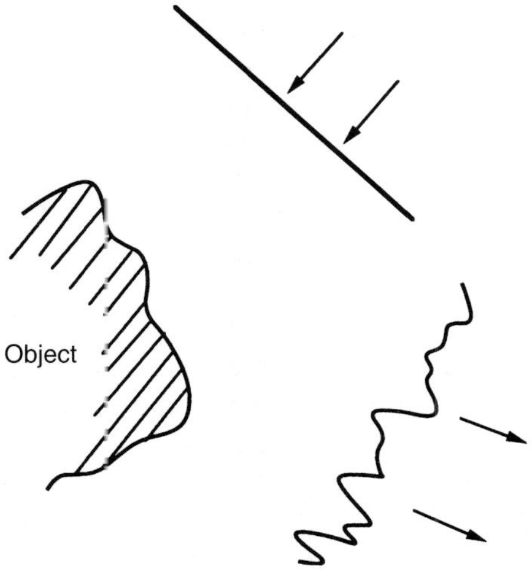


**Fig. 1** Light reflecting off an object

the effects of phase become visible. This phenomenon is, of course, the basis of all interferometry, including holography. The photographic film when developed becomes a hologram.

The hologram thus has, in an indirect way, made a record of the phase. The completion of the holographic process is the regeneration of the original wave front from the recorded pattern. We illuminate the hologram with a beam of light that, for simplicity, we take as a replica of the original reference wave. We think of the hologram now as a diffraction grating (Fig. 3), and consider that the angle of diffraction depends on the grating spacing – the closer the spacing, the greater the angle of diffraction. Then, at both the top and the bottom of the hologram, where the fringes are finest, the angle of diffraction is greatest. In the middle, where the fringes are coarser, the diffraction is at a smaller angle. Thus, the diffracted wave is a replica of the object wave that created the interference pattern when the hologram was recorded. The diffraction process is thus an inverse of the interference process, and the two processes in tandem lead back to the original wave front. Thus it is that phase can be recorded and wave fields reconstructed. This heuristic description of the holographic process basically says what mathematics says in a more quantitative and rigorous way.

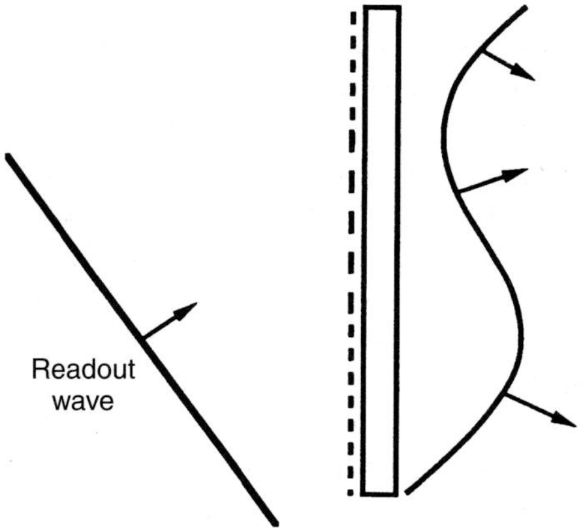

**Fig. 3** Reconstruction of the wave by the process of diffraction. Only one diffracted order is shown

Consideration of the other diffracted order of the grating leads to a diffracted wave that is conjugate to the original wave; that is, in the conjugate wave, points A and B are reversed. On the original wave, as well as in the primary wave, which is the one we have described, point A is ahead of point B, that is, the phase at A leads that at point B. In the conjugate wave, this situation is reversed. Thus, the holographic process creates not just one image, but two, known as the *primary* and *conjugate images*. The conjugation results in the image forming downstream from the hologram, thus being a real image, as opposed to the primary, which is virtual. When the hologram is read out with a collimated beam of light, the two images lie in mirror-symmetric positions about the plane of the hologram, one image forming on the source side of the hologram and the other on the observer's side. However, this symmetry disappears when the illuminating beam is not collimated, and indeed, under illumination with highly divergent or convergent beams, it is possible for the images to be both real or both virtual. Hence, to call the two images the real and virtual images of the holographic process, as is sometimes done, can be misleading, and the terms primary and conjugate are generally preferred.

The existence of the extra image has been a problem in holography, since in the original Gabor process, where the reference beam was not introduced at an

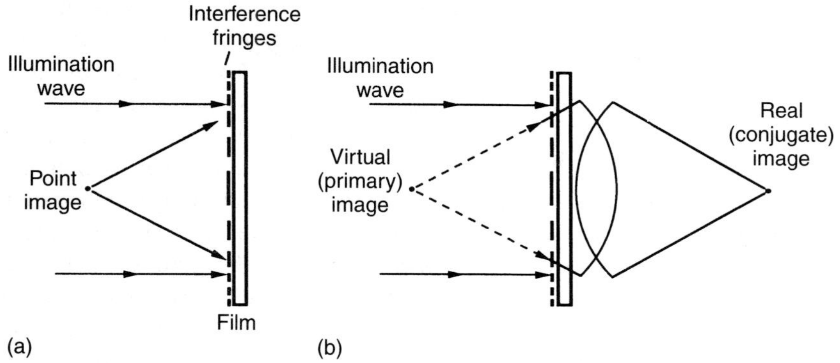


**Fig. 4** Point image hologram made in the original Gabor configuration: (a) making the hologram; and (b) the reconstruction process, showing the emergence of two images

oblique angle, the images did not separate by propagation in different directions, but instead were superimposed, so that one image could be viewed only against the background of the other, giving a noisy reconstruction.

The origin of the twin image can be explained in various ways. A simple yet revealing way is described with respect to Fig. 4. Consider an object consisting of only a single point (a small opaque dot), and let this object be placed in a collimated beam of light. Most of the light passes undisturbed around the object, falls on the film, and serves as the reference beam. Light scattered by the object point also propagates to the recording film, where the interference between the two beams creates the interference pattern shown in Fig. 4(a). In the reconstruction process, with a collimated beam, we consider as before the hologram to be like a grating, one with a spacing (or spatial frequency) that changes with position. At each position, part of the incident light is diffracted in one direction and part in the opposite, at angles that depend on the spacing. The rays are seen to cross at two locations [Fig. 4(b)], one in front of the hologram and one behind. For the latter, the rays do not actually come together; rather, when projected back, they seem to emanate from a common point, thus producing a virtual image. Thus, the diffraction process has produced both real and virtual images, and the two images fall on top of each other so that one must always observe one image with the other giving an out-of-focus background, making the image noisy.

A somewhat philosophical way of explaining the origin of the twin image is to consider that a convergent wave of the same radius of curvature produces exactly the same interference pattern as the diverging wave when made to interfere with the planar background wave. Thus, the holographic process cannot distinguish between the two waves and cannot decide which wave produced the pattern. The reconstruction process resolves the quandary by reconstructing both.

Introducing a reference beam that passes around the object and impinges on the recording plate obliquely, as we have seen, sends the two reconstructed beams in different directions, thereby solving the problem.

## 3
## Mathematical Theory

We offer a mathematical treatment that places the above viewpoint on a more rigorous basis (see also WAVE OPTICS).

The original theory, as developed by Gabor, assumed a transparent object s, which when illuminated with a coherent beam of light, that is, a monochromatic field from a point source, transmits some portion of the light without scatter and the remainder with scatter (Fig. 5). At some distance $z_0$ downstream from the object, let the field consist of a uniform component $u_0$, the background or reference wave, due to the unscattered portion of the wave, and $u$, the object beam produced by scatter from the object. The field has an irradiance $|u_0 + u|^2$. The field is recorded, typically by photographic film. If for simplicity we assume that the film has a transmittance proportional to the irradiance, thus ignoring the nonlinearity of the film and the contrast reversal that occurs upon recording, which makes the brighter parts of the field darkest on the record, we have the transmittance

$$t = I = |u_0 + u|^2 = |u_0|^2 + |u|^2 + 2u_0|u|\cos\phi, \quad (1)$$

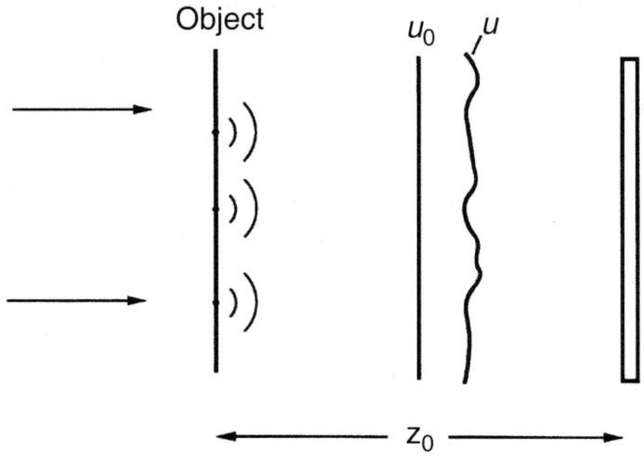


**Fig. 5** Making a hologram in the Gabor configuration. The light transmitted by the object consists of a scattered wave $u$ and an unscattered wave $u_0$

where we have dropped constants of proportionality and have assumed that $u_0$, the background wave, is uniform in amplitude and phase across the field, and that the varying part of the field, $u$, can be written in polar form as $u = |u| \exp(i\phi)$, where $|u|$ is the amplitude variation across the field and $\phi$ is the phase variation.

Upon illumination with a plane wave of monochromatic light of, for simplicity, unity amplitude, and at normal incidence, the field emanating from the hologram is

$$u_{EM} = |u_0|^2 + |u|^2 + u_0|u|e^{i\phi} + u_0|u|e^{-i\phi}, \quad (2)$$

where the final term of the film transmittance has been written in terms of the two complex components. The first term of Eq. (2) is just a uniform background. The second term is a noise term sometimes called the intermodulation product term, or sometimes the self-interference term. It is produced by interference among the various scattered waves that make up the object beam. If most of the light that passes through the object is unscattered, then $u_0$ will be much larger than $u$, and the quantity $|u|^2$ will be small and can be neglected. The term $u_0|u|e^{i\phi} = u_0 u$ is the replication of the original wave, and projects back to form a virtual image at the position occupied originally by the object. The final term is the complex conjugate of the original wave and can be written as $u_0 u^*$. If, for example, the original object were a point scatterer, then $u$ is a spherical wave centered on the object point. Similarly, $u_0 u^*$ is then a spherical wave with opposite curvature and is centered on a point lying downstream from the hologram. It thus forms a real image.

The conjugate image can be disturbing, as we have noted. Since the two image terms have the same energy, the reconstructed image will generally have a signal-to-noise ratio of unity, which is unacceptable for most applications.

The twin image of holography was a serious problem that resulted in considerable effort to remove it. Many of the early papers on holography suggested ways to eliminate the twin image. The problem is more than merely an imperfection; it is in fact rather fundamental. The complete wave front (ignoring polarization effects, which usually are of no concern) can be written as $u = |u|e^{i\phi}$, a product of an amplitude (or irradiance) part and a phase term. The ordinary photographic process records only the amplitude $|u|$, being totally insensitive to phase. The holographic process records both, but incompletely. That it is a rather fundamental defect can be appreciated by considering that, just as a complex quantity $u$ can be written in the polar form $|u|e^{i\phi}$, so it can also be written as the summation of a real term and an imaginary term, $u = |u|\cos\phi + i|u|\sin\phi$. Inspection of Eq. (1) shows that the holographic process has recorded only the real part of the field. Thus, in a basic sense the process has failed to record any more of the total wave than does conventional photography; the mix is simply different. The holographic process has recorded part of the phase, at the expense of recording only part of the amplitude. This missing information manifests itself as the twin-image problem.

Gabor had a complete solution to the problem. He proposed to make two holograms, with one having the background shifted by 90°, which can readily be done by various phase-shifting means. The second hologram would have an information-bearing term of the form $u_0|u|\sin\phi$, which is just the missing information. The two holograms would be placed in separate branches of a two-beam interferometer and upon readout, would produce two reconstructed waves that are combined by the interferometer. If the interferometer has a 90° phase shift in one branch, then the reconstructed beams would combine in what is called *quadrature*, that is, with a 90° phase shift between them, giving a complete reconstructed image wave $u_0|u|\cos\phi + iu_0|u|\sin\phi$, which is the complete field.

This solution has practical difficulties. To make a pair of phase quadrature holograms and reconstruct them in quadrature in an interferometer would require a precision of phase control rather difficult to achieve in practice. In particular, the holograms would have to be recorded on optical flats with a uniformity much better than half a wavelength of the light over the entire hologram.

It would be preferable to have the entire field on a single hologram. The Leith–Upatnieks solution [4] arises from the recognition that the recording process has an aliasing, in that a spatial frequency component, $e^{i2\pi f_1 x}$, physically manifested as a plane wave propagating at an angle $\theta$ to the normal, and another spatial frequency component, $e^{i2\pi f_1 x}$, physically manifested as a plane wave propagating at an angle $-\theta$, produce exactly the same fringe pattern when interfering with the background wave. The solution is to modulate the entire wave field $u$ onto a carrier. Each spatial frequency component of the object should be preserved, but should be shifted in spatial frequency, so that all spatial frequencies are of the same sign. Then after recording, the ambiguity would have been removed – every spatial frequency of the object wave will produce a different fringe pattern. In the reconstruction process, all of the positive spatial frequencies will belong to one image, and all negative

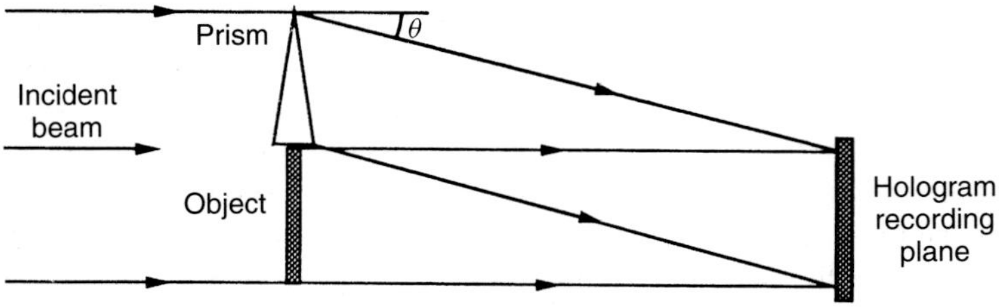


**Fig. 6** Making of off-axis reference beam hologram

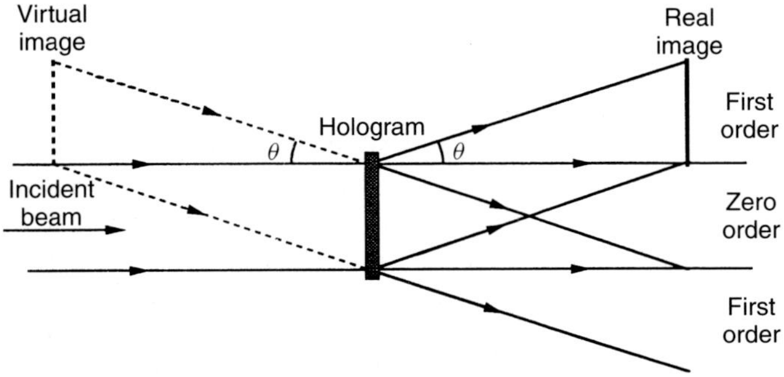


**Fig. 7** Viewing of off-axis hologram. Low-quality images occur on hologram axis. High-quality images occur in the first-order diffracted waves

spatial frequencies to the other, and the images thereby become separable.

To accomplish this objective, the reference beam is introduced obliquely, as in Fig. 6. We let the reference beam be $a_0 \exp(2\pi i f_0 x)$, which describes a wave impinging obliquely on the hologram. Equation (1) then becomes

$$t = I = |u_0|^2 + |u|^2 + 2a_0|u|\cos(2\pi f_0 x - \phi). \quad (3)$$

The final term is on a spatial carrier $f_0$. The aliasing has been removed, in that the two previously noted spatial frequency components $\exp(2\pi i f_1 x)$ and $\exp(-2\pi i f_1 x)$ now produce distinguishable fringe patterns, one of spatial frequency $f_0 + f_1$ and the other of spatial frequency $f_0 - f_1$. In the reconstruction process, the two images, primary and conjugate, of Eq. (2) become $a_0|u|\exp[i(2\pi f_0 x - \phi)]$ and $a_0|u|\exp[-i(2\pi f_0 x + \phi)]$, which signify waves propagating in different directions. If we think of the hologram as being a diffraction grating, then the two terms become respectively the $-1$ and $+1$ orders of the grating, wherein the other extraneous term, $|u_0|^2 + |u|^2$, becomes the 0th order. Thus, as shown in Fig. 7, the beams representing the various terms become fully separated after some propagation distance. Thus, the complete wave field has been recorded and perfectly recovered.

## 4
## The Rise of Modern Holography

The introduction of the off-axis or carrier-frequency technique in 1961 marked the beginning of the era of modern holography. In 1963, the technique was perfected and high-quality holographic imagery was reported, which created renewed interest in holography. Shortly thereafter [5], the formation of holograms of reflecting three-dimensional (3-D) objects, as opposed to the simple, two-dimensional transparencies that had been previously used in holography, was reported. In the reconstruction process, the observer looks into the hologram and sees an image that is a perfect likeness of the original object, fully three dimensional, with all of the normal parallax relations of real life. The observer, by moving his head, observes the same parallax relations that were observable in the original object scene. When looking from a near to a far point in the image, the observer must refocus his eyes. The observer can peer around objects to see what lies behind. The image, in short, is optically indistinguishable from the original object.

The hologram of the 3-D reflecting object, that is, the usual, real-world object, was made possible by the laser with its considerable coherence length compared with that of the mercury arc source that had been used in the prelaser era.

Coherence length is an important property of a light source used for holography and for interferometry in general (see INTERFEROMETRY). If a beam of light is split into two parts, which are then brought back together, the two beams have the possibility of producing interference effects. If the beams travel equal paths so that they are combined without a path difference being introduced, strong interference effects can be produced. If one beam travels a longer path than the other, the capability for interference diminishes. The amount of path difference that can be introduced without destroying the ability to interfere is called the *coherence length* and is determined by the spectral width of the source. A laser, being typically highly monochromatic, has a long coherence length, typically in the range of many centimeters. The mercury arc source, by comparison, has a coherence length of less than 1 mm.

Holograms of transparencies can easily be made with the mercury arc source, since the coherence length required for the basic holographic process is fairly small and can be handled very readily by the Hg arc source. However, the use of a reflecting, 3D object causes an increase of required coherence that can easily be several thousandfold. The problem is that in making a hologram of such an object, light reflected from all parts of the object must be coherent with the reference beam. If the object has a depth of 100 mm, which is fairly typical, then since the nearest point and the deepest points of the object, as well as all points in between, must interfere with the reference beam, the source must have a coherence length of the order of twice the object depth, or about 200 mm. This additional coherence length requirement, which has really nothing to do with the basic holographic process, is superimposed on the basic coherence requirement of the holographic process, which is typically about 0.1 mm. Thus, the transformation from 2-D objects to 3-D reflecting objects at once increases the needed coherence length to a level that can be met only with the laser.

Following the above events, interest in holography exploded, and holography became a research area in numerous laboratories throughout the world. Theory was developed, new ideas were conceived, and applications were sought.

In the meantime, Denisyuk in the Soviet Union described (in 1962) the volume hologram (see Sect. 7.2), in which the recording medium is sufficiently thick as to have a profound effect on the diffraction characteristics of the hologram. The Denisyuk, or volume hologram, is one of the main pillars of modern holography.

## 5 Some Early Applications

### 5.1 Imaging Through Inhomogeneities Using Holographic Phase Conjugation

The conjugate image of holography had traditionally been its greatest problem. Leith and Upatnieks [4] not only offered a practical way of avoiding the problem, but also proposed a use for the twin image: imagery through inhomogeneous media, such as highly aberrated lenses or scattering media, even as severe as frosted glass. Reduction to practice was reported in 1965 [6, 7].

Figure 8 shows a hologram-forming system in which a distorting medium has been placed between the object and the hologram. When the distorting medium is, as depicted, a diffuser such as ground glass, the object is not visible from the recording position. Similarly, from the hologram thus produced, the object cannot be imaged; the hologram reproduces only what can be seen from the recording plane. The hologram will, of course, form an image of the diffuser. Let the diffuser be described by a phase function $\exp[i\theta(x, y)]$, where $\theta$ is a noiselike function that describes the phase retardation introduced by the diffuser. The real image formed by the hologram is $\exp(-i\theta)$, since the real image is conjugate to the original object.

Placement of the original diffuser in the position where the holographic real image is formed, results in the superposition of the diffuser and its conjugate image, and the phase of one is cancelled by the complementary phase effects of the other or, mathematically, $\exp(-i\theta)\ \exp(i\theta) = 1$. The diffuser thus seems to disappear, and the object previously obscured by the diffuser is now revealed.

The process has an intriguing interpretation. As shown in Fig. 8, the readout beam retraces the reference beam path, but in reverse. The conjugate wave similarly retraces the original object beam path, but again in reverse. In the object beam, point

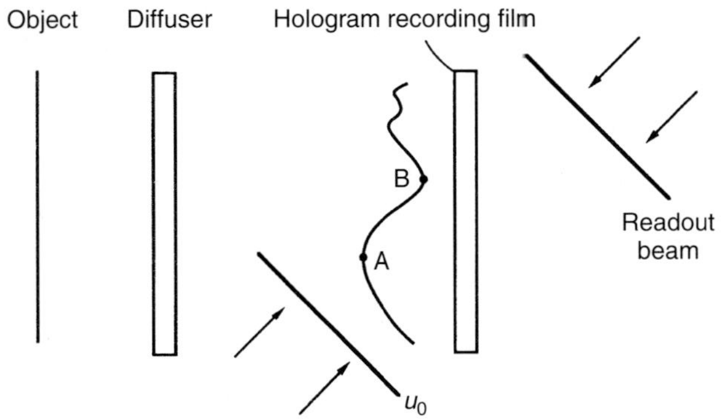


**Fig. 8** Imaging through inhomogeneity, showing the hologram being made with a reference beam $u_0$ and subsequent readout with a beam retracing the original reference beam path

B led point A in phase. The reconstruction beam has exactly the same shape, but travels in the opposite direction. Therefore, point B now lags behind point A in phase; this is the meaning of conjugation. The wave travels over the original object beam path, and at each position along its path it assumes the same shape as the object beam did at that position, until finally it arrives at the object position, whereupon the wave forms an image. The process is somewhat analogous to time reversal. In a time-reversal situation – for example, a movie running backward – curious things happen. Scrambled eggs leap out of the pan, back into the shells, which then become whole again. Here, analogously, the wave front that had been mutilated by passage through the severe inhomogeneity is made whole again. The process has thus sometimes been called *wave-front healing*.

The process languished for many years; most of the envisioned applications involved the correction in real time of inhomogeneities that were constantly changing, such as atmospheric turbulence. Photographic film, however, needs processing and thus does not operate in real time. Phase conjugation eventually became an important area of research beginning in the 1970s, with the development of light-sensitive crystals and other devices that permitted the phase conjugation to be carried out on a much faster time scale than was possible with photographic film. Phase conjugation, today, is an important area of optics, and image formation through inhomogeneities is one of the major applications.

## 5.2 Complex Spatial Filtering

Holography has greatly enhanced spatial filtering technology by making practical the construction of spatial filters with arbitrary phase and amplitude control over the filter transfer function.

Spatial filtering is an optical analog of electrical filtering. An electrical signal can be decomposed into a collection of individual frequencies, and a simple filter selectively attenuates or rejects these frequencies; a very simple example is the frequency control on an audio system, which can, for example, be adjusted to emphasize the higher-frequency components of the signal. More sophisticated filters can be designed, which can not only operate on the amplitude of the frequency components but can also alter the phase relations among these components. Similarly, optical signals, such as object and image distributions, can be decomposed into spatial frequencies, and optical spatial filtering systems can be devised to alter the frequency components.

A basic filtering system is shown in Fig. 9. An object amplitude distribution $s_0$, typically a photographic transparency, is imaged to form the light distribution $s_i$. Under coherent illumination, there occurs at the common focal plane of the two lenses a light distribution that is Fourier transform–related to the object distribution. Put another way, the various spatial frequencies that constitute the

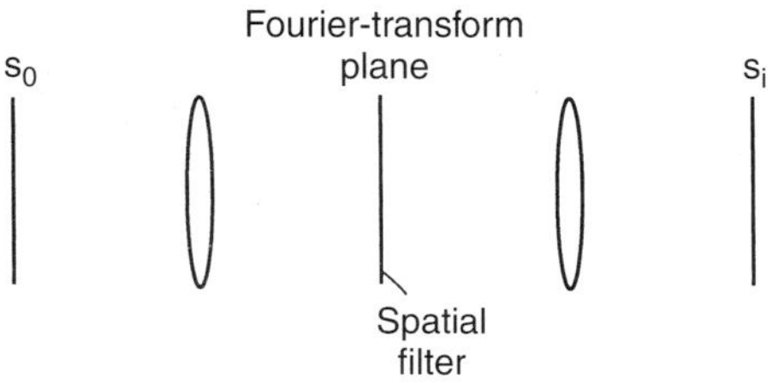


**Fig. 9** A spatial filtering system. Object $s_0$ forms an image $s_i$, and the spatial frequency spectrum is modified by an attenuating and phase-shifting structure (a spatial filter) in the Fourier transform plane

object are displayed as a spectrum in that plane. For example, if the object is a simple sinusoid on a bias, $s_0 = \frac{1}{2} + \frac{1}{2}\cos 2\pi f_0 x$, the light distribution at the common focal plane consists of three points; the point on axis represents the bias term, and a pair of off-axis points represents the spatial frequency $f_0$. More precisely, if we write $\cos 2\pi f_0 x = \frac{1}{2}\exp(i2\pi f_0 x) + \frac{1}{2}\exp(-i2\pi f_0 x)$, that is, if we divide $\cos 2\pi f_0 x$ into its positive and negative complex frequency terms, the one off-axis point of light corresponds to the positive-frequency term and the other to the negative-frequency one.

A mask that alters the light distribution at this plane, either attenuating it or shifting its phase, is a spatial filter, and its behavior is analogous to that of an electrical filter. For example, a mask consisting of a small opening on axis is a low-pass filter, because it passes only the light representing low spatial frequencies. Similarly, a stop on axis would be a high-pass filter; such a filter in microscopy produces the well-known phenomenon of dark-field illumination. Other common spatial filters give rise to schlieren imagery and phase-contrast microscopy. In analogy with electrical systems, the optical spatial filtering system can be described in terms of a transfer function $H(f_x, f_y)$, which specifies how the system passes the various spatial frequency components $(f_x, f_y)$.

The spatial filter mask can thus alter the object spatial frequency components by attenuating various spatial frequencies; this is accomplished by making the mask with an appropriate attenuation pattern. Similarly, the phase of the spectrum can be altered by making the mask with an arbitrary phase-delay pattern. Specifically, if the required transfer function is to be $H = |H|\exp(i\phi)$, with $|H|$ the amplitude part and $\phi$ the phase part, in principle, we need only to make a mask whose transmittance varies in accordance with $|H|$ and whose optical thickness varies in accordance with $\phi$; the thicker the mask, the greater the phase delay.

The historical problem has been that, although amplitude control is easy, phase control is not. In principle, one could make an arbitrary phase characteristic $\exp(i\phi)$ by generating an arbitrary thickness function on some transparent material. Such precise thickness control is possible only in some simple cases, such as the Zernike phase-contrast filter.

A hologram, on the other hand, can easily produce exactly the needed transfer function, both in amplitude and in phase [8, 9]. The required transfer function $H = |H|\exp(i\phi)$ can be produced as a hologram having the transmittance $t = s_b + |H(f_x, f_y)|\cos[2\pi\alpha f_x + \phi(f_x, f_y)]$, where $s_b$ is a bias term and where we use $f_x$ and $f_y$ as the variables, since the hologram is to be placed in a plane that displays the spatial frequency spectrum. The operation of the filter is described with reference to Fig. 10. The object $s_0$ is Fourier transformed, and the light emerging from the spatial filter mask $H$ represents the product $S_0 t$, where $S_0$ is the Fourier transform of $s_0$. In expanded form, this product is written as

$$S_0 t = S_0 s_b + \tfrac{1}{2} S_0 H \exp(2\pi i\alpha f_x) + \tfrac{1}{2} S_0 H^* \exp(-2\pi i\alpha f_x), \quad (4)$$

where the cosine term is written in exponential form.

The first term is of no interest, it simply represents a uniform filter that passes all spatial frequencies with uniform attenuation. The second term constitutes the object acted on by the complex filter $H = |H|\exp(i\phi)$ and multiplied by a factor $\exp(2\pi i\alpha f_x)$. Next, we see the usefulness

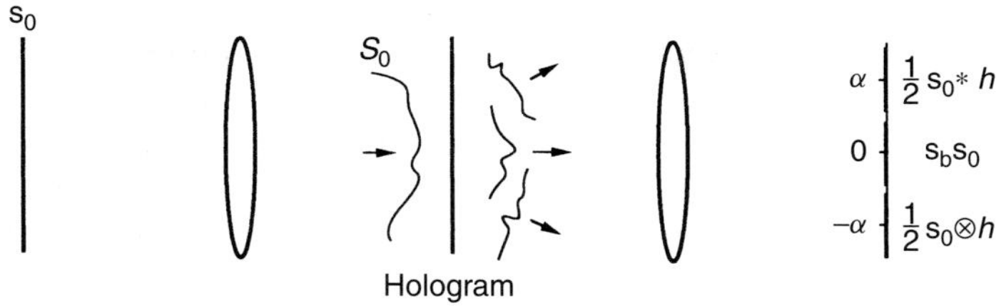


**Fig. 10** Use of a hologram as a spatial filter

of this final factor. Similarly, the final term represents the object acted upon by a conjugate filter, $H^* = |H| \exp(-i\phi)$, and multiplied by a term $\exp(-2\pi i f \alpha_x)$. Thus, the holographic process has succeeded in producing an arbitrary transfer function $H$, as well as the conjugate transfer function $H^*$. We show that these three terms are physically separated in the image plane, just as the basic terms of the conventional hologram are separated. The second lens Fourier transforms these terms, yielding an image distribution

$$I_{\mathrm{im}} = s_b s_0|_{\text{centered on axis}} + \tfrac{1}{2} s_0^* h|_{\text{centered on } x = a} + \tfrac{1}{2} s_0 \otimes h|_{\text{centered on } x = -a}, \quad (5)$$

where we have used some basic Fourier transform relations in obtaining this result. Here, * represents the well-known convolution, and ⊗ represents cross correlation. The latter is a mathematical method of comparing the similarity of two functions. Specifically, it is defined as

$$s_0 \otimes h = \iint s_0(x' + x, y' + y) \times h(x', y')\,\mathrm{d}x'\,\mathrm{d}y'. \quad (6)$$

If $s_0$ and $h$ are very similar, then the integral will have a large value when $x = y = 0$, that is, when $s_0$ and $h$ are in alignment. We say then that $s_0$ and $h$ are highly correlated.

The term $\exp(2\pi i \alpha f_x)$ has resulted in a displacement $\alpha$, a result following from a well-known theorem of Fourier transforms, namely, that linear phase shifts Fourier transform into displacements. From a physical viewpoint, the linear phase-shift terms represent wave-front tilts, and, consequently, the lens brings them to a focus at an off-axis location. If $\alpha$ is sufficiently large, the three term's will be completely separated, just as the various images are completely separated in conventional holography.

The matched filter is of special interest. A filter is said to be matched to a signal $s_0$ if its transfer function is the complex conjugate of the object Fourier transform, or $H = S_0^*$. Matched filters are used for recognizing the presence of a particular signal that may be immersed in noise or other signals and therefore not readily seen. Pattern recognition is a common application of matched filtering.

To construct a matched filter for an object $s_0$, a Fourier transform hologram is made of that object. The hologram is then used as a spatial filter, just as we have described above, with $H = S_0$. The output image is

$$i_{\mathrm{im}} s_0 s_b|_{\text{centered on axis}} + \tfrac{1}{2} s_0 * s_0|_{\text{centered on } \alpha} + \tfrac{1}{2} s_0 \otimes s_0|_{\text{centered on} -\alpha}. \quad (7)$$

The second term, generally uninteresting, is the self-convolution of the object.

The final term is the autocorrelation function of the object. It is typically a small, bright spot of light located at a position on the image corresponding to the position of the object.

Figure 11 shows experimental results using a matched filter of the letter $q$. In Fig. 11(a), we show a transparency containing an array of nine letters and numbers (alphanumerics), including $q$. The transparency is placed in the system shown in Fig. 10, with the aperture restricted so that only the letter $q$ is illuminated. The Fourier transform of $q$ [Fig. 11(b)] is formed at the frequency plane. A hologram of this distribution is formed by bringing in a reference beam. The hologram after development is placed back into its original position, and the array of nine letters is imaged through the hologram. Figure 11(c) shows an image of the correlation-forming first diffracted order. Each letter has been imaged as a light distribution that represents the cross correlation of that letter with the letter $q$. Since the letter $q$ correlates strongly with itself, we see a very strong, concentrated spot of light at the position for $q$. The other letters are considerably different in structure from $q$, and the correlation is therefore much lower. A threshold detector, adjusted to respond only when the light level rises above some predetermined value, could then automatically recognize the presence of the letter $q$.

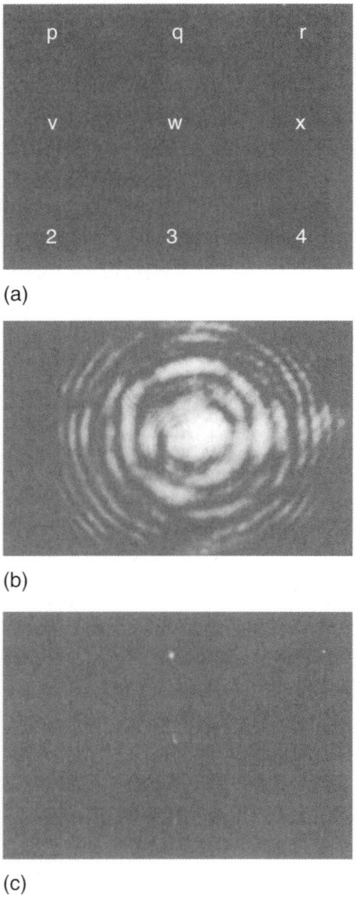


**Fig. 11** Experimental results of matched filtering: (a) object, an array of numbers and letters; (b) Fourier transform of letter $q$; and (c) image formed by imaging object through the matched filter for the letter $q$ and viewed in the correlation sideband of the filter

To give insight, a physical way of describing the operation of a matched filter is shown in Fig. 12. For simplicity, we show only the part of the filter mask. The spatial frequency spectrum $S_0$ of the object impinges on the filter mask as a wave front with some complex amplitude and phase distribution, the latter being depicted as wave-front irregularities. The wave passes through the spatial filter, where each part of the wave is phase adjusted so as to eliminate the irregularities. The wave emerges from the filter with uniform phase, essentially a plane wave; although, in general, not with uniform amplitude. This plane wave is then focused by the second lens into a small spot that is in fact the autocorrelation function we have previously described.

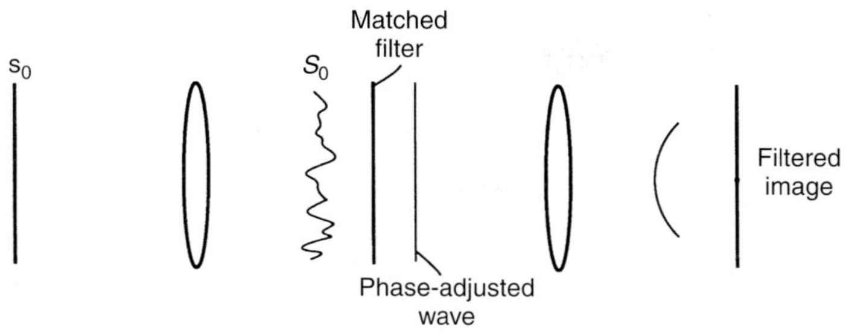


**Fig. 12** Operation of a matched filter

The use of a hologram as a complex filter and, in particular, of a Fourier transform hologram of an object as a matched filter for that object was developed at the University of Michigan Willow Run Laboratories in the period 1961–1963 under military secrecy. The initial suggestion was given by C. J. Palermo and the initial reduction to practice was by Kozma and Kelly [9], who produced a matched filter for compression of a coded pulse of the sort used in radar systems. VanderLugt [8] extended the technique to pattern recognition and character reading, and in the process made some very sophisticated matched filters.

## 5.3 Hologram Interferometry

Hologram interferometry is another major application used primarily in scientific and engineering areas. There are three principal forms of hologram interferometry: time average, double exposure, and real time.

The earliest form, time average, was first reported by Powell and Stetson [10], who made holograms of vibrating objects. The reconstructed images were found to be overlain with light and dark bands. Such an occurrence is readily explained. It is well known that holography, being an interference process between two beams, requires stability in order that the fringes remain stationary. Thus, if the object moves during the exposure, the fringes will also move and will tend to be washed out, especially those fringes originating from portions of the object where the motion is greatest. Thus, motion of the object results in dark regions on the holographic image. A pessimistic assessment would conclude that the hologram is ruined. On the other hand, one might recognize that the banded pattern superimposed on the holographic image contains valuable information about how the object moved during the exposure.

In the case of the vibrating object, the pattern overlying the object depicts the vibration amplitude. In regions where the object is stationary (the nodes of the vibration pattern), the image is bright. Other object regions reconstruct less brightly. The strength of the reconstructed image is related to the vibration amplitude by a zero-order Bessel function, which is oscillatory in nature. Hence, the image is overlaid by a fringe pattern, the fringe peaks and valleys corresponding to contours of constant amplitude (Fig. 13). Thus, inspection of the image enables the vibrational amplitude of each point on the image to be accurately determined.

Double exposure is a second form of hologram interferometry. Typically, a

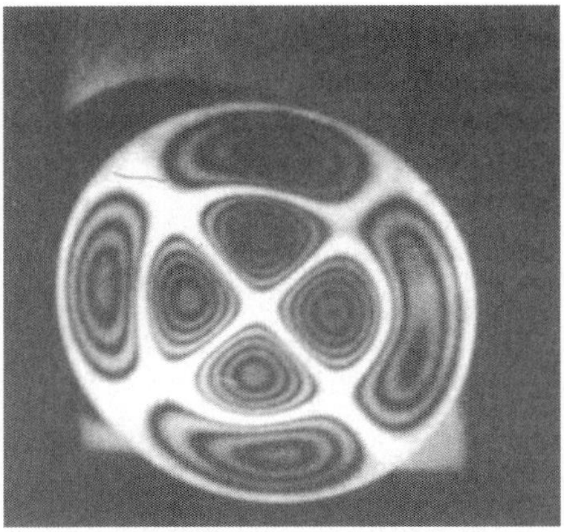

Fig. 13 Holographic image of a vibrating object, a flat plate held stationary at circumference

hologram is made of an object in an undistorted state, and then the object is distorted slightly, for example, by being stressed or heated. Then, a second exposure is made on the same photograph. On readout, two reconstruction fields are produced, one forming an image of the undistorted object, the other of the distorted object. The two wave fields interfere, forming an image with an overlay of fringes. In places where the object underwent no displacement, the two wave fields will be identical and will add in phase; the reconstruction will therefore be bright. At a position where the object shifted slightly, so that the two image waves are out of phase, destructive interference occurs and the image will be dark. Thus, the fringe pattern contains information about how the object distorted in response to the disturbance.

In the third basic form of hologram interferometry, a hologram is made of an object; then the hologram is placed back in its original position, while the object remains in its original position under continued illumination, and the reconstructed wave interferes with the object wave. The observer looking through the hologram thus sees a superposition of the object and the reconstructed image. Disturbing the object, by pressure, heating, and so on, produces a mismatch between the object wave and the reconstructed wave, causing a fringe pattern to be superimposed on the observed object–image combination. The result is akin to double-exposure hologram interferometry, except that the interference process occurs in real time, and the object can be distorted at will in many different ways. This process is appropriately called *real-time hologram interferometry*.

From the viewpoint of traditional interferometry, hologram interferometry is most remarkable. No longer is it necessary to bring two wave fields simultaneously into superposition in order to produce interference between them. The two waves can exit at different times or different places, and they can effectively be made to interfere simply by recording them holographically, thereby preserving both the amplitude and phase of each of the waves. In the context of conventional interferometry, such a process is absurd; interference could never be brought about by first recording one wave front and then at a later time and at another place, recording a second wave on the same film. Yet holography accomplishes exactly this feat.

Hologram interferometry, since its invention in 1965, has become a tool for science and engineering; its practical uses include the study of vibrations in automobile engines, the detection of defects such as cracks and debonds in airplane wings, the testing of tires, and the study of vibrational modes in musical instruments, such as violins.

Nowadays, hologram interferometry is generally carried out with a CCD (charge coupled device) camera as the recording medium, with computer reconstruction;

this method has significant advantages over photographic film. The recording can be done at lower light levels, and the hologram data can be available within a few seconds. The problems of wet film processing, including development, fixing, and drying are thus avoided. Also, with the computer, sophisticated image-processing procedures can be carried out. The CCD recording system is incorporated into an instrument consisting of a laser, beam splitter, and so on that are the usual parts of a hologram-forming system. The entire process can be carried out by persons unskilled in optics. Such systems are lightweight, highly portable, and relatively inexpensive. They have made hologram interferometer systems attractive for outside the laboratory, for example, in factory environments. The limitation of the CCD camera, which is of low resolution, is not a significant problem for hologram interferometry, since the large resolution requirements for holography are typically not present in hologram interferometry applications. The high-resolution requirements of holography arise from the recording of three-dimensional images with full parallax, and in hologram interferometry one is generally not required to form images that exhibit three dimensionality and parallax.

## 5.4
## Holographic Optical Elements

Holograms have found a use very different from the production of fascinating three-dimensional imagery. They can be used as a special kind of optical element, analogous to lenses, but operating on the principle of diffraction instead of refraction.

A hologram of a point object made with a collimated reference beam can be illuminated with a collimated beam that duplicates the reference beam, whereupon a virtual point image is formed. Alternatively, the hologram can be illuminated with a conjugate reference beam that is collimated but counterpropagating, and the hologram will form a real point image. An alternative way of viewing the reconstruction process is to consider that the hologram is acting like a lens, bringing a collimated beam to a focus. Similarly, illuminating the hologram with a duplicate of the object beam reproduces the collimated reference beam. Now the hologram is a collimating lens. Finally, the hologram can be illuminated by light reflected from an extended object, and it will form an image of this object, just as does a lens. This phenomenon is not surprising if one considers that the hologram of a point object is just a Fresnel zone-plate lens, a commonplace device described in basic texts on physical optics.

As a lens, the point-object hologram is hardly competitive with a conventional lens since it has, because of its diffractive nature, severe chromatic aberrations, and therefore it is suitable only for use with monochromatic light. Besides, the zone-plate lens has all the basic aberrations of a simple refractive lens, such as spherical aberration, coma, and so on.

However, there are many specialty niches for the holographic optical elements. Diffractive elements disperse light in a way opposite to refractive lenses; that is, red is bent the most in a diffraction process, whereas the refraction process bends violet the most. Therefore, the hologram can sometimes be used to correct the dispersion of a glass lens, producing an achromatic combination [11]. In some situations, this method can be more effective than the usual method of using glasses of different dispersiveness for color correction. The holographic lens is especially effective in combination with refractive lenses to produce a compound Fourier transforming lens in which the dispersion requirements are quite different from those of image formation. Holograms can also be used as corrector plates to remove aberrations from a lens system [12]. Diffraction gratings are nowadays often made by recording on photoresists the interference of two collimated beams of light; these are usually called *holographic gratings*, although it is questionable whether the recorded patterns of interference of two plane waves should be called *holograms*. However, if one of the two beams contains some special irregularity that is to be incorporated into the grating, then hologram would be an appropriate term.

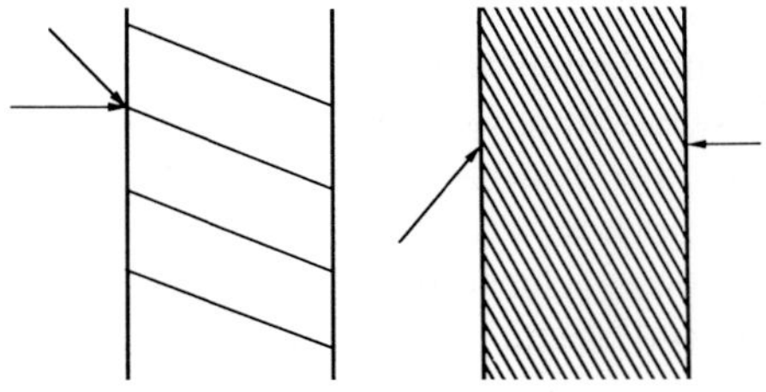


**Fig. 14** Two regimes of volume-recorded fringes: (a) two beams impinging from same side of the recording medium; (b) two beams impinging from opposite sides of the recording medium

## 6
## Volume Holography

The hologram has generally been treated as a two-dimensional structure with negligible thickness. Denisyuk [13] and van Heerden [14] departed from this view by describing holograms in which the thickness of the photographic emulsion, or other recording material, is sufficiently great to influence the diffraction properties of the hologram. When the recording material is sufficiently thick, the diffraction process is of a type known as *Bragg diffraction*.

The interference fringes formed when two light waves interfere are generally thought of as a two-dimensional pattern of light and dark lines, perhaps straight, perhaps curved. This is the way they are seen when viewed on a screen placed in the light path. However, it should be recognized that the interference typically occurs throughout the entire volume of space where the beams overlap, and the fringes are in fact three dimensional, being surfaces of brightness interspersed with surfaces of darkness. When the recording medium is thick, the recorded fringes become surfaces within the material. Figure 14 shows the two basic regimes of such recorded fringes. In the first, the two interfering beams impinge from the same side of the recording medium; in the second, from opposite sides. The fringes bisect the angle between the two beams and have a spacing that depends on the angle between the two beams. In the first regime, the fringes thus cut across the recording material and may have a spacing from very coarse to about $\lambda$, the wavelength of the illumination. In the second regime, the fringes are fine, on the order of $\lambda/2$ in spacing, and lie more nearly parallel to the emulsion surface, being exactly parallel to the surface when the beams are traveling exactly in opposite directions and normal to the surface. The criterion that distinguishes a thick recorded fringe pattern (i.e., a thick hologram) from a thin one is roughly that to be thick, the spacing between the fringes should be less than the thickness of the emulsion or other recording material. When two beams interfere with an angle of $5^\circ$ between them and are recorded by a photographic emulsion of thickness 6 μm, a typical thickness, the fringe spacing is much greater than the thickness, and the law of diffraction for thin gratings holds. At an angle of 30°, the recorded fringes will usually be sufficiently fine to produce Bragg diffraction.

Bragg diffraction can be described in various ways. A simple approach is to consider it to evolve from conventional diffraction. Figure 15 shows diffraction from a thin grating. The incident beam strikes the grating at an angle $\theta_i$ to the normal and is diffracted at an angle $\theta_r$ from the normal. The diffraction angle can be calculated from the basic consideration that the first diffracted order occurs at an angle such that the optical path from a point on the impinging wave front to a point on the diffracted wave front is one full wavelength longer (or shorter) through one slit opening than through the adjacent one, from which we readily obtain $d_1 + d_2 = \lambda$, or $\sin\theta_i + \sin\theta_r = \lambda/d$, where $d$ is the grating spacing. For the other diffracted order, the result is similar, except that one of the sine terms is negative. In either case, we require constructive interference between the light rays from the various openings. Examination of this diffraction equation shows that for a given angle of incidence, there will generally be a diffracted wave, whose angle changes with the incidence angle.

Now, suppose we let the emulsion become thick, and, for simplicity, let the fringes be normal to the emulsion surface (Fig. 16). The recorded fringes become surfaces. We consider each surface to be like a mirror that reflects some of the light incident on it, while the remainder is transmitted and undergoes partial reflection at the next surface. The diffraction process now takes on reflection properties. However, the same process that led to the previous diffraction equation still holds; the diffracted wave is still at the angle that leads to constructive interference from adjacent openings. Now, however, the openings have become surfaces that behave like mirrors, so that now the mirror law, that the angle of reflection equals the angle of incidence, must also apply. Thus we require also that $\theta_i = \theta_r$, causing the diffraction equation to become $2\sin\theta_i = \lambda/d$, which describes Bragg diffraction. According to this equation, a beam of light falling on

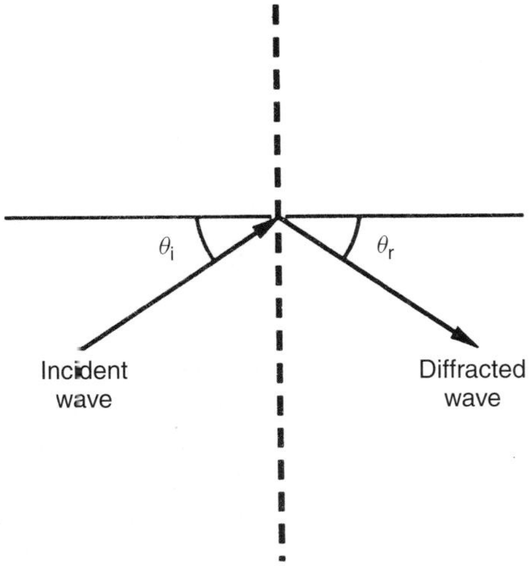


**Fig. 15** Diffraction by a grating, arbitrary angle of incidence

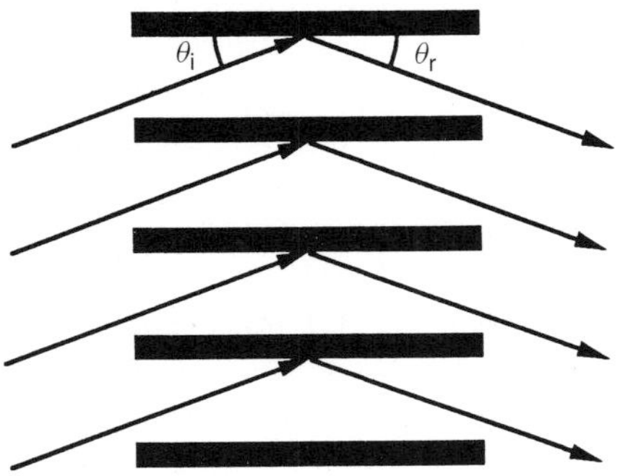


**Fig. 16** Diffraction from a thick grating (Bragg diffraction)

the diffracting structure at an angle $\theta_i$ produces a diffracted wave if the equation is satisfied, but if the angle $\theta_i$ is changed, the diffracted wave does not change direction; it simply ceases to exist since the Bragg equation will not be satisfied. Thus, a Bragg-regime hologram must be illuminated at the proper angle in order to produce a diffracted wave. This effect is readily observed with holograms illuminated with a monochromatic source such as a laser beam. At the proper angle the image is bright, but rotation of the hologram by perhaps $5^\circ$ or $10^\circ$ will result in the disappearance of the image.

Alternatively, if the hologram is illuminated by a beam of white light, then in a given direction of viewing, only a narrow band of wavelengths will satisfy the Bragg angle close enough to give diffraction. Thus, the hologram can form an acceptably sharp image when viewed with white light; the hologram simply selects a narrow band of light for forming the image, while discarding the remainder.

In order that the band $\Delta\lambda$ of diffracted light be sufficiently narrow to form an acceptably sharp image, the hologram should be made with the object and reference beams coming from opposite sides of the film. Theory shows that the spectral width of the diffracted image depends on the number of fringe surfaces that contribute to a diffracted ray, in accordance with the equation $\Delta\lambda/\lambda = 1/N$, where $N$ is the number of contributing surfaces and $\lambda$ is the midband wavelength of the diffracted light (Fig. 17). This number is greatest when the reference and object beams are introduced from opposite sides since the fringes are finer, and the incident beam encounters more partially reflecting surfaces as it propagates through the emulsion.

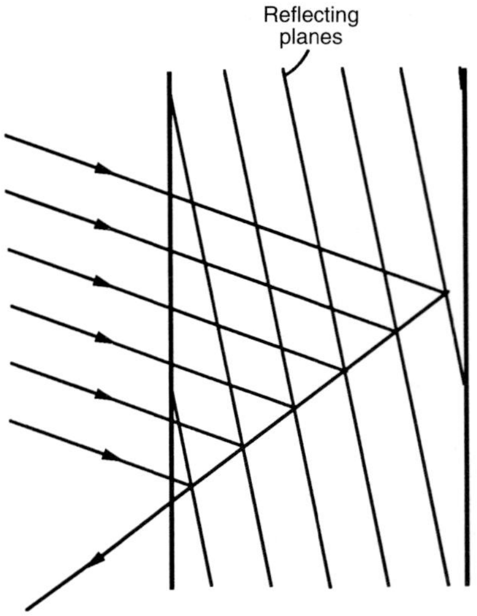


**Fig. 17** Bragg diffraction from many surfaces. The diffracted ray is formed from many incident rays

Another characteristic of the hologram made with the beams coming from opposite sides of the plate is that the image is viewed from the same side of the hologram from which the illumination comes, as seen in Fig. 15. Hence, such holograms are sometimes described as *volume reflection holograms.* More often they are known as *Denisyuk holograms.* The reflection direction of the holograms can readily be seen by noting that the fringes are almost parallel to the emulsion; hence, considered as mirrors, they reflect the light back toward the source.

The most common way to make a Denisyuk hologram is shown in Fig. 18. A beam of laser light shines through a photographic plate to illuminate an object located on the other side. Light reflected from the object then is reflected back to the plate. Thus, the same beam of light exposes the plate twice, serving as the reference beam on the first pass

**Fig. 18** Making of a Denisyuk hologram

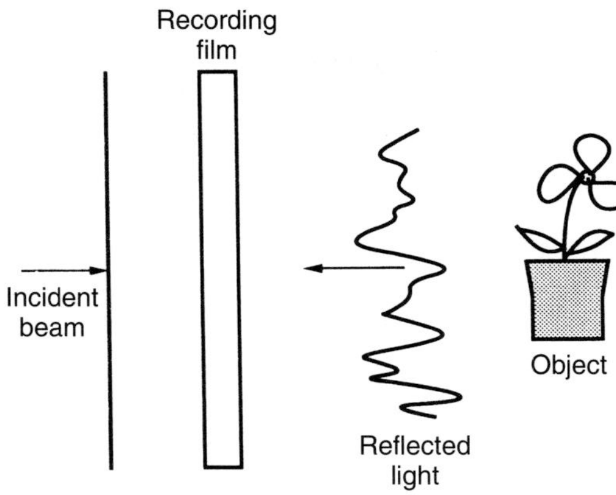


and as the object beam on the second, a configuration that is both simple and effective.

## 7
## Display Holography

### 7.1
### The Beginnings

Display holography began in 1963, when holograms of 3D, diffusely reflecting objects were first made, as opposed to simple transparencies as objects. Such holograms were the first that could be viewed without the aid of eyepieces. Images such as these had never been produced previously, and people viewed them in astonishment.

However, it took many years for display holography to develop into a viable commercial activity. There were two principal problems. First, the early holograms had to be viewed with coherent sources, such as lasers or mercury arcs. Such sources were rather expensive. Second, the holograms themselves were rather expensive, since there were no ways to mass produce them. A hologram was made either as an original or by photographically contact printing from a master. In recent years, both limitations have been overcome, with the result that display holography, in many different forms, has prospered.

### 7.2
### The Denisyuk Hologram

To be a successful display device, the hologram had to be white-light viewable. The first white-light viewable hologram, the volume reflection hologram of Denisyuk, had been described as early as 1962 [13], and high-quality Denisyuk holograms had been demonstrated by 1965. However, the early Denisyuk holograms lacked brightness, and the brightest ones were often afflicted with high levels of scattered light. The problem was one of chemistry; techniques of photographic development were required that would produce high diffraction efficiency with low noise scatter. These problems were gradually overcome over a period of many years, and today, bright, crisp, low-noise Denisyuk holograms are commonplace. A variety of recording media exist today for producing high-quality Denisyuk holograms. They can be made with very fine-grain emulsions and very special development techniques that have been perfected over many years. In addition, non–silver halide materials, such as dichromated gelatin and photopolymers, compete with silver halide and have captured a respectable percentage of the Denisyuk display hologram market. Denisyuk display holograms are generally the most expensive, since they are not easily mass produced. However, they certainly are the highest quality of display hologram. They seldom appear on covers of magazines, but rather are typically displayed in galleries and exhibitions.

The quality of the hologram is strongly dependent on the recording medium, especially for the Denisyuk holograms, since such holograms are the pinnacle of hologram perfection. The principal factors for judging hologram quality are: the brightness, or diffraction efficiency, the level of scattered light, or ambient background, and for color holograms, the degree of color saturation. Diffraction efficiency in holograms recorded on silver halide photographic plates is determined by the form of chemical processing used in the development and bleaching procedures. In general, as the diffraction efficiency is increased, the scattered light level increases, so the aim is to produce the highest diffraction efficiency in combination with the lowest light scatter. With color holography, the problem is especially critical since the blue light produces significantly higher scatter than the longer wavelengths. New photographic emulsions coming into use have silver grains smaller than the wavelength of light, and consequently provide extremely low levels of scatter, resulting in holograms of significantly higher quality that were available a decade ago. In such holograms, a portion of the image that should be black appears in these new holograms as being very black indeed.

## 7.3 The Benton Hologram

In 1969, a second kind of white-light viewable hologram, known as either the *rainbow* or *Benton hologram*, was invented [15]. The construction method has two steps (Fig. 19). First, a conventional laser transmission hologram is made. Next, a second hologram is made, using as the object the real image from the first hologram. In this process, a narrow horizontal slit is placed

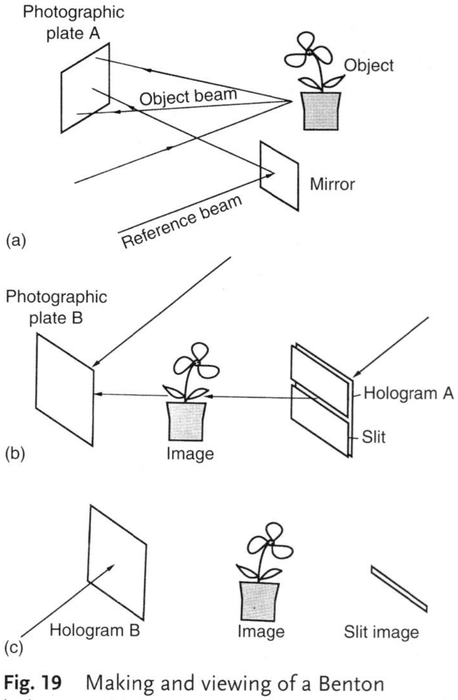


**Fig. 19** Making and viewing of a Benton hologram. (a) First step, making a conventional hologram. (b) Second step, making a second hologram from real image of the first hologram. (c) Illumination of the second hologram (with monochromatic light). An image is formed both of the object and of the slit. Illumination with polychromatic light forms a spectrum at the slit image position. An observer places his eye somewhere in the spectrum; consequently, he sees an image in monochromatic light, and the image is therefore sharp

at the first hologram, thus eliminating all parallax in the vertical plane. The second hologram, under coherent illumination, forms an image of the original object, either virtual or real, as well as a real image of the slit. To see the image, the observer must place his eye at this slit image. If he moves his head vertically, the image disappears; hence there is no vertical parallax. The image, however, can be quite bright; all the light that ordinarily would be spread over a vertical distance of perhaps 1 m at the observer's position is now confined to the slit image, and the brightness increase can be several hundredfold.

There is, however, an even more important advantage to this method: white-light viewability. If one were to illuminate the hologram with light of two different wavelengths, the lateral dispersion would result in two vertically displaced slit images, one for each color. The observer would see the entire image in one color when he positioned his eyes at one slit image and in the other color when he positioned them at the other slit image. Extrapolating this process to white-light illumination, where the slit image would be smeared into a continuous spectrum, one finds that to view the hologram the observer may position his eyes at any location on the spectrum, whereupon he will see a sharp, clear image in the color corresponding to that part of the spectrum. As he moves his head vertically, the image changes color but remains sharp and clear. If his head is not in the image plane of the slit, he will still see the entire image, but the color will change from top to bottom. Appropriately, such holograms have been called *rainbow holograms.*

In exchange for giving up vertical parallax while retaining normal horizontal parallax, one has obtained holograms that can be viewed in white light without the thick emulsion requirement. The Benton hologram thus offers a favorable tradeoff, in which the comparatively unimportant feature of vertical parallax is exchanged for the important white-light viewing capability.

## 7.4
## The Multiplex and Related Holograms

Another important development, the composite or multiplex hologram, overcomes some of the basic limitations of the more conventional holographic techniques. Here, a hologram is synthesized from a large number of ordinary photographs made in a conventional way. The pictures, made from different positions, constitute many views of the object, and they have in total all the essential information contained in a hologram. The composite method synthesizes the many pictures into a single hologram.

Composite holograms were first reported by various groups, including Pole [16], McCrickerd and George [17], and DeBettito [18]. In the next few years, many advances were made, culminating in a system perfected by Lloyd Cross and his colleagues. In the Cross method, the object is placed on a slowly rotating platform and a motion picture camera makes hundreds of pictures, just as in conventional cinematography. The pictures could then be made into a single hologram as follows. The first picture could be imaged onto a screen consisting of a piece of frosted glass. The light transmitted by the screen would be recorded as a hologram in the form of a long thin strip about 1 mm wide and perhaps 20 cm long, the purpose of the frosted glass being to direct light from all parts of the image to a recording slit that controls the width of this hologram. The actual system differs from this in that, instead of the diffuser, a large cylindrical lens at the image plane is used for directing the light to the slit; this method uses the light more efficiently and reduces laser speckle. After one frame is recorded as a hologram, the next frame is similarly projected, the recording film is moved one slit width, and a second hologram is made in an adjacent recording position, and so on.

The resulting hologram, the collection of all of the individual slit holograms, is then wrapped into an arc corresponding to the angle of rotation of the object and is illuminated as a conventional hologram. Each component hologram projects a virtual image without depth or parallax, just like the image from which the hologram was made. When an observer views the filmstrip, however, each eye looks at an image point through different holograms. The image formed on the retina of each eye is slightly different, as in a stereoscopic viewer, and consequently a three-dimensional image is perceived. As the observer moves his head laterally, different images are seen, thereby producing parallax. The Cross technique can be combined with the Benton technique, thus allowing white-light viewing. Essentially nothing is given up in so doing, since the composite holograms lack vertical parallax anyway.

In a variant of this method, the recording camera is translated laterally instead of the object rotating, resulting in what is called a *flat integral,* since the hologram is no longer curved into an arc for viewing. This flat integral can be used as a master for making a second hologram, where the image can be brought into the plane of the hologram, thus relaxing the viewing light source size requirements and giving a sharper image.

## 7.5 Construction of Commercial Display Holograms

Commercial holograms, whether for display on magazine covers, on credit cards, or in galleries, should be white light viewable. Various standard techniques have been developed for the production of commercial display holograms, all based on either the Denisyuk or Benton white-light viewing methods.

Typically, a master hologram is made as a conventional laser transmission hologram. The conjugate image may then be used as an object to produce a volume reflection hologram. The resulting hologram has an image that straddles the hologram, being partly in front of it and partly behind it, thus giving a sharper image than if the image were entirely on one side of the hologram. One of the limitations of the white-light-viewable hologram is that the image sharpness decreases with the distance of the image from the plane of the hologram.

Portrait holography can be done using this method, in which case the master hologram will have to be made with a short-pulsed laser so as to freeze object motions, since a living person cannot ever be stationary enough to be holographed, unless the exposure time is much less than 1 ms.

For inexpensive, mass-produced holograms, such as those found on magazine covers, and so on, the Benton technique is used, since Benton holograms do not require any volume effects for white-light readout and, consequently, can be readily produced by embossing methods, which are inherently surface phenomena.

To mass produce a Benton hologram, the original laser-produced master hologram is used to form a real image, which is then used as the object for making a second hologram of the Benton type. This hologram is recorded on photoresist, which when developed forms a surface relief pattern. This hologram is placed in a plating bath, where nickel is plated onto the hologram surface; when the photoresist is removed, a surface relief hologram made of nickel remains. This can be used in an embossing process, where the hologram is mounted in a press and used to stamp the surface relief pattern onto a sheet of heated plastic. Such holograms can be produced at a fast rate, thousands per hour, and the total production run can be in hundreds of thousands or even millions. These holograms can be bonded to paper by hot-stamping methods. This mass production method has made display holography a significant business.

## 8 Holography Today

Holography today is a relatively small but thriving and expanding business. Display holography is probably the leading commercial application area. Artists are employing holography as a new medium, and their works are displayed in museums and galleries. Increasingly, color holograms, as opposed to monochrome, are being seen.

The largest commercial activity is the mass production of Benton-type holograms in magazines, in advertising, for Christmas cards, for baseball cards, and so on, and as security devices on valuable objects such as credit cards and money.

Holographic optical elements are of ever-increasing importance in such applications as diffraction gratings and beam splitters. Head-up displays on fighter aircraft commonly use holograms. Holograms play a significant role in the development of all-optical or hybrid-optical electronic digital computers, an area of current intense research. Holograms have potential as memory elements in such computers, and as interconnections to connect various storage and processing elements of the computer. Other factors being equal, the more interconnections, the more powerful the computer, and holograms offer the possibility of an enormous number of interconnects in a very small volume.

Hologram interferometry continues as a solid area, with use in scientific and engineering applications. It shares its application areas with other new, laser-based types of interferometry, such as speckle interferometry and shearography.

Various other minor applications of holography exist that do not fall into these categories, but have the potential for becoming major applications. The outlook for further holographic application development is bright.

## Glossary

**Amplitude:** For a light wave, the magnitude of the electric field, or equivalently, the square root of the irradiance (or intensity).

**Bragg Diffraction:** Diffraction from a volume, or thick, grating, characterized by a diffracted beam being produced only for certain incidence angles of the illuminating beam.

**Coherence Length:** The path difference allowable between two interfering beams before interference no longer occurs.

**Coherent Light:** Light that is monochromatic and derived from a point source, or equivalently, which can be focused to a point. Laser light is highly coherent; nonlaser light (e.g., Hg arc source light) can be made moderately coherent by passing it through a filter and a pinhole.

**Composite Hologram:** A hologram formed from a number of individual subholograms, each formed from a separate image.

**Convolution:** The convolution of two functions $g(x, y)$ and $h(x, y)$ is the mathematical operation

$$g(x, y) * h(x, y) = \iint_{-\infty}^{\infty} g(\alpha, \beta) \times h(x - \alpha, y - \beta)\, d\alpha\, d\beta$$

When an object is imaged through a spatial filter, the resulting image is the convolution of the object with the point spread function $h$ of the filter, where the point spread function is the image produced by a point object.

**Cross Correlation:** Somewhat similar to a convolution, the cross correlation of two functions $g(x, y)$ and $h(x, y)$ is the mathematical operation.

$$\iint_{-\infty}^{\infty} g(\alpha, \beta) h^*(\alpha - x, \beta - y)\, d\alpha\, d\beta.$$

**Diffraction:** The tendency of light to spread out when passing through narrow apertures, or to bend when passing near an obstruction, such as a knife edge.

**Fourier Transform:** Mathematically, the Fourier transform of the spatial function $g(x, y)$ is defined as the integral

$$G(f_x, f_y) = \iint_{-\infty}^{\infty} g(x, y) \times \exp[-i2\pi(f_x x + f_y y)\, dx\, dy,$$

where $f_x, f_y$ are new variables resulting from the transformation. More physically, a decomposition of the signal $g(x, y)$ into its spatial frequency components.

**Frequency Plane:** A plane in a coherently illuminated optical system where the light distribution is Fourier transform–related to the object light distribution.

**Hologram Interferometry:** A form of holography in which multiple images (two or more) of the same object are stored on the same recording medium, and in which, on readout, interference effects occur among the various recorded images.

**Interference:** The tendency of two or more light beams in superposition to combine in a reinforcing way at some positions to form a resultant intensity greater than the sum of the two intensities individually, and to produce cancellation effects at other positions.

**Matched Filter:** An important type of filter, used for detecting the presence of signals immersed in noise, clutter, or competing signals.

**Multiplex Holography:** A particular type of composite hologram, distinguished primarily by its cylindrical shape.

**Phase:** For a signal $a \cos(2\pi f_0 \tau + \phi)$, $a$ is the amplitude, $\phi$ is the phase, and $f_0$ is the carrier frequency. The phase may be roughly described physically as the shape of the wavefront.

**Phase Conjugation:** The operation of changing a complex signal $u = a(x, y) \times \exp[i\phi(x, y)]$ into its complex conjugate $a(x, y)\exp[-i\phi(x, y)]$.

**Spatial Filter:** A mask placed in a plane in an optical imaging system where the light distribution is Fourier transform–related to the object distribution. The mask modifies the spatial frequency content of the image by attenuating or phase shifting the various spatial frequency components of the object.

**Transfer Function:** The transfer function of a system describes the way various frequency components of a signal pass through the system, that is, how much attenuation and phase shift the frequency components undergo. The system may be quite complicated, being formed from many elements, or it may be, in the simplest case, simply a filter.

## References


[1] Gabor, D. (**1948**), *Nature* **161**, 777–778.

[2] Gabor, D. (**1949**), *Proc. R. Soc.* **A197**, 454–487.

[3] Gabor, D. (**1951**), *Proc. Phys. Soc.* **B64**, 449–469.

[4] Leith, E., Upatnieks, J. (**1962**), *J. Opt. Soc. Am.* **52**, 1123–1130.

[5] Leith, E., Upatnieks, J. (**1964**), *J. Opt. Soc. Am.* **54**, 1295–1301.

[6] Leith, E., Upatnieks, J. (**1965**), *SPIE J.* **4**, 3–6.

[7] Kogelnik, H. (**1965**), *Bell Syst. Tech. J.* **44**, 2451–1455.

[8] VanderLugt, A. (**1964**), *IEEE Trans. Inform. Theory* **IT-10**, 139–145.

[9] Kozma, A., Kelly, D. (**1964**), *J. Opt. Soc. Am.* **54**, 1395.

[10] Powell, R., Stetson, K. (**1965**), *J. Opt. Soc. Am.* **55**, 1593–1597.

[11] Morris, G. M., George, N. (**1980**), *Opt. Lett.* **5**, 446–448.

[12] Veldkamp, J., Swanson, G. J. (**1983**), *Proc. SPIE* **437**, 54–59.

[13] Denisyuk, Y. (**1962**), *Sov. Phys. Dokl.* **7**, 543–545.

[14] van Heerden, P. J. (**1963**), *Appl. Opt.* **2**, 387–392.

[15] Benton, S. (**1969**), *J. Opt. Soc. Am.* **59**, 1545.

[16] Pole, R. V. (**1972**), *Appl. Phys. Lett.* **10**, 20–22.

[17] McCrickerd, J. T., George, N. (**1968**), *Appl. Phys. Lett.* **2**, 10–12.

[18] De Bitteto, D. J. (**1969**), *Appl. Opt.* **8**, 1740–1741.


## Further Reading


Abramson, N. (1981), *The Making and Evaluation of Holograms*. New York: Academic Press. Primarily deals with holographic interferometry, with emphasis on the author's own work. Full of very clever techniques.

Beiser, L. (1984), *Holographic Scanning*. New York: Wiley. Deals with a specialized application of holography.

Bjelkhagen, H. (1995), *Silver Halide Recording Materials for Holography and Its Processing*. Berlin: Springer.

Caulfield, H. J. (Ed.) (1979), *Handbook of Optical Holography*. New York: Academic Press. 32 well-known scientists have contributed to this rather thorough coverage of holography.

Collier, R. J., Burckhardt, C. B., Lin, L. H. (1971), *Optical Holography*. New York: Academic Press. A classic. Now somewhat dated, but still widely cited. For years the definitive text on holography.

Denisyuk, Yu. N. (1978), *Foundations of Holography*. St. Petersburg, Leningrad: State Optical, Order of Lenin, Order of October Revolution, S. I. Vavilov Institute. An enjoyable collection of lectures by the Soviet Union's greatest holographer.

Francon, M. (1974), *Holography*. New York: Academic Press (translated from French). A well-done basic book by a very famous French optical scientist.

Hariharan, P. (2002), *Basics of Holography*. Cambridge, UK: Cambridge University Press. A broad coverage of optical holography since its origin.

Iovine, J. (1997), *Holography for Photographers.* Boston: Focal Press.

Saxby, G. (1994), *Practical Holography*. New York: Prentice Hall.

Tonomura, A. (1999), *Electron Holography*. Berlin: Springer.

Vest, C. M. (1979), *Holographic Interferometry.* New York: Wiley. Thoroughly covers holographic interferometry.

Volkl, E. (1999), *Introduction to Electron Holography*. New York: Kluwer Academic/Plenum Publishers.

# Image Processing of Optical Data

*Maria Petrou*
*School of Electronics and Physical Sciences, University of Surrey,*
*Guildford GU2 7XE, UK*
*Phone: +44-1483-689801; Fax: +44-1483-686031; e-mail: m.petrou@surrey.ac.uk*

**Abstract**
Image processing is concerned with the processing of digital images for the purpose of image enhancement, image restoration, image compression, and feature salienation so that the processed image can be viewed by a higher-level computer vision module. These objectives are achieved by using linear and nonlinear methods. From the mathematical point of view, a digital image is a two-dimensional array of numbers; so the linear methods used are based on linear algebra. The nonlinear methods used are various types of thresholding as well as methods based on statistics and probability.

**Keywords**
image enhancement; image restoration; image compression; low-level computer vision; edge detection; image segmentation.

## 1
## Introduction

Image processing was born in the early 1920s with the introduction of the Bartlane cable picture transmission system that was used to transmit pictures for the newspapers between London and New York via the transatlantic cable [1]. The images had to be coded for the transmission and decoded and reproduced at the receiving end. Thus, *image coding* was the first topic of research under the image-processing umbrella. The other topic was the reproduction of the image so that it was visually pleasing. This was done by what we call nowadays *halftoning*: the brightness of the image is simulated using tiny different printing patterns for different levels of brightness. This problem, however, is more into the realm of image reproduction and printing, rather than image processing.

It was not before the 1960s when image processing was called to deal with other problems too: the images received from the various space probes did not look very good as they had suffered a lot of distortion during transmission. Image processing was used to improve the quality of the images. Images were enhanced and restored. *Image enhancement* means the processing attempts to improve the visual quality of the image in a **subjective** way, while *image restoration* implies the careful modelling of the degradation process and the application of techniques that try to reverse it, using **objective** criteria.

It was also in the same decade that the first computer vision systems were presented [2]: a computer vision system tries to interpret what is shown in an image. The first computer vision systems published assumed that an image had been reduced somehow into a line drawing from which meaningful objects could be identified. It was soon realized that to produce from an image a line drawing that can be easily interpreted is a major step. In fact, it is a step that bothers researchers even now. This was the start of the fourth major objective of image processing: the preparation of an image for subsequent automatic interpretation. Such a preparation almost invariably involves reduction of the information contained in an image: The contents of the image are simplified to their bare bones so that they can be easily understood and interpreted automatically. This process comes in two flavors: one either uses image processing to create a "line drawing" from the image, known as *edge map* in modern terminology, or one uses image processing to divide the image into uniform regions, a process known as *image segmentation*. Either way, image processing is used to reduce the irrelevant information contained in the image and salienate the relevant information. In the following sections we shall discuss the basic approaches to image processing, after we define what an image is.

## 2 Creation and Representation of Digital Images

An image is a two-dimensional recording of a real object or a scene. When the recording has been done with the help of optical light, we have an optical image. Alternative ways of producing an image are to use, for example, X rays, or magnetic resonance tomographs. Most people are familiar with the images they produce themselves using a photographic camera. A camera records a scene on a film. The degree of blackening of the film is proportional to the brightness of the depicted object. Development of the film involves converting the degree of blackness into brightness and results in the photographs we are familiar with. Such an image is an *analogue image*: the blackness/brightness of each point is proportional ("analogous") to the brightness of the depicted object. A *digital image* is created in a different way: the digital camera consists of many individual sensors. Each sensor records the amount of light it receives. Each sensor registers a number, the magnitude of which depends on the amount of light received by the sensor. However, these numbers are restricted to take one of relatively few possible values. For example, they may only be allowed to take value 0 or 1. In this case, we say we have a *binary* digital image. It is a common practice to allow 256 different values, represented by numbers 0, 1, 2, ..., 255. In this case, we say we have an 8-bit *grey image*. This terminology comes from the fact that any integer number between 0 and 255 can be represented by 8 bits in the memory of a computer (a bit is a digit that is either 0 or 1. There are 256 different combinations of eight 0s or 1s written next to each other). These numbers are called *pixels* from the words *picture elements*. So, a digital image consists of lots of individual numbers arranged in a rectangular grid. To view a digital image, we produce for each number a small square, the brightness of which depends on the magnitude of the number. If we print all those little squares next to each other, we have a print out of a digital image. If we have used too few sensors to record a scene, we are able to see the individual pixels that make up the image. We say then that we have *pixelation* or the *checkerboard effect*. It is possible that we did not allow the pixels to have a large range of possible values. In that case, the digital image may show some artifacts called *false contouring*.

Analogue images created by the chemical process of photography can only be improved by chemical means. The big advantage of digital images is that they are made up of individual numbers. We can therefore change the appearance of an image by simply manipulating the numbers that make it up.

A set of numbers written in an arrangement of a rectangular grid constitutes a *matrix* in the language of mathematics. So, a digital image for the mathematician is nothing more than a matrix. Mathematicians have at their disposal an arsenal of tools with which they can manipulate matrices: they come under the general title of *Linear Algebra*. So, a lot of Image processing is performed by applying linear algebra to the matrices that we call images. However, linear algebra has its limitations, and a lot of other nonlinear techniques are used.

## 3
## Application of Linear Algebra to Digital Image Processing

The basis of linear approaches in mathematics is that a complicated structure can be expressed as a sum of some simpler structures. For example, a complicated image can be represented as the sum of some trivial images. Figure 1(a) shows this pictorially. In Fig. 1(b), it is shown how a $2 \times 2$ image can be written in terms of some very simple images that consist of only black and white pixels. In Fig. 1(c), the brightness (or grey) values of each of the four pixels are replaced by numbers. The reader can verify that if they multiply each elementary image with the coefficient given in front of it and add them pixel by pixel, they will get the image on the left. Although this example concerns a trivial image consisting of only four pixels, it is directly applicable to all other sizes of images. The four trivial images on the right of the pictorial equation are appropriate for representing **any** image of size $2 \times 2$. Even though such a small image consists of only four pixels, if each pixel can take one of 256 different grey values, there are as many as $256^4$ such images! The remarkable thing is that any one of them can be represented in terms of the

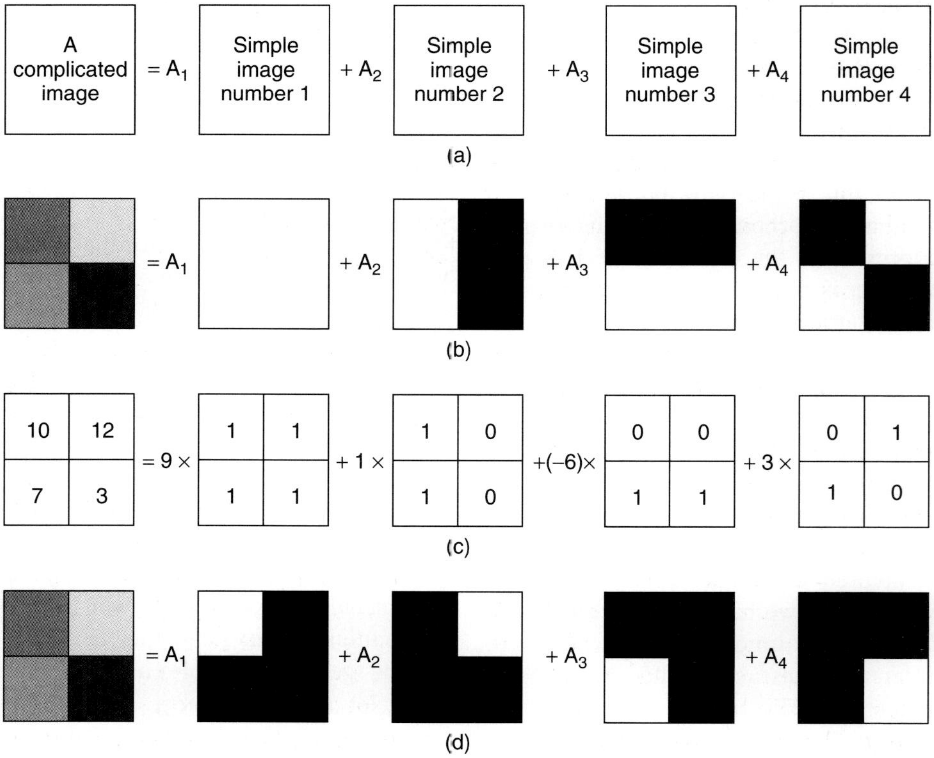


**Fig. 1** Linear image processing expands an image into the weighted sum of some elementary images. Some possible expansions of a trivial image consisting of four pixels only are shown here

same four elementary images with the appropriate choice of the four multiplying factors $A_1, A_2, A_3, A_4$. This is because the elementary images were chosen so that they form a so called *complete basis*. Writing an image as a weighted sum of the basis images is called *an expansion*. The numbers with which we multiply each basis image are the *coefficients of the expansion*. One may argue that by doing this expansion we have not gained anything: the original image was represented by four numbers (the values of its pixels), and after this expansion, we still need four numbers (the coefficients of the expansion) to represent the image. This is true, but although we have not reduced the number of numbers we use to represent the image, we have gained other benefits: the remarkable property of linear approaches is that we can manipulate the elementary basis images very easily. Then we multiply the results of this manipulation with the *same* coefficients of the expansion $A_1, A_2, A_3, A_4$ and the result we obtain will be the same as if we had manipulated the complicated original image directly.

For example, we may use as a basis images, each of which consists of a single fully bright pixel with all other pixels being black (0 brightness). It is very easy to manipulate an image that has only one pixel with nonzero value. Then by manipulating such simple images and adding the results (multiplied with the corresponding coefficients), we obtain the result we would have obtained if we were able to manipulate directly the complex image. Figure 1(d) shows such an expansion for the trivial $2 \times 2$ image.

From the above discussion, various things become apparent and they are discussed in the sections below.

### 3.1
### Linear Operators and Point Spread Functions

The trivial basis, where each basic image is black except a single pixel that is white, is used to define the *linear operators* with which we perform linear image processing. The result of applying a linear operator to an image with only one fully bright pixel is called *the point spread function* of the operator. Knowing it, we can work out the effect of the operator on **any** image.

### 3.2
### Different Bases of Image Expansion

There are many different sets of basic images in terms of which one can expand an image. Two of them are shown in Figs. 1(b, d) for the case of a $2 \times 2$ image. The most commonly used sets of basic images come from the Discrete Fourier Transform (DFT), the Haar functions, the Walsh functions and the Haddamard matrices. Other basic images also exist, for example, those defined in terms of the Discrete Cosine transform (used in JPEG coding), the Discrete Sine transform, and so on. Each such basis of images is good for representing **any** image in the world as long as it is of the particular size the basis has been designed for.

The trivial basis for expansion, consisting of elementary images with a single bright pixel each (e.g., Fig. 1(d)), is only good for defining the linear operators. It is not used for anything else because all its member basic images are equally important. The other bases, however, may be very useful for compressing images. For example, in Fig. 1(b), the first elementary image gives us some information concerning the full image. The second elementary image tells us something about the contrast between its left and right halves. The third tells us something about the image contrast between its top and bottom halves. The fourth, measures contrast in the four quadrants of the image. If we were dealing with even bigger images, we would have observed that the basis images gradually give information about smaller and smaller sections of the original image. In other words, the higher the order of the basic image, the more detailed the information it conveys. So, if we decide that we do not wish to represent an image with so much detail, we may truncate the coefficients of the expansion beyond a certain level and approximate the image with the first few basis images. In the trivial example shown in Fig. 1(c), the idea would be to set $A_4 = 0$ and reconstruct the original image by using only the first three coefficients. Then instead of needing four numbers to represent the image, we need only three (namely, $A_1$, $A_2$, and $A_3$). This is called *lossy image compression*. It is called *lossy* because we have lost information during compression, that is, the reconstructed image has fewer details than the original image. In general, it is not possible to have *lossless image compression* by using image bases that are universal, that is, good for all images of the same size.

## 3.3 Linear Image Compression

It is possible for one to construct sets of basis images, which are good for a particular image only, and not for any other image. Such a process analyses the contents of an image and writes it as the sum of some simpler images. The process is called *Singular Value Decomposition* or SVD for short. If we represent by $G$ the matrix which is our image, and by $G^T$ its transpose, we may construct matrix $GG^T$ by simple matrix multiplication. Let us denote by $\lambda_i$ and $u_i$ the eigenvalues and the eigenvectors of $GG^T$ respectively. We also construct some vectors $v_i$ by using $v_i = G^T u_i$. If we multiply column vector $u_i$ with row vector $v_i^T$, we obtain a matrix. This matrix is an elementary image of the original image $G$. It can be shown that

$$G = \sum_{i=1}^{r} \sqrt{\lambda_i} u_i v_i^T \qquad (1)$$

where $r$ is the number of nonzero eigenvalues of matrix $GG^T$. In general, $r$ is much smaller than the number of elements of $G$, so the image now is represented by fewer numbers than before. Thus, lossless compression has been achieved. We can compute the number of bits needed to represent an image of size $N \times N$ as follows: We can forget about scalars $\lambda_i$ by multiplying each vector $u_i$ with the corresponding $\sqrt{\lambda_i}$. Each vector $\sqrt{\lambda_i} u_i$ requires $N$ real numbers to be represented. The same for each vector $v_i$. If each real number requires 32 bits, the image needs $r \times 2 \times N \times 32$ bits to be represented. Alternatively, the image in its original form may require $N \times N \times 8$ bits to be represented if the image were an 8-bit grey image. We have compressed the image if $r \times 64 \times N < N \times N \times 8$, or if $8r < N$.

One may also achieve lossy compression by using only the first $k$ largest eigenvalues $\lambda_i$ and setting the rest to zero. It can be shown that in this case, the approximation of the image is in the least squares error sense.

An alternative approach to constructing a basis of elementary images is to try to construct a basis that is good for all images in a set that share the same statistical properties. For example, one may try to construct a basis of images that will be good in representing all frontal face portraits. This can be achieved by applying the so-called *Karhunen Loeve* transform, in which the statistical properties of the set of images are inferred from the statistical properties of a single available image. Such a basis can be used to approximate an image in a lossy way with accuracy that minimizes the least *mean* square error over the whole collection of images.

### 3.4 Tasks Accomplished by Linear Image Processing

Apart from lossy and lossless image compression, there are other tasks that can be achieved using linear methods. For example, one can smooth an image by removing fluctuations in the values of neighboring pixels caused by noise. One may also use linear methods to enhance the regions in an image where boundaries between different objects are depicted. The output of such an operator is the so-called *gradient image* as it depicts with brighter tones the pixels with the higher local gradient magnitude values. If the levels of noise are not high, one may sharpen an image to salienate minute details. A very difficult problem for which a solution may be obtained by linear image processing is the unblurring of a blurred image. This is usually achieved by estimating directly or indirectly the point spread function of the blurring process and by working out the point spread function of a linear operator that, if applied to the blurred image, will reverse the effect of blurring. Of course, this solution is based on the assumption that the blurring was linear in the first place.

Examples of all the above processes are shown in Fig. 2.

## 4 Nonlinear Image Processing

The linear approach to image processing is very powerful, but it has its limitations. Linear approaches are usually very popular because they are easily understood, easily implemented, and they have universal character. Nonlinear approaches tend to be isolated solutions invented to solve a particular type of problem. Nevertheless, there are some nonlinear image processing techniques that are pretty universal in character and very popular. Some of them are listed below and examples of their application are shown in Fig. 3.

### 4.1 Histogram Equalization

This is the most commonly used technique to enhance an image. It increases the contrast of an image by making the brightest pixels even brighter and the darkest pixels even darker, without altering the relative brightness of the pixels.

### 4.2 Thresholding

This is a simple operation that is used to divide the image into two regions, by using a decision criterion that looks like:

If pixel value $>$ threshold. Then, give pixel value A

If pixel value $\leq$ threshold Then, give pixel value B

**Fig. 2** Examples of linear image processing: Top: Image smoothing by reducing the noise. Second row: Salienation of the parts of the image where the brightness changes most rapidly. Third row: Image sharpening by exaggerating small details. Bottom row: Image restoration by removing the blurring caused by movement during image capture

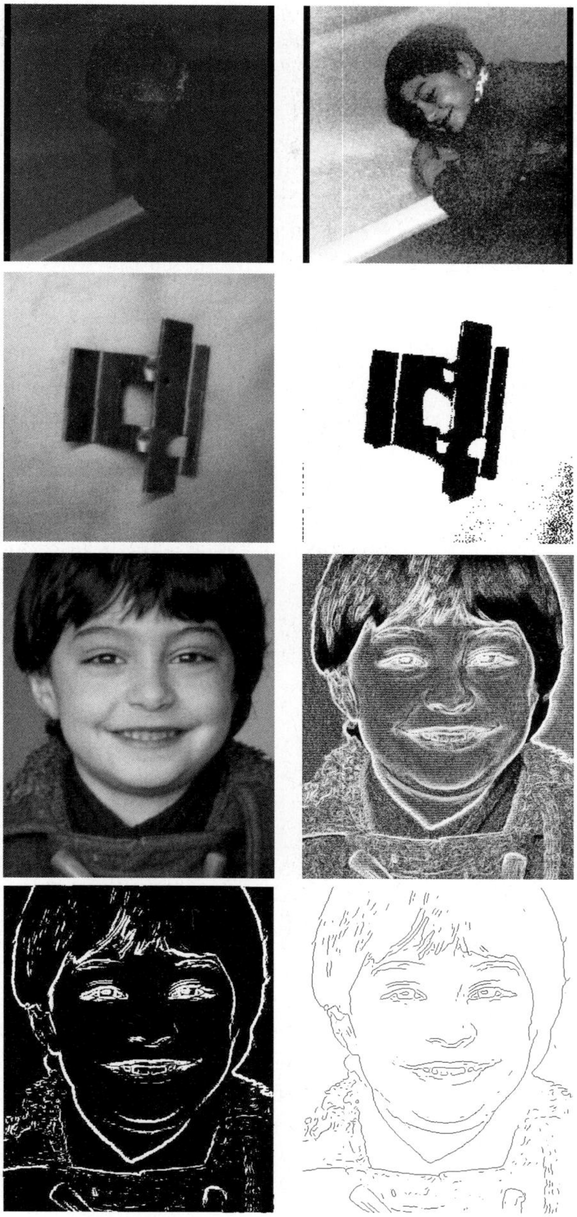

**Fig. 3** Examples of nonlinear image processing: Top: A low contrast image enhanced by histogram equalization. Second row: An image segmented into object and background by thresholding. Third row: An image and its gradient image Bottom row: The gradient image thresholded. Next to it the edge map produced by applying to the gradient image nonmaxima suppression followed by thresholding

This procedure may be applied to raw data, in order to separate an image region from another, or to an already processed image, in order to identify and flag some image features. In Fig. 3, an example is shown in which thresholding has been applied to the gradient image to isolate the pixels with the highest gradient value and produce an edge map. The result is not very good. However, a much better result can be produced if before applying the threshold, we keep only the pixels, the gradient value of which constitutes a local maximum. This process is called *nonmaxima suppression* and allows one to construct edge maps consisting of clean lines. The result is also negated, so that the edge lines are shown in black and the background in white.

### 4.3 Probabilistic Approaches

A large group of image-processing techniques have been inspired by the probability theory. These approaches attempt to decide the value that will be assigned to a pixel on the basis of probabilistic considerations. If the value they try to decide for each pixel is the "correct" grey value (because the available value of the pixel is assumed to have been corrupted), the problem they try to solve is image restoration. If the value they try to give to a pixel is just a label (i.e., an identity that puts it in a certain category), then the problem they try to solve is image segmentation: they try to divide the pixels into different categories according to some criteria. When they try to do that by saying: "Let us choose the most probable configuration of pixel values, given the image and any prior knowledge we have concerning the image", then the method is called *Maximum A Posteriori* estimate or MAP for short. In this approach, the **combination** of values assigned to all pixels is considered as a whole. Alternatively, some methods first consider each pixel **individually** and assign to it the most probable value, given its original grey value and the grey values of the other pixels, as well as any prior knowledge available on the image. Such an approach may result in global inconsistencies, as it treats each pixel as an individual and it does not consider how it interacts with and interdepends on the other pixels in the image. That is why, in a second step, such methods consider how the values they have assigned to the other pixels interact with the values assigned to the pixel under consideration, and correct their assignment accordingly. In other words, they take into consideration context. Such an approach is called *probabilistic relaxation* [3] or more generally, *constraint propagation*. Almost all probabilistic methods are iterative, that is, they start from a bad solution and gradually improve it by applying several iterations.

## 5 Color Image Processing

The bulk of image processing research has been developed for grey images. It is only in the recent years that color image processing has been developed. The main reason for this delay was the cost of acquiring and reproducing color images. A color digital image is usually represented by several matrices, not just one. Each of these matrices represents how bright each pixel is in a particular color. For example, most color sensors record images in three colors: Red, green, and blue. Then a color image can be represented by three matrices, one for each color. Each one of these matrices is called a *color band* and it can be treated like a grey image of its own right. Color in this context has to be understood as follows: the digital camera has three sets of sensors. One set is sensitive to those wavelengths of the electromagnetic spectrum that, when they reach the human retina, create the sensation of red; the other set is sensitive to those wavelengths that, when they reach the human retina, create the sensation of green, and so on. If a camera is made to produce grey images, most probably it contains sensors that are sensitive to all optical wavelengths, and so it is called *panchromatic*. Each camera may include sensors that have different sensitivities from another camera, and they certainly have different sensitivities from the human eye. So, one has to be very careful when one uses color images. There are important issues currently being researched, such as *color constancy* (the ability of the human vision system to recognize a color under different illuminations), color segmentation of images [4], color image enhancement, color edge detection, and so on. It is easy to think that one can treat each color band as a separate grey image and then combine the three results to find the output for the color image as a whole. However, some of the most commonly used methods as discussed earlier are nonlinear and therefore it is not clear how one can combine partial outputs referring to individual bands in order to produce the global output. Some biologically inspired methods have been proposed for this task [4, 5].

## 6 Concluding Remarks

Image processing deals with the process of digital images for the purpose of image improvement, image compression, and image preparation for computer vision. Image processing nowadays has a very broad spectrum of applications, in medicine, remote sensing, robotics, industrial inspection, communications, the entertainment industry, and so on. For example, image processing may be used to enhance the quality and resolution of images captured by microscopes, or by sensors onboard satellites, and it may be used to inspect products automatically as they come out of a production line. Image processing is also used in the postproduction development of films by the movie industry and in the restoration and preservation of digital archival material. It may even be used for the automatic reading of old manuscripts. For example, optical character recognition (OCR) systems are already very widespread, and they are the result of the image processing technology. Image compression finds enormous applications in teleconferencing, mobile telephony and high-definition television. It is a very active field of research, which attracts the attention of many engineers, mathematicians, and computer scientists, and it is expected to be even more widely used in the near future as high-technology approaches become more widespread, computers more accessible and much faster. There are dozens of scientific journals devoted to image processing, and dozens of books published every year. It is not possible to give a full list of the titles, but some representative texts are quoted under Further Reading.

## Glossary

**Binary Image:** An image the pixels of which take value 0 (black) or 1 (white).

**Checkerboard Effect:** The effect of being able to see the individual pixels that make up a digital image.

**Color Band:** A component of a color image, representing the brightness of the image in a particular range of wavelengths.

**Color Constancy:** The ability to recognize a color even when it is seen under different illumination conditions.

**Edge Map:** A digital representation of an image in which the places of local change in contrast are marked as thin lines.

**False Contouring:** The effect of splitting a uniform image region with gradual brightness into more than one region by forcing the image pixels to take too few grey values, that is, quantizing the brightness of an image into too few discrete levels.

**Gradient Image:** An image in which the value of each pixel is proportional to the magnitude of the local gradient of the image (i.e., the local rate of change of the image along two orthogonal directions).

**Halftoning:** A method of printing or displaying grey images by making them up using elementary binary patterns consisting of different proportions of black parts.

**Image Compression:** Representing an image with fewer bits than the original image.

**Image Enhancement:** Improving an image so that it looks subjectively better.

**Image Restoration:** Improving an image by reversing the degradation process.

**Image Segmentation:** Dividing an image into uniform regions and giving all pixels of the same region the same value.

**Image Sharpening:** Enhancing small details of an image, that is, strengthening small local image fluctuations so that they become more visible.

**Image Smoothing:** Removing the noise from an image by making neighboring pixels have similar values, that is, removing image fluctuations.

**Line Drawing:** An image representation using lines to represent the boundaries of the depicted objects.

**Linear Operator:** A process applied to an image that produces the same result; either it is applied directly to the image or to its additive components and the results are subsequently added up.

**Lossless Image Compression:** Image compression when the original image can be fully reproduced with no information being lost.

**Lossy Image Compression:** Image compression when information contained in the original image is lost.

**Nonmaxima Suppression:** Converting into background pixels all pixels in the gradient image whose values are not local maxima when compared with their neighboring pixels.

**Operator:** A mathematical process we apply to an image.

**Panchromatic Image:** An image created with a sensor that is sensitive to all optical wavelengths.

**Pixel:** A picture element. A digital image is made up of pixels, each of which carries a brightness value.

**Pixelation:** The effect of being able to see the individual pixels that make up a digital image.

**Point Spread Function:** The effect of an operator on a point light source, that is, to an image that consists of a single fully bright pixel (value 1) surrounded by black pixels (value 0).

## References

[1] Gonzalez, C., Wintz, P. (1987), *Digital Image Processing*. Reading, MA: Addison-Wesley.
[2] Roberts, L. G. (1965), in J. T. Tippet (Ed.), *Optical and Electro-Optical Information Processing*. Machine perception of three dimensional solids, Cambridge, MA: MIT Press.
[3] Christmas, W. J., Kittler, J., Petrou, M. (**1995**), *IEEE Trans. Pattern Anal. Machine Intell.* **PAMI-17**, 749–764.
[4] Mirmehdi, M., Petrou, M. (**2000**), *IEEE Trans. Pattern Anal. Machine Intell.* **PAMI-22**, 142–159.
[5] Boukouvalas, C., Petrou, M. (**2000**), *Machine Vis. Appl.* **12**, 129–136.

## Further Reading

Petrou, M., Bosdogianni, P. (2000), *Image Processing: The Fundamentals*. London: John Wiley.
Russ, J. C. (1999), *The Image Processing Handbook*. USA, Germany: CRC Press, Springer, IEEE Press.
Mitra, S. K., Sicuranza, G. L. (2001), *Nonlinear Image Processing*. San Diego: Academic Press.
Baxes, G. A. (1994), *Digital Image Processing*. New York: John Wiley.
Jain, R., Kasturi, R., Schunck, B. G. (1995), *Machine Vision*. Singapore: McGraw-Hill.
Parker, J. R. (1997), *Algorithms for Image Processing and Computer Vision*. New York: John Wiley.
Bhaskaran, V., Konstantinides, K. (1997), *Image and Video Compression Standards*. USA: Kluwer Academic Publishers.

# Imaging Detectors

## Charge-coupled Devices and CMOS Active Pixel Sensors

*Morley M. Blouke*
*Scientific Imaging Technology Inc., Tigard, OR 97223, USA*
*Phone: 503-431-7164; Fax: 503-431-7110; e-mail: mmblouke@aol.com*

**Abstract**
Charge-coupled devices (CCDs) were first introduced in 1970 with the demonstration of an eight-bit linear shift register. The concept of charge coupling was novel in the integrated circuit arena in that information was represented by discrete charge packets, which were manipulated in the device and, only at the output, converted to a voltage that could be measured. By operating in the charge domain, the charge-coupled device is a true integrated circuit; it is impossible to make such a device from a collection of discrete components.

The serious development of Active Pixel Sensors began in the early 1990s, though their heritage hails from work done in the 1960s. Active pixel sensors get their name from the nature of the structure that makes up the pixel: a charge collection site and readout circuitry are provided within each pixel.

This article provides an overview of charge-coupled devices and active pixel sensors, their design, structure, operation, and performance.

**Keywords**
charge-coupled device; active pixel sensor; CMOS-APS; visible imager; CCD.

## 1 Introduction

The movement toward providing an all solid-state replacement for the vidicon tube began in the 1960s with work done at several locations on Si-based bipolar and MOS structures, and thin-film transistors in II–VI materials. The basic problem facing these researchers was that their pixels did not integrate the optical signal. Instead, the output was simply proportional to the input intensity at the time the signal was sampled. The signal was weak and it was realized that to be effective the cell would need gain. Nevertheless, small arrays were fabricated and successfully operated [1, 2].

In 1967, it was realized by Weckler [3] that one could integrate the incident flux on the capacitance of an isolated, reverse-biased $p$–$n$ junction. Readout occurs by connecting the diode to a series resistor through a PMOSFET. A 100 × 100 array was fabricated and successfully operated in 1968 and formed the basis of the Reticon company. [4] Additional ideas on photodiode structures and readout circuitry were reported by Nobel [5] and by Chamberlain [2,6].

Charge-coupled devices (CCDs) were introduced in 1970 by Boyle and Smith [7] of Bell Laboratories. The original motivation was to find a silicon-based electronic analog for the then very successful magnetic bubble memory. In their invention, packets of charge would be the fundamental unit of information. These packets could be created, destroyed, stored, detected, and manipulated in an electronic circuit, all functions required of a memory device.

It was rapidly recognized by many workers in the field that CCDs were capable of performing a large number of electronic functions. Uses for these devices quickly spanned the areas of imaging, signal processing, display, logic functions, analog, and digital memory. By 1975, several companies were offering 16 K memory chips using CCD technology with 64 K devices in the offing. Fabrication difficulties, however, plagued their introduction, and with the continuing drive towards smaller cells and chips characteristic of the semiconductor industry, the memory houses succeeded in fabricating 64 K DRAMs before comparable CCD memories could be manufactured cheaply enough. CCD digital memories failed to succeed because they missed the window of opportunity.

It is in the area of visible imaging that CCDs have made their greatest contribution and have had their greatest impact.

The relative simplicity of the basic structure and the ease of manufacture contributed to its success. Throughout the late 1970s and 1980s, large improvements in the size and performance of CCDs were made. Two markets are largely responsible for these improvements: Astronomy and the handheld camcorder for the consumer. Astronomical applications require low-noise, high quantum efficiency (QE), high charge-transfer efficiency (CTE), and high-resolution, large-area devices. The consumer market demanded low dark current, low fixed pattern noise (FPN), high-speed, and low-defect density. Such devices are now the largest MOS chips routinely being produced and small, handheld video cameras using them are commonplace.

In addition to the advantages of solid-state construction, CCDs offer a number of other advantages for scientific applications: broad spectral bandwidth, geometrical stability and fidelity, large dynamic range, linear response, and high photometric accuracy. Today, CCDs have created as much a revolution in the area of astronomy as photographic film did 100 years ago when it was first introduced [8]. Every observatory in the world uses CCD arrays as the sensor for virtually every direct imaging and spectrographic instrument. Mosaics of 8 to over 40 large CCDs are common [9–11].

In the early 1990s, the same two markets provided the impetus to develop the complementary metal-oxide semiconductor (CMOS) sensor. NASA required low-power, compact imaging systems for the next generation space missions. The digital still camera revolution was on the verge of exploding and again, a low-power, low-cost system, a camera-on-a-chip, was required. The intent was to integrate as much functionality onto a single piece of silicon as possible. The low-cost and low-power aspects of the goal naturally pointed to CMOS processing. Attempts were made with CCDs to incorporate gate drivers and other circuitry on chip. There were occasional successes, but the task proved difficult and the results generally disappointing [12, 13]. Standard CMOS processing is generally not compatible with high-performance CCD processing.

Since the revival in the interest in MOS sensors, much development has occurred. The realization that the basic MOS pixel needed an amplifier in order to improve sensitivity and decrease noise, has given rise to the active pixel sensor (APS). Although initially large, the rapid progress in CMOS technology has allowed the miniaturization of the circuitry in the APS pixel such that its size is comparable to that of the CCD, a major factor since pixel size is often driven by the size of the available optics. Fixed pattern noise, which was one of the major problems with the early MOS sensors, has been addressed and largely solved.

Today, it is possible to acquire Video Graphics Adapter (VGA) to Ultrahigh Definition Television (UDTV) progressive scan chips with 8 to 12 bits of digitized data as the output from the chip. The chips routinely include the drivers, control logic, low-noise amplifiers, and A/D converters. Special-purpose chips are available, which perform a variety of functions, from motion detection to discrete Fourier transforms to on-chip frame differencing [14–16]. The ability to process the chip with conventional CMOS technology has opened the door to a virtually limitless space of applications. Furthermore, because the photosites in CMOS APS and CCD sensors form a regular array of sample sites, they are the natural inputting devices for computer vision applications as the information is already spatially quantized [2].

CCD and APS arrays have replaced the vidicon in all applications, with the possible exception of studio TV cameras. CCDs have been used in Stereotactic Spot Mammography systems, microscope cameras, and astronomical focal planes ranging from the Hubble Space Telescope to Keck to the amateur astronomer's telescope [17]. CMOS APS arrays are being found in applications ranging from vertex detectors in particle accelerators to professional portrait cameras to PC cameras. The digital still camera revolution is just beginning and one can purchase either a CCD-based or a CMOS APS-based multi-Megapixel camera.

It should be noted that there are other Si-based imagers, notably the charge-modulation device (CMD) and the charge-injection device or CID array. In this article, we will concentrate on CCDs and CMOS active pixel sensors.

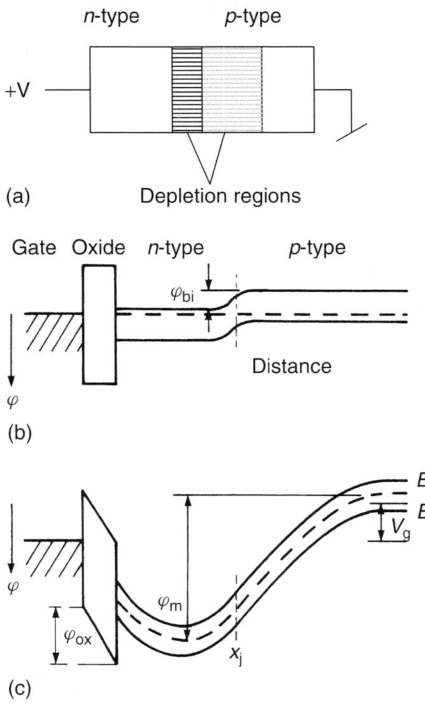


**Fig. 1** Detection devices. (a) Cross section of a *pn* junction showing depletion region. (b) Band diagram for the buried channel capacitor. $V_g = 0$. (c) Band diagram for the buried channel capacitor for the condition of positive bias on the gate ($V_g > 0$) and the channel fully depleted of electrons. Note the maximum in the potential located at $x_m$

## 2 Charge Storage

One of the necessary, fundamental properties of solid-state image sensors is the ability to locally store the photogenerated signal charges until they can be sampled and read out. In CCDs and APS devices, there are three basic structures used for this purpose: the photodiode, the photogate or MOS capacitor, and the pinned photodiode (PPD).

Figure 1 illustrates a schematic cross section of a *pn* junction. In this device, the impurities are assumed to be distributed uniformly throughout each region: $N_D$ donor atoms/cm$^3$ in the *n*-region and $N_A$ acceptors/cm$^3$ in the *p*-type region. On either side of the metallurgical junction, a region exists free of electrons ($-x_n < x < 0$, *n*-type material) or holes ($0 < x < x_p$, *p*-type side). A potential of $+V_g$ is applied to the $n+$ region. The $p$ region is grounded. For the case where $N_D \gg N_A$, the solution to Poisson's equation yields the width of the depletion region

$$w = x_n + x_p \approx x_p = \sqrt{\frac{2\varepsilon_{Si}}{qN_A}(V_g + V_{bi})} \tag{1}$$

and the associated depletion capacitance

$$C_D = \frac{dQ}{dV} = \sqrt{\frac{q\varepsilon_{Si}N_A}{2(V_g + V_{bi})}} \tag{2}$$

where $q$ is the magnitude of the charge on an electron, $\varepsilon_{Si}$ is the dielectric constant of silicon, and $V_{bi}$ is the built-in potential [18].

The fundamental building block of the CCD is the buried channel MOS capacitor. The structure of the buried channel MOS capacitor is illustrated in cross section in Fig. 1(b). In this structure, an *n*-type region is ion-implanted into the *p*-type substrate forming a *pn* junction, a distance $x_j$ below the surface. An oxide layer and then a metal or polysilicon gate deposited on the surface complete the capacitor.

If such a capacitor is biased from inversion into deep depletion (i.e., from a negative bias to a positive bias), a depletion layer of width $w$ will be formed as shown in Fig. 1(c). The potential profile of this structure can be found from an analysis of Poisson's Equation. In general, a closed form solution is not available as the impurity distribution typically cannot be described by an analytic function. There are, however, two simplified distributions that are often employed to analyze the buried channel structure. These are the box distribution and the Gaussian distribution. Although the box distribution is easier to analyze, the latter distribution more closely approximates the ion-implanted buried channel device. The solution to the box distribution can be found in the literature [19,20]. The solution to the Gaussian distribution will be used in the discussion that follows to illustrate the physics of the MOS structure.

In the Gaussian model, the buried channel is considered to be an *n*-type ion implant with a profile given by a Gaussian distribution superimposed on a more lightly doped *p*-type substrate. The impurity distribution is given by

$$\rho(x) = q\left(\sqrt{\frac{2}{\pi}}\frac{Q}{\sigma}\exp\left(-\frac{(x-x_p)^2}{2\sigma^2}\right) - N_A\right) \tag{3}$$

where $Q$ is the implant dose (ions/cm$^2$), and $N_A$ is the doping density in the substrate (ions/cm$^3$). The *n*-type region extends to the depth of the metallurgical junction, $x_j$. A peak at $x_p$ and a straggle of $\sigma$ characterize the implant. We assume that the buried channel is fully depleted and the depletion approximation is valid. The boundary conditions are: the electric displacement and potential are continuous across the silicon–silicon dioxide interface and across the *pn* junction. In addition, the potential at the gate is $\varphi(x = -t) = V_g$, and the potential and the field are 0 at the edge of the depletion region, $w$, and deeper in the substrate. Using the boundary conditions given above, the potential profile of this structure as a function of distance beneath the gate may be expressed as

$$\varphi_{ox} = \alpha x + \beta \tag{4}$$

$$\varphi_{Si}(x) = \frac{q}{\varepsilon_{Si}} \times \left(\frac{N_A x^2}{2} + Cx + D - \frac{2Q}{\sqrt{\pi}}\gamma(x)\right) \tag{5}$$

for the oxide and the silicon, respectively. In these expressions

$$\alpha = \frac{1}{t}(\beta - V_g) \tag{6}$$

$$\beta = \frac{q}{\varepsilon_{\text{Si}}}\left(D - \frac{2Q}{\sqrt{\pi}}\gamma(0)\right) \tag{7}$$

$$C = \frac{1}{w}\left(\frac{\xi}{1+\xi}\right) \times \left(\frac{2Q}{\sqrt{\pi}}\left\{(\gamma(w) - \gamma(0)) + \frac{\varepsilon_{\text{Si}}}{C_{\text{ox}}}\,\text{erf}\left(\frac{x_{\text{p}}}{\sqrt{2}\sigma}\right)\right\} - \frac{\varepsilon_{\text{Si}}}{q}V_{\text{g}} - \frac{N_{\text{A}}w^2}{2}\right) \tag{8}$$

$$D = \left(\frac{1}{1+\xi}\right)\left(\frac{2Q}{\sqrt{\pi}}\gamma(w) - \frac{N_{\text{A}}w^2}{2}\right) + \left(\frac{\xi}{1+\xi}\right)\left(\frac{2Q}{\sqrt{\pi}}\left\{\gamma(0) - \frac{\varepsilon_{\text{Si}}}{C_{\text{ox}}}\,\text{erf}\left(\frac{x_{\text{p}}}{\sqrt{2}\sigma}\right)\right\} + \frac{\varepsilon_{\text{Si}}}{q}V_{\text{g}}\right) \tag{9}$$

and

$$\gamma(x) = \int_0^x \mathrm{d}\eta \int_0^\eta \mathrm{d}\omega \exp\left(-\frac{(\omega - x_{\text{p}})^2}{2\sigma^2}\right) = (x - x_{\text{p}})\,\text{erf}\left(\frac{x - x_{\text{p}}}{\sqrt{2}\sigma}\right) + \frac{\sigma}{\sqrt{2}}\exp\left(-\frac{(x - x_{\text{p}})^2}{2\sigma^2}\right) \tag{10}$$

where $\xi = (w\varepsilon_{\text{ox}}/t\varepsilon_{\text{Si}})$, $\varepsilon_{\text{ox}}$ and $\varepsilon_{\text{Si}}$ are the dielectric constants for oxide and silicon respectively, and $t$ is the oxide thickness. Figure 2 presents calculations of the potential as a function of distance in the device for several different gate voltages. The figure illustrates one of the features of this structure that makes it so useful: the potential well has its maximum at a distance $x_{\text{m}}$ from the Si-SiO$_2$ interface and is the location where electrons will

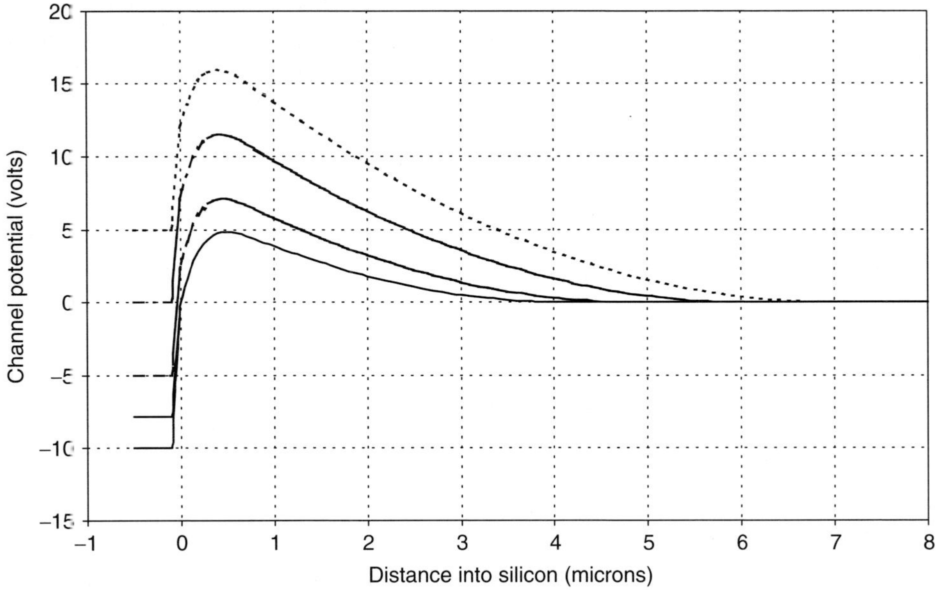


**Fig. 2** Channel potential as a function of position for $V_{\text{g}} = -10, -7.8, -5, 0, +5$ V. Note that the potential in the silicon is the same for the $-10$ and $-7.8$ V curves. The difference is in the field in the oxide. Parameters are $Q = 1.6\ \text{E}^{12}\ \text{cm}^{-2}$, $N_{\text{a}} = 5 \times 10^{14}\ \text{cm}^{-3}$, $x_{\text{p}} = 0$ cm, $x_{\text{j}} = 7 \times 10^{-5}$ cm, $t = 10^{-5}$ cm

collect. The interface contains states that can trap signal electrons leading to charge transfer problems and noise. The location away from the Si surface gives rise to the name *buried channel.*

The maximum in the potential occurs at $x = x_{\mathrm{m}}$. This position can be evaluated by setting $\mathrm{d}\varphi/\mathrm{d}x = 0$. The root of the equation gives $x_{\mathrm{m}}$

$$N_{\mathrm{A}}x_{\mathrm{m}} + C - \frac{2Q}{\sqrt{\pi}}\,\mathrm{erf}\left(\frac{(x_{\mathrm{m}} - x_{\mathrm{p}})}{\sqrt{2}\sigma}\right) = 0 \tag{11}$$

The empty potential maximum is

$$\varphi_{\max} = \left(\frac{q}{\varepsilon_{\mathrm{si}}}\right)\left(\frac{N_{\mathrm{A}}x_{\mathrm{m}}^2}{2} + Cx_{\mathrm{m}} + D - \frac{2Q}{\sqrt{\pi}}\gamma(x_{\mathrm{m}})\right) \tag{12}$$

The edge of the depletion region is given by

$$w = \frac{\varepsilon_{\mathrm{Si}}}{C_{\mathrm{ox}}} + \sqrt{\left(\frac{\varepsilon_{\mathrm{Si}}}{C_{\mathrm{ox}}}\right)^2 + \frac{2}{N_{\mathrm{A}}}\left[\frac{2Q}{\sqrt{\pi}}\left\{\gamma(0) + \frac{\varepsilon_{\mathrm{Si}}}{C_{\mathrm{ox}}} \times \left(1 - \mathrm{erf}\left(\frac{x_{\mathrm{p}}}{\sqrt{2}\sigma}\right)\right)\right\} + \frac{\varepsilon_{\mathrm{Si}}}{q}V_{\mathrm{g}}\right]} \tag{13}$$

The storage capacitance of the structure is defined as the capacitance between the gate and the signal packet. For an empty well, this capacitance is given by the series combination of the oxide capacitance and the depletion capacitance between the channel minimum and the surface.

$$C_{\mathrm{BC}} = \left(\frac{t}{\varepsilon_{\mathrm{ox}}} + \frac{x_{\mathrm{m}}}{\varepsilon_{\mathrm{Si}}}\right) \tag{14}$$

The features of the MOS capacitor that make it useful as a storage site for photo-generated charge and in the fabrication of CCDs are

1. a deep depletion region can be formed initially devoid of any carriers,
2. the depletion region acts like a potential well to which electrons are attracted,
3. the signal can be represented by the collected charge,
4. the signal charge can be generated optically or electrically, and
5. it requires a long time for carriers to be thermally generated, collected, and for the potential well to be filled, on the order of 100 s at room temperature in well-processed material.

The PPD is again an *n*-region implanted into a *p*-type substrate. An additional, very shallow *p*-type implant is placed on the surface and electrically connected to the channel stops and thus to the ground. The potential of the surface is pinned at 0 V. The potential profile within this structure can be calculated using the formalism developed for the MOS capacitor, recognizing that the potential at the surface is fixed at V = 0. The storage capacitance is given by

$$C_{\mathrm{PPD}} = \left(\frac{x_{\mathrm{m}}}{\varepsilon_{\mathrm{Si}}}\right) \tag{15}$$

This is the one-dimensional result. A more detailed analysis of the photodiode (PD), the photogate (PG), and the PPD would include capacitance associated with the periphery and the field [21].

## 2.1 Channel Potential

The charge on the diode or in the potential well may be considered as a signal and the diode or potential well can be thought of as a bucket in which signal electrons can be collected. In Fig. 1(b) and subsequent figures, the potential of the electron is positive down so that the charge fills the well from the bottom up, much as water would fill a bucket. The analogy between electrons in a CCD well and water in a bucket is an extremely powerful and effective tool in analyzing the behavior of charge in the CCD or in the APS pixels. The electrons that collect in the potential maximum can result from thermal generation in the bulk, in the depletion region, or at the surface (known as dark current); however, and most importantly, signal electrons can be intentionally introduced (e.g., electrically or optically).

Figure 3(a) illustrates the potential profile in an MOS capacitor with signal charge present. The signal collects in the middle of the channel, compensating the ionized donors in that region. The spatial distribution of the charge as a function of signal level is illustrated in Fig. 3(b). Note that for all but the largest signals the charge remains well away from the Si-$SiO_2$ interface.

Figure 4 presents the results of calculations of the potential maximum or channel potential using Eq. (12) and typical values of CCD fabrication parameters. In the figure, the channel potential is given as a function of the gate bias. It is important to note that for a given gate bias the buried channel potential beneath the gate increases for increasing implant dose and for increasing junction depth, and decreases for thinner gate dielectric. These effects have been used to make CCDs.

The second important feature illustrated in Fig. 4 is that for sufficiently negative gate biases, the channel potential curve becomes horizontal. When the surface potential is equal to the bulk potential, the surface becomes inverted (i.e., holes collect at the surface). Any further decrease in the gate voltage merely increases the concentration of holes at the surface but does not reduce the surface or the channel potential. The surface is pinned at this potential. The inversion layer in the buried channel capacitor is in good contact with the *p*-type substrate that supplies holes very rapidly.

Bearing in mind that the surface is pinned to ground potential, the potential profile in the PPD mimics the behavior of the pinned CCD.

## 2.2 Charge Detection and Collection

Signal charges are generated in the silicon by the absorption of the incident photon and the creation of an electron-hole (e-h) pair. This process of e-h creation is referred to as the *Photoelectric Effect*. If the e-h pair is created in the depletion region, the field is such as to separate the pair, collecting the electron in the potential maximum and sending the hole to the substrate. Similarly, electrons generated in the substrate can diffuse toward the potential maximum and be collected.

For the CCD, there are two techniques commonly employed in collecting the signal charge. The first is the MOS capacitor already described. The region beneath one gate is biased such that the local potential forms a potential well in which the optically generated electrons collect. This will be recognized as the basic element of the CCD pixel.

The second structure involves a PPD next to a CCD register and separated from it by a transfer gate. The diode is reverse biased by turning on the transfer gate and draining the charge in the diode into the CCD register. The channel potential will be equal to the empty channel potential as determined by Eq. (12). During integration, with the transfer gate turned off, electrons collect in the diode, decreasing

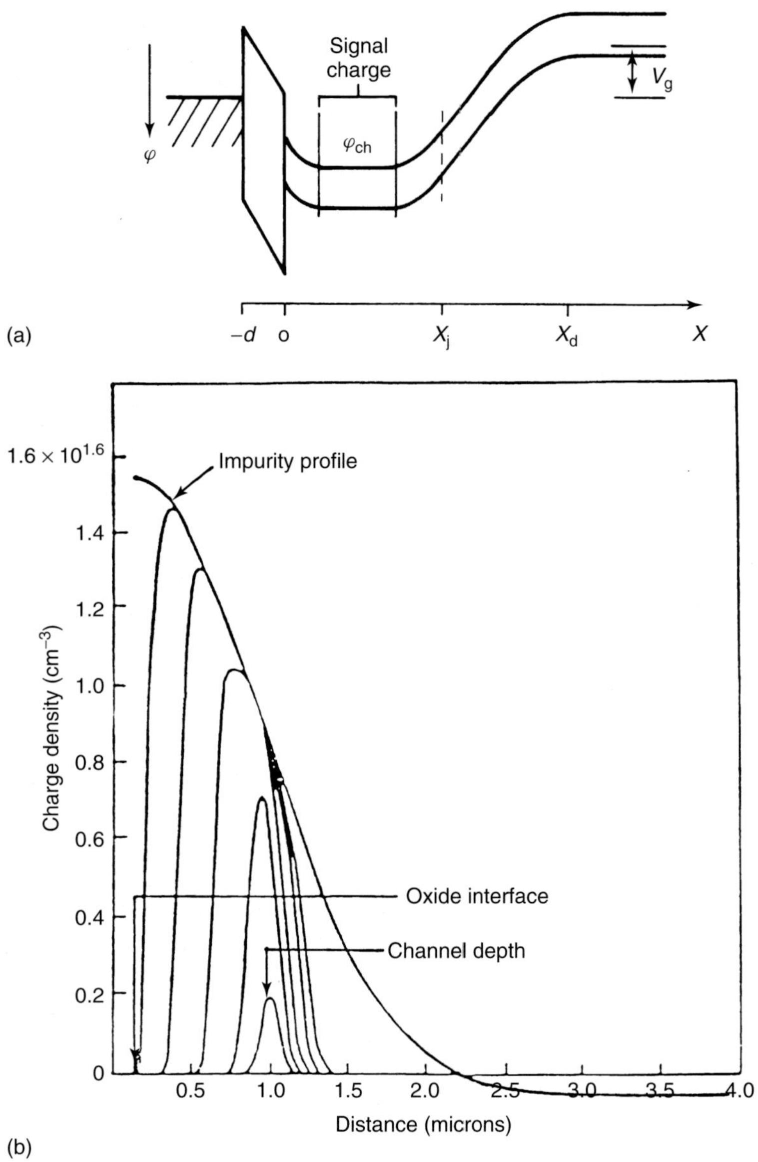


**Fig. 3** (a) Band diagram for the buried channel capacitor. $V_g > 0$, with signal charge present. (b) Calculations of the signal charge distribution in the buried channel capacitor. At low signal levels, the charge is confined to a region well away from the surface ((b) [22])

its potential. When the transfer gate is turned on again, the collected charge is dumped into the CCD register and the potential reset. This type of sensing region is commonly used with linear arrays and interline transfer (ILT) arrays.

For CMOS arrays, there are three structures: a simple photodiode in the

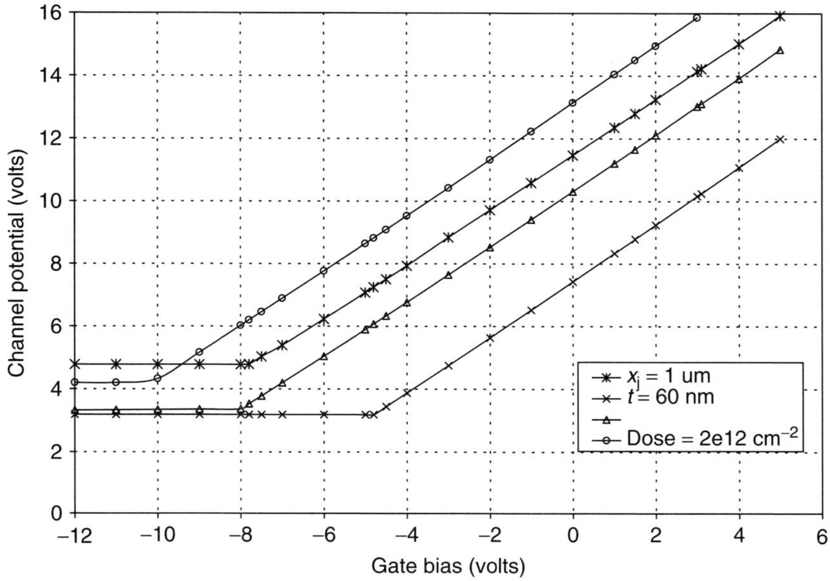


**Fig. 4** Channel potential (potential maximum) as a function of gate bias for the buried channel capacitor. The different curves represent different combinations of the fabrication parameters. Nominal parameters are $Q = 1.6\text{E}^{12}\ \text{cm}^{-2}$, $N_a = 5 \times 10^{14}\ \text{cm}^{-3}$, $x_p = 0$ cm, $x_j = 7 \times 10^{-5}$ cm, $t = 10^{-5}$ cm

*p*/*p*+ epitaxial wafer, the Photogate or MOS capacitor, similar to the CCD, or the PPD. The photodiode can be formed in a couple of ways: a simple *n*+ diffusion into the *p*/*p*+ epitaxial layer, or an *n*+ diffusion into an *n*-well. The advantage of the photodiodes is the high QE relative to the photogate, which uses polysilicon as the gate material. Most users employ the photodiode as the detector.

## 3 Charge Transfer Process

The process of moving the signal charge from one gate to a position beneath another gate is one of the major features that differentiates a CCD from an APS device. A series of MOS capacitors or gates arranged in a linear array can be used to form a simple shift register. The configuration is illustrated schematically in Fig. 5. Assume that every third gate is connected to the same electrical potential (referred to as *a clock* or *a phase*). In the figure, a triplet of gates forms the basic unit cell that can be replicated to make the CCD. The unit cell is referred to as a pixel (*pic*ture *el*ement). In the figure, the channel potential is plotted as a function of position along the array of gates; the middle gate (phase 2) of each triplet is initially biased high forming an array of potential wells.

Assume that at time $t_0$ there exists a pattern of charge packets in the wells. At some later time, $t_1$, the phase 3 clocks turn on, generating a potential well beneath the phase 3 gates. A portion of the charge now stored under the phase 2 gate will spill over into the region beneath the phase 3

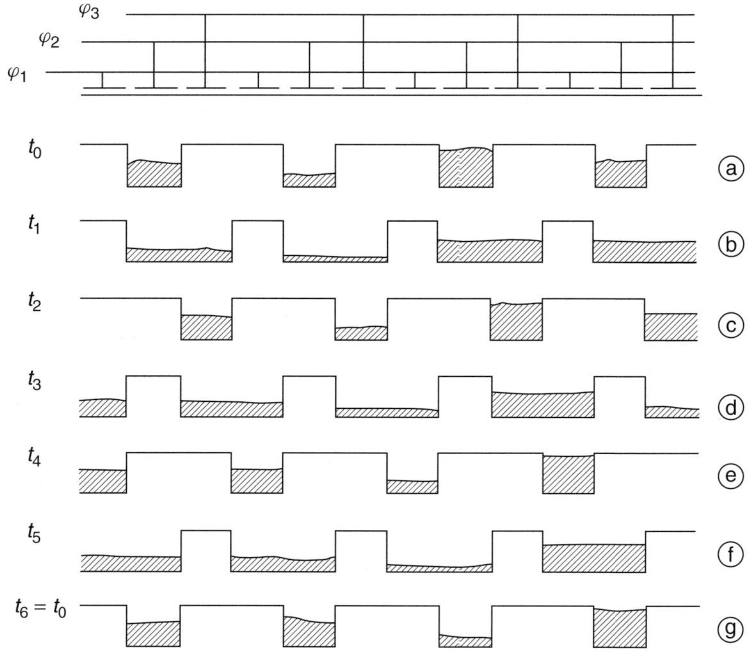


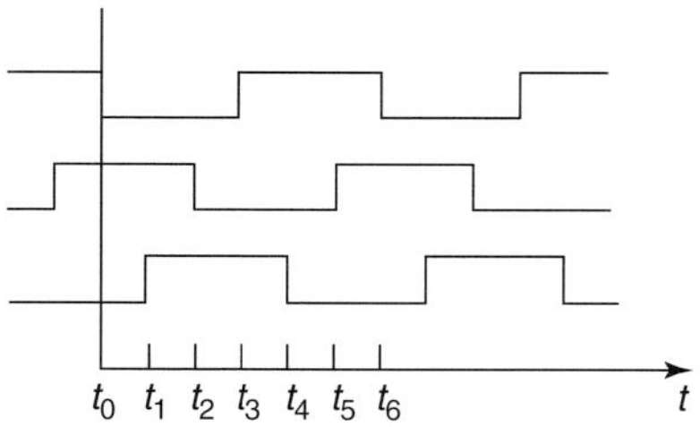


**Fig. 5** (a) – (g) Charge transfer in a three-phase serial shift register

gate and divide between the two gates. This happens for every pair of phase 2 and phase 3 gates. At $t = t_2$, the phase 2 clock turns off and all the charge now resides beneath the phase 3 gate. (See Figs. 5b and c). This sharing or coupling of the charge packet between neighboring gates gives rise to the name charge-coupled device.

The clock sequence is repeated for the phase 1 and phase 3 gates and finally the phase 1 and phase 2 gates. The result is that the charge packet has been moved or transferred from a position beneath a given phase 2 gate to a position beneath the phase 2 gate in the neighboring pixel.

Figure 5 illustrates several important and general features regarding the charge transfer process. First, there must be a well in which to collect the charge, and this well should be unique to each pixel. Secondly, at all times during the integration and transfer process, there must be potential barriers between the charge packets. Consequently, within each pixel there must exist at least one barrier and one well. Thirdly, these barriers and wells can be created and moved by simple manipulation of the clock voltages. Finally, in the illustration given above, the charge packets were moved from left to right. By simply reversing the order of any two of the clocks, it would have been possible to move the charge from right to left.

The motion of the electrons between gates is governed by three processes: (1) thermal diffusion, (2) self-induced drift, and (3) fringing fields. The fringing field is the result of the two-dimensional coupling of the electrode potentials across the gap between the gates. Both the fringing field and the self-induced field are in the sense so as to encourage charge transfer. At the beginning of the transfer process, the motion of the carriers is dominated by self-induced drift. However, once most of the charge has been transferred, the process will be governed either by the fringing fields or by thermal diffusion [23–25]. In most cases, and especially for buried channel devices, the fringing fields dominate the charge transfer process.

Although most CMOS APS sensors do not require charge to transfer a great distance, some structures do require a single transfer. These pixels generally involve more than three FETs per pixel (see Sect. 5.2.1) and use either the PPD or the Photogate detection structures. Because of the low voltages employed by CMOS arrays, the transfer of charge tends to be incomplete. The charges left behind appear as signal charge during the next frame readout resulting in lag.

## 4 Device Architecture

There are two basic architectures that are used for imaging purposes: linear arrays and area arrays. The linear shift register, as described in Sect. 3, is the basic building block of all CCD image sensors. The active imaging element can be either a photodiode or the CCD pixel itself.

### 4.1 Linear Arrays

Three of the more common implementations of linear arrays are illustrated in Fig. 6. In these structures, the photosites are reverse-biased photodiodes and typically arranged in a one-dimensional array which is matched, in the case of a CCD, to a CCD readout register. Each photodiode is coupled to a pixel in the readout register by a common transfer gate. In operation, following a suitable integration period, the charge collected in the reverse biased photosites is transferred to the empty CCD wells when the transfer gate is turned on. The charge packets are then shifted to the output amplifiers while the next line of information is integrated. In addition to the sensing sites, these arrays frequently contain pixels that provide white and dark reference signals.

To achieve high-resolution (small pixels), higher readout rates and still maintain reasonable design rules and well capacities, multiple readout registers (two and sometimes four) are used. In these cases, the device consists of parallel CCD registers on opposite sides of the photodiode array and separated from the photodiodes by a transfer gate; even numbered pixels transfer to one register while odd pixels connect with the other. The CCD registers are covered with a metal light shield to prevent stray light from contaminating the signal during the readout cycle. The registers can be clocked to two separate amplifiers or merged into

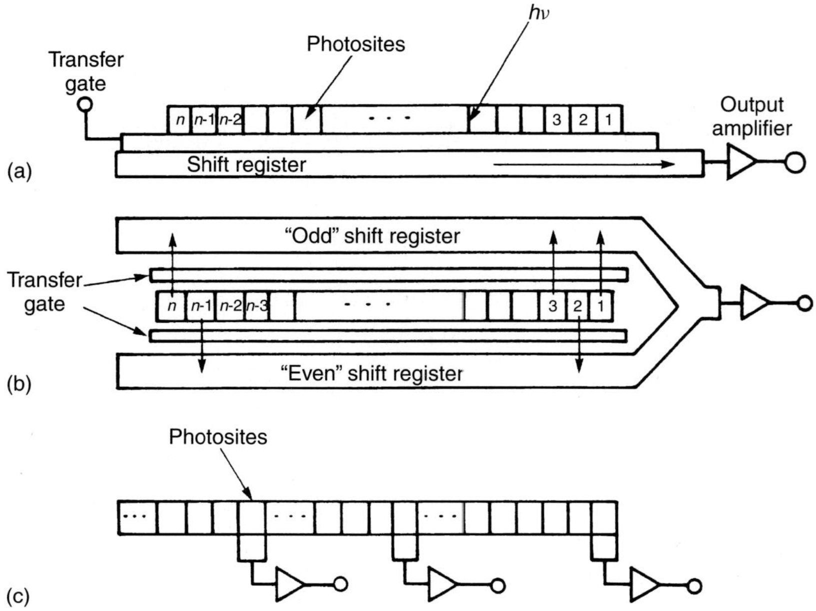


**Fig. 6** Examples of linear CCD architectures. (a) Single readout register; (b) dual readout registers with a single amplifier; and (c) output register with multiple output amplifiers

a single register and read out through a single output.

The same basic architectures apply to the linear CMOS sensors. The major difference is that between the imaging pixels and the readout shift register is an array of amplifiers that convert the detected charge to a voltage that is then sampled by the horizontal readout multiplexer [26, 27].

## 4.2 Area Arrays

### 4.2.1 CMOS Arrays

While CCD technology has benefited greatly from the development of CMOS technology (e.g., cleaner chemicals and facilities, high purity epitaxial silicon, and finer photolithography tools), it is really the APSs that have been able to make full use of the CMOS process.

**Camera-on-a-chip** Because of the ability and desire to integrate CMOS circuitry on the same die as the sensing array, APSs generally come with much of the drive electronics supplied. Figure 7 illustrates the layout of a simple APS sensor array. The central portion of the chip is dedicated to the M × N array of pixels. On either side of the array are vertical shift registers and latches for selecting the row to be read out. At the bottom of the array is a row of column amplifiers, one per column. The column amplifiers are followed by a multiplexer that sequentially (or randomly) connects the columns to an output amplifier. The analog output of the amplifier is directed to an off-chip A/D converter. Additional optional circuitry includes color and gamma correction circuitry for color arrays, on-board A/D converters, memory and digital engines

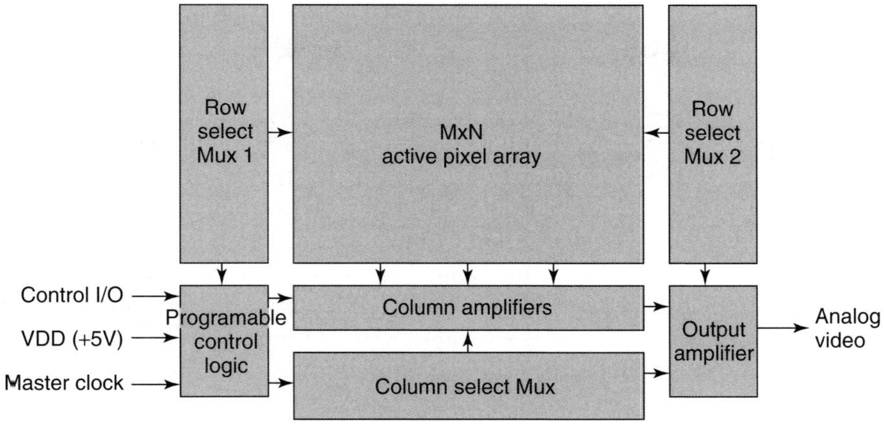


**Fig. 7** Block diagram of a simple CMOS active pixel sensor

for kernal operations, a memory array to perform binning operations, and so on.

CMOS arrays are basically full-frame (FF) arrays and, depending on the pixel type, operate in the rolling shutter or a snapshot shutter mode. There are examples of frame-transfer (FT) APSs, but they are not common [28].

**Rolling Shutter** The rolling shutter method of reading out the APS array

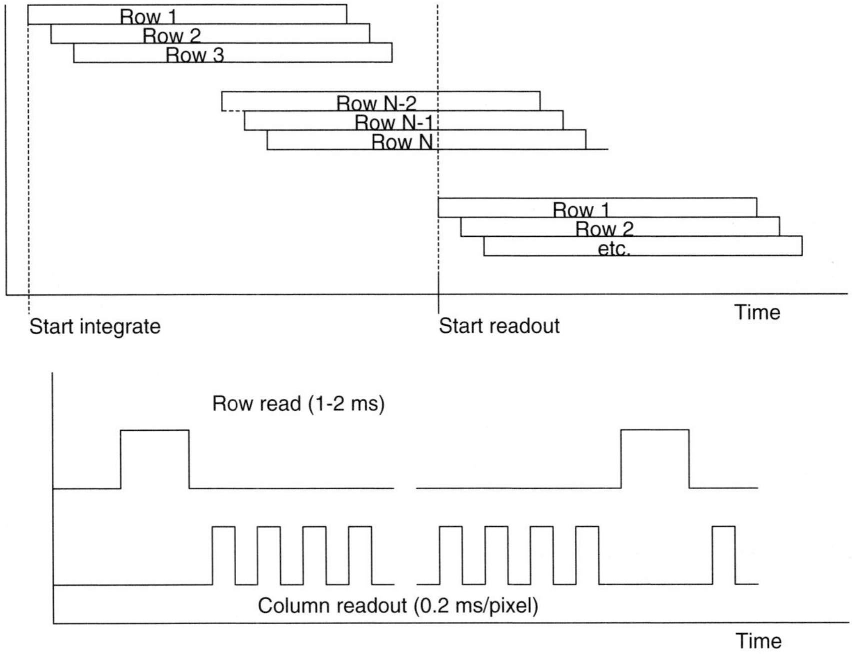


**Fig. 8** Rolling shutter readout

is illustrated in Fig. 8. Imagine that the device is being continuously exposed and readout. Following exposure, the first line is selected and the pixels read, in parallel, into the column amplifiers. The column amplifiers are sequentially addressed by the readout multiplexer, while the row is being reset. Then, row 2 is selected and the process of readout repeated. At the same time, row 1 has begun its integration cycle, which will continue until row 1 is next addressed. Following the read out of row 2, row 3 is addressed, and so on until all rows have been read out. Note that the exposure time for each row extends one line time beyond the exposure of the previous row so that the image recorded by row $n$ is at a slightly different time than row $n$-1, hence the name *rolling shutter*. This mode of operation can lead to some unusual effects if there are fast moving objects in the scene, as shown in Fig. 9(a) [29]. This method is commonly used for the three-transistor pixel.

**Snapshot Shutter** As suggested by the name, in the snapshot mode of operation the information collected by each pixel is simultaneously shifted onto a storage site within the pixel. This mode is also referred to as a *global shutter*. Integration of the next image begins again for all pixels at the same time and continues until the previously stored image is read out. Readout occurs much the same way as with the rolling shutter in that each row is sequentially addressed, and the signal sensed by the column amplifiers, which in turn are scanned and the signal detected by the output amplifier. The snapshot mode does not suffer from the same problem of capturing high-speed motion as does the rolling shutter (see Fig. 9b). In addition, because this mode requires a more complicated cell, it is possible to readout the pixels using correlated double sampling (CDS). The CDS reduces pixel-related kTC, $1/f$, and fixed-pattern noise.

**High Frame Rate Arrays** There are a number of applications that require very high frame rates and desire high resolution such as ballistics and destructive testing. Because of the highly parallel nature of the readout of the APS and the ability

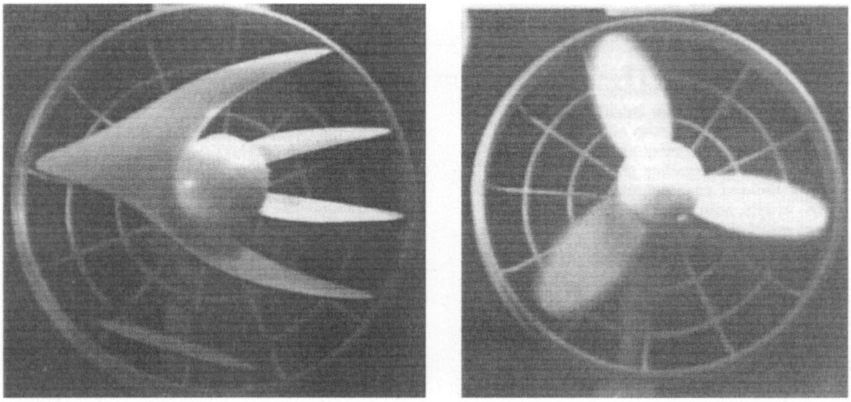

(a) (b)

**Fig. 9** Image of a fan using the (a) rolling shutter mode; and (b) the snapshot mode [29]

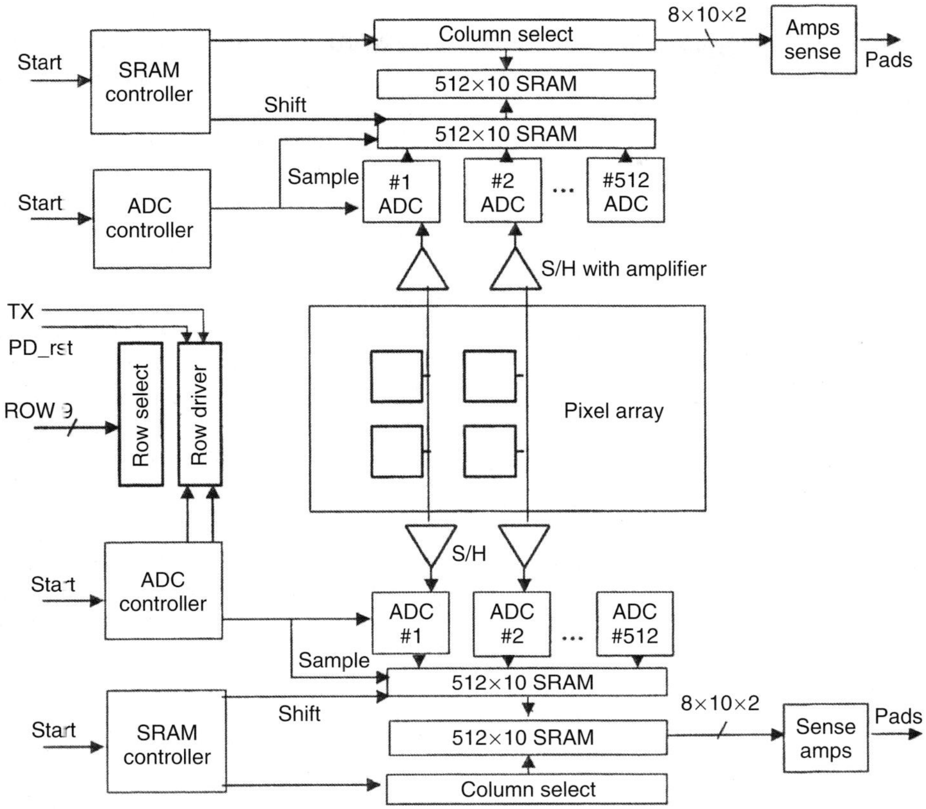


**Fig. 10** Block diagram of high frame rate, column parallel chip [31]

to add CMOS circuitry to the chip, it is possible to supply each column amplifier with its own A/D converter. This permits digitization of the pixel signals to proceed, in parallel, at the row rate rather than the pixel rate as would be the case with a single on-chip A/D.

A block diagram of such a chip is shown schematically in Fig. 10. This chip is a 512 × 512 pixel array and operates at 5000 frames per second data rate. Each column is terminated in a sample-and-hold amplifier and a 10-bit A/D converter, at the top and the bottom to increase the speed. Two 512 × 10-bit SRAMs follow the A/D converters. During digital conversion, the data are written into the first SRAM while the previous row is read out of the second SRAM. At the nominal speed of 5000 fps, this represents a data rate of 1.3 Gpix/s at 10-bit resolution [31].

**Other Arrays** The ability to integrate signal processing capabilities onto the chip provides some interesting opportunities. For example, arrays have been produced that output a discrete Cosine Transform, perform kernal operations that can detect edges or take derivatives of the scene [32], detect motion, or perform on-chip binning to trade resolution for signal-to-noise improvement.

### 4.2.2 CCD Arrays

There are three basic types of CCD area imagers: full-frame, frame-transfer, and interline transfer arrays. Figure 11 illustrates these types. The construction of each of these devices requires an array of M × N pixels.

**Full-frame CCD** The full-frame imager (Fig. 11a) can be thought of as an array of N serial shift registers oriented vertically in the column direction. Each of the shift registers is divided into M pixels. The N shift registers are placed next to each other and separated by narrow channel stop regions. The channel stops confine the charge laterally, and prevent charge collected in one pixel from communicating with the charge in the neighboring columns.

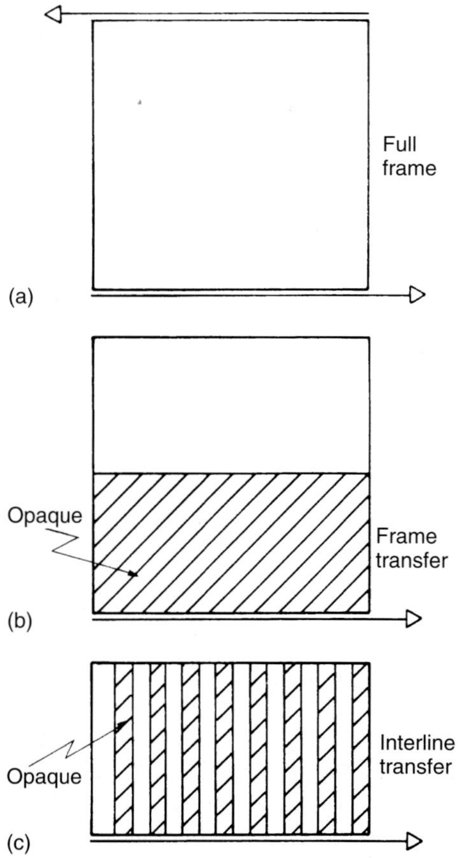


**Fig. 11** Area array imagers. (a) Full-frame imager; (b) frame-transfer device; (c) interline transfer sensor

The electrodes that define the pixels are interconnected such that all the pixels in a given row share a common set of electrodes. Busses on the edges of the array supply the electrical signals that provide the bias to the gates. Since the polysilicon gate behaves as a distributed RC delay line, there are usually busses on both sides of the array to increase the speed of operation and provide redundant biasing. During integration, all pixels collect charge under the same phase. When charge is clocked in the array, every pixel is clocked simultaneously so that rows of charge are shifted.

At the bottom of the array, there is an additional shift register oriented orthogonal to the columns. The pitch (center to center spacing) of the pixels in the horizontal register is identical to the pitch of columns. That is, each column feeds into a single pixel in the serial register. A transfer gate separates the horizontal register (or serial register) from the area array (or parallel section).

Following the integration of an image in the parallel section, the array is read out. This is accomplished by shifting one row of charge packets into the serial register. At the same time, all other rows in the parallel section shift one row towards the serial register. The information in the serial register is read out one pixel at a time. Once the first row has been detected, the next row of charge is shifted into the serial register for readout. The process continues until all M rows have been read out.

There are, of course, a number of variations to this basic design. Most of these variations have to do with trying to extract the information from the array at higher effective data rates. For multiphase CCDs, it is possible to place a serial register at the top of the array so that the array can be read out through either the top or the bottom register. In addition, it is possible to segment the parallel section into two or more separately clocked sections so that the top half of the array can be read out through the top serial register and the bottom half through the bottom serial register. Further, the serial register itself can be provided with multiple taps and amplifiers so that charge can be read out through multiple amplifiers simultaneously.

**Frame-transfer CCD** The full-frame sensor usually requires a shutter in front of the device to prevent smearing of the information in the scene during the readout operation. This feature makes the FF imager generally unsuitable for TV and high-speed imaging applications. The FT architecture is designed specifically to alleviate this problem. The basic structure is illustrated in Fig. 11(b). The device is divided into two $M \times N$ arrays. Each of the arrays can be independently clocked. The two halves of the parallel section are referred to as the *imaging array* (the upper section) and the *memory* or *storage array* (the lower section). The memory array is usually covered with an opaque aluminum layer to shield it from incident light. At the bottom of the memory section, there is a serial readout register, as with the FF device.

The scene is focused onto the imaging section of the array. Following integration, the imaging and memory arrays are clocked simultaneously, and the collected charge is moved rapidly from the imaging section into the memory section. The charge is now read out of the memory section at the desired data rate in the same manner as with the FF device, while at the same time, the clocks in the imaging section are held stationary and a new frame of data is collected. As long as the time taken to shift the image into the memory section is short compared to the integration time (<1–2%), image smearing will be acceptable. Interlacing (required for standard TV applications) is achieved by integrating the two fields that make up the FF under different phases, thus shifting the centroid of the pixel for the two fields.

**Interline Transfer CCD** The interline transfer device is similar to the FT device in that there exist imaging and memory sections. The basic organization of the chip is presented in Fig. 11(c). In this structure, unlike the FF and FT devices, the photosensitive sites are usually photodiodes. A vertical CCD shift register is positioned between each column of photodiodes. For TV applications, which require interlacing, there are two photosites for every CCD pixel in the vertical registers. The even numbered photosites (corresponding to one field) are transferred during one readout cycle, while the odd sites are transferred during the next cycle. The vertical CCD registers are covered with an Al layer so that they are optically inactive and represent the memory section of the device. The vertical registers terminate in a horizontal CCD readout register. Following the integration period, the charge collected by the photodiodes is transferred to the vertical CCD registers and transfer of the information proceeds in the same manner as with the other devices.

The advantages offered by the ILT are that it occupies approximately half the area of the FT device of the same resolution, which leads to higher wafer yields and lower manufacturing costs. It also has a higher Modulation Transfer Function (MTF, see Sect. 7.7) and it requires only one transfer before the image is shielded from the optical input, substantially reducing image smearing. Disadvantages include the fact that only half of each pixel is optically active so that the QE is lower, and the higher MTF leads to increased aliasing at spatial frequencies near Nyquist.

**Orthogonal Transfer CCD** Atmospheric turbulence causes significant degradation in astronomical images. The techniques of adaptive optics have been developed over a number of years specifically to remove the distorting effects of the atmosphere. Much of the distortion represents a tilt in the wavefront. The tilt manifests itself as a translation of the image in the focal plane. A fast tip-tilt mirror removes much of this low-order distortion. Observing the motion of a guide star in the field of view generates the feedback signals needed to drive the tip-tilt mirror.

The Orthogonal Transfer CCD (OTCCD) has been developed to perform the guiding correction directly on chip [33, 34]. Figure 12(a) illustrates the concept of the device. The chip is divided into an imaging section and a memory or guide section, similar to an FT device. The major difference is that charge packets can be moved up and down the columns as well as right and left across the columns. Figure 12(b) presents the layout of a pixel in an OTCCD. The pixel utilizes four levels of polysilicon to form the electrodes. The channel stops, which ordinarily separate columns, have been largely removed leaving posts in the four corners of each pixel. By turning off the poly 4 gates, and clocking in the sequence 1-2-3-1, charge can be made to move in the vertical direction. Conversely, holding poly 3 gates at a negative potential, clocking in the sequence 1-2-4-1 will move charge in the row direction.

In operation, the guide star is located in the guide section, while the object of interest is imaged onto the imaging section. Every 10 to 100 ms the guide section is read out and the centroid of the guide star located. The guide star's new location is then used to shift the position of the charge packets in the science field. Devices with $512 \times 512$ guiding and science fields have been successfully operated and $2\,\text{K} \times 4\,\text{K}$ devices are in the offing.

**Time Delay and Integrate CCDs** The organization of the time delay and integrate (TDI) device is similar to the FF array but it usually has a very high aspect ratio, that is, few vertical pixels and a large number of horizontal pixels. In operation, the scene is mechanically scanned across the device. The CCD wells are clocked in synchrony with the motion of the image so that a particular pixel integrates the image from one part of the scene for the entire time that it traverses the parallel section. This increases the signal-to-noise ratio by approximately the square root of the number of pixels compared to a simple linear array. To enhance resolution and MTF and to keep the charge packet in better synchronism with the motion of the image, TDI arrays frequently use multiple phases (>4) and ripple clocking. In addition, these devices often have multiple taps on the serial readout register in order to achieve higher effective data rates.

**Deep Depletion CCDs** To achieve high red response and high X-ray QE (see Sect. 7.6), some CCDs are manufactured

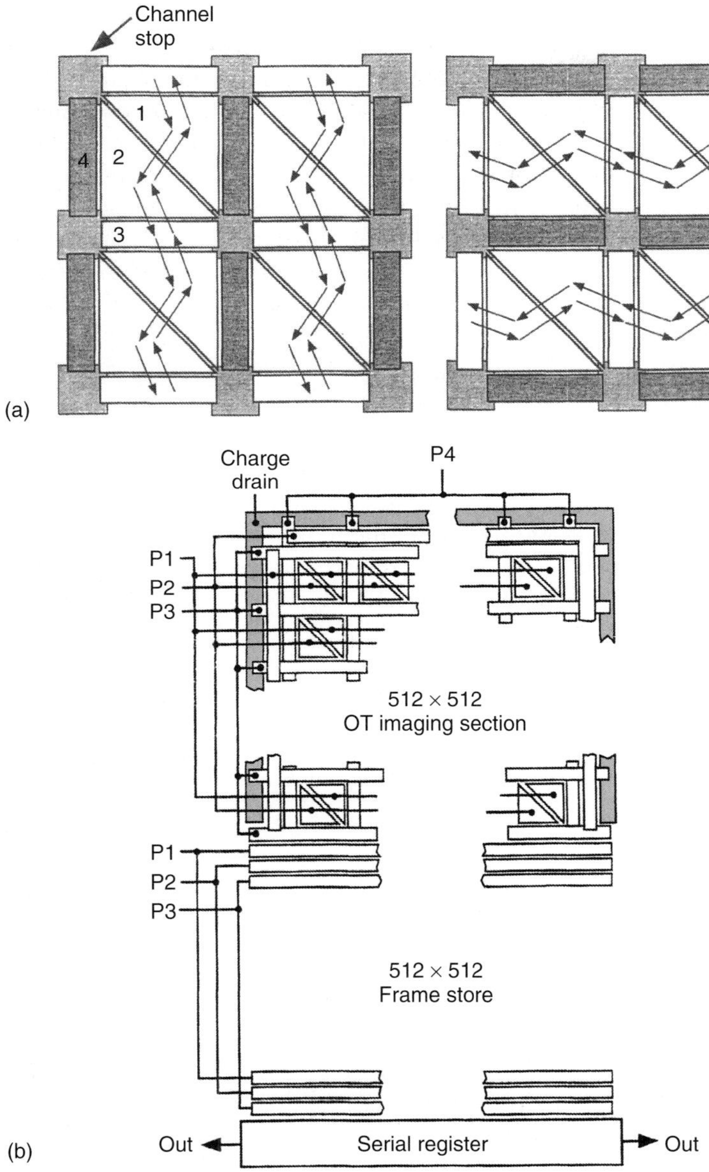


**Fig. 12** (a) Architecture of an OTCCD pixel; (b) Layout of an OTCCD chip [35, 36]

on high resistivity material [37, 38]. As noted in Eq. (13), the depletion depth, w, depends on the doping concentration of the substrate material. The devices are typically three-phase devices processed on 1000 to 1500 Ω-cm substrates and then thinned to ~30 to 50 μm. Light is incident from the back surface. The large depletion depth is desired in order to maintain the resolution and MTF of the image sensor, while at the same time providing extra material for the absorption of the photons.

The extreme in red response is the fully depleted, thinned, back-illuminated CCD [39]. This device is fabricated on 10 to 12 000 Ω-cm *n*-type material. A schematic cross section is shown in Fig. 13. It is a *p*-channel device as holes, not electrons, are the signal carriers. After processing the three-phase CCD in the conventional way, the wafer is thinned to between 200 to 300 μm. A thin layer of phosphorus-doped polysilicon is applied to the back surface to provide electrical contact to the substrate. This is followed by an antireflection (AR) coating of 60 nm Indium-Tin Oxide and 100 nm of $SiO_2$. In operation, a bias of 30 to 50 V is applied to the substrate. This voltage is sufficient to cause the depletion region to extend the full width of the thinned device (see Sect. 7.6 for performance).

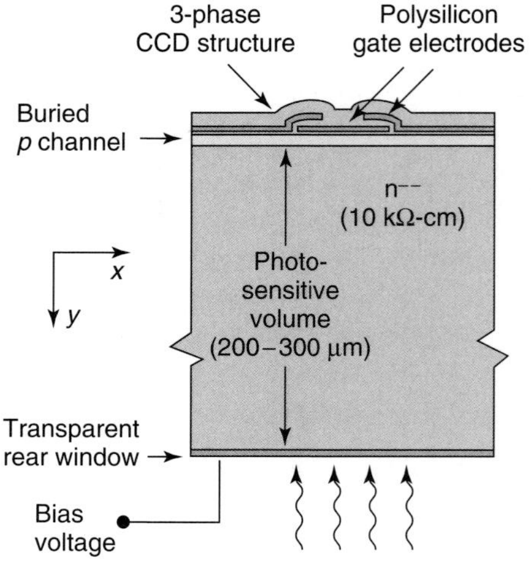


**Fig. 13** Cross section of a fully depleted CCD [39]

**CMOS/CCDs** As mentioned above, one of the long-term goals of the imager community has been to produce a camera-on-a-chip. This has largely been accomplished by the CMOS image sensors. For the CCD, this requires a marriage between the CCD process and a CMOS process. As noted in the introduction, this effort has generally not been effective. Recently, however, using Silicon-on-Insulator (SOI) technology, Suntharalingam et al. [40] have successfully fabricated a CCD and CMOS circuitry on the same wafer. The starting material is a bonded SOI wafer with a 190-nm buried oxide layer. Figure 14 shows a cross section

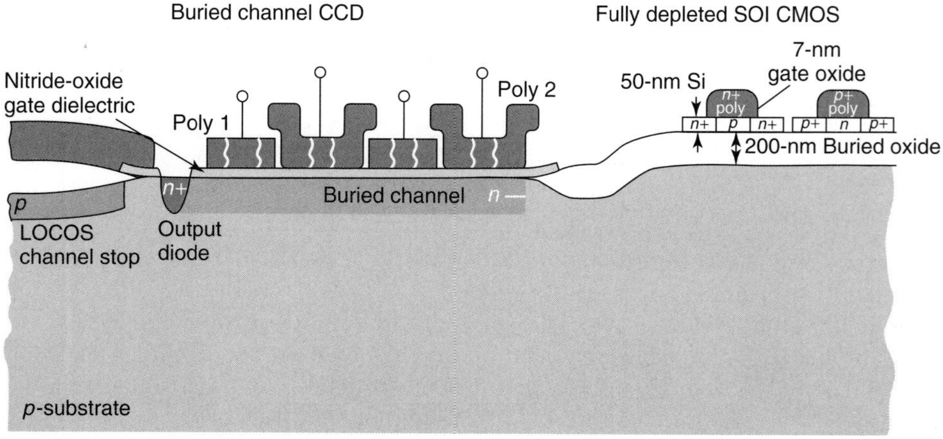


**Fig. 14** Cross section of a CCD/SOI_CMOS chip. The CCD is built on the bulk substrate, while the low-power CMOS circuitry is fabricated on the thin SOI layer [40]

illustrating the construction of such a chip. The CCD is built in the bulk part of the wafer by removing the SOI and oxide prior to forming any of the CMOS circuitry. The process uses two levels of polysilicon and three metal layers. The CCD operates with 3.3 V clocks, large full well, and excellent charge-transfer efficiency.

It should be pointed out that it is possible to build APS arrays in the same manner. The fabrication sequence requires only a minor modification of standard SOI processing. It further has the advantage of permitting lower doping of the substrate, leading to higher responsivity and lower crosstalk. One of the big advantages for both the CCD and the APS is that the wafer can be thinned resulting in 100% fill factor for the APS.

## 4.3 Color CCDs and CMOS Arrays

The FT and the ILT sensors have found widespread use in video recording and closed-circuit television (CCTV), and studio cameras. Color imaging is desired for many of these applications. A common method of providing the color is to apply red, green, and blue filters (alternatively, magenta, cyan, and yellow) directly to the devices, making certain pixels sensitive to red, green, or blue light respectively [41]. There are a number of patterns, the simplest of which may be the Bayer pattern. This pattern consists of a basic cell of $2 \times 2$ pixels with a red and a blue filter over pixels in opposite corners and green filters on the remaining two pixels. Since the eye is most sensitive to green information, there are twice as many pixels delegated to green than to either red or blue light.

After the image has been captured, the device is read out. To produce a color image, each pixel needs to have an R, a G, and a B value assigned. For a red or blue pixel, the green value is determined by linear interpolation of the four nearest-neighbor green pixels. Similarly, the red and green pixels are assigned the blue value by linear interpolation of the nearest blue signals. This creates three complete images, one each in red, blue, and green. These three images can now be manipulated to produce the required format for transmission as a color image.

## 4.4 Infrared Arrays

The spectral response of CCDs has been extended to the infrared part of the spectrum. This has been accomplished using Platinum-, Palladium- or Iridium-Silicide Schottky Barrier Diodes as the sensing element. One of the advantages of this structure is that all of the normal Si-based integrated circuit technology can be used. These devices have cut off wavelengths in the IR out to about 10 μm. The chips function as ILT CCDs where each Schottky Barrier diode is connected to a CCD pixel via a transfer gate. Although the QE is low (on the order of 10%), these devices are finding wide use with arrays of $160 \times 244$ to $1024 \times 1024$ elements being fabricated [42, 43].

For longer wavelengths, and higher QE, a hybrid approach has been utilized. In this device, a Si CCD is used both as a biasing source and as a multiplexer [44]. Both $Hg_xCd_{1-x}Te$ arrays and InSb arrays have been fabricated with up to $1\text{ K} \times 1\text{ K}$ elements. Since the late 1980s, CMOS arrays have replaced the CCD for the multiplexing function [45] and $2\text{ K} \times 2\text{ K}$ arrays are being built [46, 47].

# 5
# Pixels

## 5.1
## CCD Pixels

A number of different fabrication sequences have been developed to manufacture CCDs. These sequences utilize conventional MOS fabrication techniques and processes that are generally compatible with CMOS technology. Schematic cross sections of four of the major types of CCDs being manufactured are presented in Fig. 15.

### 5.1.1 Four Phase

The four-phase device is shown in Fig. 15(a). In this device, four gates constitute a single pixel, and there are four electrical clocks to drive the gates.

The process used to fabricate a four-phase device will serve to illustrate a representative processing sequence. Figure 16 highlights cross sections of the device at major points in the sequence. The process typically begins with a 30 to 50 Ω-cm, *p* on *p*+ epitaxial wafer. Following the growth of the oxide (45 nm) and silicon nitride (55 nm) layers (which will be the gate dielectric), the regions of the device that will represent the CCD channel and transistors are photolithographically defined. After the field implant, approximately 1 µm of $SiO_2$ (the field oxide) is thermally grown on the exposed regions of the wafer (Fig. 16b). The field implant and the field oxide combine to form channel stops that serve to isolate the active regions of the circuit from one another. The phosphorus buried channel implant is performed next, followed by the deposition of a 400-nm thick layer of doped polycrystalline silicon (polysilicon) over the wafer. This layer is photolithographically defined to form the gate electrodes (Fig. 16d). In the case of the four-phase device, the electrodes corresponding to two electrical phases are defined in this layer and alternate with one another (e.g., $\varphi 1$ and $\varphi 3$, see Fig. 15a). The electrodes are then oxidized to form an insulating layer. This insulating layer electrically isolates the first level of polysilicon from the next level of polysilicon that is then deposited. The gates for the remaining electrical phases ($\varphi 2$ and $\varphi 4$) are defined in the second layer (Fig. 16e). The source and drain regions of metal-oxide semiconductor field-effect transistors (MOSFETs) are then opened and doped with phosphorus or implanted with arsenic (Fig. 16f). A heavily phosphorus-doped reflow glass is deposited over the entire wafer to help promote good step coverage of the aluminum metallization. Contacts to the polysilicon and diodes are then opened. A 1-µm-thick layer of aluminum is deposited and patterned to form the connection between the device, the bond pads, and the outside world (Fig. 16g). Not shown in the figure are the contacts to the polysilicon gates. Finally, a 300-nm $Si_3N_4$ passivation layer is deposited and vias to the bond pads opened.

As noted above, in the process of making the four-phase device, two of the electrical phases are represented in each of the two levels of polysilicon. If, in the fabrication of the device, the polysilicon is not completely etched or removed from between the gates, two clocks will be electrically shorted. Depending on the resistivity of the short, this is frequently a fatal defect for the device.

### 5.1.2 Three Phase

Figure 15(b) illustrates the cross section of a typical three-phase register [48]. The fabrication sequence for this device is

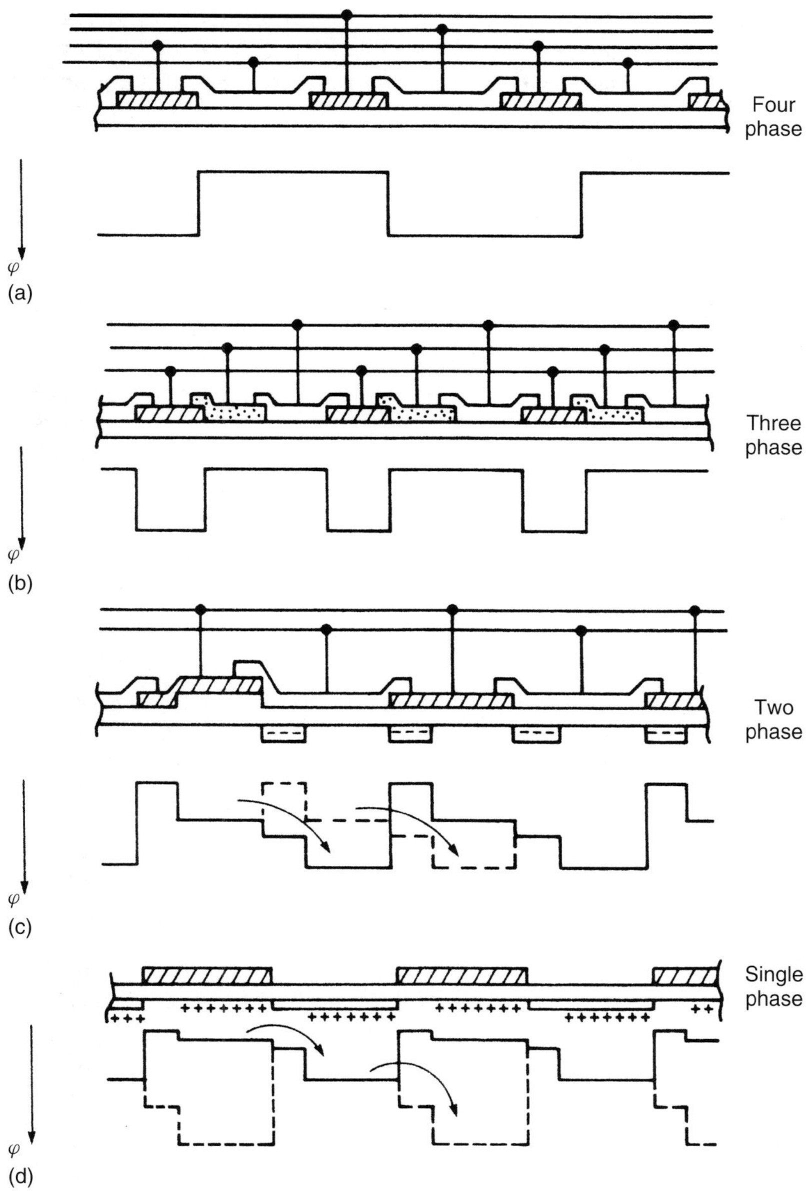


**Fig. 15** Schematic cross sections of the different CCD pixel types. (a) Four-phase device; (b) three-phase structure; (c) two-phase register; and (d) virtual-phase cross section

more complicated in that three levels of polysilicon are required: one level for each of the three gates. Again the levels of polysilicon are separated by an interlevel dielectric. If gate-to-gate shorts within a level exist because of incomplete processing, the device can still operate but will exhibit local defects in the charge

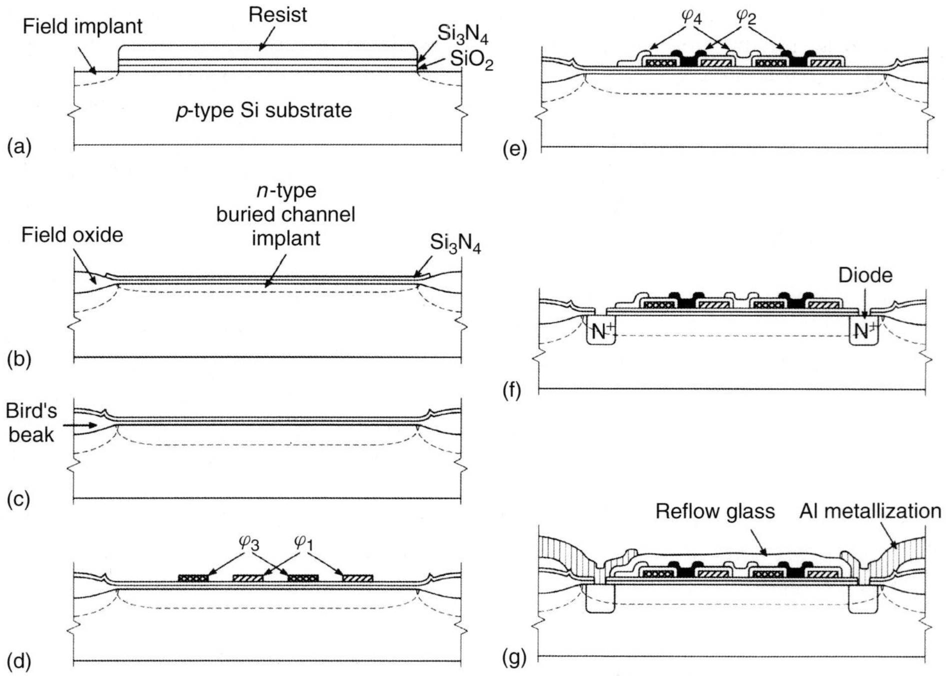


**Fig. 16** Cross section of the CCD at various stages in the fabrication process; (a). After channel definition; (b). Following growth of the field oxide; (c). Buried channel implant and drive; (d). First level polysilicon gates; (e). Second level polysilicon gates deposited and defined; (f). Diode formation; and (g). Reflow glass deposition and Al metallization

transfer process where the shorts occur. This relative lack of sensitivity to process-induced defects is a major reason why most very large area, scientific CCDs are three-phase devices. Note that as with the four-phase device, the wells and barriers are created by adjusting the clock voltages and that in both types of devices, charge packets can be made to shift in either direction.

### 5.1.3 Two Phase

There are several ways to manufacture two-phase devices, one of which is illustrated in Fig. 15(c). This process requires two layers of polysilicon and is similar to the four-phase process discussed above. The major difference is that the region beneath each gate is divided into two parts: a barrier and a well. In order to form the barrier region beneath each gate, a light boron implant is done to compensate some of the normal buried channel implant. This affects the local channel potential such that this region has a lower maximum channel potential compared to the nonboron implanted region next to it. The polysilicon gates are then deposited such that they each cover the well region and the barrier region. Each gate then controls the potential of a barrier and a well simultaneously (see Sect. 2 and Fig. 4). Good control of the photolithography and location of the implants is important in this process as misalignments can lead to undesirable potential "glitches" at the edges of the gates. These glitches tend to inhibit the transfer of charge between gates.

The two-phase device operates using two clocks that are typically 180° out of phase. With one clock biased such that the channel potential of the barrier region of the biased gate is deeper than the potential of the well region of the off gate, all the charge in that pixel will collect in the well of the biased gate. When the clocks switch states (dashed line in Fig. 15c), the charge packet will spill over the barrier and into the well of the next gate. Note that in the two-phase device the direction of charge transfer is built into the device.

The advantages offered by the two-phase structure are principally ease of fabrication and of operation since it requires only two clocks to function. Many of the CCDs built for consumer VCR and digital camera applications utilize two-phase devices and technology.

### 5.1.4 Virtual Phase

The fourth type of structure, illustrated in Fig. 15(d), is referred to as the *Virtual Phase device* [49]. The sensor is conceptually very similar to the two-phase device, but is fabricated with only one level of polysilicon. Again, every pixel consists of two regions each of which has a barrier and a well component. These regions are referred to as the *clocked* and the *virtual regions*. The clocked portion of the pixel is fabricated in a manner similar to the process described above for the two-phase device. The virtual region is processed with an additional series of implants ending with a shallow boron implant, which pins the surface potential in this region to the bulk potential. The boron dose is chosen such that the potential profile within the virtual region is fixed for all normal values of the clock voltage.

The fact that only a single level of polysilicon is required leads to high processing yields compared to the other fabrication technologies. In addition, the virtual phase device has higher QEs at short wavelengths ($\lambda < 450$ nm) than do any of the other devices when front-illuminated (see Sect. 7.6). The fact that in the virtual region, the surface potential is pinned at the bulk potential means that the channel region is isolated from the surface of the device, and hence the large contribution to the dark current that ordinarily arises because of surface states is not present (see Sect. 7.1). By operating the device such that the clocked region is also inverted during integration, the entire surface contribution to the dark current can be eliminated.

### 5.1.5 Photon-counting CCDs

The desire to perform single photon counting using a solid-state detector has led to several different approaches. A number of systems have been built using a photocathode coupled to a high voltage stage, which accelerates and focuses the photoelectrons. The accelerated electrons are detected by either a thinned CCD or a phosphor deposited on to a fiber-optic faceplate coupled to a CCD. The direct detection method suffers from radiation damage to the CCD due to the absorption of the energetic electrons. The fiber-optically coupled device experiences significant degradation in resolution (MTF, see Sect. 7.7) because the signal passes through so many interfaces.

The most promising device for the detection of single photons is based on the impact ionization that occurs in silicon in the presence of high fields. There are two versions of the device-based on three-phase technology and virtual phase technology [50, 51]. Both devices use the same operating principal although the

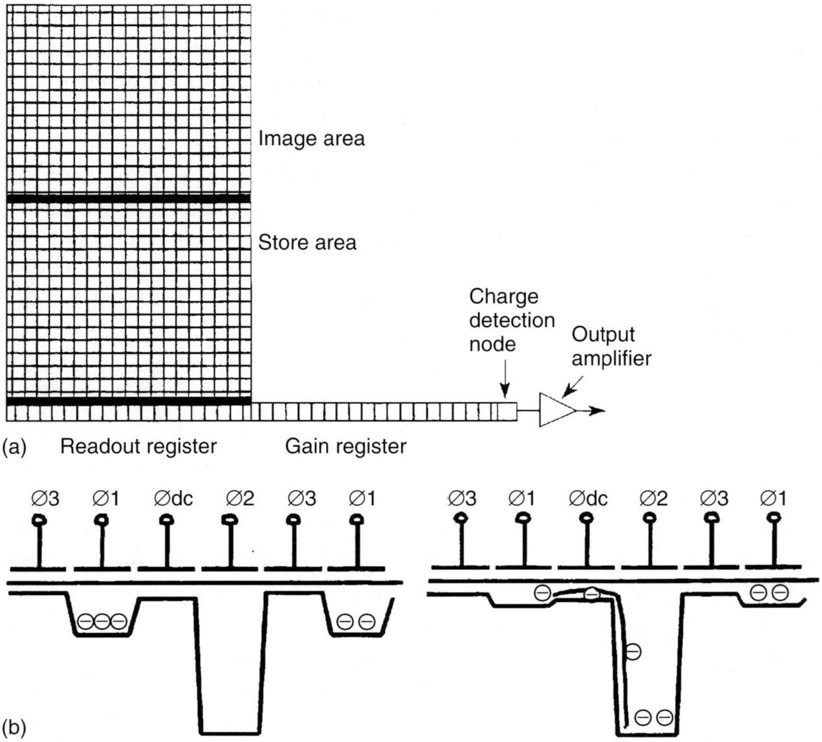


**Fig. 17** (a) Layout of a photon-counting CCD showing the location of the gain register; (b) Detail of the gain register [51]

virtual phase implementation may be more versatile. The three-phase version, the $L^3$CCD, is illustrated in Fig. 17(a). It consists of a normal FT CCD with the insertion of a gain register between the output amplifier and the normal serial register.

Each pixel in the gain register consists of four gates, three of which are clocked as in a three-phase CCD and the fourth is DC biased to a negative voltage. This is shown in Fig. 17(b). Gain is provided by avalanche multiplication. Assume phases 1 and 2 are separated by the DC gate and charge is initially stored under phase 1. When phase 2 is biased on by applying a large positive voltage to the gate, a high fringing field is created between the DC-biased gate and the phase 2 well. The DC gate allows a high fringing field to develop before the charge is transferred from phase 1 to phase 2. The voltage on phase 2 is adjusted to give an average gain per stage of $g_s < 0.015$. The overall avalanche gain for the register, $G_A$, is given by

$$G_A = (1 + g_s)^N \quad (16)$$

where $N$ is the number of stages. $N = 400$ and $g_s = 0.012$ gives an avalanche gain of $\sim$120.

The other advantage of the avalanche gain detection mode is that the noise can effectively be reduced below 1 e rms. The liability of operating in the avalanche mode is that the device behaves as though it had one half the QE [52].

## 5.2 CMOS/APS Pixels

Active pixel sensors, like CCDs, have a number of pixel structures. The simplest cell is a photodiode with a PMOS switch mentioned above [3]. It suffered from some major problems, the greatest of which were low-signal amplitudes and high noise. CCDs rapidly replaced these devices as they became available. In the early 1990s, people realized that by placing the first stage amplifier within the pixel itself, these problems could be largely ameliorated. This idea, plus the rapidly advancing CMOS technology, allowed the development of the multiple transistor cells of sizes comparable to CCD pixels. The design of these cells reflects the simplicity of construction and functionality requirements.

As mentioned earlier, one of the advantages of the APS is the ability to be fabricated using the standard CMOS process. CMOS technology refers to the integration of *n*-channel and *p*-channel MOSFETs in the same integrated circuit. Figure 18 depicts a schematic cross section of a representative CMOS structure. Illustrated in the figure are the *n*-channel and *p*-channel devices that can be built in the *p*- and *n*-wells respectively. Other important features are the silicided gates of the transistors, and the thickness (2 μm) of the epitaxial layer. The former is important because it allows higher speed operation of the device but silicided poly electrodes, like the metal lines, are opaque. The thin epi thickness means that the volume available for collecting photogenerated charges is limited. This will reduce the red sensitivity of these devices. Although the FETs occupy space, with the minimum geometry commonly at the 0.5-μm level and rapidly closing in on 0.18 μm there is room in the pixel for a few transistors.

The general CMOS process is more complex than the CCD process. Whereas the CCD process may use 10 to 15 photomasks, three levels of poly, and a single metal layer, a CMOS process can employ 25 to 35 masking steps with two layers of poly and 3 to 4 layers of metal. Common processes for making CMOS APSs use two poly layers and three layers of metal, and one poly/four metal layers.

It is of interest to minimize the number of layers used in the pixel because these layers block the light and reduce the area of the pixel available to collect light. The percentage of the pixel area sensitive to light is called the *fill factor*, and varies from ~25% to >60% depending on the light-sensing technique, design, and pixel size.

### 5.2.1 Three Transistor Cell

The most commonly used cell is the so-called *three transistor cell* or *3T cell*. The

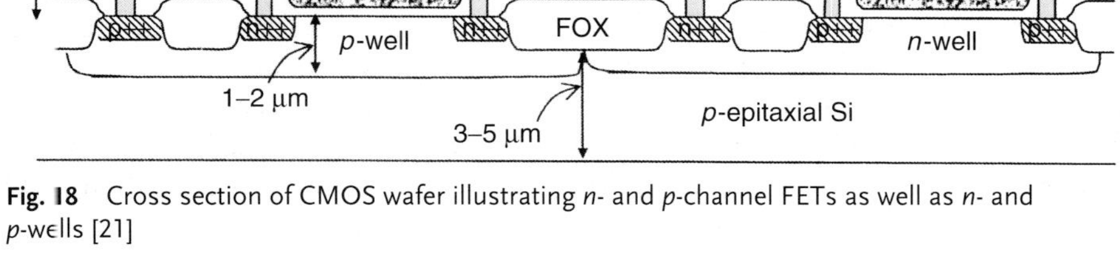


**Fig. 18** Cross section of CMOS wafer illustrating *n*- and *p*-channel FETs as well as *n*- and *p*-wells [21]

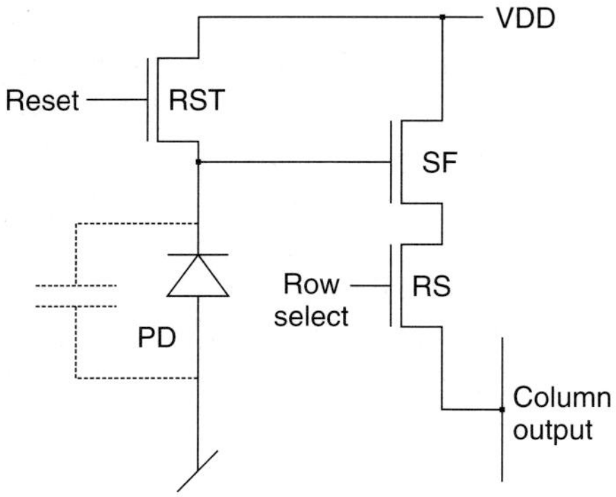


**Fig. 19** Schematic of the three transistor APS cell

pixel is shown schematically in Fig. 19. It consists of a reverse-biased photodiode as the sensing element. The photodiode is connected to the source of a reset transistor (RST), and to the gate of the source follower (SF) output transistor. The row select (RS) transistor,connects the source of the SF transistor to the column buss. Each cell has four electrical connections: the drain supply (VDD), reset, row select, and the column buss.

In operation, the photodiode is reset to the VDD voltage by momentarily turning on the reset FET. RST is then turned off and the diode allowed to discharge in proportion to the flux incident on it. After the appropriate integration time, the RS FET switches on and connects the SF FET to the load FET which is located at the end of the column and is common to all pixels in that column. The voltage across the load is sampled and is a measure of the quantity of charge collected by the diode during the integration period. Because there is no place in the cell to store charge while waiting for readout and integrating the next frame, imagers made with this pixel are read out using the rolling shutter mode.

The advantages of the 3T cell are its simplicity and higher yield, high QE, large full well, and ease of operation. Disadvantages include relatively high dark current, inability to perform correlated double sampling (CDS, see Sect. 6.3) and, thus, high noise. Since the photodiode capacitance is connected to the output diffusion, the conversion gain is sacrificed.

### 5.2.2 Four Transistor Cell

Typical four transistor cells are shown in Fig. 20. In the first example, the

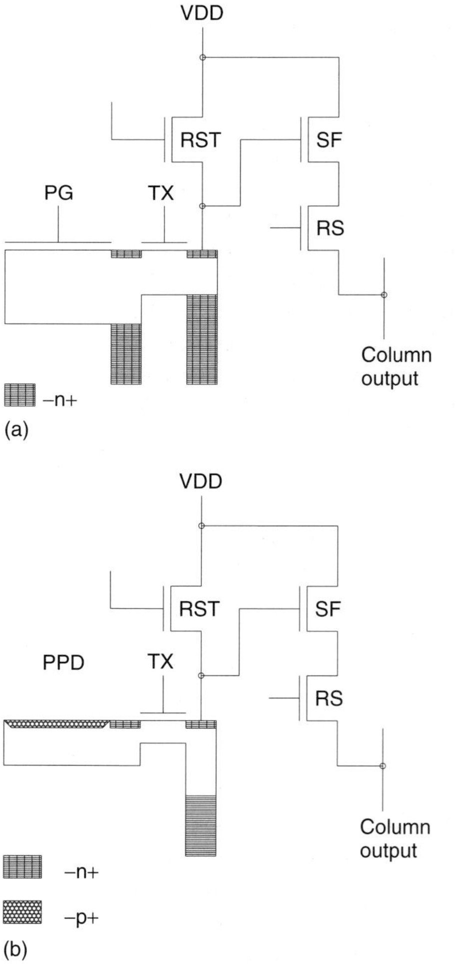


**Fig. 20** The 4T APS cell. (a) Photogate; (b) Pinned photodiode

sensing element is a photogate, PG. The second example shows a PPD. The cell contains the same FETs as the 3T cell plus an additional transistor (TX). The TX FET isolates the sensing element from the output diffusion during the integration period. The sensing element can be a photogate, or a photodiode, or a pinned photodiode.

The readout timing for the 4T cell using a photogate (PG) is shown in Fig. 20(a). The photogate is biased to VDD in order to create a well in which to collect signal charge. The TX gate is biased just above threshold voltage (0.8 V). Near the completion of the integration cycle, the RS FET turns on, connecting the SF to the column buss and the common load. As in the 3T cell, the output node is reset to VDD through the reset FET. After RST turns off, the reset level, VR, can be sampled and stored. PG is then pulsed low and the integrated signal charge transfers (*a la* CCD technology) onto the output node and the voltage, VS, across the output load is sampled. The output signal is the difference between VS and VR. This sequence is applied to each pixel in the row. Once the photogate turns off, the integration cycle begins again.

Note that the output node and all the FETs in the cell are usually covered with metal. Consequently, this cell can be used in the snapshot readout mode.

To construct the photogate structure, either two levels of polysilicon must be used with a small overlap of the PG and the TX gates. Alternatively, a single-level poly process can be used but a small gap must exist between the photogate and the TX gate. The gap is filled by an $n+$ coupling implant that aids in the transfer of charge onto the output node.

The 4T cell using the PPD is depicted in Fig. 20(b). Operation of the pixel is fundamentally the same as the PG cell except that the TX transistor is pulsed on to effect the transfer of charge from the PPD to the output node. The PPD requires some nonstandard processing in its fabrication, notably the additional $p$ and $n$ implants beneath the TX transistor. These implants have been found useful in eliminating potential pockets or barriers and promoting good charge transfer to the output node.

### 5.2.3 Other Cells

A number of pixels, including 5T and 6T cells, have been developed for special purposes. A couple of examples are illustrated in Fig. 21. The first is a 3T pixel with a logarithmic response. In this case, the 'reset' FET is biased on all the time and the current flowing through the FET is determined by the intensity of the flux falling on the diode. This performs a logarithmic transform on the voltage:

$$V_{\mathrm{OUT}} = V_o + \frac{mkT}{q} \ln\left(\frac{I}{I_o}\right) \tag{17}$$

The device is capable of a dynamic range of 6 to 7 orders of magnitude, although it requires careful calibration and is "nonlinear" at low signal levels.

The second example of a nontraditional active pixel cell is one with six FETs and two photodiodes. In this pixel, one of the photodiodes is formed by a shallow $p+$ diffusion into the $n$-well. The other junction is the $n$-well/$p$-type substrate. These two junctions are located at different depths within the Silicon. As a consequence, their spectral response is quite different: the shallow junction picks up most of the blue light while the lower junction responds more towards the red end of the spectrum.

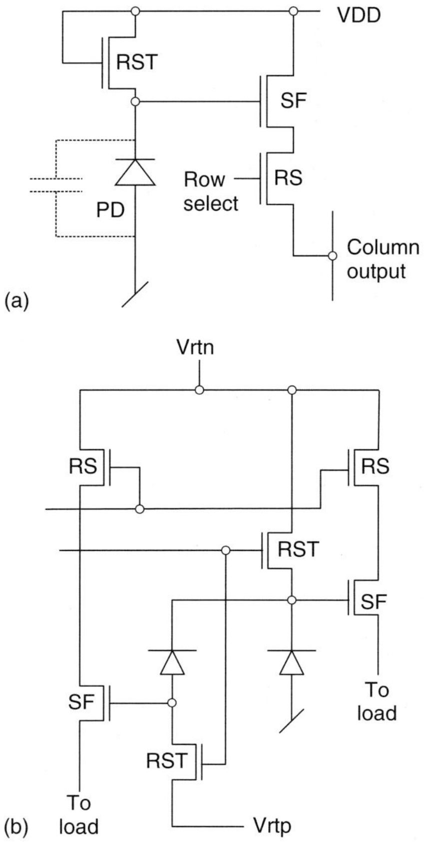


**Fig. 21** Alternative pixels. (a) 3T Cell with logarithmic response for extended dynamic range; (b) 6T cell with response to two colors

The three FETs on the right form a conventional 3T cell, while the remaining three on the left form a second 3T cell [53].

## 6 Output Amplifiers

The output of the CCD and the APS requires the detection of the quantity of charge in each charge packet. There are two common types of charge detection circuits in use today: destructive and nondestructive amplifiers. In the former amplifier circuit, the charge is sampled once and then the packet is transferred to a drain and destroyed. With the nondestructive readout the charge is sampled as it passes down the CCD channel. After sampling, the packet is transferred to the next pixel where it may be resampled. Virtually all CCDs and APSs utilize the destructive output.

### 6.1 Floating Diffusion

The most common form of readout circuit is the charge integration or floating diffusion amplifier. This is used in both the CCD and CMOS arrays. The circuit is shown schematically in Fig. 22(a).

In the case of the CCD, the channel is terminated in a diode (the output diode) that is also the source of the RST. The output diode functions as a capacitor and is biased to the voltage set by the drain of the RST. The diode is also connected to the gate of the output MOSFET. The output MOSFET is operated as a SF with either an on-chip or off-chip load.

A typical output waveform is shown in Fig. 22(b). At time $t_0$ the RST is turned on, charging the output diode to the drain voltage of the reset FET. With the diode charged and the RST turned off at $t_1$, the output voltage settles to some value. At $t_2$, a charge packet from the CCD, containing signal charge, $Q_S$, is clocked onto the diode. This results in a voltage change on the diode given by

$$\Delta V = \left(\frac{Q_S}{C_T}\right) \tag{18}$$

where $C_T$ is the total capacitance of the output node including parasitics. This voltage change is simultaneously applied to the gate of the output MOSFET and sensed as a change in the voltage across

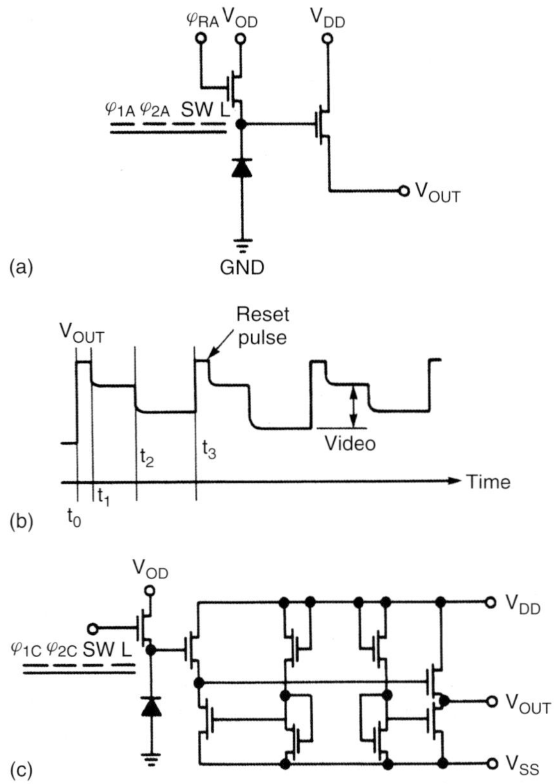


**Fig. 22** Typical floating diffusion output amplifiers. (a) Single-stage amplifier; (b) output waveform; and (c) two-stage source follower with on-chip biasing networks for the load MOSFETs. The floating diffusion amplifier is a destructive readout method

the load resistor, $R_L$. The precharge and read cycle is then repeated for the next charge packet beginning at $t_3$. Because the charge is removed by the RST in each cycle, this is a destructive readout scheme.

The gain of the SF is

$$G_{SF} = \frac{g_m R_L}{1 + g_m R_L} \tag{19}$$

where $g_m$ is the transconductance of the output MOSFET. The overall conversion gain at the output of the detection circuit is then

$$S_o = \frac{qG_{SF}}{C_T} \tag{20}$$

Typical values of the sensitivity at the output of CCD chips are 3 to 8 μV/electron. Higher values (>10–20 μV/e) can be obtained with careful design, reduction of the output FET size, and of the parasitic capacitances [35]. In APS sensors, the sensitivity can be as high as 40 to 50 μV/electron depending on the pixel technology used. This is possible partly because of the small geometries of the FETs.

In CCDs, a common variation of the simple SF circuit is the dual SF shown in Fig. 22(c). This circuit is typically used in situations in which high-speed output is desired and a large MOSFET with high current capability is needed to drive the off-chip circuitry. Most VCR and TV camera CCD chips use dual or treble SF circuits. This is unnecessary for APS chips as frequently the data are digitized on-chip.

## 6.2 Floating Gate

The second type of output circuit involves sensing the voltage change on a floating gate as the charge packet passes beneath it [54]. The voltage on the floating gate is set once a line or once a frame either by a capacitive divider consisting of a floating gate situated between a DC-biased gate above and the channel below or by a separate reset MOSFET. The voltage change detected by the floating gate is given by Eq. (18), where $C_T$ is the effective capacitance of the floating gate as seen by the signal charge. Typical sensitivities are again in the 1 to 5 μV/electron range. Once the floating gate has sensed the signal, the charge is moved along the channel to the next pixel where it may be resampled by a second floating gate amplifier. Alternatively, one can fabricate a single-stage floating gate amplifier and repeatedly clock the charge packet beneath the floating gate and sample the resulting signal many times. This technique has been employed in the "skipper amplifier" and has been used to detect signals as small as 1 electron [55].

## 6.3 CMOS Column Amplifiers

In the case of the APS array, each pixel contains a floating diffusion amplifier. The load for the in-pixel SF is located at the bottom of the column and is common to all pixels in that column. There are two configurations frequently used as the load as depicted in Fig. 23. In the first configuration, the current that flows in SF is reproduced in FET L by the current mirror. L acts as a buffer and stores the signal voltage, $V_{sig}$, for use by the following amplifier. It is crucial that

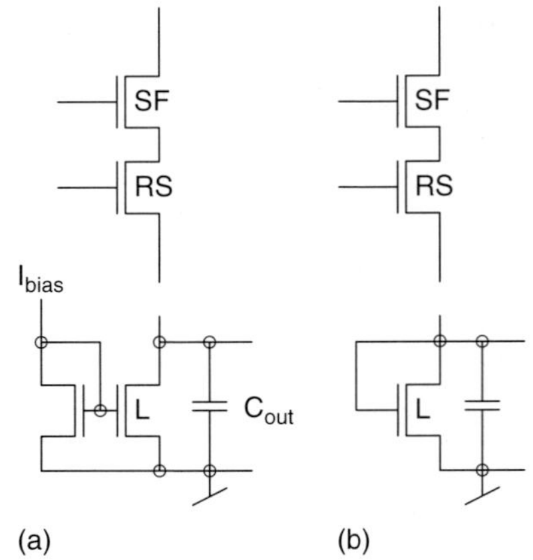


**Fig. 23** Load configurations for column amplifiers. (a) Current mirror; (b) current source

SF and L have the same dimensions in order to replicate the signal accurately. Furthermore, mismatches between the threshold voltages of $L_i$ and $L_j$ lead to differences in global offset for each column since all pixels in columns i and j pass through $L_i$ and $L_j$, respectively. This leads to a column-to-column FPN that is common to all APSs. For the configuration using the column mirror, the FPN issue is further compounded by mismatches in $V_{th}$ between $SF_i$ and $L_i$ within the column. As the separation between the two FETs increases, so also does the variation in $V_{th}$, leading to a vertical shading in a flatfield image. An example of the FPN is illustrated in Fig. 24. Because of the low output impedance, very high-speed operation is possible.

The more common approach to the SF load is shown in Fig. 23(b). In this case, the SF transistor is used as a normal SF with a load FET operated as a current source. The signal voltage, $V_{sig}$, is fed directly into the output amplifier. The FPN in this configuration is due mainly to variations in the threshold voltages of the $SF_i$ FETs. However, since these transistors are fairly

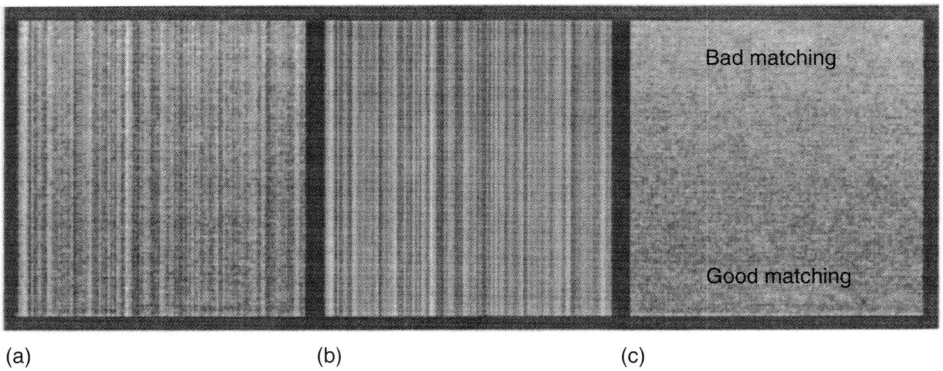


**Fig. 24** Examples of fixed-pattern noise. (a) Total noise showing pixel and column-to-column variation due to threshold voltage differences; (b) Column-to-column FPN only; (c) Noise due to pixel-to-pixel variations. Note top-to-bottom gradient [30]

close to one another the local variations are small and the FPN is improved. On the other hand, the relatively large capacitance of the column limits the speed with which these chips can be read out.

An example of an output circuit is schematically illustrated in Fig. 25. Following the column load is the column amplifier and a double delta sampling (DDS) circuit developed at JPL [56]. The column amplifier consists of two sample-and-hold circuits and is the same for each column. The output of each sample-and-hold circuit is a source follower, CSF, and a column select (CLS) FET, which couples the column amplifier to the load FET, DSL, in the DDS circuit. In operation, when the row is selected, the pixel SF is connected to the column load, and the output node is reset. At $t_1$, the FET SHR turns on and a sample of the reset level, $V_{res}$, is stored on the capacitor CR. At $t_2$, FET SHS turns on, the signal is transferred to the output node, and a sample, $V_{sig}$, is stored on capacitor CS. This process occurs in parallel for all pixels in the selected row.

The column amplifiers are now addressed sequentially through the column select FETs. Sensing the signal also proceeds in two steps. At $t_3$, with the CLS FETs turned on for column j, the CL FETs turn on and clamp the inputs to the output amplifier to $V_{CL}$. At the same time, the output of the circuit source followers (CSFs) is being stored on the capacitors CCS and CCR. Thus

$$VS(t_3) = V_{sig} - V_{thcsfS} \quad \text{and}$$
$$VR(t_3) = V_{res} - V_{thcsfR} \qquad (21)$$

where $V_{thcsfS}$ and $V_{thcsfR}$ are the threshold voltages of FETs CSFS and CSFR, respectively. At $t_4$, the clamp switches have turned off and the DDS FET has turned on, shorting CS and CR. The voltage at the gates of the CSFs is now the average of $V_{sig}$ and $V_{res}$. The voltages at the output of the CSF FETs are

$$VS(t_4) = \left(\frac{V_{sig} + V_{res}}{2}\right) - V_{thcsfS} \quad \text{and}$$
$$VR(t_4) = \left(\frac{V_{sig} + V_{res}}{2}\right) - V_{thcsfR} \qquad (22)$$

and the voltage at the output of the output amplifier, $V_{OUT}$, is:

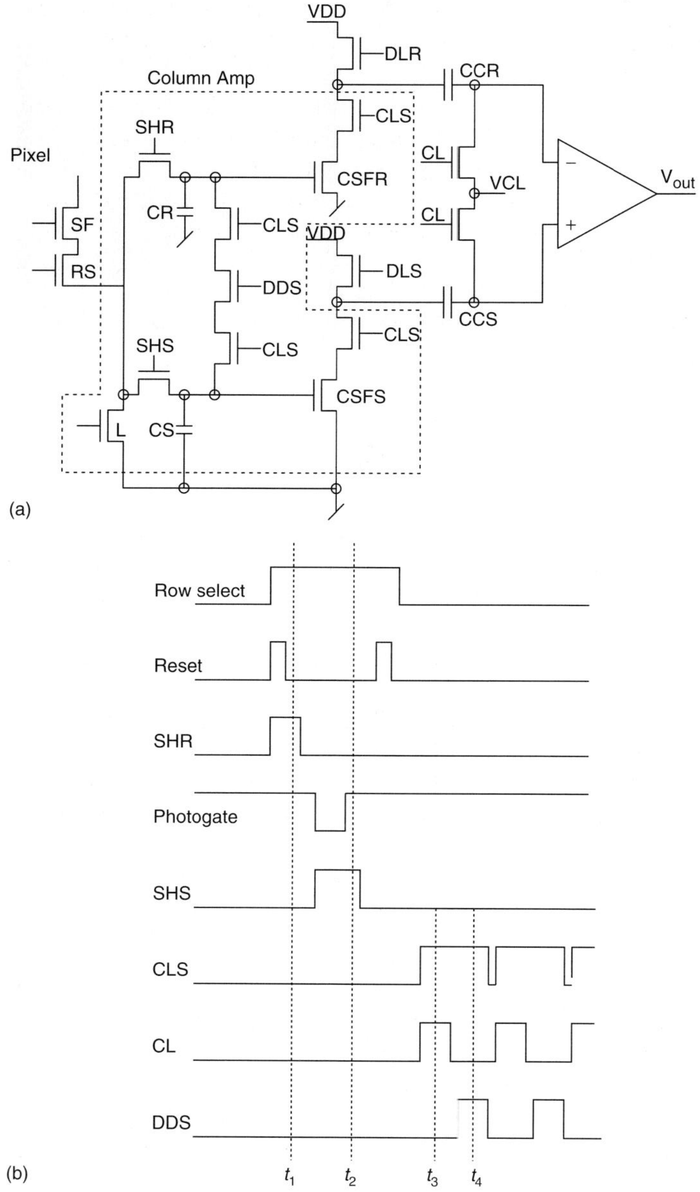


**Fig. 25** (a) Double Delta Sampling output circuit; (b) Timing for readout [21]

$$
\begin{aligned}
V_{\mathrm{OUT}} &= (VR(t_3) - VR(t_4)) \\
&\quad - (VS(t_3) - VS(t_4)) - V_{\mathrm{th}} \\
&= (V_{\mathrm{res}} - V_{\mathrm{sig}}) - V_{\mathrm{th}} \qquad (23)
\end{aligned}
$$

The DDS circuit performs CDS operation on the pixel signal. Note that the threshold voltages of the CSF FETs have disappeared and only the offset of the amplifier itself remains. The DDS circuit has proved to be very useful in removing column-to-column FPN.

## 7 Performance

There are a number of criteria by which the performance of an image sensor is measured. The major parameters of interest are the noise, dark current, charge transfer efficiency, well capacity, output sensitivity, modulation transfer function, and quantum efficiency. In this section, we will briefly discuss each parameter and indicate current state-of-the-art values where appropriate.

### 7.1 Dark Current

Dark current is the charge collected in the pixels in the absence of any intentional optical input. It arises because of the normal thermal generation processes that occur in any material at a temperature $T$. There are three major sources for dark current in CCDs and APSs. These dark current contributions are as follows:

1. the generation of electrons in surface states at the Si-$SiO_2$ interface, $J_{ss}$,
2. the generation of electrons in the depletion region of the photogate or photodiode by midgap impurity generation-recombination centers in the silicon, $J_{dep}$, and
3. the thermal generation of charge in the neutral bulk and the diffusion of that charge to the collection site, $J_b$.

Using typical values for the parameters in Table 1, it is easy to show that near room temperature $J_{ss} \gg J_{dep} \gg J_b$. In the case of CMOS sensors, there is a fourth component, namely, field assisted tunneling of carriers across the band gap in the presence of strong fields which exist in the narrow, highly doped drain regions of the small FETs [57].

It should be noted that, with the exception of tunneling, the dark current depends upon the intrinsic carrier concentration, $n_i$, and is thus exponentially dependent on

**Tab. 1** Contributions to the dark current

| *Source* | *Functional dependence* | *Typical value* |
|---|---|---|
| Surface state generation | $J_{ss} = qn_i\sigma_s v_{th} N_{ss}\pi kT/2$ | 0.2–10 nA cm$^{-2}$ |
| Generation in the depletion region | $J_{dep} = qn_i x_d/(2\tau_g)$ | 0.05–2 nA cm$^{-2}$ |
| Generation in the neutral bulk | $J_b = qn_i^2 L_n/(N_A\tau_n)$ | 1–20 pA cm$^{-2}$ |

$q = 1.6 \times 10^{-19}$ Coulomb
$k = 1.38 \times 10^{-23}$ J K$^{-1}$
$\sigma_s$ = capture cross section of interface states ($10^{-13}$–$10^{-15}$ cm$^2$)
$v_{th}$ = thermal velocity ($10^6$ cm s$^{-1}$)
$N_{ss}$ = density of interface states ($10^9$–$10^{10}$ cm$^{-2}$ eV$^{-1}$)
$\tau_g$ = generation lifetime (0.5–5 ms)
$L_n$ = electron diffusion length (0.01–0.02 cm)
$\tau_n$ = minority carrier lifetime (0.3–3 ms)
$N_A = 5 \times 10^{14}$ atoms cm$^3$

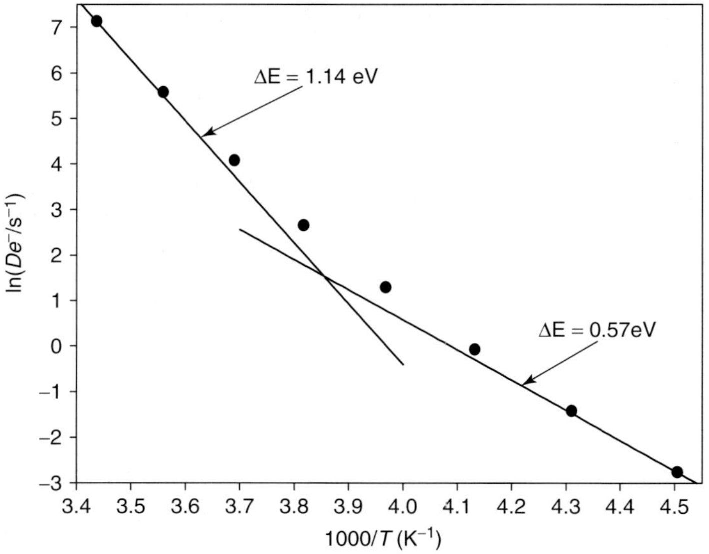


**Fig. 26** Arrhenius plot of dark current versus 1/T. Note the change in slope from $E_g/2$ to $E_g$ [58]

the temperature:

$$n_i(T) = CT^{3/2} \exp\left(-\frac{E_g}{2kT}\right) \quad (24)$$

where $C$ is a constant ($\cong 4.9 \times 10^{15}/\text{cm}^3\text{-K}^{3/2}$) $E_g$ is the energy gap, and $k$ is Boltzmann's constant. Figure 26 presents an Arrhenius plot of the dark current of a CCD sensor. Note that at high temperature the dark current varies with an activation energy of $E_a = 1.14$ eV, very close to the band gap of Silicon and implying that the dark current is dominated by diffusion from the bulk ($\sim n_i^2$). At low temperature, the variation is proportional to $E_g/2$ implying that the source is either the surface state contribution or the g-r centers that control the dark current [58]. For most scientific applications, the sensor is cooled in order to eliminate the dark current. In the case of CCDs, operating in the multipinned phase (MPP) mode, one can easily achieve an average dark current of 3 electrons/pixel/hr at −80 °C.

APSs that use the photodiode or photogate tend to have dark currents that are dominated by the surface state contribution. The dark current for photodiode-type devices is of the order 0.5 to 2 nA cm$^{-2}$ while for photogate pixels it is ~100 pA cm$^{-2}$, at room temperature.

#### 7.1.1 MPP Operation

The Multipinned Phase or MPP is an important CCD variation and is designed to specifically capitalize on the ability to reduce the dark current by integrating with the surface inverted [59]. Under conditions of inversion, the surface states are filled with holes and any electron that is generated quickly recombines with a free hole.

Figure 27(a) shows a schematic cross section of a typical three-phase MPP device. Two of the phases are fabricated as normally made. A light $p$-type implant is implanted into the CCD channel prior to formation of the third gate.

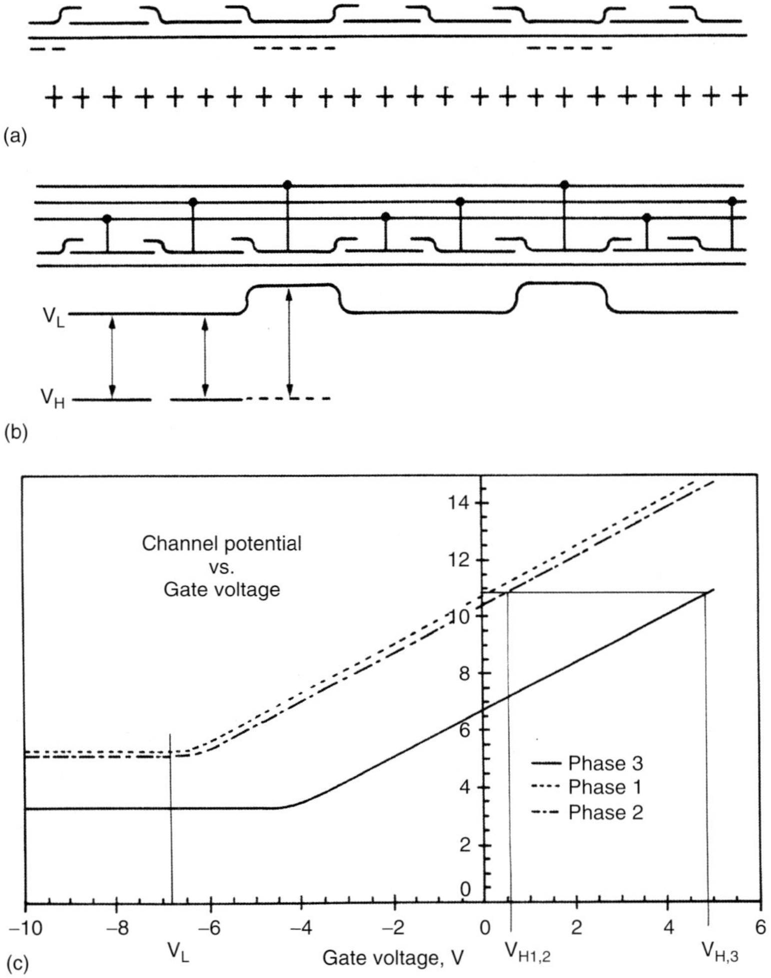


**Fig. 27** (a) Illustration of a three-phase Multipinned Phase (MPP) device. The extra *p*-type MPP implant is under the phase three electrodes. (b) With all the electrodes biased into inversion, the potential under phase three gates forms a barrier and confines the signal charge to a pixel. (c) Channel potential as a function of gate voltage for an MPP CCD

This implant shifts the channel potential under the third phase relative to the first and second phases. Measured curves of channel potential as a function of gate voltage are shown in Fig. 27(c). The upper curves represent the empty channel potential for first and second phases, while the lower curve represents the potential for the third phase. During integration, all three phases are biased into inversion. As seen from the channel potential curves, a barrier between pixels exists since the potential under the phase one and phase two gates is deeper than that of the phase three gate. Charge transfer occurs in the normal fashion; one simply must ensure that the channel potential under phase three is higher when phase three is on than the off potential of phases one and two. A similar technique works for the four-phase device. MPP CCDs are currently available with dark currents in the 1 to 10 pA cm$^{-2}$ range.

## 7.2 Charge Transfer Efficiency

The charge transfer efficiency (CTE) is fundamentally a CCD related parameter. One of the motivations for the development of the APS is the desire to eliminate the problems associated with the movement of charge packets long distances in large devices. The CTE is the parameter that describes the ability of the device to efficiently transfer charge from one pixel to its neighbor in a CCD. Charge transfer does come into play with the APS when moving charge from the PG or the PD onto the floating diffusion, a one-time transfer.

In area arrays, the charge must move first in the column direction and then in the serial direction toward the output amplifier. Consequently, there are two CTEs that are measured: parallel and serial CTEs, respectively.

Two general techniques are used to measure the CTE of CCDs. These involve evaluating the performance of the device when (1) a single pulse is inserted at various points in the device, or (2) a string of unit-sized pulses or a step function is used. The major difference between the two methods is that in one case the system has come to a steady-state condition with the string of ones being input and in the other technique we are dealing with an impulse response.

Detailed analysis of the first technique yields an expression for the charge transfer efficiency

$$\varepsilon = \left(1 - \frac{\sum\limits_{i>m} S_i}{S_o}\right)^{1/N} \tag{25}$$

where $\sum S_i$ is the sum of all the charge in the pixels following the pulse at pixel m, $S_o = S_m + \sum S_i$ represents the size of the initial charge packet and $N$ is the number of pixels through which the target pixel was transferred. This process assumes proportional losses and that none of the charge is lost. The CTE, $\varepsilon$, and the inefficiency are simply related by $\varepsilon = 1 - \eta$. Note that with this technique, the excess charges in the pixels following the target pixel in both the serial and the parallel directions can be examined simultaneously. Since the size of the charge packet is well known, the method provides an absolute method of measuring the CTE.

A very effective method of implementing this technique is to use an X-ray photon as the source of the charge [60]. When the X-ray photon is absorbed, it produces an energetic electron, which, in turn, creates a cloud of electron–hole pairs in its immediate neighborhood. The size of this cloud is such that frequently the entire charge packet is collected in one pixel. This technique has the additional beneficial effect of producing a charge packet of statistically well-known size. For example, the $Fe^{55}$, $K_\alpha$ X-ray photon at 5.9 keV generates about 1620 electrons/photon. Figure 28(a) presents a histogram of X-ray events. Note the well-defined peak at 1620 electrons and the smaller peak at higher energy that represents the $Fe^{55}$ $K_\beta$ photon at 6.4 keV. The size of the charge packet (i.e., 1620 electrons) and the full width at half maximum (FWHM) of

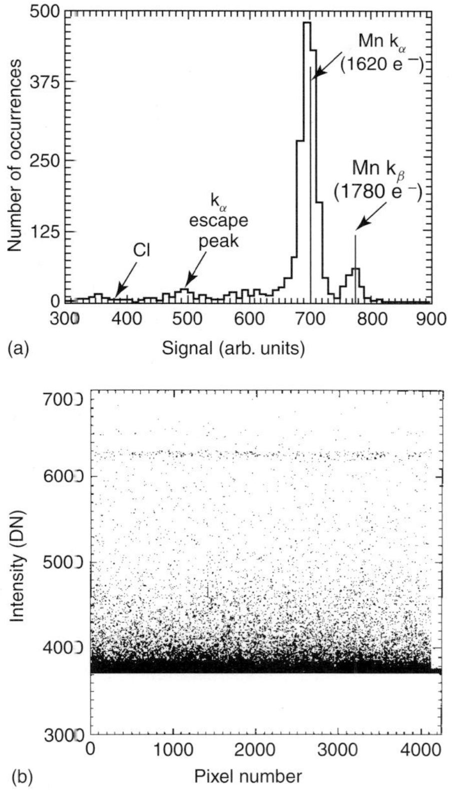


**Fig. 28** (a) X-ray histogram generated by 2250 individual events using an Fe55 source. Note the excellent separation between the $K_\alpha$ and the $K_\beta$ peaks [61]. (b) Column plot of Fe55 events for a 2 K × 4 K CCD indicating the high CTE

the distribution are indicative of both the low noise as well as the excellent CTE of this device. Figure 28(b) presents a column stack plot for a 2 K × 4 K imager. The high CTE is evidenced by the horizontal nature of the stacked $K_\alpha$ X-ray events, implying that the size of the X-ray event in the last row is as large as the X-ray event in the first row.

Very high CTEs are required in order to maintain the integrity of the charge in a packet in the large devices such as the 2048 × 2048 arrays. CTEs of >0.999999 per pixel are routinely available.

In APS sensors that use the 4T cell, the signal charge must be transferred from the sensing site to the output node. This transfer is identical to the process that occurs in CCDs. However, because of the limited voltage swing available, high CTE is frequently not achieved.

## 7.3 Linearity

The transfer function of an imaging system can be represented by a power series that relates the output signal to the input signal and the system offset. The system offset represents the signal that is measured in the absence of any intentional input. In the simpler cases, the expression takes the form

$$S_{\text{out}} = KS_{\text{in}}^{\gamma} + S_{\text{dark}} + S_{\text{off}} \qquad (26)$$

where $S_{\text{in}}$ is the input signal (photons), $S_{\text{out}}$ is the output signal (electrons), $K$ is the proportionality constant, $\gamma$ is the contrast, $S_{\text{dark}}$ is the dark signal, and $S_{\text{off}}$ is the system offset. For a linear device, $\gamma = 1$.

### 7.3.1 CCD

The linearity of a CCD depends on a number of factors. Principal among these factors is the charge detection circuit. Imagers depend upon optical input, which is a very linear process in these devices. At the output of the CCD, the capacitance of the floating diffusion and the $g_m$ of the output MOSFET are voltage dependent. As long as the voltage swing at the output node is not too large, this process can also be very linear. Measurements on a number of typical scientific type CCDs have shown $\gamma = 1.000 \pm 0.0002$, making these devices some of the most linear optical transducers known [62].

Because the detection process is so linear, and because the arrival of photons is governed by Poisson statistics, it is possible to show that the conversion gain is given as the reciprocal of the slope of the graph of mean signal versus variance [20]. That is,

$$C_g^{-1} = \frac{N^2}{S_{out}} \tag{27}$$

where both the noise $N$ and $S_{out}$ are in units of ADU. $C_g$ then has the units of e/ADU. Note that it is not necessary to calibrate the electronics between the CCD and the computer in order to calculate the conversion gain.

The linearity is an extremely important property for scientific applications where accurate measurement of the incident photon flux is desired. The linear response allows one to calibrate each pixel with a computationally simple algorithm:

$$Q_i = \frac{\mu_c}{a_i}(S_i - b_i) \tag{28}$$

where $Q_i$ is the corrected signal and $S_i$ is the measured signal in the $i$th pixel, respectively. $\mu_c$ is the mean value of the calibration frame, $a_i$ normalizes the response of the $i^{th}$ pixel (e.g., corrects for differences in QE), and $b_i$ corrects for differences in the offset (e.g., dark current, bias charge). $\mu_c$, $a_i$, $b_i$ are all determined from flat-field frames or dark frames.

#### 7.3.2 APS Linearity

The linearity problem is more acute in the APS. The source is again the output amplifier plus the photodiode or photogate. The problem is that the capacitances of these junctions are voltage dependent (see Eq. 2). While in the CCD, the output node is biased to 13 to 15 V and a full well signal produces a 1 to 2 V swing on the output node, in the APS the same 1 V swing on the output diode is a substantial fraction of the 3.3 V bias.

Figure 29 presents data of the response of a CMOS sensor as a function of photon flux. The nonlinearity manifests itself as the curvature at high signal levels. Nonlinearities are reported for the various detectors ranging from $< \sim 1\%$ for the PG detector to 2 to 3% for the photodiode. These values for the nonlinearity are quoted for signals <80% full well.

### 7.4 Noise

Ideally, for an imager, the noise should be dominated by the arrival statistics of photons in the scene. The total noise, $N_T$, can be expressed as

$$N_T = \sqrt{\eta \Phi A_{pix} \tau_{int} + N_{FPN}^2 + N_{dark} + N_{read}^2 + \frac{C_g^2}{12}} \tag{29}$$

where $N_{FPN}$ is the FPN, $N_{dark}$ is the average number of electrons in the dark current, $N_{read}$ is the read noise associated with the on-chip amplifier, and $C_g^2/12$ is the quantization noise associated with the digitizer.

Figure 30 shows a typical curve of noise as a function of signal for a high-performance CCD illuminated uniformly at increasing exposure levels. At the very low signal levels, the measured noise is dominated by the thermal and 1/f noise associated with the output MOSFET. At very large signals the noise is linearly related to the signal, and represents an FPN due to variations in the pixel response. At intermediate levels, the noise is dominated by photon statistics as indicated by the line of slope 1/2. Because of the linearity and stability of the CCD, it is possible to accurately normalize the pixel

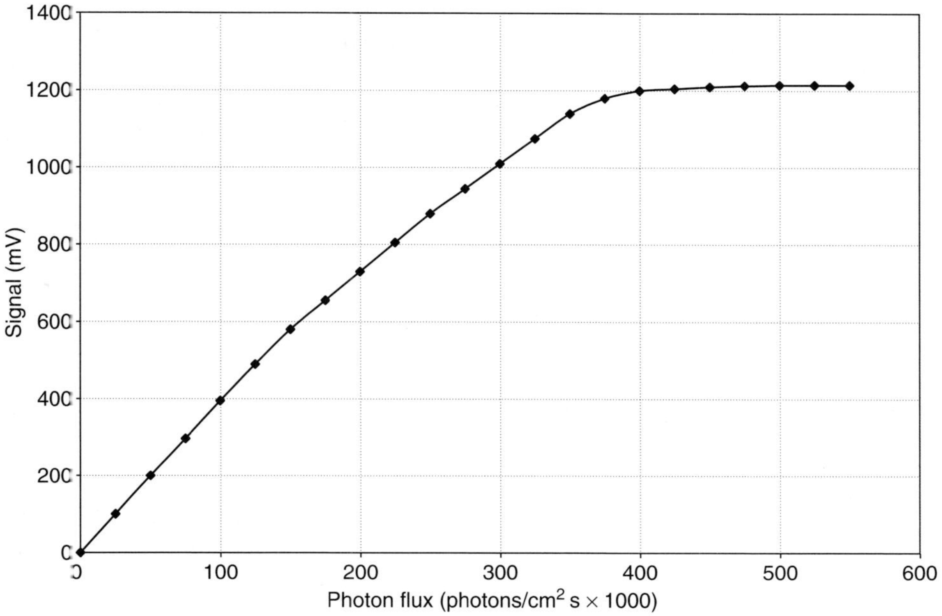


**Fig. 29** Output signal as a function of input photon flux for an APS

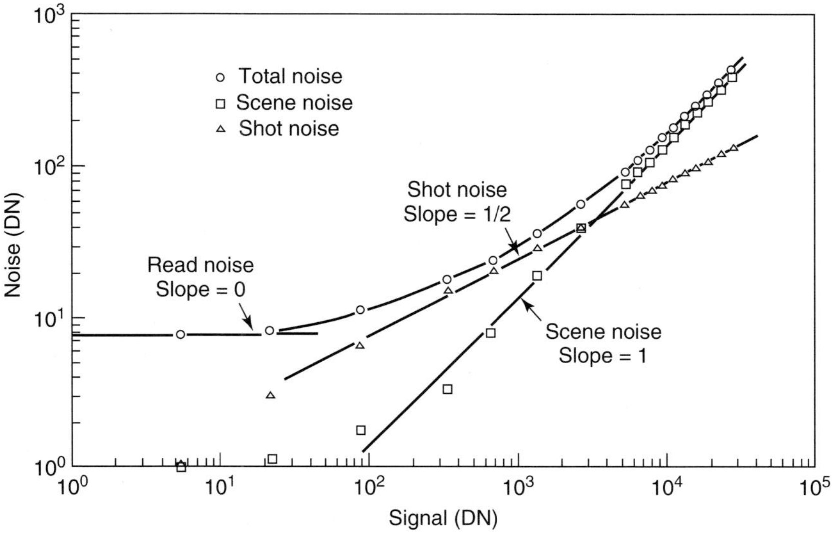


**Fig. 30** Typical graph of noise as a function of signal. The curve is divided into three regimes: read noise, signal shot noise, and pixel-to-pixel fixed-pattern noise. Note that with careful calibration and renormalization, the device is signal shot-noise-limited over most of the dynamic range of operation [62]

response variations with the result that the device is indeed photon-noise-limited over most of its dynamic range. To be photon-noise-limited is the best performance one can achieve.

### 7.4.1 Input Noise

Optically, input charge also has a noise associated with it since the arrival of photons follows Poisson statistics. As mentioned above, this noise is proportional to the square root of the number of collected charges. There is also an FPN due to optical input. This fixed pattern arises due to pixel-to-pixel variations in the QE and/or collecting area of each pixel. Noise of this type is dominant at high signal levels but can be removed by flat fielding (see Sect. 7.3.1).

The third source of input noise is the thermally generated dark current collected during the integration and readout cycles. As with optical generation, the dark current is also governed by Poisson statistics and produces a shot noise proportional to the square root of the average dark current. This produces additional FPN, which can easily be removed by subsequent image processing or simply cooling the device.

### 7.4.2 CCD Readout Noise

The major source of noise in modern CCDs is the charge detection and readout circuit. This consists of two parts. During the resetting operation, the floating diffusion is charged through the resistive channel of the reset MOSFET to a reference voltage [63]. This charging operation introduces an uncertainty in the voltage of the floating diffusion, which may be expressed as an rms equivalent number of electrons,

$$N_{kTC} = \frac{\sqrt{kTC_T}}{q} \tag{30}$$

where $C_T$ is the total capacitance attached to the floating diffusion. For a typical device, this can be ~100 electrons. CDS techniques can remove virtually all this noise [64]. For CCD systems not utilizing CDS circuits, the kTC noise is the dominant source.

The remaining source of noise is the noise of the MOSFET itself. In a typical scientific CCD application, the source of the output MOSFET is connected to an off-chip load resistor and the circuit operated as a source follower. The signal is amplified and passes through a CDS circuit before being digitized. Assuming a dual slope integrator type CDS circuit, the rms read noise in electrons, $N_r$, is given by the expression

$$N_{\mathrm{FET}} = \left(\frac{2}{S_o}\right) \times \sqrt{\left\{\int_0^{\infty} S_V(f)\frac{\sin^4(\pi f\tau)}{(\pi f\tau)^2}\,\mathrm{d}f\right\}} \tag{31}$$

where $S_o$ is the output sensitivity of the chip, $S_V(f)$ is the power spectral density of the voltage noise of the MOSFET (in $V^2$/Hz), $H_{CDS}(f) = \sin^4(\pi f\tau)/(\pi f\tau)^2$ is the transfer function of the dual slope integrator type CDS circuit, and $\tau$ is the integration time of the CDS circuit. A $1/f$ component and a white noise floor usually characterize the noise-power spectral density of the MOSFET:

$$S_V(f) = \left(\frac{8kT}{3g_m}\right)\left(\left(\frac{f_c}{f}\right)^m + 1\right) \tag{32}$$

where m is the exponent of the $1/f$ component of the noise and $f_c$ is the corner frequency. (m can take on values between 0.7 and 1.5, and $f_c$ typically lies somewhere between a few kHz and a few MHz depending on the fabrication process and operating conditions of the

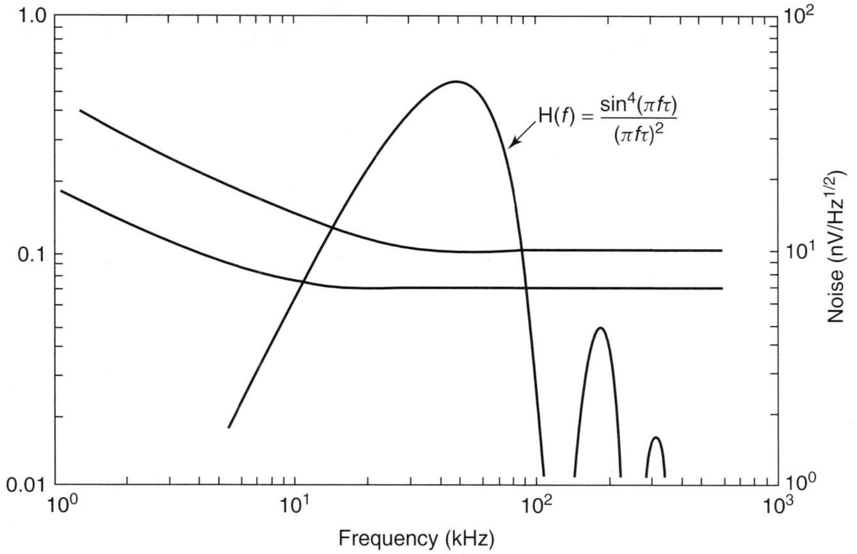


**Fig. 31** Noise spectra of typical scientific CCD output amplifiers and the transfer curve of a dual slope CDS circuit, assuming an integration time of $\tau = 8$ µsec in Eq. (31). The two noise spectra yield calculated values of the rms noise of 2.6 and 4.2 e

MOSFET.) In addition to removing the reset noise, the CDS circuit is also effective in removing MOSFET $1/f$ noise. Fig. 31 presents a graph of the CDS transfer function and noise spectra from typical MOSFETs. Read noise as low as 1.6 to 2 rms electrons at 20 kpix $s^{-1}$ and $<10$ e at 1 Mpix $s^{-1}$ data rates is currently being measured on high-performance, slow-scan devices without resorting to multiple sampling techniques.

### 7.4.3 **APS Noise**

The noise sources present in an APS sensor are essentially the same as with a CCD. At high signal levels, noise is dominated by the shot noise on the incident flux. At low-signal levels, the noise is dominated by the pixel-related read noise, FPN due to the pixel-to-pixel threshold variations noted above and noise related to the signal chain.

For the 4T photodiode cell, resetting the floating diffusion can take place in two modes: hard and soft reset. The hard reset mode leads to the usual $kTC$ noise. In this mode, the RST is turned on such that the FET is operating in the saturation region and the node is reset to VDD. In the soft reset mode, the reset FET is operated in weak inversion and the resetting operation occurs through the leakage current flow. The node is reset to (VDD-$V_{th}$). Since the current is essentially one directional, the node does not come to equilibrium with the sea of electrons in the photodiode as is the case with a hard reset. The noise in the case of the soft reset is

$$N_R = \frac{\sqrt{kTC}}{2q} \tag{33}$$

The downside of the soft reset mode is that there are serious linearity problems at low signal levels. In either case, this noise is effectively removed by the CDS of the column amplifiers. Note, however, that it is not possible to take an independent sample of the floating diffusion for the PD based pixel because the PD and the floating diffusion are one and the same. Thus, the $kTC$ contribution to the noise remains for pixels based on the 3T cell.

Although the threshold voltage variations discussed above give rise to FPN, the pixel-related FPN is also largely removed by CDS in the column amplifiers. The column-to-column FPN is also removed by the DDS circuit discussed above. Typical values of the FPN are 0.1 to 0.4% of the saturation or full well signal.

The remaining noise source to be considered is the signal chain itself. It can be shown that the input referred noise is given by the expression [21, 65]

$$N^2 = \frac{2kT}{C_{CS}}\frac{[(1-G)C_g + C_{fd}](C_g + C_{fd})}{G} + 2(C_g + C_{fd})^2 S_{1/f}(SF)\ln(1+\pi^2 f_c\tau) \quad (34)$$

where $C_{CS}$ is the sample-and-hold capacitor in the column amplifier (see Fig. 25), $G$ is the gain of the pixel SF, $C_g$ is the parasitic gate-to-source capacitance of the pixel SF, $C_{fd}$ is the capacitance of the floating diffusion, and the term $S_{1/f}(\mathrm{SF})\ln(1+\pi^2 f_c\tau)$ represents the $1/f$ noise contribution of the SF transistor. $f_c$ is the corner frequency and $\tau$ is the time between samples. The first term represents the white-noise term, and can be recognized as the $kTC$ noise associated with resetting the sample-and-hold capacitor by the SF FET in the pixel. Typical noise for APS sensors is in the 10 to 30 e range for 30-fps operation. Recent results on a 1280 × 1024 4T cell device exhibited 1.4 e noise at a data rate of 6 fps or ~8 Mpix $s^{-1}$ [66].

#### 7.4.4 Noise versus Readout Rate

It is of interest to compare the behavior of the noise of CCDs and CMOS APS arrays as a function of the readout rate. For CCDs, the low signal-level noise is primarily determined by the noise characteristics of the output amplifier, assuming that white noise dominates the spectrum. Then, as the readout rate increases, the noise will increase as the square root of the bandwidth. On the other hand, noise in APS sensors is dominated by kTC noise which is frequency independent. Figure 32 presents data of noise as a function of readout rate for several CMOS and CCD arrays. Beyond approximately 10 to 20 Mpix $s^{-1}$ the CMOS arrays will exhibit better performance.

### 7.5 Well Capacity and Dynamic Range

The well capacity measures the quantity of charge that can be contained in a single pixel before that charge is inadvertently shared with neighboring pixels (blooming). It is also the measure of the saturation signal.

The size of the well is defined by the voltage difference between the well and the neighboring barrier; however, the maximum voltage difference is set by the requirement that the full well charge be confined to the buried channel and restricted from interacting with the surface. No simple expression has been developed for the buried channel device, assuming an arbitrary impurity profile; however, analysis using the box distribution gives an approximate expression,

$$N_{FW} = \left(N_D x_j - \sqrt{\frac{2\varepsilon_{Si} N_A \Delta V}{q}}\right) A_{pix} \quad (35)$$

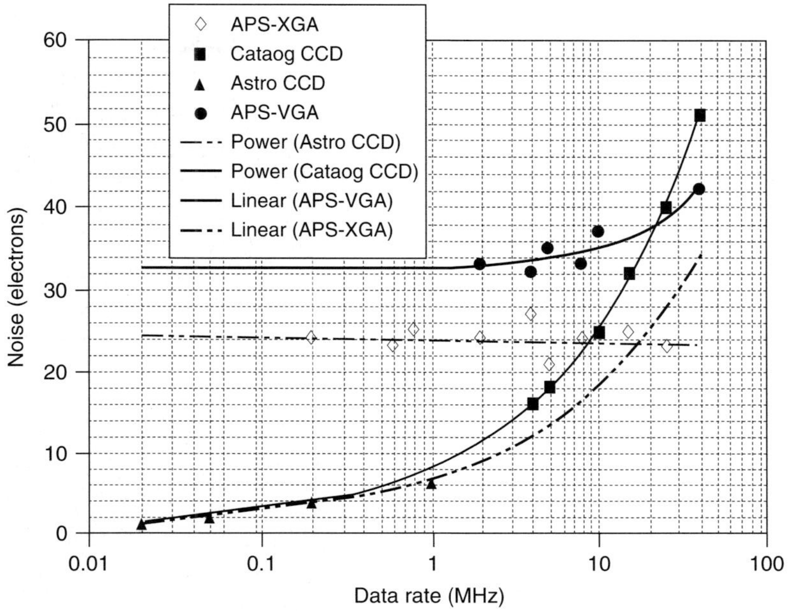


**Fig. 32** Noise as a function of readout frequency for CCD and CMOS APS arrays

A good approximation for buried channel devices is $N_{FW} \sim 10^3$ electrons per $\mu m^2$ of pixel area: a well capacity of about 400 ke for a $20 \times 20\text{-}\mu m^2$ pixel. There is, of course, substantial variation in reported well capacity among devices depending on the details of the design, fabrication process and measurement technique. Well capacities as high as 3 ke/$\mu m^2$ have been reported. However, given the same design rules and process, a four-phase device should have a somewhat higher well capacity than either a three-phase or two-phase device.

The dynamic range for CCDs and APSs is usually defined as the ratio of the maximum well capacity to the minimum readout noise. The typical dynamic range for TV-type imagers is on the order of 60 dB, but for very high-quality scientific chips, the dynamic range can exceed 100 dB or 100 000. The magnitude of this extraordinary range is captured in the series of photos presented in Fig. 33. Film has a dynamic range of perhaps 100 to 200. To be able to reproduce all of the data that exist in a single frame of CCD data requires multiple photographic images. Either the brightest or the dimmest objects can be seen but not both without seriously underexposing or overexposing the film. In addition, because of the linearity of the CCD, it is possible to perform photometry at the 0.5% level throughout this dynamic range.

For APSs, the full well is frequently limited by the voltage swing of the signal chain rather than by the capacity of the PG or PD to hold charge. For example, for a 3.3 V process, $\Delta V_{max} \sim 1.2$ V for a PD and a PG-based pixel and $\sim$0.5 V for a PPD. Conversion gains are $\sim$5 $\mu$V/e for the PD and $\sim$20 $\mu$V/e for the PG and PPD structures. This implies a maximum well capacity of $\sim$240 000 e for the PD, 60 000 e for the PG device, and $\sim$25 000 e for the

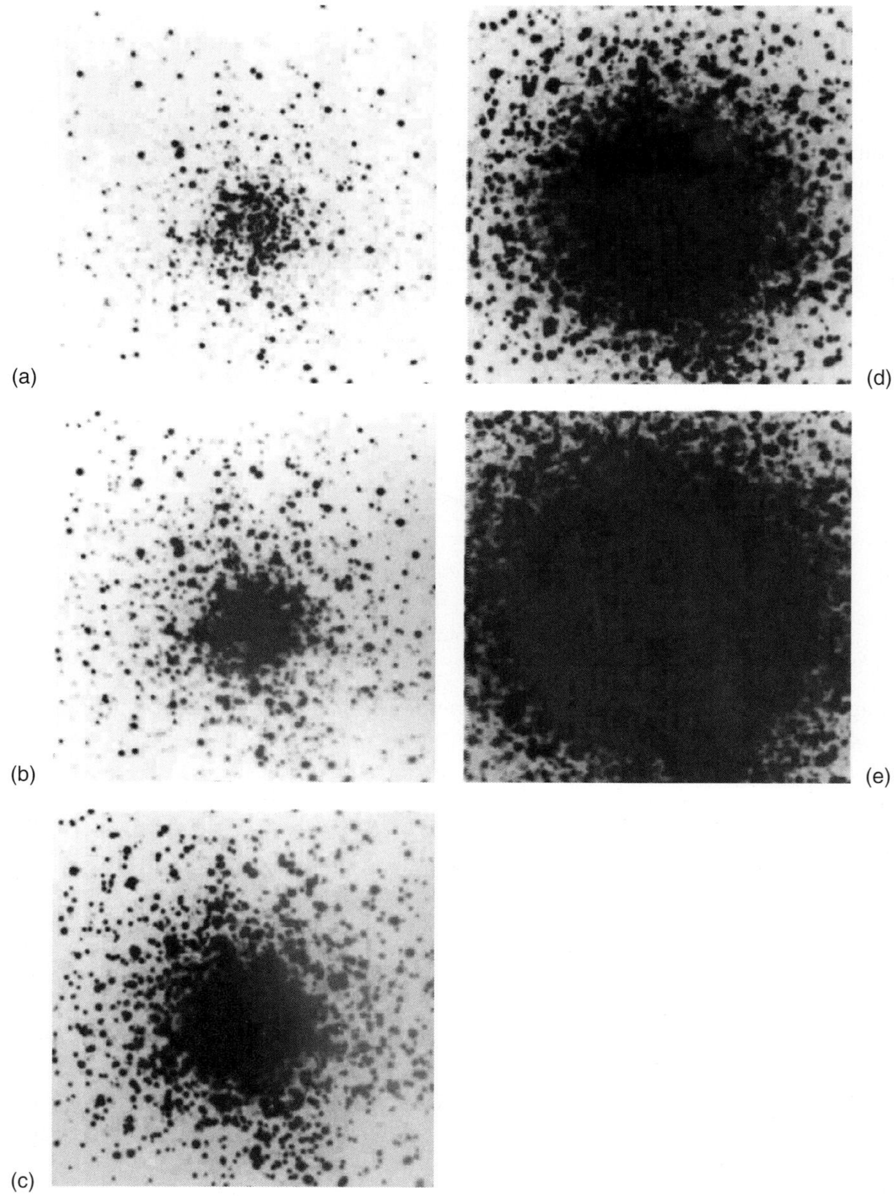


**Fig. 33** This series of photographs illustrates the enormous dynamic range of the CCD. In (a), only the very brightest stars in the globular cluster 47 Tucana can be seen. This represents the largest signals in the CCD data frame. As one goes lower into the data (b–e) displaying fainter and fainter objects, the bright objects saturate the film. It requires at least eight photos to reproduce all the data that exist in the single CCD exposure. (J. Kristian, Carnegie Observatories)

PPD. These estimates pertain to pixels that are roughly 15 times the minimum process feature size.

### 7.5.1 Antiblooming

Blooming occurs when the input photon flux creates more photoelectrons than the pixel can handle. The excess charge must go somewhere. In CCDs, the natural place is for the excess charge to spill over the barrier separating pixels in the same column and start filling the neighboring pixels. Blooming thus occurs up and down columns. (Blooming does not occur across columns as the barrier presented by the channel stop is too high.)

To prevent blooming in CCDs, several techniques have been created. These fall into two categories: lateral and vertical antiblooming. The two forms are illustrated in Fig. 34. The lateral antiblooming structure is depicted in Fig. 34(a). The channel stops are photolithographically patterned and implanted with an $n+$ diode. The presence of the biased diode within the channel stop allows one to adjust the height of the potential barrier presented by the channel stop to be just below the barrier of the phase separating pixels. Excess charge spills laterally over the channel stop into the antiblooming drain and is removed from the chip.

Vertical antiblooming is illustrated in Fig. 34(b). In this structure, one begins with an $n$-type substrate. A $p$-type implant is performed followed by a second $n$-type

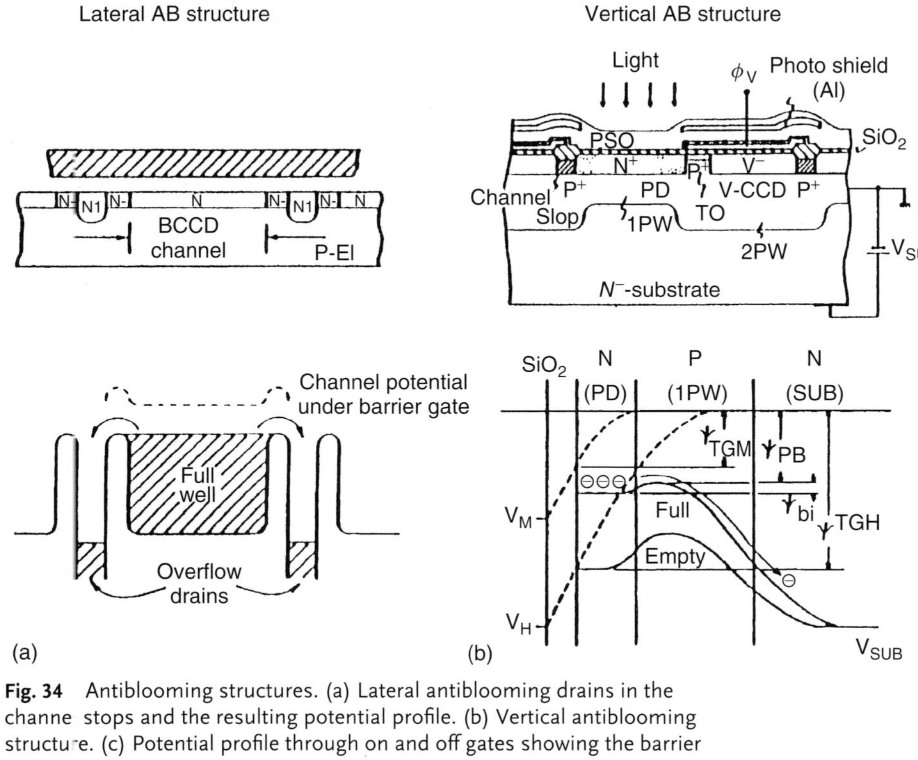


**Fig. 34** Antiblooming structures. (a) Lateral antiblooming drains in the channel stops and the resulting potential profile. (b) Vertical antiblooming structure. (c) Potential profile through on and off gates showing the barrier to the substrate

implant to form the buried channel. An analysis similar to the one given in Sect. 2.1 yields a potential profile as shown in Fig. 34(c). The basic idea is to adjust the clock voltages such that the empty channel of the barrier phase is slightly higher than the potential barrier between the well and the substrate. Excess charges then spill harmlessly into the substrate. One additional advantage of this structure is that it is possible to bias the substrate such that the barrier between the CCD wells and the substrate disappears. Until this voltage is released, no charge will collect in the pixels. Thus, this vertical antiblooming structure also acts as an electronic shutter.

In APS devices, lateral antiblooming comes essentially for free in the 4T pixels. As long as the barrier presented by the RST is lower than TX, the excess charge will spill over the TX barrier into the floating diffusion and into the VDD supply.

## 7.6 Spectral Response

One of the extraordinary performance parameters that make the Si-based CCDs and CMOS APS arrays so unique and useful is the ability of these devices to efficiently detect photons over an extremely wide range of wavelengths. Figure 35 illustrates the measured QE of a thinned CCD from 0.1 nm in the soft X ray to 1100 nm in the near IR [67]. The remarkable feature of the curve is that the measured efficiency is greater than 50% over most of this spectral region. A thinned APS array should show similar performance.

The QE is defined as

$$\eta_{\mathrm{QE}} = \frac{N}{\Phi A_{\mathrm{pix}} \tau_{\mathrm{int}} N_{\mathrm{ex}}} \tag{36}$$

$N_{\mathrm{ex}}$ is the expected average number of electrons generated per incident photon

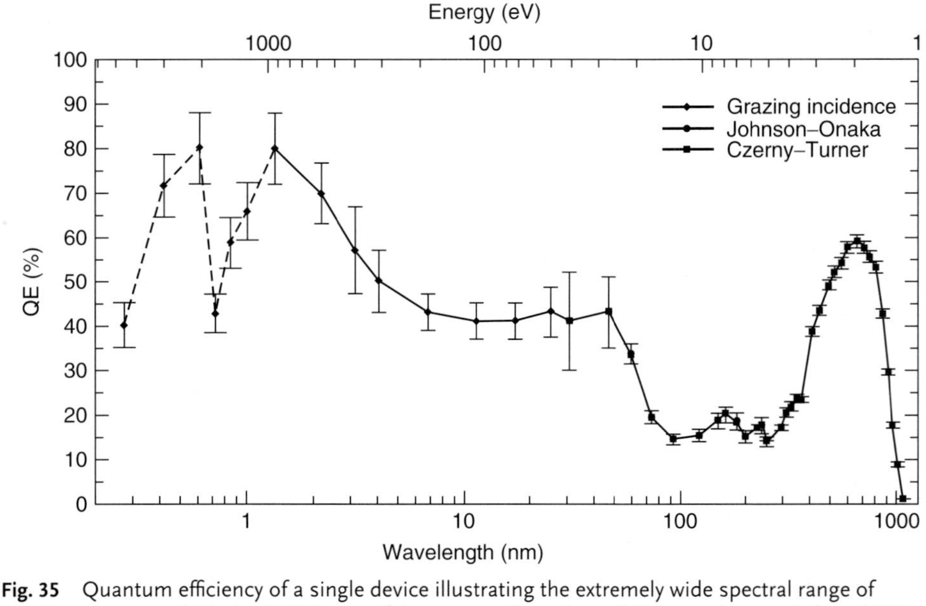


**Fig. 35** Quantum efficiency of a single device illustrating the extremely wide spectral range of wavelengths over which the CCD has useful response, from the soft X ray to the near infrared [67]

(hν/3.65). With this definition, the maximum QE is 100% at all wavelengths.

### 7.6.1 Front-illumination

Virtually all CMOS sensors and most CCDs operate in the front-illuminated mode, that is, where the photon flux is incident on the circuit side of the device. The spectral response of a Si image sensor depends on a number of factors. The dominant parameters are the absorption coefficient of silicon, which is small for long wavelengths but rises to a maximum of about $2 \times 10^6\ \text{cm}^{-1}$ near 230 nm (corresponding to an absorption length of 5 nm) and the quantity of material available for collection of the photogenerated carriers (the thickness of the epitaxial layer).

Examples of the measured QE for selected front-illuminated CCD and APS devices is presented in Fig. 36(a). The curve for the triple poly process represents the response that front-illuminated devices using all polysilicon gates can achieve. There are several features to be observed. First, for wavelengths shorter than ~450 nm the QE drops precipitously and below 400 nm the device sensitivity is essentially nil. This is because, polysilicon, like silicon, absorbs blue light very strongly and thus little of the short-wavelength light makes it through the gate to the CCD wells. Secondly, the response is not a smooth function of wavelength but rather has significant peaks and valleys in the response. This is a result of the fact that the layers that make up the circuit side of the device act as a thin-film interference filter. The virtual phase device shows significantly higher response than the triple poly device. The higher response is largely due to the fact that half of each pixel in a virtual phase device is not covered by a polysilicon gate.

An alternative to removing the polysilicon electrode is to use a conducting, transparent material as the gate electrode. Indium-Tin Oxide (ITO) is such a material and has been used to make two-phase CCDs with one electrode replaced with ITO [68]. The QE of this device is the top curve presented in Fig. 36(a). Again, only half the pixel is sensitive to short-wavelength light since the other gate is polysilicon. To increase the collection efficiency of each pixel, a cylindrical microlens is placed over each row, focusing the incident photons on to the ITO portion of the pixel.

The QE curves for PD and PG type APS devices are also presented in the figure. Like the virtual phase device, the PD-based pixel has significant blue response because the blue photons can enter the device without encountering any absorbing materials. It will be noted, however, that the response is lower than the VP device. This is due to several factors. First, the fill factor (the fraction of the pixel exposed to light) is about 60% for this device. This is due to the metal lines and silicided poly gates, which make up the circuitry of the active pixel and block the incoming photons. Secondly, because the epi thickness is thinner, the volume of silicon available for generating signal electrons is smaller than in the CCD pixel. This effect is most pronounced in the red part of the spectrum. The bottom curve in the figure is the QE curve for the PG structure.

Another technique to extend the response of devices into the blue and ultraviolet region is to coat the packaged device with an ultraviolet downconverting phosphor such as coronene, liumogen, or a thin plastic film containing laser dyes [69, 70]. These compounds strongly absorb ultraviolet photons and reemit a

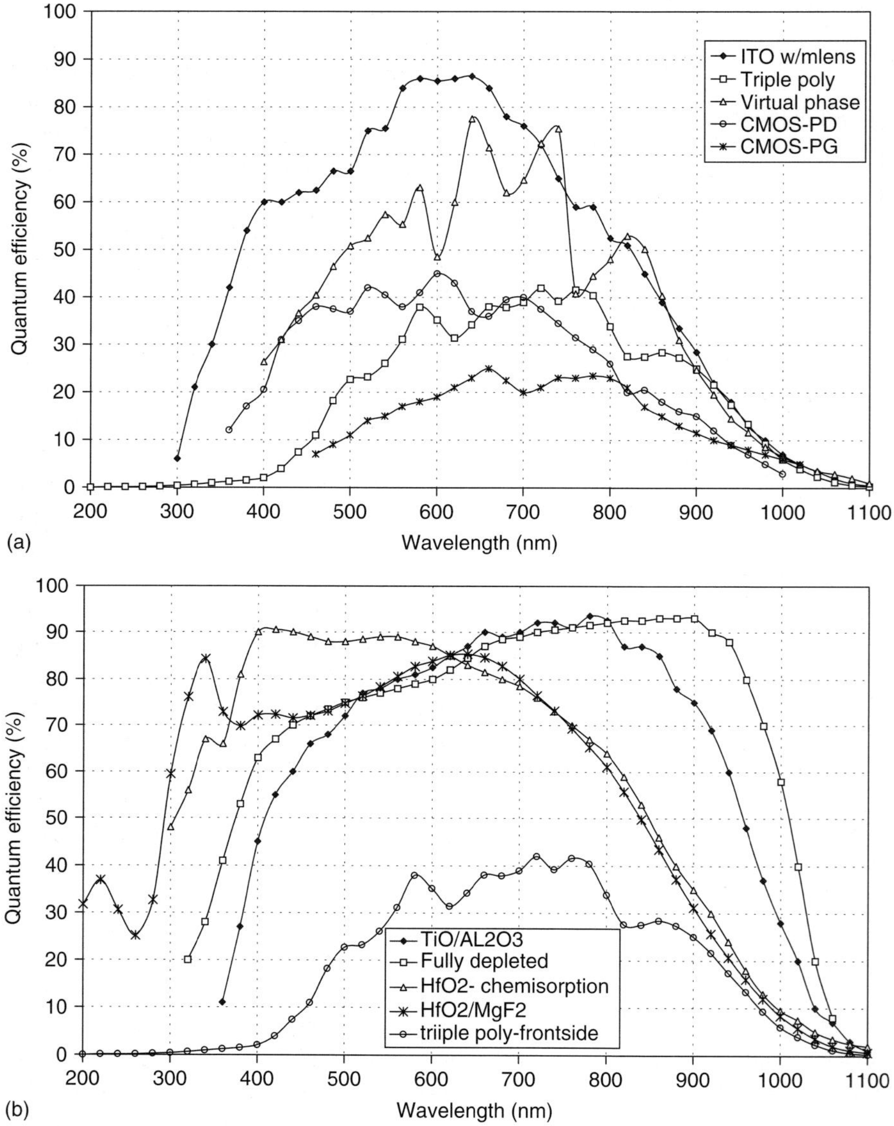


**Fig. 36** Quantum-efficiency curves for selected devices. (a) Front-illuminated devices including CCDs and CMOS APS chips. (b) Back-illuminated devices

photon between 450 and 600 nm where the CCD is very sensitive. These coatings have produced devices with overall QEs of between 20 and 30% in the blue and UV down to wavelengths of 50 nm. This technique has been used successfully with both front- and back-illuminated devices.

### 7.6.2 **Back Illumination**

This method produces devices that have the highest QEs at virtually all

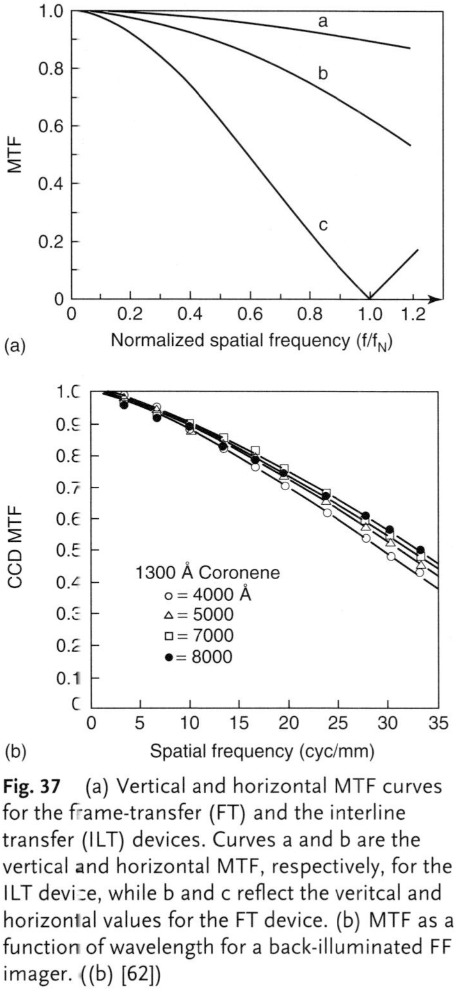


**Fig. 37** (a) Vertical and horizontal MTF curves for the frame-transfer (FT) and the interline transfer (ILT) devices. Curves a and b are the vertical and horizontal MTF, respectively, for the ILT device, while b and c reflect the veritcal and horizontal values for the FT device. (b) MTF as a function of wavelength for a back-illuminated FF imager. ((b) [62])

wavelengths. Although the process is beginning to find application to CMOS APS devices, it has been used extensively with high-performance CCDs and will be discussed in that context. In this technique, the device is fabricated, as usual, on the 525- to 625-μm thick, epitaxial silicon wafer. After completing the device, the Si substrate underlying the active area is thinned to between 10 and 20 microns. Several methods of thinning are employed: whole wafer thinning where the device is supported by an underlying rigid substrate, or window thinning where only the active area of the device is thinned and a thick Si rim surrounding the active area supports the thin membrane. Light is incident from the thinned side of the array so that the electrode structure does not have to be traversed by the signal photons (back-illuminated mode). Photons that are absorbed produce electrons that diffuse to and are collected in the CCD wells.

The key to the successful implementation of this method is the treatment of the back surface following thinning [59, 71, 72]. Figure 36(b) compares spectral response data for the triple poly, front-illuminated device and several back-illuminated devices with different, thin-film AR coatings applied to them. At short wavelengths, the index of refraction of silicon increases to a value in excess of 4. Consequently, for these wavelengths one desires an AR coating material with a high index. $HfO_2/MgF_2$ is frequently used for shorter wavelength coatings. $SiO_2$ and $TiO_2/Al_2O_3$ have also been employed. As may be seen from the figure, QEs in excess of 90% are possible.

Part of the reason for the high red response of the $TiO_2/Al_2O_3$ device shown in Fig. 36(b) is the fact that this device is 52-μm thick, much thicker than the normal 12 to 15 μm common for a thinned device. The response of the 200-μm fully depleted device is even higher at the long-wavelength extreme of the spectrum. The thicker device provides more material for the red photons to be absorbed [39].

## 7.7 Modulation Transfer Function

The spatial resolution of an imager can be characterized by the number of pixels in the array or, in more analytic terms, by means of the modulation transfer function (MTF). The MTF quantifies the ability of the device to reproduce a spatially varying input signal. Device resolution is degraded by several factors: lateral diffusion of photogenerated carriers, the discrete nature of the pixels, charge transfer inefficiency, and nonuniform aperture response function due to spatial variation in the transmission in the electrode structure within the pixel. The latter term is important because it can give rise to significant aliasing. Note also that for two-dimensional arrays, there are two MTFs corresponding generally to the row and column directions.

TDI arrays have additional components of the MTF related to the relative motion of the image with respect to the moving charge packet and the number of clock phases that define a pixel.

Each of the above effects can be characterized in terms of an MTF as summarized in Table 2. The overall MTF of the imager at spatial frequency $f$ ($= 1/\lambda_s$) is the product of the individual contributions [73]. Figure 37(a) presents curves of the MTF as a function of normalized spatial frequency for the ILT, the FT imagers, and an APS. The curves are based on the aperture response function typical of a discrete sampling system

$$MTF = \sin c\left(\frac{\pi\,\Delta x}{p}\frac{f}{f_N}\right) \quad (37)$$

where $\sin c\,x = (\sin x)/x$, and $f_N$ is the Nyquist spatial frequency. $\Delta x/p$ is the fraction of the pixel exposed to the scene in that direction (the exposed length of the pixel may be larger in one direction than in the orthogonal direction). In the interlace mode, for the FF and the FT devices, $\Delta x = p$ in both the vertical and horizontal directions, while $\Delta x = p/2$ for the ILT sensor. For both devices, $f_N = 1/(2p)$ in the horizontal direction and $1/p$ in the vertical. As a result, the horizontal and vertical MTF of the ILT device are given by curves a and b respectively, and by curves b and c for

**Tab. 2** Components of the modulation transfer function

| ***Source*** | ***Functional dependence*** |
|---|---|
| Aperture response | $MTF^a_{Ap}(k) = \sin c(kp/2) = \sin c\{(\pi\,\Delta x/p)(f/f_N)\}$ |
| Lateral diffusion | $MTF_d(k) = (1 - (\exp(-\alpha L_d))/(1+\alpha L))/(1 - (\exp(-\alpha L_d))/(1+\alpha L_d))$ |
| Charge transfer | $MTF_\eta(k) = \exp[-n\eta(1-\cos(k\pi))]$ |
| Velocity mismatch | $MTF_v(k) = \sin c(k\pi(\Delta v/v)/2)$ |
| Phase sharing | $MTF_\phi(k) = \sin c(k\pi n\phi/[2(n\phi - 1)])$ |

$\alpha$ = Absorption coefficient of Si at optical wavelength $\lambda$ ($cm^{-1}$).
$L_d$ = diffusion length of electrons in Si (cm).
$L = (k^2 + 1/L_d^2)^{-1/2}$
$\Delta v$ = velocity mismatch between motion of charge and image on TDI sensor.
$n\phi$ = number of clock phases in TDI sensor.
$k$ = spatial angular frequency, $2\pi/\lambda_s$.
$\Delta x$ = fraction of pixel exposed to scene.
$p$ = pixel pitch.
f = spatial frequency.
$f_N$ = Nyquist frequency.
$MTF^a$ = Modulation transfer function.

the FF and FT imager. Because the APS pixel is covered in metal on the edges, $\Delta x/p$ is less than 1 in both the vertical as well as the horizontal directions. The MTF for the APS device follows curve b in both directions. The liability of the higher MTF of the ILT and the APS devices is that spatial frequencies in the scene greater than the Nyquist frequency will be detected and displayed as lower frequency signals (i.e., significant aliasing can occur).

The wavelength dependence of the MTF is illustrated in Fig. 37(b) for a thinned, back-illuminated CCD with a UV downconverting phosphor. The wavelength dependence is a direct result of the longer absorption length for long-wavelength photons and the ability of photogenerated carriers to diffuse laterally in a field-free region before they are collected in the CCD wells [74–76]. For front-illuminated devices the long-wavelength photons give poorer performance, whereas, as illustrated in the figure, for the back-illuminated devices it is the short-wavelength photons that will generally yield lower MTF.

## 8 Summary

The solid-state image sensor has made enormous progress since its initial introduction in 1967 as the Reticon, both in terms of physical size, performance, and capability. The motivating force behind the impressive development of these sensors was the desire for a low-power, light-weight, all solid-state replacement for the vidicon camera. Inexpensive devices, capable of capturing a TV format color image on a single chip are routinely available. These solid-state devices are now the preferred image sensors for virtually all consumer, industrial, as well as scientific imaging applications, especially when low light levels are involved.

The performance of CCDs and APSs has benefited greatly from the drive in the CMOS industry toward small devices and submicron geometries. Not only have the processing chemicals and the basic silicon material been dramatically improved but the handling and cleanliness of the fabrication facilities and equipment have undergone substantial modernization. CCDs are exquisitely sensitive to defects, both in their effect on dark current and the transfer efficiency. In the past 20 years, the improvements in semiconductor manufacture have allowed the dark current in CCDs to decrease by over three orders of magnitude from a value of $>10\ nA/cm^2$ to $<10\ pA/cm^2$ today. Charge transfer efficiencies now routinely exceed 0.999995. The development of very quiet buried channel MOSFETs has permitted read noise performance of slow-scan, scientific sensors to achieve the 1–2 electron level.

The original three-phase, 8-bit CCD was demonstrated using a single level of Al metallization to form the gates [77]. Since then, the technology has evolved towards multilevel gate structures and single, two, three, and four-phase architectures. Chip size has evolved to where scientific arrays of 2K × 2K, 2K × 4K, and 4K × 4K pixels have been fabricated [55, 78]. These 2K × 2K devices, fabricated as a single device on each 100-mm wafer, represent the first true wafer scale integrated circuits. Though still expensive, the technology has now progressed to the point where thinned 2K × 4K and 4K × 4K devices are routinely available.

Large focal planes using these sensors are now being built. Using the thinned 2K × 4K chip as a building block, several

8K × 8K focal planes are now in use at several telescopes around the world. An 8K × 12K focal plane using red-sensitive chips and an array of 40 of these chips are now in use at the Canada-France-Hawaii Telescope on Mauna Kea. There are plans to fabricate a focal plane with ~1500 1K × 1K chips (either CCDs or APSs) for the Large Synoptic Survey Telescope.

The processing uniformity and capabilities of today's manufacturing facilities have made this possible. With CTE > 0.999995, QE > 80 to 90% over the visible, noise at the 2-electron level, and cooled dark current at 3 electron/pixel/h, back-illuminated CCDs have become the nearly perfect detector. The L3CCD and the Impactron are both going to allow an all solid-state photon-counting system to be built.

Over the past decade, APSs have also made great strides in performance. Not only is the QE now comparable to that of some front-illuminated CCDs, but so also is the noise. This makes the CMOS device competitive with the CCD for many low and midlevel applications. In addition, resolution has increased to meet the size of the largest CCDs. A 13.89 Mpix color sensor is currently being produced as a replacement for 35-mm film in SLR cameras [79]. It is now possible to obtain a 3936 × 2196, 60-fps, progressive scan, UDTV chip that outputs 10-bit digitized data at 518 Mpix $s^{-1}$ data rate [80]. The remaining parts of the system, for example, image compression circuitry, memory, and the liquid crystal display (LCD) driver are yet to be added to complete the chip.

Many robotic systems can benefit from some preprocessing of the data before data analysis can begin or transferring to a human operator. Some of these operations include level shifting, edge enhancement, frame-to-frame subtraction, motion detection, and pattern recognition. With even moderate frame rates, the number of arithmetic operations that can be required to analyze the data in the digital domain could easily exceed $10^8$ per second [81]. The ability to integrate signal processing capabilities directly onto the chip gives CMOS sensors a very bright future.

## Glossary

**Aliasing:** The phenomenon that occurs in sampled data systems in which high-frequency components ($f > f_N$) are transformed and interpreted as low frequency signals ($f < f_N$). $f_N$ is the Nyquist frequency, which is the highest frequency that can be correctly interpreted and reproduced by a sampled data system. It takes two samples to reproduce the sine wave at Nyquist.

**Accumulation:** The condition where the majority carrier density at the surface (electrons in *n*-type material) is larger than in the neutral bulk.

**Blooming:** The spilling of charge from one pixel into neighboring pixels when the target pixel is overexposed or its well capacity is exceeded. Blooming usually occurs in the column direction.

**Box Distribution:** A simplified model of the impurity distribution in a buried channel device. The *n*-type region is assumed to be uniformly doped to the metallurgical junction and the *p*-type material is uniform thereafter.

**Buried Channel:** The region of maximum potential where signal charge resides in a buried channel capacitor or CCD.

Because of the doping profile, the buried channel is physically located at a distance below the Si-$SiO_2$ interface, which isolates the signal charge from the interface states.

**Channel Potential:** The maximum potential in the semiconductor region of an MOS capacitor or a pinned photodiode.

**Charge-transfer Efficiency:** The measure of the ability of the CCD to retain the integrity of a charge packet when transferring the charge from one pixel to the next.

**Clock:** A periodic signal applied to the CCD gates to manipulate and move charge packets within the circuit.

**Depletion Approximation:** Simplifying assumptions that describe the charges present in a semiconductor. Specifically, the minority-carrier density is assumed to be so small it need not be considered, the majority-carrier density is negligible within and up to the edge of the depletion region and is equal to the impurity density outside that region.

**Depletion Region:** A region within a semiconductor from which mobile carriers have been excluded.

**Dynamic Range:** The ratio of the well capacity to the read noise.

**Gate:** One of multiple electrodes that form FETs or the MOS capacitor. The gate is used to control the potential in the semiconducting region below it.

**Gaussian Distribution:** An approximation to the distribution of *n*-type impurities in the channel region of a buried channel device as a result of an implant and subsequent high-temperature processing.

**Integration:** The process where photons are imaged onto a sensor and the photogenerated charge is collected in the photosites. During integration, the clocks are typically held at a constant voltage.

**Inversion:** The condition where the concentration of minority carriers at the surface (holes in *n*-type material) is larger than the concentration of majority carriers in the neutral bulk.

**Output Sensitivity:** The change in voltage measured at the output of the device amplifier in response to a unit charge at the input.

**Phase:** A clock voltage used to drive a CCD gate.

**Pinning:** The condition of a CCD gate in a buried channel device such that the surface is inverted. The inversion charge connects the surface to ground potential and shields the underlying channel from any further decrease in the gate voltage.

**Pitch:** The center-to-center spacing between neighboring pixels.

**Pixel:** The fundamental unit cell of the CCD. A pixel is also sometimes referred to as a *pel.*

**Read Noise:** The noise measured at the output in the absence of any signal.

**Trap:** A defect site in the device that can capture and retain signal charge carriers. These traps can occur at the silicon–silicon dioxide interface or in the channel or neutral bulk of the device.

**Well Capacity:** The maximum quantity of charge that a pixel can contain without communicating with its neighbors.

## References


[1] Schuster, M. A., Strull, G. (**1966**), *IEEE Trans. Electron Devices* **ED-13**, 907–912.

[2] Fossum, E. R. (**1997**), *IEEE Trans. Electron Devices* **ED-44**, 1689–1698.

[3] Weckler, G. P. (**1967**), *IEEE J. Solid-State Circuits* **SC2**, 65–71.

[4] Dyke, R., Weckler, G. (**1968**), *IEEE Trans. Electron Devices* **ED-15**, 196–201.

[5] Nobel, P. (**1968**), *IEEE Trans. Electron Devices* **ED-15**, 202–209.

[6] Chamberlain, S. G. (**1969**), *IEEE J. Solid-State Circuits* **SC-4**, 333–342.

[7] Boyle, W. S., Smith, G. E. (**1970**), *Bell Syst. Tech. J.* **49**, 587–593.

[8] Kristian, J., Blouke, M. M. (**1982**), *Sci. Am.* **247**, 66–74.

[9] Boulade, O. (1998), in J. W. Beletic, P. Amico (Eds.), *Optical Detectors For Astronomy*. Dordrecht: Kluwer Academic Publishers, pp. 203–207.

[10] Wolfe, T., Armandroff, T., Blouke, M. M., Rector, T., Reed, R., Saha, A., Schommer, R., Smith, C., Smith, R., Walker, A. (2000), in M. M. Blouke, N. Sampat, G. M. Williams, T. Yeh (Eds.), *Sensors, Cameras, and Systems for Scientific, Industrial, and Digital Photography Applications*. Proc. SPIE Vol. 3965. Bellingham: SPIE Press, pp. 80–91.

[11] Starr, B. M., Cuillandre, J.-C., Isani, S., Luppino, G. (2000), in P. Amico, J. W. Beletic (Eds.), *Optical Detectors For Astronomy II*. Dordrecht: Kluwer Academic Publishers, pp. 165–179.

[12] Dawson, R., Preisig, J., Carnes, J., Pridgen, R. (**1977**), *RCA Rev.* **38**, 406–435.

[13] Ong, D. (**1981**), *IEEE Trans. Electron Devices* **ED-28**, 6–12.

[14] Ma, S., Chen, L. (**1999**), *IEEE J. Solid-State Circuits* **SC-34**, 1415–1418.

[15] Dickinson, A., Ackland, B., Eid, E.-S., Inglis, D., Fossum, E. R. (**1995**), *ISSCC Dig. Tech. Papers* pp. 226–227.

[16] Kawahito, S., Yoshida, M., Sasaki, M., Umehara, K., Tadokkra, Y., Murata, K., Doushou, S., Matsuzawa, A. (**1997**), *ISSCC Dig. Tech. Papers*, pp. 184–185.

[17] Sirianni, M., Clampin, M., Hartig, G. F., Ford, H. C., Golimowski, D. A., Illingworth, G., Sullivan, P., Blouke, M. M., Lesser, M., Burmester, B., Schrein, R., Kimble, R. (2000), in M. Iye, A. E. Moorwood (Eds.), *Optical and IR Telescope Instrumentation and Detectors*. Proc. SPIE Vol. 4008, Bellingham: SPIE Press, pp. 669–674.

[18] Muller, R. S., Kamins, T. I. (1977), *Device Electronics for Integrated Circuits*, (2nd ed.), New York: John Wiley & Sons, Chap. 4.

[19] Blouke, M. M. (1992), Charge-coupled devices, *Encyclopedia of Applied Physics*, Vol 3. New York: VCH Publishers, pp. 241–272.

[20] Janesick, J. R. (2001), *Scientific Charge-coupled Devices*, Bellingham: SPIE Press, pp. 86–87.

[21] Pain, B. (2000), *Fundamentals of CMOS Digital Image Sensors*, SPIE Short Course Notes, Bellingham: SPIE Press.

[22] Mohsen, A. M., Tompsett, M. F. (**1974**), *IEEE Trans. Electron Devices* **ED-21**, 701–711.

[23] Carnes, J. E., Kosonocky, W. F., Ramberg, E. G. (**1972**), *IEEE Trans. Electron Devices* **ED-19**, 798–808.

[24] Kim, C. K., Lenzlinger, M. (**1971**), *J. Appl. Phys.* **42**, 3586–3594.

[25] Lee, H.-S., Heller, L. G. (**1972**), *IEEE Trans. Electron Devices* **ED-19**, 1270–1279.

[26] Fowler, B., Balicki, J., How, D., Mims, s., Canfield, J., Godfrey, M. (2003), *Proc. IEEE Workshop on Charge Coupled Devices and Advanced Image Sensors.*

[27] Lehmann, M., Richter, M., Cavallier, L., Lustenberer, F., Blanc, N. (2003), *Proc. IEEE Workshop on Charge Coupled Devices and Advanced Image Sensors.*

[28] Zhou, Z., Pain, B., Fossum, E. R. (**1997**), *IEEE Trans. Electron Devices* **ED-44**, 1764–1768.

[29] Wäny, M., Israel, G. P. (**2003**), *IEEE Trans. Electron Devices* **ED-50**, 57–62.

[30] Goy, J., Courtois, B., Karam, J. M., Pressecq, F. (2001, 2002), M. M. Blouke, J. Canosa, N. Sampat (Eds.), *Sensors and Camera Systems for Scientific, Industrial, and Digital Photography Applications II*. Proc. SPIE Vol. 4306, Bellingham: SPIE Press, pp. 93–99.

[31] Krymski, A., Tu, N., *IEEE Trans. Electron Devices* **ED-50**, 136–143.

[32] Huppertz, J., Kneip, T., Schwartz, M., Hosticka, B. (1999), *Proc. IEEE Workshop on Charge Coupled Devices and Advanced Image Sensors*, pp. 207–210.

[33] Burke, B. E., Reich, R. K., Savoye, E. D., Tonry, J. L. (**1994**), *IEEE Trans. Electron Devices* **ED-41**, 2482–2484.
[34] Tonry, J. L., Burke, B. E., Schecter, P. L. (**1997**), *Pubs. Astron. Soc. Pacific* **109**, 1154–1164.
[35] Burke, B. E., Gregory, J. A., Bautz, M. W., Prigochin, G. Y., Kissel, S. E., Kosicki, B. B., Loomis, A. H., Young, D. J. (**1997a**), *IEEE Trans. Electron Devices* **ED-44**, 1633–1642.
[36] Burke, B. E., Tonry, J. L. (1997b), in M. M. Blouke (Ed.), *Solid State Sensor Arrays Development and Applications*. Proc. SPIE Vol. 3019, Bellingham: SPIE Press, pp. 233–240.
[37] Burke, B. E., Mountain, R. W., Harrison, D. C., Bautz, M. W., Doty, J. P., Ricker, G. R., Daniels, P. J. (**1991**), *IEEE Trans. Electron Devices* **ED-38**, 1069–1076.
[38] Lumb, D., Eggel, H., Laine, R., Peacock, A. (1996), *EUV, x-ray, and gamma ray instrumentation for Astronomy VII*, Proc. SPIE Vol. 2808, pp. 326–337.
[39] Holland, S. E., Groom, D. E., Palaio, N. P., Stover, R. J., Wei, M. (**2002**), *IEEE Trans. Electron Devices* ED-50, pp. 225–238.
[40] Suntharalingam, V., Burke, B. E., Burns, J. A., Cooper, M. J., Keast, C. L. (2000), in M. M. Blouke, N. Sampat, G. M. Williams, T. Yeh (Eds.), *Sensors and Camera Systems for Scientific, Industrial, and Digital Photography Applications*, Proc. SPIE Vol. 3965, Bellingham: SPIE Press, pp. 246–253.
[41] Parulski, K. A. (**1985**), *IEEE Trans. Electron Devices* **ED-32**, 1381–1389.
[42] Kosonocky, W. F., Shallcross, F. V., Villani, T. S., Groppe, J. V. (**1985**), *IEEE Trans. Electron Devices* **ED-32**, 1564–1573.
[43] Kimata, M., Denda, M., Yutani, N., Ivade, S., Tsubouchi, N. (1987), *Proc. 1987 IEEE Int. Solid-State Circuits Conf.*, pp. 110–111.
[44] Felix, P., Moulin, M., Munier, B., Portman, J., Reboul, J.-P. (**1980**), *IEEE Trans. Electron Devices* **ED-27**, 175–188.
[45] Bailey, R. B., Kozlowski, L. J., Chen, J., Bui, D. Q., Vural, K., Edwall, D. D., Gil, R. V., Vanderwyck, A. N., Gertner, E. R., Gubala, M. N. (**1991**), *IEEE Trans. Electron Devices* **ED-38**, 1104–1109.
[46] Ando, K. J., Love, P. J., Lum, N. A., Gulbransen, D. J., Hoffman, A. W., Corrales, E., Mills, R. E., Murray, M. E. (2003), in P. Amico, J. W. Beletic (Eds.), *Optical Detectors For Astronomy III*. Dordrecht: Kluwer Academic Publishers.
[47] Kozlowski, L. J., Loose, M., Bai, Y., Luo, J., Xue, S., Hughes, G. W., Vural, K. (2003), in P. Amico, J. W. Beletic (Eds.), *Optical Detectors For Astronomy III*. Dordrecht: Kluwer Academic Publishers.
[48] Bertram, W. J., Mohsen, A. M., Morris, F. J., Sealer, D. A., Sequin, C. H., Tompsett, M. F. (**1974**), *IEEE Trans. Electron Devices* **ED-21**, 758–767.
[49] Hynecek, J. (**1981**), *IEEE Trans. Electron Devices* **ED28**, 483–493.
[50] Jerram, P., Poole, P., Bell, R., Bowring, S., Spencer, S., Hazelwood, M., Moody, I., Catlett, N., Heyes, P. (2001), M. M. Blouke, J. Canosa, N. Sampat (Eds.), *Sensors and Camera Systems for Scientific, Industrial, and Digital Photography Applications II*. Proc. SPIE Vol. 4306, Bellingham: SPIE Press, pp. 178–186.
[51] Hynecek, J. (**2001**), *IEEE Trans. Electron Devices* **ED48**, 2238–2241.
[52] Mackay, C. D., Tubbs, R. N., Bell, R., Burt, D., Jerram, P., Moody, I. (2001), M. M. Blouke, J. Canosa, N. Sampat (Eds.), *Sensors and Camera Systems for Scientific, Industrial, and Digital Photography Applications II*, Proc. SPIE Vol. 4306, Bellingham: SPIE Press, pp. 289–298.
[53] Findlater, K. M., Renshaw, D., Hurwitz, J. E. D., Henderson, R. K., Purcell, M. D., Smith, S. G., Bailey, T. E. R. (**2003**), *IEEE Trans. Electron Devices* **ED-50**, 32–42.
[54] Wen, D. D. (**1974**), *IEEE Trans. Solid-State Circuits* **SC-9**, 410–414.
[55] Janesick, J. R., Elliott, T., Dingizian, A., Bredthauer, R., Chandler, C., Westphal, J. A., Gunn, J. E. (1990), in G. H. Jacoby (Ed.), *CCDs in Astronomy*. San Francisco: Astro. Soc. Pac. Conf. Proc., pp. 18–39.
[56] Nixon, R. H., Kemeny, S. E., Staller, C. O., Fossum, E. R. (1995), in M. M. Blouke (Ed.), *Charge-coupled Devices and Solid-state Optical Sensors V*. Proc. SPIE Vol. 2415, Bellingham: SPIE Press, pp. 117–123.
[57] Loukianova, N. V., Folkerts, H. O., Maas, J. P. V., Verbugt, D. W. E., Mienop, A. J., Hoekstra, W., Roks, E., Theuwissen, A. J. P. (**2003**), *IEEE Trans. Electron Devices* **Ed-50**, 77–83.
[58] Widenhorn, R., Blouke, M. M., Weber, A., Rust, A., Bodegom, E. (2002), in M. M.

Blouke, J. Canosa, N. Sampat (Eds.), *Sensors, Cameras, and Systems for Scientific, Industrial, and Digital Photography Applications III*. Proc. SPIE Vol. 4669, Bellingham: SPIE Press, pp. 193–201.

[59] Janesick, J. R., Elliott, S. T., Fraschetti, G., Collins, S., Blouke, M. M., Corrie, B. L. (1989), in M. M. Blouke, D. Pophal (Eds.), *Optical Sensors and Electronic Photography*. Proc. SPIE Vol. 1071, Bellingham: SPIE Press, pp. 153–169.

[60] Janesick, J. R., Elliott, T., Collins, S., Blouke, M. M., Freeman, J. (**1987**), *Opt. Eng.* **26**, 692–714.

[61] Janesick, J. R., Elliott, T., Daud, T., McCarthy, J., Blouke, M. M. (1985), in K. N. Prettyjohns, E. L. Dereniak (Eds.), *Solid State Imaging Arrays*. Proc. SPIE Vol. 570, Bellingham: SPIE Press, pp. 46–79.

[62] Blouke, M. M., Janesick, J. R., Elliott, T., Hall, J. E., Cowens, M. W., May, P. J. (**1983**), *Opt. Eng.* **22**, 607–614.

[63] Carnes, J. E., Kosonocky, W. F. (**1972a**), *RCA Rev.* **33**, 327–343.

[64] White, M. H., Lampe, D. R., Blaha, F. C., Mack, I. A. (**1974**), *IEEE Trans. Solid-State Circuits* **SC-9**, 1–12.

[65] Yadid-Pecht, O., Mansoorian, B., Fossum, E. R., Pain, B. (1997), in M. M. Blouke (Ed.), *Solid State Sensor Arrays: Development and Applications*. Proc. SPIE Vol. 3019, Bellingham: SPIE Press, pp. 125–136.

[66] Krymski, A., Khaliullin, N., Rhodes, H. (2003b), *IEEE Workshop on Charge Coupled Devices and Advanced Image Sensors*, pp. 1–6.

[67] Poletto, L., Boscolo, A., Pelizzo, M. G., Placentino, L., Tondello, G. (1997), EUV, X-ray, and gamma ray Instrumentation for Astronomy, Proc. SPIE Vol. 3114, Bellingham: SPIE Press, pp. 617–624.

[68] Ciccarelli, A., Davis, B., Des Jardin, W., Doan, H., Meisenzahl, E., Pace, L., Putnam, G., Shephered, J., Stevens, E., Summa, J., Wetzel, K. (2002), in M. M. Blouke, J. Canosa N. Sampat (Eds.), *Sensors and Camera Systems for Scientific, Industrial, and Digital Photography Applications III*. Proc. SPIE Vol. 4669, Bellingham: SPIE Press, pp. 153–160.

[69] Blouke, M. M., Cowens, M. W., Hall, J. E., Westphal, J. A., Christensen, A. B. (**1980**), *Appl. Opt.* **19**, 3318–3321.

[70] Cullum, M., Dieries, S., D'Odorico, S., Reiss, R. (**1985**), *Astron. Astrophys.* **153**, L1–L3.

[71] Blouke, M. M. (1990), in M. B. Denton (Ed.), *International Conference on Scientific Optical Imaging*. Proc. SPIE Vol. 1439, Bellingham: SPIE Press, pp. 136–143.

[72] Lesser, M. P. (1993), in M. M. Blouke (Ed.), *Charge-coupled Devices and Solid-state Optical Sensors III*. Proc. SPIE Vol. 1900, Bellingham: SPIE Press, pp. 219–227.

[73] Chamberlain, S. G., Harper, D. H. (**1978**), *IEEE Trans. Electron Devices* **ED-25**, 145–154.

[74] Blouke, M. M., Robinson, D. A. (**1982**), *IEEE Trans. Electron Devices* **ED-28**, 251–256.

[75] Stevens, E. G., Lavine, J. P. (**1994**), *IEEE Trans. Electron Devices* **ED-41**, 1753–1760.

[76] Shcherback, I., Yadid-Pecht, O. (**2001**), *IEEE Trans. Electron Devices* **ED-48**, 2710–2715.

[77] Amelio, G. F., Tompsett, M. F., Smith, G. E. (**1970**), *Bell Syst. Tech. J.* **49**, 593–600.

[78] Blouke, M. M., Heidtmann, D. L., Corrie, B. L., Lust, M. L., Janesick, J. R. (1985), in K. N. Prettyjohns, E. L. Dereniak (Eds.), *Solid State Imaging Arrays*. Proc. SPIE Vol. 570, Bellingham: SPIE Press, pp. 82–88.

[79] Meynants, G., Dierickx, B., Alaerts, A., Uwaarts, D., Cos, S., Scheffer, D. (2003), *Proc. IEEE Workshop on Charge Coupled Devices and Advanced Image Sensors.*

[80] Iversen, S. (2003), *Proc. IEEE Workshop on Charge Coupled Devices and Advanced Image Sensors.*

[81] Fossum, E. R. (**1987**), *Opt. Eng.* **26**, 916–922.

## Further Reading

Amico, P., Beletic, J. W. (Eds.) (2000), *Optical Detectors for Astronomy II*. Dordrecht: Kluwer Academic Publishers, pp. 187–200.

Amico, P., Beletic, J. W. (Eds.) (2002), *Optical Detectors for Astronomy III*. Dordrecht: Kluwer Academic Publishers, pp. 187–200.

Ashikawa, M., Kosonocky, W. F. (Eds.) (**1985**), *IEEE Trans. Electron Devices* **ED-32**.

Beletic, J. W., Amico, P. (Eds.) (1998), *Optical Detectors for Astronomy*. Dordrecht: Kluwer Academic Publishers.

Beynon, J. D. E., Lamb, D. R. (Eds.) (1980), *Charge-Coupled Devices and their Applications*. London: McGraw-Hill.

Blouke, M. M., Canosa, J., Sampat, N. (Eds.) (2001). *Sensors and Camera Systems for Scientific, Industrial, and Digital Photography Applications II*. Proc. SPIE 4306 (2001).

Blouke, M. M., Canosa, J., Sampat, N. (Eds.), *Sensors and Camera Systems for Scientific, Industrial, and Digital Photography Applications III*. Proc. SPIE 4669 (2002).

Blouke, M. M., Canosa, J., Sampat, N. (Eds.), *Sensors and Camera Systems for Scientific, Industrial, and Digital Photography Applications IV*. Proc. SPIE 5017 (2003).

Blouke, M. M., Sampat, N., Williams, G. M., Yeh, T. (Eds.) (2000), *Sensors and Camera Systems for Scientific, Industrial, and Digital Photography Applications*. Proc. SPIE 3965 (2000).

Buil, C. (1991), *CCD Astronomy*. Richmond: Willmann-Bell.

Buss, D. D., Tompsett, M. F. (Eds.) (**1976**), *IEEE Trans. Electron Devices* **ED-23**. Feb. Special Issue on CCDs.

Chamberlain, S. G., Kuhn, M. (Eds.) (**1978**), *IEEE Trans. Electron Devices* **ED-25**. Feb. Special Issue on CCDs.

Fossum, E. R., Teranishi, N., Theuwissen, A. J. P., Hynecek, J. (Eds.) (**2003**), *IEEE Trans. Electron Devices* **ED-50**. Jan. Special Issue on Image Sensors.

Howes, M. J., Morgan, D. V. (Eds.) (1979), *Charge-coupled Devices and Systems*. New York: John Wiley & Sons.

Janesick, J. R. (Ed.) (**1987**), *Opt. Eng.* **26**(8).

Janesick, J. R. (Ed.) (**1987**), *Opt. Eng.* **26**(9).

Janesick, J. R. (Ed.)(**1987**), *Opt. Eng.* **26**(10).

Janesick, J. R. (2001), *Scientific, Charge-Coupled Devices*. Bellingham: SPIE Press.

Kosonocky, W. F., Okuto, Y., Theuwissen, A. J. P. (Eds.) (**1991**), *IEEE Trans. Electron Devices* **ED-38**. Feb. Special Issue on Image Sensors.

Kosonocky, W. F., Fossum, E. R., Stancampiano, C. V., Teranishi, N., Theuwissen, A. J. P., (**1997**), *IEEE Trans. Electron Devices* **ED-38**, Oct. Special Issue on Solid State Image Sensors.

Melen, R., Buss, D. (Eds.) (1977), *Charge-coupled Devices: Technology and Applications*. New York: IEEE Press.

Pain, B. (2000), Fundamentals of CMOS Digital Image Sensors, SPIE short course notes.

Theuwisssen, A. J. P. (1995), *Solid-state Imaging with Charge-coupled Devices*. Dordrecht: Kluwer Academic Publishers.

# Information Processing, Optical

*Bahram Javidi*
*Electrical and Computer Engineering Department, University of Connecticut, USA*
*Phone: 860-486-2867; Fax: 860-486-2447; e-mail: bahram@engr.uconn.edu*

*Joseph L. Horner*
*RL/ERCP, Air Force Research Laboratory, Hanscom Air Force Base, Bedford, MA, USA*
*Phone: 781-377-3841; Fax: 781-377-2836; e-mail: joseph.horner@hanscom.af.mil*

**Abstract**
This article presents a brief review of the fundamentals of optical information processing. We discuss optical information processing techniques, materials and devices for optical processing, optical pattern recognition, optical security, optical neural computing, and optical memory. Optical image processing algorithms and architectures as well as basic hardware concepts such as the fundamentals of optical spatial light modulators are reviewed.

**Keywords**
optical information processing; optical imaging; optical pattern recognition; optical security; optical spatial light modulators; optical neural computing; optical memory.

## 1 Introduction

One of the most important applications of optics and optoelectronics is in information technology. Processing of information with optics [1–4] offers many advantages and capabilities, including high speed, large-volume data handling, compactness, low power consumption, and ruggedness. Electronic machines are mostly serial processors while optical systems are inherently parallel processing systems [5, 6]. When one is dealing with large volumes of data, this feature offers a powerful alternative to electronic computing. Presently, compact custom-made optical hardware can process two-dimensional arrays of data of up to a million pixels at kilohertz frame rates, which is substantially higher than their electronics counterpart.

In the past, most applications of optical-processing systems have been for military hardware because of high cost and performance demands. With recent advances in optical devices and components [7] such as optics memory and display, optical processing systems have become more attractive for commercial applications. However, the progress in the field still seems to be of an evolutionary nature. Progress is needed in the development of reliable active optical devices and materials to realize low-cost optical systems.

This article presents the fundamentals of coherent optical signal processing. We discuss optical information-processing techniques, optical pattern recognition, optical neural computing, and materials and devices for optical processing. Signal-processing algorithms and architectures in optics as well as basic hardware concepts such as the fundamentals of optical spatial light modulators are covered. Each section begins with a review of the basic concepts and follows with a discussion and examples of recent advances in the field. A complete bibliography on the fundamentals of each topic is also included to assist the reader.

## 2
## Fourier-transform Property of Lenses

An attractive feature of an optical system is that the Fourier transform of an image can be generated in the space domain [8]. As a result, various types of filtering operations for information processing can be implemented. It is well known that the complex amplitude of the light distribution in the back focal plane of a lens is the two-dimensional complex Fourier transform of the transmittance in the front focal plane of the lens [1]. For convenience the lens is modeled as a thin spherical lens. By a thin lens, it is meant that a ray of light entering at coordinates $(x, y)$ on one face emerges at approximately the same coordinates on the opposite face. In other words, there is negligible translation of light within the lens volume. Therefore, a thin lens simply introduces a phase shift (delay) to an incident wave front by an amount proportional to both the thickness of the lens at each point and the index of refraction of the lens material $n$ (for the wavelength $\lambda$). The lens transmittance function $t_1(x, y)$ can be expressed as [1, 7]

$$t_1(x, y) = \exp\left[j\frac{2\pi}{\lambda}n\Delta_0\right] \times \exp\left[-j\frac{\pi}{\lambda f}(x^2 + y^2)\right], \quad (1)$$

where $\Delta_0$, and $f$ are the maximum thickness and the focal length of the lens, respectively. Equation (1) is valid only under the paraxial approximation, that is, only the portions of the wave front that are near the lens axis are delayed in phase according to Eq. (1). The complex amplitude of the light distribution in the back focal plane of a lens [see Fig. 1(a)], $E(\alpha, \beta;\lambda)$, in terms of that in the front plane, $e(x, y;\lambda)$, is

$$E(\alpha, \beta; \lambda) = K \iint e(x, y; \lambda) \times \exp\left[-j\frac{2\pi}{\lambda f}(\alpha x + \beta y)\right] \mathrm{d}x\, \mathrm{d}y, \quad (2)$$

where a time-dependent factor $\exp[j\omega t]$ has been dropped for convenience; $K$ is a constant under coherent illumination; $(x, y)$ are the coordinates in the input plane, which is the front focal plane of the lens; $(\alpha, \beta)$ are the coordinates in the Fourier plane, which is the back focal plane of the lens; and $\lambda$ is the wavelength of the illuminating light. This Fourier-transform relationship is the fundamental relationship in the analysis of the optical processors. If we place a second lens L2 behind the Fourier plane as shown in Fig. 1(b), the light distribution at the back focal plane of L2 is the double Fourier transform of the input field. Here, we have assumed that lenses L1 and L2 have the same focal lengths. Different image-processing operations can be achieved by placing a spatial filter at the Fourier plane. For example, by placing an opaque spot at the origin of the Fourier

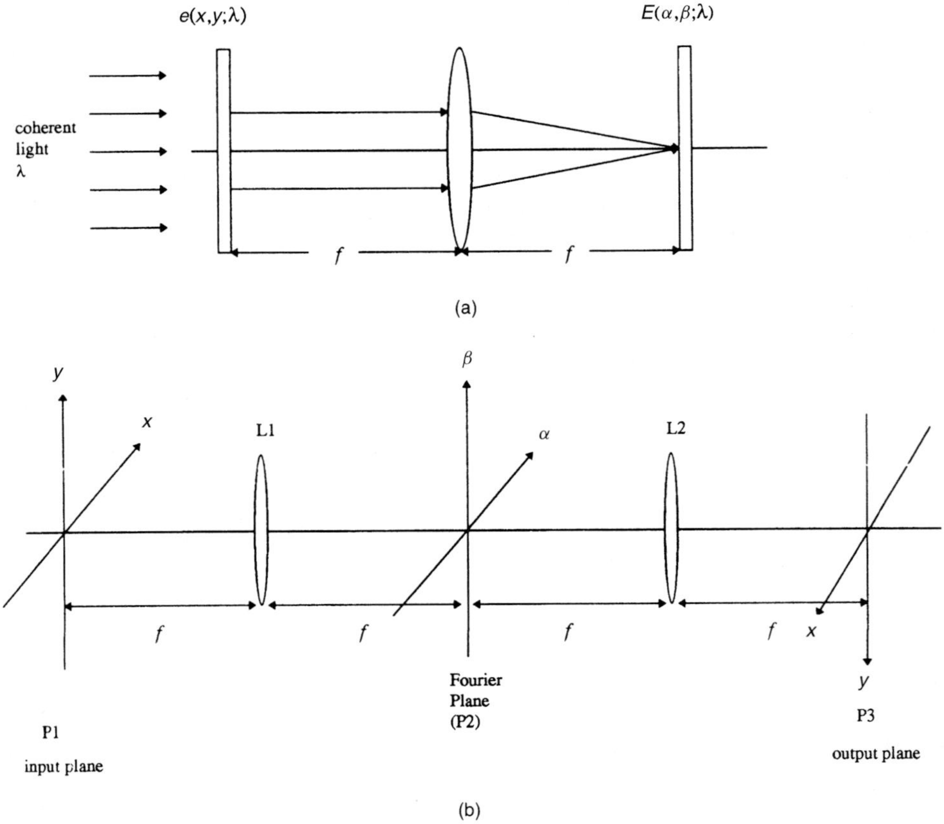


**Fig. 1** (a) The Fourier-transform property of a thin lens; (b) A $4f$ optical system performs the Fourier transform twice. Plane $(\alpha, \beta)$ is the Fourier plane, where a filter function can be inserted to execute different image-processing operations

plane, we can block the low or zero spatial frequencies of the input signal, thus generating a high-pass filtered version of the input field. Similarly, for more sophisticated types of image processing, a complex spatial filter $F(\alpha, \beta)$, where $F(\alpha, \beta)$ is the Fourier transform of $f(x, y)$, can be inserted at the Fourier plane, resulting in a light field of $F(\alpha, \beta)E(\alpha, \beta)$ leaving the filter plane. Therefore, at the output plane P3, we obtain the Fourier transform $F(\alpha, \beta)E(\alpha, \beta)$, which is equivalent to the convolution of the input signal $e(x, y)$ with the filter function $f(x, y)$. By proper choice of the spatial filter, numerous signal-processing operations can be performed, and since it is two dimensional, it lends itself naturally to image processing.

## 3
## Filters

### 3.1
### Correlation and Matched Filtering

One of the early achievements in the field was the invention of the optical matched filter correlator by [9]. The matched filter, developed during World War II for extracting radar returns from noisy background and detector noise, gives the optimum theoretical response of a system when the signal is corrupted by additive overlapping Gaussian. In the derivation of the matched filter, optimum is defined as maximizing the signal-to-noise ratio (SNR), which is defined as the ratio of the output signal peak to the root mean square of the output noise. The definition of "optimum" and the fact that the noise overlaps or blankets the target or the signal is very important [10]. If different criteria are used, the matched filter is no longer optimum. This is discussed in more detail later.

The procedure to synthesize an optical matched spatial filter by a holographic technique is shown in Fig. 2 (see also HOLOGRAPHY). It turns out that the matched filter designed for detecting a specific image or target in the presence of white noise [4] is just the target itself, in the spatial domain. In the Fourier domain it is equal to the complex conjugate of the Fourier transform of the target. It is possible to produce the matched spatial filter of a reference signal $s(x, y)$ at the filter plane P2. Referring to Fig. 2, the reference signal $s(x, y)$ is inserted at the input plane P1. The light distribution $S(\alpha, \beta)$ at the filter plane P2 is the Fourier transform of the input reference signal $s(x, y)$:

$$S(\alpha, \beta; \lambda) = K \iint s(x, y) \times \exp\left[-j\frac{2\pi}{\lambda f}(\alpha x + \beta y)\right] \mathrm{d}x\,\mathrm{d}y, \tag{3}$$

where $(x, y)$ and $(\alpha, \beta)$ are the coordinates in planes P1 and P2, respectively; $f$ is the focal length of the transform lens L; and $K$ is a complex constant. A plane-wave reference beam of uniform amplitude incident on the plane P2 at an angle $\theta$ with respect to the optical axis can be written as

$$R(\alpha, \beta) = c \exp\left(-j\frac{2\pi}{\lambda}\sin\theta \cdot \beta\right), \tag{4}$$

where $\sin\theta/\lambda$ is the spatial frequency corresponding to the plane wave with wavelength $\lambda$ [1]. To produce the matched

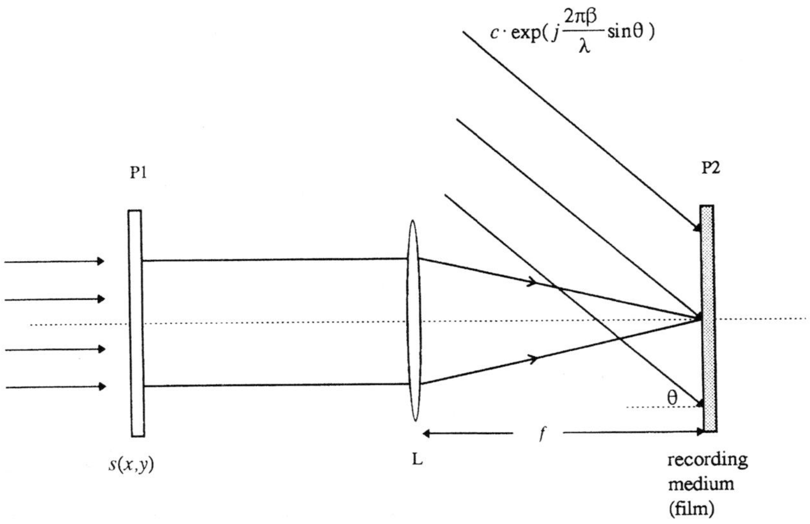


**Fig. 2** An interferometric technique for synthesizing an optical spatial filter

filter at the Fourier plane for detecting the reference signal $s(x, y)$, a conventional holographic technique is used to record the interference patterns of the reference-signal transform $S(\alpha, \beta)$ with the reference beam $R(\alpha, \beta)$. This can be done by placing a detector such as a high-resolution photographic film at plane P2. The total intensity distribution at the filter plane $I(\alpha, \beta)$ is obtained as

$$I(\alpha, \beta) = |R(\alpha, \beta) + S(\alpha, \beta)|^2. \quad (5)$$

The film is then developed to produce a transmittance function $H(\alpha, \beta)$.

For simplicity, we assume that $H(\alpha, \beta)$ is proportional to the intensity of the distribution $I(\alpha, \beta)$. Under this condition, the filter transmittance function can be written as

$$\begin{aligned} H(\alpha, \beta) &= c^2 + |S(\alpha, \beta)|^2 \\ &\quad + cS(\alpha, \beta) \exp\left(j2\pi \frac{\sin\theta}{\lambda}\beta\right) \\ &\quad + cS^*(\alpha, \beta) \exp\left(-j2\pi \frac{\sin\theta}{\lambda}\beta\right). \end{aligned} \quad (6)$$

The first and second terms of Eq. (6) represent the on-axis terms, since they are not modulated by a spatial carrier term. The last term in Eq. (6) is the desired matched spatial filter of $s(x, y)$, which is proportional to the complex conjugate of the reference signal's spectrum $S^*(\alpha, \beta)$. The third term produces the convolution term.

Referring to Fig. 1(b), if the matched spatial filter described above is placed at the Fourier plane, and an arbitrary signal $g(x, y)$ is inserted at the input plane, then the complex amplitude of the light leaving the filter plane is the product of the filter's transmittance function and the input signal's spectrum; that is,

$$\begin{aligned} t(\alpha, \beta) &= G(\alpha, \beta) S^*(\alpha, \beta) \\ &\quad \times \exp\left(-j\frac{2\pi\beta}{\lambda}\sin\theta\right), \end{aligned} \quad (7)$$

where $t(\alpha, \beta)$ is the light leaving plane P2, and $G(\alpha, \beta)$ is the Fourier transform of the input signal $g(x, y)$. For simplicity, we have considered only the matched spatial filter term of the filter's transmittance function, because the spatial frequency sin $\theta/\lambda$ spatially separates the correlation term from the other terms at the output plane P3. Plane P2 is located at the front focal plane of lens L2, as shown in Fig. 1(b). Therefore, the light pattern in plane P3 is proportional to the Fourier transform of $t(\alpha, \beta)$ and is given by

$$\begin{aligned} r_c(x, y) &= \iint g(\xi, \eta) \\ &\quad \times S^*(\xi - x, \eta - y + f\sin\theta)\, d\xi\, d\eta, \end{aligned} \quad (8)$$

where $r_c(x, y)$ is the light distribution at the output plane.

We note that $r_c(x, y)$ is the cross-correlation between the input signal $g(x, y)$ and the reference signal $s(x, y)$ that is diffracted in the vicinity of $(0, f\sin\theta)$ at the output plane. If the input signal is equivalent to the reference signal $s(x, y)$, then the autocorrelation of the reference signal is obtained at the output plane.

## 3.2 Spatial Filter Development

The holographic matched filter discussed above was a breakthrough in its time, but being described by a complex mathematical function, it is difficult to implement on real-time devices like spatial light modulators. In 1984 the phase-only filter [11] was introduced. This approach makes use of the fact that when we look at the phase and amplitude information of a signal in the Fourier transform plane, the phase information is vastly more important than the amplitude information. Interesting examples of reconstructing two-dimensional signals (images) from the phase information were shown by [12]. This concept was first applied to the matched filter [11] with very good results. That is, if we represent the Fourier transform of a signal $r(x, y)$ by

$$\mathcal{F}(r(x, y)) = R(\alpha, \beta) \exp[j\phi_R(\alpha, \beta)], \quad (9)$$

where $\mathcal{F}(.)$ represents the Fourier transform and $R(\alpha, \beta)$ and $\phi_R(\alpha, \beta)$ are the amplitude and the phase of the Fourier transform of the signal $r(x, y)$, the transmittance function of the matched filter for this signal is

$$H(\alpha, \beta) = R(\alpha, \beta) \exp[-j\phi_R(\alpha, \beta)], \quad (10)$$

and that of the phase-only filter (POF) is

$$H_{\mathrm{POF}}(\alpha, \beta) = \exp[-j\phi_R(\alpha, \beta)]. \quad (11)$$

The advantage of the POF as compared with the matched filter is that it produces a very narrow correlation peak. This means that if there are multiple targets in the input, their correlation signals will not overlap and cause false alarms. Also, in an optical system, the POF has the advantage of passing all the light incident on it, since it is a pure phase element. The holographic implementation of the matched filter typically puts only a fraction of a percent of the light into the correlation plane [13]. Figure 3 shows a comparison of the matched filter, the phase-only filter, and the further developed filter called the *binary phase-only filter*, which we discuss next.

### 3.3 Binary Phase-only Filters (BPOFs)

In the POF, the amplitude information was discarded, leaving only the phase information. A further simplification is possible by forcing the phase function to assume one of two values, 0 or $\pi$ [14, 15]. If we think of each regime, phase and amplitude, as being represented by a one-byte word, the BPOF represents a reduction of 94% in the information of the signal target. It is surprising that the filter works at all, yet, as Fig. 3 shows, it works almost as well as the POF, except that the noise background is a little higher. That is the trade-off. Using a nonlinear Fourier-transform method of analysis, it was shown that the BPOF could be represented as a sum of infinite harmonic terms, the first term of which is just the term for the POF [16, 17]. That is,

$$\begin{aligned} H_{\mathrm{BPOF}}(\alpha, \beta) = {} & k_1 \exp[-j\phi_R(\alpha, \beta)] \\ & + k_2 \exp[j\phi_R(\alpha, \beta)] \\ & + \sum \text{other harmonics}, \end{aligned} \quad (12)$$

where $k_1$ and $k_2$ are constants. The second term is the POF for an inverted-target signal. Therefore, the BPOF cannot distinguish the difference between an object and its inverted version. Whether or not this is a problem depends on the application. For example, if we are trying to find tanks in a battlefield, this would not be a disadvantage. In practice the BPOF, like the POF, can be implemented in a real system in real time on a single phase-modulating spatial

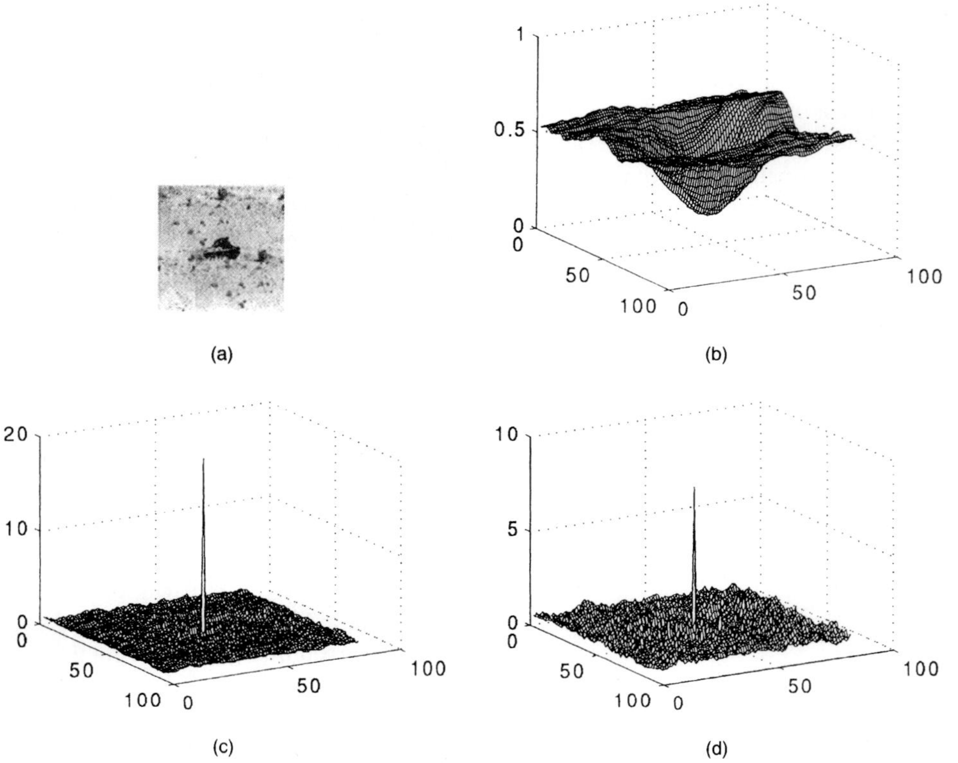


**Fig. 3** The comparison of the classical matched filter, the phase-only matched filter, and the binary phase-only filter. A tank image is the target to be detected. (a) The input scene; (b) Correlation output of the classical matched filter; (c) Correlation output of the POF; (d) Correlation output of the BPOF

light modulator (SLM), which is now commercially available [18].

### 3.4
### Optimum Filtering for Spatially Nonoverlapping Target and Scene Noise

The matched filtering [9, 19] and its variations [6, 11, 15, 16] are well-established techniques for optical pattern-recognition applications. The matched filter function is derived to optimize the conventional SNR defined in Sec. 2, under the condition that the additive input noise is overlapping with the target and is, at least in the wide sense, stationary. In many pattern-recognition applications, however, the input scene noise does not overlap the target (it is then sometimes called *disjoint noise*). This means that the target is in the foreground and blocks the scene noise. For this class of problems, the matched filter and the optimum filter derived under the assumption of overlapping input target and scene noise may not perform well [10].

Now, we consider that an input image $s(t)$ (one-dimensional notation is used for simplicity) is real and contains a target $r(t)$ in the presence of random noise that includes nonoverlapping scene noise and distortion noise on the target (the second and third terms in the following equation, respectively):

$$s(t,\tau) = r(t-\tau) + n_b(t) \times [w_0(t) - w_r(t-\tau)] + n_r(t)w_r(t-\tau),$$

where $w_0(t)$ and $w_r(t)$ are window functions of the input image and the target, respectively. The window $w_0(t)$ is defined to be 0 outside the input image and 1 inside. The target window $w_r(t)$ is defined to be 1 wherever the target exists and 0 otherwise. The quantity $\tau$ is a random variable that indicates the random location of the target in the input scene, and $n_b(t)$ and $n_r(t)$ are noise for modeling the background noise and the distortion on the target, respectively.

One solution for this problem is using multiple-hypothesis testing to design an optimum receiver for the disjoint input target and scene noise [20]. It is shown that for a noise-free target, the optimum receiver is similar to a correlator normalized by the input scene energy within the target window. In addition, given that the target is noise free and the scene noise probability-density function is bounded, then the actual scene noise statistics becomes irrelevant to the detection process.

Another solution is the optimum-filter approach [21] for detecting targets in spatially disjoint scene noise. The filter is designed by maximizing a new performance metric, peak-to-output energy (POE), that is defined as the ratio of the square of the expected value of the output signal at the target location to the expected value of the average output signal energy. So the filter produces a sharp output signal at the target location with a low output-noise floor.

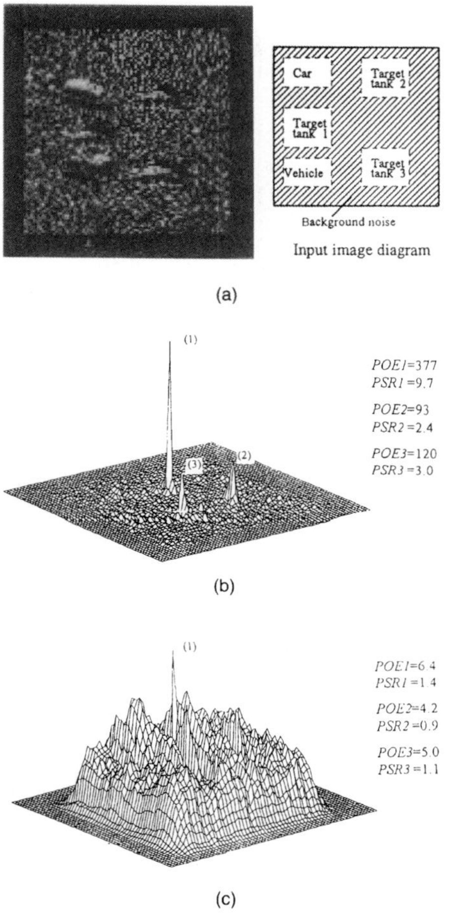


**Fig. 4** The performance of the optimum filter for spatially disjoint target and scene noise [21]. (a) The input scene where three target tanks and two objects (a car and a vehicle) are embedded in white Gaussian-distributed background noise with mean of $m_b = 0.4$ and standard deviation of $\sigma_b = 0.3$. Target tank 1 is identical to the reference tank used in the filter design. Target tank 2 is rotated by 4°. Target tank 3 is scaled up by 10%. The target distortion noise parameters are chosen as $m_r = 0$ and $\sigma_r = 0.2$ in the filter design; (b) Correlation output of the classical matched filter; (c) Correlation output of the optimum filter

We provide a test result of the optimum filter to show its performance. Three target tanks and two objects (a car and a vehicle) are embedded in white Gaussian-distributed background noise with mean of $m_b = 0.4$ and standard deviation of $\sigma_b = 0.3$ [Fig. 4(a)]. Target tank 1 is identical to the reference tank used in the filter design. Target tank 2 is rotated by 4°. Target tank 3 is scaled up by 10%. The target-distortion noise parameters are chosen as mean $m_r = 0$ and standard deviation $\sigma_r = 0.2$ in the filter design. The filter output is plotted in Fig. 4(b) and compared with the output of the conventional matched filter in Fig. 4(c). Here the peak-to-sidelobe ratio (PSR) is defined as the correlation peak intensity to the largest noise peak intensity at the output plane.

## 4
## Correlations

### 4.1
### Classical Joint-transform Correlators

Another architecture for correlation of objects is the joint-transform correlator (JTC) [1, 22], as shown in Fig. 5. Here the reference function $r(x, y)$ and an unknown input object $s(x, y)$ are presented together in the input plane, and their combined or joint Fourier transform is produced in the focal plane behind the first lens. If the joint Fourier transform is recorded on a square-law or energy detector such as photographic film and a second Fourier

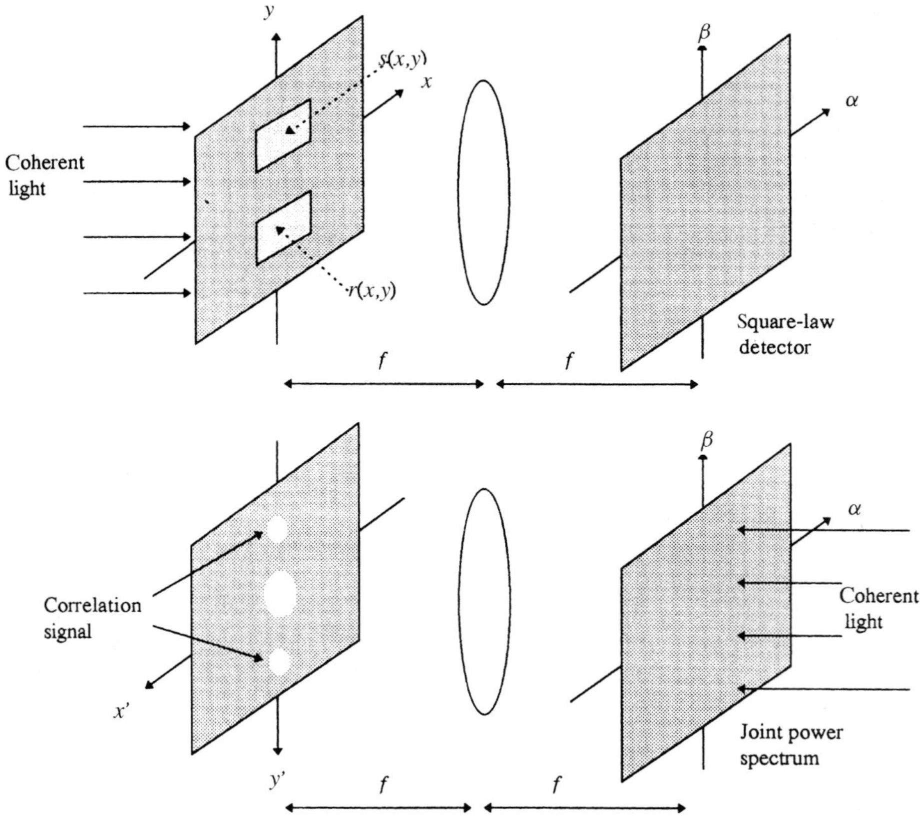


Fig. 5 Joint-transform correlator – an architecture for correlation of objects

transform is taken, a correlation of the two objects can be realized. The main advantage of the joint-transform correlator is that both the input signal and the reference signal are Fourier-transformed simultaneously, and the interference between the transforms is achieved in one single step. Thus, the need for making filters is eliminated. The input images can be displayed on a spatial light modulator (SLM) for real-time operation. A disadvantage of the JTC is that the space–bandwidth product (equivalent to the total number of image pixel elements) available in the input plane must be shared by the input and reference signals as well as a space between these signals to ensure separation of correlation patterns from the zero-frequency or dc terms at the output plane.

The implementation of the joint-transform correlator using a spatial light modulator is shown in Fig. 6 [23]. Plane P1 is the input plane that contains the reference signal $r(x + x_0, y)$ and the input signal $s(x - x_0, y)$. The amplitude of the light distribution at the back focal plane of the transform lens FTL1 is the sum of the Fourier transforms of the two input image functions; that is,

$$I(\alpha, \beta) = S(\alpha, \beta)\exp[j\phi_S(\alpha, \beta)] \times \exp(-jx_0\alpha) + R(\alpha, \beta) \times \exp[j\phi_R(\alpha, \beta)]\exp(jx_0\alpha), \tag{13}$$

where $(\alpha, \beta)$ are the spatial-frequency coordinates and $S(\alpha, \beta)\exp[j\phi_S(\alpha, \beta)]$ and $R(\alpha, \beta) \times \exp[j\phi_R(\alpha, \beta)]$ correspond to the Fourier transforms of the input and reference signals $s(x, y)$ and $r(x, y)$, respectively. The Fourier-transform intensity distribution at plane P2 can be written as

$$\begin{aligned} E(\alpha, \beta) = |I(\alpha, \beta)|^2 = S^2(\alpha, \beta) + R^2(\alpha, \beta) \\ + S(\alpha, \beta)\exp[j\phi_S(\alpha, \beta)]R(\alpha, \beta) \\ \times \exp[-j\phi_R(\alpha, \beta)]\exp(-j2x_0\alpha) \\ + S(\alpha, \beta)\exp[-j\phi_S(\alpha, \beta)]R(\alpha, \beta) \\ \times \exp[j\phi_R(\alpha, \beta)]\exp(j2x_0\alpha), \end{aligned} \tag{14}$$

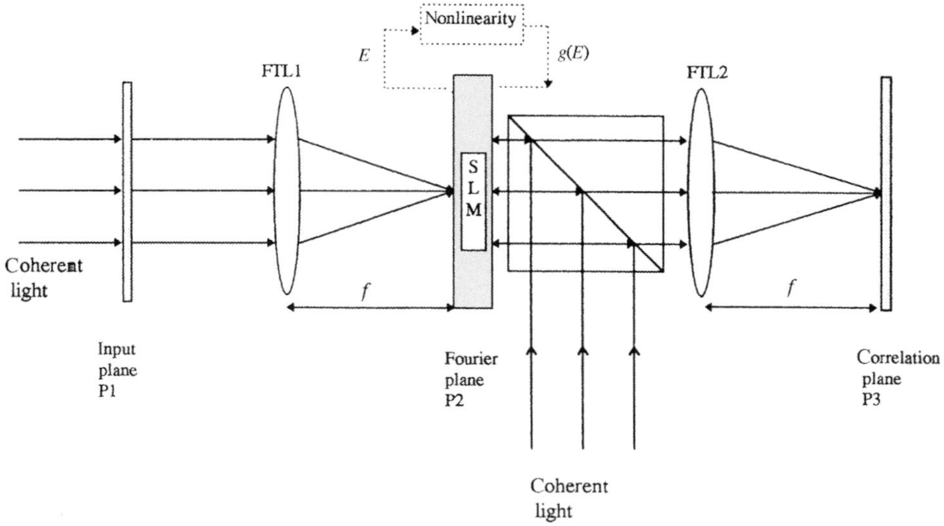


**Fig. 6** Nonlinear joint-transform correlator [4]

where we assume that $E(\alpha, \beta)$ has been normalized.

For the linear or classical joint-transform correlator, the inverse Fourier transform (or the Fourier transform with coordinates reversed) of Eq. (14) will produce the following correlation signals at the output plane:

$$h(x', y') = R_{11}(x', y') + R_{22}(x', y') + R_{12}(x' - 2x_0, y') + R_{21}(x' + 2x_0, y'), \quad (15)$$

where

$$R_{21}(x', y') = R_{12}(-x', -y') = \iint s(\xi, \zeta)r(\xi - x', \zeta - y')\, d\xi\, d\zeta,$$

$$R_{11}(x', y') = \iint s(\xi, \zeta)s(\xi - x', \zeta - y')\, d\xi\, d\zeta,$$

$$R_{22}(x', y') = \iint r(\xi, \zeta)r(\xi - x', \zeta - y')\, d\xi\, d\zeta.$$

The first two terms of Eq. (15) are autocorrelation terms. The terms of interest are the third and fourth terms, which are the cross-correlations of the reference signal with the input signal. Here $\xi$ and $\zeta$ are just dummy variables.

### 4.2 Nonlinear Joint-transform Correlators

More recently, nonlinearities were introduced into the JTC. The binary JTC is obtained by binarizing the joint power spectrum into two values [24]. It has been shown that in terms of discrimination a binary joint-transform correlator has superior performance compared with that of the conventional linear JTC. Later, the binary JTC was generalized to nonlinear JTCs [24–27], and thus a family of correlators and $k$th-power–law nonlinear JTCs, including the conventional JTC for $k = 1$ and the binary JTC for $k = 0$, were introduced. Here, $k$ represents the severity of the nonlinearity of the transformation of $\mathrm{sgn}(E_m)|E_m|^k$, where $E_m$ is the modified joint power spectrum ($E_m = E - S^2 - R^2$) and sgn(.) is the signum function; $k = 1$ corresponds to the linear case, and $k = 0$, to the binary case. It was also shown that JTCs with $k = 1$ correspond to classical matched filters, and JTC's with $k = 0$ correspond to inverse filters. Theoretical and experimental studies have shown that the nonlinear joint-transform correlator can produce very good correlation performance. The nonlinear JTC can use the nonlinearity of a nonlinear device such as a SLM at the Fourier plane to alter the Fourier-transform interference intensity. It has been shown that compared with the classical correlator, the compression type of nonlinear joint-transform correlator ($k < 1$) provides higher peak intensity, larger peak-to-sidelobe ratio, narrower correlation width, and better correlation sensitivity.

In Fig. 6, the joint-transform correlator is implemented with use of a liquid-crystal light valve (LCLV) as an optically addressed SLM at the Fourier plane. The Fourier-transform interference pattern, Eq. (14), is displayed at the input of the LCLV to obtain the intensity of the Fourier-transform interference. For the nonlinear JTC, the LCLV nonlinearly transforms the joint power spectrum according to the nonlinear characteristics of the device. The transformed joint power spectrum can be considered as the output of a nonlinear system. The nonlinear characteristic of the nonlinear device is denoted by $g(E)$, where $E$ is the Fourier transform of the interference intensity. An expression for the nonlinearly transformed interference intensity can be obtained by an approach similar to the analysis of nonlinear systems using the transform method of communication theory [17, 25]. Let the Fourier transform of the nonlinear characteristic of the nonlinear device be defined by

$$G(\omega) = \int_{-\infty}^{\infty} g(E) \exp(-j\omega E)\, \mathrm{d}E. \quad (16)$$

The output of the nonlinear device can be written as

$$g(E) = \sum_{\nu=0}^{\infty} H_\nu[R(\alpha, \beta), S(\alpha, \beta)] \times \cos[2\nu x_0\alpha + \nu\phi_S(\alpha, \beta) - \nu\phi_R(\alpha, \beta)], \quad (17)$$

where

$$H_\nu[R(\alpha, \beta), S(\alpha, \beta)] = \frac{\varepsilon_\nu}{2\pi} j^\nu \int G(\omega) \times \exp\{j\omega[R^2(\alpha, \beta) + S^2(\alpha, \beta)]\} \times J_\nu(2\omega R(\alpha, \beta) S(\alpha, \beta))\, \mathrm{d}\omega.$$

Here,

$$\varepsilon_\nu = \begin{cases} 1, & \nu = 0, \\ 2, & \nu > 0, \end{cases}$$

and $J_\nu$ is a Bessel function of the first kind, order $\nu$.

It can be seen from Eq. (17) that for $\nu = 1$, the nonlinear system has preserved the phase of the cross-correlation term $\phi_S(\alpha, \beta) - \phi_R(\alpha, \beta)$, and only the amplitude is affected. This explains the good correlation properties of the first-order correlation signal at the output plane. For a $k$th-power–law nonlinearity, that is,

$$g(E) = |E|^k \mathrm{sgn}(E - \nu_T),$$

where $\nu_T = R^2 + S^2$, Eq. (17) becomes

$$g_k(E) = \sum_{\nu=1}^{\infty} \frac{\varepsilon_\nu \Gamma(k+1)[R(\alpha, \beta) S(\alpha, \beta)]^k}{[\Gamma(1 - 1/2(\nu - k)) \times \Gamma(1 + 1/2(\nu + k))]} \times \cos[2\nu x_0\alpha + \nu\phi_S(\alpha, \beta) - \nu\phi_R(\alpha, \beta)], \quad (18)$$

where $\Gamma$ is the gamma function, and $k$ is the severity of the nonlinearity; $k = 1$ corresponds to a linear device, and $k = 0$ corresponds to a hard-clipping nonlinearity.

It can be seen from the above equation that each harmonic term is phase modulated by $\nu$ times the phase difference of the input-signal and the reference-signal Fourier transforms, and the higher-order correlation signals are diffracted to $2\nu x_0$. The correct phase information of the joint power spectrum is obtained for the first-order harmonic term ($\nu = 1$); that is,

$$g_{1k}(E) = \frac{2\Gamma(k+1)[R(\alpha, \beta) S(\alpha, \beta)]^k}{\Gamma(1 - 1/2(\nu - k))\Gamma(1 + 1/2(\nu + k))} \times \cos[2x_0\alpha + \phi_S(\alpha, \beta) - \phi_R(\alpha, \beta)]. \quad (19)$$

It is noted that varying the severity of the nonlinearity $k$ will produce correlation signals with different characteristics. For highly nonlinear transformations (small $k$), the high spatial frequencies are emphasized, and the correlation becomes more sensitive in discrimination.

## 5 Spatial Light Modulators (SLMs)

Spatial light modulators are very important optical components of optical information-processing systems. They are used in optical computing systems, programmable optical interconnects, optical neural nets, and optical pattern-recognition systems. The SLM input is either a time-dependent electrical signal or a light distribution such as an image. SLMs with electrical or optical input are called *electrically addressed* or *optically addressed SLMs*, respectively. The SLM modulates the amplitude and/or phase, or polarization, of the read-out light beam as a function of the input signal, which can be an optical image or an electrical signal. Figure 7 shows an example of an optically addressed SLM. The writing light $A_i(x, y)$ is incident on the input of the SLM. In general, $A_i(x, y)$ is a two-dimensional spatially varying amplitude distribution, imaged from a display monitor onto the SLM. The output light distribution $A_0(x, y)$ is a function of the input light amplitude $A_i(x, y)$. The details of the component pieces are discussed in Sec. 5.4 (Liquid-Crystal Light Valve).

Various SLMs differ in addressing methods and the modulating materials used. In Fig. 8, the input light is converted to an electric field distribution by a photoconductor. The electric field is related to the input light intensity (optically addressed SLM). The electric field can also be directly applied by use of transparent conductive electrodes (electrically addressed SLM). This electric field modifies the properties of the electro-optic or modulating material. For example, it may change the optical refractive index of the modulating material.

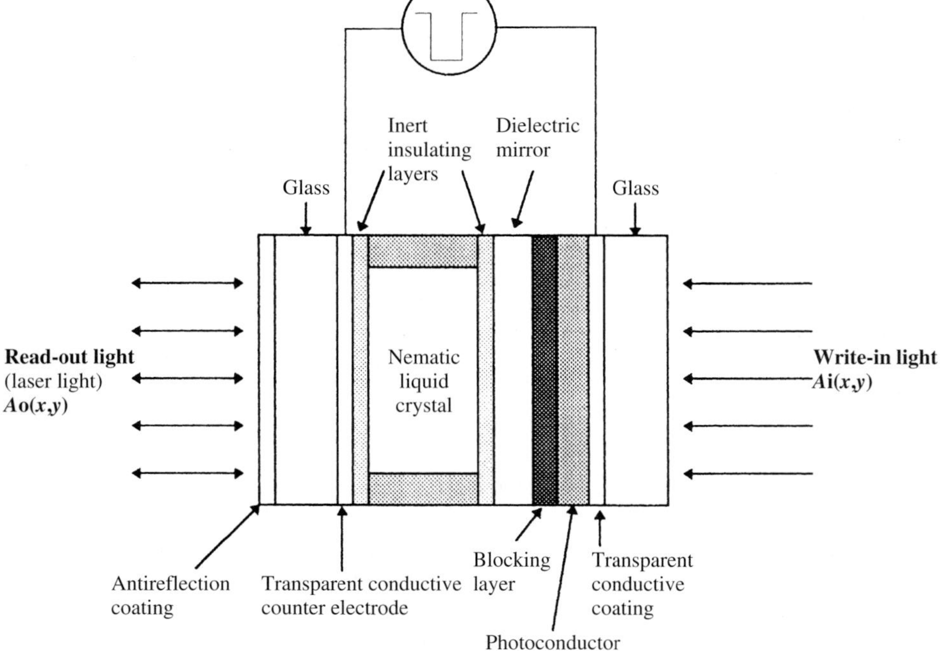


**Fig. 7** Structure of the Hughes liquid-crystal light valve

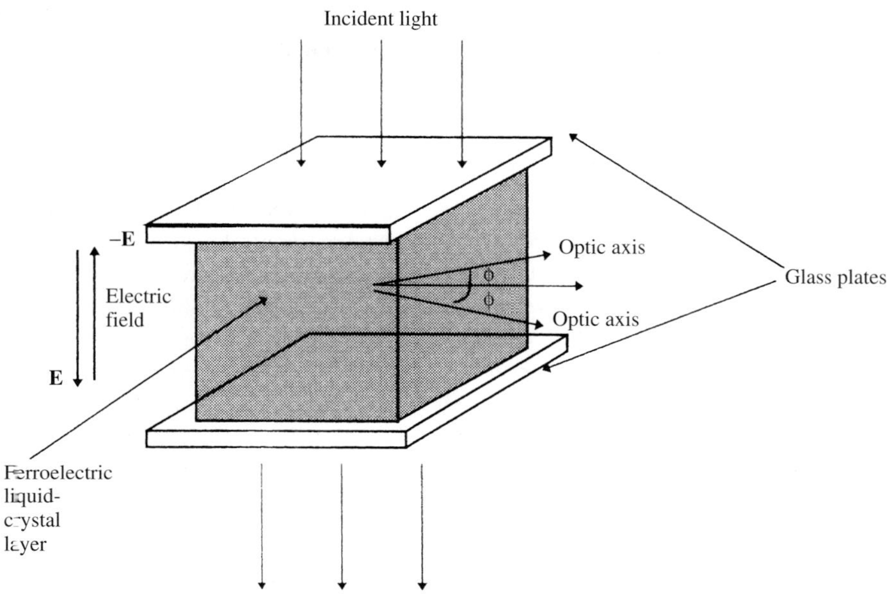


**Fig. 8** Ferroelectric liquid-crystal device

The read-out light beam is modulated by the modulating element and reflected back to create the read-out image. Some modulating properties are electro-optic effect, molecular alignment by electric field that exists in liquid crystals, photorefractive effect, electrostatic deformation, and acousto-optic effect. The electro-optic effect, which is the change in index of refraction of the medium as a function of the applied electric field, is used in a number of SLMs, such as the Pockels read-out optical modulator. The SLMs can also operate in the transmissive mode. A SLM operating in complex amplitude mode has an output

$$A_0(x, y) = \hat{A}_0(x, y) \exp[jP_0(x, y)], \quad (20)$$

where $P_0(.)$ represents the phase. The phase-only SLM output is $\exp[jP_0(x, y)]$, with $\hat{A}_0(x, y)$ a constant. The amplitude-only SLM output is $\hat{A}_0(x, y)$, with $\exp[j(P_0(x, y)]$ being constant.

## 5.1 Capabilities of SLMs

Some of the capabilities of SLMs are listed here:

1. *Image storage*: This is useful for optical memory, optical database/knowledge-base processors, optical pattern recognition, and neural networks.
2. *Incoherent to coherent light conversion*: This is used for converting real scenes illuminated under natural or other incoherent light into a coherent image. The real scene is imaged onto a SLM which is read out by coherent light. The coherent image can then be processed by optical signal-processing techniques.
3. *Real-time holography*: The interference generated between the object beam and the reference beam can be positioned on the SLM and the holographic pattern therefore displayed on the SLM. However, the space–bandwidth

product (SBP) of the typical SLM limits the effectiveness of this method.

4. *Fourier-plane spatial filtering*: Spatial filters can be displayed on SLMs in the Fourier plane. In this case, the input image Fourier transformed by a lens is multiplied by the filter on the Fourier-plane SLM. An additional Fourier transformation using a lens will produce the convolution between the input image and the impulse response of the filter. This can be used in optical spatial filtering, pattern recognition, and neural nets. If the filter is generated electronically by a computer, a SLM with electrical-to-optical conversion is used to display the filter.
5. *Nonlinear transformation and thresholding*: SLMs can be used to transform an image nonlinearly or binarize an image. This property is also useful for logic operations and switching in digital optical computing. In information processing, nonlinear characteristics of the SLM can be used in nonlinear filtering and nonlinear signal processing. Contrast reversal can be performed using SLMs.
6. *Image amplification*: SLMs are used in projection systems for image amplification. An object is imaged onto the SLM. It is read out from the SLM to provide a brighter image.

## 5.2 SLM Performance Considerations

There are many considerations in designing and using a SLM for optical processing. Frame rate determines how fast an image written on the SLM can be updated. Spatial resolution is a measure of how fine an image displayed on the SLM can be. Space–bandwidth product is a measure of the number of pixels available. Dynamic range is the number of gray levels that can be represented by a pixel. Contrast ratio, a measure of the ability of an SLM to block the light, is the ratio of the maximum and minimum output light levels. Flatness of the mirrors or windows of the SLM to a fraction of a wavelength of the light is important for optical processing where phase information is critical. Nonlinear input–output characteristics of the SLM are often considered for the specific image-processing applications. The exposure sensitivity and read-out light efficiency define the light budget of the system. SLMs also have write-in and read-out wavelength range, and electrical driving-signal power requirements.

## 5.3 Photorefractive Materials and Devices

Photorefractive materials can store optical images using variations in the index of refraction through the electro-optic effect. A photorefractive material, upon exposure to a light beam or an image, generates free charges, such as electrons or holes. These charges are redistributed and fall into trapping sites, thereby creating a charge distribution. This space-charge field produces a spatially dependent electric field that changes the index of refraction of the material through the electro-optic effect. The variations in the index of refraction result in the refraction or diffraction of light. The image stored in the photorefractive material can be read out by an optical beam. For a 1D signal with no applied field, the change in the index of refraction $\Delta n(x)$ as a function of the input intensity $I(x)$ is

$$\Delta n(x) = \frac{-K\Delta I(x)}{I(x)}, \qquad (21)$$

where $K$ is a constant dependent on the electro-optic coefficient, refractive index of the material, temperature, and electron mobility. Examples of photorefractive materials are lithium niobate ($LiNbO_3$), barium titanate ($BaTiO_3$), and bismuth silicon oxide (BSO). Photorefractive materials are used in optical storage and memory, real-time optical information processing, neural networks, holography, distortion compensation, and phase conjugation.

Photorefractive devices can be used to generate spatial filters or holograms in real time and have been used for an optical lock-in amplifier/detector. An image $I(x, y)$ is spatially mixed with a reference beam, and their interference intensity is recorded in a photo-refractive device. The interference intensity produces changes in the refractive index $\Delta n(x, y)$ that are stored in the form of a volume phase hologram. When the device is illuminated by the reference wave, the object beam $I(x, y)$ is reconstructed. For spatial filtering, the Fourier transform of $I(x, y)$ is stored in the device as a filter function.

An interesting property of photorefractive crystals is real-time holographic wave mixing. The interference between a reference beam and an object beam produces a grating that couples them, and energy can be exchanged. Each wave reflects from the grating and can add to the other beam. As a result, one wave may be amplified and take energy away from the other beam. The first complete theoretical treatment of this was given by Kogelnik [28]. Table 1 shows a number of available SLMs and their characteristics (optically addressed). In the following subsections, we discuss some of the SLMs in detail.

### 5.4 Liquid-crystal Light Valves

A liquid-crystal light valve (LCLV) is a device commonly used as a spatial light modulator to convert an incoherent light signal to a coherent light signal for optical-processing applications. Figure 7 shows an example of a multilayer sandwich structure of the LCLV (Hughes H-4060). The device consists of a nematic liquid-crystal

**Tab. 1** A number of available optically addressed SLMs and their characteristics

| | *Microchannel SLM* | *PROM* | *LC* | *Deformable mirror* |
|---|---|---|---|---|
| Modulating element | $LiNbO_3$ crystal | $Bi_{12}SiO_{20}$ | CdS; Si; photoconductor | Micromirror |
| Write light | X ray to 1 mm | Blue | Visible | UV to long IR |
| Read-out | Blue to 4 mm | Red | Visible | |
| Sensitivity | 2–30 mJ $cm^{-2}$ | 10 mJ $cm^{-2}$ | 1–5 mJ $cm^{-2}$ | |
| Spatial resolution | 10 line pairs/mm | 5 line pairs/mm | 20–40 line pairs/mm | 50 line pairs/mm |
| Frame rate | 5–60 Hz | 10–30 Hz | 40 Hz (nematic)<br>MHz (ferroelectric) | 4 kHz |
| Operating voltage | 3 kV | 4 kV | 5–10 V | Low, <12 V |
| Storage time | Days | 1 h in dark | | |
| Contrast ratio | 1000 : 1 | 1000 : 1 | 40 : 1 (nematic) | 40 : 1 |

layer, a photoconductor (CdS or silicon), a dielectric mirror that provides the high reflectivity required, and a light-blocking layer used to isolate optically the reading light from the writing light. The optical birefringence of the liquid-crystal layer is electrically controlled by the voltage applied on the layer through a pair of transparent conductive electrodes. The operation of the LCLV can be explained by a hybrid field-effect mode [29]. The voltage applied on the device is primarily divided between the photoconductor and the liquid-crystal layer. With changes of the voltage drop across the photoconductor, the voltage drop across the crystal layer changes. The liquid crystal used is a nematic type with positive dielectric anisotropy and homogeneously aligned with 45° twist. In the "off" state (no writing light), the voltage drop on the liquid-crystal layer is very small. The linearly polarized reading light will be reflected from the dielectric mirror and the polarization state of the reflected light leaving the LCLV is unchanged. In the "on" state, the resistance of the photoconductor decreases proportionally to the light intensity of the writing light. The voltage across the liquid-crystal layer then increases coordinately. This causes the liquid-crystal molecular orientation to tilt toward the homeotropic alignment. The polarization orientation of the reading light exiting the liquid crystal will rotate by a certain angle from its original polarization direction. The degree of the rotation of the polarization is proportional to the voltage drop across the liquid-crystal layer and, therefore, proportional to the intensity of the writing light. Therefore the LCLV can modulate the polarization direction of the coherent reading light proportionally to the intensity of the writing light. By passing the reading light reflected from the LCLV through an analyzer, the modulation on the polarization orientation can be easily converted to an amplitude modulation. When the writing-light intensity has spatial variation, different portions of the photoconductor generate different voltage drops across the different portions of the liquid-crystal layer. As a result, spatial variation of the polarization of the reading light is achieved.

The input–output characteristic of a LCLV can be controlled electronically by changing the frequency, wave form, and voltage of the power supply of the LCLV [23, 30–32]. Some other LCLVs, such as a wire-grid-mirror LCLV [33, 34], have interesting and useful nonlinear input–output characteristics. The resolution of the LCLV is better than 40 line pairs/mm (a wire-grid-mirror LCLV has a lower resolution about 10 line pairs/mm because of the wire-grid structure of the mirror). The frame speed of the LCLV can vary from several frames per second to a few hundred frames per second. The speed of the LCLV depends on the response time of the twisted liquid crystal and the photoconductors. Usually, a silicon photoconductor is faster than a CdS, and a binary LCLV is faster than a gray-scale LCLV. The contrast of the LCLV is better than 100 : 1.

### 5.5 Liquid-crystal Television Displays (LCTVs)

Liquid-crystal devices are widely used in small television sets, television projectors, and lap-top computers. These displays have been used in the optical signal-processing community for the last several years because of their low cost and commercial availability [35]. The liquid-crystal displays used in LCTVs were not originally designed for coherent optical systems. Their optical quality is not ideal for a coherent system mainly because of the phase variation of materials and the surfaces of the devices, which are less critical in an incoherent optical system than a coherent one. However, recent experiments show that the LCTV is still a good device in the applications where cost is an important factor and an electrically addressed device is needed.

The liquid-crystal display in a LCTV consists of a 90° twisted liquid-crystal layer sandwiched between two polarizers that have parallel polarization direction. The electric field can be applied to the liquid crystal through two transparent conductive electrodes on the two sides of the liquid-crystal layer. The transparent electrodes are pixelated and can be electrically addressed. When there is no electric field applied, the orientation of the input light is rotated by 90° from one side of the liquid-crystal layer to the other side, and the result is no light passing through because the two polarizers are parallel. When an electric field is applied, the twist and the tilt of the liquid-crystal molecules are altered depending on the voltage across the liquid-crystal layer. As a result, a fraction of the light passing through the liquid-crystal layer will retain the same polarization as the input light and, therefore, will pass through the second polarizer. The fraction of the light that passes through the display is proportional to the voltage applied to the liquid-crystal layer.

Phase modulation and complex modulation can also be obtained by properly setting up the polarization axes of the polarizer and analyzer [36].

Liquid-crystal displays used in projector-type LCTVs usually have more pixels ($300 \times 300$ or more) and smaller pixel size than those in conventional LCTVs. The contrast of the display is about 10 : 1.

## 5.6 Ferroelectric Liquid-crystal Spatial Light Modulators

The surface-stabilized ferroelectric liquid-crystal (FLC) device [37] contains a FLC layer between two glass plates. The FLC has two optical axis orientations, which are parallel to the glass surface and have the tilt angles $\phi = 22.5^\circ$ and $\phi = -22.5^\circ$ corresponding to the sign of the electric field applied across the FLC layer, as shown in Fig. 8. Since the FLCs have a strong interaction with light (the material birefringence is typically 0.1 to 0.2), with proper choice of the thickness of the FLC layer it functions like an electrically switchable half-wave plate for a specific wavelength of the incident light. When the device is placed between a crossed polarizer and analyzer, and has one of the optical axes of the material aligned with the polarizer, the incident light will not change its polarization on passing through the FLC. Thus, the light will be extinguished by the analyzer at the output. If the electric field is reversed, the optical axis of the material is also rotated by $2\phi = 45^\circ$. This results in a 90° rotation of the polarization of the incident light. Thus, the light passing through FLC is transmitted by the analyzer.

Ferroelectric liquid-crystal spatial light modulators are binary devices and have high speed. They have small pixel size and can be fabricated with very high resolution.

## 5.7 Deformable-mirror Spatial Light Modulators

A deformable-mirror spatial light modulator consists of a number of movable mechanical mirror elements that can be electrically addressed by an underlying array of MOS transistors. The materials used for a deformable-mirror spatial light modulator can be categorized as elastomers, cantilever beams, and membranes. The elastomer requires high address voltage. It is not good for high-density integration. The cantilever-beam approach does not require high voltage (<12 V) and can be easily fabricated on silicon [38]. It was invented by Texas Instruments Corp.

Each deformable mirror element consists of a small mirror that is supported by the polysilicon lines that also serve as the base for transistor gates. The structure of a mirror element is shown in Fig. 9. There is an air gap between the mirror and the silicon substrate. When the addressed gate is on, the mirror is deflected toward the silicon substrate according to the voltage drop across the air gap. If the mirror is undeflected and illuminated by light, it acts just as a conventional mirror. If the mirror is deflected, the direction of the reflected light will be different from the direction when the mirror is undeflected. If the mirror is adjusted to reflect the light along the optical axis of the system when the mirror is in the "on" state, and off the optical axis of the optical system when the mirror is in the "off" state, the light can be turned "on" and "off" by using an aperture function to block out the off-axis light.

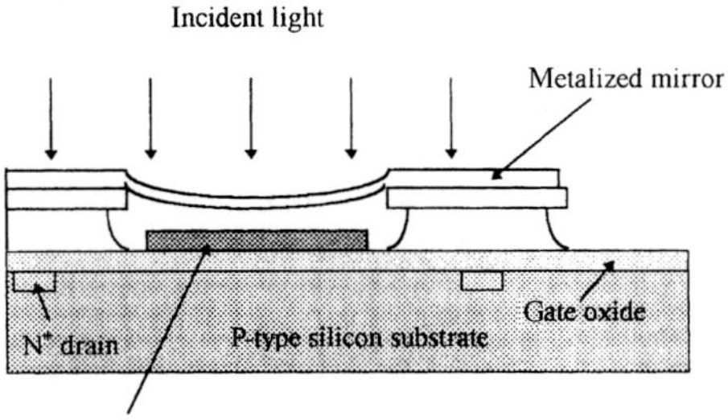


**Fig. 9** Deformable-mirror element

The deformable-mirror device has small pixel size, about $15 \times 15\,\mu m^2$, and can be fabricated as large as $\sim 10^6$ pixels in a two-dimensional array. It has a high contrast. Since it is a mirror-type modulator, it works at a wide range of wavelengths and has an almost pure phase-modulating characteristic. Texas Instruments Corp. is using it in their high-definition TV (HDTV) system with good results.

## 6 Acousto-optic Signal Processing

So far we have been discussing two-dimensional optical processors, which use spatial light modulators to impress the information of the light beam. Another class of systems exist which are basically one dimensional and use ultrasound or acousto-optical principles to perform their signal processing.

We will first consider the basic acousto-optic (AO) modulator, the device that impresses an electrical signal on a beam of light. This is shown in Fig. 10. The transducer, which launches an acoustic wave into the medium above it in the

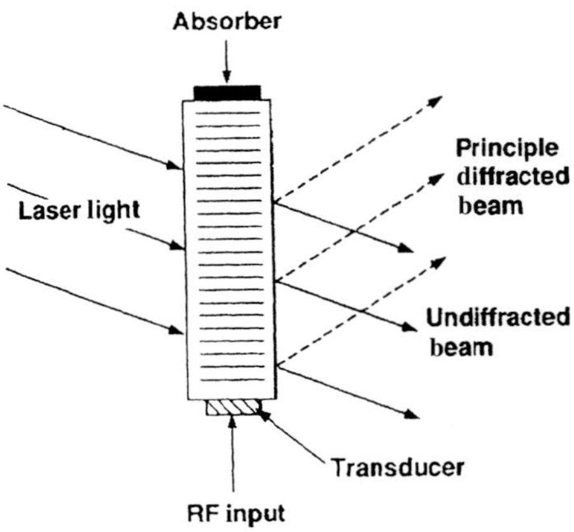


**Fig. 10** Acoustic-optic modulator

figure, is typically a piezoelectric crystal. The acoustic wave induces compression and rarefaction into the medium, thus modulating its index of refraction. Hence, the medium becomes in effect a moving phase grating that diffracts the light incident on the device from the left. The upper end of the device usually contains an absorbent material to quench the incident wave and prevent reflections and standing waves that could impair the operation of the device. Note that the device is inherently one dimensional, in contrast to a SLM, which is two dimensional. The number of distinct signal elements (so-called *space–bandwidth product*) in currently available devices is about 2000.

The AO modulator can be used in a system to correlate two electrical signals. There are two basic configurations for this; the time-integrating (TI) correlator and the space-integrating (SI) correlator. We will discuss the TI correlator first, shown in Fig. 11. One input signal modulates the light-emitting diode (LED) shown in the far left side of the figure. The other electrical signal is fed into the AO modulator as shown. The resultant multiplication of the two signals emerging from the right side of the AO cell is

$$s_1\left(t - \frac{x}{v}\right)^* s_2(t), \qquad (22)$$

where $x$ is measured from the central optical axis and $v$ is the velocity of the acoustic wave in the AO cell. The lens $L_1$ takes a spatial Fourier transform of this in order to provide a means of blocking the zero-order undiffracted light from the AO cell. The second lens, $L_2$, takes a second Fourier transform and restores the spatial signal minus the dc. The light is detected by a detector, such as a CCD detector, and if it is integrated for a time period $T$, the desired correlation is achieved,

$$\int_0^T s_1\left(t - \frac{x}{v}\right) s_2(t)\,dt. \qquad (23)$$

The other architecture commonly used is the space-integrating system, shown in Fig. 12. A LED, through a collimating lens, illuminates the AO cell as before. In this case the two signals are fed in to opposite ends of the AO cell. This results in the two propagating waves passing each other and in effect doing the shifting required in the correlation operation. The light again goes through a Fourier-transform lens where the undiffracted zero-order light is blocked to improve performance by cutting down on stray light. Lens $L_2$ retransforms the light and focuses it down on to the detector. The instantaneous detector output gives a temporal profile of the correlation signal. If just the value of the correlation peak is desired, the detector can be fed to an electronic peak detecting circuit.

There are many variations of the basic schemes we have discussed here, and the reader is referred for these to Berg and Lee [39], Korpel [40], and Xu and Stroud [41]. So far we have been limited

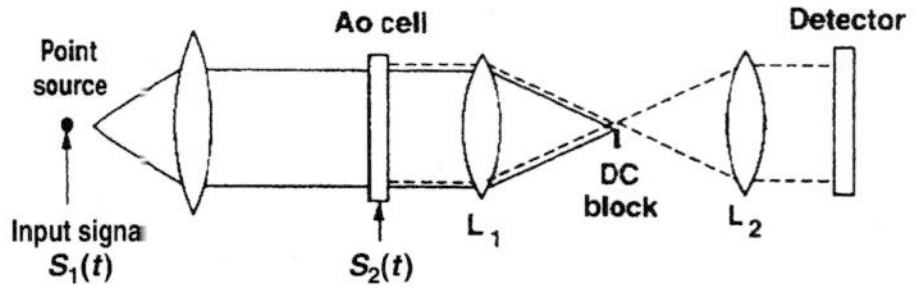


**Fig. 11** Time-integrating correlator using an acoustic-optic modulator

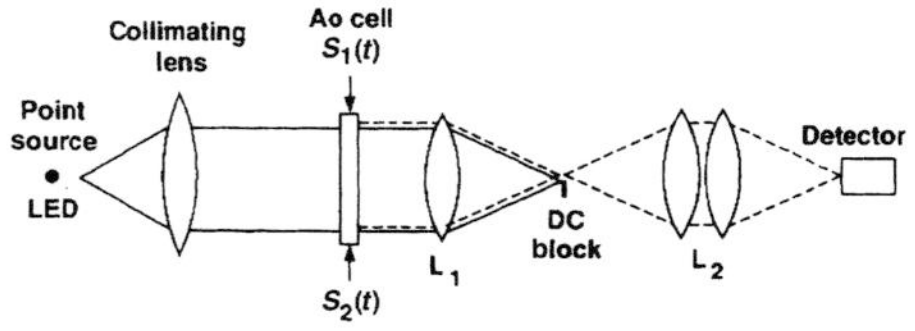


**Fig. 12** Space-integrating correlator using an acoustic-optic modulator

to one-dimensional signals. A clever architecture was proposed by Psaltis [42] that effectively makes the AO correlator a two-dimensional system capable of correlating images and, therefore, pattern recognition and target tracking. An impressive functional system using this scheme has been built and demonstrated by a group at Sandia National Laboratories [43].

Comparing the two architectures, the space-integrating systems have single-element detectors, good dynamic range and unlimited time-delay ranges, and wide bandwidths. The time-integrating systems have limited time-delay ranges, require a large detector, that is, many pixels, but generally have a larger processing gain.

The simplest application of the technology is the so-called *channelizer*. It consists of a light source, a Fourier-transform lens, and an AO cell. It is simply an optical analog spectrum analyzer. It takes an rf input, from a receiver for instance, and separates it into the various component parts, optically.

## 7 Optical Interconnections

Interconnections are a major communication problem in high-speed, high-density, VLSI digital, and electronic circuits. Optical interconnections (*q.v.*) are intended to complement and/or replace conventional electrical interconnections, such as wires, cables, or connectors on integrated circuits. Optical interconnections have definite advantages over electronic interconnections in terms of speed and density, and can remedy these problems. Free-space optical interconnections can be three dimensional, and beams can pass through one another without cross talk. In contrast, electronic interconnections are planar and cannot intersect without adequate insulation and some degree of cross talk. Thus, optics can offer advantages in terms of the number of interconnections in a given area. Time delays are also a problem in VLSI circuits. Photons travel at the speed of light, and the velocity is independent of the number of interconnections. On the other hand, in electrical transmission, delay times increase with the number of fan outs. In addition, the coupling in electronic interconnections is affected by the bandwidth of the data carried by the connection. This is not the case in optics.

Optical interconnections can be realized by a variety of methods and components, including free-space optics, optical fibers, optical wave guides, optical holograms, and holographic devices. In free-space optics, simple optical components (mirrors, lenses, and gratings) can be used to realize the interconnections by mapping points between the object and image plane. The motivation for using free-space optics for interconnections is the high space–bandwidth–product handling capability of optics. For example, a million points in an area of a square millimeter in the object plane are imaged/connected to a million points in the image plane using a lens. An electronic counterpart would require a million connections in a square millimeter with special care taken to ensure that the connections are properly insulated to avoid short circuits. Optical components may be used for mapping input beams to fan in, fan out, shift, shuffle, etc. For fan in, a lens focuses the input beam onto its focal point. For fan out, a source placed at the focal point of a lens is simply collimated. Shuffle and shift of beams are accomplished by holographic and diffractive optical elements. In general, an optical mask is designed using computer-generated holograms [44] to perform the interconnections. This mask may be considered as containing a large number of phase gratings. Each grating has a different spatial frequency, which deflects or tilts the beam at a given angle. Thus, arbitrary optical interconnection maps can be created. Computer-generated holograms may have very high spatial frequencies, up to thousands of lines per millimeter. A phase grating has a complex amplitude transmittance $\exp[-j2\pi(f_x x + f_y y)]$, where $f_x$ and $f_y$ are the spatial frequencies in the $x$ and $y$ directions. When a light beam illuminates the grating, it is modulated by the grating, which causes the beam to tilt by an angle proportional to the grating spatial frequency. By changes in the grating frequency, the tilt angle of the illuminating beam is controlled. Any arbitrary interconnection pattern may be obtained by a computer-generated hologram that contains a number of grating segments with different spatial frequencies. These grating segments tilt the illuminating beam to create the desired interconnection.

Dynamic or reconfigurable interconnections may be obtained by writing the grating or the hologram onto a spatial light modulator. Currently, the available spatial light modulators have a limited space–bandwidth product, which limits the number of interconnection points that can be achieved. Reconfigurable interconnections have been accomplished by writing the gratings into photorefractive materials.

## 8 Optical Memory

In this section, we briefly discuss the significance of optical memory and its application to optical information processing and computing. Optical storage is currently being used in many commercial applications for audio and video recordings, electronic photography, and computer memories. Optical memory is compact, easily removable, and inexpensive. The current storage density of an optical disk is one bit/$\mu m^2$. With the development of short-wavelength semiconductor lasers, the storage density is likely to increase. In general it is read-only memory (see DATA STORAGE, OPTICAL).

The true potential of optical memory is its parallel access capability that may prove advantageous over magnetic storage. By illuminating a disk by a broad optical beam, many stored bits are accessed in parallel. As shown in Fig. 13, the illuminated bits on the disk are imaged onto a detector array and converted into electronic signals to be utilized by the computer. Given the commercially available optical components and devices, 100 000 channels can be accessed in parallel. The access time to a block of data is a few tens of milliseconds. At present, most optical disks have been of the write-once, read-many (worm) type. The challenge is to develop an erasable – read/write many – optical disk. New materials are now making this possible.

The optical disk can be used to store binary data as 2D blocks for analog image processing. Computer-generated holographic techniques can be utilized to store images in a binary format. The Fourier transform of the data is written in the holographic format, which has the advantage of robustness to misalignments between the disk and the detector array. The data are retrieved by illuminating the hologram on the optical disk, which reconstructs the stored images. The parallel-access optical memory has numerous applications in

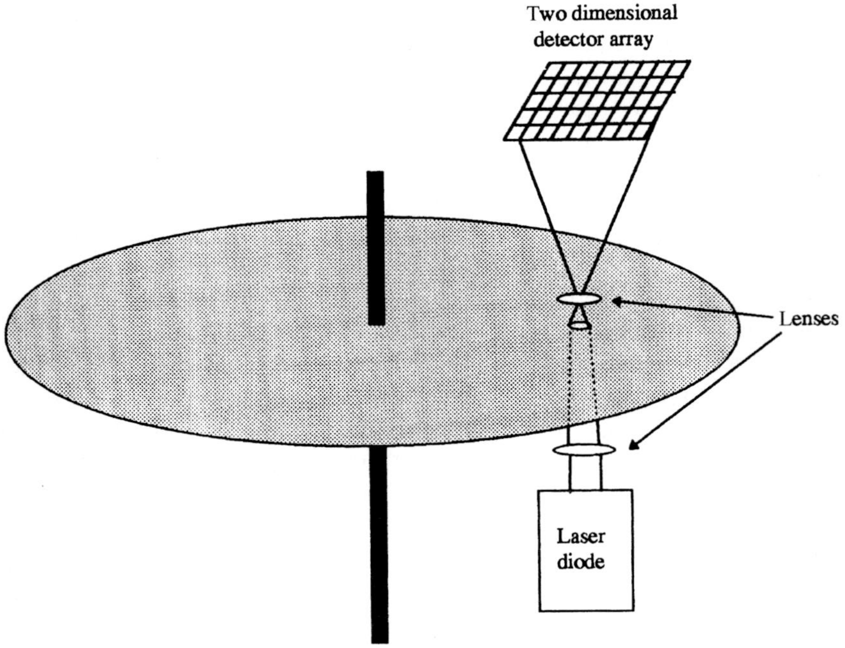


**Fig. 13** Optic memory accessible in parallel

image processing, database management, and networks. It provides the capability to access a large volume of data rapidly. Parallel-access optical memory is attractive in pattern recognition and associative memory by recording the weights or reference patterns. The pattern to be inspected/searched is displayed on a spatial light modulator to obtain the product between the input pattern and the weights. The product is imaged on a detector, normalized according to the intensity of the input pattern, and is maximum when the input pattern matches the data illuminated on the disk. This process is iterated by rotating the disk and illuminating various portions of the disk to search the entire data to obtain the maximum output. This system is attractive because it provides processing of a large volume of stored data in one disk revolution. The disadvantage of holographic encoding is that it is inefficient in its use of space–bandwidth product: only about one third of the space–bandwidth product is used for the actual information.

Additional improvements in the storage capacity of optical memory can be accomplished by utilizing thick-medium optical disks. The data are stored holographically by interfering the data (such as images) and a plane-wave reference beam. The image is retrieved by reilluminating the hologram on the optical memory with a similar reference beam. In a thick-medium hologram, the image is reconstructed only if the Bragg condition is satisfied, which means that the angle of the reconstruction beam should be equal to the angle of the reference beam used to synthesize the hologram. This property allows the storage of multiple images in the same hologram using angular multiplexing, and retrieving or accessing the images using the proper angle of illumination. The 3D disk is mechanically rotated to scan the entire volume disk. The recording materials are similar to those used in photoreflective recording, such as lithium niobate, and can provide up to $10^{12}$ pixels for a 5-cm-radius disk. Using acoustic-optic deflectors, the access time to each stored datum in a hologram is 10 µs, which makes the total access time 10 s. Rapid advances have taken place in photorefractive holographic materials such as the commercially available photopolymers and in optoelectronic devices, and the need for large–storage-capacity memory has stimulated much interest in research and development of optical memory. Parallel access and fast data-transfer rates seem to be the key to likely success of optical memory.

The applications remain diverse, ranging from pattern recognition, neural networks, and image processing to database management. The volume optical disk or the 3D disk provides additional storage capacity and makes optical storage a very attractive alternative to its nonoptical counterparts.

## 9 Neural Networks

### 9.1 Overview of Neural Networks

Artificial neural networks (*q.v.*) (NN), which are also referred to as neuromorphic systems, parallel-distributed processing models, and connectionist machines, are intended to provide humanlike performance by mimicking biological neural systems. They are used in image processing, signal processing, and pattern recognition. They are characterized by massive interconnection of simple computational elements, or nodes, called *neurons*. Neurons are nonlinear and typically analog, and can have a slow response, typically several hundred hertz. A neuron produces an output by nonlinearly transforming a sum of $N$ inputs as shown in Fig. 14(a),

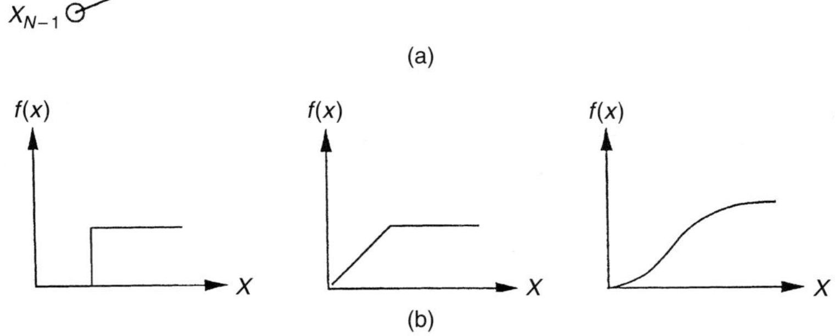


**Fig. 14** (a) A neuron produces an output ($Y$) by nonlinearly transforming a sum of $N$ inputs ($X_0 - X_{N-1}$); (b) Three types of neuron nonlinearity

where $f(.)$ represents the nonlinear characteristic of the neuron and $w_i$ is the weight of the interconnection. Three types of neuron nonlinearity are shown in Fig. 14(b). Neural nets provide many computational benefits. The information is stored in the interconnections ($w_i$). Training or learning changes the interconnection weights.

Because of their large degree of parallelism and massive interconnection capability, NN provide fault tolerance, since damaging a few nodes will not affect the overall performance significantly. Neural networks do not require complete knowledge of the statistical models of the signals to be processed, and instead they utilize available training data. Neural systems are best for problems with no clear algorithmic solutions.

Neural nets are characterized by the network topology, neuron input/output characteristics, and learning rules. Learning plays an important role in the performance of the NN. The ability to adapt the weights is essential in applications such as pattern recognition where the underlying statistics are not available, and the new inputs are continuously changing. Conventional statistical techniques are not adaptive. For classification, a neural-network algorithm is used to compute matching values between the input and the stored data and then to select the class that generates the minimum value. A probabilistic model is used to compute the likelihood or probability that the input belongs to a certain class. If Gaussian distribution is used, simple solutions can be obtained.

Neural-net classifiers may outperform conventional statistical techniques when the underlying distribution of data is generated by nonlinear processes and is strongly non-Gaussian. Neural-net classifiers contain more than one stage. The output of the first stage exhibits the degree of matching between the input and the weights stored in the network. The maximum of these values is enhanced, and the outputs are forwarded to the second stage. This provides a strong output corresponding to the most likely class. If supervision is provided, this information can be used to adapt the weights of the network using a learning rule, which will improve the performance of the system by reducing the probability of error.

Neural-net–based classifiers are used in three important applications. The first is to identify the class of an input pattern when the input is partially obscured or distorted. This has applications in pattern recognition and classification. The second is in vector quantization and data compression, which are useful in signal and image processing. They are used to reduce the amount of data to be processed without losing the important information. The third is signal and image processing.

### 9.2 Perceptron

The perceptron learning rule can be implemented in both single-layer and multilayer networks. Figure 15(a) is a single-layer perceptron with a single output that classifies an input into two classes, A and B. The decision boundary for this network is a hyperplane that divides the space representation of the input. For example, if the input vector has only two elements, the decision boundary is a line, as shown in Fig. 15(b). The output is the inner product of the inputs and the weights adjusted by a threshold and binarized. An output

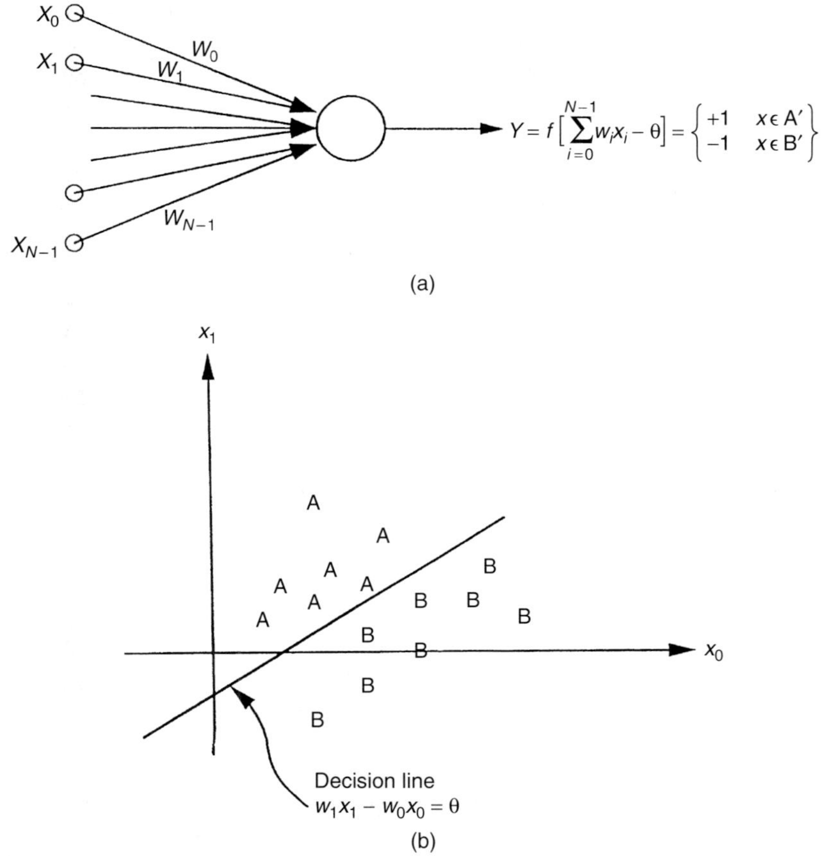


**Fig. 15** (a) A single-layer perceptron network with a single output; (b) The decision boundary for the network that separates two classes, A and B. The input vector has two elements, and the decision boundary is a line

of +1 represents class A, and −1 represents class B. The connection weights can be adopted using the following learning rule:

$$w_i(t-1) = w_i(t) + \alpha[d(t) - y(t)]x(t), \tag{24}$$

with

$$d(t) = \begin{cases} +1, x \in \mathrm{A}, \\ -1, x \in \mathrm{B}, \end{cases} \tag{25}$$

where $0 < \alpha < 1$ is a positive gain, $d(t)$ is the desired response, $x(t)$ is the input, $y(t)$ is the actual response, and $w_i(t)$ is the weight. If the inputs from the two classes are separable such that they are on opposite sides of a hyperplane, then the perceptron classifier works successfully by placing the decision boundary between the two classes. In this algorithm, the statistical properties of the input are not known. Instead, the procedure focuses on the error between the ideal output and the actual output, which represents the overlap between the different classes. Thus, the algorithm works well when the input statistics are non-Gaussian, generated by nonlinear processes.

### 9.3
### Multilayer Perceptron

When the classes cannot be represented by hyperplane decision boundaries, and are separated by complex decision surfaces, a multilayer perceptron is needed. A multilayer perceptron is a feed-forward network that consists of an input layer, an output layer, and as many hidden layers as needed. A two-layer perceptron is shown in Fig. 16. The nonlinearities used within the nodes of the multilayer perceptron provide the capability to generate the complex decision boundaries.

The back-propagation algorithm is used to train the multilayer perceptron. It is an iterative algorithm designed to minimize the mean square error (MSE) between the desired output and the actual output of a multilayer network. Thus, for each training input $x$, a desired output $d$ is specified, and continuous differentiable nonlinearities are used in the network. The actual outputs $y_i$ are calculated using the input, weights, and nonlinearities. The weights are adapted to minimize the mean square difference

$$\mathrm{MSE} = n^{-1}\sum_{i=0}^{n-1}(d - y_i)^2. \qquad (26)$$

The number of nodes, the number of hidden layers, and the thresholds need to be set. This poses a practical difficulty. There is no general way to resolve these issues, and the solutions are application dependent. Thresholds can be set using training.

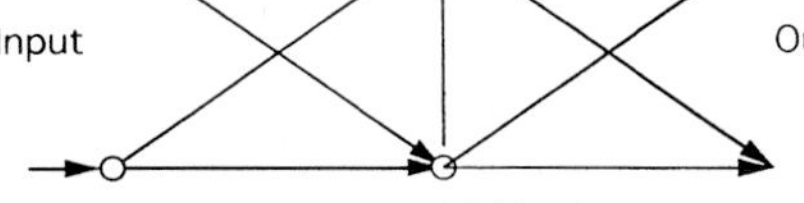


**Fig. 16** A two-layer perceptron network

### 9.4
### Associative Memory

An associative-memory processor stores signals or patterns in the memory [45]. It is capable of producing an output that is a reproduction of the stored input pattern in response to an input that is a partially obscured or distorted version of one of the stored patterns. For one-dimensional signals, the patterns are stored in a matrix $W$. The output of the associative memory is the input vector multiplied by the matrix $W$ followed by nonlinear transformations. The associative processors include autoassociative memory and heteroassociative memory. In autoassociative memory, the output recollection is the same as the stored pattern associated with the input. If $x$ is one of the stored patterns, and $x'$ is a distorted version of $x$, then the output is $x$. An important property of associative memory is fault tolerance, which means being able to produce the original pattern in response to an input that is a noisy or partially obscured version of a stored pattern. There are also applications to content-addressable memory. When the associative memory is used for pattern recognition, the output is compared to the stored signals to determine if there is a match. The heteroassociative memory produces outputs that are arbitrarily associated to a given input $x$.

The Hopfield associative memory, which is used with binary inputs such as black and white images or binary data, does not perform as well when continuous gray-scale input values are used. The network contains $N$ nodes with hard limiting nonlinearity $f[.]$ and binary outputs. The output is fed back to the input. The weights $t_{ij}$ are fixed using the $M$ associative signals $x^k$:

$$t_{ij} = \begin{cases} \sum_{k=0}^{M-1} x_i^k x_j^k, & i \neq j, \\ 0, \; i = j; \end{cases} \tag{27}$$

$0 \leq i, j \leq M - 1$. The output at the time $t$ is $\mu_i$ and at time $t + 1$ is

$$\mu_i(t + 1) = f\left[\sum_{i=0}^{n-1} t_{ij}\mu_i\right]. \tag{28}$$

The input is an unknown pattern $x'$ and $\mu_i(0) = x'$. The process is repeated until the output remains unchanged. The output is forced to match one of the patterns $x^k$ stored in the weight. The network converges to a correct solution if the output is the correct version of the distorted input. Graded nonlinearities improve the performance of the Hopfield network. The Hopfield network has many major limitations. First, the storage capacity is limited. If many patterns are stored, the network may converge to a false memory pattern different from all stored patterns, which will produce a no-match output when used as a classifier. This problem can be remedied if the patterns are generated randomly and the number of classes $M$ is less than 0.15 times the number of the input nodes $N$. The second limitation is that the network may not converge to a correct solution if the stored patterns are too similar to one another. In this case, the stored patterns are considered to be unstable. This problem can be remedied by orthogonalizing the patterns before storing them in the network [45].

## 9.5 Vector-matrix Architectures

Psaltis and Farhat [46, 47] proposed an optical implementation of the Hopfield [48] network, as shown in Fig. 17. The Hopfield network is a single-layer network with feedback. It can be used as an autoassociative memory. In this network, neurons accept the input and present the output, and each neuron is connected to all other neurons via the interconnection weights. The weights form a matrix that is called the *memory matrix*. The matrix is generated by summing all the outer products of the exemplars, which are used to represent the pattern to be retrieved or recognized, in the learning process (an outer product of two vectors is calculated by the matrix multiplication of one column vector with transpose of the second column vector). The diagonal elements of the matrix are set equal to zero to prevent the connection of each node to itself. In the recall process, when an unknown input vector is presented to the network, the output is obtained by performing the matrix multiplication of the input vector and the memory matrix, which is a summation of the stored exemplars weighted by the inner product between the input and the corresponding

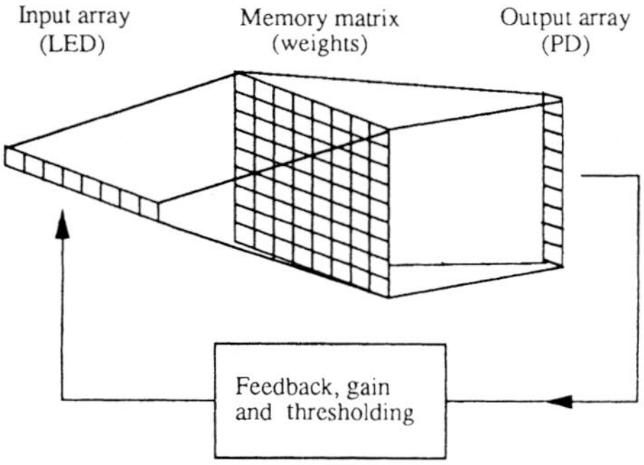


**Fig. 17** Scheme of an optical vector-matrix multiplier with nonlinear feedback

stored exemplars. The iterations are repeated until the output converges to a stable state, which is one of the stored exemplars that has the fewest different bits from the input. The Hopfield network has limited storage capacity, and only binary patterns can be stored and recalled.

In the architecture shown in Fig. 17, a LED array is used to represent the input array, a photodiode (PD) array is used to represent the output array, and a programmable spatial light modulator is used to store the weights. An anamorphic lens is used between the input and the weight mask to perform the multiplication of the input and the memory matrix, and another anamorphic lens is used between the weight mask and the output array to carry out the summation of the multiplication results in the row direction and to generate the output. When an unknown input is imposed on to the network, the product of the input and the memory matrix is obtained at the output array, and the output is fed back optically to the input through thresholding and gain. Thus, the Hopfield network is implemented by this architecture.

A number of novel optical implementations for autoassociative modules providing versatile adaptive learning capabilities were proposed by Fisher et al. [49]. These modules can be configured into larger multiple-module architectures.

Paek et al. [50] described mathematically how all the features of the Hopfield model and higher-order models of neural networks can be derived automatically without any biological assumptions. They also derived the computational circuits for implementing these neural-network models.

## 9.6 Optical-correlator Neural Nets

Optical correlators are inherently two-dimensional systems. They can perform some operations on two-dimensional images. As noted in Sec. 9.5, Psaltis and Farhat [46] proposed using optical correlators to implement the Hopfield model for two-dimensional images. An autoassociative content-addressable memory using the optical correlator is shown in Fig. 18 [51]. Originally, the Hopfield network was based on one-dimensional vectors, and outer products of the vectors were used. To implement the outer product between two-dimensional images, spatial-frequency multiplexing is introduced by using two optical correlators. The first correlator is used to obtain the cross-correlations between the input image and the stored images. The 2D images are stored holographically by means of the Fourier-transform hologram. The hologram is recorded on a thermoplastic plate. Each image is recorded with a different reference beam angle, so-called *angular multiplexing*. In the retrieval process, the input is imaged onto a liquid-crystal light valve (LCLV) (Sec. 5.4). An argon laser is used to read out the input image from the LCLV. Lens L1 is used to perform the Fourier transform of the input image. On the back focal plane of lens L1, the Fourier transform of the input image is multiplied by the transmittance of the hologram, which records the complex conjugate of the Fourier transforms of the stored images. The product is then passed through another Fourier-transform lens L2. On the focal plane of lens L2, a pinhole array is placed to sample the correlation signals between the input image and the stored images recorded on the hologram. The separation between

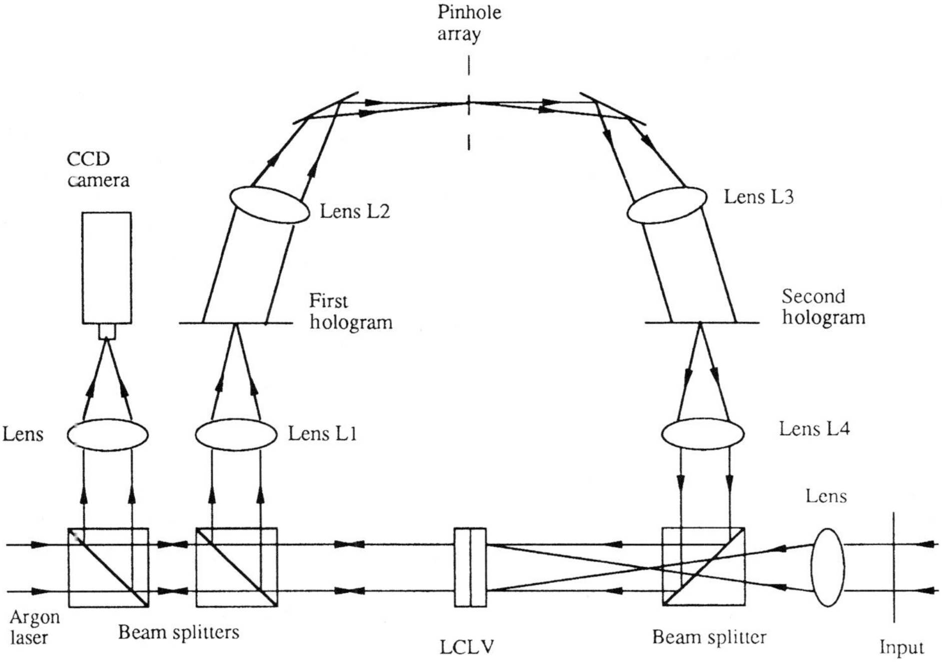


**Fig. 18** Schematic diagram of a double-correlator neural network with optical loop

the pinholes is determined by the focal length of L2 and the reference beam angles used for recording the images on the hologram. In the second correlator architecture, the output light of the pinhole array illuminates another hologram that is similar to the first one. The output at the back focal plane of lens L4 is a weighted superposition of the images stored in the second hologram, with the weights proportional to the correlations of the input and the images stored in the first hologram. By doing this, the outer-product model with associative property is realized for two-dimensional images. The same LCLV is used to obtain the output and feed it back into the first correlator to form the iteration loop. The iterations are repeated until the output is stable. A facial image retrieval test with four people's images stored on the hologram and with partial and distorted images as inputs was shown [51].

Paek et al. [52] proposed a holographic memory, shown in Fig. 19. A photorefractive crystal, lithium niobate ($LiNbO_3$), is used to store multiple holograms. An argon laser with $\lambda = 514.5$ nm is used to record the holograms. A low-threshold electronically pumped vertical-cavity surface-emitting microlaser diode array (VCSEL) with $\lambda = 960$ nm is used to retrieve the images stored in the holograms. The angular separation of the diodes is 0.04° to ensure the angular separation for selective read out. The system has a $10^4$-bits/s access speed.

Khoury et al. [53] proposed using the binary phase-only filter [11] to realize an associative memory. A $2f$ correlator architecture [54] was used instead of a $4f$ correlator. Because the binary phase-only

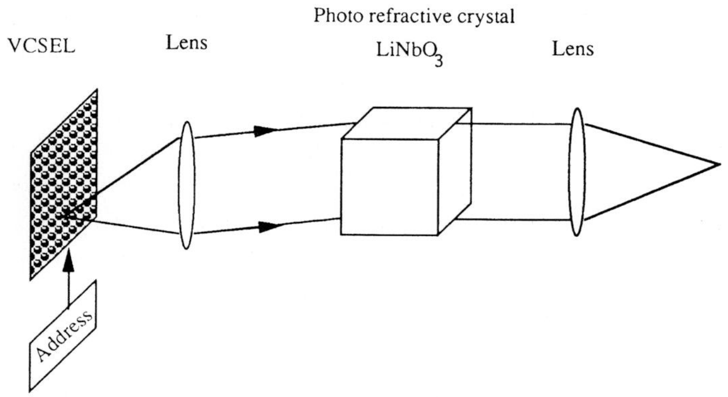


**Fig. 19** A holographic memory system

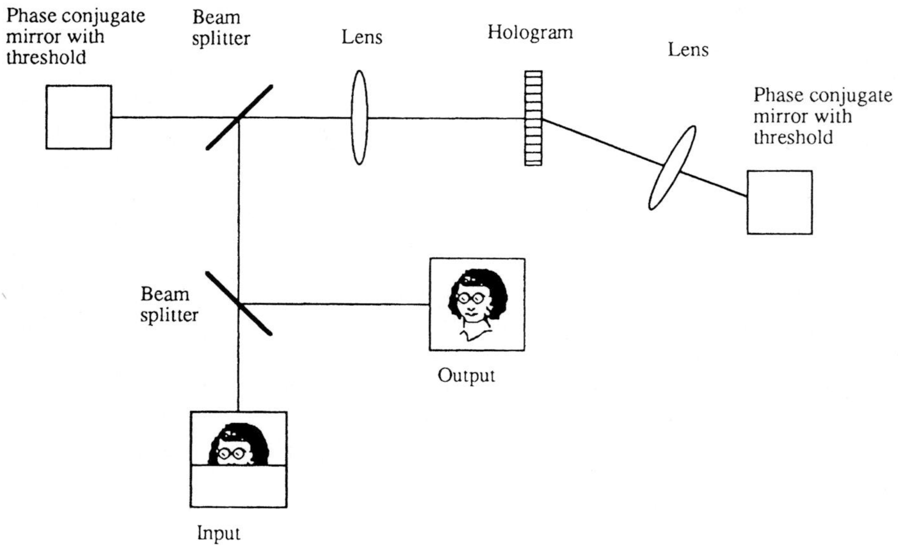


**Fig. 20** Schematic diagram of a nonlinear holographic associative memory

filter provides sharp autocorrelation, no feedback or nonlinearity was employed. A mirror was used in the correlation plane to reflect the correlation output back to the binary phase-only filter to restore the stored image.

### 9.7 Hologram-based Neural Nets

Owechko et al. [55] proposed an autoassociative memory, shown in Fig. 20. This network uses the outer-product neural-network model. The desired images are stored in Fourier-transform holograms with different reference beams in different angles. A thermoplastic film is used to record the hologram. An input is Fourier transformed by the first lens, and the Fourier transform of the input is multiplied by the hologram that records the conjugate of the Fourier transforms of the stored images. The correlations between the input and the stored images are obtained on the focal back plane of the second lens. A phase-conjugate mirror is used to reflect the correlation signal with a conjugated phase back to the hologram. This way, the hologram acts as a memory that, with the illumination of the correlation signals, generates a weighted superposition of the stored images. This newly formed image is then reflected back by another phase-conjugate mirror as a new input. Thus, the two phase-conjugate mirrors provide a resonator cavity for feedback, and all the stored images oscillate inside the cavity. With the threshold and the nonlinear reflectivity of the phase-conjugate mirror, the system will converge to the strongest correlated stored image while the other images will vanish.

A single-image experiment is performed by storing a portrait in the hologram and retrieving it from a partial version of the original image.

Paek and Lehmen [56] realized a holographic associative memory that is capable of identifying individual words and inserting word breaks into a concatenated word string. The architecture is quite similar to the two-correlator architecture described above, except that electronics was used at the correlation plane to find the correlation peaks and stretch them in the horizontal direction. The stretched correlation peaks were forwarded to the second correlator to restore the desired words with proper space between them.

## 9.8
## Hughes Programmable Multilayer Optical Neural Net

In a hybrid optoelectronic system [57], shown in Fig. 21, two liquid-crystal light valves are used to form the oscillation cavity, and a photorefractive crystal is used to store the volume phase holograms. It is a modification of the system described in the previous section [55]. Computers are used to realize the feedback loops. It is claimed that $7 \times 10^5$ neurons and $7 \times 10^7$ interconnections can be achieved with this system. The neuron update rate can be $10^7$ neurons/s, and the data rate can be $2 \times 10^9$ interconnects/s.

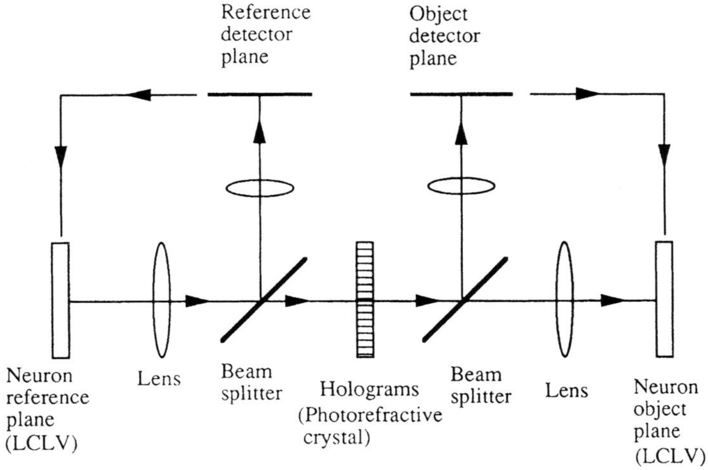


**Fig. 21** Schematic diagram of a programmable optical neural network

### 9.9 Learning in Multilayer Neural Nets

A handwritten-character recognition system was built by Psaltis and Quio [58]. It was realized by a multilayer optical neural network, shown in Fig. 22. A rotating mirror was used to change the reference beam in 26 different directions, two $LiNbO_3$ photorefractive crystals were used to represent the input layer and the hidden layer, and a CCD camera was used to represent the output layer. The learning method proposed by Kanerva [59] was used to train the system. The weights are initially assigned random values and are updated with the new inputs. Each input character has 100 pixels, the hidden layer contains $10^5$ units, and the output layer has 26 units that represent one of the 26 letters of the alphabet. In tests, 104 patterns were used to train the system, and a test set of 520 patterns was used. The error probability was 44%. It is claimed that the system has a rate of $10^{12}$ multiplications/s.

## 10 Conclusions

We have tried to present a balanced and comprehensive overview of the field of optical signal processing/computing as it has evolved from its beginnings in the early 1960s to today. It has broadened considerably in the intervening four decades – starting out with image enhancement using optical high-pass and low-pass filtering and pattern recognition with the optical correlator. Within optics it is a very large field. For example, the Optical Society of America publishes a separate monthly research journal on information processing alone. Many of the innovative algorithms developed in the context of optical processing are also implementable on a digital computer and perform extremely well compared with various algorithms developed by the digital community. Improvements are occurring rapidly in the basic building block for optical computing and processing, namely, the spatial light modulator.

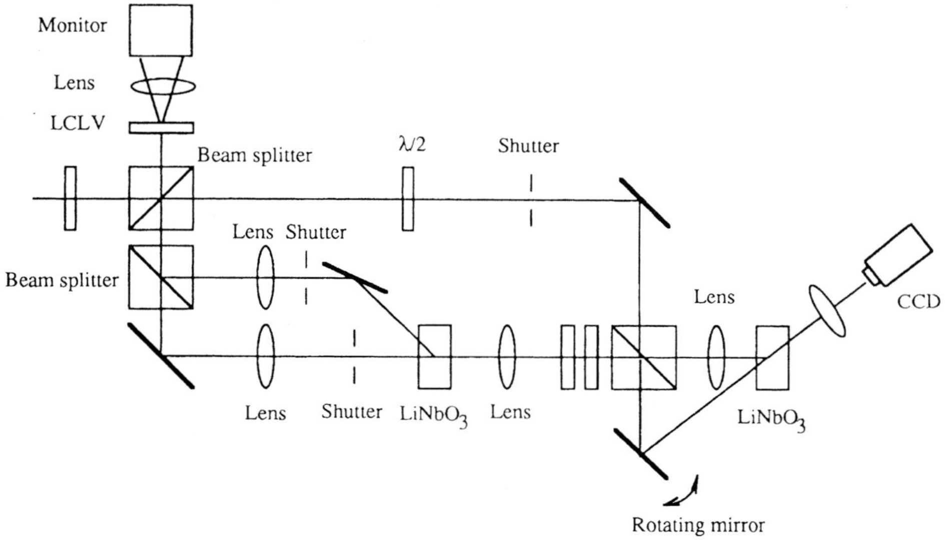


**Fig. 22** A two-layer optical network for character recognition

Nearly half a dozen companies are pursuing this technology, which stimulates new advances in applications and materials such as liquid crystals. New areas of application are opening all the time, such as the use of this technology for law enforcement, security, and anticounterfeiting. Although government-funded research is undergoing a transition at this time, with the end of the cold war, we feel the field has a bright and continuing future.

There are new research activities in the field of optical information processing, including information security [60–70] and three-dimensional image processing [71–80]. The new optical security techniques can enhance electronic information security against anti-counterfeiting and copying. The new three-dimensional image processing provides additional performance gains compared with electronic 2D image processing.

## Acknowledgment

We acknowledge the support of Rome Laboratory at Hanscom Air Force Base in preparing this article.

## References


[1] Goodman, J. W. (1968), *Introduction to Fourier Optics*. New York: McGraw-Hill.

[2] Guenther, R. D. (1990), *Modern Optics*. New York: Wiley.

[3] McAulay, A. D. (1991), *Optical Computer Architecture*. New York: Wiley.

[4] Javidi, B., Horner, J. L. (1994), *Real Time Optical Information Processing*. Boston: Academic Press.

[5] Lohmann, A. U., Brown, B. R. (**1966**), *Appl. Opt.* **5**, 967.

[6] Flannery, D. L., Horner, J. L. (**1989**), *Proc. IEEE* **77**, 1511.

[7] Saleh, B. E. A. (1991), *Fundamentals of Photonics*. New York: Wiley.

[8] O'Neill, E. L. (1963), *Introduction to Statistical Optics*. Reading, MA: Addison-Wesley.

[9] VanderLugt, A. (**1964**), *IEEE Trans. Inf. Theory* **IT-10**, 139.

[10] Javidi, B., Wang, J. (**1992**), *Appl. Opt.* **31**, 6826.

[11] Horner, J. L., Gianino, D. (**1984**), *Appl. Opt.* **23**, 812.

[12] Oppenheim, A., Lim, J. (**1981**), *Proc. IEEE* **69**, 529.

[13] Horner, J. L. (**1982**), *Appl. Opt.* **21**, 4511.

[14] Psaltis, D., Paek, E., Vankatesh, S. (**1984**), *Opt. Eng.* **23**, 698.

[15] Horner, J. L., Leger, J. R. (**1985**), *Appl. Opt.* **24**, 609–611.

[16] Javidi, B. (**1990**), *Appl. Opt.* **29**, 1215.

[17] Horner, J. L., Javidi, B., Wang, J. (**1992**), *Opt. Commun.* **91**, 189.

[18] Lu, K., Saleh, B. (**1990**), *Opt. Eng.* **29**, 240.

[19] Turin, J. L. (**1960**), *IRE Trans. Inf. Theory* **IT-6**, 311.

[20] Javidi, B., Refregier, P., Willett, P. K. (**1993**), *Opt. Lett.* **18**, 1160.

[21] Javidi, B., Wang, J. (**1994**), *J. Opt. Soc. Am.* **11**, 2604.

[22] Weaver, C. S., Goodman, J. W. (**1966**), *Appl. Opt.* **5**, 1248.

[23] Javidi, B., Tang, Q., Gregory, D., Hudson, T. (**1991**), *Appl. Opt.* **30**, 1772.

[24] Javidi, B., Kuo, C. (**1988**), *Appl. Opt.* **27**, 663.

[25] Javidi, B. (**1989**), *Appl. Opt.* **28**, 2358.

[26] Javidi, B., Horner, J. L. (**1989a**), *Appl. Opt.* **28**, 1027.

[27] Javidi, B., Horner, J. L. (**1989b**), *Opt. Eng.* **28**, 837.

[28] Kogelnik, H. (**1969**), *Bell Syst. Tech. J.* **48**, 2909.

[29] Bleha, W. P., Lipton, T., Wiener-Avnear, E., Grinberg, J., Reif, P. G., Casasent, D., Brown, H. B., Markevitch, B. V. (**1978**), *Opt. Eng.* **17**, 371.

[30] Efron, U., Grinberg, J., Braatz, P. O., Little, M. J., Reif, P. G., Schwartz, R. N. (**1985**), *J. Appl. Phys.* **57**, 1356.

[31] Warde, C., Fisher, A. (1987), in J. L. Horner (Ed.), *Optical Signal Processing*. New York: Academic Press.

[32] Javidi, B., Zhang, G. (**1992**), *Opt. Eng.* **31**, 934.

[33] Efron, U., Au, A., Bak, C. S., Goodwin, N. W., Reif, P. G., Garvin, H. L., Byles, W., Owechko, Y., Welkowsky, M. S. (1989), in B. Javidi (Ed.), *Optical Information Processing Systems and Architectures,* SPIE Proceedings No. 1151, Bellingham, WA: SPIE, pp. 591–606.
[34] Javidi, B., Zhang, G., Fazlollahi, A. H., Efron, U. (**1994**), *Appl. Opt.* **33**, 2834.
[35] Young, M. (**1986**), *Appl. Opt.* **25**, 1024.
[36] Kirsch, J. C., Gregory, D. A., Thie, M. W., Jones, B. K. (**1992**), *Opt. Eng.* **31**, 963.
[37] Johnson, K. M., Moddel, G. (**1989**), *Appl. Opt.* **28**, 4888.
[38] Hornbeck, L. J. (**1983**), *IEEE Trans. Electron Devices* **ED30**, 539.
[39] Berg, N., Lee, J. (1983), *Acousto-Optic Signal Processing.* New York: Marcel Dekker.
[40] Korpel, A. (1988), *Acousto-Optics.* New York: Marcel Dekker.
[41] Xu, J., Stroud, R. (1992), *Acoustic-Optic Devices: Principles, Design, and Applications.* New York: Wiley.
[42] Psaltis, D. (**1984**), *Proc. IEEE* **72**, 962.
[43] Molley, P. A., Stalker, K. T. (**1990**), *Opt. Eng.* **29**, 1073.
[44] Dallas, W. J. (1980), Computer-generated holograms, in B. R. Frieden (Ed.), *The Computer in Optical Research, Topics in Applied Physics,* Vol. 41, Berlin: Springer, pp. 291–366.
[45] Lippmann, R. P. (**1987**), *IEEE ASSP Mag.* **4**(2), 4–22.
[46] Psaltis, D., Farhat, N. H. (**1985**), *Opt. Lett.* **10**, 98.
[47] Farhat, N. H., Psaltis, D., Prata, A., Paek, E. (**1985**), *Appl. Opt.* **24**, 1469.
[48] Hopfield, J. J. (**1982**), *Proc. Natl. Acad. Sci. U.S.A.* **79**, 2554.
[49] Fisher, A., Lippincott, W., Lee, J. (**1987**), *Appl. Opt.* **26**, 5039.
[50] Paek, E., Liao, P., Gharavi, H. (**1992**), *Opt. Eng.* **31**, 986.
[51] Hsu, K. -Y., Li, H. -Y., Psaltis, D. (**1990**), *Proc. IEEE* **78**, 1637.
[52] Paek, E. G., Wullert II, J. R., Jain, M., Lehmen, A. V., Scherer, A., Harbison, J., Florez, L. T., Yoo, H. J., Martin, R., Jewell, J. L., Lee, Y. H. (**1990**), *Opt. Lett.* **15**, 341.
[53] Khoury, J., Kane, J. S., Hemmer, P., Woods, C. (**1992**), *Appl. Opt.* **31**, 1818.
[54] Fielding, K. H., Horner, J. L., Makekeau, C. (**1990**), *Appl. Opt.* **29**, 4332.
[55] Owechko, Y., Dunning, G. J., Marom, E., Soffer, B. H. (**1987**), *Appl. Opt.* **26**, 1900.
[56] Paek, E., Lehmen, A. (**1989**), *Opt. Eng.* **28**, 519.
[57] Owechko, Y., Soffer, B. H. (1988), Programmable Multi-layer Optical Neural Networks with Asymmetrical Interconnection Weights, *Proc. IEEE International Conference on Neural Networks,* p. 385.
[58] Psaltis, D., Qiuo, Y. (**1990**), *Opt. Photon. News* **1**, 17.
[59] Kanerva, P. (1986), in J. S. Denker (Ed.) *Neural Networks for Computing.* New York: American Institute of Physics.
[60] Javidi, B. (Ed.) (2001), *Smart Imaging Systems.* Bellingham, Washington: SPIE Press, The International Society for Optical Engineering.
[61] Javidi, B., Sergent, A., Zhang, G. (**1997**), *Opt. Eng.* **36**, 935–942.
[62] Goudail, F., Bollaro, F., Refregier, P., Javidi, B. (**1998**), *J. Opt. Soc. Am. A (JOSA A)* **15**, 2629–2638.
[63] Towghi, N., Javidi, B., Luo, Z. (**1999**), *J. Opt. Soc. Am. A (Optics, Image Science and Vision)* **16**(8), 1915–1927.
[64] Matoba, O., Javidi, B. (**1999**), *Opt. Lett.* **24**, 762–765.
[65] Matoba, O., Javidi, B. (**2000**), *Appl. Opt. IP* **39**, 2975–2981.
[66] Rosen, J., Javidi, B. (**2001**), *Appl. Opt. IP* **40**(20), 3346–3351.
[67] Javidi, B. (**1997**), *Phys. Today* **50**(3), 27–32.
[68] Javidi, B. (**1997**), *Opt. Photon. News* **8**, p. 28.
[69] Javidi, B., Horner, J. L. (**1994**), *Opt. Eng.* **33**(6), 1752–1756.
[70] Refregier, P., Javidi, B. (**1995**), *Opt. Lett.* **20**, 767–769.
[71] Javidi, B. (Ed.) (2002), *Image Recognition and Classification: Algorithms, Systems, and Applications.* New York: Marcel-Dekker.
[72] Javidi, B., Okano, F. (Eds.) (2002), *Three Dimensional Television, Video, and Display Technologies.* Berlin: Springer Verlag.
[73] Javidi, B., Tajahuerce, E. (**2000**), *Opt. Lett.* **25**, 610–612.
[74] Tajahuerce, E., Javidi, B. (**2000**), *Appl. Opt.* **39**, 6595–6601.
[75] Frauel, Y., Castro, A., Javidi, B. (**2001**), *Appl. Opt.* **40**(20), 3387–3393.
[76] Tajahuerce, E., Matoba, O., Javidi, B. (**2001**), *Appl. Opt.* **40**(23), 3877–3886.

[77] Tajahuerce, E., Matoba, O., Javidi, B. (**2001**), *Appl. Opt.* **40**(20), 3318–3325.
[78] Frauel, Y., Javidi, B. (**2001**), *Opt. Lett.* **26**(19), 1478–1480.
[79] Jang, J. S., Javidi, B. (**2002**), *Opt. Lett.* **27**(5), 324–326.
[80] Shin, Y., Javidi, B. (**2002**), *Appl. Opt.* **41**(14), 2644–2649.

# Interconnections, Optical

*Takao Matsumoto*
*NTT Transmission Systems Laboratories, Kanagawa, Japan*

*Jean Trewhella*
*IBM T.J. Watson Research Center, Yorktown Heights, NY 10598, USA*
*Phone: (914) 945-2786; Fax: (914) 945-1974; e-mail: jeanmh@us.ibm.com*

**Abstract**
Although progress has been made in optical interconnections, electrical interconnections continue to dominate most of the hierarchical interconnect layers that comprise computer systems. (The delay in the industry's move to optics is largely due to the continued improvements in electrical links, most notably the addition of equalization schemes [1].) The challenges faced by electrical interconnections [2] are described along with a presentation of the potential benefits optical interconnections afford. Commercial success for optical interconnections has largely been in rack-to-rack links, where the distances are great enough to require the low loss advantages of optical transmission. Further work on optical schemes for each layer of interconnection hierarchy, rack to rack, board to board, and intraboard is presented. These schemes can be broken into two general types of guided wave solutions that use fibers or waveguides as transmission media and free-space solutions, which require only the optical elements. Advantages and examples of links using each type are described. It is shown that free-space optical interconnections board to board can reduce wiring congestion dramatically over electrical or guided wave approaches, which force all interconnections to pass between the boards through the backplane. The density achievable can be further improved by employing 3D interconnection arrangements which are discussed in the intraboard examples. Optical interconnection will gradually penetrate from the outer levels of the systems to the inner as the technology matures, but all of these interesting possibilities will remain possibilities unless future areas of study include focus on low cost, small size for high density and high reliability of optical interconnections and devices.

**Keywords**
interconnections, optical; interconnections, electrical; waveguide; optical bus; free-space optical interconnection; integrated optics; optical backplane; multiprocessor systems; parallel processor systems.



## 1 Introduction

Signal processing in telecommunication systems, computers, and sensors has so far been performed by using electrical circuits. This is because functions such as signal transmission, distribution, connection, conversion, switching, memory, and so on, can be done more easily with electronics than with other schemes such as mechanics, optics, bionics, and so on. Another reason is that small-size, high-speed, and high-reliability electronic circuits have become available at relatively low cost as a result of the rapid progress of semiconductor technology.

The advancement of semiconductor technology has also made it possible to highly integrate electronic circuits. A very large scale integrated (VLSI) circuit, for example, contains more than 20 000 000 transistors in a single chip. However, it is still very important to reduce the circuit size in various types of electronic instruments and systems, especially computers. Thus, it is now believed that, if only electronics continues to be used in circuits, the performance of such instruments and systems will be severely limited. Performance is already being degraded as a result of problems such as signal delay and wiring congestion in electrical interconnections.

On the other hand, with the progress of optical-fiber transmission technology in the past 20 years, several kinds of optical devices and components have been developed that have the potential to replace electronic devices and components. Several problems inherent in electrical interconnections can be solved by using optical interconnections, which can be implemented through the use of electronic/optic (E/O) and optic/electronic (O/E) converting devices, and optical-transmission media [3]. In future, the signal processing will greatly depend on the potential of optical interconnections.

## 2
## Limit of Electrical Interconnection

The operating speed and the logical configuration of digital signal-processing circuits need to be enhanced to improve the circuit performance. Figure 1 shows the relationship between the clock frequency and the number of gates required for various signal-processing systems. State-of-the-art VLSI fabrication technology is also plotted therein. The number of interconnections between the gates can be assumed to be nearly proportional to the number of gates. In high-speed optical-transmission systems, the number of interconnections may be small, but the system hardware is required to operate at as high as over 10 Gbps. Supercomputers may have lower operation speeds than optical-transmission systems, but they require a larger number of interconnections than the regular computer systems.

When high performance is required in such systems, the limitations of electrical interconnections become clear. The limitations are evident in the following areas:

1. *Signal delay and clock skew*: Microstrip lines, that are stripe-shaped conductors implemented on a substrate and which are widely used as electrical interconnections between functional devices, give rise to signal delay as a result of the dielectric constant of the substrate material. In the currently popular alumina substrate, for example, the delay is 100 ps $cm^{-1}$, which is about three times as large as the free-space delay. On the other hand, when circuit integration enables the dimensions of a circuit to be reduced, the interconnection delay becomes larger than the gate delay. As a simple model, we assume

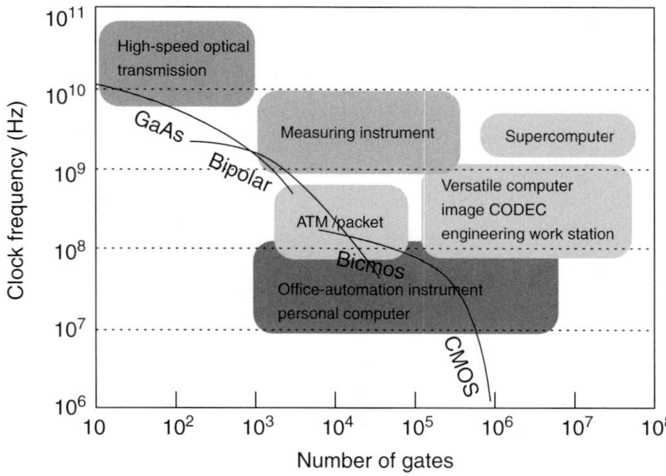


**Fig. 1** Relationship between clock frequency and number of gates required for various signal-processing systems

that all the dimensions of a VLSI chip are scaled down by the factor $k$ [3]. Since the capacitance of a fixed interconnection scales down by $k$ and the resistance scales up by $k$, the resistance–capacitance (RC) time constant and the interconnection delay do not change. On the other hand, gate delays decrease with the scaling because the channel length is decreased by $k$. Thus, interconnection delay becomes the prime chip–speed-limiting factor. In digital signal-processing hardware, signals and clock should arrive at a gate without any relative delay, since all the circuits operate synchronously according to a common clock. However, the path-length difference between signal and clock causes a relative delay and so breaks the synchronization. This phenomenon, which limits the speed of the processing hardware, is called *clock skew*. A technology to decrease the interconnection delay is needed to decrease the effect of clock skew.

2. *Signal reflection*: In electrical interconnections, characteristic impedance ($Z_0$) is defined for a transmission line of the interconnection. When an electrical device having an impedance $Z \neq Z_0$ is attached to a transmission-line end, the impedance mismatching causes signal reflection. Even if the impedance is matched, signal reflection will occur as a result of stray capacitance and inductance produced by discontinuity in the physical structure of the conductors. Moreover, if an interconnection link is relatively long in comparison with the signal bit length, the effects of signal reflection become very noticeable because of the delayed multireflection. The bus-interconnection scheme, in which processors and memories are interconnected with a common transmission line, is widely employed in computer systems. The transmission line includes a series of taps, to which the lines from the processors or memories are connected. Impedance mismatching occurs in taps as a matter of course, and this causes signal reflection.

3. *Cross talk and noise*: As electrical interconnections become more congested, the spacing between interconnection lines decreases and the length of adjacent-line coupling increases. This causes cross talk in the interconnection lines. Some kind of shielding must be devised between coupled lines as well as coupled circuits to avoid such cross talk.

Electromagnetic interference (EMI) from the environment induces noise and so degrades the electrical circuit performance. Therefore, anti-EMI countermeasures need to be taken when circuits are to be located near power-supplying units or switching units.

## 3 Potential of Optical Interconnection

Optical interconnection is based on the propagation of an electromagnetic wave at very high frequency (several THz). Therefore, it is free from the physical restraints due to the electric current or voltage inherent in electrical interconnections. The advantages of optical interconnections are the following:

1. *Low loss and wide bandwidth*: Transmission loss is one of the most important factors that determine the performance of interconnections. Optical interconnections have an extremely low loss characteristic when they are implemented by using optical fibers or free space. Optical-fiber loss around the 1.55-μm wavelength is only 0.1 dB km$^{-1}$. Even if there is a loss problem, present-day optical-fiber amplifier technology can

compensate for it. When the effects of diffraction and reflection are ignored, free-space transmission loss is virtually nil, and when optical waveguides are implemented in substrates, the loss is as low as 0.1 dB $cm^{-1}$ at 1.55 microns wavelength and 0.01 dB $cm^{-1}$ at 0.85 microns wavelength [4]. Optical interconnections make it possible to utilize the extremely wide bandwidths of transmission media. For example, the wavelength region of 1.5 to 1.6 μm, where optical fibers have low-loss characteristics, corresponds to a frequency region of 13 THz. Optical wavelength-division multiplexing technique simultaneously uses multiple channels of different wavelengths in a single transmission medium, and thus enables the users to utilize a very wide wavelength region.

2. *Small cross talk and small reflection*: Since the concept of "current" is not applicable to optical interconnections, the effects of cross talk and noise due to electromagnetic induction can be ignored. Cross talk in optical interconnections can occur if optical signals are leaking from one channel to another, and this problem can be effectively suppressed by carefully designing the optical interconnections in hardware. Since the refractive-index difference between any two materials normally used in optical interconnections is small, the reflection arising at a joint is very small. Multireflection in a transmission medium is therefore considerably suppressed.

3. *Electrical isolation*: Since optical interconnections eliminate the need for conductors in signal transmission, transmitters can be electrically isolated from receivers. Furthermore, optical interconnections block the surges of high voltage and/or large current generated by thunder or by malfunctions in the system. As a result, voltage difference can be ignored in designing systems, and system reliability is improved.

4. *Nonphysical links*: Optical free-space interconnections enable transmitters to be separated from receivers with no need for the physical links that are essential in the electrically interconnected systems. This means that we can directly connect them without using a detour circuit, and thus can decrease signal-propagation delay and wiring congestion. Since air flow is not obstructed in the optical free-space interconnection region, devices and components can be easily cooled.

5. *Bundle interconnections and distributed interconnections*: One of the functions of a lens is to map two-dimensional (2-D) information from one plane to another. This function makes it possible to optically interconnect two groups of 2-D distributed multichannel ports facing each other. Bundles of multiple channels can be easily guided by serially aligned lenses. A light wave radiated from a source can illuminate a wide surface. The radiation is suitable for distributing optical signals to multiple detectors. Fanout can be increased with no need for a sophisticated layout design.

6. *Crossing of links*: Optical beams do not interfere with each other even if they are crossed. Therefore, optical free-space interconnection links can be freely chosen without worrying about whether they cross or not. This is a big advantage in designing interconnection hardware since the hardware configuration can be simple and flexible.

The above-mentioned issues are appropriate in general cases of optical interconnections. In a specific case where a peculiar device is to be implemented, we indicate other possible advantages.

Compactness and high-speed operation of optical-interconnection networks were discussed by Kiamilev et al. [5].

As a matter of course, optical interconnections have some disadvantages that should be taken into account in hardware design. The issues are the following:

1. *Necessity of E/O and O/E converters*: Since VLSIs are operated by electrical signals, E/O and O/E converters are necessary to implement optical interconnections. The dimensions and power consumption of E/O and O/E converters should be reduced.
2. *Limitation of miniaturization*: Miniaturization of interconnection circuits while keeping low-loss throughput is difficult, because of the limitation on the fabrication accuracy and the inherent characteristics of optical waveguides. Bending and scattering losses of optical interconnections are unignorable problems.
3. *Power loss due to multiple signal dropping*: In electrical interconnections, the high-impedance coupling technique can considerably suppress the signal power loss at a coupling point. In optical interconnections, such a technique is unavailable, and multiple signal dropping at serial taps along an interconnection link causes considerable power loss.

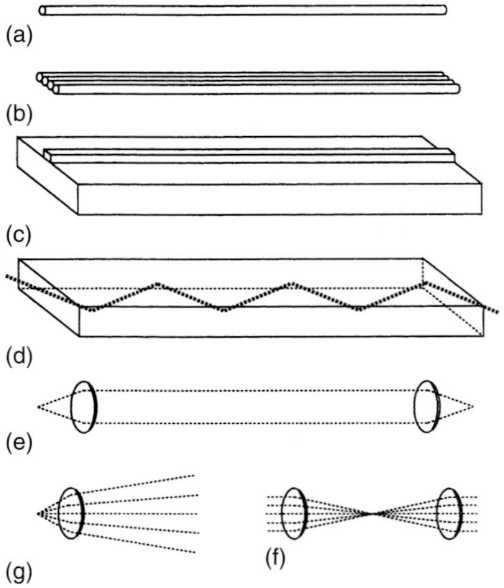


**Fig. 2** Images of optical-interconnection schemes: (a) optical fiber; (b) ribbon fiber; (c) stripe waveguide; (d) slab waveguide; (e) point-to-point free-space beam; (f) distributed free-space beam; and (g) bundle free-space beam

## 4 Optical-interconnection Schemes

Optical interconnections can be classified into three schemes in terms of the transmission medium employed, that is, optical fiber, optical waveguide, and free space. Typical images of these schemes are shown in Fig. 2.

An optical-fiber interconnection scheme can utilize the optical-transmission technology that has already been developed for public telecommunication systems. Since optical interconnection is usually employed in a specific closed and small region, specifications such as wavelength, fiber parameter, transmission speed, and so on, can be chosen more freely than in public telecommunication systems. Today, the mature single-mode fiber technology is useful in a wide variety of applications. Multimode fibers, however, make it possible to reduce interconnection hardware costs because of the ease with which they can be connected. Parallel signal transmission can be easily performed by using a ribbon fiber, that is composed of plural fibers contacting each other side by side. Since an optical ribbon fiber is arranged in a 1-D array structure, optical sources and detectors should employ the same structure. Optoelectronic integrated-circuit (OEIC) technology, which integrates optical and electrical semiconductor devices in a chip, is suitable for fabricating such array devices.

There are two ways of formulating optical-waveguide interconnection schemes: One way employs stripe waveguides and the other slab waveguides. In the former, it is easy to couple light waves to optical devices such as laser diodes, photodetectors, modulators, couplers, and so on, since the light waves are confined to a small section of the waveguide. Therefore, optical-stripe waveguides are indispensable for the realization of optical integrated circuits. With present-day fabrication technology, the substrates used for optical integrated circuits are small, for example, less than several inches in diameter. Accordingly, the stripe-waveguide interconnection method will be applied to relatively short-length interconnection links.

In the slab-waveguide method, an optical beam is totally reflected at the two surfaces of a slab, that is, the top and bottom surfaces, and propagates along the waveguide. Since a light wave is easily diffused in a slab, performance degradation factors such as cross talk and signal loss occur. In order to suppress the signal diffusion, some means of focusing light waves should be fabricated on the waveguide surfaces. A Fresnel lens, which can be fabricated in a small size by using ordinary etching or micromachine technology, is suitable for this application.

Optical free-space interconnection schemes include several methods with different beam shapes. Figure 2 shows three examples:

1. a point-to-point interconnection method, which interconnects a pair of optical sources and detectors,
2. a distributed interconnection method, which distributes an optical signal from an optical source to multiple detectors, and
3. a bundle interconnection method, which interconnects multiple optical sources and detectors with one-by-one correspondence in a lump.

The first method is used for general interconnections, the second for broadcasting of signals, and the third for transmitting multichannel signals or image signals.

Optical free-space interconnection schemes are advantageous in that optical devices can easily be inserted in the middle of an interconnection. Optical beams can be routed and/or switched by inserting mirrors, beam controllers, polarization controllers, polarization prisms, filters, and so on. Therefore, a flexible network can be implemented by optical free-space beams.

## 5
## Optical-interconnection Hierarchy

The progress of integrated-circuit technology has drastically reduced the size of electrical circuits, which created an interconnection hierarchy in electrical instruments and systems. This is especially true in computers, where interconnections can be classified into several levels because of the sophisticated hardware in which digital signals are processed and transmitted (Fig. 3).

First, in the case of large-scale computers occupying a wide setting space, optical interconnection between the racks is required. Since supercomputers and parallel-processing computers handle a large volume of data with a high-processing speed, the interconnections between their racks must have a large throughput. Since signals are usually transmitted in byte units (or multiple-byte units) in computers, parallel signal transmission is

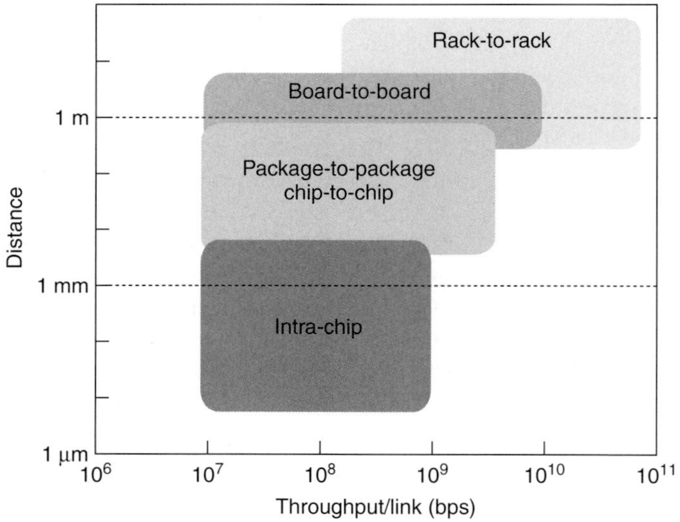


**Fig. 3** Relationship between distance and throughput for various hierarchical levels of interconnection

widely employed. When parallel multiple signals are multiplexed into a serial signal, a higher-speed transmission link is necessary to maintain the system performance. If single-mode fiber transmission is employed, link capacity can be over 10 Gbps.

Next, we have board-to-board interconnection, that is used for signal transmission within a single rack. This interconnection corresponds to the backplane interconnection used in conventional systems. When a backplane is used, interconnection-link detours give rise to problems such as signal delay, electromagnetic induction, signal reflection, and wiring congestion. However, when optical fibers or waveguides are used for board-to-board interconnection, the effects of electromagnetic induction and signal reflection are reduced. If optical free-space interconnection is employed for the direct links between adjacent boards, the backplane interconnection can be removed, and so signal delay and wiring congestion can be reduced.

A third classification is package-to-package or chip-to-chip interconnection. This is used for signal transmission between the VLSI packages or chips accommodated on a board. In the former case, any one of three schemes, that is, optical fiber, waveguide, and free space, can be employed. However, in the latter case, where link length is quite small, optical-waveguide and free-space schemes are employed.

Finally, intrachip interconnection is used in the smallest regions. When we reduce the size of an electrical circuit in order to improve its high-speed characteristics, the total performance of the circuit is mainly determined by the interconnections. Since the chip area of VLSIs is growing because of the progress of fabrication technology, the link length of intrachip interconnections is increasing. Therefore, optical intrachip interconnection may be useful for realizing advanced VLSIs.

## 6 Rack-to-rack Optical Interconnection

Rack-to-rack interconnection technology is the most advanced of the technologies employed in the above-mentioned levels of optical interconnection. This is because the optical-fiber transmission technology, which must be mainly used in rack-to-rack interconnection, is already developed and in use in the commercial field of telecommunications. The keys to making rack-to-rack optical interconnection more useful for practical applications are reducing the cost and size of E/O and O/E modules [6] and increasing the number of channels. To meet these requirements, parallel transmission techniques employing E/O and O/E array modules and ribbon fibers are being widely researched and developed [7].

There is a rack-to-rack optical-interconnection link using a 12-channel ribbon fiber [8]. In this link, the transmitter and receiver modules have an emitter-coupled-logic interface, and the modules contain an InGaAsP/InP light-emitting diode (LED) array and an InGaAs/InP photodiode (PD) array, respectively. A 200-Mbps transmission over 1 km has been reported at a signal wavelength of 1.3 μm.

Another example employing a ribbon fiber is a six-channel link comprising a laser diode (LD) array and transmitting a 1-Gbps signal per channel [9]. This link features the use of a silicon "opto-hybrid" submount, which carries all the optoelectronic components, fibers, and active and passive electronic devices. Solder bumps and grooves on the submount are used for mounting LD array chips and ribbon fibers, respectively. This structure effectively reduces the size and cost of the E/O and O/E modules, and power consumption can also be effectively reduced through LD bias-current reduction.

Figure 4 shows the structure of a four-channel transmitter/receiver module employing LD and PD arrays [10]. Transmitter and receiver OEICs fabricated in GaAs chips are mounted on a substrate and coupled to silica waveguides fabricated on the substrate by using a bonding technique. Preliminary measurements carried out at a data rate of 1 Gbps revealed that good coupling efficiency was obtained at a low coupling loss of between 3.9 and 4.2 dB.

In general, space is more efficiently used in 2-D arrays than in 1-D arrays. Optics is especially suitable for the bundle transmission and processing of 2-D information, as was mentioned in Sect. 4. A 2-D parallel transmission link of 36 channels has been reported, comprising a surface emitting a $2 \times 18$ LD array, a $2 \times 18$ PD array, and a multifiber array. 500-Mbps transmission at a wavelength of 0.85 μm was confirmed on using it [11].

## 7 Board-to-board Optical Interconnection

There are two methods of carrying out board-to-board optical interconnection, that is, the adoption of optical-transmission technology to backplane interconnections. One method is simply to replace the electrical interconnections in conventional systems with optical interconnections, and the other is to connect E/O and O/E modules on adjacent boards directly without detouring through the backplane. The former method uses optical fibers, waveguides, and free-space beams, and the latter only free-space beams.

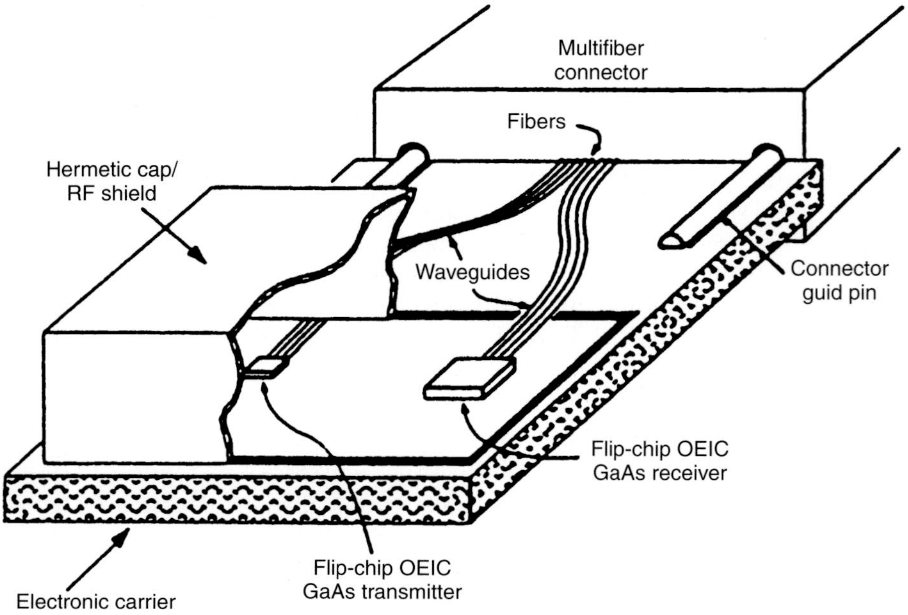


**Fig. 4** Structure of the four-channel transmitter/receiver module [10]

An experimental parallel computer system employing optical backplane interconnection has been reported by a group at Nippon Telegraph and Telephone Corp. [12]. In this system, electrical signals from every board are converted to optical beams, which are then emitted vertically to the boards along the backplane (Fig. 5). The beams are emitted parallel to each other to prevent them from interfering with each other. A receiver module is used to pick up an optical beam, which is then guided through the optical fiber to an O/E module on a board. By changing the position of the receiver module, the optical beam to be guided to a board is manually selected. A 144-channel × 144-channel optical switch is realized in the backplane section. An experiment on electrically controlled optical-beam switching was performed using a liquid-crystal polarization control device with a 2-D array structure.

Figure 6 shows an example employing a D-fiber backplane [13]. The D fiber has a core exposed to the external free space. The cross section of the core is one-half of a conventional one and is in the shape of the letter "D." By bringing the flat planes of two fiber cores into contact, it becomes possible to optically couple the fibers. In the rack system, since D fibers are laid on boards as well as on a backplane, the boards can be optically connected and disconnected merely by inserting and removing them. Through an experiment on a rack accommodating boards, it was confirmed that the coupling loss of fibers was −15 dB and that the system ran for over nine months.

Figures 7 and 8 show examples of backplane interconnection using optical waveguides [14, 15]. In Fig. 7, a glass substrate is used as a backplane and E/O and O/E modules on every board are optically coupled to the glass substrate. A light wave emitted

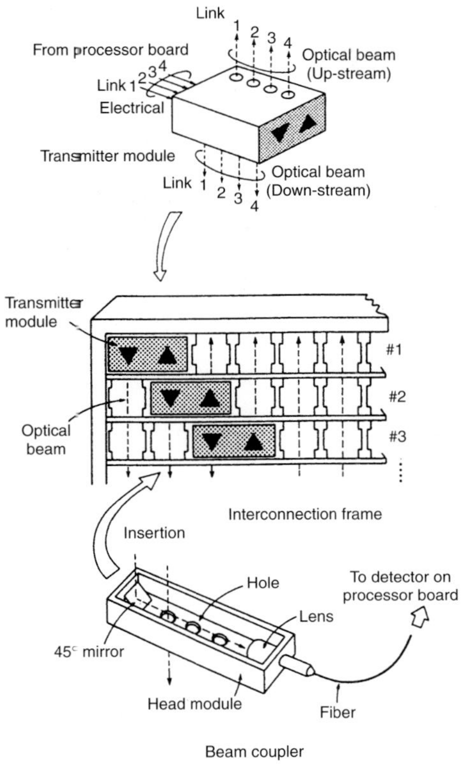


**Fig. 5** Optical backplane of a parallel computer system employing fiber and beam interconnection [12]

from a LD propagates in the waveguide in a zigzag path. Holographic optical elements are fabricated on the glass substrate and used for collimating and routing optical beams. Thus, an optical signal transmitted from one board can be distributed to a number of different boards. An experiment using nine boards was carried out where a 200-Mbps optical clock signal was transmitted from each board and distributed to the other eight boards.

In Fig. 8, 1-D arrays of rod lenses are serially aligned so that the arrays function as optical waveguides on a backplane. A LED (or LD) emits an unmodulated optical beam from one end of each rod lens train. Optical transparent modulators and detectors are attached to one of the edges of each board, and the edges are inserted in a slot between the rod lens arrays. By coupling an optical beam with a modulator and a detector, signals can be transmitted between any two boards. It was theoretically shown that an optical bundle beam of 3-mm outer diameter can carry up to 21 channels.

Direct optical interconnection between adjacent boards without detouring through a backplane can reduce both signal delay and wiring congestion. Figure 9 shows a model of a 3-D mesh multiprocessor system [16]. It is composed of $n$ stacked processor boards, on each of which $n^2$ processing units are arranged and interconnected in a 2-D square mesh. It is assumed that the backplane is perpendicular to the stacked processor boards. The signal delay between two adjacent processing units for the intraboard interconnection and that for the board-to-board interconnection are assumed to be $T$ and $\alpha T$, respectively. The minimum delay between an arbitrary pair of processing units using a direct free-space optical interconnection and that between two units using the backplane interconnection were calculated under the assumption that the routing time for relaying signals at each processing unit is ignored. Distribution of the signal delay for $n = 10$ is shown in Fig. 10. The center values and the widths of the curve for the free-space interconnection are smaller than those for the backplane interconnection for both the values of $\alpha$. This demonstrates that the direct board-to-board free-space optical-interconnection scheme is effective in reducing signal delay and clock skew.

In the model shown in Fig. 9, each board has $8n$ unused ports, which are allocated

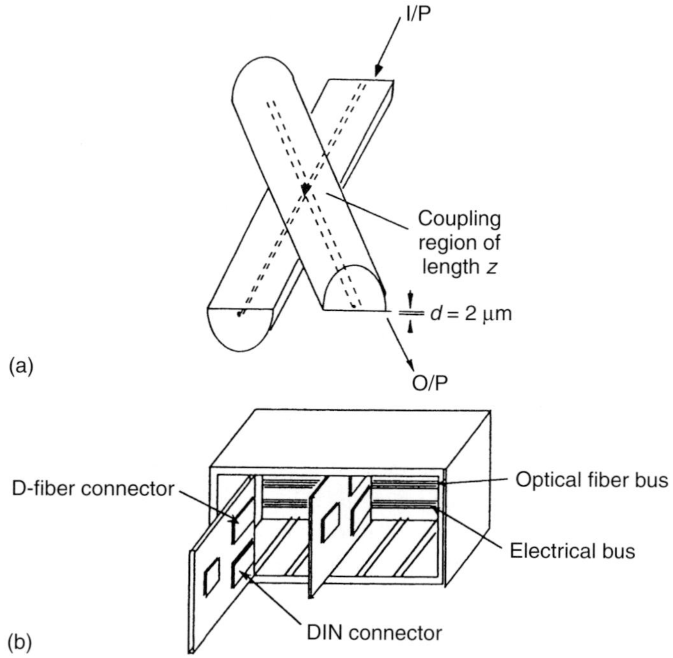


**Fig. 6** Backplane interconnection employing D fibers: (a) a D-fiber cross point; (b) boards inserted in a rack [13]

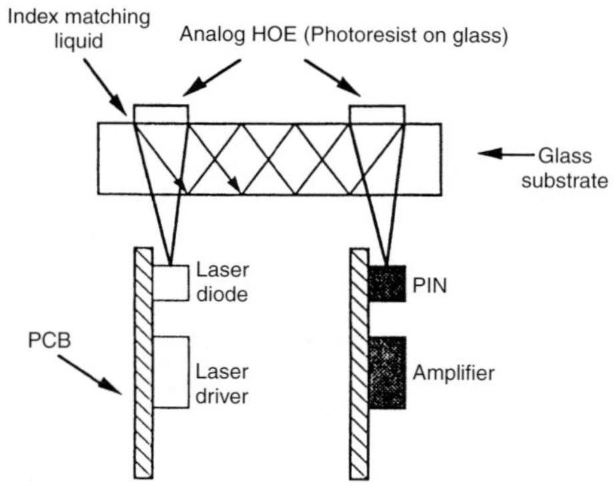


**Fig. 7** Schematic representation of a backplane interconnection using a glass substrate [14]

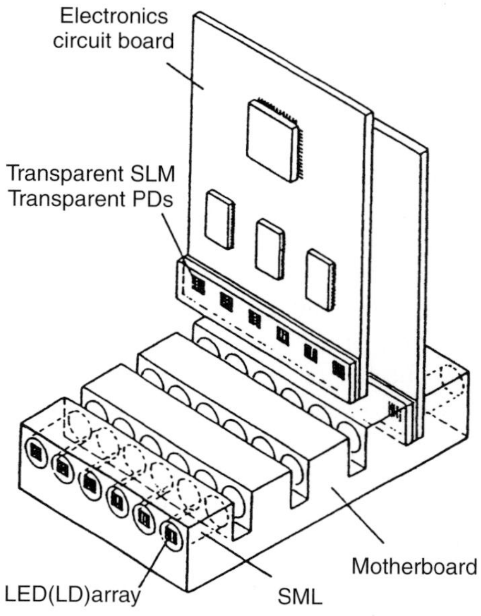


**Fig. 8** Schematic representation of a backplane interconnection using rod lens arrays [15]

on the edges. When the system is extended to a larger system by incorporating with other systems, the unused ports are connected to the added systems through a backplane. Thus, we assume that each board uses $8n$ backplane-interconnection

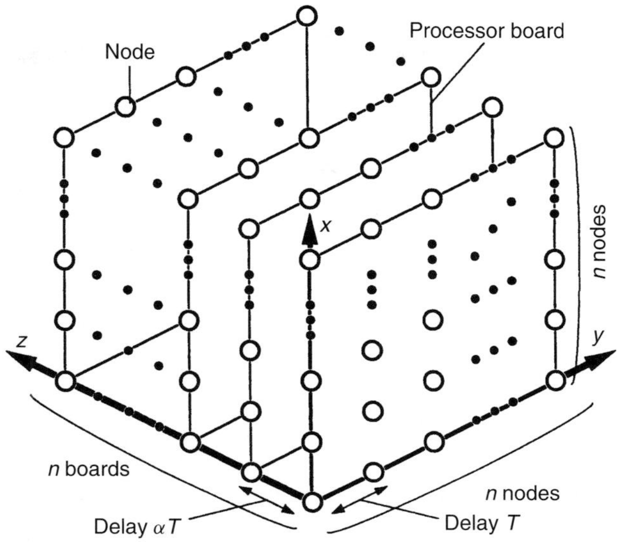


**Fig. 9** 3-D mesh multiprocessor system model [16]

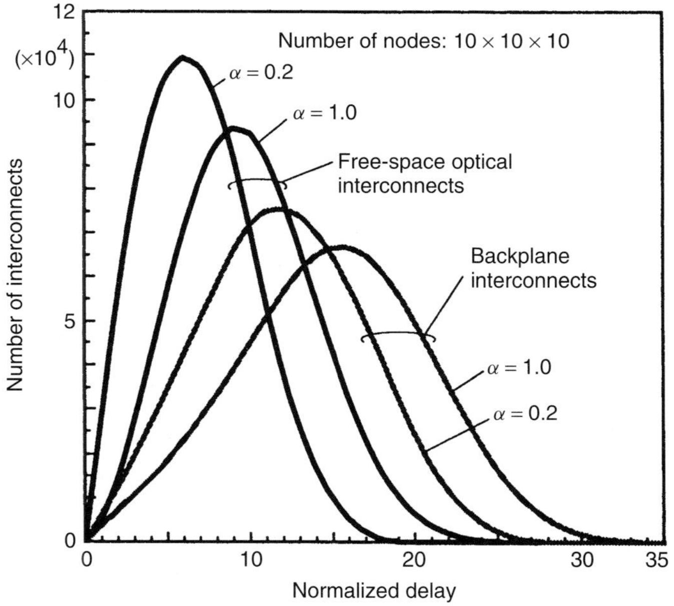


**Fig. 10** Distribution of normalized signal delay [16]

links for system extension, no matter which interconnection scheme is employed. In the backplane-interconnection scheme, board-to-board connection is performed only through the backplane, and each board needs $4n^2$ backplane-interconnection links to form a 3-D mesh network. Therefore, the total number of links between the backplane and a board is $4n^2 + 8n$. On the other hand, a free-space optical-interconnection scheme requires only $8n$ links. As shown in Fig. 11, it is apparent that optical free-space interconnection can drastically reduce wiring congestion on a backplane.

Figure 12 shows an example of a multiprocessor system with board-to-board free space optical interconnections distributed over the processor boards [16]. The system, named *COSINE-III,* consists of 64 processing units interconnected in a 3-D mesh network. Each processing unit has six bidirectional links, four of which are electrically implemented on a board and two of which are optically implemented between adjacent boards. A differential signal-transmission method is employed to suppress the effect of received optical fluctuations caused by board displacement and that of environmental light leakage. Optical sources and detectors are attached to the boards with stiffeners to avoid misalignment due to board warp. The optical source is an LED with a wavelength of 0.82 μm. System operation was confirmed by the running of a parallel-processing program.

**Fig. 12** Outside view of the 3-D mesh multiprocessor system COSINE-III. One side panel is removed [16]

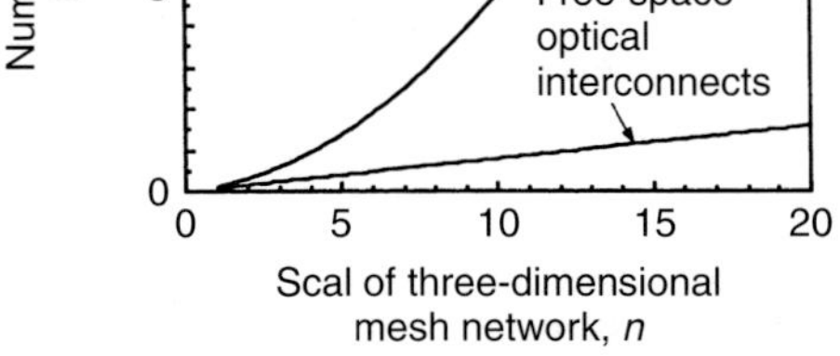


**Fig. 11** Relaxation of wiring congestion by employing optical free-space interconnection [16]

## 8
## Intraboard Optical Interconnection

The concept of intraboard interconnection includes package-to-package, chip-to-chip, and intrachip interconnections. When electrical interconnection is used, intraboard links are laid in a 2-D arrangement. However, when optical interconnection is employed, a 3-D arrangement becomes possible by adding optical free-space links perpendicular to the 2-D arrangement. Although several types of 2-D and 3-D arrangements for use in optical intraboard interconnections have been proposed, most of them are still in the concept

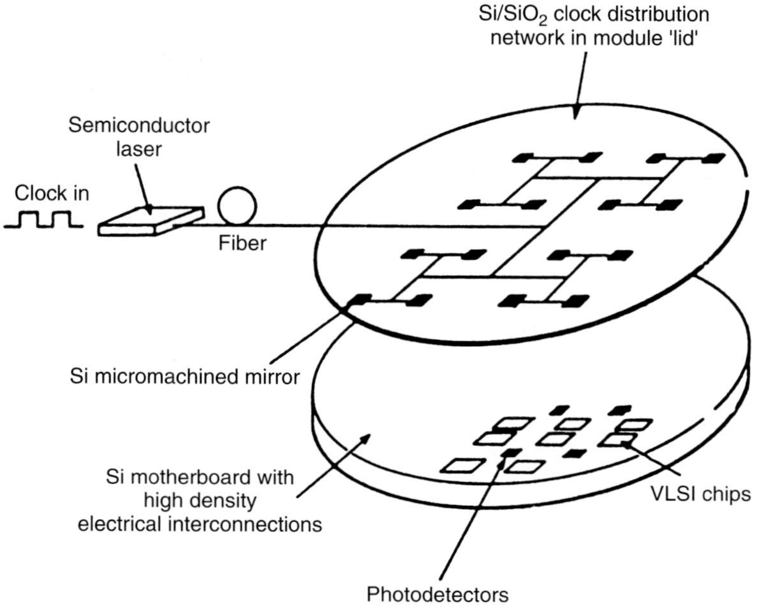


**Fig. 13** Optical clock signal distribution scheme using waveguides and beam deflection [17]

phase, and so their feasibility has not yet been estimated from the viewpoint of practical application.

A layout that distributes a clock signal over a VLSI chip by using optical waveguides and beams is shown in Fig. 13 [17]. In this layout, a silica waveguide on a silicon substrate is divided into multiple waveguides in the shape of a H tree. An optical clock signal is coupled to a micromachined mirror allocated at every output port of the waveguide, and deflected onto a receiver on the circuit board. The clock delay is the same in all of the branches because of the H-tree structure. Link fanout can be increased by repeating the division of the branches.

When passive optical waveguides are coupled to active optical devices, the optical-interconnection performance is improved. Figure 14 shows an example of such an optical-waveguide circuit, which

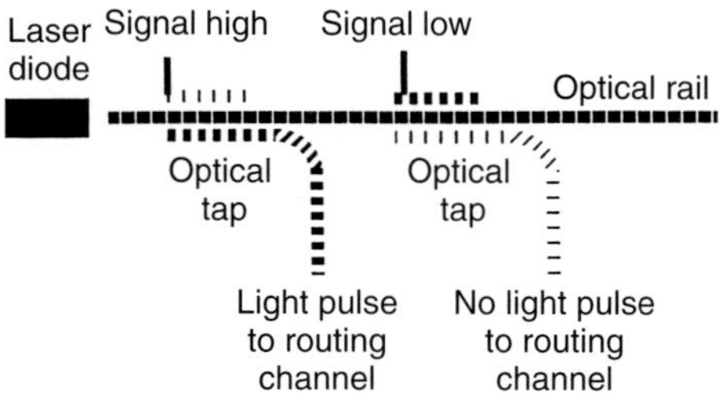


**Fig. 14** Schematic representation of optical interconnection composed of waveguides and active taps [18]

is composed of a waveguide and an active tap [18]. The active tap picks up a portion of the light wave transmitted through the waveguide, and modulates the light wave according to a supplied control signal. Because of the thermal effect of optical sources, the number of optical interconnections on a chip is usually limited. When we adopt the arrangement shown in Fig. 14, the number of interconnections can be considerably increased, since a common optical source is used.

Figure 15 shows an example in which optical-waveguide interconnection is applied to a multichip module [19]. The interconnection network of the module is composed of two parts, a silicon substrate on which VLSI chips are mounted and electrical interconnections are implemented, and a silicon template on which optical sources and detectors are mounted and optical-waveguide interconnections are implemented. Since optical interconnections are suitable for long-path interconnections, they constitute an overlay network for long-path signal transmission in a multichip module.

Figure 16 shows an example of optical interconnection based on planar optics [20]. A light wave emitted by a microlaser is transmitted from one chip to the next through the repetition of reflection in the substrate. The reflection of the light wave is performed by diffractive optical elements and mirrors.

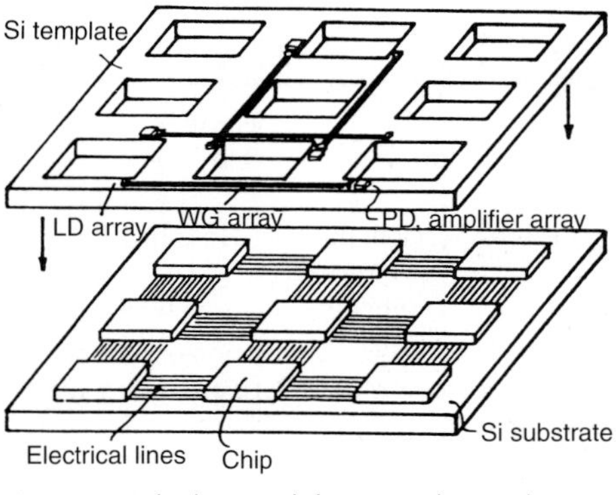


**Fig. 15** Multichip module using electrical interconnection on a substrate and optical interconnection on a template [19]

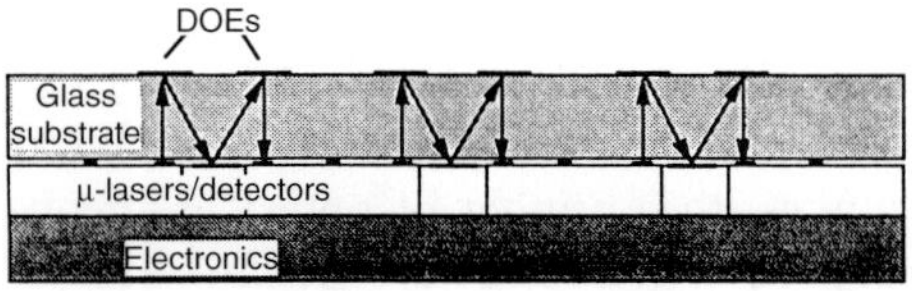


**Fig. 16** Planar optics for chip-to-chip interconnection [20]

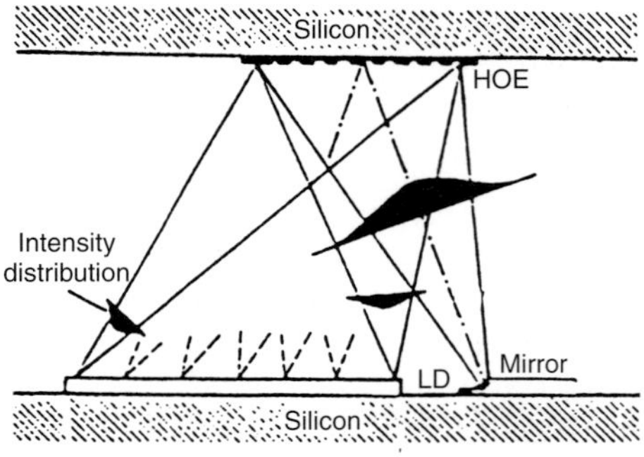


**Fig. 17** Optical clock signal distribution through beam reflection on a package ceiling [21]

A clock-distributing method using optical free-space interconnection has been proposed and is shown in Fig. 17 [21]. In this method, light waves emitted by a LD illuminate a hologram device attached to the ceiling of the package. The hologram device then distributes the light waves and focuses them to the detectors on a VLSI chip. The number and the directions of the distributed light waves can be changed by modifying the design of the hologram device. A preliminary experiment of the method has been carried out.

## 9 Optical-interconnection Devices

Optical-interconnection devices can be classified into two categories, one being the devices needed to realize optical-interconnection links and the other being the devices fabricated by employing optical-interconnection technology. Since the devices in the former category have already been discussed in terms of their structure and application, they will not be further described here.

A number of devices can be fabricated by employing optical-interconnection technology; one of them is the artificial neural network chip shown in Fig. 18 [22]. This chip is an integrated device in a hybrid structure with a 32 line-shaped LED array, a 32 × 32 interconnection matrix mask, and a 32 line-shaped photodetector (PD) array. The stripes of the LED array are arranged to be perpendicular to those of the PD array. This chip has a crossbar-interconnection function, in which the weightings at the cross points can be determined by the transparency of the elements in the matrix mask. An artificial neural network was realized by combining this chip with an electrical circuit having sigmoidal characteristics. An alphabet-recognition experiment was carried out on the network, and the results obtained were verified by computer simulation.

In a parallel-processor system, a common memory significantly increases the overall speed and throughput in computations. Figure 19 shows the cross-sectional structure of a 3-D optically coupled common memory having a multilayered structure [23]. The figure shows a case with only two layers, each of which is composed of memory cells, LEDs, and photoconductors (PCs). An LED and a PC, which are on different layers and adjacent to each other, are optically coupled in order to transmit signals between the memory cells. LEDs and PCs are produced by using the heteroepitaxial growth of GaAs on silicon after completing the silicon device fabrication process. Computer simulation showed that the device will have a speed of 32 Gbps.

## 10
## Systems and Applications

As was described in Sect. 6, rack-to-rack optical interconnection can utilize the technologies developed for telecommunication

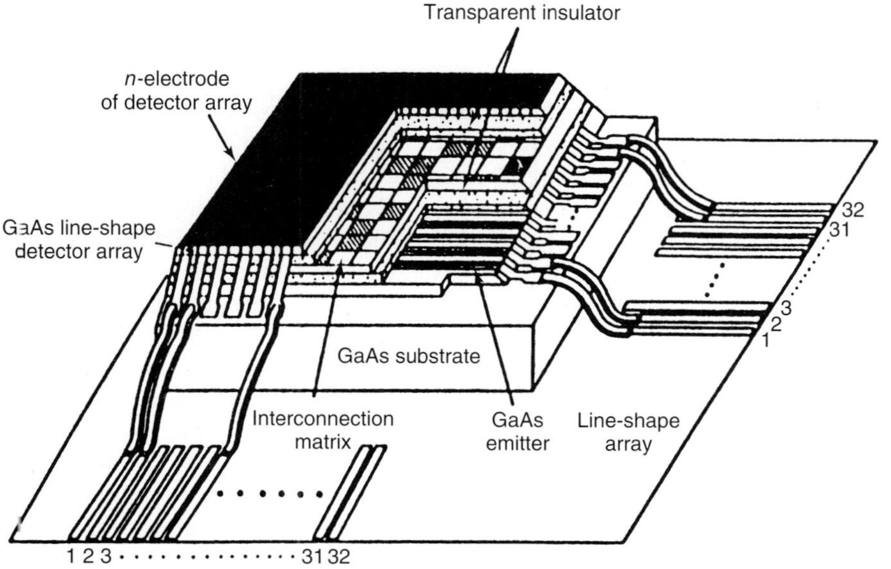


Fig. 18 Schematic view of a neural network chip employing optical interconnection [22]

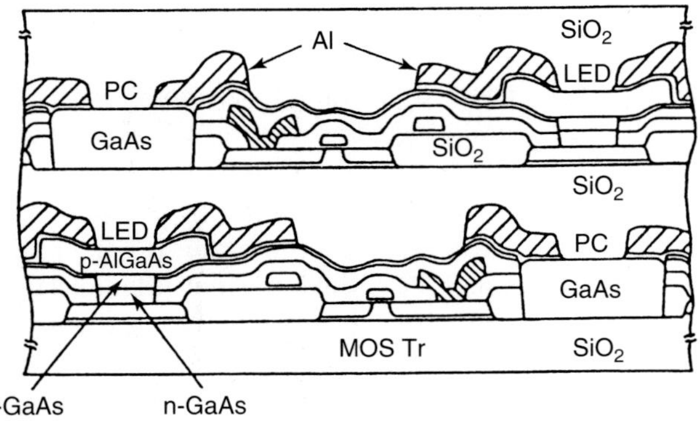


**Fig. 19** Cross-sectional view of 3-D optically coupled common memory [23]

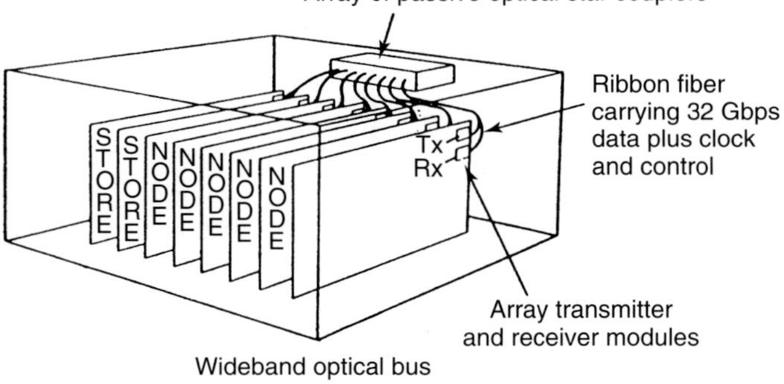


**Fig. 20** Multifiber bus using a ribbon-fiber star coupler [9]

systems. Although reduction of cost and size of O/E and E/O modules is important and should be continuously striven for, the key technologies required for such interconnection can essentially be regarded as mature technologies.

On the other hand, the optical-interconnection technologies to be applied inside of system frames are still premature, and so the number of systems employing them is very small. Figure 20 shows a system image in which multiple boards in a rack are optically coupled via an array of optical star couplers [9]. An optical star coupler is a device composed of multiple optical couplers and used to mix the signals in multiple optical fibers. In this system, 8-bit parallel signals transmitted from each board are distributed to other boards through ribbon fibers. A six-channel interconnection at 700 Mbps per channel has been reported. The transmitter and receiver modules comprise a silicon opto-hybrid submount described in Sect. 6.

Figure 21 shows an extended version of a parallel computer system called *COSINE-III,* which was described in Sect. 7 [16]. Since optical beams are transmitted and received on both sides of the system unit, it is easy to combine multiple system units serially. This combination makes it possible to extend the scale of the system.

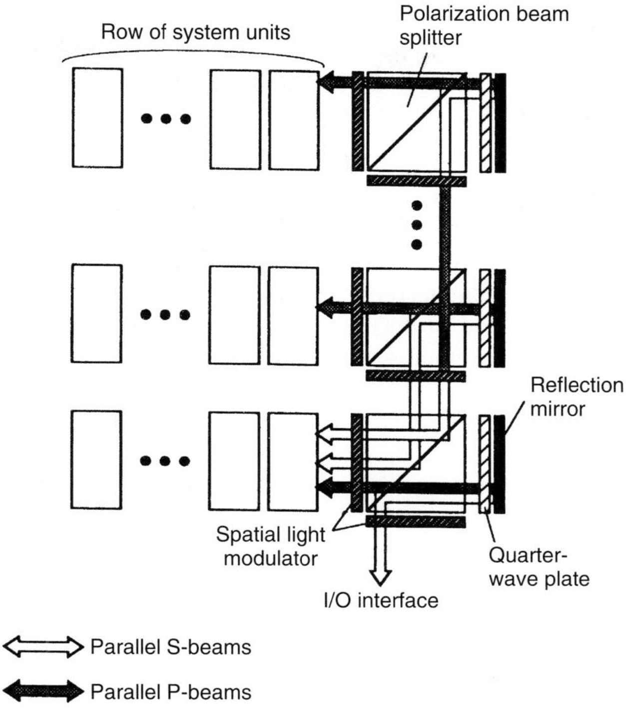


**Fig. 21** Schematically represented system extension of the COSINE-III multiprocessor system [16]

When optical-beam switching sections are inserted in the optical links, the network of the system becomes flexible. Liquid-crystal spatial modulators are suitable for the multichannel optical-beam switching.

Figure 22 shows a multistage optical space-division switching system, in which optical-fiber interconnection is used to link the switching stages [24]. Each board accommodates $LiNbO_3$ matrix switches and short fibers terminated by optical connectors. The newly developed component that joins the boards contains a fiber and connectors, and is axially flexible. A test system with a capacity of 128 channels has been developed, and has been confirmed to operate successfully for 600 Mbps signal switching at a 1.3-μm wavelength. An orthogonal arrangement of the matrix switch

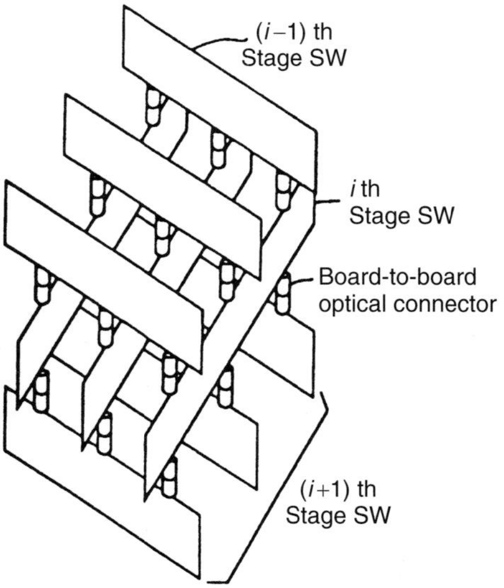


**Fig. 22** Optical space-division switching system using board-to-board optical interconnection [24]

boards can simplify the complicated interconnection network of the switching system. Such an arrangement straightens all the links between the boards.

## 11
## Future Prospects

Because of the technical trend toward high-speed and multichannel signal processing in computers as well as communications network nodes, the electrical interconnections used inside and outside of instruments will be gradually replaced by optical interconnections. This will be accelerated by the active research currently being carried out worldwide.

Optical interconnection includes various hierarchical levels as shown in Fig. 3. In the higher levels, since the size of the units to be linked is relatively large, the size of the devices and components is not necessarily minute. Therefore, conventional technologies can be easily applied thereto. However, the development of high-speed, multichannel interconnection links for practical use is strongly needed.

On the other hand, in the lower hierarchical levels of optical interconnection, the scale of circuit should be small, and thus optical-waveguide and integrated-circuit technologies are required to fabricate them. Therefore, novel fabrication technologies and hardware architecture should be aggressively researched. With the progress of these technologies, optical interconnection will gradually penetrate from the higher levels to the lower levels.

The main subjects for future research in the area of optical-interconnection devices are as follows:

1. array implementation of E/O and O/E modules,
2. optical integrated-circuit and waveguide technology,
3. hologram technology,
4. 2-D device technology,
5. OEIC technology, and
6. multiquantum-well laser technology.

These technologies will be advanced by combining research work to be done in the fields of optics and electronics.

**Tab. 1** Technological subjects for realizing optical multidimensional circuits

| *Technological subjects* \ *Circuit type* | *Zero-dimensional (Device unit)* | *One-dimensional* | *Two-dimensional* | *Three-dimensional* |
|---|---|---|---|---|
| Bandwidth, sensitivity, output power, noise | ↕ | ↕ | ↕ | ↕ |
| Cross talk, multichannel, density, uniformity | | ↕ | ↕ | ↕ |
| Surface emission/detection, congestion of input/output ports | | | ↕ | ↕ |
| Congestion of internal interconnects, thermal radiation | | | | ↕ |

As was described in Sects. 7 and 8, optical interconnection is suitable for implementing 3-D circuits. Table 1 shows the problems to be solved for the implementation of optical multidimensional circuits [25]. As the number of dimensions increases, the number of problems also increases.

## References

[1] de Araujo, D. N., Diepenbrock, J., Cases, M., Pham, N. (2003), Transmitter and Channel Equalization for High-Speed Server Interconnects, *IEEE 12th Topical Meeting on Electrical Performance of Electronic Packaging Princeton*, NJ, USA 27–29 Oct. 2003.

[2] Cutler, R. (**2001**), Challenges for achieving 10 Gbps copper backplanes, *Electron. Eng. (UK)* **73**(898), 41–42.

[3] Goodman, J. W., Leonberger, F. J., Kung, S.-Y., Athale, R. A. (**1984**), *Proc. IEEE* **72**, 850–866.

[4] Eldada, L., Shacklette, L. W. (**2000**), Advances in polymer integrated optics, *IEEE J. Sel. Top. Quantum Electron (USA)* **6**(1), 54–68.

[5] Kiamilev, F. E., Marchand, P., Krishnamoorthy, A. V., Esener, S. C., Lee, S. H. (**1991**), *IEEE J. Lightwave Technol.* **9**, 1674–1692.

[6] Trewhella, J., Johnson, G. W., Hogan, W. K., Karst, D. L. (**2003**), Evolution of optical subassemblies in IBM data communication transceivers, *IBM J. Res. Dev.* **47**(2/3), 251–258.

[7] Kuchta, D. (2003), 120 Gb/s VCSEL based Parallel Optical Transmitter and Custom 120 Gb/s Testing Station, *Proceedings of the 16th Annual Meeting of the IEEE Lasers and Electro Optics Society*, Tucson, AZ, October 26–30, 2003.

[8] Ota, Y., Swartz, R. G. (**1991**), *IEEE Mag. Lightwave Telecommun. Syst.* **2**, 24–32.

[9] Parker, J. W., Ayliffe, P. J., Clapp, T. V., Ceear, M. C., Harrison, P. M., Peall, R. G. (**1992**), *Electron. Lett.* **28**, 801–803.

[10] Jackson, K. P., Flint, E. B., Cina, M. F., Lacey, D., Trewhella, J. M. (1992), *Electronic Components and Technology '92 Digest*, pp. 93–97.

[11] Hasnain, G., Novotny, R. A., Wyann, J. D., Leibenguth, R. (1992), *CLEO '92 Digest*, pp. 170–171.

[12] Matsumoto, T., Sakano, T., Noguchi, K., Sawabe, T. (1990), *ICCD '90 Digest*, pp. 426–429.

[13] Mackenzie, F., Hodgkinson, T. G., Cassidy, S. A., Healey, P. (**1992**), *Opt. Quantum Electron.* **24**, S491–S504.

[14] Sebillotte, C. (1990), in G. Arjavalingam, J. Pazaris (Eds.), *Microelectronic Interconnects and Packages: Optical and Electrical Technologies*, SPIE Proceedings Vol. 1389. Bellingham, WA: SPIE, pp. 600–611.

[15] Hamanaka, K. (1991), *Optical Computing '91 Digest*, pp. 32–35.

[16] Sakano, T., Matsumoto, T., Noguchi, K. (1993), *ICCD '93 Digest*, pp. 278–283.

[17] Parker, J. W. (1990), *Optical Computing '90 Digest*, pp. 408–413.

[18] Lytel, R., Ticknor, A. J., Van Eck, T. E., Lipscomb, G. F. (1991), *Photonics Switching '91 Digest*, pp. 218–221.

[19] Karstensen, H., Schneider, H. W., Staudt, A., Zarschizky, H., Gerndt, C., Klement, E., Tischer, H. (1990), in H. Bartelt (Ed.), *Optical Interconnections and Networks*, SPIE Proceedings Vol. 1281. Bellingham, WA: SPIE, pp. 23–32.

[20] Jahns, J., Lee, Y. H., Jewell, J. L. (1990), *Optical Computing '90 Digest*, pp. 164–166.

[21] Zarschizky, H., Karstensen, H., Gerndt, Ch., Klement, E., Schneider, H. W. (1990), in G. Arjavalingam, J. Pazaris (Eds.), *Microelectronic Interconnects and Packages: Optical and Electrical Technologies*, SPIE Proceedings Vol. 1389. Bellingham, WA: SPIE, pp. 484–495.

[22] Nitta, Y., Ohta, J., Mitsunaga, K., Kyuma, K. (1990), *Optical Computing '90 Digest*, pp. 319–320.

[23] Koyanagi, M., Takata, H., Mori, H., Iba, J. (**1990**), *IEEE J. Solid-State Circuits* **25**, 109–116.

[24] Sawano, T., Matsuda, K., Suzuki, S., Fujiwara, M. (1993), *Photonics Switching '93 Digest*, pp. 56–59.

[25] Matsumoto, T., Noguchi, K., Koga, M. (**1992**), *Trans. IEICE Jpn.* **J75-C-I**, 223–234.

## Further Reading

Feldman, M. R., Esener, S. C., Guest, C. C., Lee, S. H. (**1988**), *Appl. Opt.* **27**, 1742–1751.

Haugen, P. R., Rychnovsky, S., Husain, A., Hutcheson, L. D. (**1986**), *Opt. Eng.* **25**, 1076–1085.

Miller, D. A. B. (**1989**), *Opt. Lett.* **14**, 146–148.

Crow, J. (2002), Intramachine communications, *Handbook of Fiber Optic Data Communication.* San Diego: Academic Press, ISBN 0-12-207891-8, pp. 379–433.

Chen, R. T. (**2000**), Fully Embedded Board-Level Guided-Wave Optoelectronic Interconnects, *Proc. IEEE* **88**(6), 780–793.

Eldada, L. (2001), Telecom optical componentry: Past, present, future, *Fiber Optic Components, Subsystems, and Systems for Telecommunications.* Proc. SPIE 4604, Bellingham, WA: SPIE, pp. 1–15.

Sakano, T., Uchiyama, K., Shake, I., Morioka, T., Hagimoto, K. (**2002**), Large dispersion tolerance optical signal transmission system based on temporal imaging, *Opt. Lett. (USA)* **27**(8), 583–585.

# Interferometry

*Parameswaran Hariharan*
*School of Physics, University of Sydney, Australia*
*Phone: (612) 9413 7159; Fax: (612) 9413 7200; e-mail: hariharan_optics@hotmail.com*

*Katherine Creath*
*Optineering, Tucson, Arizona, USA*
*Phone: (520) 882-2950; Fax: (520) 882-6976; e-mail: kcreath@ieee.org*

**Abstract**
This article reviews the field of interferometry. It begins by outlining the fundamentals of two-beam and multiple-beam interference. The rest of the article discusses the applications of interferometry for measurement of length, optical testing, fringe analysis, interference microscopy, interferometric sensors, interference spectroscopy, nonlinear interferometers, interferometric imaging, space-time and gravitation, holographic interferometry, Moiré techniques, and speckle interferometry. Each section provides a referenced (and cross-referenced) overview of the application area.

**Keywords**
interferometry; interference; interferometers; optical testing; optical metrology; nondestructive testing.

## 1 Introduction

Optical interferometry uses the phenomenon of interference between light waves to make extremely accurate measurements. The interference pattern contains, in addition to information on the optical paths traversed by the waves, information on the spectral content of the light and its spatial distribution over the source.

Young was the first to state the principle of interference and demonstrate that the summation of two rays of light could give rise to darkness, but the father of optical interferometry was undoubtedly Michelson. Michelson's contributions to interferometry, from 1880 to 1930, dominated the field to such an extent that optical interferometry was regarded for many years as a closed chapter. However, the last four decades have seen an explosive growth of interest in interferometry due to several new developments.

The most important of these was the development of the laser, which made available, for the first time, an intense source of light with a remarkably high degree of spatial and temporal coherence. Lasers have removed most of the limitations imposed by conventional light sources and have made possible many new techniques, including nonlinear interferometry.

Another development that has revolutionized interferometry has been the application of electronic techniques. The use of photoelectric detector arrays and digital computers has made possible direct measurements of the optical path difference at an array of points covering an interference pattern, with very high accuracy, in a very short time.

Light scattered from a moving particle has its frequency shifted by an amount proportional to the component of its velocity in a direction determined by the directions of illumination and viewing.

Lasers have made it possible to measure this frequency shift and, hence, the velocity of the particles, by detecting the beats produced by mixing the scattered light and the original laser beam.

Another major advance has been the use of single-mode optical fibers to build analogs of conventional two-beam interferometers. Since very long optical paths can be accommodated in a small space, fiber interferometers are now used widely as rotation sensors. In addition, since the length of the optical path in such a fiber changes with pressure or temperature, fiber interferometers have found many applications as sensors for a number of physical quantities.

In the field of stellar interferometry, it is now possible to combine images from widely spaced arrays of large telescopes to obtain extremely high resolution. Interferometry is also being applied to the detection of gravitational waves from black holes and supernovae.

Holography (see HOLOGRAPHY) is a completely new method of imaging based on optical interference. Holographic interferometry has made it possible to map the displacements of a rough surface with an accuracy of a few nanometers, and even to make interferometric comparisons of two stored wavefronts that existed at different times. Holographic interferometry and a related technique, speckle interferometry (see SPECKLE AND SPECKLE METROLOGY), are now used widely in industry for nondestructive testing and structural analysis.

The applications outlined above provide a glimpse of the many areas of optics that use interferometry. This article is meant to provide an overview. More detail is available in the cross-referenced articles as well as in the lists of works cited and further reading.

## 2 Interference and Coherence

When two light waves are superimposed, the resultant intensity depends on whether they reinforce or cancel each other. This is the well-known phenomenon of interference (see WAVE OPTICS).

If, at any point, the complex amplitudes of two light waves, derived from the same monochromatic point source and polarized in the same plane, are $A_1 = a_1 \exp(-i\phi_1)$ and $A_2 = a_2 \exp(-i\phi_2)$, the intensity (or the irradiance, units W m$^{-2}$) at this point is

$$\begin{aligned} I &= |A_1 + A_2|^2 \\ &= I_1 + I_2 + 2(I_1 I_2)^{1/2} \cos(\phi_1 - \phi_2), \end{aligned} \tag{1}$$

where $I_1$ and $I_2$ are the intensities due to the two waves acting separately and $\phi_1 - \phi_2$ is the difference in their phases.

The visibility $V$ of the interference fringes is defined by the relation

$$\begin{aligned} \mathrm{V} &= \frac{I_{\mathrm{max}} - I_{\mathrm{min}}}{I_{\mathrm{max}} + I_{\mathrm{min}}} \\ &= \frac{2(I_1 I_2)^{1/2}}{I_1 + I_2}. \end{aligned} \tag{2}$$

Interference effects can be observed quite easily by viewing a transparent plate illuminated by a point source of monochromatic light. In this case, interference takes place between the waves reflected from the front and back surfaces of the plate.

For a ray incident at an angle $\theta_1$ on a plane-parallel plate (thickness $d$, refractive index $n$) and refracted within the plate at an angle $\theta_2$, the optical path difference between the two reflected rays is

$$\Delta p = 2nd\cos\theta_2 + \frac{\lambda}{2}, \tag{3}$$

since an additional phase shift of $\pi$ is introduced by reflection at one of the surfaces. The interference fringes are circles centered on the normal to the plate (fringes of equal inclination, or Haidinger fringes).

With a collimated beam, the interference fringes are contours of equal optical thickness (Fizeau fringes). The variations in the phase difference observed can represent variations in the thickness or the refractive index of the plate. A polished flat surface can be compared with a reference flat surface, by placing them in contact and observing the fringes of equal thickness formed in the air film between them. Introduction of a small tilt between the test and reference surfaces produces a set of almost straight and parallel fringes. Any deviations of the test surface from a plane are seen as a departure of the fringes from straight lines. The errors of the test surface can then be evaluated, as shown in Fig. 1, by measuring the maximum deviation ($\Delta x$) of a fringe from a straight line as well as the spacing between successive fringes ($x$). Each fringe corresponds to a change in the optical path difference of half a wavelength.

## 2.1
## Localization of Fringes

An extended monochromatic source can be considered as an array of independent point sources. Since the light waves from these sources take different paths to the point where interference is observed, the elementary interference patterns produced by any two of them will not, in general, coincide. Interference fringes are then observed with maximum contrast only in a particular region (the region of localization).

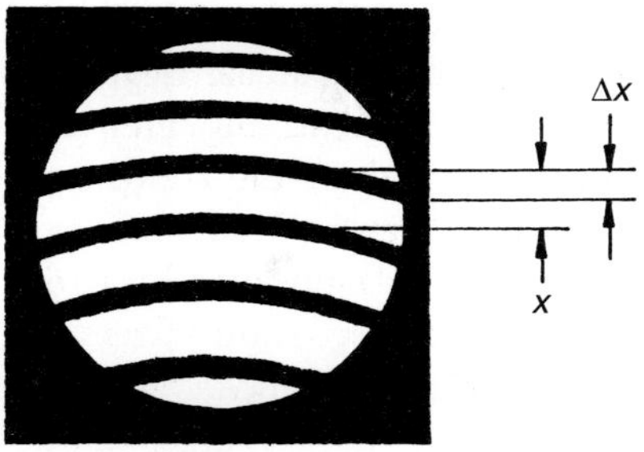


**Fig. 1** Evaluation of the errors of a polished flat test surface by interference

With a plane-parallel plate, the interference fringes are localized at infinity. With a wedged thin film, and near-normal incidence, the interference fringes are localized in the wedge.

## 2.2
## Coherence

A more detailed analysis [1] shows that the interference effects observed depend on the degree of correlation between the wave fields at the point of observation. The intensity in the interference pattern is given by the relation

$$I = I_1 + I_2 + 2(I_1 I_2)^{1/2}|\gamma| \times \cos(\arg\gamma + 2\pi\nu\tau), \qquad (4)$$

where $I_1$ and $I_2$ are the intensities of the two beams, $\nu$ is the frequency of the radiation, $\tau$ is the mean time delay between the arrival of the two beams and $\gamma$ is the (*complex*) degree of coherence between the wave fields.

With two beams of equal intensity, the visibility of the interference fringes is equal to $|\gamma|$, with a maximum value of 1 when the correlation between the wave fields is complete.

The correlation between the fields at any two points, when the difference between the optical paths to the source is small enough for effects due to the spectral bandwidth of the light to be neglected, is a measure of the spatial coherence of the light. If the size of the source and the separation of the two points are very small compared to their distance from the source, it can be shown that the complex degree of coherence is given by the normalized two-dimensional Fourier transform of the intensity distribution over the source (see FOURIER AND OTHER TRANSFORM METHODS).

Similarly, the correlation between the fields at the same point at different times is a measure of the temporal coherence of the light and is related to its spectral bandwidth. With a point source (or when interference takes place between corresponding elements of the original wavefront), the visibility of the fringes as a function of the delay is the Fourier transform of the source spectrum.

To make this analysis complete, we must also take into account the polarization effects. In general, for maximum visibility, the beams must start in the same state of polarization (see POLARIZED LIGHT, BASIC CONCEPTS OF) and interfere in the same state of polarization. For natural (unpolarized) light, the optical path difference must be the same for all polarizations. The effects of deviations from these conditions, which can be quite complex, have been discussed by [2].

## 3
## Two-beam Interferometers

Two methods are used to obtain two beams from a common source.

In wavefront division, two beams are isolated from separate areas of the primary wavefront. This technique was used in Young's experiment and in the Rayleigh interferometer.

More commonly, two beams are derived from the same portion of the primary wavefront (amplitude division) using a beam splitter (a transparent plate coated with a partially reflecting film), a diffraction grating or a polarizing prism.

### 3.1
### The Michelson Interferometer

In the Michelson interferometer, a single beam splitter is used, as shown in Fig. 2, to divide and recombine the beams. However, to obtain interference fringes with white light, the two optical paths must contain the same thickness of glass. Accordingly, a compensating plate (of the same thickness and the same material as the beam splitter) is introduced in one beam.

The interference pattern observed is similar to that produced in a plate ($n = 1$) bounded by one mirror and the image of the other mirror produced by reflection from the beam splitter. With an extended source, the interference fringes are circles localized at infinity (fringes of equal

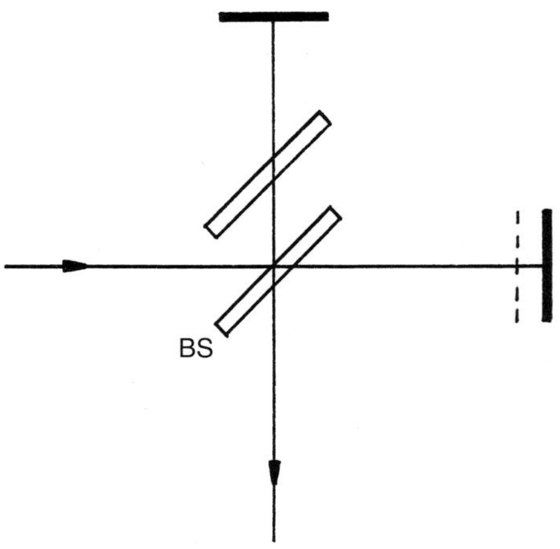


**Fig. 2** The Michelson interferometer

inclination). With collimated light (the Twyman–Green interferometer), straight, parallel fringes of equal thickness (Fizeau fringes) are obtained.

### 3.2 The Mach–Zehnder Interferometer

As shown in Fig. 3, the Mach–Zehnder interferometer (MZI) uses two beam splitters to divide and recombine the beams.

The MZI has the advantage that each optical path is traversed only once. In addition, with an extended source, the region of localization of the fringes can be made to coincide with the test section. The MZI has been used widely to map local variations of the refractive index in wind tunnels, flames, and plasmas.

A variant, the Jamin interferometer, along with the Rayleigh interferometer, is commonly used to measure the refractive index of gases and mixtures of gases. Accurate measurements of the refractive index of air are a prerequisite for interferometric measurements of length (see Sect. 5).

### 3.3 The Sagnac Interferometer

In one form of the Sagnac interferometer, as shown in Fig. 4, the two beams traverse exactly the same path in opposite directions. However, with an odd number of reflections in the path, the wavefronts are laterally inverted with respect to each other.

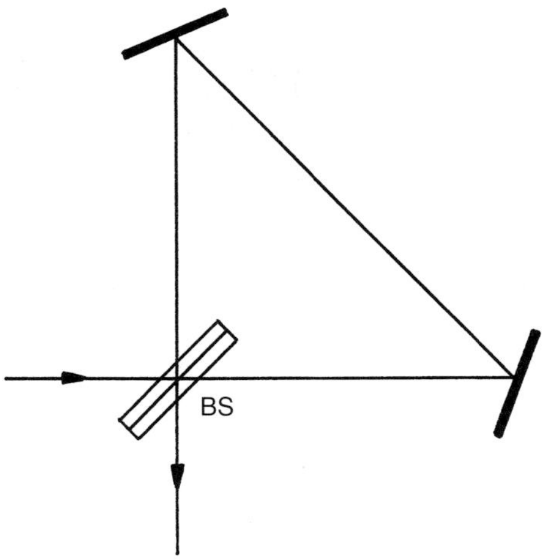


**Fig. 4** The Sagnac interferometer

Since the optical paths traversed by the two beams are always very nearly equal, fringes can be obtained easily with an extended, white-light source. Modified forms of the Sagnac interferometer are used for rotation sensing, since the rotation of the interferometer with an angular velocity $\Omega$ about an axis making an angle $\theta$ with the normal to the plane of the beams introduces an optical path difference

$$\Delta p = \left(\frac{4\Omega A}{c}\right)\cos\theta, \qquad (5)$$

between the two beams, where $A$ is the area enclosed by the beams and $c$ is the speed of light.

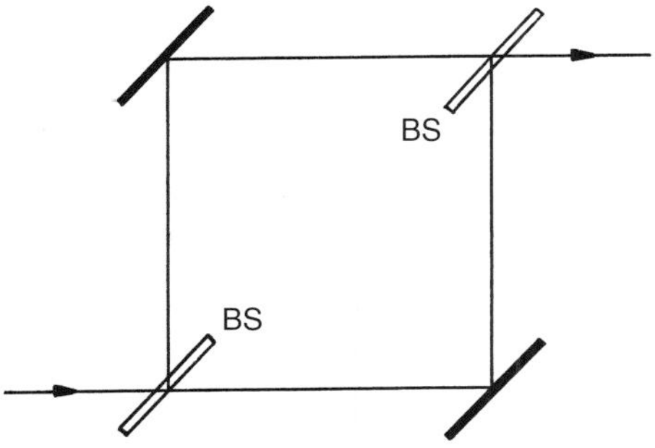


**Fig. 3** The Mach–Zehnder interferometer

### 4 Multiple-beam Interference

With two highly reflecting surfaces, we have to take into account the effects of multiply reflected beams (see WAVE OPTICS). The intensity in the interference pattern formed by the transmission is

$$I_T(\phi) = \frac{T^2}{1 + R^2 - 2R\cos\phi}, \quad (6)$$

where $R$ and $T$ are, respectively, the reflectance and transmittance of the surfaces and $\phi = (4\pi/\lambda)nd\cos\theta_2$. As the reflectance $R$ increases, the intensity at the minima decreases, and the bright fringes become sharper.

The finesse, defined as the ratio of the separation of adjacent fringes to the full width at half maximum (FWHM) of the fringes (the separation of points at which the intensity is equal to half its maximum value), is

$$F = \frac{\pi R^{1/2}}{1 - R}. \quad (7)$$

The interference fringes formed by the reflected beams are complementary to those obtained by transmission.

### 4.1 Fringes of Equal Chromatic Order

With a white-light source, interference fringes cannot be seen for optical path differences greater than a few micrometers. However, if the reflected light is examined with a spectroscope, the spectrum will be crossed by dark bands corresponding to interference minima. With a thin film (thickness $d$, refractive index $n$), if $\lambda_1$ and $\lambda_2$ are the wavelengths corresponding to adjacent dark bands, we have

$$d = \frac{\lambda_1\lambda_2}{2n|\lambda_2 - \lambda_1|}. \quad (8)$$

With two highly reflecting surfaces enclosing a thin film, very sharp fringes of equal chromatic order (FECO fringes) can be obtained, using a white-light source and a spectrograph. FECO fringes permit measurements with a precision of $\lambda/500$.

A major application of FECO fringes has been to study the microstructure of surfaces [3, 4]. However, the test surface must be coated with a highly reflective coating.

## 5 Measurement of Length

One of the earliest applications of interferometry was in measurements of lengths. Because of the limited distance over which interference fringes could be observed with conventional light sources, Michelson had to perform a laborious series of comparisons to measure the number of wavelengths of a spectral line in the standard meter. The extremely narrow spectral bandwidth of light from a laser has led to the development of a number of interferometric techniques for direct measurements of large distances. The values obtained for the optical path length are divided by the value of the refractive index of air, under the conditions of measurement, to obtain the true length.

### 5.1 Electronic Fringe Counting

If an additional phase difference of $\pi/2$ is introduced between the beams in one half of the field, two detectors can provide signals in quadrature to drive a bidirectional counter. These signals can also be processed to obtain an estimate of the fractional interference order [5].

### 5.2 Heterodyne Interferometry

In the Hewlett–Packard interferometer [6], a He-Ne laser is forced to oscillate

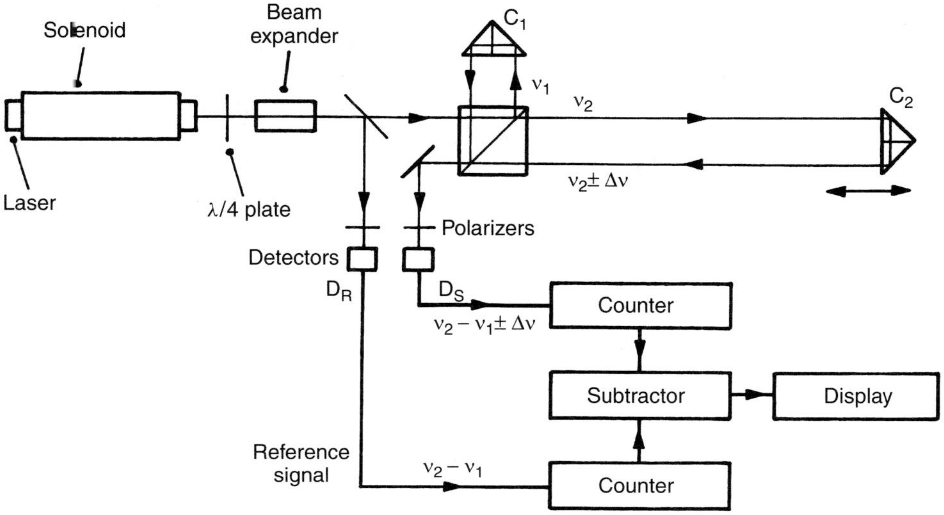


**Fig. 5** Fringe-counting interferometer using a two-frequency laser (after Dukes, J. N., Gordon, G. B. (**1970**), *Hewlett-Packard J.* **21**, 2–8 [6]. © Hewlett–Packard Company. Reproduced with permission)

simultaneously at two frequencies separated by about 2 MHz by applying a longitudinal magnetic field. As shown in Fig. 5, these two frequencies that have opposite circular polarizations pass through a $\lambda/4$ plate that converts them to orthogonal linear polarizations.

A polarizing beam splitter reflects one frequency to a fixed cube-corner, while the other is transmitted to a movable cube-corner. Both frequencies return along a common axis and, after passing through a polarizer set at 45°, are incident on a photodetector. The beat frequencies from this detector and a reference detector go to a differential counter. If one of the cube-corners is moved, the net count gives the change in the optical path in wavelengths.

Very small changes in length can be measured by heterodyne interferometry. In one technique [7], a small frequency shift is introduced between the two beams, typically by means of a pair of acousto-optic modulators operated at slightly different frequencies. The output from a detector viewing the interference pattern contains a component at the difference frequency, and the phase of this heterodyne signal corresponds to the phase difference between the interfering beams.

In another technique, the two mirrors of a Fabry–Perot interferometer are attached to the two ends of the sample, and the wavelength of a laser is locked to a transmission peak [8]. A change in the separation of the mirrors results in a change in the wavelength of the laser and, hence, in its frequency. These changes can be measured with high precision by mixing the beam from the laser with the beam from a reference laser, and measuring the beat frequency.

## 5.3 Two-wavelength Interferometry

If an interferometer is illuminated simultaneously with two wavelengths $\lambda_1$ and $\lambda_2$, the envelope of the fringes yields the interference pattern that can be obtained with a synthetic wavelength

$$\lambda_s = \frac{\lambda_1 \lambda_2}{|\lambda_1 - \lambda_2|}. \quad (9)$$

One way to implement this technique is with a $CO_2$ laser, which is switched rapidly between two wavelengths, as one of the mirrors of an interferometer is moved over the distance to be measured. The output signal from a photodetector is then processed to obtain the phase difference at any point [9].

## 5.4 Frequency-modulation Interferometry

Absolute measurements of distance can be made with a semiconductor laser by sweeping its frequency linearly with time [10]. If the optical path difference between the two beams in the interferometer is $L$, one beam reaches the detector with a time delay $L/c$, and they interfere to yield a beat signal with a frequency

$$f = \left(\frac{L}{c}\right)\left(\frac{\mathrm{d}f}{\mathrm{d}t}\right), \quad (10)$$

where $\mathrm{d}f/\mathrm{d}t$ is the rate at which the laser frequency varies with time.

## 5.5 Laser-feedback Interferometry

If, as shown in Fig. 6, a small fraction of the output of a laser is fed back to it by an external mirror, the output of the laser varies cyclically with the position of the mirror [11]. A displacement of the mirror by half a wavelength corresponds to one cycle of modulation.

A very simple laser-feedback interferometer can be set up with a single-mode semiconductor laser. An increased measurement range and higher accuracy can be obtained by mounting the mirror on a piezoelectric translator, and using an active feedback loop to hold the optical path constant [12].

# 6 Optical Testing

A major application of interferometry is in testing optical components and optical systems (see OPTICAL METROLOGY).

## 6.1 Flat Surfaces

The Fizeau interferometer (see Fig. 7) is used widely to compare a polished flat surface with a standard flat surface without placing them in contact and risking damage to the surfaces. Measurements are made on the fringes of equal thickness formed with collimated light in the air space separating the two surfaces.

To determine absolute flatness, it is possible to use a liquid surface as a reference [13]; however, a more often-used method is to test a set of three nominally flat surfaces in pairs. Errors of each of the three surfaces can be evaluated using this technique without the need for a known standard flat surface [14–16].

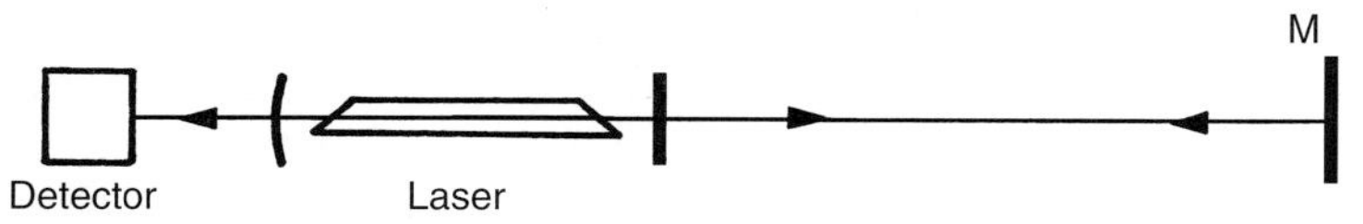


**Fig. 6** Laser-feedback interferometer

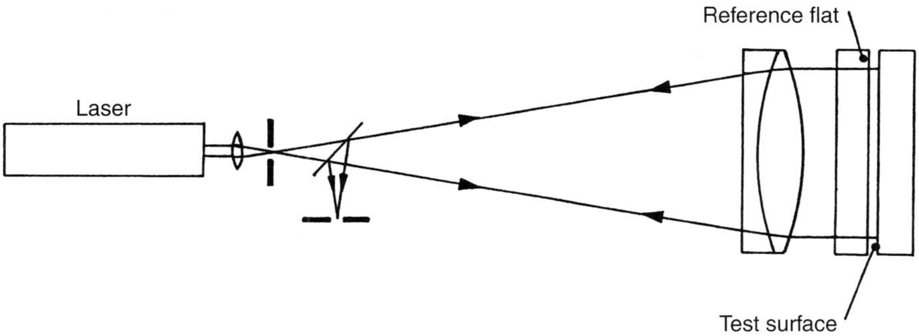


**Fig. 7** Fizeau interferometer used to test flat surfaces

6.2
## Homogeneity

The homogeneity of a material can be checked by preparing a plane-parallel sample and placing it in the test path of the interferometer. The effects of surface imperfections and systematic errors can be minimized by submerging the sample in a refractive-index matching oil and making measurements with and without the sample in the test path [17].

6.3
## Concave and Convex Surfaces

The Fizeau and Twyman–Green interferometers can be used to test concave and convex surfaces [18, 19]. Typical test configurations for curved optical surfaces are shown in Fig. 8.

6.4
## Prisms

Figure 9 shows a test configuration for a 60° prism. With a prism having a roof

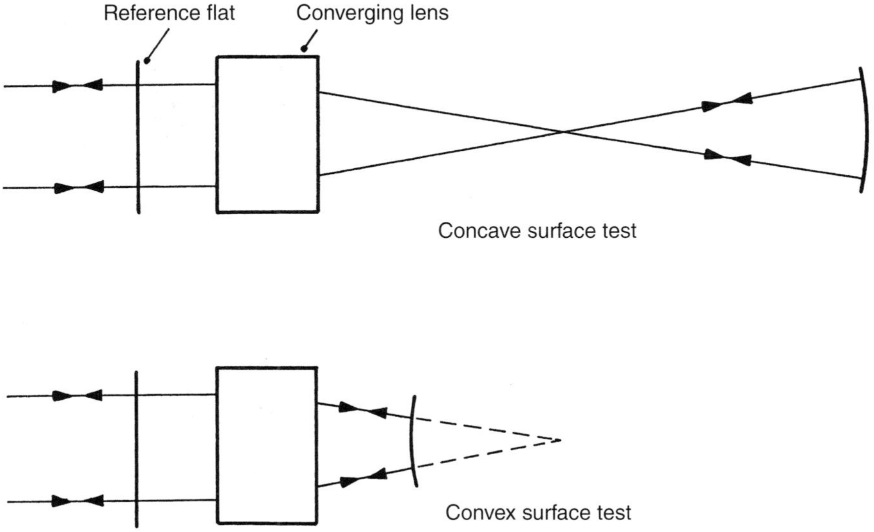


**Fig. 8** Fizeau interferometer used to test concave and convex surfaces

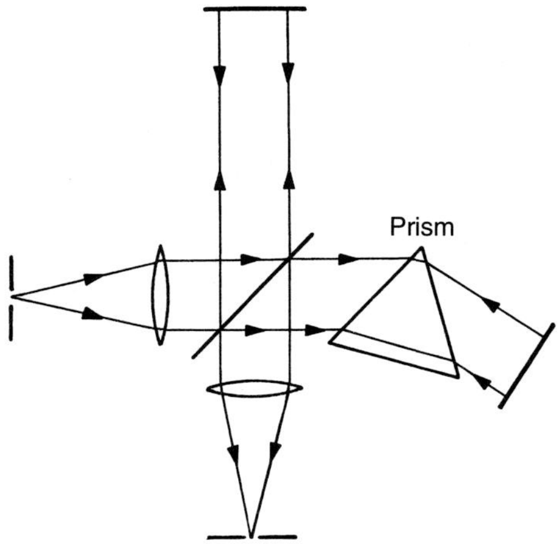


**Fig. 9** Twyman–Green interferometer used to test a prism

angle of 90°, the beam is retro-reflected back through the system.

## 6.5
## Aspheric Surfaces

Problems can arise in testing an aspheric surface against a spherical reference wavefront because the fringes in some parts of the resulting interferogram may be too closely spaced to be resolved. One way to solve this problem is to use a compensating null-lens [20]; another is to use a computer-generated hologram to produce a reference wavefront matching the desired aspheric wavefront [21, 22]. Shearing interferometry is yet another way to reduce the number of fringes in the interferogram [23, 24].

## 6.6
## Optically Rough Surfaces

One way to test fine ground surfaces, before they are polished, is by infrared interferometry with a $CO_2$ laser at a wavelength of 10.6 μm [25]. A simpler alternative, with nominally flat surfaces, is to use oblique incidence [26]. Another means of measuring these surfaces uses scanning white-light techniques similar to those described in Sect. 8.3.

## 6.7
## Shearing Interferometers

Shearing interferometers, in which interference takes place between two images of the test wavefront, have the advantage that they eliminate the need for a reference surface of the same dimensions as the test surface [23, 24]. With a lateral shear, as shown in Fig. 10(a), the two images undergo a mutual lateral displacement. If the shear is small, the wavefront aberrations can be obtained by integrating the phase data from two interferograms with orthogonal directions of shear. With a radial shear, as shown in Fig. 10(b), one of the images is contracted or expanded with respect to the

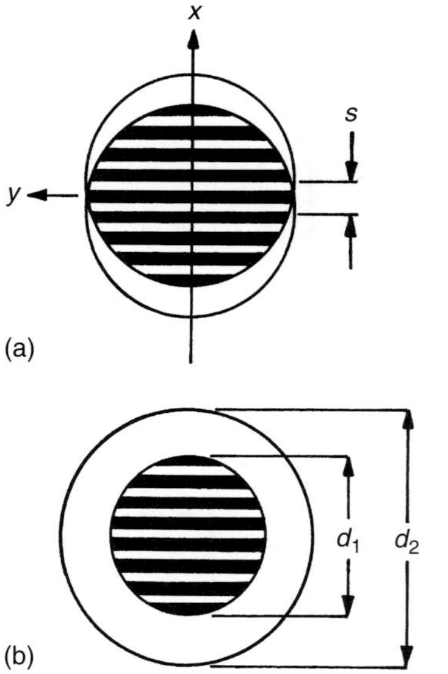


**Fig. 10** Images of the test wavefront in (a) lateral and (b) radial shearing interferometers

other. If the diameter of one image is less than (say) 0.3 of the other, the interferogram obtained is very similar to that obtained with a Fizeau or Twyman–Green interferometer.

Other forms of shear, such as rotational, inverting or folding shears, can also be used for specific applications.

### 6.8 The Point-diffraction Interferometer

As shown in Fig. 11, the point-diffraction interferometer [27] consists of a pinhole in a partially transmitting film placed at the focus of the test wavefront.

The interference pattern formed by the test wavefront, which is transmitted by the film, and a spherical reference wave produced by diffraction at the pinhole corresponds to a contour map of the wavefront aberrations. Both wavefronts traverse the same path, making this compact interferometer insensitive to vibrations.

## 7 Fringe Analysis

High-accuracy information can be extracted from the fringe pattern, including the calculation of aberration coefficients (see OPTICAL ABERRATIONS; [28]), by using an electronic camera interfaced with a computer to measure and process the intensity distribution in the interference pattern. Several methods are available for this purpose (see [29]).

### 7.1 Fringe Tracking and Fourier Analysis

Early approaches to fringe analysis were based on fringe tracking [30]. In order to analyze a single fringe pattern, it is desirable to introduce a tilt between the interfering wavefronts so that a large number of nominally straight fringes are obtained. The shape of the fringe will be modified by the errors of the test wavefront. Fourier analysis of the fringes

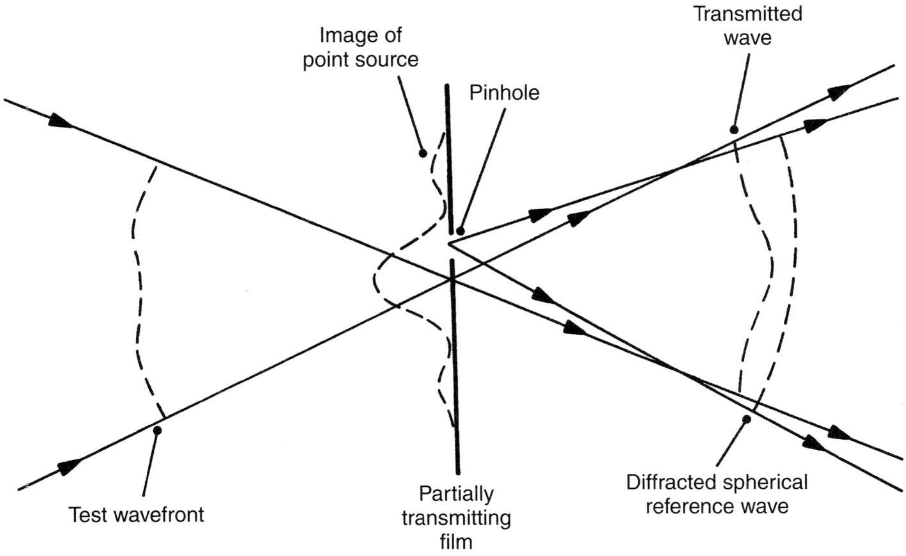


**Fig. 11** The point-diffraction interferometer [27]

can then determine the test wavefront deviation [31–33].

## 7.2 Phase-shifting Interferometry

Direct measurement of the phase difference between the beams at a uniformly spaced array of points offers many advantages. In order to determine the phase of the wavefront at each data point, at least three interferograms are required. The phase difference between the interfering beams is usually varied linearly with time, and the intensity signal is integrated at each point over a number of equal phase segments covering one period of the sinusoidal output signal. This technique is often simplified by adjusting the phase difference in equal steps.

The most common way to accomplish the phase shift between the object and reference beams is by changing the optical path difference between the beams through a shift of the reference mirror along the optical axis. Other ways include tilting a glass plate, moving a grating, frequency-shifting, or rotating a half-wave plate or analyzer. Typically, intensity information from four or five interferograms are used to calculate the original phase difference between the wavefronts on a point-by-point basis [34, 35]. The repeatability of measurements is around $\lambda/1000$.

Because the phase-calculation algorithm utilizes an arctangent function, which does not yield any information on the integral interference order, it is necessary to use a phase unwrapping procedure to detect changes in the integral interference order and remove discontinuities in the retrieved phase [29, 36].

Figure 12 shows a three-dimensional plot of the errors of a flat surface produced by an interferometer using a digital phase-measurement system.

Normally, to implement such a phase unwrapping procedure, it is necessary to have at least two measurements per fringe spacing; this constraint, limits the phase gradients that can be measured. However, techniques are available that can be used in special situations, with some *a priori*

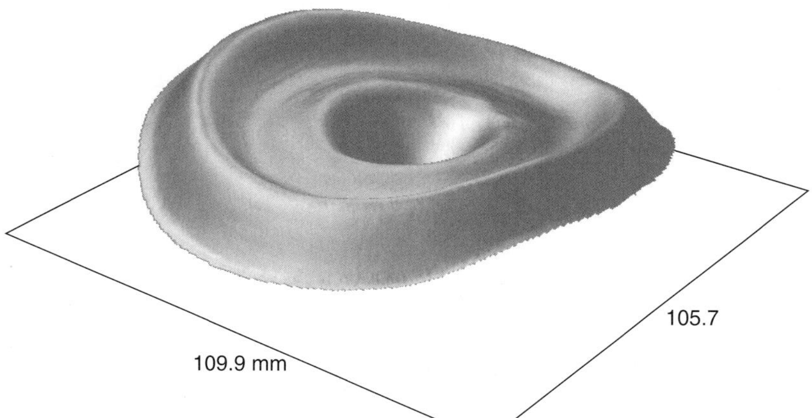


**Fig. 12** Three-dimensional plot of the errors of a flat surface (326 nm Peak-to-Valley) obtained with a phase-measurement interferometer (Courtesy of Veeco Instruments Inc.)

knowledge about the test surface, to work around this limitation [37].

### 7.3 Determining Aberrations

With most optical systems, it is then convenient to express the deviations of the test wavefront as a linear combination of Zernike circular polynomials in the form

$$W(\rho,\theta) = \sum_{k=0}^{n}\sum_{l=0}^{k}\rho^{k} \times (A_{kl}\cos l\theta + B_{kl}\sin l\theta), \qquad (11)$$

where $\rho$ and $\theta$ are polar coordinates over the pupil, and $(k-l)$ is an even number. If the optical path differences at a suitably chosen array of points are known, the coefficients $A_{kl}$ and $B_{kl}$ can be calculated from a set of linear equations [38, 39].

## 8 Interference Microscopy

Interference microscopy provides a noncontact method for studies of surfaces as well as a method for studying living cells without the need to stain them.

Two-beam interference microscopes have been described using optical systems similar to the Fizeau and Michelson interferometers. For high magnifications, a suitable configuration is that described by Linnik [40] in which a beam splitter directs the light onto two identical objectives; one beam is incident on the test surface, while the other is directed to the reference mirror.

### 8.1 The Mirau Interferometer

The Mirau interferometer permits a very compact optical arrangement. As shown schematically in Fig. 13, light from an illuminator is incident through the microscope objective on a beam splitter. The transmitted beam falls on the test surface, while the reflected beam falls on an aluminized spot on a reference surface. The two reflected beams are recombined at the same beam splitter and return through the objective.

As shown in Fig. 14, very accurate measurements of surface profiles can be made using phase shifting. With a rough surface, the data can be processed to obtain the rms surface roughness and the autocovariance function of the surface [41].

### 8.2 The Nomarski Interferometer

Common-path interference microscopes use polarizing elements to split and recombine the beams [42]. In the Nomarski interferometer (see Fig. 15), two polarizing

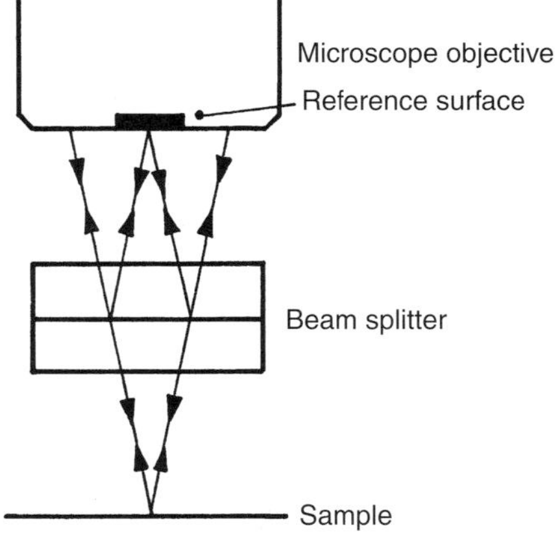


**Fig. 13** The Mirau interferometer

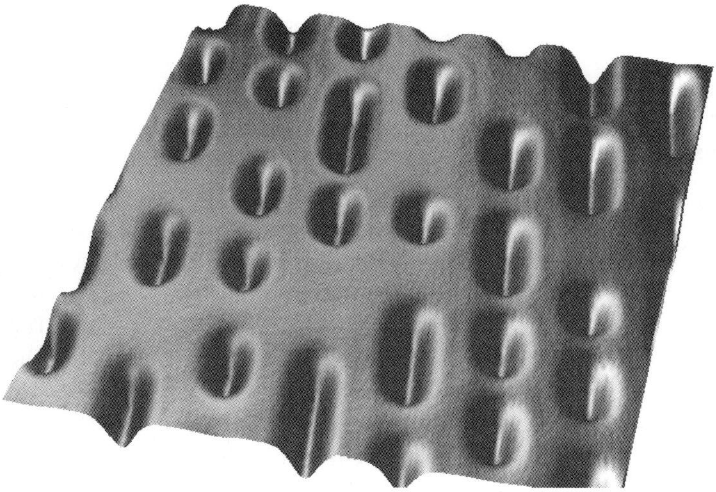

**Fig. 14** Pits (approximately 90 nm deep) on the surface of a mass-replicated CD-ROM. 11 μm × 13 μm field of view (Courtesy of Veeco Instruments Inc.)

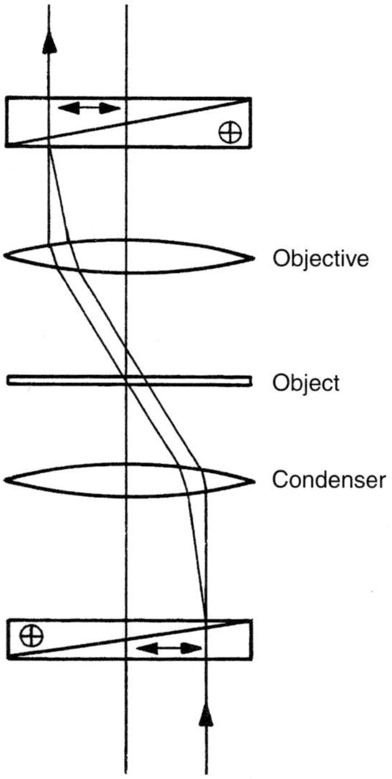


**Fig. 15** The Nomarski interferometer

prisms (see MICROSCOPY) introduce a lateral shear between the two beams.

With small isolated objects, two images are seen covered with fringes that map the phase changes introduced by the object. With larger objects, the interference pattern is a measure of the phase gradients, revealing edges and local defects.

The use of phase-shifting techniques to extract quantitative information from the interference pattern has been described by [43].

## 8.3
## White-light Interferometry

With monochromatic light, ambiguities can arise at discontinuities and steps producing a change in the optical path difference greater than a wavelength. This problem can be overcome by using a broadband (white-light) source. When the surface is scanned in height and the corresponding variations in intensity at each point are recorded, the height position corresponding to equal optical paths at which the visibility of the fringes

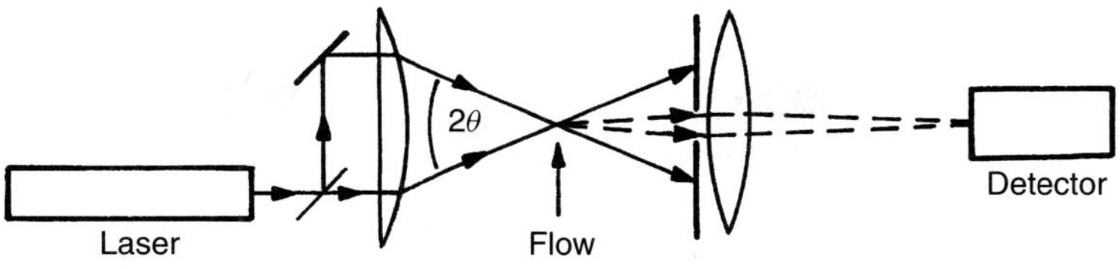


**Fig. 16** Laser–Doppler interferometer for measurements of flow velocities

is a maximum yields the height of the surface at that point.

The method most commonly used to recover the fringe visibility function from the fringe intensity is by digital filtering [44]. More recent techniques combine phase-shifting techniques with signal demodulation [45, 46]. More detail on these techniques is available in OPTICAL METROLOGY.

Another technique that can be used with white light is spectrally resolved interferometry [47, 48]. A spectroscope is used to analyze the light from each point on the interferogram. The optical path difference between the beams at this point can then be obtained from the intensity distribution in the resulting channeled spectrum (see Sect. 4.1).

White-light interferometry techniques used for biomedical applications are generally referred to as optical coherence tomography or coherence radar (BIOMEDICAL IMAGING TECHNIQUES; [49, 50]).

## 9
## Interferometric Sensors

Interferometers can be used as sensors for several physical quantities.

### 9.1
### Laser–Doppler Interferometry

Laser–Doppler interferometry [51] makes use of the fact that light scattered by a moving particle has its frequency shifted.

In the arrangement shown in Fig. 16, two intersecting laser beams making angles $\pm\theta$ with the direction of observation are used to illuminate the test field.

The frequency of the beat signal observed is

$$\Delta\nu = \frac{2v \sin\theta}{\lambda}, \tag{12}$$

where $v$ is the component of the velocity of the particle in the plane of the beams at right angles to the direction of observation.

Simultaneous measurements of the velocity components along orthogonal directions can be made by using two pairs of illuminating beams (with different wavelengths) in orthogonal planes.

It is also possible to use a self-mixing configuration for velocimetry, in which the light reflected from the moving object is mixed with the light in the laser cavity. A very compact system has been described by [52] using a laser diode operated near threshold with an external cavity to ensure single-frequency operation.

Very small vibration amplitudes can be measured by attaching one of the mirrors in an interferometer to the vibrating object. If the reflected beam is made to interfere with a reference beam with a fixed-frequency offset, the time-varying output contains, in addition to a component at the offset frequency (the carrier), two sidebands [53]. The vibration amplitude is given by the relation

$$a = \left(\frac{\lambda}{2\pi}\right)\left(\frac{I_S}{I_C}\right), \qquad (13)$$

where $I_C$ and $I_S$ are, respectively, the power in the carrier and the sidebands.

## 9.2 Fiber Interferometers

Since the length of the optical path in an optical fiber changes when it is stretched, or when its temperature changes, interferometers in which the beams propagate in single-mode optical fibers (see FIBER OPTICS and SENSORS, OPTICAL) can be used as sensors for a number of physical quantities [54]. High sensitivity can be obtained, because it is possible to have very long, noise-free paths in a very small space.

In the interferometer shown in Fig. 17, light from a laser diode is focused on the end of a single-mode fiber and optical fiber couplers are used to divide and recombine the beams. Fiber stretchers are used to shift and modulate the phase of the reference beam. The output goes to a pair of photodetectors, and measurements are made with a heterodyne system or a phase-tracking system [55].

It is also possible, as shown in Fig. 18, to use a length of a birefringent single-mode fiber, in a configuration similar to a Fizeau interferometer, as a temperature-sensing element [56].

The outputs from the two detectors are processed to give the phase retardation between the waves reflected from the front and rear ends of the fiber. Changes in temperature of 0.0005 °C can be detected with a 1-cm-long sensing element.

Measurements of electric and magnetic fields can also be made with fiber interferometers by bonding the fiber sensor to a piezoelectric or magnetostrictive element. In addition, it is possible to multiplex several optical fiber sensors in a single system to make measurements of various quantities at a single location or even at different locations (see [57, 58]).

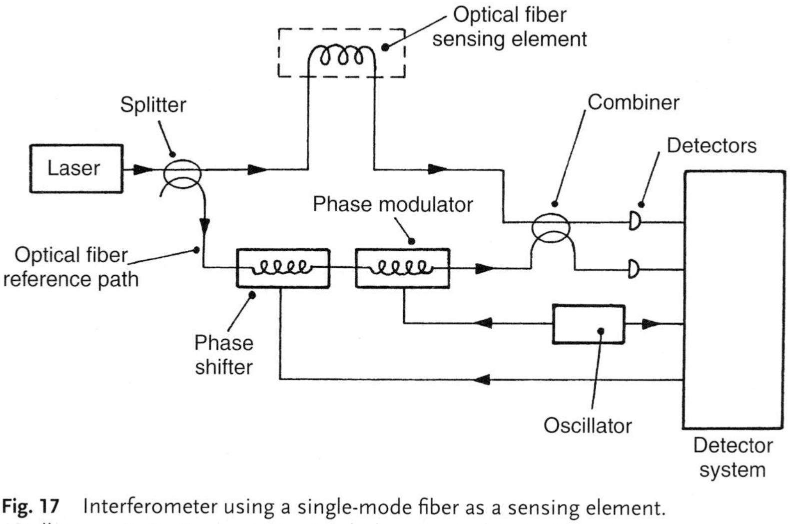


**Fig. 17** Interferometer using a single-mode fiber as a sensing element. (Giallorenzi, T. G., Bucaro, J. A., Dandridge, A., Sigel, Jr G. H., Cole, J. H., Rashleigh, S. C., Priest, R. G. (**1982**), *IEEE J. Quantum Electron.* **QE-18**, 626–665 [55] © IEEE, 1982. Reproduced with permission.)

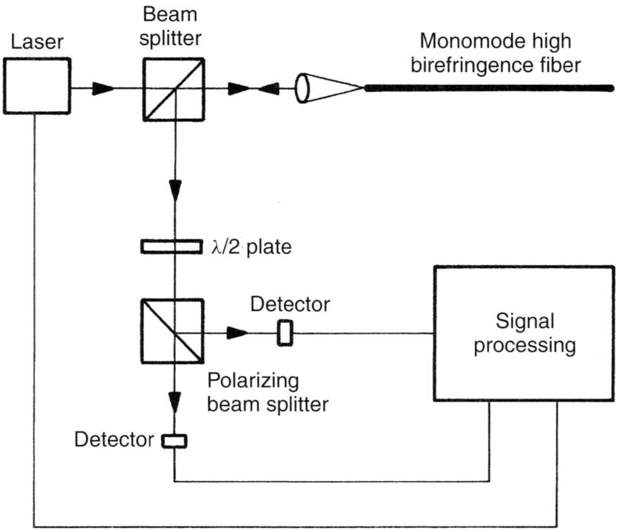


**Fig. 18** Interferometer using a birefringent single-mode fiber as a sensing element [56]

### 9.3 Rotation Sensing

Another application of fiber interferometers is in rotation sensing [59, 60]. The configuration used, in which the two waves traverse a closed multiturn loop in opposite directions, is the equivalent of a Sagnac interferometer (see Sect. 3.3). A typical system is shown in Fig. 19 [61].

## 10 Interference Spectroscopy

Interferometric techniques are now used widely in high-resolution spectroscopy (see SPECTROSCOPY, LASER) because they offer, in addition to higher resolution, a higher throughput.

### 10.1 Etendue of an Interferometer

The throughput of an optical system is proportional to a quantity known as its etendue (see OPTICAL RADIATION SOURCES AND STANDARDS).

In the optical system shown in Fig. 20, the effective areas $A_S$ and $A_D$ of the source and detector are images of each other.

The etendue of the system is

$$E = A_S\Omega_S = A_D\Omega_D, \qquad (14)$$

where $\Omega_S$ is the solid angle subtended by the lens $L_S$ at the source and $\Omega_D$ is the solid angle subtended by the lens $L_D$ at the detector.

Since the etendue of a conventional spectroscope is limited by the entrance slit, a much higher etendue can be obtained with an interferometer.

### 10.2 The Fabry–Perot Interferometer

The Fabry–Perot interferometer (FPI) [62] uses multiple-beam interference between two flat, parallel surfaces coated with

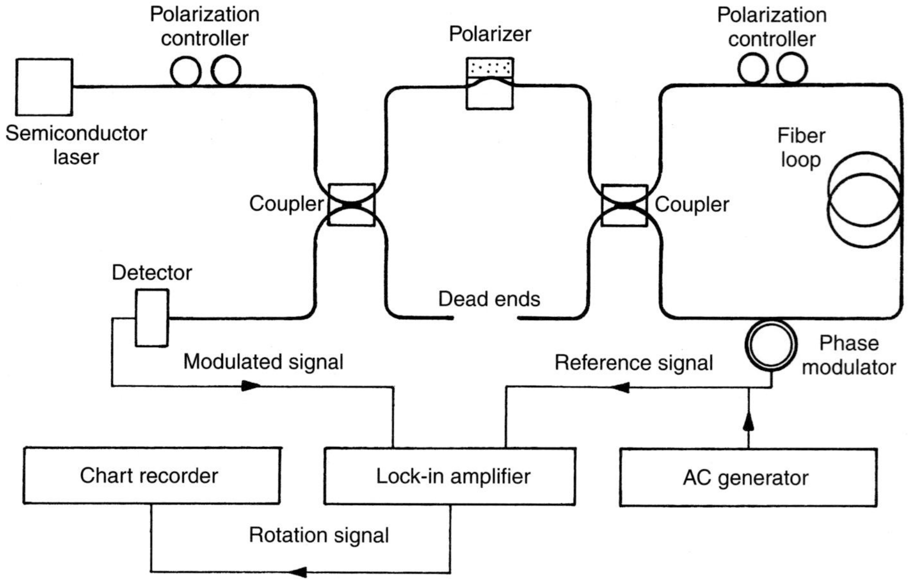


**Fig. 19** Fiber interferometer for rotation sensing [61]

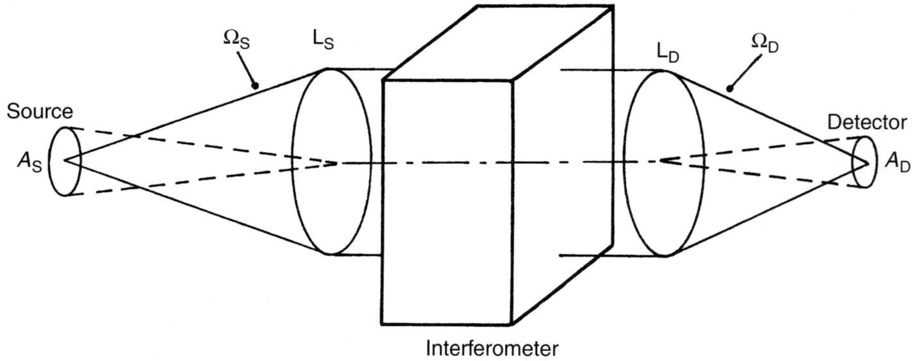


**Fig. 20** Etendue of an interferometer

semitransparent, highly reflecting coatings. With a fixed spacing $d$, any single wavelength produces a system of sharp, bright rings (fringes of equal inclination), defined by Eq. (6), centered on the normal to the surfaces.

For any angle of incidence, with a broadband source, it also follows from Eq. (6) that the separation of successive intensity maxima corresponds to a wavelength difference $\lambda^2/2nd$. This wavelength difference, known as the free spectral range (FSR), is the range of wavelengths that can be handled by the FPI without successive interference orders overlapping.

The resolving power of the FPI is obtained by dividing the FSR by the finesse (see Eq. 7) and is given by the relation

$$R = \frac{\lambda}{\Delta\lambda} = \left(\frac{2nd}{\lambda}\right)\left[\frac{\pi R^{1/2}}{1-R}\right], \quad (15)$$

where $\Delta\lambda$ is the half-width of the peaks, and $R$ is the reflectance of the surfaces.

One way to overcome the limited FSR of the FPI is by imaging the fringes onto the slit of a spectrometer, but this procedure limits the etendue of the system. A better way is to use two or more FPIs, with different values of $d$, in series.

Another important characteristic of an FPI is the contrast factor, defined by the ratio of the intensities of the maxima and minima, which is

$$C = \left[\frac{1+R}{1-R}\right]^2. \quad (16)$$

For applications such as Brillouin spectroscopy, in which a weak spectrum line may be masked by the background due to a neighboring strong spectral line, a much higher contrast factor can be obtained by passing the light several times through the same FPI [63].

A much higher throughput can be obtained with the confocal Fabry–Perot interferometer shown in Fig. 21 that uses two spherical mirrors separated by a distance equal to their radius of curvature, so that their foci coincide. Since the optical path difference is independent of the angle of incidence, a uniform field is obtained. An extended source can be used, and the transmitted intensity is recorded as the separation of the plates is varied [64].

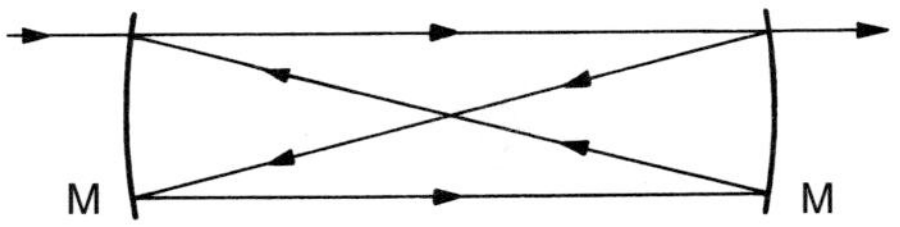


**Fig. 21** Confocal Fabry–Perot interferometer

## 10.3 Wavelength Measurements

Accurate measurements of the wavelength of the output from a tunable laser, such as a dye laser, can be made with an interferometric wavelength meter.

In the dynamic wavelength meter shown in Fig. 22, a beam from the dye laser as well as a beam from a reference laser, whose wavelength is known, traverse the same two paths. The wavelength of the dye laser is determined by counting fringes simultaneously at both wavelengths, as the end reflector is moved [65].

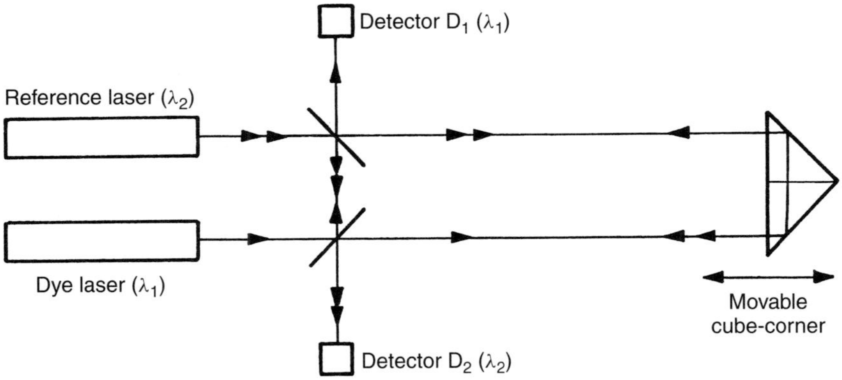


**Fig. 22** Dynamic wavelength meter [65]

In a simpler arrangement [66], the fringes of equal thickness formed in a wedged air film are imaged on a linear detector array. The spacing of the fringes is used to evaluate the integral interference order, and their position to determine the fractional interference order.

### 10.4
### Laser Linewidth

The extremely small spectral bandwidth of the output from a laser can be measured by mixing, as shown in Fig. 23, light from the laser with a reference beam from the same laser that has undergone a frequency shift and a delay [67].

### 10.5
### Fourier-transform Spectroscopy

Major applications of Fourier-transform spectroscopy include measurements of infrared absorption spectra as well as emission spectra from faint astronomical objects (see SPECTROMETERS, ULTRAVIOLET AND VISIBLE LIGHT and SPECTROMETERS, INFRARED).

With a scanning spectrometer, the total time of observation $T$ is divided between, say, $m$ elements of the spectrum. Since in the infrared, the main source of noise is the detector, the signal-to-noise ($S/N$) ratio is reduced by a factor $m^{1/2}$. This reduction in the $S/N$ ratio can be avoided by varying the optical path difference in an interferometer linearly with time, in which case each element of the spectrum generates an output modulated at a frequency that is inversely proportional to its wavelength. It is then possible to record all these signals simultaneously (or, in other words, to multiplex them) and then, by taking the Fourier transform of the recording (see FOURIER AND OTHER TRANSFORM METHODS), to recover the spectrum [68–70].

As shown in Fig. 24, a Fourier-transform spectrometer is basically a Michelson interferometer illuminated with an approximately collimated beam from the source whose spectrum is to be

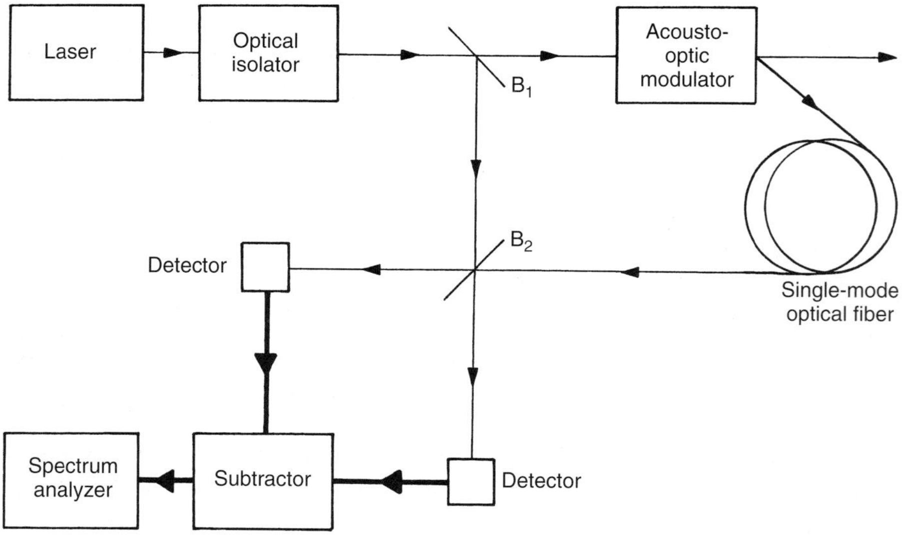


**Fig. 23** Measurement of laser linewidth by heterodyne interferometry [67]

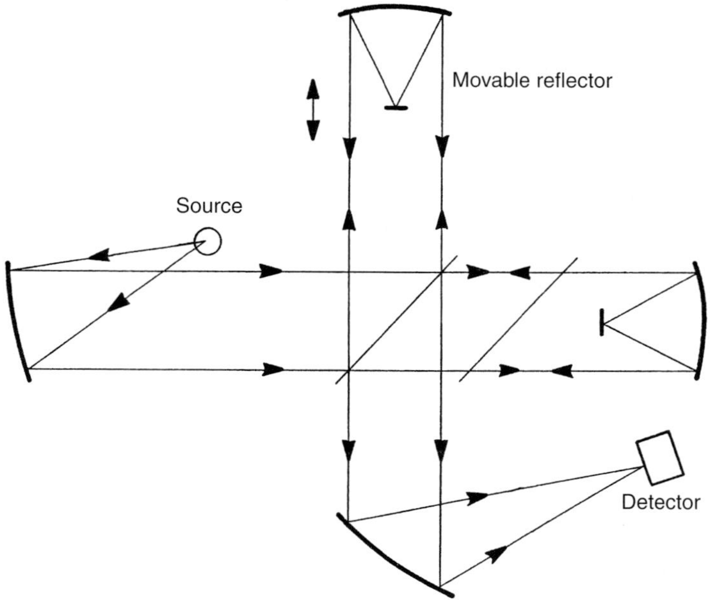


**Fig. 24** Fourier-transform spectrometer

recorded. The mirrors are often replaced by "cat's eye" reflectors to minimize problems due to tilting as they are moved.

The interferogram is then sampled at a number of equally spaced points. To avoid ambiguities (aliasing) the change in the optical path difference between samples must be less than half the shortest wavelength in the spectrum. Finally, the spectrum is computed using the fast Fourier-transform algorithm [71]. Errors in the computed spectrum can be reduced by a process called apodization, where the interference signal is multiplied by a symmetrical weighting function whose value decreases gradually with the optical path difference.

## 11
## Nonlinear Interferometers

The high light intensity available with pulsed lasers has opened up completely new areas of interferometry based on the use of nonlinear optical materials (see NONLINEAR OPTICS).

### 11.1
### Second-harmonic Interferometers

Second-harmonic interferometers produce a fringe pattern corresponding to the phase difference between two second-harmonic waves generated at different points in the optical path from the original wave at the fundamental frequency.

Figure 25 is a schematic of an interferometer using two frequency-doubling crystals that can be considered as an analog of the Mach–Zehnder interferometer [72].

In this interferometer, the infrared beam from a *Q*-switched Nd:YAG laser ($\lambda_1 = 1.06\ \mu m$) is incident on a frequency-doubling crystal, and the green ($\lambda_2 =$

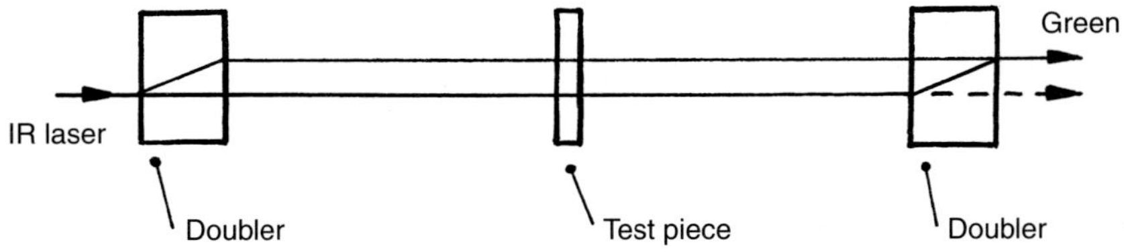


**Fig. 25** Second-harmonic interferometer using two frequency-doubling crystals

0.53 μm) and infrared beams emerging from this crystal pass through the test piece. At the second crystal, the infrared beam undergoes frequency doubling to produce a second green beam that interferes with the one produced at the first crystal. The interference order at any point is

$$N(x, y) = \frac{(n_2 - n_1)\,\mathrm{d}(x, y)}{\lambda_2}, \qquad (17)$$

where $\mathrm{d}(x, y)$ is the thickness of the test specimen at any point $(x, y)$ and $n_1$ and $n_2$ are its refractive indices for infrared and green light, respectively.

Other types of interferometers that are analogs of the Fizeau, Twyman, and point-diffraction interferometers have also been described [73, 74].

## 11.2 Phase-conjugate Interferometers

In a phase-conjugate interferometer, the test wavefront is made to interfere with its conjugate, eliminating the need for a reference wave and doubling the sensitivity.

In the phase-conjugate interferometer shown in Fig. 26, which can be regarded as an analog of the Fizeau interferometer, a partially reflecting mirror is placed in front of a single crystal of barium titanate, which functions as a self-pumped phase-conjugate mirror [75, 76].

An interferometer in which both mirrors have been replaced by a single-phase conjugator has the unique property that the field of view is normally completely dark and is unaffected by misalignment or by air currents. However, because of the delay in the response of the phase

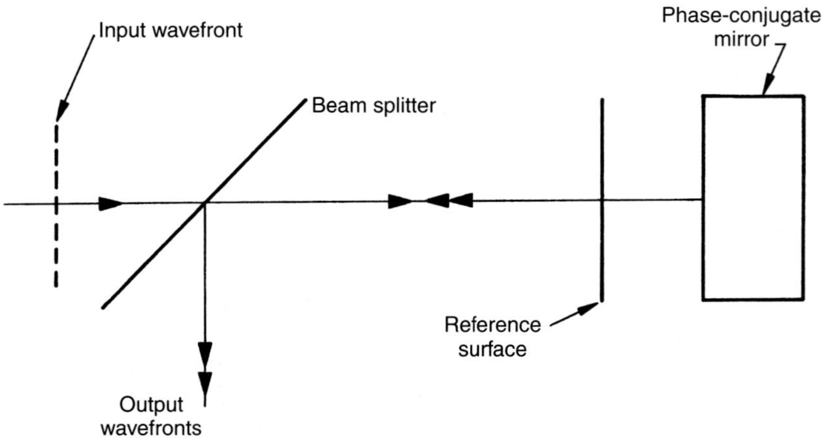


**Fig. 26** Phase-conjugate interferometer [76]

conjugator, any sudden local change in the optical path produces a bright spot that slowly fades away [77].

### 11.3 Measurement of Nonlinear Susceptibilities

A modified Twyman–Green interferometer can be used for measurements of nonlinear susceptibilities. A system that can be used to measure the relative phase shift between two-phase conjugators, as well as the ratio of their susceptibilities, and yields high sensitivity, even with weak signals, has been described by [78].

## 12 Interferometric Imaging

Interferometric imaging started with the development of techniques to measure the diameters of stars that could not be resolved with conventional telescopes (see ASTRONOMICAL TELESCOPES AND INSTRUMENTATION).

Michelson's stellar interferometer [79] used the fact that the angular diameter of a star can be calculated from observations of the visibility of the fringes in an interferometer using light from the star reaching the surface of the earth at two points separated by a known distance.

If we assume the star to be a uniform circular source with an angular diameter $2\alpha$, and $D$ is the separation of two mirrors receiving the light from the star and feeding it to a telescope, as shown in Fig. 27 the visibility of the fringes is

$$V = \frac{2J_1(u)}{u}, \tag{18}$$

where $u = 2\pi\alpha D/\lambda$ and $J_1$ is a first-order Bessel function. The fringe visibility drops

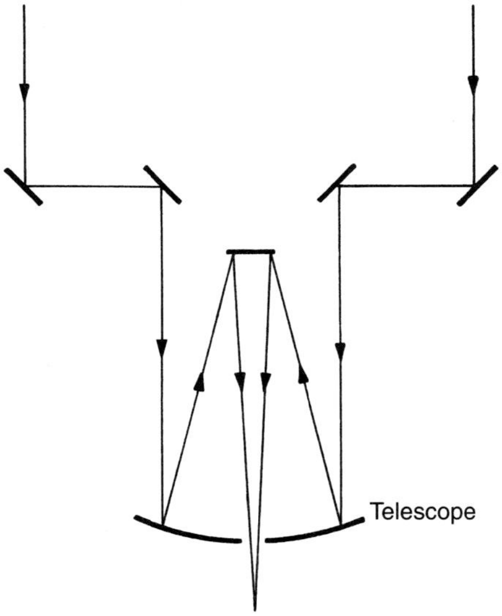


**Fig. 27** Michelson's stellar interferometer

to zero when

$$D = \frac{1.22\lambda}{2\alpha}. \tag{19}$$

Measurements with Michelson's stellar interferometer over baselines longer than 6 m presented serious difficulties because of the difficulty of maintaining the optical path difference between the beams stable and small enough not to affect the visibility of the fringes. However, modern detection, control, and data handling techniques have made possible a new version of Michelson's stellar interferometer [80] designed to make measurements over baselines up to 640 m.

### 12.1 The Intensity Interferometer

The problem of maintaining the equality of the two paths was minimized in the intensity interferometer, which used measurements of the degree of correlation between the fluctuations in the outputs of two photodetectors at the foci of two large light collectors separated by a variable distance [81, 82].

The actual instrument used light collectors operated with separations up to 188 m. With a bandwidth of 100 MHz it was only necessary to equalize the two optical paths to within 30 cm, but, because of the narrow bandwidth, measurements could only be made on 32 of the brightest stars.

## 12.2
## Heterodyne Stellar Interferometers

In these instruments, as shown in Fig. 28, light from the star is received by two telescopes and mixed with light from a laser at two photodiodes. The resulting heterodyne signals are multiplied in a correlator. The output signal is a measure

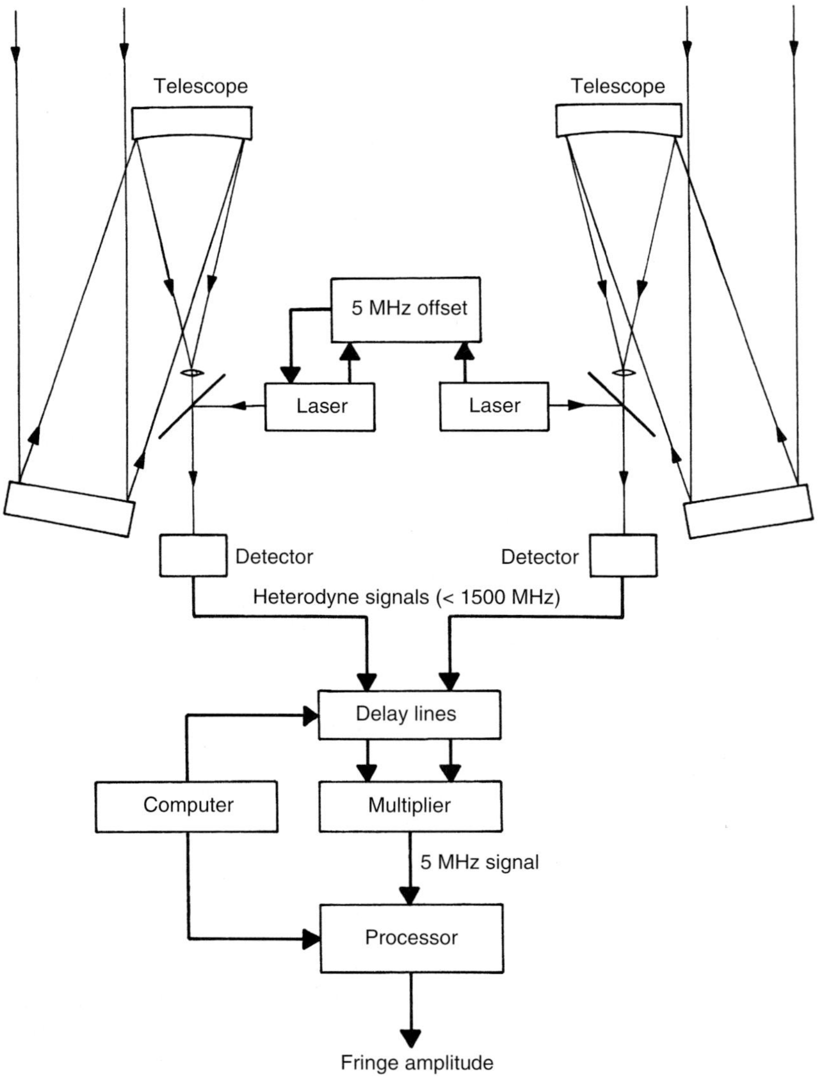


**Fig. 28** Schematic of a heterodyne stellar interferometer [83]

of the degree of coherence of the wave fields at the two photo detectors [83].

As with the intensity interferometer, this technique only requires the two optical paths to be equalized to within a few centimeters; however, the sensitivity is higher, since it is proportional to the product of the intensities of the laser and the star. An infrared interferometer comprising two telescopes with an aperture of 1.65 m has been constructed, capable of yielding an angular resolution of 0.001 second of arc [84]. Larger telescopes are planned (see ASTRONOMICAL TELESCOPES AND INSTRUMENTATION).

## 12.3 Stellar Speckle Interferometry

Stellar speckle interferometry makes use of the fact that, due to local inhomogeneities in the earth's atmosphere, the image of a star produced by a large telescope, when observed under high magnification, has a speckle structure [85] (see also ASTRONOMICAL TELESCOPES AND INSTRUMENTATION). However, individual speckles have dimensions corresponding to a diffraction-limited image of the star. Reference [86] showed that a high-resolution image could be extracted from a number of such speckled images recorded with sufficiently short exposures to freeze the speckles.

While the angular resolution that can be obtained by speckle interferometry is limited by the aperture of the telescope, it has been applied successfully to a number of problems, including the study of close double stars.

## 12.4 Telescope Arrays

The ultimate objective would be the ability to produce high-resolution images of stars. Unfortunately, with a two-element interferometer, it is only possible to obtain information on the fringe amplitude because the value of the phase is affected by instrumental and atmospheric effects. However, with a triangular array, the closure phase is determined only by the coherence function. As the number of elements increases, the image becomes better constrained [87].

Some images have already been obtained from a large, multielement interferometer (the Cambridge Optical Aperture Synthesis Telescope) [88] and several other telescope arrays are nearing completion (see ASTRONOMICAL TELESCOPES AND INSTRUMENTATION).

# 13 Space-time and Gravitation

Michelson's classical experiment to test the hypothesis of a stationary ether showed an effect that was less than one-tenth of that expected. This experiment was repeated by [89], with a much higher degree of accuracy, by locking the frequency of a He-Ne laser to a resonance of a thermally isolated Fabry–Perot interferometer mounted along with it on a rotating horizontal granite slab. When the frequency of this laser was compared to that of a stationary, frequency-stabilized laser, the frequency shifts were found to be less than 1 part in $10^6$ of those expected with a stationary ether.

## 13.1 Gravitational Waves

It follows from the general theory of relativity that binary systems of neutron stars, collapsing supernovae, and black holes should be the sources of gravitational waves.

Because of the transverse quadrupolar nature of a gravitational wave, the local distortion of space-time due to it stretches space in a direction normal to the direction of propagation of the wave, and shrinks it along the orthogonal direction. This local strain could, therefore, be measured by a Michelson interferometer in which the beam splitter and the end mirrors are attached to separate, freely suspended masses [90, 91].

Theoretical estimates of the intensity of gravitational radiation due to various possible events, suggest that a sensitivity to strain of the order of $10^{-21}$ over a bandwidth of a kilohertz would be needed. This would require an interferometer with unrealistically long arms.

One way to obtain a substantial increase in sensitivity is, as shown in Fig. 29, by using two identical Fabry–Perot cavities, with their mirrors mounted on freely suspended test masses, as the arms of the interferometer. The frequency of the laser is locked to a transmission peak of one interferometer, while the optical path length in the other is continually adjusted so that its peak transmittance is also at this frequency [92].

A further increase in sensitivity can be obtained by recycling the available light. Since the interferometer is normally adjusted so that observations are made on a dark fringe, to avoid overloading the detector, most of the light is returned toward the source and is lost. If this light is reflected back into the interferometer in the right phase, by an extra mirror placed in the input beam, the amount of light traversing the arms of the interferometer can be increased substantially [93].

Two other techniques that can be combined with these techniques for obtaining even higher sensitivity are signal recycling [94] and resonant side-band extraction [95].

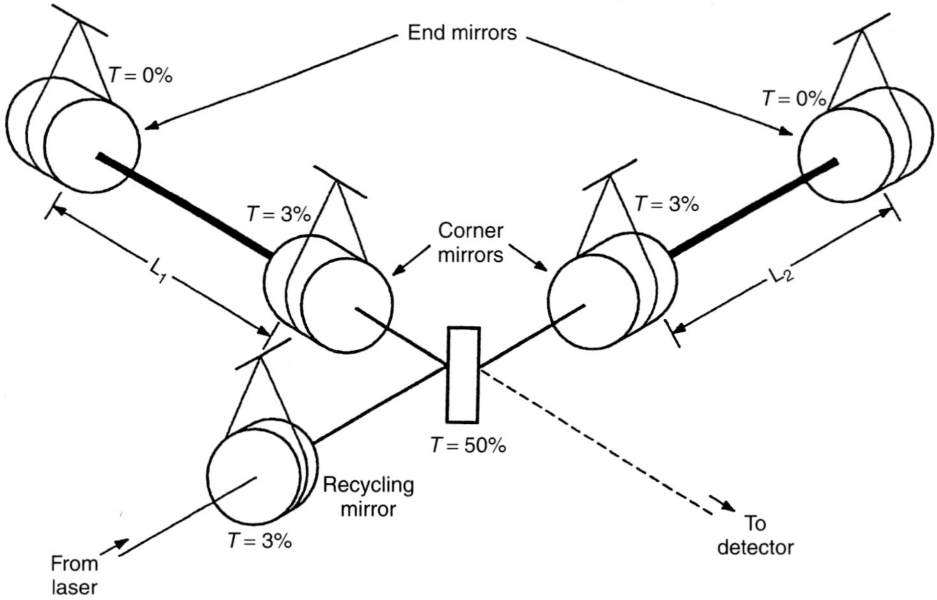


**Fig. 29** Schematic of an interferometric gravitational wave detector [91]

### 13.2 LIGO

The laser interferometer gravitational observatory (LIGO) project [96, 97] involves the construction of three laser interferometers with arms up to 4-km long, two at one site and the third at another site separated from the first by almost 3000 km. The test masses and the optical paths in these interferometers are housed in a vacuum. Correlating the outputs of the three interferometers should make it possible to distinguish the signals due to gravitational waves from the bursts of instrumental and environmental noise.

### 13.3 Limits to Measurement

The limit to measurements of such small displacements is ultimately related to the number of photons $n$ that pass through the interferometer in the measurement time. The resulting uncertainty in measurements of the phase difference between the beams has been shown to be [98]

$$\Delta\phi \geq \frac{1}{2\sqrt{n}}, \tag{20}$$

and is known as the standard quantum limit (SQL).

## 14 Holographic Interferometry

Holography (see HOLOGRAPHY and OPTICAL TECHNIQUES FOR MECHANICAL MEASUREMENT) makes it possible to store and reconstruct a perfect three-dimensional image of an object. The reconstructed wave can then be made to interfere with the wave generated by the object to produce fringes that contour, in real time, any changes in the shape of the object. Alternatively, two holograms can be recorded with the object in two different states and the wavefronts reconstructed by these two holograms can be made to interfere.

Since holographic interferometry makes it possible to measure, with very high precision, changes in the shape of objects with rough surfaces, it has found many applications including nondestructive testing and vibration analysis [99–101].

### 14.1 Strain Analysis

The phase difference at any point $(x, y)$ in the interferogram is given by the relation (see Fig. 30)

$$\Delta\phi = \mathbf{L}(x, y) \cdot (\mathbf{k}_1 - \mathbf{k}_2) = \mathbf{L}(x, y) \cdot \mathbf{K}, \tag{21}$$

where $\mathbf{L}(x, y)$ is the vector displacement of the corresponding point on the surface of the object, $\mathbf{k}_1$ and $\mathbf{k}_2$ are the propagation vectors of the incident and scattered light, and $\mathbf{K} = \mathbf{k}_1 - \mathbf{k}_2$ is known as the sensitivity vector [102].

To evaluate the vector displacements (out-of-plane and in-plane), it is convenient to use a single direction of observation and

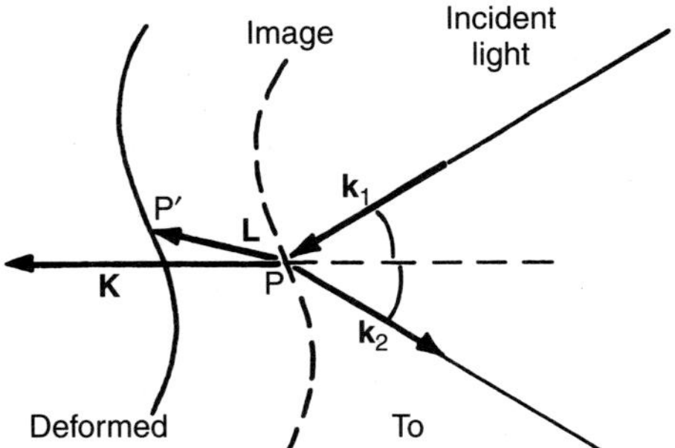


**Fig. 30** Phase difference produced by a displacement of the object

record four holograms with the object illuminated from two different angles in the vertical plane and two different angles in the horizontal plane. Phase shifting is used to obtain the phase differences at a network of points [103]. These values can then be used, along with information on the shape of the object, to obtain the strains.

### 14.2
### Vibration Analysis

One way to study the vibrating objects is to record a hologram of the vibrating object with an exposure time that is much longer than the period of the vibration [104]. The intensity at any point $(x, y)$ in the image is then given by the relation

$$I(x, y) = I_0(x, y) J_0[\mathbf{K} \cdot \mathbf{L}(x, y)], \qquad (22)$$

where $I_0(x, y)$ is the intensity with the stationary object, $J_0$ is a zero-order Bessel function and $\mathbf{L}(x, y)$ is the amplitude of vibration of the object. The fringes observed (time-average fringes) are contours of equal vibration amplitude, with the dark fringes corresponding to the zeros of the Bessel function.

Alternatively, a hologram can be recorded of the stationary object, and the real-time interference pattern obtained with the vibrating object can be viewed using stroboscopic illumination. A brighter image can be obtained by recording the hologram with stroboscopic illumination, synchronized with the vibration cycle, and viewing the interference fringes formed with the stationary object, using continuous illumination. Phase-shifting techniques can then be used to map the instantaneous displacement of the vibrating object (see Fig. 31) [105].

### 14.3
### Contouring

The simplest method of contouring an object is by recording two holograms with the object illuminated from slightly different angles. More commonly used techniques are two-wavelength contouring and two-refractive-index contouring.

In two-wavelength contouring [106], a telecentric lens system is used to image

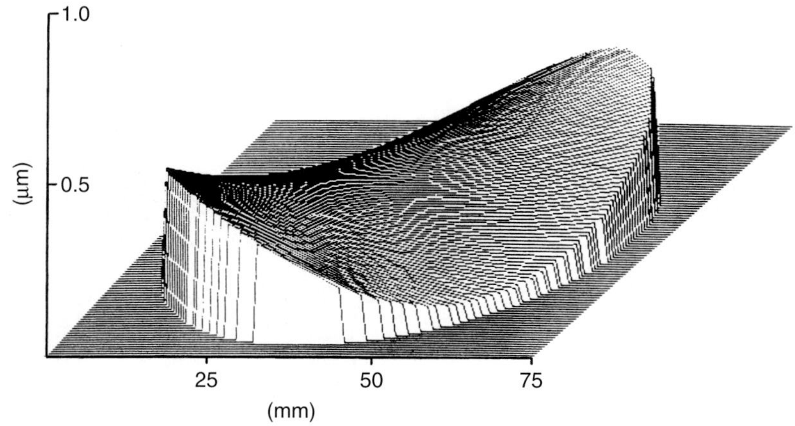


**Fig. 31** Three-dimensional plot of the instantaneous displacement of a metal plate vibrating at 231 Hz obtained by stroboscopic holographic interferometry using phase shifting[105]

the object on the hologram plane and exposures are made with two different wavelengths, $\lambda_1$ and $\lambda_2$. When the hologram is illuminated with one of the wavelengths (say $\lambda_2$), fringes are seen contouring the reconstructed image, separated by an increment of height

$$|\delta z| = \frac{\lambda_1 \lambda_2}{2(\lambda_1 - \lambda_2)} \tag{23}$$

In two-refractive-index contouring [107], the object is placed in a cell with a glass window and imaged by a telecentric system.

Two holograms are recorded on a plate placed near the stop of the telecentric system with the cell filled with liquids having refractive indices $n_1$ and $n_2$, respectively. Contours are obtained with a spacing

$$|\delta z| = \frac{\lambda}{2(n_1 - n_2)}. \tag{24}$$

Digital phase-shifting techniques can be used with both these methods of holographic contouring. Figure 32 shows a three-dimensional plot of a wear mark on a flat surface obtained by phase shifting, using the two-refractive-index technique [108].

## 15 Moiré Techniques

Moiré techniques complement holographic interferometry and can be used where a contour interval greater than 10 μm is required [109] (see also OPTICAL TECHNIQUES FOR MECHANICAL MEASUREMENT).

A simple way to obtain Moiré fringes is to project interference fringes (or a grating) onto the object and view it through a grating of approximately the same spacing. The contour interval is determined by the fringe (grating) spacing and the angle between the illumination and viewing directions. Phase shifting is possible by shifting one grating or the projected fringes.

### 15.1 Grating Interferometry

The in-plane displacements of nearly flat objects can be measured with

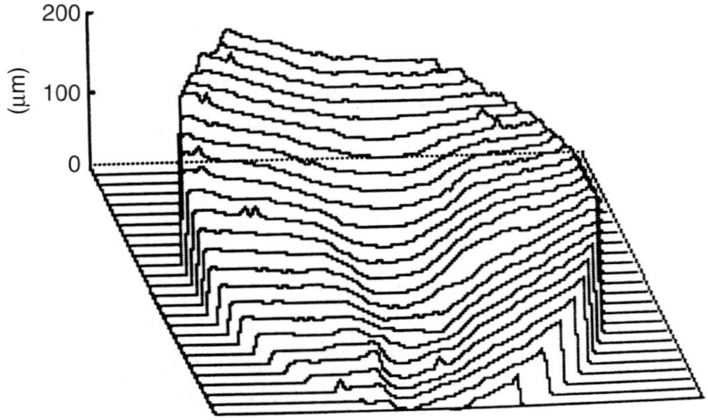


**Fig. 32** Three-dimensional plot of a wear mark on a flat surface obtained by phase shifting, using the two-refractive-index technique [108]

submicrometer sensitivity by grating interferometry [109, 110].

A reflection grating is attached to the object under test with a suitable adhesive and illuminated by two coherent beams symmetrical to the grating normal. The interference pattern produced by the two diffracted beams reflected from the grating yields a map of the in-plane displacements. Polarization techniques can be used for phase shifting.

## 16
## Speckle Interferometry

The image of an object illuminated by a laser is covered with a stationary granular pattern known as a speckle pattern (see SPECKLE AND SPECKLE METROLOGY). Speckle interferometry [111] utilizes interference between the speckled image of an object illuminated by a laser and a reference beam derived from the same laser. Any change in the shape of the object results in local changes in the intensity distribution in the speckle pattern. If two photographs of the speckled image are superimposed, fringes are obtained, corresponding to the degree of correlation of the two speckle patterns that contour the changes in shape of the surface [112].

Speckle interferometry can be a very simple way of measuring the in-plane displacements using an optical system in which the surface is illuminated by two beams making equal but opposite angles to the normal.

### 16.1
### Electronic Speckle Pattern Interferometry (ESPI)

Measurements can be made at video rates using a TV camera interfaced to a computer [111, 113]. As shown in Fig. 33, the object is imaged on a CCD array along with a coaxial reference beam. This technique was originally known as electronic speckle pattern interferometry (ESPI).

If an image of the object in its original state is subtracted from an image of the object at a later stage, regions in which the speckle pattern has not changed, corresponding to the condition

$$\mathbf{K} \cdot \mathbf{L}(x, y) = 2m\pi, \qquad (25)$$

where $m$ is an integer, appear dark, while regions where the pattern has changed are covered with bright speckles [114, 115].

This technique is also known as TV holography.

### 16.2
### Phase-shifting Speckle Interferometry

Each speckle can be regarded as an individual interference pattern and the

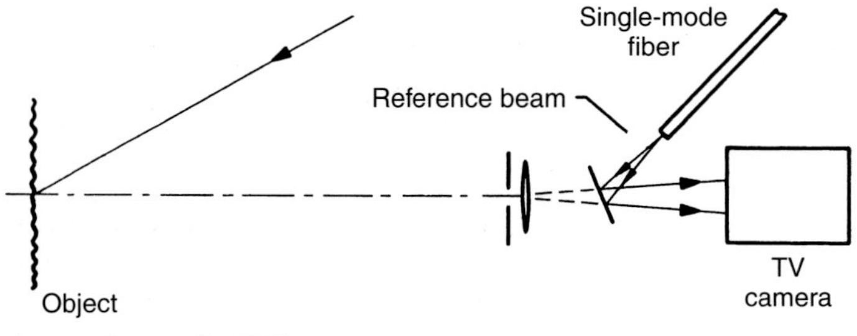


Fig. 33 System for ESPI

phase difference between the two beams at this point can be measured by phase shifting, with the object before and after an applied stress. The result of subtracting the second set of values from the first is then a contour map of the deformation of the object [116, 117].

Further developments in speckle interferometry with phase-measurement techniques, high-resolution CCD arrays, and real-time processing have created many techniques encompassing what is now most commonly known as digital holography (see OPTICAL TECHNIQUES FOR MECHANICAL MEASUREMENT and HOLOGRAPHY).

## 16.3 Vibrating Objects

If the period of the vibration is small compared to the exposure time or the scan time of the camera, the contrast of the speckles at any point is given by the expression

$$C = \frac{\{1 + 2\alpha J_0^2[\mathbf{K} \cdot \mathbf{L}(x, y)]\}^{1/2}}{1 + \alpha}, \qquad (26)$$

where $\alpha$ is the ratio of the intensities of the reference beam and the object beam, **K** is the sensitivity vector and $\mathbf{L}(x, y)$ is the vibration amplitude at that point. Regions corresponding to the zeros of the $J_0$ Bessel function appear as dark fringes.

Phase-shifting techniques can also be applied to the analysis of vibrations [118].

## Glossary

**Beam splitter:** An optical element that divides a single beam of light into two beams of the same wave form.

**Coherence:** A complex quantity whose magnitude denotes the correlation between two wave fields; its phase denotes the effective phase difference between them.

**Degree of Coherence:** The value of the coherence expressed as a fraction of that for complete correlation between the wave fields.

**Fringes of Equal Inclination:** Interference fringes created from two collinear interfering beams having wavefronts with different radii of curvature.

**Fringes of Equal Thickness:** Interference fringes created with collimated beams when the optical path difference depends only on the thickness and refractive index.

**Interference Order:** The number of wavelengths in the optical path difference between two interfering beams.

**Interferogram:** The varying part of an interference pattern, after subtracting any uniform background.

**Moiré Fringes:** Relatively coarse fringes produced by the superposition of two fine fringe patterns with slightly different spacings or orientations.

**Optical Path Difference:** The difference in the optical path length between two interfering beams.

**Optical Path Length:** The product of the refractive index of the medium traversed by a beam and the length of the path in the medium.

**Optical Phase:** The resultant phase of a light beam after allowing for changes due to the optical path traversed and reflection at any surfaces.

**Phase Shifting:** A technique that shifts the phase of one interfering beam relative to the other in order to determine optical path difference from the intensity in an interference fringe pattern.

**Region of Localization:** The region in which interference fringes are observed with maximum contrast.

## References


[1] Born, M., Wolf, E. (1999), *Principles of Optics*. Cambridge: Cambridge University Press.

[2] Leonhardt, K. (1981), *Optische Interferenzen*. Stuttgart: Wissenschaftliche Vergesselschaft.

[3] Tolansky, S. (1970), *Multiple Beam Interferometry of Surfaces and Thin Films*. Oxford: Oxford University Press.

[4] Bennett, J. M., Mattson, L. (1989), *Introduction to Surface Roughness and Scattering*. Washington, DC: Optical Society of America.

[5] Smythe, R., Moore, R. (**1984**), *Opt. Eng.* **23**, 361–364.

[6] Dukes, J. N., Gordon, G. B. (**1970**), *Hewlett-Packard J.* **21**, 2–8.

[7] Crane, R. (**1969**), *Appl. Opt.* **8**, 538–542.

[8] Jacobs, S. F., Shough, D. (**1981**), *Appl. Opt.* **20**, 3461–3463

[9] Matsumoto, H. (**1986**), *Appl. Opt.* **25**, 493–498.

[10] Kubota, T., Nara, M., Yoshino, T. (**1987**), *Opt. Lett.* **12**, 310–312.

[11] Ashby, D. E. T. F., Jephcott, D. F. (**1963**), *Appl. Phys. Lett.* **3**, 13–16.

[12] Yoshino, T., Nara, M., Mnatzakanian, S., Lee, B. S., Strand, T. C. (**1987**), *Appl. Opt.* **26**, 892–897.

[13] Bünnagel, R. (**1956**), *Z. Angew. Phys.* **8**, 342–350.

[14] Schultz, G., Schwider, J. (1976), in E. Wolf (Ed.), *Progress in Optics*, Vol. 13. Amsterdam: North Holland, pp. 93–167.

[15] Fritz, B. S. (**1984**), *Opt. Eng.* **23**, 379–383.

[16] Hariharan, P. (**1998**), *Proc. SPIE* **3479**, 2–13.

[17] Schwider, J. (1990), in E. Wolf (Ed.), *Progress in Optics*, Vol. 29. Amsterdam: North Holland, pp. 271–359.

[18] Mantravadi, M. V. (1992a), in D. Malacara (Ed.), *Optical Shop Testing*. New York: John Wiley, pp. 1–49.

[19] Malacara, D. (1992a), in D. Malacara (Ed.), *Optical Shop Testing*. New York: John Wiley, pp. 51–94.

[20] Offner, A., Malacara, D. (1992), in D. Malacara (Ed.), *Optical Shop Testing*. New York: John Wiley, pp. 427–454.

[21] Loomis, J. S. (**1980**), *Opt. Eng.* **19**(5), 679–685.

[22] Creath, K., Wyant, J. C. (1992), in D. Malacara (Ed.), *Optical Shop Testing*, (2nd ed.), New York: John Wiley & Sons, pp. 599–652.

[23] Mantravadi, M. V. (1992b), in D. Malacara (Ed.), *Optical Shop Testing*. New York: John Wiley, pp. 123–172.

[24] Malacara, D. (1992b), in D. Malacara (Ed.), *Optical Shop Testing*. New York: John Wiley, pp. 173–206.

[25] Kwon, O., Wyant, J. C., Hayslett, C. R. (**1980**), *Appl. Opt.* **19**, 1862–1869.

[26] Birch, K. G. (**1973**), *J. Phys. E: Sci. Instrum.* **6**, 1045–1048.

[27] Smartt, R. N., Steel, W. H. (**1975**), *Jpn. J. Appl. Phys.* **14**(Suppl. 14-1), 351–356.

[28] Wyant, J. C., Creath, K. (1992), in R. R. Shannon, J. C. Wyant (Eds.), *Applied Optics and Optical Engineering*, Vol. XI. San Diego: Academic Press, pp. 1–54.

[29] Robinson, D. W., Reid, G. T. (Eds.) (1993), *Interferogram Analysis: Digital Processing Techniques for Fringe Pattern Measurement*. London: IOP Publishing.

[30] Yatagai, T. (1993), in D. W. Robinson, G. T. Reid (Eds.), *Interferogram Analysis: Digital Fringe Pattern Measurement Techniques*. Bristol: IOP Publishing, pp. 72–93.

[31] Takeda, M., Ina, H., Kobayashi, S. (**1982**), *J. Opt. Soc. Am.* **72**, 156–160.

[32] Takeda, M. (**1990**), *Ind. Metrol.* **1**, 79–99.

[33] Kujawinska, M. (1993), in D. W. Robinson, G. T. Reid (Eds.), *Interferogram Analysis: Digital Fringe Pattern Measurement Techniques*. Bristol: IOP Publishing, pp. 141–193.

[34] Creath, K. (1988), in E. Wolf (Ed.), *Progress in Optics*, Vol. XXVI. Amsterdam: Elsevier, pp. 349–393.

[35] Creath, K. (1993), D. W. Robinson, G. T. Reid (Eds.), *Interferogram Analysis: Digital Fringe Pattern Measurement Techniques*. Bristol: IOP Publishing, pp. 94–140.

[36] Ghiglia, D. C., Pritt, M. D. (1998), *Two-Dimensional Phase Unwrapping: Theory, Algorithms, and Software*. New York: John Wiley.

[37] Greivenkamp, J. E., Bruning J. H. (1992), in D. Malacara (Ed.), *Optical Shop Testing*. New York: John Wiley, pp. 501–598.

[38] Malacara, D., DeVore, S. L. (1992), in D. Malacara (Ed.), *Optical Shop Testing*. New York: John Wiley, pp. 455–499.

[39] Malacara, D., Servin, M., Malacara, Z. (1998), *Interferogram Analysis for Optical Testing*. New York: Marcel Dekker.

[40] Linnik, V. P. (**1933**), *C. R. Acad. Sci. URSS* **1**, 18.

[41] Wyant, J. C., Creath, K. (**Nov 1985**), *Laser Focus/Electro Opt.* 118–132.

[42] Françon, M., Mallick, S. (1971), *Polarization Interferometers: Applications in Microscopy and Macroscopy*. London: Wiley-Interscience.

[43] Cogswell, C. J., Smith, N. I., Larkin, K. G., Hariharan, P. (**1997**), *Proc. SPIE* **2984**, 72–81.

[44] Lee, B. S., Strand, T. C. (**1990**), *Appl. Opt.* **29**, 3784–3788.

[45] Hariharan, P., Roy, M. (**1994**), *J. Mod. Opt.* **41**, 2197–2201.

[46] Larkin, K. G. (**1996**), *J. Opt. Soc. Am. A.* **17**, 832–843.

[47] Calatroni, J., Guerrero, A. L., Sainz, C., Escalona, R. (**1996**), *Opt. Laser Technol.* **28**, 485–489.

[48] Sandoz, P., Tribillon, G., Perrin, H. (**1996**), *J. Mod. Opt.* **43**, 701–708.

[49] Schmit, J. M. (**1999**), *IEEE J. Selected Top. Quantum Electron.* **5**(2), 1205–1215.

[50] Häusler, G., Lindner, M. W. (**1998**), *J. Biomed. Opt.* **3**(1), 21–31.

[51] Durst, F., Melling, A., Whitelaw, J. H. (1976), *Principles and Practice of Laser-Doppler Anemometry*. London: Academic Press.

[52] de Groot, P. J., Gallatin, G. M. (**1989**), *Opt. Lett.* **14**, 165–167.

[53] Puschert, W. (**1974**), *Opt. Commun.* **10**, 357–361.

[54] Culshaw, B. (1984), *Optical Fibre Sensing and Signal Processing*. London: Peregrinus.

[55] Giallorenzi, T. G., Bucaro, J. A., Dandridge, A., Sigel, Jr G. H., Cole, J. H., Rashleigh, S. C., Priest, R. G. (**1982**), *IEEE J. Quantum Electron.* **QE-18**, 626–665

[56] Leilabady, P. A., Jones, J. D. C., Corke, M., Jackson, D. A. (**1986**), *J. Phys. E: Sci. Instrum.* **19**, 143–146.

[57] Farahi, F., Gerges, A. S., Jones, J. D. C., Jackson, D. A. (**1988a**), *Electron. Lett.* **24**, 54–55.

[58] Farahi, F., Jones, J. D. C., Jackson, D. A. (**1988b**), *Electron. Lett.* **24**, 409–410.

[59] Vali, V., Shorthill, R. W. (**1976**), *Appl. Opt.* **15**, 1099–1100.

[60] Ezekiel, S., Arditty, H. J. (Eds.) (1982), *Fiber-Optic Rotation Sensors and Related Technologies*. New York: Springer-Verlag.

[61] Bergh, R. A., Lefevre, H. C., Shaw, H. J. (**1981**), *Opt. Lett.* **6**, 502–504.

[62] Vaughan, J. M. (1989), *The Fabry-Perot Interferometer*. Bristol: Adam Hilger.

[63] Sandercock, J. R.(**1970**), *Opt. Commun.* **2**, 73–76.

[64] Fork, R. L., Herriott, D. R., Kogelnik, H. (**1964**), *Appl. Opt.* **3**, 1471–1484.

[65] Kowalski, F. V., Hawkins, R. T., Schawlow, A. L. (**1976**), *J. Opt. Soc. Am.* **66**, 965–966.

[66] Snyder, J. J. (1977), in J. L. Hall, J. L. Carsten (Eds.), *Laser Spectroscopy III*. Berlin: Springer-Verlag, pp. 419–420.

[67] Abitbol, C., Gallion, P., Nakajima, H., Chabran, C. (**1984**), *J. Opt. (Paris)* **15**, 411–418.

[68] Vanasse, G. A., Sakai, H. (1967), in E. Wolf (Ed.), *Progress in Optics*, Vol. 6. Amsterdam: North Holland, pp. 261–330.

[69] Bell, R. J. (1972), *Introductory Fourier Transform Spectroscopy*. New York: Academic Press.

[70] Chamberlain, J. (1979), *The Principles of Interferometric Spectroscopy*. Chichester: Wiley.

[71] Kauppinen, J., Partanen, J. (2001), *Fourier Transforms in Spectroscopy,*. New York: John Wiley.

[72] Hopf, F. A., Tomita, A., Al-Jumaily, G. (**1980**), *Opt. Lett.* **5**, 386–388.

[73] Hopf, F. A., Cervantes, M. (**1982**), *Appl. Opt.* **21**, 668–677.

[74] Liepmann, T. W., Hopf, F. A. (**1985**), *Appl. Opt.* **24**, 1485–1488.

[75] Feinberg, J. Hellwarth, R. W. (**1981**), *J. Opt. Soc. Am.* **71**(12), 1563.

[76] Gauthier, D. J., Boyd, R. W., Jungquist, R. K., Lisson, J. B., Voci, L. L. (**1989**), *Opt. Lett.* **14**, 323–325.
[77] Anderson, D. Z., Lininger, D. M., Feinberg, J. (**1987**), *Opt. Lett.* **12**, 123–125.
[78] Saltiel, S. M., Van Wonterghem, B., Rentzepis, P. M. (**1989**), *Opt. Lett.* **14**, 183–185.
[79] Michelson, A. A., Pease, F. G. (**1921**), *Astrophys. J.* **52**, 249–259.
[80] Davis, J., Tango, W. J., Booth, A. J., Minard, R. A., Owens, S. M., Shobbrook, R. R. (**1994**), *Proc. SPIE* **2200**, 231–241.
[81] Hanbury Brown, R., Twiss, R. Q. (**1954**), *Philos. Mag.* **19**, 10–11.
[82] Hanbury Brown, R. (1974), *The Intensity Interferometer*. Cambridge: Cambridge University Press.
[83] Johnson, M. A., Betz, A. L., Townes, C. H. (**1974**), *Phys. Rev. Lett.* **33**, 1617–1620.
[84] Bester, M., Danchi, W. C., Townes, C. H. (**1990**), *Proc. SPIE* **1237**, 40–48.
[85] Dainty, J. C. (Ed.) (1984), *Laser Speckle and Related Phenomena*, Topics in Applied Physics, Vol. 9, (2nd ed.), Berlin: Springer.
[86] Labeyrie, A. (1976), in E. Wolf (Ed.), *Progress in Optics*, Vol. 14, Amsterdam: North Holland, pp. 49–87.
[87] Baldwin, J. E., Haniff, C. A., Mackay, C. D., Warner, P. J. (**1986**), *Nature* **320**, 595–597.
[88] Baldwin, J. E., Beckett, M. G., Boysen, R. C., Burns, D., Buscher, D. F., Cox, G. C., Haniff, C. A., Mackay, C. D., Nightingale, N. S., Rogers, J., Scheuer, P. A. G., Scott, T. R., Tuthill, P. G., Warner, P. J., Wilson, D. M. A., Wilson, R. W. (**1996**), *Astron. Astrophys.* **306**, L13–L16.
[89] Brillet, A., Hall, J. L. (**1979**), *Phys. Rev. Lett.* **42**, 549–552.
[90] Saulson, P. R. (1994), *Fundamentals of Interferometric Gravitational Wave Detectors*. Singapore: World Scientific.
[91] Weiss, R. (**1999**), *Rev. Mod. Phys.* **71**, S187–S196.
[92] Drever, R. W. P. (1983), in N. Deruelle, T. Piran (Eds.), *Gravitational Radiation*. Amsterdam: North Holland, pp. 321–338.
[93] Strain, K. A., Meers, B. J. (**1991**), *Phys. Rev. Lett.* **66**, 1391–1394.
[94] Meers, B. J. (**1988**), *Phys. Rev. D* **38**, 2317–2326.
[95] Mizuno, J., Strain, K. A., Nelson, P. G., Chen, J. M., Schilling, R., Rüdiger, A., Winkler, W., Danzmann, K. (**1993**), *Phys. Lett. A* **175**, 273–276.
[96] Abramovici, A., Althouse, W. E., Drever, R. W. P., Gürsel, Y., Kawamura, S., Raab, F. J., Shoemaker, D., Sievers, L., Spero, R. E., Thorne, K. S., Vogt, R. E., Weiss, R., Whitcomb, S. E., Zucker, M. E. (**1992**), *Science* **256**, 325–333.
[97] Barish, B. C., Weiss, R. (**Oct 1999**), *Phys. Today* 44–50.
[98] Caves, C. M. (**1980**), *Phys. Rev. Lett.* **45**, 75–79.
[99] Vest, C. M. (1979), *Holographic Interferometry*. New York: John Wiley.
[100] Rastogi, P. K. (Ed.) (1994). *Holographic Interferometry*. Berlin: Springer-Verlag.
[101] Kreis, T. (1996), *Holographic Interferometry: Principles and Methods*. Berlin: Akademie-Verlag.
[102] Ennos, A. E. (**1968**), *J. Phys. E: Sci. Instrum.* **1**, 731–734.
[103] Hariharan, P., Oreb, B. F., Brown, N. (**1983**), *Appl. Opt.* **22**, 876–880.
[104] Powell, R. L., Stetson, K. A. (**1965**), *J. Opt. Soc. Am.* **55**, 1593–1598.
[105] Hariharan, P., Oreb, B. F. (**1986**), *Opt. Commun.* **59**, 83–86.
[106] Haines, K. A., Hildebrand, B. P. (**1965**), *Phys. Lett.* **19**, 10–11.
[107] Tsuruta, T., Shiotake, N., Tsujiuchi, J., Matsuda, K. (**1967**), *Jpn. J. Appl. Phys.* **6**, 661–662.
[108] Hariharan, P., Oreb, B. F. (**1984**), *Opt. Commun.* **51**, 142–144.
[109] Patorski, K. (1993), *Handbook of the Moiré Fringe Technique*. Amsterdam: Elsevier.
[110] Post, D., Han, D., Ifju, P. (1994), *High Sensitivity Moiré*. New York: Springer.
[111] Jones, R., Wykes, C. (1989), *Holographic and Speckle Interferometry*. Cambridge: Cambridge University Press.
[112] Leendertz, J. A. (**1970**), *J. Phys. E: Sci. Instrum.* **3**, 214–218.
[113] Lokberg, O. J., Slettemoen, G. A. (1987), in R. R. Shannon, J. C. Wyant (Eds.), *Applied Optics and Optical Engineering*, Vol. X. San Diego: Academic Press, pp. 455–504.
[114] Butters, J. N., Leendertz, J. A. (**1971**), *Opt. Laser Technol.* **3**, 26–30.
[115] Macovski, A., Ramsey, S. D., Schaefer, I. F. (**1971**), *Appl. Opt.* **10**, 2722–2727.
[116] Creath, K. (**1985**), *Appl. Opt.* **24**, 3053–3058.

[117] Robinson, D. W., Williams, D. C. (**1986**), *Opt. Commun.* **57**, 26–30.

[118] Stetson, K. A., Brohinsky, W. R. (**1988**), *J. Opt. Soc. Am.* **5**, 1472–1476.

## Further Reading

Adrian, R. J. (Ed.) (1993), *Laser Doppler Velocimetry*, Vol. MS78. Bellingham: SPIE.

Brown, G. M. (Ed.) (2000), *Modern Interferometry*, Selected SPIE Papers on CD-ROM, Vol. 15. Bellingham: SPIE.

Hariharan, P. (1985), *Optical Interferometry*. Sydney: Academic Press.

Hariharan, P. (Ed.) (1991), *Interferometry*, Vol. MS28. Bellingham: SPIE.

Hariharan, P. (1992), *Basics of Interferometry*. San Diego: Academic Press.

Hariharan, P. (1995), *Optical Holography*. Cambridge: Cambridge University Press.

Hariharan, P., Malacara-Hernandez, D. (Eds.) (1995), *Interference, Interferometry and Interferometric Metrology*, Vol. MS110. Bellingham: SPIE.

Lawson, P. R. (Ed.) (1997), *Long Baseline Stellar Interferometry*, Vol. MS139. Bellingham: SPIE.

Meinischmidt, P., Hinsch, K. D., von Ossietzky, C., Sirohi, R. S. (Eds.) (1996), *Electronic Speckle Pattern Interferometry: Principles and Practice*, Vol. MS132. Bellingham: SPIE.

Sirohi, R. S., Hinsch, K. D., von Ossietzky, C. (Eds.) (1998), *Holographic Interferometry: Principles and Techniques*, Vol. MS144. Bellingham: SPIE.

Steel, W. H. (1983), *Interferometry*. Cambridge: Cambridge University Press.

Udd, E. (Ed.) (1991), *Fiber Optic Sensors: An Introduction for Engineers and Scientists*. New York: John Wiley & Sons.

Udd, E., Tatam, R. P. (Eds.) (1994), *Interferometry '94: Interferometric Fiber Sensing: 16–20 May 1994, Warsaw, Poland*, SPIE Proceedings No. 2341, Bellingham: SPIE.

# Laser Cooling and Trapping of Neutral Atoms

*Harold J. Metcalf*
*Department of Physics, State University of New York, Stony Brook, N.Y. 11794-3800, USA*
*Phone: (631) 632-8185; Fax: (631) 632-8176; e-mail: hmetcalf@notes.cc.sunysb.edu*

*Peter van der Straten*
*Debye Institute, Department of Atomic and Interface Physics, Utrecht University, 3508 TA Utrecht, The Netherlands*
*Phone: 31-30-2532846; Fax: 31-30-2537468; e-mail: P.vanderStraten@phys.uu.nl*

**Abstract**
This article presents a review of some of the principal techniques of laser cooling and trapping that have been developed during the past 20 years. Its approach is primarily experimental, but its quantitative descriptions are consistent in notation with most of the theoretical literature.

**Keywords**
laser cooling; atom trapping; optical lattice; Bose–Einstein condensation.

# 1 Introduction

The combination of laser cooling and atom trapping has produced astounding new tools for atomic physicists [1]. These experiments require the exchange of momentum between atoms and an optical field, usually at a nearly resonant frequency. The energy of light $\hbar\omega$ changes the internal energy of the atom, and the angular momentum $\hbar$ changes the orbital angular momentum $\ell$ of the atom, as described by the well-known selection rule $\Delta\ell = \pm 1$. By contrast, the linear momentum of light $p = \hbar\omega/c = \hbar k$ cannot change the internal atomic degrees of freedom, and therefore must change the momentum of the atoms in the laboratory frame. The force resulting from this momentum exchange between the light field and the atoms can be used in many ways to control atomic motion, and is the subject of this article.

## 1.1 Temperature and Entropy

The idea of "temperature" in laser cooling requires some careful discussion and disclaimers. In thermodynamics, temperature is carefully defined as a parameter of the state of a closed system in thermal equilibrium with its surroundings. This, of course, requires that there be thermal contact, that is, heat exchange, with the environment. In laser cooling, this is clearly not the case because a sample of atoms is always absorbing and scattering light. Furthermore, there is essentially no heat exchange (the light cannot be considered as heat even though it is indeed a form of energy). Thus, the system may very well be in a steady state situation, but certainly not in thermal equilibrium, so the assignment of a thermodynamic "temperature" is completely inappropriate.

Nevertheless, it is convenient to use the label of temperature to describe an atomic sample whose average kinetic energy $\langle E_k \rangle$ has been reduced by the laser light, and this is written simply as $k_B T/2 = \langle E_k \rangle$, where $k_B$ is the Boltzmann's constant (for the case of one dimension, 1-D). It must be remembered that this temperature assignment is absolutely inadequate for atomic samples that do not have a Maxwell–Boltzmann velocity distribution, whether or not they are in thermal contact with the environment; there are infinitely many velocity distributions that have the same value of $\langle E_k \rangle$ but are so different from one another that characterizing them by the same "temperature" is a severe error. (In the special case where there is a true damping force, $F \propto -v$, and where the diffusion in momentum space is a constant independent of momentum, solutions of the Fokker–Planck equation can be found analytically and can lead to a Maxwell–Boltzmann distribution that does indeed have a temperature.)

Since laser cooling decreases the temperature of a sample of atoms, there is less disorder and therefore less entropy. This seems to conflict with the second law of thermodynamics, which requires the entropy of a closed system to always increase with time. The explanation lies in the consideration of the fact that in laser cooling, the atoms do not form a closed system. Instead, there is always a flow of laser light with low entropy into the system and fluorescence with high entropy out of it. The decrease of entropy of the atoms is accompanied by a much larger increase in entropy of the light field. Entropy considerations for a laser beam are far from trivial, but recently it has been shown that the entropy lost by the atoms is many orders of magnitude smaller than the entropy gained by the light field.

### 1.2
### Phase Space Density

The phase space density $\rho(\vec{r}, \vec{p}, t)$ is defined as the probability that a single particle is in a region $d\vec{r}$ around $\vec{r}$ and has momentum $d\vec{p}$ around $\vec{p}$ at time $t$. In classical mechanics, $\rho(\vec{r}, \vec{p}, t)$ is just the sum of the $\rho(\vec{r}, \vec{p}, t)$ values of each of the $N$ particles in the system divided by $N$. Since the phase space density is a probability, it is always positive and can be normalized over the six-dimensional volume spanned by position $\vec{r}$ and momentum $\vec{p}$. For a gas of cold atoms, it is convenient to choose the elementary volume for $\rho(\vec{r}, \vec{p}, t)$ to be $\hbar^3$, so it becomes the dimensionless quantity

$$\rho_\phi = n\lambda_{\mathrm{deB}}^3, \tag{1}$$

where $\lambda_{\mathrm{deB}}$ is the deBroglie wavelength of the atoms in the sample and $n$ is their spatial density.

The Liouville theorem requires that $\rho_\phi$ cannot be increased by using conservative forces. For instance, in light optics one can focus a parallel beam of light with a lens to a small spot. However, that simply produces a high density of light rays in the focus in exchange for the momentum part of $\rho_\phi$ because the beam entering the lens is parallel but the light rays are divergent at the focus.

For classical particles, the same principle applies. By increasing the strength of the trapping potential of particles in a trap, one can increase the density of the atoms in the trap but at the same time, the compression of the sample results in a temperature increase, leaving the phase space density unchanged.

In order to increase the phase space density of an atomic sample, it is necessary to use a force that is not conservative, such as a velocity-dependent force. In laser cooling, the force on the atoms can be a damping force, that is, always directed opposite to the atomic velocity, so that the momentum part of $\rho_\phi$ increases. This process arises from the irreversible nature of spontaneous emission.

## 2
## Optical Forces on Neutral Atoms

The usual form of electromagnetic forces is given by $\vec{F} = q(\vec{\mathcal{E}} + \vec{v} \times \vec{B})$, but for neutral atoms, $q = 0$. The next order of force is the dipole term, but this also vanishes because neutral atoms have no inherent dipole moment. However, a dipole moment can be induced by a field, and this is most efficient if the field is alternating near the atomic resonance frequency. Since these frequencies are typically in the optical range, dipole moments are efficiently induced by shining nearly resonant light on the atoms.

If the light is absorbed, the atom makes a transition to the excited state, and the subsequent return to the ground state can be either by spontaneous or by stimulated emission. The nature of the optical force that arises from these two different processes is quite different and will be described separately.

The spontaneous emission case is different from the familiar quantum-mechanical calculations using state vectors to describe the system. Spontaneous emission causes the state of the system to evolve from a pure state into a mixed state and so the density matrix is needed to describe it. Spontaneous emission is an essential ingredient for the dissipative nature of the optical forces.

## 2.1 Radiative Optical Forces

In the simplest case–the absorption of well-directed light from a laser beam–the momentum exchange between the light field and the atoms results in a force

$$\vec{F} = \frac{d\vec{p}}{dt} = \hbar\vec{k}\gamma_p, \qquad (2)$$

where $\gamma_p$ is the excitation rate of the atoms. The absorption leaves the atoms in their excited state, and if the light intensity is low enough that they are much more likely to return to the ground state by spontaneous emission than by stimulated emission, the resulting fluorescent light carries off momentum $\hbar k$ in a random direction. The momentum exchange from the fluorescence averages zero, so the net total force is given by Eq. (2).

The excitation rate $\gamma_p$ depends on the laser detuning from atomic resonance $\delta \equiv \omega_l - \omega_a$, where $\omega_l$ is the laser frequency and $\omega_a$ is the atomic resonance frequency. This detuning is measured in the atomic rest frame, and it is necessary that the Doppler-shifted laser frequency in the rest frame of the moving atoms be used to calculate $\gamma_p$. In Sect. 2.3.3, we find that $\gamma_p$ for a two-level atom is given by the Lorentzian

$$\gamma_p = \frac{s_0\gamma/2}{1 + s_0 + [2(\delta + \omega_D)/\gamma]^2}, \qquad (3)$$

where $\gamma \equiv 1/\tau$ is an angular frequency corresponding to the natural decay rate of the excited state. Here, $s_0 = I/I_s$ is the ratio of the light intensity $I$ to the saturation intensity $I_s \equiv \pi hc/3\lambda^3\tau$, which is a few mW cm$^{-2}$ for typical atomic transitions. The Doppler shift seen by the moving atoms is $\omega_D = -\vec{k}\cdot\vec{v}$ (note that $\vec{k}$ opposite to $\vec{v}$ produces a positive Doppler shift for the atoms). The force is thus velocity-dependent and the experimenters' task is to exploit this dependence to the desired goal, for example, optical friction for laser cooling.

The maximum attainable deceleration is obtained for high intensities of light. High-intensity light can produce faster absorption, but it also causes equally fast stimulated emission; the combination produces neither deceleration nor cooling. The momentum transfer to the atoms by stimulated emission is in the opposite direction to what it was in absorption, resulting in a net transfer of zero momentum. At high intensity, Eq. (3) shows saturation of $\gamma_p$ at $\gamma/2$, and since the force is given by Eq. (2), the deceleration saturates at a value $\vec{a}_{max} = \hbar\vec{k}\gamma/2M$.

## 2.2 Dipole Optical Forces

While detuning $|\delta| \gg \gamma$, spontaneous emission may be much less frequent than stimulated emission, unlike the case of the dissipative radiative force that is necessary for laser cooling, given by Eqs. (2) and (3). In this case, absorption is most often followed by stimulated emission, and seems to produce zero momentum transfer because the stimulated light has the same momentum as the exciting light. However, if the optical field has beams with at least two different $\vec{k}$-vectors present, such as in counterpropagating beams, absorption from one beam followed by stimulated emission into the other indeed produces a nonzero momentum exchange. The result is called the dipole force, and is reversible and hence conservative, so it cannot be used for laser cooling.

The dipole force is more easily calculated from an energy picture than from a momentum picture. The force then derives from the gradient of the potential of an atom in an inhomogeneous light field, which is appropriate because the force is conservative. The potential arises from the shift of the atomic energy levels in the light field, appropriately called the "light shift", and is found by direct solution of the Schrödinger equation for a two-level atom in a monochromatic plane wave. After making both the dipole and rotating wave approximations, the Hamiltonian can be written as

$$\mathcal{H} = \frac{\hbar}{2}\begin{bmatrix} -2\delta & \Omega \\ \Omega^* & 0 \end{bmatrix} \tag{4}$$

where the Rabi frequency is $|\Omega| = \gamma\sqrt{s_0/2}$ for a single traveling laser beam. Solution of Eq. (4) for its eigenvalues provides the dressed state energies that are light-shifted by

$$\omega_{\mathrm{ls}} = \frac{\left[\sqrt{|\Omega|^2+\delta^2}-\delta\right]}{2}. \tag{5}$$

For sufficiently large detuning $|\delta| \gg |\Omega|$, approximation of Eq. (5) leads to $\omega_{\mathrm{ls}} \approx |\Omega|^2/4\delta = \gamma^2 s_0/8\delta$.

In a standing wave in 1-D with $|\delta| \gg |\Omega|$, the light shift $\omega_{\mathrm{ls}}$ varies sinusoidally from node to antinode. When $\delta$ is sufficiently large, the spontaneous emission rate may be negligible compared with that of stimulated emission, so that $\hbar\omega_{\mathrm{ls}}$ may be treated as a potential $U$. The resulting dipole force is

$$\vec{F} = -\vec{\nabla}U = -\frac{\hbar\gamma^2}{8\delta I_s}\vec{\nabla}I, \tag{6}$$

where $I$ is the total intensity distribution of the standing-wave light field of period $\lambda/2$. For such a standing wave, the optical electric field (and the Rabi frequency) at the antinodes is double that of each traveling wave that composes it, and so the total intensity $I_{\mathrm{max}}$ at the antinodes is four times that of the single traveling wave.

## 2.3 Density Matrix Description of Optical Forces

### 2.3.1 Introduction

Use of the density matrix $\rho$ for pure states provides an alternative description to the more familiar one that uses wave functions and operators but adds nothing new. Its equation of motion is $i\hbar(\mathrm{d}\rho/\mathrm{d}t) = [\mathcal{H}, \rho]$, and can be derived directly from the Schrödinger equation. Moreover, it is a straightforward exercise to show that the expectation value of any operator $\mathcal{A}$ that represents an observable is $\langle\mathcal{A}\rangle = \mathrm{tr}\,(\rho\mathcal{A})$.

Application of the Ehrenfest theorem gives the expectation value of the force as $\langle F\rangle = -\mathrm{tr}\,(\rho\nabla\mathcal{H})$. Beginning with the two-level atom Hamiltonian of Eq. (4), we find the force in 1-D to be

$$\langle F\rangle = \hbar\left(\frac{\partial\Omega}{\partial z}\rho_{eg}^* + \frac{\partial\Omega^*}{\partial z}\rho_{eg}\right). \tag{7}$$

Thus, $\langle F\rangle$ depends only on the off-diagonal elements $\rho_{eg} = \rho_{ge}^*$, terms that are called *the optical coherences.*

### 2.3.2 Open Systems and the Dissipative Force

The real value of the density matrix formalism for atom-light interactions is its ability to deal with open systems. By not including the fluorescent light that is lost from an atom-laser system undergoing cooling, a serious omission is being made in the discussion above. That is, the closed system of atom plus laser light that can be described by Schrödinger wave functions and is thus in a pure state, undergoes evolution to a "mixed" state by virtue of the spontaneous emission. This omission can be rectified by simple ad hoc additions to the equation of motion, and the result is called the optical Bloch equations (OBE). These are written explicitly as

$$\begin{aligned}
\frac{d\rho_{gg}}{dt} &= +\gamma\rho_{ee} + \frac{i}{2}(\Omega^*\tilde{\rho}_{eg} - \Omega\tilde{\rho}_{ge}) \\
\frac{d\rho_{ee}}{dt} &= -\gamma\rho_{ee} + \frac{i}{2}(\Omega\tilde{\rho}_{ge} - \Omega^*\tilde{\rho}_{eg}) \\
\frac{d\tilde{\rho}_{ge}}{dt} &= -\left(\frac{\gamma}{2} + i\delta\right)\tilde{\rho}_{ge} + \frac{i}{2}\Omega^*(\rho_{ee} - \rho_{gg}) \\
\frac{d\tilde{\rho}_{eg}}{dt} &= -\left(\frac{\gamma}{2} - i\delta\right)\tilde{\rho}_{eg} + \frac{i}{2}\Omega(\rho_{gg} - \rho_{ee}),
\end{aligned} \tag{8}$$

where $\tilde{\rho}_{eg} \equiv \rho_{eg}e^{-i\delta t}$ for the coherences.

In these equations, the terms proportional to the spontaneous decay rate $\gamma$ have been put in "by hand", that is, they have been introduced into the OBEs to account for the effects of spontaneous emission. The spontaneous emission is irreversible and accounts for the dissipation of the cooling process. For the ground state, the decay of the excited state leads to an increase of its population $\rho_{gg}$ proportional to $\gamma\rho_{ee}$, whereas for the excited state, it leads to a decrease of $\rho_{ee}$, also proportional to $\gamma\rho_{ee}$. These equations have to be solved in order to evaluate the optical force on the atoms.

### 2.3.3 Solution of the OBEs in Steady State

In most cases, the laser light is applied for a period long compared to the typical evolution times of atom-light interaction, that is, the lifetime of the excited state $\tau = 1/\gamma$. Thus, only the steady state solution of the OBEs have to be considered, and these are found by setting the time derivatives in Eq. (8) to zero. Then the probability $\rho_{ee}$ to be in the excited state is found to be

$$\rho_{ee} = \frac{\gamma_p}{\gamma} = \frac{s_0/2}{1 + s_0 + (2\delta/\gamma)^2} = \frac{s/2}{1+s}, \tag{9}$$

where $s \equiv s_0/[1 + (2\delta/\gamma)^2]$ is the off-resonance saturation parameter. The excited-state population $\rho_{ee}$ increases linearly with the saturation parameter $s$ for small values of $s$, but for $s$ of the order of unity, the probability starts to saturate to a value of 1/2. The detuning dependence of $\gamma_p$ (see Eq. 3) showing this saturation for various values of $s_0$ is depicted in Fig. 1.

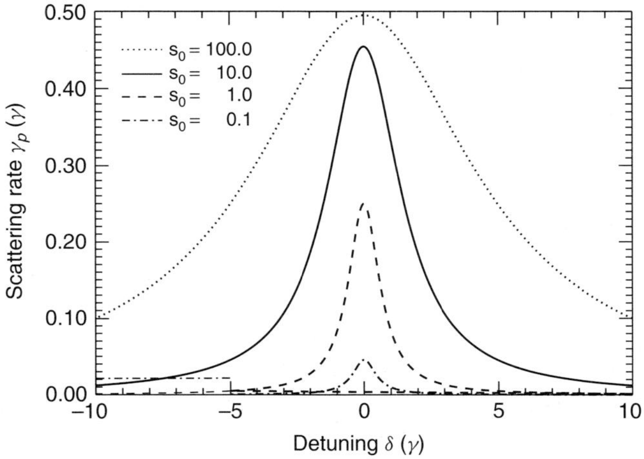


**Fig. 1** Excitation rate $\gamma_p$ as a function of the detuning $\delta$ for several values of the saturation parameter $s_0$. Note that for $s_0 > 1$, the line profiles start to broaden substantially from power broadening

#### 2.3.4 Radiative and Dipole Forces

Some insight into these forces emerges by expressing the gradient of the Rabi frequency of Eq. (7) in terms of a real and imaginary part so that $(\partial\Omega/\partial z) = (q_r + iq_i)\Omega$. Then Eq. (7) becomes

$$F = \hbar q_r(\Omega\rho_{eg}^* + \Omega^*\rho_{eg}) + i\hbar q_i(\Omega\rho_{eg}^* - \Omega^*\rho_{eg}) \qquad (10)$$

Thus, the first term of the force is related to the dispersive part of the atom-light interaction, whereas the second term is related to the absorptive part of the atom-light interaction.

To appreciate the utility of the separation of $\nabla\Omega$ into real and imaginary parts, consider the interaction of atoms with a traveling plane wave $\mathcal{E}(z) = \mathcal{E}_0(e^{i(kz-\omega t)} + \text{c.c.})/2$. In this case, $q_r = 0$ and $q_i = k$, and so the force is caused only by absorption. The force is given by $F_{sp} = \hbar k\gamma\rho_{ee}$ and is the radiative force of Eqs. (2) and (3).

For the case of counterpropagating plane waves, there is a standing wave whose electric field is $\mathcal{E}(z) = \mathcal{E}_0\cos(kz)(e^{-i\omega t} + \text{c.c.})$. Thus, $q_r = -k\tan(kz)$ and $q_i = 0$, so there is only the dispersive part of the force, given by

$$F_{dip} = \frac{2\hbar k\delta s_0 \sin 2kz}{1 + 4s_0\cos^2 kz + (2\delta/\gamma)^2}. \qquad (11)$$

This replaces Eq. (6) for the dipole force and removes the restriction $|\delta| \gg |\Omega|$, thereby including saturation effects. Even though the average of this force over a wavelength vanishes, it can be used to trap atoms in a region smaller than the wavelength of the light.

#### 2.3.5 Force on Moving Atoms

In order to show how these forces can be used to cool atoms, one has to consider the force on moving atoms. For the case of atomic velocities that are small compared with $\gamma/k$, the motion can be treated as a small perturbation in the atomic evolution that occurs on the time scale $1/\gamma$. Then the first-order result is given by

$$\frac{d\Omega}{dt} = \frac{\partial\Omega}{\partial t} + v\frac{\partial\Omega}{\partial z} = \frac{\partial\Omega}{\partial t} + v(q_r + iq_i)\Omega. \qquad (12)$$

For the case of atoms moving in a standing wave, this results in the same damping force as Eq. (13) below.

## 3 Laser Cooling

### 3.1 Slowing Atomic Beams

Among the earliest laser cooling experiments was the deceleration of atoms in a beam [2]. The authors exploited the Doppler shift to make the momentum exchange (hence the force) velocity dependent. It worked by directing a laser beam opposite an atomic beam so that the atoms could absorb light, and hence momentum $\hbar k$, very many times along their paths through the apparatus as shown in Fig. 2 [2, 3]. Of course, excited-state atoms cannot absorb light efficiently from the laser that excited them, so between absorptions they must return to the ground state by spontaneous decay, accompanied by the emission of fluorescent light. The spatial symmetry of the emitted fluorescence results in an average of zero net momentum transfer from many such fluorescence events. Thus, the net force on the atoms is in the direction of the laser beam, and the maximum deceleration is limited by the spontaneous emission rate $\gamma$.

Since the maximum deceleration $\vec{a}_{max} = \hbar\vec{k}\gamma/2M$ is fixed by atomic parameters, it is straightforward to calculate the minimum

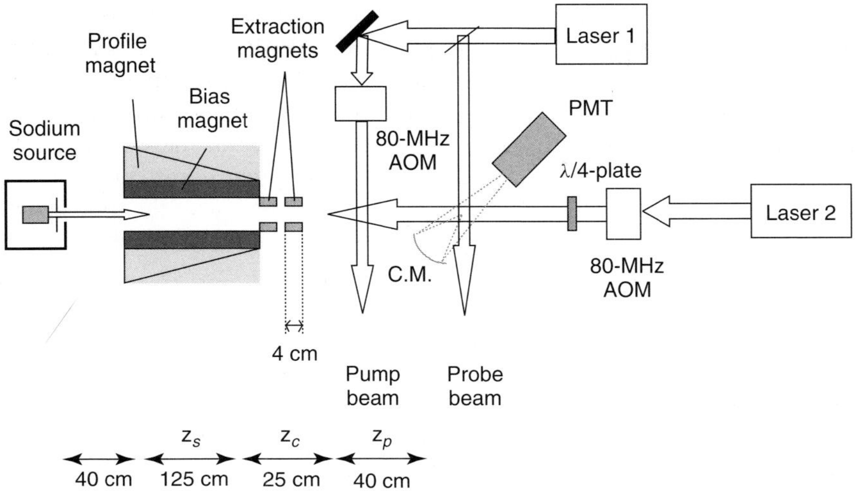


**Fig. 2** Schematic diagram of apparatus for beam slowing. The tapered magnetic field is produced by layers of varying length on the solenoid

stopping length $L_{\text{min}}$ and time $t_{\text{min}}$ for the rms velocity of atoms $\bar{v} = 2\sqrt{k_B T/M}$ at the source temperature. The result is $L_{\text{min}} = \bar{v}^2/2a_{\text{max}}$ and $t_{\text{min}} = \bar{v}/a_{\text{max}}$. In Table 1 are some of the parameters for slowing a few atomic species of interest from the peak of the thermal velocity distribution.

Maximizing the scattering rate $\gamma_p$ requires $\delta = -\omega_D$ in Eq. (3). If $\delta$ is chosen for a particular atomic velocity in the beam, then as the atoms slow down, their changing Doppler shift will take them out of resonance. They will eventually cease deceleration after their Doppler shift has been decreased by a few times the power-broadened width $\gamma' = \gamma\sqrt{1+s_0}$, corresponding to $\Delta v$ of a few times $v_c = \gamma'/k$. Although this $\Delta v$ of a few $\text{m s}^{-1}$ is considerably larger than the typical atomic recoil velocity $v_r = \hbar k/M$ of a few $\text{cm s}^{-1}$, it is still only a small fraction of the average thermal velocity $\bar{v}$ of the atoms, such that significant further cooling or deceleration cannot be accomplished.

**Tab. 1** Parameters of interest for slowing various atoms. The stopping length $L_{\text{min}}$ and time $t_{\text{min}}$ are minimum values. The oven temperature $T_{\text{oven}}$ that determines the peak velocity is chosen to give a vapor pressure of 1 Torr. Special cases are H at 1000 K for dissociation of $H_2$ into atoms, and He in the metastable triplet state, for which two rows are shown: one for a 4-K source and another for the typical discharge temperature

| *Atom* | *$T_{oven}$ [K]* | *$\bar{v}$ [m s$^{-1}$]* | *$L_{min}$ [m]* | *$t_{min}$ [ms]* |
|---|---|---|---|---|
| H | 1000 | 5000 | 0.012 | 0.005 |
| He* | 4 | 158 | 0.03 | 0.34 |
| He* | 650 | 2013 | 4.4 | 4.4 |
| Li | 1017 | 2051 | 1.15 | 1.12 |
| Na | 712 | 876 | 0.42 | 0.96 |
| K | 617 | 626 | 0.77 | 2.45 |
| Rb | 568 | 402 | 0.75 | 3.72 |
| Cs | 544 | 319 | 0.93 | 5.82 |

In order to achieve deceleration that changes the atomic speeds by hundreds of $\text{m s}^{-1}$, it is necessary to maintain $(\delta + \omega_D) \ll \gamma$ by compensating such large changes of the Doppler shift. This can be done by changing $\omega_D$ through the angular dependence of $\vec{k} \cdot \vec{v}$, or changing $\delta$ either via $\omega_l$ or $\omega_a$. The two most common methods for maintaining this resonance are sweeping the laser frequency $\omega_l$ along with the changing $\omega_D$ of the decelerating atoms [4–6], or by spatially varying the atomic resonance frequency with an inhomogeneous d.c magnetic field to keep the decelerating atoms in resonance with the fixed frequency laser [2, 3, 7].

The use of a spatially varying magnetic field to tune the atomic levels along the beam path was the first method to succeed in slowing atoms [2, 3]. It works as long as the Zeeman shifts of the ground and excited states are different so that the resonant frequency is shifted. The field can be tailored to provide the appropriate Doppler shift along the moving atom's path. A solenoid that can produce such a spatially varying field has layers of decreasing lengths. The technical problem of extracting the beam of slow atoms from the end of the solenoid can be simplified by reversing the field gradient and choosing a transition whose frequency decreases with increasing field [9].

For alkali atoms such as sodium, a time-of-flight (TOF) method can be used to measure the velocity distribution of atoms in the beam [8]. It employs two additional beams labeled pump and probe from Laser 1 as shown in Fig. 2. Because these beams cross the atomic beam at 90°, $\omega_D = -\vec{k} \cdot \vec{v} = 0$, and they excite atoms at all velocities. The pump beam is tuned to excite and empty a selected ground hyperfine state (hfs), and it transfers more than 98% of the population as the atoms pass through its 0.5 mm width. To measure the velocity distribution of atoms in the selected hfs, this pump laser beam is interrupted for a period of $\Delta t = 10 - 50\ \mu s$ with an acoustic optical modulator (AOM). A pulse of atoms in the selected hfs passes the pump region and travels to the probe beam. The time dependence of the fluorescence induced by the probe laser, tuned to excite the selected hfs, gives the time of arrival, and this signal is readily converted to a velocity distribution. Figure 3 shows the measured velocity distribution of the atoms slowed by Laser 2.

## 3.2 Optical Molasses

### 3.2.1 Doppler Cooling

A different kind of radiative force arises in low intensity, counterpropagating light beams that form a weak standing wave. It is straightforward to calculate the radiative force on atoms moving in such a standing wave using Eq. (3). In the low intensity case where stimulated emission is not important, the forces from the two light beams are simply added to give $\vec{F}_{OM} =$

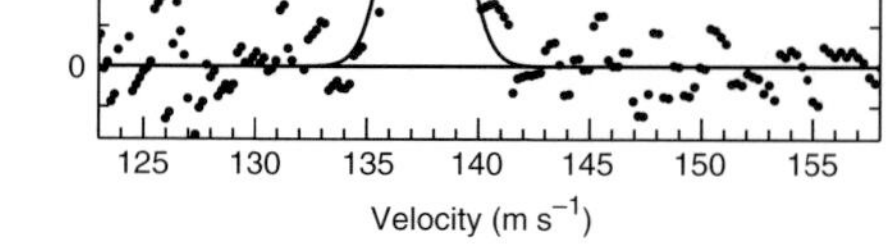


**Fig. 3** The velocity distribution measured with the TOF method. The experimental width of approximately $\frac{1}{6}(\gamma/k)$ is shown by the dashed vertical lines between the arrows. The Gaussian fit through the data yields an FWHM (full width at half maximum) of 2.97 m s$^{-1}$ (figure taken from Molenaar, P. A., vander Straten, P., Heideman, H. G. M., Metcalf, H. (**1997**), *Phys. Rev. A* **55**, 605–614)

$\vec{F}_+ + \vec{F}_-$, where $\vec{F}_\pm$ are found from Eqs. (2) and (3) Then the sum of the two forces is

$$\vec{F}_{\mathrm{OM}} \cong \frac{8\hbar k^2 \delta s_0 \vec{v}}{\gamma(1 + s_0 + (2\delta/\gamma)^2)^2} \equiv -\beta\vec{v}, \tag{13}$$

where terms of order $(kv/\gamma)^4$ and higher have been neglected. The force is proportional to velocity for small enough velocities, resulting in viscous damping for $\delta < 0$ [10, 11] that gives this technique the name "optical molasses" (OM).

These forces are plotted in Fig. 4. For $\delta < 0$, this force opposes the velocity and therefore viscously damps the atomic motion. The force $\vec{F}_{\mathrm{OM}}$ has maxima near $v \approx \pm\gamma\sqrt{s_0 + 1}/2k$ and decreases rapidly for larger velocities.

### 3.2.2 Doppler Cooling Limit

If there were no other influence on the atomic motion, all atoms would quickly decelerate to $v = 0$ and the sample would reach $T = 0$, a clearly unphysical result. In laser cooling and related aspects of optical control of atomic motion, the forces arise because of the exchange of momentum between the atoms and the laser field. These necessarily discrete steps of size $\hbar k$ constitute a heating mechanism that must be considered.

Since the atomic momentum changes by $\hbar k$, their kinetic energy changes on an average by at least the recoil energy $E_{\mathrm{r}} = \hbar^2 k^2/2M = \hbar\omega_{\mathrm{r}}$. This means that the average frequency of each absorption is at least $\omega_{\mathrm{abs}} = \omega_{\mathrm{a}} + \omega_{\mathrm{r}}$. Similarly, the energy $\hbar\omega_{\mathrm{a}}$ available from each spontaneous decay must be shared between the outgoing light and the kinetic energy of the atom recoiling with momentum $\hbar k$. Thus, the average frequency of each emission is $\omega_{\mathrm{emit}} = \omega_{\mathrm{a}} - \omega_{\mathrm{r}}$. Therefore, the light field loses an average energy of $\hbar(\omega_{\mathrm{abs}} - \omega_{\mathrm{emit}}) = 2\hbar\omega_{\mathrm{r}}$ for each scattering event. This loss occurs at a rate of $2\gamma_p$ (two beams), and the energy is converted to atomic kinetic energy because the atoms

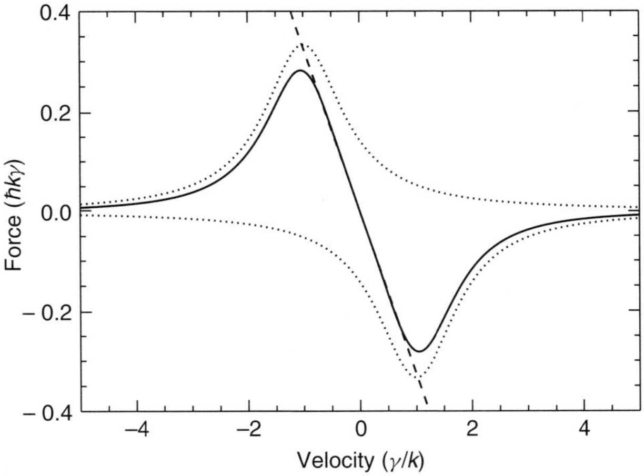


**Fig. 4** Velocity dependence of the optical damping forces for 1-D optical molasses. The two dotted traces show the force from each beam, and the solid curve is their sum. The straight line shows how this force mimics a pure damping force over a restricted velocity range. These are calculated for $s_0 = 2$ and $\delta = -\gamma$, so there is some power broadening evident

recoil from each event. The atomic sample is thereby heated because these recoils are in random directions.

The competition between this heating and the damping force of Eq. (13) results in a nonzero kinetic energy in steady state, where the rates of heating and cooling are equal. Equating the cooling rate, $\vec{F}_{\mathrm{OM}} \cdot \vec{v}$, to the heating rate, $4\hbar\omega_{\mathrm{r}}\gamma_p$, we find the steady state kinetic energy to be $(\hbar\gamma/8)(2|\delta|/\gamma + \gamma/2|\delta|)$. This result is dependent on $|\delta|$, and has a minimum at $2|\delta|/\gamma = 1$, whence $\delta = -\gamma/2$. The temperature found from the kinetic energy is then $T_{\mathrm{D}} = \hbar\gamma/2k_{\mathrm{B}}$, where $T_{\mathrm{D}}$ is called the Doppler temperature or the Doppler cooling limit. For ordinary atomic transitions, $T_{\mathrm{D}}$ is typically below 1 mK.

Another instructive way to determine $T_{\mathrm{D}}$ is to note that the average momentum transfer of many spontaneous emissions is zero, but the rms scatter of these about zero is finite. One can imagine these decays as causing a random walk in momentum space, similar to Brownian motion in real space, with step size $\hbar k$ and step frequency $2\gamma_p$, where the factor of 2 arises because of the two beams. The random walk results in an evolution of the momentum distribution as described by the Fokker–Planck equation, and can be used for a more formal treatment of laser cooling. It results in diffusion in momentum space with diffusion coefficient $D_0 \equiv 2(\Delta p)^2/\Delta t = 4\gamma_p(\hbar k)^2$. Then the steady state temperature is given by $k_{\mathrm{B}}T = D_0/\beta$. This turns out to be $\hbar\gamma/2$ as above for the case $s_0 \ll 1$ when $\delta = -\gamma/2$. This remarkable result predicts that the final temperature of atoms in OM is independent of the optical wavelength, atomic mass, and laser intensity (as long as it is not too large).

### 3.2.3 Atomic Beam Collimation – One-dimensional Optical Molasses – Beam Brightening

When an atomic beam crosses a 1-D OM as shown in Fig. 5, the transverse motion of the atoms is quickly damped while the longitudinal component is essentially unchanged. This transverse cooling of

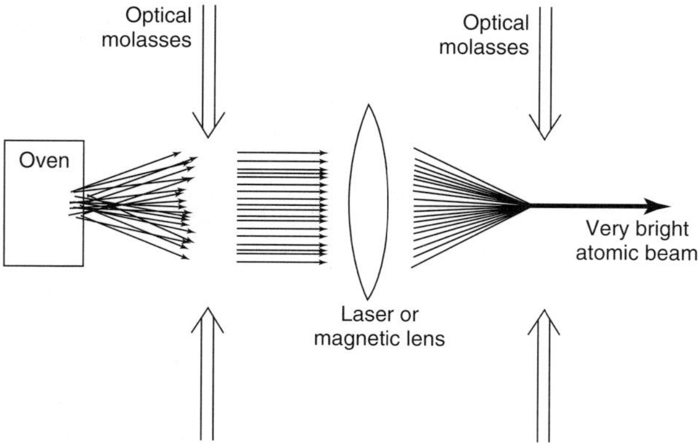


**Fig. 5** Scheme for optical brightening of an atomic beam. First, the transverse velocity components of the atoms are damped out by an optical molasses, then the atoms are focused to a spot, and finally the atoms are recollimated in a second optical molasses (figure taken from Sheehy, B., Shang, S. Q., van der Straten, P., Metcalf, H. (**1990**), *Chem. Phys.* **145**, 317–325)

an atomic beam is an example of a method that can actually increase its brightness (atoms/s-sr-cm$^2$) because such active collimation uses dissipative forces to compress the phase space volume occupied by the atoms. By contrast, the usual focusing or collimation techniques for light beams and most particle beams is restricted to selection by apertures or conservative forces that preserve the phase space density of atoms in the beam.

This velocity compression at low intensity in one dimension can be easily estimated for two-level atoms to be about $v_c/v_D = \sqrt{\gamma/\omega_r} \equiv \sqrt{1/\varepsilon}$. Here $v_D$ is the velocity associated with the Doppler limit for laser cooling discussed above: $v_D = \sqrt{\hbar\gamma/2M}$. For Rb, $v_D = 12\ \mathrm{cm\ s^{-1}}$, $v_c = \gamma/k \simeq 4.6\ \mathrm{m\ s^{-1}}$, $\omega_r \simeq 2\pi \times 3.8$ kHz, and $1/\varepsilon \simeq 1600$. (The parameter $\varepsilon$ characterizes optical forces on atoms.) Including two transverse directions along with the longitudinal slowing and cooling discussed above, the decrease in three-dimensional 3-D phase space volume for laser cooling of an Rb atomic beam from the momentum contribution alone can exceed $10^6$. Clearly optical techniques can create atomic beams enormously more times intense than ordinary thermal beams and also many orders of magnitude brighter.

### 3.2.4 Experiments in Three-dimensional Optical Molasses

By using three intersecting orthogonal pairs of oppositely directed beams, the movement of atoms in the intersection region can be severely restricted in all 3-D, and many atoms can thereby be collected and cooled in a small volume.

Even though atoms can be collected and cooled in the intersection region, it is important to stress that this is *not* a trap (see Sect. 4 below), that is, atoms that wander away from the center experience no force directing them back. They are allowed to diffuse freely and even escape, as long as there is enough time for their very slow diffusive movement to allow them to reach the edge of the region of intersection of the laser beams. Since the atomic velocities are randomized during the damping time $M/\beta = 2/\omega_r$, atoms execute a random walk in position space with a step size of $2v_D/\omega_r = \lambda/(\pi\sqrt{2\varepsilon}) \cong$ few μm. To diffuse a distance of 1 cm requires about $10^7$ steps or about 30 s [13, 14].

In 1985, the group at Bell Labs was the first to observe 3-D OM [11]. Preliminary measurements of the average kinetic energy of the atoms were done by blinking off the laser beams for a fixed interval. Comparison of the brightness of the fluorescence before and after the turnoff was used to calculate the fraction of atoms that left the region while it was in the dark. The dependence of this fraction on the duration of the dark interval was used to estimate the velocity distribution and hence the temperature. This method, which is usually referred to as *release and recapture,* is specifically designed to measure the temperature of the atoms, since the usual way of measuring temperatures cannot be applied to an atomic cloud of a few million atoms. The result was consistent with $T_D$ as calculated from the Doppler theory, as described in Sect. 3.2.2.

Later a more sensitive ballistic technique was devised at NIST that showed the astounding result that the temperature of the atoms in OM was very much lower than $T_D$ [15]. These experiments also found that OM was less sensitive to perturbations and more tolerant of alignment errors than was predicted by Doppler theory. For example, if the intensities of the two counterpropagating laser beams forming an OM were unequal, then the force on the atoms at rest would not vanish, but the force on the atoms with some nonzero drift velocity *would* vanish. This drift velocity can be easily calculated by using unequal intensities $s_{0+}$ and $s_{0-}$, to derive an analog of Eq. (13). Thus, atoms would drift out of an OM, and the calculated rate would be much faster than observed by deliberately unbalancing the beams in the experiments [16].

## 3.3
## Cooling Below the Doppler Limit

### 3.3.1 Introduction

It was an enormous surprise to observe that the ballistically measured temperature of the Na atoms was as much as 10 times *lower* than $T_D = 240\ \mu K$ [15], the temperature minimum calculated from theory. This breaching of the Doppler limit forced the development of an entirely new picture of OM that accounts for the fact that in 3-D, a two-level picture of atomic structure is inadequate. The multilevel structure of atomic states, and optical pumping among these sublevels, must be considered in the description of 3-D OM.

In response to these surprising measurements of temperatures below $T_D$, two groups developed a model of laser cooling that could explain the lower temperatures [17, 18]. The key feature of this model that distinguishes it from the earlier picture is the inclusion of the multiplicity of sublevels that make up an atomic state (e.g., Zeeman and hfs). The dynamics of optically pumping the moving atoms among these sublevels provides the new mechanism for producing ultralow temperatures [19].

The dominant feature of these models is the nonadiabatic response of moving atoms to the light field. Atoms at rest in a steady state have ground-state orientations caused by optical pumping processes that distribute the populations over the different ground-state sublevels. In the presence of polarization gradients, these orientations reflect the local light field. In the low-light-intensity regime, the orientation of stationary atoms is completely determined by the ground-state distribution; the optical coherences and the excited-state population follow the ground-state distribution adiabatically.

For atoms moving in a light field that varies in space, optical pumping acts to adjust the atomic orientation to the changing conditions of the light field. In a weak pumping process, the orientation of moving atoms always lags behind the orientation that would exist for stationary atoms. It is this phenomenon of nonadiabatic following that is the essential feature of the new cooling process.

Production of spatially dependent optical pumping processes can be achieved in several different ways. As an example, consider two counterpropagating laser beams that have orthogonal polarizations, as discussed below. The superposition of the two beams results in a light field having a polarization that varies on the wavelength scale along the direction of the laser beams. Laser cooling by such a light field is called polarization gradient cooling. In a 3-D OM, the transverse wave character of light requires that the light field always has polarization gradients.

### 3.3.2 Linear ⊥ Linear Polarization Gradient Cooling

One of the most instructive models for discussion of sub-Doppler laser cooling was introduced in Ref. [17] and very well described in Ref. [19]. If the polarizations of two counterpropagating laser beams are identical, the two beams interfere and produce a standing wave. When the two beams have orthogonal linear polarizations (same frequency $\omega_l$) with their $\hat{\mathcal{E}}$ vectors perpendicular (e.g., $\hat{x}$ and $\hat{y}$), the configuration is called lin ⊥ lin or lin-perp-lin. Then the total field is the sum of the two counterpropagating beams given by

$$\vec{\mathcal{E}} = \mathcal{E}_0\,\hat{x}\cos(\omega_l t - kz) + \mathcal{E}_0\hat{y}\cos(\omega_l t + kz)$$
$$= \mathcal{E}_0[(\hat{x} + \hat{y})\cos\omega_l t\cos kz + (\hat{x} - \hat{y})\sin\omega_l t\sin kz]. \quad (14)$$

At the origin, where $z = 0$, this becomes

$$\vec{\mathcal{E}} = \mathcal{E}_0(\hat{x} + \hat{y})\cos\omega_l t, \quad (15)$$

which corresponds to linearly polarized light at an angle $+\pi/4$ to the $x$-axis. The amplitude of this field is $\sqrt{2}\mathcal{E}_0$. Similarly, for $z = \lambda/4$, where $kz = \pi/2$, the field is also linearly polarized but at an angle $-\pi/4$ to the $x$-axis.

Between these two points, at $z = \lambda/8$, where $kz = \pi/4$, the total field is

$$\vec{\mathcal{E}} = \mathcal{E}_0\left[\hat{x}\sin\left(\omega_l t + \frac{\pi}{4}\right) - \hat{y}\cos\left(\omega_l t + \frac{\pi}{4}\right)\right]. \quad (16)$$

Since the $\hat{x}$ and $\hat{y}$ components have sine and cosine temporal dependence, they are $\pi/2$ out of phase, and so Eq. (16) represents circularly polarized light rotating about the $z$-axis in the negative sense. Similarly, at $z = 3\lambda/8$ where $kz = 3\pi/4$, the polarization is circular but in the positive sense. Thus, in this lin ⊥ lin scheme the polarization cycles from linear to circular to orthogonal linear to opposite circular in the space of only half a wavelength of light, as shown in Fig. 6. It truly has a very strong polarization gradient.

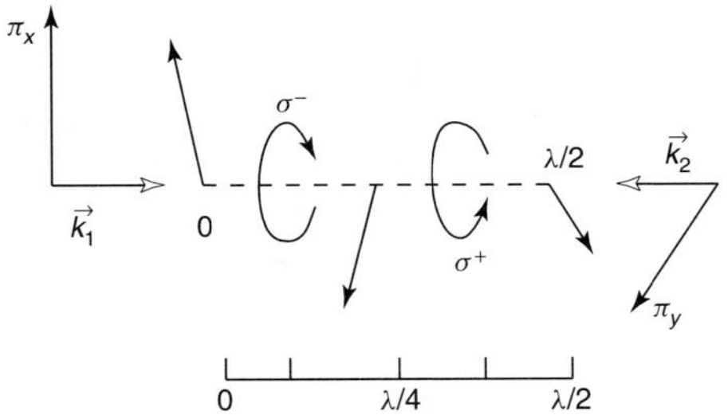


**Fig. 6** Polarization gradient field for the lin ⊥ lin configuration

Since the coupling of the different states of multilevel atoms to the light field depends on its polarization, atoms moving in a polarization gradient will be coupled differently at different positions, and this will have important consequences for laser cooling. For the $J_g = 1/2 \rightarrow J_e = 3/2$ transition (one of the simplest transitions that show sub-Doppler cooling [20]), the optical pumping process in purely $\sigma^+$ light drives the ground-state population to the $M_g = +1/2$ sublevel. This optical pumping occurs because absorption always produces $\Delta M = +1$ transitions, whereas the subsequent spontaneous emission produces $\Delta M = \pm 1$, 0. Thus, the average $\Delta M \geq 0$ for each scattering event. For $\sigma^-$-light, the population is pumped toward the $M_g = -1/2$ sublevel. Thus, atoms traveling through only a half wavelength in the light field, need to readjust their population completely from $M_g = +1/2$ to $M_g = -1/2$ and back again.

The interaction between nearly resonant light and atoms not only drives transitions between atomic energy levels but also shifts their energies. This light shift of the atomic energy levels, discussed in Sect. 2.2, plays a crucial role in this scheme of sub-Doppler cooling, and the changing polarization has a strong influence on the light shifts. In the low-intensity limit of two laser beams, each of intensity $s_0 I_s$, the light shifts $\Delta E_g$ of the ground magnetic substates are given by a slight variation of the approximation to Eq. (5) that accounts for the multilevel structure of the atoms. We write

$$\Delta E_g = \frac{\hbar s_0 C_{ge}^2 \gamma^2}{8\delta}, \qquad (17)$$

where $C_{ge}$ is the Clebsch–Gordan coefficient that describes the coupling between the particular levels of the atom and the light field.

In the present case of orthogonal linear polarizations and $J = 1/2 \rightarrow 3/2$, the light shift for the magnetic substate $M_g = 1/2$ is three times larger than that of the $M_g = -1/2$ substate when the light field is completely $\sigma^+$. On the other hand, when an atom moves to a place where the light field is $\sigma^-$, the shift of $M_g = -1/2$ is three times larger. So, in this case, the optical pumping discussed above causes a larger population to be there in the state with the larger light shift. This is generally true for any transition $J_g \rightarrow J_e = J_g + 1$. A schematic diagram showing the populations and light shifts for this particular case of negative detuning is illustrated in Fig. 7.

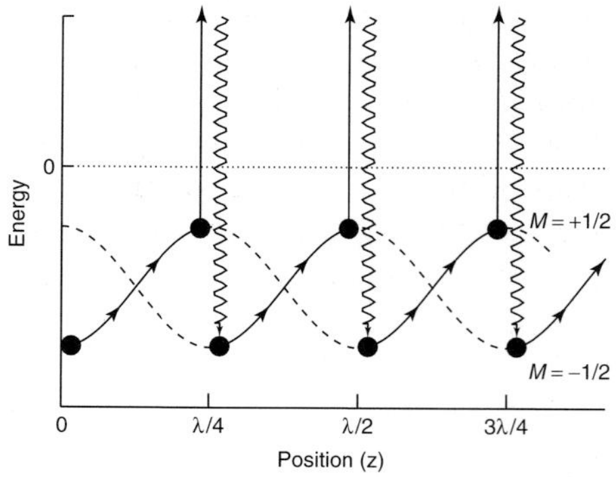


**Fig. 7** The spatial dependence of the light shifts of the ground-state sublevels of the $J = 1/2 \Leftrightarrow 3/2$ transition for the case of lin ⊥ lin polarization configuration. The arrows show the path followed by atoms being cooled in this arrangement. Atoms starting at $z = 0$ in the $M_g = +1/2$ sublevel must climb the potential hill as they approach the $z = \lambda/4$ point where the light becomes $\sigma^-$ polarized, and they are optically pumped to the $M_g = -1/2$ sublevel. Then they must begin climbing another hill toward the $z = \lambda/2$ point where the light is $\sigma^+$ polarized and they are optically pumped back to the $M_g = +1/2$ sublevel. The process repeats until the atomic kinetic energy is too small to climb the next hill. Each optical pumping event results in the absorption of light at a frequency lower than emission, thus dissipating energy to the radiation field

### 3.3.3 Origin of the Damping Force

To discuss the origin of the cooling process in this polarization gradient scheme, consider atoms with a velocity $v$ at a position where the light is $\sigma^+$-polarized, as shown at the lower left of Fig. 7. The light optically pumps such atoms to the strongly negative light-shifted $M_g = +1/2$ state. In moving through the light field, atoms must increase their potential energy (climb a hill) because the polarization of the light is changing and the state $M_g = 1/2$ becomes less strongly coupled to the light field. After traveling a distance $\lambda/4$, atoms arrive at a position where the light field is $\sigma^-$-polarized, and are optically pumped to $M_g = -1/2$, which is now lower than the $M_g = 1/2$ state. Again, the moving atoms are at the bottom of a hill and start to climb. In climbing the hills, the kinetic energy is converted to potential energy, and in the optical pumping process, the potential energy is radiated away because the spontaneous emission is at a higher frequency than the absorption (see Fig. 7). Thus, atoms seem to be always climbing hills and losing energy in the process. This process brings to mind a Greek myth, and is thus called "Sisyphus laser cooling".

The cooling process described above is effective over a limited range of atomic velocities. The force is maximum for atoms that undergo one optical pumping process while traveling over a distance $\lambda/4$. Slower atoms will not reach the hilltop before the pumping process occurs and faster atoms will travel a longer distance before being pumped toward the other sublevel, so $\Delta E/\Delta z$ is smaller. In both cases, the energy loss is smaller and therefore the cooling process less efficient. Nevertheless, the damping constant $\beta$ for this process is much larger than for Doppler cooling, and therefore the final steady state temperature is lower [17, 19].

In the experiments of Ref. [21], the temperature was measured in a 3-D molasses under various configurations of the polarization. Temperatures were measured by a ballistic technique, in which the flight time of the released atoms was measured as they fell through a probe a few centimeters below the molasses region. The lowest temperature obtained was 3 μK, which is a factor 40 below the Doppler temperature and a factor 15 above the recoil temperature of Cs.

### 3.3.4 The Limits of Sisyphus Laser Cooling

The extension of the kind of thinking about cooling limits in the case of Doppler cooling to the case of the sub-Doppler processes must be done with some care, because a naive application of similar ideas would lead to an arbitrarily low final temperature. In the derivation in Sect. 3.2.2, it is explicitly assumed that each scattering event changes the atomic momentum $p$ by an amount that is a small fraction of $p$ and this clearly fails when the velocity is reduced to the region of $v_r \equiv \hbar k/M$.

This limitation of the minimum steady state value of the average kinetic energy to a few times $E_r \equiv k_B T_r/2 = M v_r^2/2$ is intuitively comforting for two reasons. First, one might expect that the last spontaneous emission in a cooling process would leave atoms with a residual momentum of the order of $\hbar k$, since there is no control over its direction. Thus, the randomness associated with this would put a lower limit of $v_{min} \sim v_r$ on such cooling. Second, the polarization gradient cooling mechanism described above requires that atoms be localizable within the scale of $\sim \lambda/2\pi$ in order to be subject to only a single polarization in the spatially inhomogeneous light field. The uncertainty principle then requires that these atoms have a momentum spread of at least $\hbar k$.

The recoil limit discussed here has been surpassed by evaporative cooling of trapped atoms [22] and two different optical cooling methods, neither of which can be described in simple notions. One of these uses optical pumping into a velocity-selective dark state [23]. The other one uses carefully chosen, counterpropagating, laser pulses to induce velocity-selective Raman transitions, and is called *Raman cooling* [24].

## 4 Traps for Neutral Atoms

In order to confine any object, it is necessary to exchange kinetic for potential energy in the trapping field, and in neutral atom traps, the potential energy must be stored as internal atomic energy. Thus, practical traps for ground-state neutral atoms are necessarily very shallow compared with thermal energy because the energy-level shifts that result from convenient size fields are typically considerably smaller than $k_B T$ for $T = 1$ K. Neutral atom trapping therefore depends on substantial cooling of a thermal atomic sample, and is often connected with the cooling process. In most practical cases, atoms are loaded from magneto-optical traps (MOTs) in which they have been efficiently accumulated and cooled to mK temperatures (see Sect. 4.3), or from optical molasses, in which they have been optically cooled to μK temperatures (see Sect. 3.2).

The small depth of typical neutral atom traps dictates stringent vacuum requirements because an atom cannot remain trapped after a collision with a thermal energy background gas molecule. Since these atoms are vulnerable targets for thermal energy background gas, the mean free time between collisions *must* exceed the desired trapping time. The cross section for destructive collisions is quite large because even a gentle collision (i.e., large impact parameter) can impart enough energy to eject an atom from a trap. At pressure $P$ sufficiently low to be of practical interest, the trapping time is $\sim (10^{-8}/P)$ s, where $P$ is in Torr.

## 4.1 Dipole Force Optical Traps

### 4.1.1 Single-beam Optical Traps for Two-level Atoms

The simplest imaginable optical trap consists of a single, strongly focused Gaussian laser beam (see Fig. 8) [25, 26] whose intensity at the focus varies transversely with $r$ as

$$I(r) = I_0 e^{-2r^2/w_0^2}, \tag{18}$$

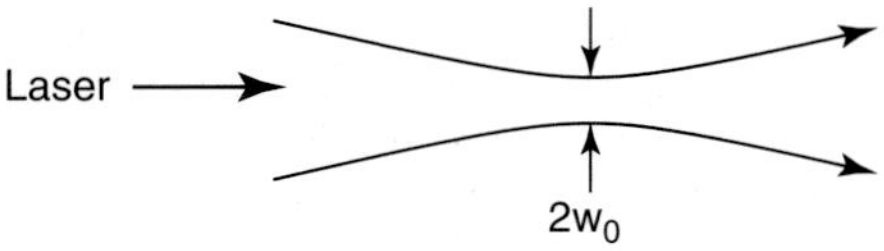


**Fig. 8** A single focused laser beam produces the simplest type of optical trap

where $w_0$ is the beam waist size. Such a trap has a well-studied and important macroscopic classical analog in a phenomenon called *optical tweezers* [27–29].

With the laser light tuned below resonance ($\delta < 0$), the ground-state light shift is negative everywhere, but largest at the center of the Gaussian beam waist. Ground-state atoms, therefore, experience a force attracting them toward this center, given by the gradient of the light shift, which is found from Eq. (5), and for $\delta/\gamma \gg s_0$ is given by Eq. (6). For the Gaussian beam, this transverse force at the waist is harmonic for small $r$ and is given by

$$F \simeq \frac{\hbar\gamma^2}{4\delta}\frac{I_0}{I_s}\frac{r}{w_0^2}e^{-2r^2/w_0^2}. \tag{19}$$

In the longitudinal direction, there is also an attractive force but it is more complicated and depends on the details of the focusing. Thus, this trap produces an attractive force on the atoms in three dimensions.

Although it may appear that the trap does not confine atoms longitudinally because of the radiation pressure along the laser beam direction, careful choice of the laser parameters can indeed produce trapping in 3-D. This can be accomplished because the radiation pressure force decreases as $1/\delta^2$ (see Eqs. 2 and 3), but by contrast, the light shift and hence the dipole force decreases only as $1/\delta$ for $\delta \gg \Omega$ (see Eq. 5). If $|\delta|$ is chosen to be sufficiently large, atoms spend very little time in the untrapped (actually repelled), excited state because its population is proportional to $1/\delta^2$. Thus, a sufficiently large value of $|\delta|$ produces longitudinal confinement and also maintains the atomic population primarily in the trapped ground state.

The first optical trap was demonstrated in Na with light detuned below the D-lines [26]. With 220 mW of dye laser light tuned about 650 GHz below the atomic transition and focused to an ~10 μm waist, the trap depth was about $15\hbar\gamma$ corresponding to 7 mK.

Single-beam dipole force traps can be made with the light detuned by a significant fraction of its frequency from the atomic transition. Such a far-off-resonance trap (FORT) has been developed for Rb atoms using light detuned by nearly 10% to the red of the $D_1$ transition at $\lambda = 795$ nm [30]. Between 0.5 and 1 W of power was focused to a spot about 10 μm in size, resulting in a trap 6 mK deep where the light-scattering rate was only a few hundreds per second. The trap lifetime was more than half a second.

There is a qualitative difference when the trapping light is detuned by a large fraction of the optical frequency. In one such case, Nd : YAG light at $\lambda = 1064$ nm was used to trap Na whose nearest transition is at $\lambda = 596$ nm [31]. In a more extreme case, a trap using $\lambda = 10.6$ μm light from a $CO_2$ laser has been used to trap Cs whose optical transition is at a frequency ~12 times higher ($\lambda = 852$ nm) [32]. For such large values of $|\delta|$, calculations of the trapping force cannot exploit the rotating wave approximations as was done for Eqs. (4) and (5), and the atomic behavior is similar to that in a DC field. It is important to remember that for an electrostatic trap Earnshaw's theorem precludes a field maximum, but that in this case there is indeed a local 3-D intensity maximum of the focused light because it is not a static field.

#### 4.1.2 Blue-detuned Optical Traps

One of the principal disadvantages of the optical traps discussed above is that the negative detuning attracts atoms to the region of highest light intensity. This may result in significant spontaneous emission unless the detuning is a large fraction of the optical frequency such as the Nd : YAG laser trap [31] or the $CO_2$ laser trap [32]. More important in some cases is that the trap relies on Stark shifting of the atomic energy levels by an amount equal to the trap depth, and this severely compromises the capabilities for precision spectroscopy in a trap [33].

Attracting atoms to the region of *lowest* intensity would ameliorate both these concerns, but such a trap requires positive detuning (blue), and an optical configuration having a dark central region. One of the first experimental efforts at a blue detuned trap used the repulsive dipole force to support Na atoms that were otherwise confined by gravity in an optical "cup" [34]. Two rather flat, parallel beams detuned by 25% of the atomic resonance frequency were directed horizontally and oriented to form a V-shaped trough. Their Gaussian beam waists formed a region ≃1 mm long where the potential was deepest, and hence provided confinement along their propagation direction as shown in Fig. 9. The beams were the $\lambda = 514$ nm and $\lambda = 488$ nm from an argon laser, and

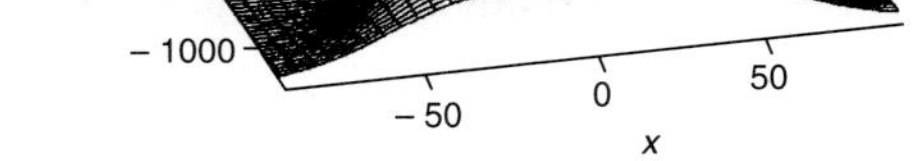


**Fig. 9** The light intensity experienced by an atom located in a plane 30 μm above the beam waists of two quasi-focused sheets of light traveling parallel and arranged to form a V-shaped trough. The *x* and *y* dimensions are in μm (figure taken from Davidson, N., Lee, H. J., Adams, C. S., Kasevich, M., Chu, S. (**1995**), *Phys. Rev. Lett.* **74**, 1311–1314)

the choice of the two frequencies was not simply to exploit the full power of the multiline Ar laser, but also to avoid the spatial interference that would result from the use of a single frequency.

Obviously, a hollow laser beam would also satisfy the requirement for a blue-detuned trap, but conventional textbook wisdom shows that such a beam is not an eigenmode of a laser resonator [35]. Some lasers can make hollow beams, but these are illusions because they consist of rapid oscillations between the $TEM_{01}$ and $TEM_{10}$ modes of the cavity. Nevertheless, Maxwell's equations permit the propagation of such beams, and in the recent past there have been studies of the LaGuerre–Gaussian modes that constitute them [36–38]. The several ways of generating such hollow beams have been tried by many experimental groups and include phase and amplitude holograms, hollow waveguides, axicons or related cylindrical prisms, stressing fibers, and simply mixing the $TEM_{01}$ and $TEM_{10}$ modes with appropriate cylindrical lenses.

An interesting experiment has been performed using the ideas of Sisyphus cooling (see Sect. 3.3) with evanescent waves combined with a hollow beam formed with an axicon [39]. In the previously reported experiments with atoms bouncing under gravity from an evanescent wave field [40, 41], they were usually lost to horizontal motion for several reasons, including slight tilting of the surface, surface roughness, horizontal motion associated with their residual motion, and horizontal ejection by the Gaussian profile of the evanescent wave laser beam. The authors of Ref. [39] simply confined their atoms in the horizontal direction by surrounding them with a wall of blue-detuned light in the form of a vertical hollow beam. Their gravito-optical surface trap cooled Cs atoms to $\simeq 3$ μK at a density of $\simeq 3 \times 10^{10}/\text{cm}^3$ in a sample whose $1/e$ height in the gravitational field was only 19 μm. Simple ballistics gives a frequency of 450 bounces per second, and the $\simeq$6-s lifetime (limited only by background gas collisions) corresponds to several thousand bounces. However, at such low energies, the deBroglie wavelength of the atoms is $\simeq 1/4$ μm, and the atomic motion is no longer accurately described classically, but requires deBroglie wave methods.

## 4.2 Magnetic Traps

### 4.2.1 Introduction

Magnetic trapping of neutral atoms is well suited for use in very many areas, including high-resolution spectroscopy, collision studies, Bose–Einstein condensation (BEC), and atom optics. Although ion trapping, laser cooling of trapped ions, and trapped ion spectroscopy were known for many years [42], it was only in 1985 that neutral atoms were first trapped [43]. Such experiments offer the capability of the spectroscopic ideal of an isolated atom at rest, in the dark, available for interaction with electromagnetic field probes.

Because trapping requires the exchange of kinetic energy for potential energy, the atomic energy levels will necessarily shift as the atoms move in the trap. These shifts can severely affect the precision of spectroscopic measurements. Since one of the potential applications of trapped atoms is in high-resolution spectroscopy, such inevitable shifts must be carefully considered.

### 4.2.2 Magnetic Confinement

The Stern–Gerlach experiment in 1924 first demonstrated the mechanical action of inhomogeneous magnetic fields on neutral atoms having magnetic moments, and the basic phenomenon was subsequently developed and refined. An atom with a magnetic moment $\vec{\mu}$ can be confined by an inhomogeneous magnetic field because of an interaction between the moment and the field. This produces a force given by

$$\vec{F} = \vec{\nabla}(\vec{\mu} \cdot \vec{B}) \qquad (20)$$

since $E = -\vec{\mu} \cdot \vec{B}$. Several different magnetic traps with varying geometries that exploit the force of Eq. (20) have been studied in some detail in the literature. The general features of the magnetic fields of a large class of possible traps has been presented [44].

W. Paul originally suggested a quadrupole trap composed of two identical coils carrying opposite currents (see Fig. 10). This trap clearly has a single center in which the field is zero, and is the simplest of all possible magnetic traps. When the coils are separated by 1.25 times their radius, such a trap has equal depth in the radial ($x$-$y$ plane) and longitudinal ($z$-axis) directions [44]. Its experimental simplicity makes it most attractive, both because of ease of construction and of optical access to the interior. Such a trap was used in the first neutral atom trapping experiments at NIST on laser-cooled Na atoms for times exceeding 1 s, and that time was limited only by background gas pressure [43].

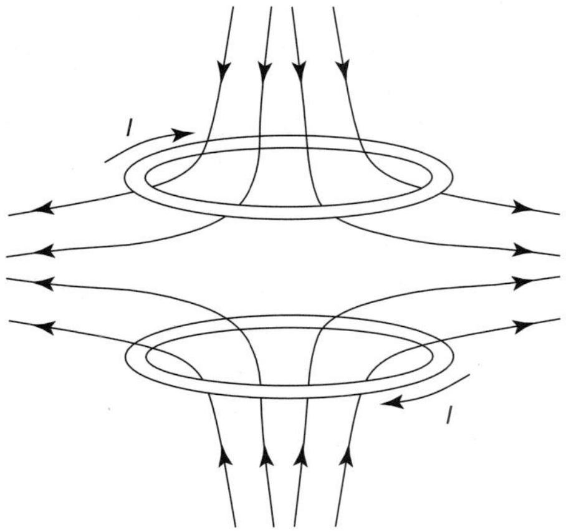


**Fig. 10** Schematic diagram of the coil configuration used in the quadrupole trap and the resultant magnetic field lines. Because the currents in the two coils are in opposite directions, there is a $|\vec{B}| = 0$ point at the center

The magnitude of the field is zero at the center of this trap, and increases in all directions as

$$B = \nabla B\sqrt{\rho^2 + 4z^2}, \qquad (21)$$

where $\rho^2 \equiv x^2 + y^2$ and the field gradient is constant (see Ref. [44]). The field gradient is constant along any line through the origin, but has different values in different polar directions because of the '4' in Eq. (21). Therefore, the force of Eq. (20) that confines the atoms in the trap is neither harmonic nor central, and orbital angular momentum is not conserved.

The requisite field for the quadrupole trap can also be provided in two dimensions by four straight currents as indicated in Fig. 11. The field is translationally invariant along the direction parallel to the currents, so a trap cannot be made this way without additional fields. These are provided by end coils that close the trap, as shown.

Although there are very many different kinds of magnetic traps for neutral particles, this particular one has played a special

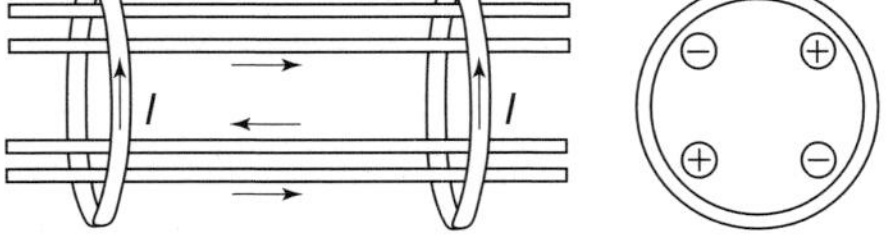


**Fig. 11** The Ioffe trap has four straight current elements that form a linear quadrupole field. The axial confinement is accomplished with end coils as shown. These fields can be achieved with many different current configurations as long as the geometry is preserved

role. There are certain conditions required for trapped atoms not to be ejected in a region of zero field such as occurs at the center of a quadrupole trap (see Sects. 4.2.3 and 4.2.4). This problem is not easily cured; the Ioffe trap has been used in many of the BEC experiments because it has $|\vec{B}| \neq 0$ everywhere.

### 4.2.3 Classical Motion of Atoms in a Quadrupole Trap

Because of the dependence of the trapping force on the angle between the field and the atomic moment (see Eq. 20), the orientation of the magnetic moment with respect to the field must be preserved as the atoms move about in the trap. Otherwise, the atoms may be ejected instead of being confined by the fields of the trap. This requires velocities low enough to ensure that the interaction between the atomic moment $\vec{\mu}$ and the field $\vec{B}$ is adiabatic especially when the atom's path passes through a region where the field magnitude is small and therefore the energy separation between the trapping and nontrapping states is small. This is especially critical at the low temperatures of the BEC experiments. Therefore energy considerations that focus only on the trap depth are not sufficient to determine the stability of a neutral atom trap; orbit and/or quantum state calculations and their consequences must also be considered.

For the two-coil quadrupole magnetic trap of Fig. 10, stable circular orbits of radius $\rho_1$ in the $z = 0$ plane can be found classically by setting $\mu \nabla B = M v^2/\rho_1$, so that $v = \sqrt{\rho_1 a}$, where $a \equiv \mu \nabla B / M$ is the centripetal acceleration supplied by the field gradient (cylindrical coordinates are appropriate). Such orbits have an angular frequency of $\omega_T = \sqrt{a/\rho_1}$. For traps of a few centimeter size and a few hundred Gauss depth, $a \sim 250$ m s$^{-2}$, and the fastest trappable atoms in circular orbits have $v_{\max} \sim 1$ m s$^{-1}$ so $\omega_T/2\pi \sim 20$ Hz. Because of the anharmonicity of the potential, the orbital frequencies depend on the orbit size, but in general, atoms in lower-energy orbits have higher frequencies.

For the quadrupole trap to work, the atomic magnetic moments must be oriented with $\vec{\mu} \cdot \vec{B} < 0$ so that they are repelled from regions of strong fields. This orientation must be preserved while the atoms move around in the trap even though the trap fields change directions in a very complicated way. The condition for adiabatic motion can be written as $\omega_Z \gg |\mathrm{d}B/\mathrm{d}t|/B$, where $\omega_Z = \mu B/\hbar$ is the Larmor precession rate in the field.

Since $v/\rho_1 = v \nabla B / B = |\mathrm{d}B/\mathrm{d}t|/B$ for a uniform field gradient, the adiabaticity condition is

$$\omega_Z \gg \omega_T. \tag{22}$$

More general orbits must satisfy a similar condition. For the two-coil quadrupole trap, the adiabaticity condition can be easily calculated. Using $v = \sqrt{\rho_1 a}$ for circular orbits in the $z = 0$ plane, the adiabatic condition for a practical trap ($\nabla B \sim 1$ T/m) requires $\rho_1 \gg (\hbar^2/M^2 a)^{1/3} \sim 1$ μm as well as $v \gg (\hbar a/M)^{1/3} \sim 1$ cm s$^{-1}$. Note that violation of these conditions (i.e., $v \sim 1$ cm s$^{-1}$ in a trap with $\nabla B \sim 1$ T/m) results in the onset of quantum dynamics for the motion (deBroglie wavelength $\approx$ orbit size).

Since the nonadiabatic region of the trap is so small (less than $10^{-18}$ m$^3$ compared with typical sizes of $\sim$2 cm, corresponding to $10^{-5}$ m$^3$), nearly all the orbits of most atoms are restricted to regions where they are adiabatic. Therefore, most of such laser-cooled atoms stay trapped for many thousands of orbits corresponding to several minutes. At laboratory vacuum chamber pressures of typically $10^{-10}$ torr, the mean free time between collisions that can eject trapped atoms is ~2 min, so the transitions caused by nonadiabatic motion are not likely to be observable in atoms that are optically cooled.

### 4.2.4 Quantum Motion in a Trap

Since laser and evaporative cooling have the capability to cool atoms to energies where their deBroglie wavelengths are on the scale of the orbit size, the motional dynamics must be described in terms of quantum mechanical variables and suitable wave functions. Quantization of the motion leads to discrete bound states within the trap having $\vec{\mu} \cdot \vec{B} < 0$, and also a continuum of unbound states having $\vec{\mu} \cdot \vec{B}$ with opposite sign.

Studying the behavior of extremely slow (cold) atoms in the two-coil quadrupole trap begins with a heuristic quantization of the orbital angular momentum using $Mr^2\omega_T = n\hbar$ for circular orbits. The energy levels are then given by

$$E_n = \frac{3}{2}E_1 n^{2/3}, \quad \text{where}$$
$$E_1 = (Ma^2\hbar^2)^{1/3} \sim h \times 5\,\text{kHz}, \qquad (23)$$

For velocities of optically cooled atoms of a few cm $s^{-1}$, $n \sim 10 - 100$. It is readily found that $\omega_Z = n\omega_T$, so that the adiabatic condition of Eq. (22) is satisfied only for $n \gg 1$.

These large-$n$ bound states have small matrix elements coupling them to the unbound continuum states [45]. This can be understood classically since they spend most of their time in a stronger field, and thus satisfy the condition of adiabaticity of the orbital motion relative to the Larmor precession. In this case, the separation of the rapid precession from the slower orbital motion is reminiscent of the Born–Oppenheimer approximation for molecules.

On the other hand, the small-$n$ states, whose orbits are confined to a region near the origin where the field is small, have much larger coupling to the continuum states. Thus, they are rapidly ejected from the trap. The transitions to unbound states resulting from the coupling of the motion with the trapping fields are called *Majorana spin flips*, and effectively constitute a "hole" at the bottom of the trap. The evaporative cooling process used to produce very cold, dense samples reduces the average total energy of the trapped atoms sufficiently that the orbits are confined to regions near the origin and so, such losses dominate [44, 45].

There have been different solutions to this problem of Majorana losses for confinement of ultracold atoms for the BEC experiments. In the JILA-experiment, the hole was rotated by rotating the magnetic field, and thus, the atoms do not spend sufficient time in the hole to make a spin flip. In the MIT experiment, the hole was plugged by using a focused laser beam tuned to the blue side of atomic resonance, which expelled the atoms from the center of the magnetic trap. In the Rice experiment, the atoms were trapped in an Ioffe trap, which has a nonzero field minimum. Most BEC experiments are now using the Ioffe trap solution.

## 4.3 Magneto-optical Traps

### 4.3.1 Introduction

The most widely used trap for neutral atoms is a hybrid employing both optical and magnetic fields to make a magneto-optical trap (MOT), first demonstrated in

1987 [46]. The operation of an MOT depends on both inhomogeneous magnetic fields and radiative selection rules to exploit both optical pumping and the strong radiative force [46, 47]. The radiative interaction provides cooling that helps in loading the trap and enables very easy operation. MOT is a very robust trap that does not depend on precise balancing of the counterpropagating laser beams or on a very high degree of polarization.

The magnetic field gradients are modest and have the convenient feature that the configuration is the same as the quadrupole magnetic traps discussed in Sect. 4.2.2. Appropriate fields can readily be achieved with simple, air-cooled coils. The trap is easy to construct because it can be operated with a room-temperature cell in which alkali atoms are captured from the vapor. Furthermore, low-cost diode lasers can be used to produce the light appropriate for many atoms, so the MOT has become one of the least expensive ways to make atomic samples with temperatures below 1 mK.

Trapping in an MOT works by optical pumping of slowly moving atoms in a linearly inhomogeneous magnetic field $B = B(z)$ (see Eq. 21), such as that formed by a magnetic quadrupole field. Atomic transitions with the simple scheme of $J_g = 0 \rightarrow J_e = 1$ have three Zeeman components in a magnetic field, excited by each of three polarizations, whose frequencies tune with the field (and therefore with position) as shown in Fig. 12 for 1-D. Two counterpropagating laser beams of opposite circular polarization, each detuned below the zero-field atomic resonance by $\delta$, are incident as shown.

Because of the Zeeman shift, the excited state $M_e = +1$ is shifted up for $B > 0$, whereas the state with $M_e = -1$ is shifted down. At position $z'$ in Fig. 12,

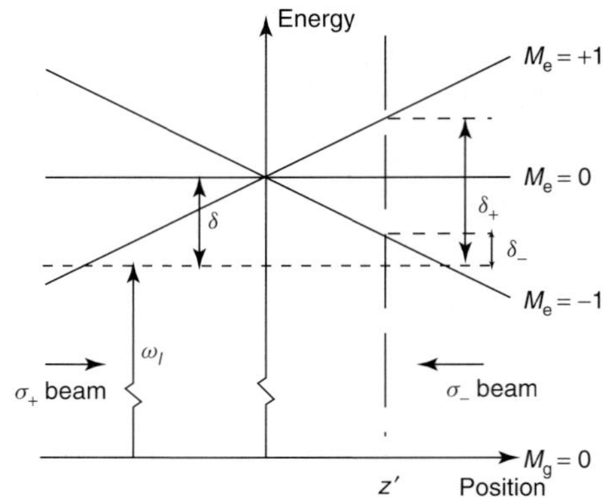


**Fig. 12** Arrangement for an MOT in 1-D. The horizontal dashed line represents the laser frequency seen by an atom at rest in the center of the trap. Because of the Zeeman shifts of the atomic transition frequencies in the inhomogeneous magnetic field, atoms at $z = z'$ are closer to resonance with the $\sigma^-$ laser beam than with the $\sigma^+$ beam, and are therefore driven toward the center of the trap

the magnetic field, therefore, tunes the $\Delta M = -1$ transition closer to resonance and the $\Delta M = +1$ transition further out of resonance. If the polarization of the laser beam incident from the right is chosen to be $\sigma^-$ and correspondingly $\sigma^+$ for the other beam, then more light is scattered from the $\sigma^-$ beam than from the $\sigma^+$ beam. Thus, the atoms are driven toward the center of the trap where the magnetic field is zero. On the other side of the center of the trap, the roles of the $M_e = \pm 1$ states are reversed and now more light is scattered from the $\sigma^+$ beam, again driving the atoms toward the center.

So far, the discussion has been limited to the motion of atoms in 1-D. However, the MOT scheme can easily be extended to 3-D by using six instead of two laser beams. Furthermore, even though very few atomic species have transitions as simple as $J_g = 0 \rightarrow J_e = 1$, the scheme works for any $J_g \rightarrow J_e = J_g + 1$ transition. Atoms that scatter mainly from the $\sigma^+$

laser beam will be optically pumped toward the $M_g = +J_g$ substate, which forms a closed system with the $M_e = +J_e$ substate.

### 4.3.2 Cooling and Compressing Atoms in an MOT

For a description of the motion of atoms in an MOT, consider the radiative force in the low intensity limit (see Eqs. 2 and 3). The total force on the atoms is given by $\vec{F} = \vec{F}_+ + \vec{F}_-$, where $\vec{F}_\pm$ can be found from Eqs. (2) and (3), and the detuning $\delta_\pm$ for each laser beam is given by $\delta_\pm = \delta \mp \vec{k} \cdot \vec{v} \pm \mu' B/\hbar$. Here, $\mu' \equiv (g_e M_e - g_g M_g)\mu_B$ is the effective magnetic moment for the transition used. Note that the Doppler shift $\omega_D \equiv -\vec{k} \cdot \vec{v}$ and the Zeeman shift $\omega_Z = \mu' B/\hbar$ both have opposite signs for opposite beams.

The situation is analogous to the velocity damping in an OM from the Doppler effect as discussed in Sec. 3.2, but here the effect also operates in position space, whereas for molasses it operates only in velocity space. Since the laser light is detuned below the atomic resonance in both cases, compression and cooling of the atoms is obtained simultaneously in an MOT.

When both the Doppler and Zeeman shifts are small compared to the detuning $\delta$, the denominator of the force can be expanded as for Eq. (13) and the result becomes

$$\vec{F} = -\beta\vec{v} - \kappa\vec{r}, \qquad (24)$$

where the damping coefficient $\beta$ is defined in Eq. (13). The spring constant $\kappa$ arises from the similar dependence of $\vec{F}$ on the Doppler and Zeeman shifts, and is given by $\kappa = \mu' \beta \nabla B/\hbar k$

The force of Eq. (24) leads to damped harmonic motion of the atoms, where the damping rate is given by $\Gamma_{MOT} = \beta/M$ and the oscillation frequency $\omega_{MOT} = \sqrt{\kappa/M}$. For magnetic field gradients $\nabla B \approx 0.1\ \mathrm{T\,m^{-1}}$, the oscillation frequency is typically a few kHz, and this is much smaller than the damping rate that is typically a few hundred kHz. Thus, the motion is overdamped, with a characteristic restoring time to the center of the trap of $2\Gamma_{MOT}/\omega_{MOT}^2 \approx$ several milliseconds for typical values of detuning and intensity of the lasers.

### 4.3.3 Capturing Atoms in an MOT

Although the approximations that lead to Eq. (24) for force hold for slow atoms near the origin, they do not apply for the capture of fast atoms far from the origin. In the capture process, the Doppler and Zeeman shifts are no longer small compared to the detuning, so the effects of the position and velocity can no longer be disentangled. However, the full expression for the force still applies and the trajectories of the atoms can be calculated by numerical integration of the equation of motion [48].

The capture velocity of an MOT is serendipitously enhanced because atoms traveling across it experience a decreasing magnetic field just as in beam deceleration described in Sect. 3.1. This enables resonance over an extended distance and velocity range because the changing Doppler shift of decelerating atoms can be compensated by the changing Zeeman shift as atoms move in the inhomogeneous magnetic field. Of course, it will work this way only if the field gradient $\nabla B$ does not demand an acceleration larger than the maximum acceleration $a_{max}$. Thus, atoms are subject to the optical force over a distance that can be as long as the trap size, and can therefore be slowed considerably.

The very large velocity capture range $v_{cap}$ of an MOT can be estimated by using $F_{max} = \hbar k\gamma/2$ and choosing a maximum size of a few centimeters for the beam diameters. Thus, the energy change can be as large as a few K, corresponding to $v_{cap} \sim 100$ m s$^{-1}$ [47]. The number of atoms in a vapor with velocities below $v_{cap}$ in the Boltzmann distribution scales as $v_{cap}^4$, and there are enough slow atoms to fall within the large MOT capture range even at room temperature, because a few K includes $10^{-4}$ of the atoms.

#### 4.3.4 Variations on the MOT Technique

Because of the wide range of applications of this most versatile kind of atom trap, a number of careful studies of its properties have been made [47, 49–56], and several variations have been developed. One of these is designed to overcome the density limits achievable in an MOT. In the simplest picture, loading additional atoms into an MOT produces a higher atomic density because the size of the trapped sample is fixed. However, the density cannot increase without limit as more atoms are added. The atomic density is limited to $\sim 10^{11}$ cm$^{-3}$ because the fluorescent light emitted by some trapped atoms is absorbed by others.

One way to overcome this limit is to have much less light in the center of the MOT than at the sides. Simply lowering the laser power is not effective in reducing the fluorescence because it will also reduce the capture rate and trap depth. But those advantageous properties can be preserved while reducing fluorescence from atoms at the center if the light intensity is low only in the center.

The repumping process for the alkali atoms provides an ideal way of implementing this idea [57]. If the repumping light is tailored to have zero intensity at the center, then atoms trapped near the center of the MOT are optically pumped into the "wrong" hfs state and stop fluorescing. They drift freely in the "dark" at low speed through the center of the MOT until they emerge on the other side, into the region where light of both frequencies is present and begin absorbing again. Then they feel the trapping force and are driven back into the "dark" center of the trap. Such an MOT has been operated at MIT [57] with densities close to $10^{12}/\mathrm{cm}^3$, and the limitations are now from collisions in the ground state rather than from multiple light scattering and excited-state collisions.

## 5 Optical Lattices

### 5.1 Quantum States of Motion

As the techniques of laser cooling advanced from a laboratory curiosity to a tool for new problems, the emphasis shifted from attaining the lowest possible steady state temperatures to the study of elementary processes, especially the quantum mechanical description of the atomic motion. In the completely classical description of laser cooling, atoms were assumed to have a well-defined position and momentum that could be known simultaneously with arbitrary precision. However, when atoms are moving sufficiently slowly that their deBroglie wavelength precludes their localization to less than $\lambda/2\pi$, these descriptions fail and a quantum mechanical description is required. Such exotic behavior for the motion of whole atoms, as opposed to electrons in the atoms, had not been considered before the advent of laser cooling simply because it was too far out of the range of ordinary experiments. A series of experiments in the early 1990s provided dramatic evidence for these new quantum states of motion of neutral atoms, and led to the debut of deBroglie wave atom optics.

The quantum description of atomic motion requires that the energy of such motion be included in the Hamiltonian. The total Hamiltonian for atoms moving in a light field would then be given by

$$\mathcal{H} = \mathcal{H}_{\text{atom}} + \mathcal{H}_{\text{rad}} + \mathcal{H}_{\text{int}} + \mathcal{H}_{\text{kin}}, \quad (25)$$

where $\mathcal{H}_{\text{atom}}$ describes the motion of the atomic electrons and gives the internal atomic energy levels, $\mathcal{H}_{\text{rad}}$ is the energy of the radiation field and is of no concern here because the field is not quantized, $\mathcal{H}_{\text{int}}$ describes the excitation of atoms by the light field and the concomitant light shifts, and $\mathcal{H}_{\text{kin}}$ is the kinetic energy operator of the motion of the center of mass of the atoms. This Hamiltonian has eigenstates of not only the internal energy levels and the atom-laser interaction that connects them, but also that of the kinetic energy operator $\mathcal{H}_{\text{kin}} \equiv P^2/2M$. These eigenstates will therefore be labeled by quantum numbers of the atomic states as well as the center of mass momentum $p$. An atom in the ground state, $|g; p\rangle$, has an energy $E_g + p^2/2M$, that can take on a range of values.

In 1968, V.S. Letokhov [58] suggested that it is possible to confine atoms in the wavelength-size regions of a standing wave by means of the dipole force that arises from the light shift. This was first accomplished in 1987 in 1-D with an atomic beam traversing an intense standing wave [59]. Since then, the study of atoms confined to wavelength-size potential wells has become an important topic in optical control of atomic motion because it opens up configurations previously accessible only in condensed matter physics using crystals.

The limits of laser cooling discussed in Sect. 3.3.4 suggest that atomic momenta can be reduced to a "few" times $\hbar k$. This means that their deBroglie wavelengths are equal to the optical wavelengths divided by a "few". If the depth of the optical potential wells is high enough to contain such very slow atoms, then their motion in potential wells of size $\lambda/2$ must be described quantum mechanically, since they are confined to a space of size comparable to their deBroglie wavelengths. Thus, they do not oscillate in the sinusoidal wells as classical localizable particles, but instead occupy discrete, quantum mechanical bound states [60], as shown in the lower part of Fig. 13.

The basic ideas of the quantum mechanical motion of particles in a periodic potential were laid out in the 1930s with the Kronig–Penney model and Bloch's theorem, and optical lattices offer important opportunities for their study. For example,

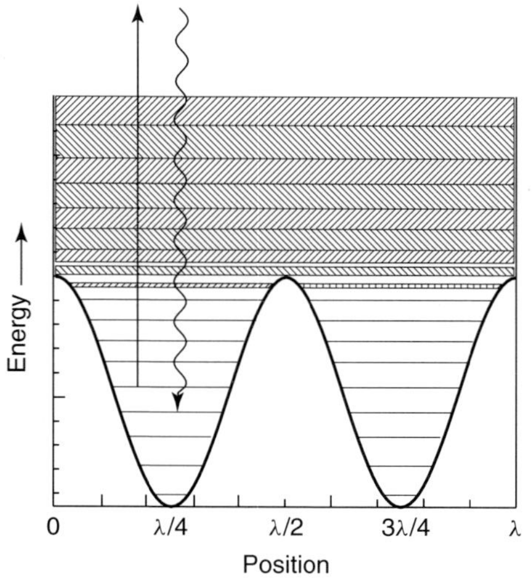


**Fig. 13** Energy levels of atoms moving in the periodic potential of the light shift in a standing wave. There are discrete bound states deep in the wells that broaden at higher energy, and become bands separated by forbidden energies above the tops of the wells. Under conditions appropriate to laser cooling, optical pumping among these states favors populating the lowest ones as indicated schematically by the arrows

these lattices can be made essentially free of defects with only moderate care in spatially filtering the laser beams to assure a single transverse mode structure. Furthermore, the shape of the potential is exactly known and does not depend on the effect of the crystal field or the ionic energy level scheme. Finally, the laser parameters can be varied to modify the depth of the potential wells without changing the lattice vectors, and the lattice vectors can be changed independently by redirecting the laser beams. The simplest optical lattice to consider is a 1-D pair of counterpropagating beams of the same polarization, as was used in the first experiment [59].

Of course, such tiny traps are usually very shallow, so loading them requires cooling to the μK regime. Even atoms whose energy exceeds the trap depth must be described as quantum mechanical particles moving in a periodic potential that display energy band structure [60]. Such effects have been observed in very careful experiments.

Because of the transverse nature of light, any mixture of beams with different $\vec{k}$-vectors necessarily produces a spatially periodic, inhomogeneous light field. The importance of the "egg-crate" array of potential wells arises because the associated atomic light shifts can easily be comparable to the very low average atomic kinetic energy of laser-cooled atoms. A typical example projected against two dimensions is shown in Fig. 14.

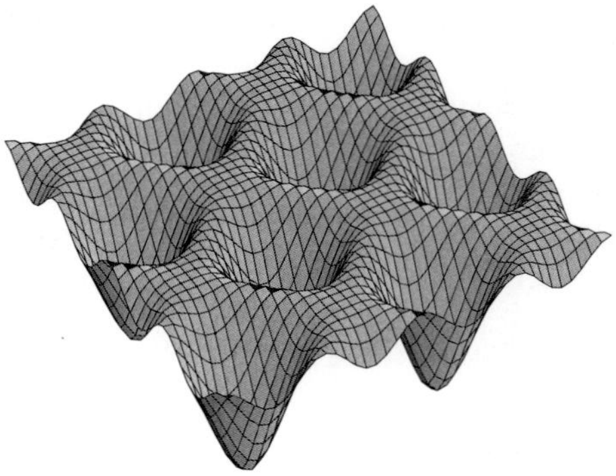

**Fig. 14** The "egg-crate" potential of an optical lattice shown in two dimensions. The potential wells are separated by $\lambda/2$

Atoms trapped in wavelength-sized spaces occupy vibrational levels similar to those of molecules. The optical spectrum can show Raman-like sidebands that result from transitions among the quantized vibrational levels [61, 62] as shown in Fig. 15. These quantum states of atomic motion can also be observed by stimulated emission [62, 63] and by direct RF spectroscopy [64, 65].

## 5.2
## Properties of 3-D Lattices

The name "optical lattice" is used rather than optical crystal because the filling fraction of the lattice sites is typically only a few percent (as of 1999). The limit arises because the loading of atoms into the lattice is typically done from a sample of trapped and cooled atoms, such as an MOT for atom collection, followed by an OM for laser cooling. The atomic density in such experiments is limited by collisions and multiple light scattering to a few times $10^{11}$ cm$^{-3}$. Since the density of lattice sites of size $\lambda/2$ is a few times $10^{13}$ cm$^{-3}$, the filling fraction is necessarily small. With the advent of experiments that load atoms directly into a lattice from a BEC, the filling factor can be increased to 100%, and in some cases it may be possible to load more than one atom per lattice site [66, 67].

In 1993 a very clever scheme was described [68]. It was realized that an $n$-dimensional lattice could be created by only $n+1$ traveling waves rather than

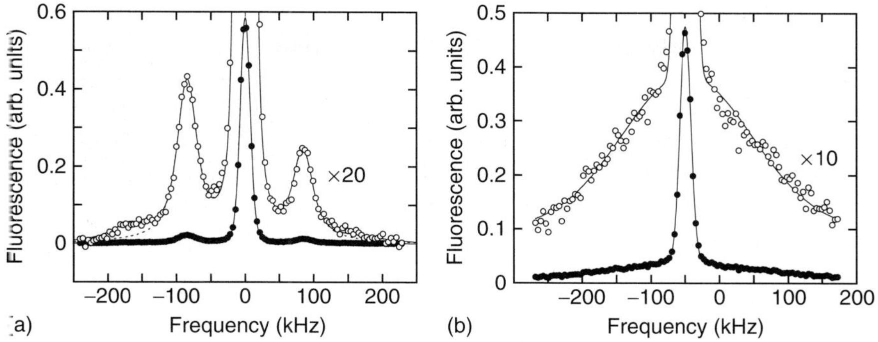


**Fig. 15** (a) Fluorescence spectrum in a 1-D lin ⊥ lin optical molasses. Atoms are first captured and cooled in an MOT, and then the MOT light beams are switched off leaving a pair of lin ⊥ lin because of spontaneous emission of the atoms to the same vibrational state from which they are excited, whereas the sideband on the left (right) is due to spontaneous emission to a vibrational state with one vibrational quantum number lower (higher) (see Fig. 13). The presence of these sidebands is a direct proof of the existence of the band structure. (b) Same as (a) except that the 1-D molasses is $\sigma^+ - \sigma^-$, which has no spatially dependent light shift and hence no vibrational states (figure taken from Jessen, P. S., Gerz, C., Lett, P. D., Phillips, W. D., Rolston, S. L., Spreeuw, R. J. C., Westbrook, C. I. (**1992**), *Phys. Rev. Lett.* **69**, 49–52)

$2n$. The real benefit of this scheme is that in case of phase instabilities in the laser beams, the interference pattern is only shifted in space, but the interference pattern itself is not changed. Instead of producing optical wells in 2-D with four beams (two standing waves), these authors used only three. The $\vec{k}$-vectors of the coplanar beams were separated by $2\pi/3$, and they were all linearly polarized in their common plane (not parallel to one another). The same immunity to vibrations was established for a 3-D optical lattice by using only four beams arranged in a quasi-tetrahedral configuration. The three linearly polarized beams of the 2-D arrangement described above were directed out of the plane toward a common vertex, and a fourth circularly polarized beam was added. All four beams were polarized in the same plane [68]. The authors showed that this configuration produced the desired potential wells in 3-D.

## 5.3
## Spectroscopy in 3-D Lattices

The group at NIST developed a new method that superposed a weak probe beam of light directly from the laser upon some of the fluorescent light from the atoms in a 3-D OM, and directed the light from these combined sources onto on a fast photodetector [70]. The resulting beat signal carried information about the Doppler shifts of the atoms in the optical lattices [61]. These Doppler shifts were expected to be in the sub-MHz range for atoms with the previously measured 50-μK temperatures. The observed features confirmed the quantum nature of the motion of atoms in the wavelength-size potential wells (see Fig. 15) [15].

The NIST group also studied atoms loaded into an optical lattice using of laser light from the spatially ordered array [71]. They cut off the laser beams that formed the lattice, and before the atoms had time to move away from their positions, they pulsed on a probe laser beam at the Bragg angle appropriate for one of the sets of lattice planes. The Bragg diffraction not only enhanced the reflection of the probe beam by a factor of $10^5$, but by varying the time between the shut-off of the lattice and turn-on of the probe, they could also measure the "temperature" of the atoms in the lattice. The reduction of the amplitude of the Bragg-scattered beam with time provided some measure of the diffusion of the atoms away from the lattice sites, much like the Debye–Waller factorDebye–Waller factor in X-ray diffraction.

## 5.4 Quantum Transport in Optical Lattices

In the 1930s, Bloch realized that applying a uniform force to a particle in a periodic potential would not accelerate it beyond a certain speed, but instead would result in Bragg reflection when its deBroglie wavelength became equal to the lattice period. Thus, an electric field applied to a conductor could not accelerate electrons to a speed faster than that corresponding to the edge of a Brillouin zone, and that at longer times the particles would execute oscillatory motion. Ever since then, experimentalists have tried to observe these Bloch oscillations in increasingly pure and/or defect-free crystals.

Atoms moving in optical lattices are ideally suited for such an experiment, as was beautifully demonstrated in 1996 [69]. The authors loaded a 1-D lattice with atoms from a 3-D molasses, further narrowed the velocity distribution, and then instead of applying a constant force, simply changed the frequency of one of the beams of the 1-D lattice with respect to the other in a controlled way, thereby creating an accelerating lattice. Seen from the atomic reference frame, this was the equivalent of a constant force trying to accelerate them. After a variable time $t_a$, the 1-D lattice beams were shut off and the measured atomic velocity distribution showed beautiful Bloch oscillations as a function of $t_a$. The centroid of the very narrow velocity distribution was seen to shift in velocity space at a constant rate until it reached

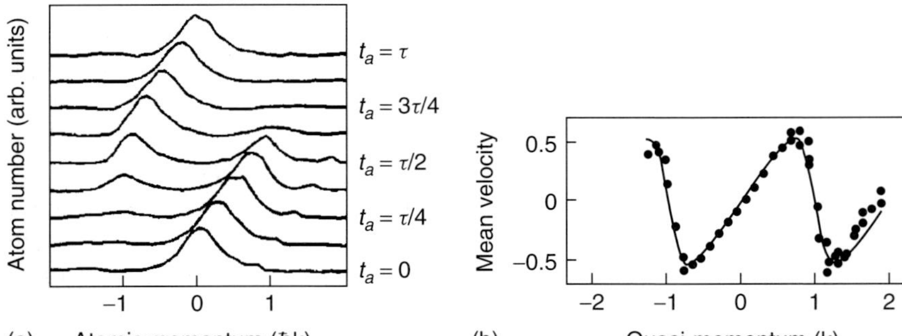


**Fig. 16** Plot of the measured velocity distribution *verses.* time in the accelerated 1-D lattice. (a) Atoms in a 1-D lattice are accelerated for a fixed potential depth for a certain time $t_a$ and the momentum of the atoms after the acceleration is measured. The atoms accelerate only to the edge of the Brillouin zone where the velocity is $+v_r$, and then the velocity distribution appears at $-v_r$. (b) Mean velocity of the atoms as a function of the quasi-momentum, that is, the force times the acceleration time (figure taken from Ben Dahan, M., Peik, E., Reichel, J., Castin, Y., Salomon, C. (**1996**), *Phys. Rev. Lett.* **76**, 4508–4511)

$v_r = \hbar k/M$, and then it vanished and reappeared at $-v_r$ as shown in Fig. 16. The shape of the "dispersion curve" allowed measurement of the "effective mass" of the atoms bound in the lattice.

## 6 Bose–Einstein Condensation

### 6.1 Introduction

In the 1920s, Bose and Einstein predicted that for sufficiently high phase space density, $\rho_\phi \sim 1$ (see Sect. 1.2), a gas of atoms undergoes a phase transition that is now called Bose–Einstein condensation. It took 70 years before BEC could be unambiguously observed in a dilute gas. From the advent of laser cooling and trapping, it became clear that this method could be instrumental in achieving BEC.

BEC is another manifestation of quantization of atomic motion. It occurs in the absence of resonant light, and its onset is characterized by cooling to the point where the atomic deBroglie wavelengths are comparable to the interatomic spacing. This is in contrast with the topics discussed in Sect. 5 where the atoms were in an optical field and their deBroglie wavelengths were comparable to the optical wavelength $\lambda$.

Laser cooling alone is inherently incapable of achieving $\rho_\phi \sim 1$. This is easily seen from the recoil limit of Sect. 3.3.4 that limits $\lambda_{deB}$ to $\lambda$/"few". Since the cross section for optical absorption near resonance is $\sigma \sim \lambda^2$ near $\rho_\phi \sim 1$, this limit of $\lambda_{deB} \sim \lambda$/"few" results in the penetration depth of the cooling light into the sample being smaller than $\lambda$. Thus, the sample would have to be extremely small and contain only a few atoms, hardly a system suitable for investigation.

Temperatures lower than the recoil limit are readily achieved by evaporative cooling, and so all BEC experiments employ it in their final phase. Evaporative cooling is inherently different from the other cooling processes discussed in Sect. 3, and hence discussed here separately.

Since the first observations in 1995, BEC has been the subject of intense investigation, both theoretical and experimental. No attempt is made in this article to even address, much less cover, the very rich range of physical phenomena that have been unveiled by these studies. Instead, we focus on the methods to achieve $\rho_\phi \sim 1$ and BEC.

### 6.2 Evaporative Cooling

Evaporative cooling is based on the preferential removal of those atoms from a confined sample with an energy higher than the average energy followed by a rethermalization of the remaining gas by elastic collisions. Although evaporation is a process that occurs in nature, it was applied to atom cooling for the first time in 1988 [72].

One way to think about evaporative cooling is to consider cooling of a container of hot liquid. Since the most energetic molecules evaporate from the liquid and leave the container, the remaining molecules obtain a lower temperature and are cooled. Furthermore, it requires the evaporation of only a small fraction of the liquid to cool it by a considerable amount.

Evaporative cooling works by removing the higher-energy atoms as suggested schematically in Fig. 17. Those that remain have much lower average energy (temperature) and so they occupy a smaller volume near the bottom of the

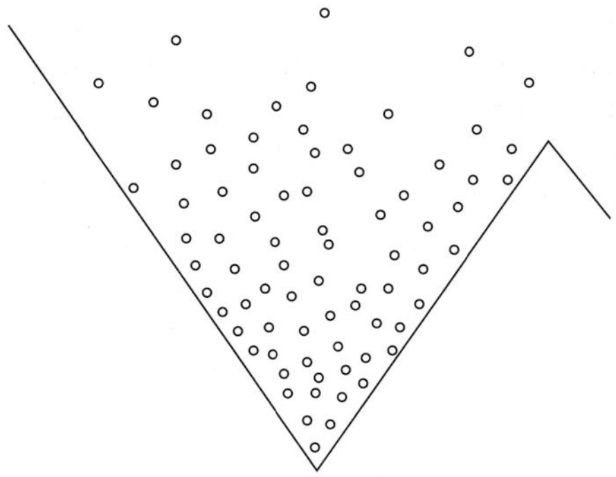

**Fig. 17** Principle of the evaporation technique. Once the trap depth is lowered, atoms with energy above the trap depth can escape and the remaining atoms reach a lower temperature

trap, thereby increasing their density. For trapped atoms, it can be achieved by lowering the depth of the trap, thereby allowing the atoms with energies higher than the trap depth to escape, as discussed first by Hess [73]. Elastic collisions in the trap then lead to a rethermalization of the gas. This technique was first employed for evaporative cooling of hydrogen [72, 74–76]. Since both the temperature and the volume decrease, $\rho_\phi$ increases.

Recently, more refined techniques have been developed. For example, to sustain the cooling process the trap depth can be lowered continuously, achieving a continuous decrease in temperature. Such a process is called *forced evaporation* and is discussed in Sect. 6.3 below.

### 6.2.1 Simple Model

This section describes a simple model of evaporative cooling. Since such cooling is not achieved for single atoms but for the whole ensemble, an atomic description of the cooling process must be replaced by thermodynamic methods. These methods are completely different from the rest of the material in this article, and will therefore remain rather elementary.

Several models have been developed to describe this process, but we present here the simplest one [77] because of its pedagogical value [22]. In this model, the trap depth is lowered in one single step and the effect on the thermodynamic quantities, such as temperature, density, and volume, is calculated. The process can be repeated and the effects of multiple steps added up cumulatively.

In such models of evaporative cooling, the following assumptions are made:

1. The gas behaves sufficiently ergodically, that is, the distribution of atoms in phase space (both position and momentum) depends only on the energy of the atoms and the nature of the trap.
2. The gas is assumed to begin the process with $\rho_\phi \ll 1$ (far from the BEC transition point), and so it is described by classical statistics.
3. Even though $\rho_\phi \ll 1$, the gas is cold enough that the atomic scattering is pure (s-wave) quantum mechanical, that is, the temperature is sufficiently low that all higher partial waves do not contribute to the cross section. Furthermore, the cross section for elastic scattering is energy-independent and is given by $\sigma = 8\pi a^2$, where $a$ is the scattering length. It is also assumed that the ratio of elastic to inelastic collision rates is sufficiently large that the elastic collisions dominate.
4. Evaporation preserves the thermal nature of the distribution, that is, the thermalization is much faster than the rate of cooling.
5. Atoms that escape from the trap neither collide with the remaining atoms nor exchange energy with them. This is called *full evaporation*.

### 6.2.2 Application of the Simple Model

The first step in applying this simple model is to characterize the trap by calculating how the volume of a trapped sample of atoms changes with temperature $T$. Consider a trapping potential that can be expressed as a power law given by

$$U(x, y, z) = \varepsilon_1 \left|\frac{x}{a_1}\right|^{s_1} + \varepsilon_2 \left|\frac{y}{a_2}\right|^{s_2} + \varepsilon_3 \left|\frac{z}{a_3}\right|^{s_3}, \qquad (26)$$

where $a_j$ is a characteristic length and $s_j$ the power, for a certain direction $j$. Then the volume occupied by trapped atoms scales as $V \propto T^{\xi}$ [78], where

$$\xi \equiv \frac{1}{s_1} + \frac{1}{s_2} + \frac{1}{s_3}. \qquad (27)$$

Thus, the effect of the potential on the volume of the trapped sample for a given temperature can be reduced to a single parameter $\xi$. This parameter is independent of how the occupied volume is defined, since many different definitions lead to the same scaling. When a gas is held in a 3-D box with infinitely high walls, then $s_1 = s_2 = s_3 = \infty$ and $\xi = 0$, which means that $V$ is independent of $T$, as expected. For a harmonic potential in 3-D, $\xi = 3/2$; for a linear potential in 2-D, $\xi = 2$; and for a linear potential in 3-D, $\xi = 3$.

The evaporative cooling model itself [77] starts with a sample of $N$ atoms in volume $V$ having a temperature $T$ held in an infinitely deep trap. The strategy for using the model is to choose a finite quantity $\eta$, and then (1) lower the trap depth to a value $\eta k_B T$, (2) allow for a thermalization of the sample by collisions, and (3) determine the change in $\rho_\phi$.

Only two parameters are needed to completely determine all the thermodynamic quantities for this process (the values after the process are denoted by a prime). One of these is $\nu \equiv N'/N$, the fraction of atoms remaining in the trap after the cooling. The other is $\gamma$ (This $\gamma$ is not to be confused with the natural width of the excited state.), a measure of the decrease in temperature caused by the release of hot atoms and subsequent cooling, modified by $\nu$, and defined as

$$\gamma \equiv \frac{\log(T'/T)}{\log(N'/N)} = \frac{\log(T'/T)}{\log \nu}. \qquad (28)$$

This yields a power-law dependence for the decrease in temperature caused by the loss of the evaporated particles, that is, $T' = T\nu^{\gamma}$. The dependence of the other thermodynamic quantities on the parameters $\nu$ and $\gamma$ can then be calculated.

The scaling of $N' = N\nu$, $T' = T\nu^{\gamma}$, and $V' = V\nu^{\gamma\xi}$ can provide the scaling of all the other thermodynamic quantities of interest by using the definitions for the density $n = N/V$, the phase space density $\rho_\phi = n\lambda_{\mathrm{deB}}^3 \propto nT^{-3/2}$, and the elastic collision rate $k_{\mathrm{el}} \equiv n\sigma v \propto nT^{1/2}$. The results are given in Table 2. For a given value of $\eta$, the scaling of all quantities depends only on $\gamma$. Note that for successive steps $j$, $\nu$ has to be replaced with $\nu^j$.

**Tab. 2** Exponent $q$ for the scaling of the thermodynamic quantities $X' = X\nu^q$ with the reduction $\nu$ of the number of atoms in the trap

| *Thermodynamic variable* | *Symbol* | *Exponent q* |
|---|---|---|
| Number of atoms | $N$ | 1 |
| Temperature | $T$ | $\gamma$ |
| Volume | $V$ | $\gamma\xi$ |
| Density | $n$ | $1 - \gamma\xi$ |
| Phase space density | $\rho$ | $1 - \gamma(\xi + 3/2)$ |
| Collision rate | $k$ | $1 - \gamma(\xi - 1/2)$ |

The fraction of atoms remaining is fully determined by the final trap depth $\eta$ for a given potential characterized by the trap parameter $\xi$. In order to determine the change of the temperature in the cooling process, it is necessary to consider in detail the distribution of the atoms in the trap, and this is discussed more fully in Refs. [1, 22, 77].

#### 6.2.3 Speed of Evaporation

So far, the speed of the evaporative cooling process has not been considered. If the trap depth is ramped down too quickly, the thermalization process does not have time to run its course and the process becomes less efficient. On the other hand, if the trap depth is ramped down too slowly, the loss of particles by inelastic collisions becomes important, thereby making the evaporation inefficient.

The speed of evaporation can be found from the principle of detailed balance [22]. Its application shows that the ratio of the evaporation time and the elastic collision time is

$$\frac{\tau_{\mathrm{ev}}}{\tau_{\mathrm{el}}} = \frac{\sqrt{2}e^{\eta}}{\eta}. \tag{29}$$

Note that this ratio increases exponentially with $\eta$.

Experimental results show that $\sim$2.7 elastic collisions are necessary to rethermalize the gas [79]. In order to model the rethermalization process, Luiten et al. [80] have discussed a model based on the Boltzmann equation where the evolution of the phase space density $\rho(\vec{r}, \vec{p}, t)$ is calculated. This evolution is not only caused by the trapping potential, but also by collisions between the particles. Only elastic collisions, whose cross section is given by $\sigma = 8\pi a^2$ with $a$ as the scattering length, are considered. This leads to the Boltzmann equation [81].

#### 6.2.4 Limiting Temperature

In the models discussed so far, only elastic collisions have been considered, that is, collisions where kinetic energy is redistributed between the partners. However, if part of the internal energy of the colliding partners is exchanged with their kinetic energy in the collision, then it is inelastic. Inelastic collisions can cause problems for two reasons: (1) the internal energy released can cause the atoms to heat up and (2) the atoms can change their internal states, and the new states may no longer be trapped. In each case, such collisions can lead to trap loss and are therefore not desirable.

Apart from collisions with the background gas and three-body recombination, there are two inelastic processes that are important for evaporative cooling of alkali atoms: dipolar relaxation and spin relaxation. The collision rate $nk_{\mathrm{dip}}$ for them at low energies is independent of velocity [82]. Since the elastic collision rate is given by $k_{\mathrm{el}} = n\sigma v_{\mathrm{rel}}$, the ratio of good (= elastic) to bad (= relaxation) collisions goes down when the temperature does. This limits the temperature to a value $T_e$ near which the ratio between good and bad collisions becomes unity, and $T_e$ is given by

$$k_{\mathrm{B}} T_e = \frac{\pi M k_{\mathrm{dip}}^2}{16\sigma^2}. \tag{30}$$

The limiting temperature for the alkalis is of the order of 1 nK, depending on the values of $\sigma$ and $k_{\mathrm{dip}}$.

In practice, however, this ratio has to be considerably larger than unity, and so the practical limit for evaporative cooling occurs when the ratio is $\sim 10^3$ [22]. In the model of Ref. [80], the authors discuss different strategies for evaporative cooling. Even for the strategy of the lowest temperature, the final temperature is higher than $T_e$.

The collision rate between atoms with one in the excited state (S + P collisions) is also much larger at low temperatures than the rate for such collisions with both atoms in the ground state (S + S collisions). Since S + P collisions are generally inelastic and the inelastic energy exchange generally leads to a heating of the atoms, increasing the density increases the loss of cold atoms. To achieve BEC, resonant light should therefore be avoided, and thus laser cooling is not suitable for achieving BEC.

### 6.3 Forced Evaporative Cooling

In all the earliest experiments that achieved BEC, the evaporative cooling was "forced" by inducing rf transitions to magnetic sublevels that are not bound in the magnetic trap. Atoms with the highest energies can access regions of the trap where the magnetic field is stronger, and thus their Zeeman shifts would be larger. A correspondingly high-frequency rf field would cause only these most energetic atoms to undergo transitions to states that are not trapped, and in doing so, the departing atoms carry away more than the average energy. Thus, a slow sweep of the rf frequency from high to low would continuously shave off the high-energy tail of the energy distribution, and thereby continuously drive the temperature lower and the phase space density higher. The results of evaporative cooling from the first three groups that have obtained BEC have shown that using this rf shaving technique, it is much easier to select high-energy atoms and waste them than it is to cool them.

For the evaporation of the atoms, it is important that atoms with an energy above the cutoff are expelled from the trap. By using RF-evaporation, one can expel the atoms in all three dimensions equally and thus obtain a true 3-D evaporation. In the case of the time orbiting potential (TOP) trap, the atoms are evaporated along the outer side of the cloud that is exposed to the highest magnetic field on the average. This is a cylinder along the direction of rotation axis of the magnetic field and thus the evaporation takes place in 2-D.

Once the energy of the atoms becomes very small, the atoms sag because of gravity and the outer shell of the cloud is no longer at a constant magnetic field. Atoms at the bottom of the trap have the highest energy and thus the evaporation becomes 1-D. In case of harmonic confinement, $U_{\text{trap}} = U''z^2/2$, the equipotential surface is at $z \approx \sqrt{2\eta k_B T/U''}$. Now, the gravitational energy is given by $U_{grav} = mgz$ and thus the limiting temperature for 1-D evaporation to take place is given by [22]

$$k_B T < \frac{2\eta(mg)^2}{\mu B''} \tag{31}$$

For a curvature of $B'' = 500\ \text{T m}^{-2}$, the limiting temperature becomes 1 μK for $^7$Li, 10 μK for $^{23}$Na, and 150 μK for $^{87}$Rb. Below this temperature, evaporation becomes less efficient.

In the three experiments that obtained BEC for the first time in 1995, the problem of this "gravitational sag" was not known, but it did not prevent the experimentalist from observing BEC. The solution used in those experiments was because of the light mass ($^7$Li), tight confinement ($^{23}$Na), and TOP trap ($^{87}$Rb). In the last case, the axis of rotation is in the $z$-direction and thus the evaporation always remains 2-D. Table 3 shows typical values of $\rho_\phi$ for various situations.

**Tab. 3** Typical numbers for the phase space density as obtained in the experiments aimed at achieving BEC. The different stages of cooling and trapping the atoms are discussed in the text

| *Stages* | *T* | $\lambda_{deB}$ | *n* | $\rho_\phi$ |
|---|---|---|---|---|
| Oven | 300 °C | 0.02 nm | $10^{10}$ cm$^{-3}$ | $10^{-16}$ |
| Slowing | 30 mK | 2 nm | $10^{8}$ cm$^{-3}$ | $10^{-12}$ |
| Pre-cooling | 1 mK | 10 nm | $10^{9}$ cm$^{-3}$ | $10^{-9}$ |
| Trapping | 1 mK | 10 nm | $10^{12}$ cm$^{-3}$ | $10^{-6}$ |
| Cooling | 1 μK | 0.3 μm | $10^{11}$ cm$^{-3}$ | $3 \times 10^{-3}$ |
| Evaporation | 70 nK | 1 μm | $10^{12}$ cm$^{-3}$ | 2.612 |

## 7 Conclusion

In this article, we have reviewed some of the fundamentals of optical control of atomic motion. The reader is cautioned that this is by no means an exhaustive review of the field, and that many important and current topics have been omitted. Much of the material here was taken from our recent textbook [1], and the reader is encouraged to consult that source for the origin of many of the formulas presented in the present text, as well as for further reading and more detailed references to the literature.

## Glossary

**Atomic Beam Slowing:** Using laser light to slow down the velocities of atoms in a beam.

**Bad Collisions:** Inelastic scattering of atoms leading to loss of atoms from the trap during evaporation.

**Beam Brightening:** Increasing the brightness of an atomic beam by using laser cooling.

**Bloch Oscillations:** The oscillatory motion of particles moving through a periodic potential, predicted in the 1930s by Bloch for electrons in solid state.

**Bose–Einstein Condensation (BEC):** New state of matter theoretically described in the 1920s by Bose and Einstein and experimentally first observed in dilute gases in 1996, where the interatomic spacing is smaller than the deBroglie wavelength of the atoms.

**Brightness:** The number of atoms emitted from a source or in a beam per unit of time, per unit of solid angle, and per unit of area of the source.

**Capture Range:** The range in velocity, where the optical forces are effective.

**Damping Force:** The optical force on atoms that leads to damping of their velocity.

**Dark-spot Magneto-optical Trap:** Magneto-optical trap, where in the center of the trap the atoms are not kept in the cycling transition, which reduces the loss of atoms due to inelastic collisions.

**Dipole Force Trap:** Trap using the dipole optical force to trap atoms.

**Dipole Forces:** The optical force on the atom caused by the gradient of the light shift of the atom.

**Doppler Cooling Limit:** Limit of cooling atoms given by the Doppler theory.

**Doppler Cooling Techniques:** Using the Doppler shift and the detuning of the light from atomic resonance in order to cool atoms.

**Doppler Theory:** Theory describing the cooling of two-level atoms by exploiting the Doppler shift of the atoms in combination with the detuning of the laser light from resonance.

**Ehrenfest Theorem:** Theorem by Ehrenfest making the correspondence between classical relations, like $F = -\mathrm{grad}V$, and their quantum-mechanical counterpart, in this case $\langle F \rangle = -\mathrm{grad}\langle H \rangle$.

**Entropy:** Thermodynamic measure of the disorder in the system.

**Evaporative Cooling:** The preferential removal of high-energy atoms from a gas sample, thereby reducing the temperature of the remaining atoms.

**Far-off-resonance Traps:** Dipole force trap, where the light is far detuned from resonance, which minimizes the scattering of light due to spontaneous emission.

**Fokker-Planck Equation:** Equation describing the evolution of the atomic momentum distribution under the combined influence of force and diffusion.

**Good Collisions:** Elastic scattering of atoms leading to the thermalization of the atoms during evaluation.

**Laser Cooling:** Using the interaction between laser light and atoms to reduce the average kinetic energy of the atoms.

**lin ⊥ lin Cooling:** Polarization gradient cooling, where the two beams have perpendicular, linear polarization.

**Magnetic Traps:** Traps where atoms are trapped due to their magnetic moment by an inhomogeneous magnetic field.

**Magneto-optical Trap (MOT):** Combination of light forces and inhomogeneous magnetic field, leading to cooling and trapping of atoms.

**Majorana Losses:** Loss of atoms from magnetic traps, where the atoms in the center of the trap undergo a motion-induced transition to an untrapped state.

**Multilevel Atoms:** Atoms that have degenerate magnetic substates in ground and excited state, which are coupled by laser light.

**Optical Bloch Equations (OBE):** Relations describing the evolution of the state vector of the internal state of the atom due to the interaction with laser light.

**Optical Forces:** The force induced on the atom by laser light.

**Optical Lattice:** The periodic trapping potential for atoms created by the interference of laser beams.

**Optical Molasses:** Viscous damping of the atomic velocities by the use of counterpropagating laser beams.

**Phase Space Density:** Probability of finding a particle in a certain region of space, with a certain range of velocities.

**Polarization Gradient Cooling:** The use of polarization gradients to cool atoms below the Doppler limit.

**Polarization Gradient:** The interference pattern that occurs when two counterpropagating laser beams have different polarization. The resulting polarization of the light field is not constant in space, that is, shows a gradient.

**Quadrupole Trap:** The simplest magnetic trap, where the magnetic field is generated by two separated identical coils carrying opposite currents.

**Radiative Forces:** The optical force on the atom caused by the scattering of light.

**Recoil Limit:** Limit in laser cooling determined by the recoil by one photon on the atom.

**Saturation Intensity:** Intensity of the light where the rate for spontaneous emission and the rate for stimulated emission are equal.

**Saturation Parameter:** Ratio between the intensity of the laser light and the saturation intensity.

**$\sigma^+$-$\sigma^-$ Cooling:** Polarization gradient cooling, where the two beams have opposite, circular polarization.

**Sisyphus Cooling:** The continuous transfer of kinetic energy to potential energy in laser cooling, where the potential energy is radiated away by spontaneously emitted photons. A special case of Sisyphus cooling is polarization gradient cooling to sub-Doppler temperature.

**Spontaneous Emission:** The return of the atom from the excited state to the ground state by the emission of one photon in a random direction.

**Stimulated Emission:** The return of the atom from the excited state to the ground state by the emission of one photon in the direction of the laser light.

**Sub-Doppler Cooling:** Using laser cooling techniques beyond the Doppler techniques to cool atoms below the Doppler limit.

**s-wave Scattering:** Scattering of atoms at low temperatures, where only the lowest-order partial wave is effective.

**Temperature:** A measure of the average kinetic energy of the atoms, that is, $k_B T/2 = \langle E_k \rangle$.

**Time-of-flight (TOF) Method:** Technique to determine the flight time of atoms over a well-defined flight path in order to detect the atomic velocity.

**Time-orbiting Potential (TOP) trap:** Magnetic quadrupole trap, where the zero of the magnetic field oscillates in space in order to reduce trap loss due to Majorana flips.

**Two-level Atoms:** Hypothetical atoms that only have one nondegenerate ground state and one nondegenerate excited state, which are resonant with the laser light.

## References

[1] Metcalf, H. J., vander Straten, P. (1999), *Laser Cooling and Trapping.* New York: Springer-Verlag.
[2] Phillips, W., Metcalf, H. (**1982**), *Phys. Rev. Lett.* **48**, 596–599.
[3] Prodan, J., Phillips, W., Metcalf, H. (**1982**), *Phys. Rev. Lett.* **49**, 1149–1153.
[4] Prodan, J., Phillips, W. (**1984**), *Prog. Quant. Electron.* **8**, 231–235.
[5] Ertmer, W., Blatt, R., Hall, J. L., Zhu, M. (**1985**), *Phys. Rev. Lett.* **54**, 996–999.
[6] Watts, R., Wieman, C. (**1986**), *Opt. Lett.* **11**, 291–293.
[7] Bagnato, V., Lafyatis, G., Martin, A., Raab, E., Ahmad-Bitar, R., Pritchard, D. (**1987**), *Phys. Rev. Lett.* **58**, 2194–2197.
[8] Molenaar, P. A., vander Straten, P., Heideman, H. G. M., Metcalf, H. (**1997**), *Phys. Rev. A* **55**, 605–614.
[9] Barrett, T. E., Dapore-Schwartz, S. W., Ray, M. D., Lafyatis, G. P. (**1991**), *Phys. Rev. Lett.* **67**, 3483–3487.

[10] Dalibard, J., Phillips, W. (**1985**), *Bull. Am. Phys. Soc.* **30**, 748.

[11] Chu, S., Hollberg, L., Bjorkholm, J., Cable, A., Ashkin, A. (**1985**), *Phys. Rev. Lett.* **55**, 48–51.

[12] Sheehy, B., Shang, S. Q., vander Straten, P. Metcalf, H. (**1990**), *Chem. Phys.* **145**, 317–325.

[13] Gould, P., Lett, P., Phillips, W. D. (1987), New measurement with optical molasses, in W. Persson, S. Svanberg, (Eds.), *Laser Spectroscopy VIII*, Berlin: Springer, pp. 64–67.

[14] Hodapp, T., Gerz, C., Westbrook, C., Futrlehner, C., Phillips, W. (**1992**), *Bull. Am. Phys. Soc.* **37**, 1139.

[15] Lett, P., Watts, R., Westbrook, C., Phillips, W., Gould, P., Metcalf, H. (**1988**), *Phys. Rev. Lett.* **61**, 169–172.

[16] Lett, P. D., Watts, R. N., Tanner, C. E., Rolston, S. L., Phillips, W. D., Westbrook, C. I. (**1989**), *J. Opt. Soc. Am. B* **6**, 2084–2107.

[17] Dalibard, J., Cohen-Tannoudji, C. (**1989**), *J. Opt. Soc. Am. B* **6**, 2023–2045.

[18] Ungar, P. J., Weiss, D. S., Chu, S., Riis, E. (**1989**), *J. Opt. Soc. Am. B* **6**, 2058–2071.

[19] Cohen-Tannoudji, C., Phillips, W. D. (**1990**), *Phys. Today* **43**, 33–40.

[20] Gupta, R., Padua, S., Xie, C., Batelaan, H., Metcalf, H. (**1994**), *J. Opt. Soc. Am. B* **11**, 537–541.

[21] Salomon, C., Dalibard, J., Phillips, W. D., Clairon, A., Guellati, S. (**1990**), *Europhys. Lett.* **12**, 683–688.

[22] Ketterle, W., Vandruten, N. J. (**1996**), *Adv. At. Mol. Opt. Phys.* **37**, 181–236.

[23] Aspect, A., Arimondo, E., Kaiser, R., Vansteenkiste, N., Cohen-Tannoudji, C. (**1988**), *Phys. Rev. Lett.* **61**, 826–829.

[24] Kasevich, M., Chu, S. (**1992**), *Phys. Rev. Lett.* **69**, 1741–1744.

[25] Ashkin, A. (**1970**), *Phys. Rev. Lett.* **24**, 156–159.

[26] Chu, S., Bjorkholm, J., Ashkin, A., Cable, A. (**1986**), *Phys. Rev. Lett.* **57**, 314–317.

[27] Ashkin, A. (**1980**), *Science* **210**, 1081–1088.

[28] Ashkin, A., Dziedzic, J. M. (**1985**), *Phys. Rev. Lett.* **54**, 1245–1248.

[29] Ashkin, A., Dziedzic, J. M. (**1987**), *Science* **235**, 1517–1520.

[30] Miller, J. D., Cline, R. A., Heinzen, D. J. (**1993**), *Phys. Rev. A* **47**, R4567–R4570.

[31] Adams, C. S., Lee, H. J., Davidson, N., Kasevich, M., Chu, S. (**1995**), *Phys. Rev. Lett.* **74**, 3577–3580.

[32] Takekoshi, T., Knize, R. J. (**1996**), *Opt. Lett.* **21**, 77–79.

[33] Metcalf, H., Phillips, W. (**1986**), *Metrologia* **22**, 271–278.

[34] Davidson, N., Lee, H. J., Adams, C. S., Kasevich, M., Chu, S. (**1995**), *Phys. Rev. Lett.* **74**, 1311–1314.

[35] Siegman, A. (1986), *Lasers*. Mill Valley: University Sciences.

[36] Simpson, N., Dholakia, K., Allen, L., Padgett, M. (**1997**), *Opt. Lett.* **22**, 52–54.

[37] McGloin, D., Simpson, N., Padgett, M. (**1998**), *App. Opt.* **37**, 469–472.

[38] Beijersbergen, M. (1996), Phase Singularities in Optical Beams. Ph.D. Thesis, University Leiden, Leiden, The Netherlands.

[39] Ovchinnikov, Yu. B., Manek, I. Grimm, R. (**1997**), *Phys. Rev. Lett.* **79**, 2225–2228.

[40] Aminoff, C. G., Steane, A. M., Bouyer, P., Desbiolles, P., Dalibard, J., Cohen-Tannoudji, C. (**1993**), *Phys. Rev. Lett.* **71**, 3083–3086.

[41] Kasevich, M. A., Weiss, D. S., Chu, S. (**1990**), *Opt. Lett.* **15**, 607–609.

[42] Wineland, D., Itano, W., Bergquist, J., Bollinger, J. (1985), Trapped Ions and Laser Cooling. NBS Technical Note 1086. Washington, DC: US Govt. Printing Office.

[43] Migdall, A., Prodan, J., Phillips, W., Bergeman, T., Metcalf, H. (**1985**), *Phys. Rev. Lett.* **54**, 2596.

[44] Bergeman, T., Erez, G., Metcalf, H. (**1987**), *Phys. Rev. A* **35**, 1535.

[45] Bergeman, T. H., Balazs, N. L., Metcalf, H., Mcnicholl, P., Kycia, J. (**1989**), *J. Opt. Soc. Am. B* **6**, 2249–2256.

[46] Raab, E., Prentiss, M., Cable, A., Chu, S., Pritchard, D. (**1987**), *Phys. Rev. Lett.* **59**, 2631–2634.

[47] Metcalf, H. (**1989**), *J. Opt. Soc. Am. B* **6**, 2206–2210.

[48] Molenaar, P. (1995), Photoassociative Reactions of Laser-Cooled Sodium. Ph.D. Thesis, Utrecht University, Utrecht, The Netherlands.

[49] Cornell, E. A., Monroe, C., Wieman, C. E. (**1991**), *Phys. Rev. Lett.* **67**, 2439–2442.

[50] Steane, A. M., Foot, C. J. (**1991**), *Europhys. Lett.* **14**, 231–236.

[51] Steane, A. M., Chowdhury, M., Foot, C. J. (**1992**), *J. Opt. Soc. Am. B* **9**, 2142–2158.

[52] Wallace, C. D., Dinneen, T. P., Tan, K. Y. N., Kumarakrishnan, A., Gould, P. L.,

Javanainen, J. (**1994**), *J. Opt. Soc. Am. B* **5**, 703–711.

[53] Walker, T., Sesko, D., Wieman, C. (**1990**), *Phys. Rev. Lett.* **64**, 408–411.

[54] Sesko, D. W., Walker, T. G., Wieman, C. E. (**1991**), *J. Opt. Soc. Am. B* **8**, 946–958.

[55] Gibble, K. E., Kasapi, S., Chu, S. (**1992**), *Opt. Lett.* **17**, 526–528.

[56] Lindquist, K., Stephens, M., Wieman, C. (**1992**), *Phys. Rev. A* **46**, 4082–4090.

[57] Ketterle, W., Davis, K. B., Joffe, M. A., Martin, A., Pritchard, D. E. (**1993**), *Phys. Rev. Lett.* **70**, 2253–2256.

[58] Letokhov, V. S. (**1968**), *JETP Lett.* **7**, 272.

[59] Salomon, C., Dalibard, J., Aspect, A., Metcalf, H., Cohen-Tannoudji, C. (**1987**), *Phys. Rev. Lett.* **59**, 1659–1662.

[60] Castin, Y., Dalibard, J. (**1991**), *Europhys. Lett.* **14**, 761–766.

[61] Jessen, P. S., Gerz, C., Lett, P. D., Phillips, W. D., Rolston, S. L., Spreeuw, R. J. C., Westbrook, C. I. (**1992**), *Phys. Rev. Lett.* **69**, 49–52.

[62] Verkerk, P., Lounis, B., Salomon, C., Cohen-Tannoudji, C., Courtois, J. Y., Grynberg, G. (**1992**), *Phys. Rev. Lett.* **68**, 3861–3864.

[63] Lounis, B., Verkerk, P., Courtois, J. Y., Salomon, C., Grynberg, G. (**1993**), 1-D *Europhys. Lett.* **21**, 13–17.

[64] Gupta, R., Padua, S., Xie, C., Batelaan, H., Bergeman, T., Metcalf, H. (**1992**), *Bull. Am. Phys. Soc.* **37**, 1139.

[65] Gupta, R., Padua, S., Bergeman, T., Metcalf, H. (1993), Search for motional quantization of laser-cooled atoms, in E. Arimondo, W. Phillips, F. Strumia (Eds.), *Laser Manipulation of Atoms and Ions, Proceedings of Fermi School CXVIII, Varenna*, Amsterdam: North Holland.

[66] Anderson, B. P., Kasevich, M. A. (**1998**), *Nature* **282**, 1686–1689.

[67] Greiner, M., Mandel, O., Esslinger, T., Hänsch, T. W., Bloch, I. (**2002**), *Nature* **415**, 39–44.

[68] Grynberg, G., Lounis, B., Verkerk, P., Courtois, J. Y., Salomon, C. (**1993**), *Phys. Rev. Lett.* **70**, 2249–2252.

[69] Ben Dahan, M., Peik, E., Reichel, J., Castin, Y., Salomon C. (**1996**), *Phys. Rev. Lett.* **76**, 4508–4511.

[70] Westbrook, C. I., Watts, R. N., Tanner, C. E., Rolston, S. L., Phillips, W. D., Lett, P. D., Gould, P. L. (**1990**), *Phys. Rev. Lett.* **65**, 33–36.

[71] Birkl, G., Gatzke, M., Deutsch, I. H., Rolston, S. L., Phillips, W. D. (**1995**), *Phys. Rev. Lett.* **75**, 2823–2826.

[72] Masuhara, N., Doyle, J. M., Sandberg, J. C., Kleppner, D., Greytak, T. J., Hess, H. F., Kochanski, G. P. (**1988**), *Phys. Rev. Lett.* **61**, 935.

[73] Hess, H. F. (**1986**), *Phys. Rev. B* **34**, 3476.

[74] Doyle, J. M., Sandberg, J. C., Yu, I. A., Cesar, C. L., Kleppner, D., Greytak, T. J. (**1991**), *Phys. Rev. Lett.* **67**, 603.

[75] Luiten, O. J., Werij, H. G. C., Setija, I. D., Reynolds, M. W., Hijmans, T. W., Walraven, J. T. M. (**1993**), *Phys. Rev. Lett.* **70**, 544–547.

[76] Setija, I. D., Werij, H. G. C., Luiten, O. J., Reynolds, M. W., Hijmans, T. W., Walraven, J. T. M. (**1993**), *Phys. Rev. Lett.* **70**, 2257–2260.

[77] Davis, K. B., Mewes, M. O., Ketterle, W. (**1995**), *App. Phys. B* **60**, 155–159.

[78] Bagnato, V., Pritchard, D. E., Kleppner, D. (**1987**), *Phys. Rev. A* **35**, 4354.

[79] Monroe, C., Cornell, E., Sackett, C., Myatt, C., Wieman, C. (**1993**), *Phys. Rev. Lett.* **70**, 414.

[80] Luiten, O. J., Reynolds, M. W., Walraven, J. T. M. (**1996**), *Phys. Rev. A* **53**, 381.

[81] Huang, K. (1963), *Statistical Mechanics*. New York: Wiley.

[82] Landau, L. D., Lifshitz, E. M. (1958), *Quantum Mechanics (Non-Relativistic Theory)*. Oxford: Pergamon Press.

# Laser Isotope Separation

*Vladilen Letokhov and Evgeny Ryabov*
*Institute of Spectroscopy of Russian Academy of Sciences, 142190 Troitsk, Moscow Region, Russia*
*Phone: +7(095) 334 0579; Fax: +7(095) 334 0886; e-mail: letokhov@isan.troitsk.ru; ryabov@isan.troitsk.ru*

**Abstract**
This paper gives in a concise form, an idea of the principles of various laser isotope separation methods, the current status of studies in this field, and industrial applications of laser isotope separation techniques.

**Keywords**
separation; isotope; photoionization; photodissociation; laser; multiphoton.

## 1
## Introduction

Most atoms have several naturally occurring isotopes. The different isotopes of some specified elements have the same number of protons, but different number of neutrons. The isotopes are widely used in modern medicine, science and technology, as well as in power production. Though the history of isotope separation methods is more than half a century old, they continue to be further developed on the basis of the latest achievements of science and technology. In the 1970s, a new stage started in the development of high-efficiency isotope separation techniques using laser radiation and its fundamentally new capabilities to affect atoms and molecules.

All the isotope separation methods known to date [1, 2] can be divided into four groups: (1) electromagnetic, (2) kinetic, (3) physicochemical, and (4) optical or laser separation techniques. The latter are based on the isotope shift in the absorption spectra of atoms and molecules, which allows one to excite these particles by laser radiation in an isotope-selective fashion and then exert the desired influence on the chemical and physical processes involving them and depending materially on their degree of excitation.

The first successful experiment on separation of isotopes by means of light was performed by Kuhn and Martin [3] who excited phosgene molecules ($COCl_2$) by the 281.618-nm spectral line of aluminum spark. Later, a similar method was used to separate Hg isotopes in small quantities and also isotopes of some other elements. Even these first experiments demonstrated the high enrichment factor typical of the optical method in a one-cycle separation process. However, it was only the advent of laser sources that made it possible to realize all the potential advantages of the optical isotope separation method.

Laser radiation possesses unique characteristics from the standpoint of optical isotope separation. It can be highly monochromatic and can have a high intensity, quite sufficient to saturate the absorbing transition, that is, to excite a substantial portion of the absorbing atoms or molecules. Its duration can be much shorter than the lifetime of excited states in atoms and molecules, and its space coherence is high enough to allow the laser beam to be collimated to irradiate extended volumes of matter.

The existing laser isotope separation (LIS) methods can be classified [4] in accordance with the type of process implemented with the atoms or molecules excited in an isotope-selective fashion (Fig. 1). A common feature of all these methods is the isotope-selective excitation of an electronic or a vibrational level in a quantum transition showing a distinct isotope shift. Thereafter, several entirely different approaches to the organization of the separation process are possible [5, 6].

One can organize a chemical reaction between the atom or molecule subject to excitation with a suitable partner before the excited particle starts to relax to its ground unexcited state or gives its excitation to some particle of undesirable isotopic composition upon inevitable collisions (see Fig. 1). Therein lies the *photochemical approach*. At the root of the *photophysical approach* is the subsequent laser-induced dissociation or ionization of the particles preexcited in an isotope-selective manner (see Fig. 1). In both approaches, there are formed new products enriched with the desired isotope, which strongly differ from the original substances. The latter circumstance makes these enriched products easy

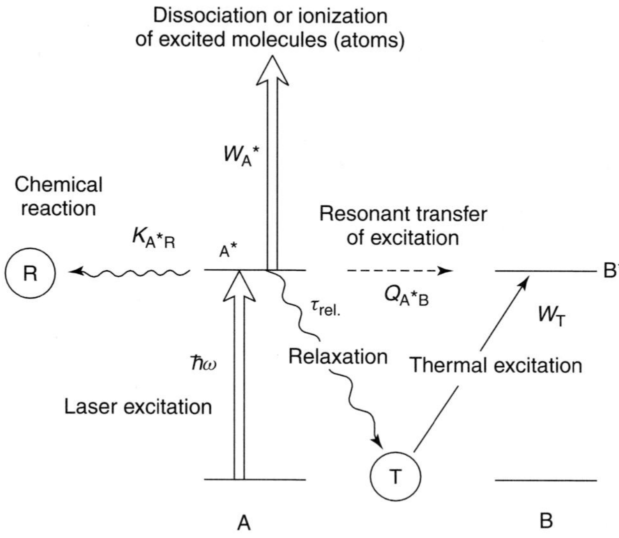


**Fig. 1** General scheme of selective photoexcitation of particles A by laser light and classification of the processes responsible for the loss and conservation of excitation selectivity

to extract from the mixture by the standard methods.

The photochemical approach has, as yet, not found any wide application. This is due to the difficulty of finding a suitable chemical reaction whose rate $K_{A^*R}$ substantially exceeds both the relaxation rate $\tau_{rel.}^{-1}$ of the preexcited particles and the rate $Q_{A^*B}$ with which excitation is transferred in a resonance fashion to particles containing the other isotope. (Here and below, the asterisk is used to denote the excited particles.) It is only when the condition

$$K_{A^*R} \gg Q_{A^*B}, \tau_{rel.}^{-1} \qquad (1)$$

is satisfied that one can expect high isotopic selectivity and efficiency of the photochemical process.

The high intensity of visible, UV, and IR-pulsed laser radiation made it possible to develop different photophysical methods of affecting atoms and molecules in an isotope-selective manner. These include the method of resonance stepwise photoionization of atoms [7–9] (this method was given the acronym AVLIS, from Atomic Vapor Laser Isotope Separation), the method of resonance stepwise IR-UV photodissociation of molecules [9, 10], and the method of resonance multiple-photon dissociation (MPD) of polyatomic molecules by IR radiation [11, 12] (the latter two methods were given the acronym MLIS, from Molecular Laser Isotope Separation). It was precisely these methods that formed the basis for the LIS techniques that are being practically implemented today. In effecting the photoionization of the atom, or the photodissociation of the molecule of interest by an additional laser pulse (Fig. 1), it is necessary that the rate $W$ of these processes be higher than the excitation transfer rate $Q_{A^*B}$ and the relaxation rate $\tau_{rel.}^{-1}$. As the induced photoionization and photodissociation rates are proportional to the radiation intensity

$I$, the condition for high selectivity and efficiency of the photophysical ionization and dissociation processes,

$$W_{\mathrm{ph.i.}}(I),\ W_{\mathrm{ph.d.}}(I) \gg Q_{\mathrm{A^*B}},\ \tau_{\mathrm{rel.}}^{-1}, \qquad (2)$$

can always be satisfied. These considerations equally apply to other photophysical ionization and dissociation processes, specifically to the multistep photoionization of atoms and MPD of molecules, wherein the particle being excited absorbs not two, but much more laser photons. Of course, in all cases it is necessary to meet the requirement that the isotopic selectivity of the process must not be lost in the secondary processes involving the ionization or dissociation products, and this is much easier with the products of photoprocesses than with selectively excited particles.

## 2 Advantages of Laser Isotope Separation

The potential advantages of the laser methods for separating isotopes over the traditional techniques are as follows [4, 5, 13].

1. *The high selectivity of the elementary separation act*: The separation factor (selectivity) $\alpha$ is defined as the ratio between the relative yields $\beta_i$ and $\beta_j$ of the two isotopes ($\alpha = \beta_i/\beta_j$). In the majority of the traditional techniques, the separation factor $\alpha$ in a single stage is very small, close to unity [1, 2], and so to attain a high degree of separation, $K$, one has to use a great many ($n \gg 1$) separation stages:

$$K = \alpha^n \gg 1 \qquad (3)$$

The laser methods of isotope-selective ionization of atoms and dissociation of molecules can, in principle, provide for $\alpha \gg 1$. This makes it possible to diminish the number of separation stages many times, reducing it in many cases to one separation stage.

2. *Low expenditures of energy*: For the traditional methods, the expenditure of energy per atom of the objective isotope is usually 10 to $10^4$ times higher than for the laser methods [4, 5].

3. *The possibility of separating the desired isotope* without affecting the other isotopes of the same element. When the element of interest has several isotopes of intermediate mass (for example, $^{17}$O or $^{40}$K), one can selectively excite the atoms or molecules of the desired isotopic species and separate them without affecting in any substantial manner the particles of other isotopic species by using tunable laser radiation of appropriate frequency.

4. *Noncontact separation capability*, that is, the possibility of implementing the separation process by means of laser radiation alone. The absence in the laser methods of the traditional separating elements (plates, membranes, etc.) minimizes the contact between the product being enriched and the surface of the separating plant and hence the contamination of the latter. This is especially important, for example, in separating radioactive isotopes.

5. *Versatility*: With rare exception, the laser methods can be used with a sufficiently high efficiency to separate isotopes of any elements, be they light, medium weight, or heavy, as distinct from the traditional methods whose efficiency depends on the mass of the isotope being enriched.

6. *Short start-up time*: With the laser methods, steady state operation is attained practically at once, whereas in the traditional

methods, the start-up period may last a few months [1, 2, 13].

## 3
## Photoionization Isotope Separation Method (AVLIS)

The method of isotope-selective photoionization of atoms [8] is, in principle, suitable for separating isotopes of any element, but it is more expedient to use it for those elements that are difficult to incorporate in suitable molecular compounds. In all cases, the substance being irradiated is an atomic beam in a vacuum chamber.

The isotope-selective photoionization of atoms can be carried into effect by several schemes whose common feature is the preliminary isotope-selective laser excitation of one or several levels in sequence in the atoms of interest, followed by the ionization of the excited atoms by means of an auxiliary laser. Some most widespread schemes of selective ionization of atoms are presented in Fig. 2. To excite atoms of a certain isotopic species, use is usually made of a single pulse of laser radiation with a narrow, well-controlled spectrum tuned in to the absorption line of the objective isotope. In some cases, the excited atoms are directly ionized by an additional UV laser pulse in a transition to the continuum (see Fig. 2a), but most frequently one more step of excitation to a higher state is used (see Fig. 2b) in order that ionization can be effected by means of the more accessible visible lasers, even in the case of atoms with an ionization potential of $I = 6–8$ eV.

The direct photoionization in a transition to the continuum has the material disadvantage that the ionization cross section $\sigma_{\text{ph.i.}}$ is small in comparison with the resonance excitation cross section $\sigma_{\text{res.}}$ ($\sigma_{\text{ph.i.}}/\sigma_{\text{res.}} = 10^{-4} \div 10^{-6}$), which makes the process difficult to put into practice. For this reason, the following more efficient methods of ionization of excited atoms have been developed and put to use:

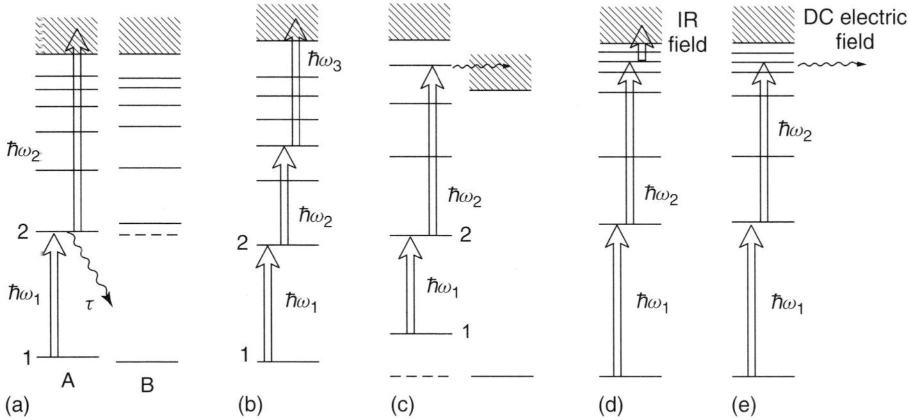


Fig. 2 Schemes of selective stepwise photoionization of atoms by laser radiation: (a) two-step photoionization; (b) three-step photoionization; (c) stepwise selective excitation of an autoionization state; (d) stepwise selective excitation of a Rydberg state and its subsequent photoionization by an IR laser; (e) stepwise selective excitation of a Rydberg state and its subsequent ionization by a DC electric field

1. Final-stage excitation of a narrow autoionization state that decays spontaneously into the continuum (see Fig. 2c).
2. IR laser excitation of a state lying comparatively close to the ionization limit in atoms featuring an increased ionization cross section (see Fig. 2d).
3. Excitation of a high-lying (Rydberg) state near the ionization limit that can easily be ionized by a DC field pulse (see Fig. 2d).

The above three methods enable one to effectively ionize selectively excited isotopic atoms of any element in the periodic table that is of interest to isotope separation by the atomic scheme. The separation factor $\alpha$ here can reach $10^2$ to $10^5$ and more.

Apart from separating isotopes, the selective laser ionization technique can be used to separate atoms with different isomeric nuclei [14, 15]. This technique is also being successfully used for separating isobars, that is, atoms of different elements having the same mass [16].

Laser isotope separation based on the stepwise ionization of atoms seems versatile and selective enough, but at the same time it is a sufficiently complex process of separating substances at the atomic level. Compared to the processes based on the photodissociation of molecules (described later in the text), it suffers from the following drawbacks: the need to vaporize the substance, which sometimes requires a high temperature, the need for the laser pulses to have a high repetition frequency because the atoms rapidly leave the irradiation region, and the need to excite atoms from many different ground states if it is necessary to completely extract the desired isotope with the original substance being vaporized only once.

## 4 Photodissociation Isotope Separation Techniques (MLIS)

Isotopes of any element in the periodic table that can be incorporated into a suitable molecular compound at a vapor pressure of 10 to 1000 Pa and normal temperature are amenable to separation by the method of isotope-selective photodissociation of molecules by laser radiation, that is, by the MLIS technique. Many molecular compounds have a well-resolvable isotope shift in their IR-absorption spectra. Therefore, to separate isotopes incorporated in such compounds, use can be made of dissociation techniques employing selective vibrational excitation. A common feature of all such techniques is the isotope-selective one- and multiple-photon excitation of vibrational levels, followed by the dissociation of the vibrationally excited molecules, either by UV radiation exciting an electronic state in the molecules [9, 10], or by intense IR radiation, with the molecules residing in their electronic ground state [11, 12].

### 4.1 IR-UV Photodissociation

The isotope-selective two-step photodissociation of molecules consists in the pulsed isotope-selective excitation of a vibrational state in the molecules by laser radiation and subsequent pulsed photodissociation of the vibrationally excited molecules alone. The second stage must be done before the preexcited molecules transfer their excitation to molecules of a different isotopic composition, that is, before the isotopic selectivity of excitation on the first stage is lost (Fig. 3a). The selective two-step photodissociation of molecules is possible if their excitation is accompanied

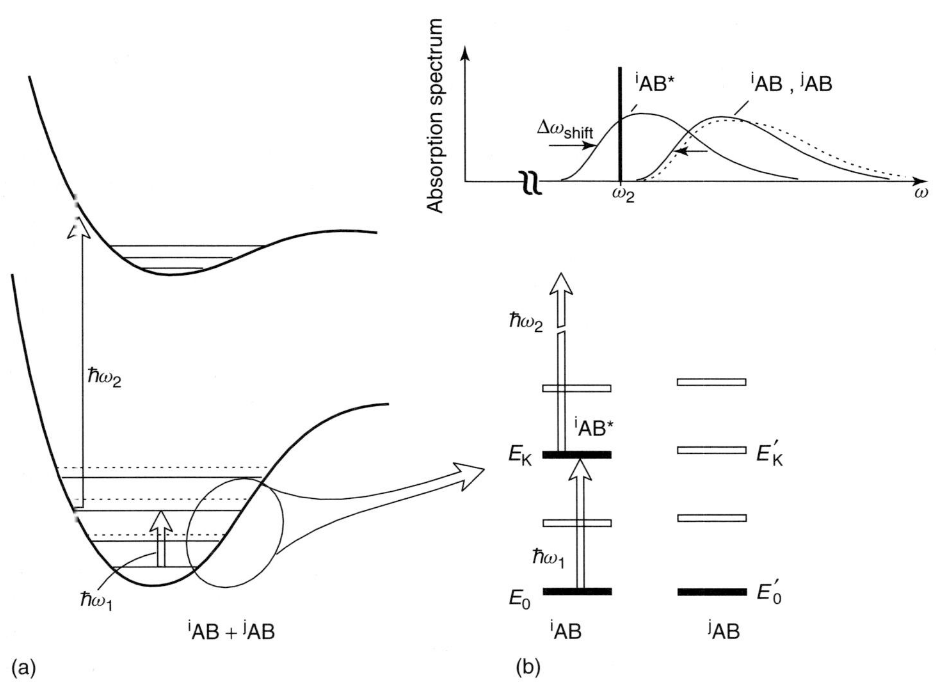


**Fig. 3** (a) Schematic diagram of the IR-UV isotope-selective photodissociation of molecules $^i$AB in a mixture with molecules $^j$AB of different isotopic composition via an intermediate vibrational state (for example, $\nu = 2$); (b) Absorption spectrum

by a shift of their photoabsorption band, leading to their photodissociation. In that case, molecules of the desired isotopic composition, selectively excited by a laser pulse of frequency $\omega_1$, can be photodissociated by a second laser pulse of frequency $\omega_2$ selected to fall within the shift region where the ratio between the absorption coefficients of the excited and unexcited molecules is maximum (Fig. 3b).

The first experiment on isotope separation by the IR-UV two-step isotope-selective photodissociation was performed in 1972 [17] with $^{14}NH_3$ and $^{15}NH_3$ molecules. The isotope-selective excitation of several vibrational levels in a polyatomic molecule through multiple-photon absorption of intense IR radiation can provide for a much greater shift of its UV absorption band toward the long-wavelength region than in the case of one-photon absorption. This extends the possibilities of using the IR–UV photodissociation method for isotope separation purposes. The method was studied in detail in experiments with $^{12}CF_3I$ and $^{13}CF_3I$ molecules [18].

### 4.2
### Multiple-photon IR Photodissociation

In addition to the photodissociation of molecules via excited vibrational and electronic states under the effect of IR–UV radiation, there can also take place their isotope-selective photodissociation in the electronic ground state on absorption of a large number (20–50) of IR photons. This phenomenon was for the first time observed in 1974 at the Institute of Spectroscopy of the USSR Academy of Sciences in separating the $^{10}B$ and $^{11}B$ isotopes in the $BCl_3$ molecule [11]. Thereupon successful experiments were performed on separating $^{32}S$ and $^{34}S$ isotopes in the $SF_6$ molecule [12]. Later on, the effect of isotope-selective IR MPD was demonstrated for a few tens of various isotopes in around a hundred different molecules [19]. The selectivity $\alpha$ ranged from a few units in molecules with isotopes of heavy elements ($UF_6$) to some $10^4$ for isotopes of light elements (hydrogen, deuterium).

The photodissociation of molecules in the electronic ground state under the effect of a resonance IR radiation pulse is a universal phenomenon for polyatomic molecules. Such molecules are characterized by a great number of vibrational states interacting with one another because of anharmonicity, the interaction growing stronger as the vibrational energy level of the molecule is increased. As a result of this interaction, starting with some threshold energy $E_{\text{onset}}$, the vibrational motion becomes stochastic, so that a vibrational quasicontinuum is formed (for details, see, e.g., [6, 19]). The spectra of vibrational transitions in the quasicontinuum become sufficiently broad to provide for the possibility of quasiresonance excitation of the molecules up to the dissociation limit and higher, where they dissociate. Figure 4 schematically illustrates the main characteristics of a vibrationally excited molecule that are essential for its isotope-selective IR MPD.

Studies into the evolution of the vibrational spectra of polyatomic molecules subject to increasing vibrational excitation

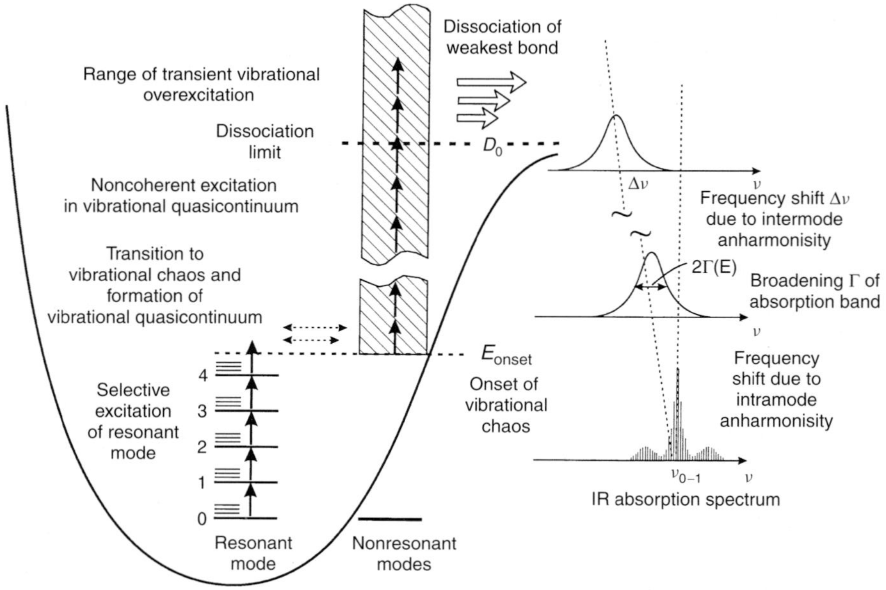


**Fig. 4** Simplified model of the multiple-photon excitation and dissociation of a polyatomic molecule by an intense IR field: (left) selective excitation of the resonance mode and subsequent quasiresonance excitation in the quasicontinuum above its boundary $E_{\text{onset}}$; (right) evolution of the IR absorption spectra near the fundamental mode frequency with increasing vibrational energy

helped develop various IR dissociation schemes capable of a maximum isotopic selectivity (separation factor) and dissociation yield. For example, MPD in a two-frequency IR field, with the isotope-selective excitation and dissociation functions being separated, [20] and dissociation of comparatively simple polyatomic molecules with a high-lying quasicontinuum boundary in a multiple-frequency IR field can provide for a separation factor as high as $10^3$ to $10^4$ [21].

One more method of multiple-frequency IR dissociation of molecules is based on the excitation of overtones of high-frequency vibrations (C–H, Si–H, etc.) in them as a result of one-photon absorption and the subsequent IR MPD of these overtone preexcited (OP) molecules [22]. By virtue of the high selectivity of the first step, this method can provide for high dissociation selectivity. To illustrate, in the OP IR MPD of $^{13}CF_3H$ molecules preexcited to the second C–H stretch overtone in [22], the separation factor $\alpha(13/12)$ reached around 9000.

Note in conclusion that apart from molecules, one can use the Van der Waals clusters for isotope separation purposes. Specifically, such clusters are formed when molecules are made to escape into a vacuum in a supersonic manner [23]. To dissociate them, absorption of a single IR photon usually proves sufficient. The dissociation products – the molecules containing the target isotope – escape from a collimated beam, and this fact can be used for isotope separation purposes [24]. This approach seems very attractive. However, the scaling of this process to the level of industrial facility presents certain difficulties because of the small scattering angle of the photodissociation fragments.

## 5 Development of Industrial LIS Technology

### 5.1 The Scaling of the LIS Process

The above-considered LIS methods based on the photoionization of atoms and photodissociation of molecules (the AVLIS and the MLIS approach respectively) are capable of very high elementary separation act parameters (ESAP), namely, the yield of the desired isotope, $\beta_i$, and the separation factor (selectivity), $\alpha = \beta_i/\beta_j$. However, going over from tabletop laboratory experiments to industrial technology has required a great deal of work on the scaling of the process. To develop an efficient industrial LIS technology, it is necessary that the following two initial conditions be satisfied.

1. Availability of a laser radiation source with an average power from 1 kW to 1 MW, depending on the required output capacity.
2. Realization of sufficiently high ESAP for the given working substance with the desired isotope.

Besides, when scaling an LIS process, it is necessary to retain the ESAP unchanged, when irradiating the working substance with high-repetition-rate pulses of high average power, to provide for high radiation utilization factor, and to implement an optimal process scheme, including the extraction and subsequent processing of the enriched products. All these problems have been solved, to some degree, when developing the AVLIS and MLIS technologies.

### 5.2 The Scaling of the AVLIS Technology

In the AVLIS approach, the separation process consists of three stages:

1. Vaporization of the starting material and obtaining a flow of atomic vapor of the necessary density and volume.
2. Selective photoionization of atoms with the desired isotope by laser radiation.
3. Extraction and collection of the ions formed on special collectors.

The selectivity of multiple-step photoionization of atoms may be very high: $\alpha \geq 10^2$ to $10^5$. But this is only true of single, noninteracting atoms. To attain high output capacity in the industrial implementation of the AVLIS approach requires using increased atomic vapor densities. This substantially reduces the separation factor reached in practice. A number of negative factors limit the density of the atomic vapor and the size of the irradiation region to the following typical values: vapor density $\leq 10^{14}$ cm$^{-3}$; cross-sectional dimension of the irradiation region – a few centimeters (mainly because of resonance charge exchange); length of the irradiation region – a few tens of meters.

The AVLIS approach also places a number of constraints on the laser radiation source. The characteristic lifetime of excited atomic states, as regards their spontaneous decay, ranges from $10^{-8}$ to $10^{-7}$ s. Therefore, the laser pulse duration should not be longer than a few tens of nanoseconds. The characteristic velocity of particles in atomic vapors is around $10^5$ cm s$^{-1}$. For this reason, to maximize the "interception" of the atoms crossing the laser beam, the laser pulse repetition frequency should be a few tens of kilohertz.

In developing industrial AVLIS technology, primary attention was concentrated on the separation of uranium isotopes. These works were performed in the United States [25], France [26], Japan [27], and some other countries. Native uranium contains two isotopes – $^{235}$U (0.7%) and $^{238}$U (99.3%). For uranium to be suitable for use as a fuel in nuclear power engineering, it is necessary to raise its $^{235}$U content to some 3.2%. The ionization potential of atomic uranium equals 6.19 eV, and so a three-step photoionization scheme using visible radiation quanta was selected for the purpose. To vaporize metallic uranium, use was made of a beam of accelerated electrons. Dye lasers were used as a radiation source. These were pumped by copper-vapor lasers that provided for the necessary 20- to 30-ns pulse duration at a repetition frequency of around $10^4$ Hz. In the first half of the 1980s, an experimental separation plant with a potential output of around $10^3$ kg of low-enrichment uranium per year was built at Lawrence Livermore National Laboratory (LLNL) in the United States. The production of a few kilograms of such uranium was reported. Early in the 1990s, the "First Plant Prototype" was started and tested at LLNL. The total power output of the dye lasers amounted to 24 kW. The possibility was discussed of building a full-scale plant with an output capacity of up to 1700 tons of 3 to 5% enriched uranium per year. However, in 1999, work on the AVLIS technology at LLNL was announced suspended. Further progress in the development of this technology, as applied to uranium, will apparently be possible when the switch takes place to the new generation of laser radiation sources based on wholly solid-state systems.

As for the separation of the isotopes of other elements, LLNL reported experiments on the industrial-scale extraction of the $^{157}$Gd isotope. There is also a report [28] on the development of an industrial plant for the commercial production of the $^{168}$Y isotope for medical applications. The output capacity of the plant is around a gram per month.

## 5.3
## The Scaling of the MLIS Technology

In the MLIS approach, the isotope of interest is incorporated in a polyatomic molecule. As compared with the AVLIS technology, this opens up additional possibilities for selecting a starting working compound suitable for the existing IR lasers. The most serious work on the scaling of the MLIS technology was done in separating isotopes of uranium and carbon.

### 5.3.1 Separating Uranium Isotopes

The only molecular uranium compound suitable for using the MLIS approach is the $UF_6$ molecule. The highest-frequency vibration, $\nu_3$, in this molecule that is active in IR absorption lies in the region of 16 μm. The isotope shift in this vibration between $^{235}UF_6$ and $^{238}UF_6$ is comparatively small, that is, $\Delta\nu_{is} \approx 0.6\ cm^{-1}$. Therefore, to resolve the isotopic structure of IR transitions, use is made of gas-dynamic cooling of $UF_6$ to temperatures $\lesssim$100 K. To effect isotope-selective IR excitation of $UF_6$, special lasers tunable in the region of 16 μm were developed.

Work on the development of the MLIS technology was done in the United States, Japan, France, and some other countries. Subject to investigation were two approaches – two-step IR + UV photodissociation and IR MPD. The first approach was developed at Los Alamos National Laboratory (USA). But for a number of reasons this project gave way to the AVLIS project and was closed early in the 1980s. The approach based on the purely infrared excitation and dissociation of $UF_6$ seems to be more promising, both economically and technologically. Use can be made of both two- and three-frequency MPD. By appropriately selecting the laser radiation parameters, one can attain the necessary selectivity $\alpha$ ($^{235}U/^{238}U$) $\geq$4, which is, in principle, sufficient for the production of nuclear fuel. The development of an industrial prototype of a uranium isotope separation plant based on this MLIS version was reported.

### 5.3.2 Carbon-isotope Separation

Two stable carbon isotopes – $^{13}C$ and $^{12}C$ – exist in nature in concentrations of around 1.1 and 98.9% respectively. The $^{13}C$ isotope is widely used in medicine for the noninvasive diagnosing of a number of diseases and also in biological investigations. The $CF_2HCl$ (Freon-22) molecule proved most suitable for large-scale laser separation of carbon isotopes. The IR MPD of $CF_2HCl$ by $CO_2$ laser radiation gives rise to $CF_2$ fragments that thereupon combine together to form the $C_2F_4$ molecule. The $CF_2HCl$ molecule provides for high ESAP values at moderate laser radiation fluence, $\Phi = 5-10\ J\ cm^{-2}$. The IR MPD yield of the objective isotope $^{13}C$ amounts to 1 to 10%. The selectivity $\alpha(^{13}C/^{12}C)$ of the process can be as high as $10^2$ to $10^3$. What is important is that the working pressure of $CF_2HCl$ gas may in that case be as high as a few kilopascals. Another important circumstance is that Freon-22 is a sufficiently low-priced product manufactured on a mass scale. Once the $^{13}C$ isotope is extracted, it is returned to the manufacturer, which solves the problem of utilization of the used working compound and substantially improves the economic characteristics of the LIS process.

In Russia (Institute of Spectroscopy of the Russian Academy of Sciences, Kurchatov Institute of Atomic Energy), an industrial carbon LIS technology was developed on the basis of the IR MPD of $CF_2HCl$ and a plant was built by the Gas-Oil Co. in the city of Kaliningrad for commercial production of the $^{13}C$ isotope using this technology. With this technology, the separation reactor is placed inside the cavity of a high-pulse-repetition-rate (HPRR) TEA $CO_2$ laser. Such an arrangement is equivalent to a multiple-pass cell and provides for high radiation utilization factor. The working substance is a mixture of $CF_2HCl$ and $N_2$. Nitrogen is used as a buffer gas to protect the optical elements and improve selectivity (see e.g., [19]). A block diagram of one of the modules of this plant is presented in Fig. 5. The laser separation block includes an HPRR TEA $CO_2$ laser and a separation reactor wherein the working mixture is irradiated. The parameters of this $CO_2$ laser are as follows: pulse repetition rate up to 600 pps, pulse energy up to 3 J, and average power up to 1.8 kW. The irradiated mixture is fed to the rectification block where the enriched tetrafluoroethylene is extracted. In the conversion block, this $C_2F_4$ is converted into $CO_2$ that is delivered to customers. The plant consists of three such modules with a total output of up to 1.5 g of 40% $^{13}C$ per hour. If necessary, further enrichment to 99% can be effected by conventional methods or by means of a second laser enrichment stage.

## 6
## Conclusion

In the course of the past 30 years of intense research and development work, the main LIS methods in the atomic (AVLIS) and molecular (MLIS) versions have been found and studied in detail.

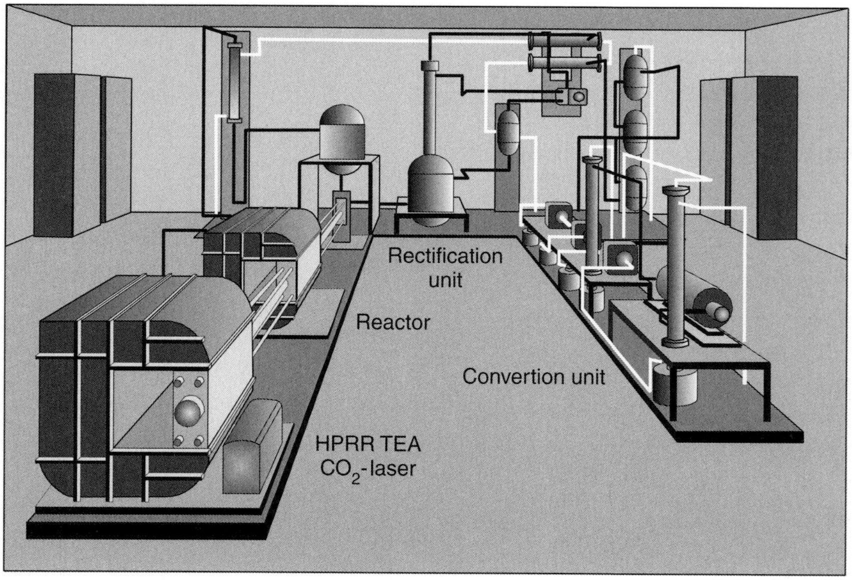


**Fig. 5** Block diagram of one module of the laser carbon-isotope separation plant

Those methods have been selected among them, which can be used for the industrial production of isotopes on various scales, in quantities of both up to a kilogram for scientific and medical applications and up to a few tons for nuclear power plants. The large-scale production of isotopes requires highly reliable, efficient, and cost-effective lasers with an average energy from a few kilowatts to a few hundreds of kilowatts. Among such lasers are $CO_2$-gas lasers that are being successfully used in the MLIS of $^{13}C/^{12}C$. The AVLIS of uranium isotopes, for example, requires tunable visible lasers with a high pulse repetition frequency (over 10 kHz). At the first stage, use was made, for this purpose, of dye lasers pumped by Cu-vapor laser pulses, which made the AVLIS of uranium isotopes little probable because of the high maintenance costs of such lasers in industrial production. However, the development of semiconductor laser arrays for pumping solid-state lasers is progressing so swiftly that it seems quite realistic that the AVLIS technology based on this new, much more consummate laser equipment will be implemented on an industrial scale in the near future, as the consumption of isotopic fissile materials grows higher.

## Glossary

**AVLIS:** Atomic Vapor Laser Isotope Separation.

**Dissociation:** Separation of a molecule into atoms or group of atoms.

**Ionization:** Loss of an electron by the neutral atom or molecule.

**IR MPD:** Infrared Multiple Photon Dissociation.

**Isotope:** One of two or more atoms having the same atomic number but differing in atomic weight.

**MLIS:**
Molecular Laser Isotope Separation.

## References


[1] London, H. (1961), *Separation of Isotopes.* London: G. Newnes.

[2] Chemla, M., Perre, J. (1974), *La Separation des Isotopes.* Paris: Press Univ. de France.

[3] Kuhn, W., Martin, H. (**1933**), *Z. Phys., Chem. Bd.* **21**, 93.

[4] Letokhov, V. S. (**1973**), *Science* **180**(4085), 451–458.

[5] Letokhov, V. S., Moore, C. B. (1977), in C. B. Moore (Ed.), *Chemical and Biochemical Applications of Lasers,* Vol. III. Orlando: Academic Press, Chap. 1, pp. 1–165.

[6] Letokhov, V. S. (1983), *Nonlinear Laser Chemistry. Multiple Photon Excitation.* Berlin: Springer-Verlag.

[7] Letokhov, V. S. (1970), Soviet Patent No. 7,846,79 (appl. on March, published on August 1, 1980).

[8] Ambartzumian, R. V., Kalinin, V. P., Letokhov, V. S. (**1971**), *Sov. Phys. JETP Lett.* **13**, 217–219.

[9] Letokhov, V. S., Ambartzumian, R. V. (**1971**), *IEEE J.* **QE-7**, 305–306.

[10] Letokhov, V. S. (1970), Soviet Patent No. 7,846,80 (appl. on March 30, published on August 1, 1980).

[11] Ambartzumian, R. V., Letokhov, V. S., Ryabov, E. A., Chekalin, N. V. (**1974**), *Sov. Phys. JETP Lett.* **20**, 273–275.

[12] Ambartzumian, R. V., Gorokhov, Y. A., Letokhov, V. S., Makarov, G. N. (**1975**), *Sov. Phys. JETP Lett.* **21**, 171–174.

[13] Villani, S. (Ed.) (1979), *Uranium Enrichment, Topics in Applied Physics,* Vol. 10, Berlin: Springer-Verlag.

[14] Letokhov, V. S. (**1973**), *Opt. Commun.* **7**, 59–60.

[15] Mishin, V. I., Sekatskii, S. K., Fedoseyev, V. N., Buyanov, N. B., Letokhov, V. S., Barzakh, A. E., Denisov, V. I., Dernyatin, A. G.,

Ivanov, V. S., Chubukov, I. Ya., Alkazov, G. D. (**1987**), *Opt. Commun.* **61**, 383–386.

[16] Mishin, V. I., Fedoseyev, V. N., Kluge, H.-J., Letokhov, V. S., Ravn, H. L., Scheerer, F., Shirakabe, Y., Sundell, S., Tengblad, O., ISOLDE Collaboration (**1993**), *Nucl. Instrum. Methods Phys. Res.* **B73**, 550–560.

[17] Ambartzumian, R. V., Letokhov, V. S., Makarov, G. N., Puretzkii, A. A. (**1972**), *Sov. Phys. JETP Lett.* **15**, 501–503.

[18] Knyazev, I. N., Kudriavtsev, Yu. A., Kuzmina, N. P., Letokhov, V. S., Sarkisian, A. S. (**1978**), *Appl. Phys.* **17**, 427–429.

[19] Bagratashvili, V. N., Letokhov, V. S., Makarov, A. A., Ryabov, E. A. (1985), *Multiple Photon Infrared Laser Photophysics and Photochemistry*. Chur: Harwood Academic Publications.

[20] Ambartzumian, R. V., Gorokhov, Y. A., Letokhov, V. S., Makarov, G. N., Puretzkii, A. A. (**1975**), *Sov. Phys. JETP Lett.* **22**, 177–178.

[21] Evseev, A. V., Letokhov, V. S., Puretzkii, A. A. (**1985**), *Appl. Phys. B* **B36**, 93–103.

[22] Boyarkin, O. V., Kowalczyk, M., Rizzo, T. R. (**2003**), *J. Chem. Phys.* **118**, 93–103.

[23] Hagena, O. F., Obert, W. (**1972**), *J. Chem. Phys.* **56**, 1793–1802.

[24] Zellweger, J., Phillippoz, J., Melinon, P., Monot, R., van den Bergh, H. (**1984**), *Phys. Rev. Lett.* **52**, 522–525.

[25] Lawrence Livermore National Laboratory (**2000**), *Sci. Technol. Rev.*, May issue, 13–17.

[26] Camarcat, N., Lafon, A., Perves, J. P., Rosengard, A. (1993), *Laser Isotope Separation*, SPIE 1859, pp. 14–23.

[27] Marioka, K. (1993), *Laser Isotope Separation*, SPIE 1859, pp. 2–13.

[28] Yakovlenko, S. I. (**1988**), *Quant. Electron.* **25**, 971–987.

## Further Reading

Lupo, W. D., Quack, M. (**1987**), *Chem. Rev.* **87**, 181–216.

Lyman, J. L. (1987), in L. J. Radziemski, R. W. Solarz, J. A. Paisner (Eds.), *Laser Spectroscopy and Its Applications*. New York, Basel: Marcel Dekker, pp. 417–505.

Paisner, J. A. (**1988**), *Appl. Phys.* **B46**, 253–260.

# Laser Physics

*John J. Zayhowski*
*MIT Lincoln Laboratory, Lexington, MA 02420–9108, USA*
*Phone: 781-981-0701; Fax: 781-981-0602; e-mail: Zayhowski@LL.MIT.EDU*

*Paul L. Kelley*
*Department of Electrical Engineering, Tufts University, Medford, MA 02155, USA*
*e-mail: pkelley@tufts.edu*

**Abstract**
This article concerns the physics of lasers, including population inversion, pumping processes, stimulated emission, and resonators. Important laser characteristics such as narrow spectral linewidth, high spatial collimation, and ultrashort pulses are discussed. A review is given of several types of laser media, laser dynamics, and modes of laser operation. Methods for control of laser output and practical considerations in laser design are considered. Finally, there is a discussion of nonlinear optical techniques for frequency conversion of laser radiation.

**Keywords**
laser; coherent radiation; laser, solid-state; laser, gas; laser, semiconductor; laser dynamics; rate-equation model; tunable lasers; mode-locked lasers; optical amplifiers; frequency conversion.

## 1 Introduction

The laser is a device that generates coherent, highly directional electromagnetic radiation somewhere in the wavelength range from submillimeter through X-ray. Lasers can operate at a single wavelength (and frequency) or, when mode locked, on a large number of frequencies. The word *laser* is an acronym for "light amplification by stimulated emission of radiation." The principle of operation of lasers is similar to that of the maser, which is somewhat arbitrarily defined as a device operating in the range from the radio or microwave region down to millimeter wavelengths. Since the first laser was operated in 1960, the laser has come to play an important role through its revolutionary impact on applied optical technology, including fiber-optical communications and optical data storage.

## 2 Basic Theory of Operation

### 2.1 Population Inversion and Stimulated Emission

Quantum theory shows that matter can exist only in certain allowed energy levels or states. In thermal equilibrium, lower-energy states of matter are preferentially populated, with an occupation probability proportional to $e^{-E/kT}$, where $E$ is the state energy, $T$ the temperature, and $k$ the Boltzmann constant. An excited state can decay spontaneously (i.e., with only zero-point electromagnetic radiation present) to a lower-energy state, emitting a quantum or wave packet of electromagnetic radiation (photon) with transition frequency $\nu = \Delta E/h$, where $\Delta E$ is the energy difference between the two states and $h$ is Planck's constant. In the presence of radiation at frequency $\nu$, a transition from the upper state to the lower state can be induced, with the simultaneous emission of a photon in phase (coherent) with the stimulating radiation. This stimulated emission process is the reverse of the absorption process. If matter can be forced out of thermal equilibrium to a sufficient degree, so that the upper state has a higher population than the lower state (population inversion), more stimulated emission than absorption occurs, leading to coherent growth (amplification or gain) of the electromagnetic wave at the transition frequency.

Laser transitions in the optical region are most often electric dipole in character. In the dipole approximation to the Hamiltonian, the transitions arise from a term of the form $er\mathcal{E}$, where $e$ is the electronic charge, $r$ is the quantum-mechanical coordinate operator defined relative to the center of coordinates of the material system (such as an atom or molecule), and $\mathcal{E}$ is the electric field of the optical wave at the center of coordinates. The transition rate and gain cross-section are proportional to the square of this interaction term. The transition-matrix element of the coordinate operator between upper and lower laser levels ranges from about one hundredth of a Bohr radius ($\approx 0.5 \times 10^{-8}$ cm), for vibrational transitions in molecules and for local-field-induced transitions of rare earths in solids, to several hundreds of Bohr radii, for highly excited Rydberg atoms.

A laser generally consists of three components: (1) an active medium with energy levels that can be selectively populated, (2) a pump to produce population inversion between some of these energy levels, and (usually) (3) a resonant electromagnetic cavity that contains the active medium and provides feedback to maintain the coherence of the electromagnetic field (see Fig. 1). In a continuously operating laser, coherent radiation will build up in the cavity to the level required to balance the stimulated emission and cavity losses (see Sect. 5.3). The system is then said to be lasing, and radiation is emitted in a direction defined by the cavity.

### 2.2 Pumping and Relaxation Processes

A material system can become excited and displaced from normal thermal equilibrium when driven by processes

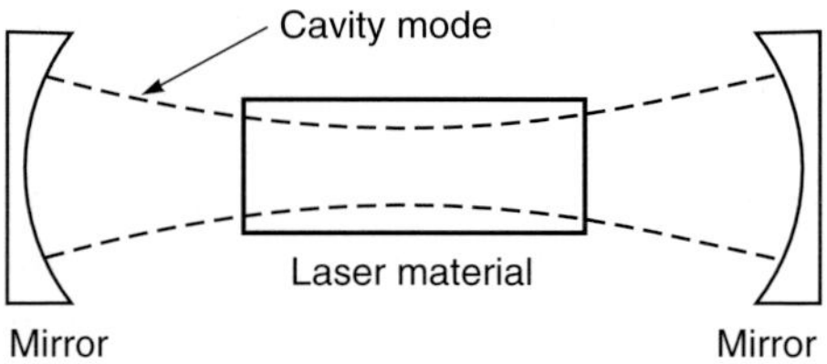


**Fig. 1** Simplified schematic of a laser oscillator. The mirrors at the ends of the laser form an open resonator. Stable modes that consist of electromagnetic waves that travel back and forth in the resonator are amplified by the active laser material. In the radiative steady state, the gain due to amplification balances the loss due to intracavity absorption, mirror reflection losses, and diffraction beyond the edges of the mirrors. The pumping system is not shown, nor are ancillary intracavity elements that are often used for temporal (including frequency selection) and spatial control of the laser output. Usually, one of the mirrors is partially transmitting so that some of the highly directional radiation leaves the cavity through the mirror. The dashed lines are approximately characteristic of the transverse extent of the lowest-order transverse mode

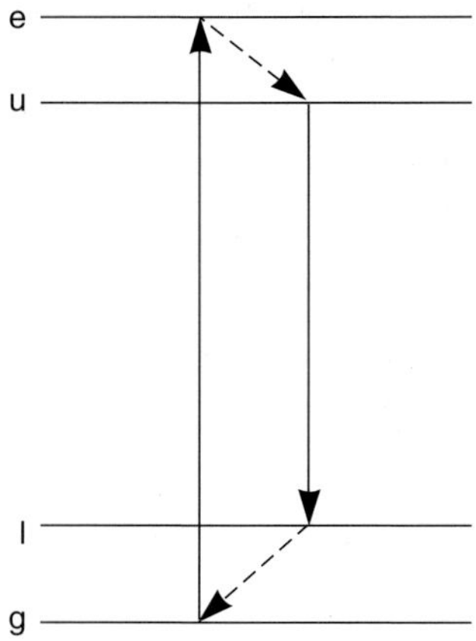


**Fig. 2** Schematic representation of a four-level system. Population is pumped from g to e and laser operation occurs on the transition between u and l

such as chemical reactions or under sufficiently strong external influence. External influences include electron beams and optical fields that selectively excite energy levels of the material. Applied voltages can create electrical currents, also resulting in disequilibrium. Disequilibration, if carried out by a sufficiently selective process, can result in population inversion and laser operation. This "pumping" can be carried out continuously, with single pulses, or with multiple pulses of excitation. The inversion and its duration depend on the relaxation rates for the different energy levels and the degrees of freedom of the system, as well as on the rate of stimulated emission. The energy-level scheme of the laser plays an important role in obtaining inversion; in Sect. 5.1 we will discuss the difference in the operation of three-level and four-level lasers.

Let us try to understand pumping and relaxation in an "ideal" four-level laser with the aid of Fig. 2. The pumping process, indicated by the upward arrow, is assumed to excite the system from the lowest energy level, denoted by g for ground state, to the highest level, denoted by e for excited state. Pumping might occur in a variety of ways, one of which could be through radiative excitation using light whose frequency coincides with the transition frequency between g and e. The state e is assumed to relax to the upper laser level u. The population of the upper laser level is radiatively transferred, either through spontaneous or stimulated emission, to the lower laser level l. Finally, the lower laser level can either relax to the ground state or absorb the laser radiation and repopulate the upper laser level. Several conclusions concerning optimum operation can be made from this model. First, the relaxation rates from e to u and from l to g should be as rapid as possible in order to maintain the maximum population inversion between u and l.

Second, the pumping rate between g and e should be sufficiently rapid to overcome the spontaneous emission from u to l. Third, the thermal equilibrium population of l should be as small as possible. Fourth, decay of e to any level other than u should be as slow as possible (for optical pumping, e can decay radiatively to g) and the nonradiative decay of u should be slow. For radiative pumping, it is advantageous to have e distinct from u and not to have radiative decay of u to g; for other types of pumping this advantage is not obvious.

Not every practical laser satisfies this "ideal" model. The Cu-vapor laser violates the first conclusion listed in the last paragraph, since the lower laser level has a very slow decay rate and the laser self-terminates owing to filling of this level with concomitant reduction in the population inversion. Nevertheless, the Cu-vapor laser operates with fair efficiency as a powerful repetitively pulsed laser.

## 2.3 Resonators and Cavity Modes

An important aspect of the laser involves the design of resonators to accommodate the characteristics of the active medium and the diffractive properties of radiation, and at the same time meet requirements such as low angular beam divergence and high efficiency. The electromagnetic field in a resonator has well-defined modes that have patterns both transverse to and along the cavity axis. Waveguiding with index-of-refraction profiles or reflecting walls can be usefully employed in some cases (most notably in semiconductor and fiber lasers), not only to confine the radiation to the amplifying medium, but also to force the laser to operate in a single transverse mode. More frequently, however, laser resonators are open in the sense that the transverse structure is defined only by axial mirrors or lenses (see Fig. 1). Open resonators formed with convergent optics ("stable" resonators) generally have the lowest diffraction losses, while planar resonators have higher losses, and resonators formed with divergent optics ("unstable" resonators) have the highest losses. Figure 3 shows a few of the lowest-order transverse mode distributions for a stable resonator.

There is generally one transverse mode of a cavity that has the largest net gain (product of the amplifier gain and the transmittance of the remainder of the cavity). This is the transverse mode that oscillates first, and is typically the lowest-order ($TEM_{00}$) mode (see Fig. 3). Single-transverse-mode lasers, in particular, lasers operating in the $TEM_{00}$ mode, have nearly optimal "spatial brightness" in the sense that the beam divergence is near the minimum value for the spot size of the laser on the output mirror; such a laser is said to have diffraction-limited

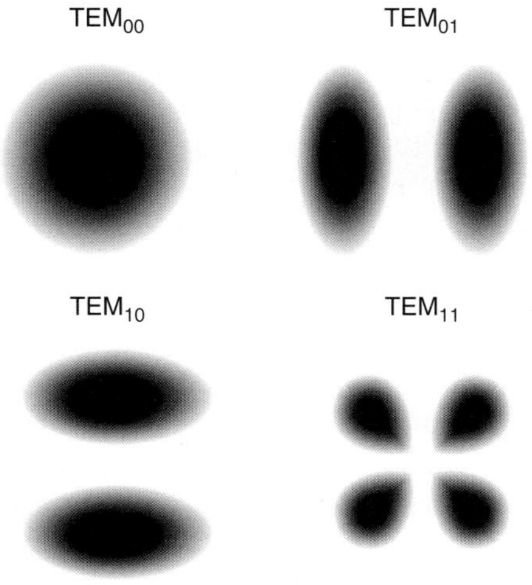


**Fig. 3** Lowest-order transverse modes for a "stable" resonator with square symmetry. The $TEM_{mn}$ notation indicates that the modes have nearly transverse electric and magnetic fields with $m$ nodes vertically and $n$ nodes horizontally

output. The optimal resonator for a particular laser (e.g., "stable" or "unstable") is determined by the geometry of the gain medium, the desired cavity length, and the single-pass gain.

Cavity modes have an axial periodicity that is determined by the cavity length. The frequency spacing between the axial (longitudinal) modes is the inverse of the round-trip time for radiation in the cavity. The gain in a laser is peaked at a transition frequency determined by the energy levels of the active medium and laser operation tends to occur at the axial-mode frequency (or frequencies) closest to the gain peak.

## 3 Important Characteristics of Laser Radiation

### 3.1 Linewidth – Spectral Brightness

Lasers can have very narrow linewidths and, therefore, very high spectral brightness (power per unit spectral interval). Linewidths as narrow as a fraction of a hertz have been obtained (1-Hz stability corresponds to a fractional stability of about $2 \times 10^{-15}$ for visible lasers).

The first step in achieving narrow-line operation is to design the laser to operate in a single longitudinal and transverse mode (see Sect. 2.3). Once single-mode operation has been obtained, the linewidth of the output is determined by technical and fundamental noise. Technical noise arises from sources that can be controlled, such as power-supply fluctuations, variations in the thermal environment, environmental vibrations, and so on. Fundamental noise arises from sources that cannot be eliminated, such as spontaneous emission and fundamental thermal fluctuations. The random effect of noise causes the laser frequency to drift. The influence of noise can often be reduced by adjusting controllable parameters such as cavity lifetime. In addition, the frequency drift can be reduced by measuring the frequency of the laser and providing feedback to adjust the cavity mode frequency to compensate for the drift. The frequency can be measured in several ways: by using a stable Fabry-Pérot etalon, by comparing the laser frequency with a narrow resonance in an atom or molecule, or by comparing the frequency with a nearby frequency reference.

### 3.2 Intensity and Directionality (Angular Confinement) – Spatial Brightness

As discussed in Sect. 2.3, lasers operating solely in the lowest-order transverse mode ($TEM_{00}$) have optimal spatial brightness and are described as operating in the diffraction limit. In this case, for propagation of initially collimated radiation in a nonconfining medium, the product $A\Omega$, where $A$ is the laser output aperture and $\Omega$ is the solid angle into which the power is confined at long distances, has a minimum value of order $\lambda^2$, where $\lambda = c/\nu$ is the wavelength of the radiation and $c$ is the velocity of light. This aperture–solid-angle product is invariant to transformations of the laser beam such as beam expansion. Because of this diffraction-limited character, single-mode laser beams can maintain high intensities over long distances and have a high degree of directionality (small solid angle).

### 3.3 Short Pulses – Temporal Brightness

Short pulses of laser radiation can be made in a variety of ways including: gain switching, $Q$ switching, cavity dumping, mode-locking, and nonlinear frequency broadening and pulse compression. These techniques are described in following sections. To obtain the shortest possible pulses, mode locking is used to obtain pulses in the picosecond (1 ps $= 10^{-12}$ s) to femtosecond (1 fs $= 10^{-15}$ s) range, often followed by frequency broadening and pulse compression. Pulses as short as 4.5 fs have been obtained at a center wavelength of 800 nm, corresponding to less than 2 cycles of light, and very high peak powers are possible. To obtain even shorter pulses, nonlinear optical techniques are used to frequency convert the output of mode-locked lasers into the VUV and soft X-ray spectral regions, where pulses as short as 100 attoseconds (1 as $= 10^{-18}$ s) have been reported.

High temporal brightness (high peak power) competes with high spectral brightness in the sense that the shortness of a pulse is limited by the bandwidth of the radiation. Said another way, for a fixed average power in a continuous train of bandwidth-limited pulses, peak power cannot be increased without reducing spectral brightness.

## 4 Types of Lasers

A rich variety of physical systems have been exploited to produce laser radiation over a five-decade range of wavelengths. We can only briefly describe here some of the most significant of these. More detail can be found in the articles on specific types of lasers.

### 4.1 Solid-state Lasers

The first laser used a rod of ruby (single-crystal $Al_2O_3$ with the ion $Cr^{3+}$ substituted for a small fraction of the $Al^{3+}$ ions) as the active medium. In this laser, a xenon flashlamp pumps the chromium ions from their ground state to a broad band of states, from which they rapidly decay nonradiatively to a long-lived state at an energy of about 14 422 cm$^{-1}$ (693.4 nm) above the ground state. (In spectroscopy and laser physics, energy $E$ and frequency $\nu$ are often given in terms of wave number $1/\lambda$, where $E = h\nu = hc/\lambda$.) It is the narrow ($\approx 11$ cm$^{-1}$) emission line from this level back to the ground state that gives rise to ruby laser emission. When lasing occurs, the emission narrows in frequency width to less than 1 cm$^{-1}$, and leaves the ruby rod with an angular spread of a few milliradians.

The ruby laser is a three-level system (see Fig. 4 and Sect. 5.1), and therefore requires depopulation of the ground state by more than 50% to obtain population inversion. For this reason, pumping of the laser requires a high-intensity source, and continuous-wave (cw) operation of the system is difficult to achieve. In the free-running mode, as opposed to the $Q$-switched mode (see below and Sect. 6.2), the output radiation of the ruby laser fluctuates rapidly over a time of about 1 ms and, for a 1-cm-diameter × 10-cm-long laser rod, is emitted in pulses of about 1 J.

The high threshold pump-power requirement of a three-level laser is greatly relaxed in a four-level system. In the four-level system, radiative transitions do not terminate on the ground state (see Fig. 2). If the final state is not significantly populated at the operating temperature, then population inversion can be maintained with only moderate pump power and such lasers can usually be operated on a continuous basis.

One type of four-level solid-state laser is based solely on the electronic levels

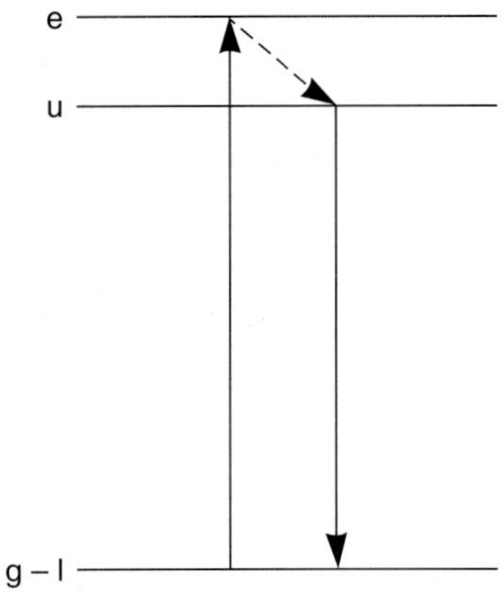


**Fig. 4** Schematic representation of a three-level system. Note that g and l are now the same level

of ions in a crystal or other solid-state host, with the laser transition terminating on an excited electronic level of the ion. The rare-earth ions $Nd^{3+}$, $Ho^{3+}$, $Er^{3+}$, and $Tm^{3+}$, which can also be operated as quasi-three-level systems (with the lower laser level partially occupied), are the most frequently used active ions in such lasers. Extensively developed lasers of this type are the Nd-glass laser and the Nd:YAG ($Y_3Al_5O_{12}$) laser, both of which have their most important transitions in the vicinity of 1.06 μm. Nd:YAG lasers have produced in excess of 5 kW in continuous operation, several joules in low-repetition-rate (10 Hz) pulsed operation, and about 10 mJ in high-repetition-rate (1 kHz) pulsed operation. In order to increase efficiency and decrease heat loading in the solid-state laser medium, Nd:YAG lasers have been pumped with arrays of semiconductor diode lasers. Large Nd-glass lasers have produced tens of joules at repetition rates of several hertz, and several kilojoules at very low repetition rates (≈1 per hour).

Rare-earth-doped glass fiber lasers are of use in optical communications. Single-mode $Er^{3+}$ fiber lasers at 1.55 μm (a quasi-three-level transition) have given output powers on the order of 10 W. At power levels on the order of 100 mW, they are in use as amplifiers in long-distance fiber transmission.

The availability of high-brightness diode lasers has made some quasi-three-level solid-state systems very attractive for continuous, as well as pulsed, operation. The rare-earth ion $Yb^{3+}$, which operates as a quasi-three-level system at 1.03 μm, has been found to be very suitable for diode laser pumping because of its relatively broad pump band. $Yb^{3+}$ also has a paucity of low-lying electronic states, resulting in no absorption of pump light by ions in the excited state. The small energy difference between the excitation photon and the emitted photon leads to low thermal loading, but comes with the disadvantage of significant thermal occupation of the lower laser level at room temperature, and the Yb-ions must be pumped sufficiently hard to overcome the consequent absorption. The relatively small amount of heat generated in Yb-doped materials is attractive for scaling to high average powers, and Yb:YAG lasers have been operated with more than 1 kW of continuous output, making them competitive with more extensively developed Nd:YAG systems. Double-clad fiber lasers using Yb have been operated with output powers over 400 W. In the double-clad structure, the laser radiation is confined to a single-mode inner core while the multimode pump radiation travels through both the inner core and a larger multimode region. The shape of the multimode region can be tailored to the shape of the pump beam. For instance, a rectangular region can be used to accommodate the shape of a pump-diode array.

A second type of four-level solid-state laser involves fewer electronic levels of the active ion. Instead, laser transitions are employed that terminate on an excited vibrational level of the ion in the host lattice. The basic optical pumping–emission cycle is similar to that of a molecular dye laser (see Sect. 4.3). One important class of these solid-state lasers uses transition-metal ions such as $Cr^{3+}$, $Cr^{4+}$, $Ti^{3+}$, $Ni^{2+}$, and $Co^{2+}$. As with the ruby ($Al_2O_3$) laser, the alexandrite ($BeAl_2O_4$) laser employs the $Cr^{3+}$ ion. In the latter case, however, the laser transition can terminate on a variety of final vibrational states. As a result, this laser is tunable from 700 to 818 nm, and has pulsed output energies similar to the ruby laser. The Ti:$Al_2O_3$ laser has even broader tunability, covering a range from 660 nm to beyond 1.1 μm. It has operated continuously at power levels up to 17 W with Ar-laser pumping, and pulsed at energies of hundreds of millijoules when pumped by doubled Nd:YAG laser radiation.

Lasers based on color centers in alkali–halide crystals operate on a principle similar to dye lasers (see Sect. 4.3) and transition-metal lasers. Using different types of *F*-centers in various alkali halides, wavelength coverage over a range from 0.82 to 3.3 μm can be obtained, with continuous output powers ranging from tens of milliwatts to over a watt. Stability of *F*-centers can be a problem, and low-temperature storage is required for several of these lasers.

The energy stored in a pumped solid-state laser medium can be delivered as a giant pulse in a time much shorter than the spontaneous lifetime of the upper laser energy level through the use of *Q* switching (quality-factor switching). This technique (see Sect. 6.2), which was first used in the ruby laser, has been applied to many laser systems. In the case of the Nd:YAG laser, *Q*-switched pulses are typically on the order of 10 ns in duration and can be obtained with the use of a saturable absorber, an electro-optic element, or an acousto-optic element as the switch in the laser cavity.

Because solid-state lasers have spectrally broad gain regions, many equally spaced longitudinal cavity modes can lie within the gain bandwidth when cavity lengths are on the order of tens of centimeters. As a result, mode locking (see Sect. 6.5) can be used to obtain a train of high-intensity, short pulses. Mode locking has been applied to many solid-state laser systems such as Nd:YAG (20-ps pulses) and Ti:$Al_2O_3$ (5-fs pulses). Mode locking has also been used in many other types of laser (see, e.g., Sect. 4.3 on dye lasers).

## 4.2 Gas Lasers

A number of methods can be used to produce population inversion in gaseous media. Inversion can exist between some of the energy levels of the constituents in a gas discharge (electrical discharge in a gas). The first such system, demonstrated not long after the announcement of the ruby laser, was the He-Ne laser, now a standard item in optics laboratories. This system makes use of a discharge in He at a pressure of about 1 torr, with an admixture of Ne at about 0.1 torr. The discharge excites He atoms to their first excited level, about 160 000 $cm^{-1}$ above their ground state. This excitation is readily transferred by collisions to a Ne atomic level with nearly the same excitation energy (resonant transfer). These excited states decay radiatively to lower-energy Ne states, giving rise to continuous laser emission in the red at 15 820 $cm^{-1}$ (632.8 nm) with an output power in the range of $10^{-2}$ W. Other transitions produce strong emission at 8680 $cm^{-1}$ (1.15 μm) and 2957 $cm^{-1}$ (3.39 μm). In a pure Ne discharge, excitation would occur to many Ne levels, and population inversion would not occur as effectively.

An important gas-discharge laser is based on the energy levels of the argon ion ($Ar^+$). Through a complex series of steps, argon-ion–electron collisions in the discharge lead to population inversion and lasing at a number of frequencies near 20 500 $cm^{-1}$ (488 nm). Continuous output at power levels of tens of watts in the blue–green make this device especially useful as a spectroscopic source in Raman scattering, and for pumping continuously operating tunable dye and Ti:$Al_2O_3$ lasers.

Other intense laser sources arise from atomic transitions of metal ions in a pulsed He discharge (the Cu- and Cd-vapor lasers, for example). Gas-discharge lasers have also been made to operate in the UV, but special problems arise in this frequency range; these are discussed below in Sect. 4.5.

Gas lasers with output at longer wavelengths make use of the vibration–rotational energy levels of molecules. In addition to the electronic-state energy-level structure characteristic of atoms, there is vibration–rotational structure associated with the relative motion of the nuclei. The spacing of vibrational energy levels corresponds to frequencies in the infrared. It is in this region of the electromagnetic spectrum that molecular gas-discharge lasers are especially important.

The most efficient and powerful of the molecular gas-discharge lasers is the $CO_2$ laser. One version of this laser makes use of a dilute mixture of $CO_2$ in an $N_2$ discharge. The $N_2$ molecules are excited by collisions with electrons to their first excited vibrational state, from which the excitation is resonantly transferred by molecular collisions to preferentially excite $CO_2$ molecules to a particular vibrational state. These molecules in turn undergo radiative transitions to lower vibrational levels. The presence of the rotational structure gives rise to a cluster of many lines that can lase, grouped near frequencies of 944 $cm^{-1}$ (10.6 μm) and 1042 $cm^{-1}$ (9.6 μm). The electric-discharge $CO_2$ laser is quite efficient (better than 10% electrical power converted to laser power) and is capable of producing continuous-output powers of greater than 1 kW. Other important molecular gas-discharge lasers make use of vibration–rotational or pure rotational transitions of $H_2O$, CO, and HCN, and produce emission at (78 μm, 119 μm), 5.3 μm, and (337 μm, 311 μm), respectively.

A number of other methods of generating laser radiation using molecular or atomic energy levels in gases have been devised. Powerful pulses of laser radiation at 1.315 μm from excited iodine atoms have been produced by flash photolysis (UV photodecomposition) of $CH_3I$ (methyl iodide) and, using energy transfer from chemically generated $O_2(^1\Delta)$, kilowatts of continuous power have been obtained. In the gas-dynamic laser, a nonthermal distribution of molecular vibrational energy levels is produced by the rapid expansion of a hot gas through a nozzle. This method has produced continuous emission of tens of kilowatts at 10.6 μm from $CO_2$ gas. In the chemical laser, two reacting molecular species in a gas produce a product that is left in an excited vibration–rotation state and returns to the ground state radiatively. An example is the HF (DF) laser, which lases in the 2.5- to 3.5-μm region when $H_2$ ($D_2$) and $F_2$ gases combine chemically. Other types of lasers produced by excitation of gases are TEA (transversely excited atmosphere) lasers, e-beam (electron-beam excited) lasers, and UV-preionized, electric-discharge lasers.

$CO_2$ laser radiation has been used to pump other gases, yielding far-infrared emission. If there is a coincidence between a $CO_2$ laser line and a vibration–rotational transition in another gas, an excited level can be populated directly or by collisional transfer of excitation. Pure rotational transitions can take place radiatively to an unpopulated level, producing far-infrared laser radiation. Such gases as $NH_3$ (output at 291 μm) and $CH_3OH$ (output at 164 and 205.3 μm) have been made to lase by this method.

## 4.3
## Dye Lasers

Many organic dyes, when illuminated with visible or UV radiation, fluoresce strongly at lower frequencies. This so-called Stokes shift of the fluorescence can be understood in terms of the change of equilibrium internuclear position with electronic excitation, the rapid vibrational relaxation within an electronic state, and the Franck-Condon principle, which states that an electronic transition in a molecule takes place so rapidly that the nuclear coordinates can be regarded as nearly fixed. An optical pumping–fluorescence cycle for a molecular system is indicated schematically in Fig. 5(a). After absorption of a photon and nonradiative vibrational cascading to the lowest vibrational level, the excited electronic state can decay by fluorescence at a lower frequency, and the molecule returns to the ground state by nonradiative vibrational cascading. Since the final state in the radiative transition process is unoccupied, the pumping requirements for such a laser system need not be too severe. Discrete vibrational structure is washed out in a liquid, but the general outline of

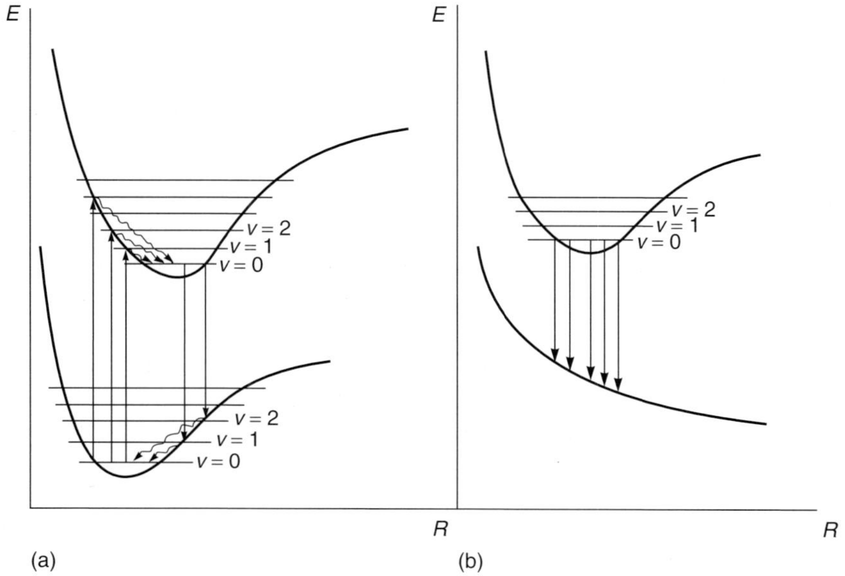


**Fig. 5** Light emission from molecular systems (a) bound–bound system and (b) bound–free system

the cycle indicated in Fig. 5(a) is still preserved in dye fluorescence in a liquid. This process is virtually identical to that occurring in tunable transition-metal solid-state lasers such as $Ti^{3+}$:$Al_2O_3$, $Cr^{3+}$:$BeAl_2O_3$, and $Cr^{4+}$:$Mg_2SiO_4$. In the solid, the substitutional ion's coordinates with respect to neighboring host ions play the same role as the internuclear coordinates in the dye molecules.

A problem that arises with the use of dyes in a laser is the existence of long-lived electron spin-triplet states into which the usual excited electron spin-singlet state can relax nonradiatively. (For rhodamine 6G, a commonly used dye, this nonradiative relaxation time is $\approx 10^{-7}$ s.) This process interferes with laser action, but can be largely circumvented by circulating the dye solution through the laser cavity. Dye lasers can be flashlamp pumped, or pumped with radiation from Ar ion, frequency-doubled Nd:YAG, or $N_2$ lasers. Dye systems fluoresce over a wide band of frequencies from the near-infrared through the visible, and are well adapted for use in a cw laser, with wavelength tuning over as much as 40 nm for a single dye. A battery of dyes placed in optical cavity structures gives tunable laser radiation over a range from roughly 10 000 $cm^{-1}$ (1 μm) to 25 000 $cm^{-1}$ (400 nm).

Because of the broad bandwidth of dye fluorescence lines and, in the case of pulsed dye lasers, the low $Q$ of the resonant structure, the dye laser emission line is fairly broad. It can, however, be greatly narrowed without much loss in output power through the use of a diffraction grating in place of one of the mirrors in the cavity. Even narrower linewidths may be obtained by using intracavity frequency-selective elements (etalons). Single-longitudinal-mode operation can be obtained in both pulsed and cw dye lasers. Dye-jet fluctuations normally limit the stability of cw dye lasers, although stabilities of hundreds of kilohertz are readily obtained, and stabilities of a fraction of a hertz have been achieved. Thus, the dye laser became an important source of radiation for spectroscopy, but recently the tunable solid-state laser has proven more practical.

Since the dye laser has a spectrally broad gain region and many equally spaced longitudinal cavity modes falling within the gain bandwidth, it is well suited for mode-locked operation (see Sect. 6.5). Using continuous mode-locked dye lasers, pulse trains having pulses of less than 10-fs duration have been obtained. These short pulses are used for studies of fast electronic processes in solids and organic molecules.

## 4.4 Semiconductor Lasers

In 1962, it was reported that a forward-biased semiconductor diode of GaAs radiated efficiently at about 11 800 $cm^{-1}$ (850 nm). In the following year, a number of groups reported the observation of laser emission in this frequency region from suitably prepared diode structures. While these semiconductor devices are solids, they are differentiated from those lasers involving optically active ions in ionic hosts on account of their markedly different physical and technological characteristics.

To understand how semiconductor lasers function, it is necessary to consider the nature of the electronic energy states in a semiconductor. A periodic crystal has bands of allowed energy levels separated by forbidden energy gaps. In an intrinsic semiconductor at low temperatures, there are just enough electrons present to fill the uppermost-occupied energy band (valence band), leaving the next higher band

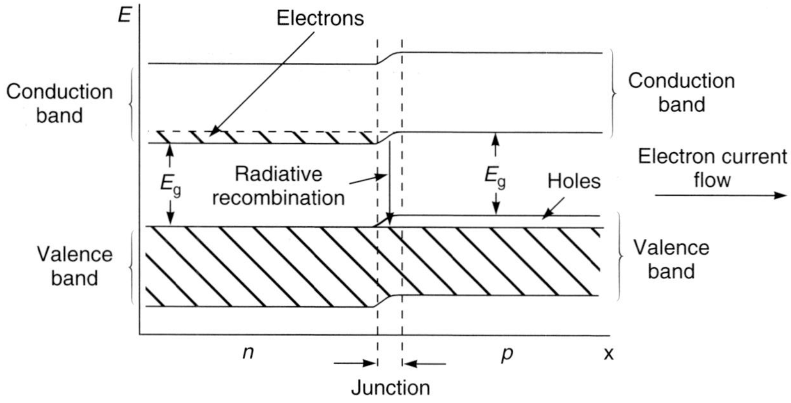


**Fig. 6** Schematic showing the emission of radiation from a forward-biased light-emitting diode

(conduction band) empty. In an *n*-type semiconductor, impurity atoms (donors) are present that contribute electrons to the conduction band; in a *p*-type semiconductor, there are impurity atoms (acceptors) present that can bind electrons, leaving behind missing electrons (holes) in the valence band.

Figure 6 shows a schematic of a $p$–$n$ junction, fabricated by forming *p*- and *n*-type semiconductor layers in intimate contact. When a voltage is applied in the forward direction, electrons are injected from the *n* region into the depletion region of the junction (a region of the junction about 1 μm thick). At the same time, holes are injected from the *p* region. As electrons drop into hole states, they may emit radiation (electron–hole recombination radiation) at a frequency in the vicinity of the energy gap ($\nu = E_g/h$). When the injection current density is sufficiently high, population inversion and gain will be induced.

A typical semiconductor diode-laser heterostructure is shown schematically in Fig. 7. The dielectric film serves to guide the current into a narrow stripe region in order to concentrate the electron–hole pair density. The $x$ coordinate shown in Fig. 6 corresponds to the vertical direction in Fig. 7.

In many diode (or injection) lasers, a cavity structure (typically a fraction of a millimeter in dimensions) is provided by plane-parallel, cleaved facets at right angles to the junction plane. Laser emission is perpendicular to the cleaved facets, once the diode injection current reaches a threshold value. Multimode behavior often occurs in the plane of the junction. Since stimulated emission occurs in a narrow area near the junction (a few to several hundred micrometers in the plane of the junction and a fraction of a micrometer perpendicular to the junction), the angular spread of emitted radiation is fairly large, as expected from consideration of diffraction.

The technology of diode lasers has undergone considerable development, with the primary goals of achieving room-temperature operation, low thresholds, high output powers, improved mode quality, wavelength diversity, and long lifetimes. Progress has included improvements in electrical and optical confinement, and closer coupling of the gain

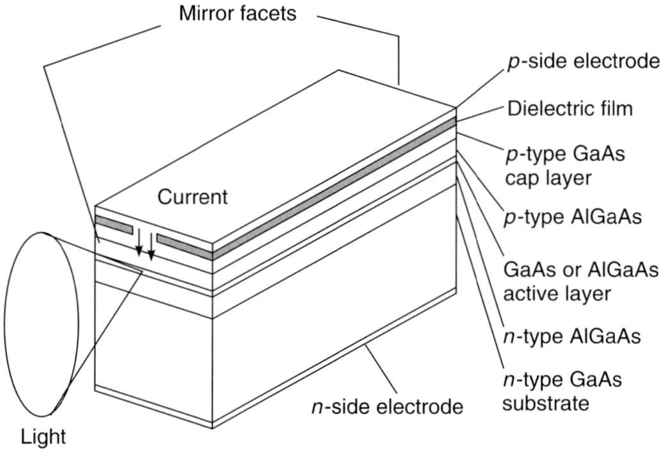


**Fig. 7** Typical semiconductor diode-laser heterostructure. The material compositions shown produce lasers operating in the 750- to 850-nm wavelength range. The device length is typically 250 μm, the width 50 to 100 μm, and the height on the order of 50 μm. The epitaxial layers (i.e., excluding the substrate and the contacts) contribute only a few micrometers to the height. For thermal and electrical connection, the device is soldered to an electrically conducting heat sink, and an electrical lead is attached to the topside of the device. The width of the emitting region is 10 to 20 μm, while the height is approximately 0.5 μm; this causes the beam to diverge rapidly in the vertical direction

region to the heat sink. Much of the improvement has involved advances in materials growth, including molecular-beam epitaxy (MBE) and metal-organic chemical vapor deposition (MOCVD). These growth techniques have resulted in material of exceptionally high quality as well as structures with quantum confinement of carriers (electrons and holes).

Quantum-well diode lasers have exceptional optical and electrical properties including very low laser thresholds. By reducing the volume of the active region through the use of quantum wells, the amount of current required to bleach the absorption of the semiconductor (the transparency current) is reduced, with a commensurate reduction in threshold. In addition, the quantum confinement of carriers in the active region increases their radiative gain cross section. In strained quantum well devices, the density of states in the valence band is reduced, leading to a further reduction in the transparency current and reducing parasitic effects such as Auger recombination, which is important at long wavelengths.

In 1969 (before the advent of quantum-well devices), room-temperature continuous operation was achieved in a GaAlAs double-heterostructure laser. This laser structure consists of a small region of GaAs sandwiched between *p*- and *n*-type layers of the wider-bandgap alloy $Al_xGa_{1-x}As$ ($x < 1$). With further development, device lifetimes of tens of years were obtained, with output powers in the tens of milliwatts. These improvements opened up applications of considerable significance – in particular, fiber-optical communication and optical data storage. GaAlAs diode lasers have also been operated continuously at room temperature with output powers in the range of several watts. Linear arrays of diodes, in the form of multiple stripes in a 1-cm bar, have given a total output of over 50 W. In addition, electrical-to-optical power conversion efficiencies of greater than 50% have been obtained.

Continuous laser operation at room temperature has also been obtained in the quaternary alloy system InGaAsP. Because laser operation is much less sensitive to dislocation effects in this system, long-lifetime devices were readily achieved. When operating at 1.3 and 1.55 μm, these lasers are matched to low-loss (<1 db $km^{-1}$), low-dispersion fiber optics, and are currently used in high-data-rate, long-distance communication.

InGaAsSb/AlGaAsSb quantum-well lasers have produced 1 W of continuous output (5 W pulsed) at wavelengths as long as 2.5 μm, at near room temperature. Cooled to liquid-nitrogen temperature, this material system can be used to produce laser output at wavelengths as long as 4 μm.

The compositionally tuned lead-salt lasers ($Pb_xSn_{1-x}Te$, $PbS_xSe_{1-x}$) operate at cryogenic temperatures. As with other semiconductor lasers, changes in composition change $E_g$ and, therefore, the frequency of laser emission. Since $E_g$ and hence the dielectric constant at near-band gap wavelengths are sensitive functions of temperature in these small gap semiconductor systems, cavity-mode frequencies and gain peaks are tunable, giving rise to temperature-tunable laser output. These tunable sources have been used extensively for high-resolution infrared spectroscopy in the 5- to 20-μm region.

Quantum-cascade lasers offer an alternative to lead-salt lasers for operation at mid-infrared wavelengths and longer. The active region of a quantum-cascade laser features several epitaxially grown layers of semiconductor. The device generates radiation based on electronic transitions that occur in the stacked layers. The thickness of the layers, rather than the fundamental bandgap $E_g$ of the materials, determines the frequency of the emitted radiation. Quantum-cascade lasers have been operated at wavelengths from 3.4 to 67 μm, with the potential to extend the range from 1 to 100 μm. Continuous operation with output powers in excess of 0.5 W has been obtained at liquid-nitrogen temperature, and pulsed operation with peak powers approaching 1 W has been demonstrated at room temperature. These systems have also demonstrated tunability in excess of 150 nm.

In the visible region, using the quaternary alloy AlGaInP, continuous room-temperature operation has been obtained at 635 nm and pulsed operation at 603 nm; continuous liquid-nitrogen-temperature operation has been achieved at 583.6 nm, which is in the yellow region of the spectrum. Initial attempts to obtain operation in the blue–green spectral region focused on $Zn_{1-x}Cd_xSe/ZnS_ySe_{1-y}$ quantum-well heterostructure diodes. These lasers have been operated continuously at room temperature, but have not yet demonstrated acceptable lifetimes for commercial applications. Operation of SiC diode lasers at 403 nm, in the blue–violet spectral region, has faced a similar problem. More recently, heterostructure diode lasers based on InAlGaN alloys have demonstrated in excess of 10 000 h lifetime for continuous room-temperature operation at wavelengths as short as 375 nm, and this system has the potential to be tailored for outputs throughout the visible and near-ultraviolet.

Operating at an output power of a few milliwatts, InAlGaN diode lasers are now in use in CD/DVD recorders.

VCSELs (vertical-cavity surface-emitting lasers) have become an important class of semiconductor lasers. As with cleaved-cavity semiconductor lasers, these devices are constructed in layers using MBE or MOCVD. The peculiarity of the VCSEL is that some of the layers are used to form distributed Bragg mirrors on either side of the junction region. Radiation builds up in the direction normal to the junction instead of in the plane of the junction. Careful cleaving does not have to be done and the output beams are circularly symmetric.

Semiconductor lasers with one of the radiation feedback elements external to the semiconductor material are also technologically significant. Wavelength-tuned external-cavity lasers are flexible sources for optical communications and other applications. Both conventional cleaved-cavity diodes and VCSELs have been used to make external-cavity lasers. In order to reduce the effect of the internal cavity (formed by the interface of the semiconductor to air), very good antireflection coatings must be used, having reflectivities of the order 0.0001.

Semiconductor optical amplifiers (SOAs) are being developed to provide high-power outputs and booster amplification in fiber-communication systems. Tapered-waveguide designs are used to prevent filamentation or "hot spots" at the end of the amplifier.

In addition to injection pumping of semiconductor lasers, a number of other methods not requiring the fabrication of a junction have been used successfully. These include electron-beam pumping and optical pumping.

## 4.5 UV and X-ray Lasers

To produce stimulated emission in the UV and X-ray region, special problems must be addressed. Short lifetimes of inverted populations become important, since spontaneous radiative lifetime varies with frequency as $\nu^{-3}$. For example, the nitrogen gas-discharge laser, which radiates in the near UV (337 nm), can only operate in a pulsed mode, and must be pumped by a powerful intermittent source. Another problem is associated with the difficulty of devising resonant structures, since the reflectivity of materials becomes very small in the vacuum-UV and X-ray regions. A different conceptual scheme must frequently be used in these spectral regions, involving directionally amplified spontaneous emission from an inverted population (superradiance), where the directional amplification is achieved by the geometry of the pumped region.

The nitrogen and hydrogen discharge lasers make use of radiative transitions between two bound electronic levels (bound–bound transitions). On the other hand, there are molecular systems (excimers) in which radiative transitions occur between a bound excited state and a free or very weakly bound ground state (for bound–free transitions – see Fig. 5b). Xe and Kr form the excited molecular states $Xe_2^*$ and $Kr_2^*$, although the diatomic molecules are unstable in their ground states. At high pressure these gases, when pumped by powerful electron-beam sources, emit superradiantly at 172 and 145.7 nm, respectively. At somewhat longer wavelengths, rare-gas–halide excimers, such as $ArF^*$ at 248 nm, $KrF^*$ at 193 nm, and $XeCl^*$ at 308 nm, have been operated both by electron-beam pumping and by transversely excited discharge.

These lasers can produce multijoule pulses at 100-Hz repetition rates and are commercially available.

X-ray lasers at wavelengths shorter than a few tens of nanometers present a more difficult challenge. Nevertheless, several groups have achieved laser operation in this region. A multijoule visible laser can directly pump the gain medium through inversion produced in a laser-generated plasma, or indirectly by using the X-rays from a laser plasma to pump a separate X-ray laser medium. The first approach has been used to achieve operation near 21 nm in Se XXV, which has a $1s^2 2s^2 2p^6$ Ne-like electron configuration. Current research involves efforts to improve the very low efficiency of X-ray lasers as well as studies of potential new X-ray laser systems.

### 4.6
### Free-electron Lasers

Relativistic electrons traveling in a periodically alternating transverse magnetic field (wiggler) can be stimulated to give up radiation to a copropagating electromagnetic field of wavelength $\lambda = \lambda_w/2\gamma^2$, where $\lambda_w$ is the wiggler period and $\gamma$ is the ratio of the electron energy to its rest-mass energy. This phenomenon can be pictured as stimulated inverse Compton scattering from the electromagnetic field of the wiggler seen in the electron rest frame. To obtain efficient conversion over reasonable distances with high output power, the electron-beam must be very monoenergetic, constitute a high current, and have very low angular divergence. Efficient operation also requires either the use of an electron storage ring or a tapered-period wiggler. Free-electron lasers have been operated at several wavelengths ranging from the visible to millimeter waves. While free-electron lasers do not have the wavelength restrictions imposed on other lasers, and do not have the inhomogeneities characteristic of many laser media, they require electron beams of a current level and quality that are difficult to achieve simultaneously. Further, the electron-beam and wiggler requirements tend to become more demanding as wavelength decreases. These systems have the large sizes that are associated with relativistic electron-beam sources.

## 5
## Laser Dynamics

### 5.1
### Rate-equation Model

Many of the properties of the laser can be determined from a rate-equation model for the population of the laser levels and the photon number in the laser cavity. The rate equations provide a simple and intuitive, yet accurate, picture of the behavior of many lasers. In the most simplified form, the increase in photon number within the laser cavity is balanced by the decrease in the population difference between the upper and lower laser levels. In addition, the population difference increases on account of the pumping process, while the photon number decreases because of absorption, diffraction of the beam out of the cavity, and transmission through the mirrors.

The rate-equation model can be derived as an approximation to the fundamental equations relating the electromagnetic field, the material polarization, and the populations. The validity of the rate equations requires that the polarization can be accurately approximated by assuming that it instantaneously follows the field; this is a situation that applies to most lasers. In order to describe the problem in terms of total population and total photon number within the laser cavity, it is necessary that the gain of the laser be small during one pass through the cavity and that the laser operate in a single longitudinal mode (see Sect. 5.6).

As stated above, the number of photons $q$ within the laser cavity is affected by two types of events: the emission of a photon by the gain medium ($\dot{q}_{\mathrm{em}}$) and the escape of a photon from the cavity or absorption by unpumped transitions ($\dot{q}_{\mathrm{dec}}$). Photon emission can be either stimulated ($\dot{q}_{\mathrm{stim}}$) or spontaneous ($\dot{q}_{\mathrm{spon}}$). Once a laser is above threshold, the stimulated-emission rate is much greater than the spontaneous rate and, to first order, spontaneous emission can be ignored. We will return to the issue of spontaneous emission in the next section when we discuss the buildup of a laser from noise.

The stimulated-emission rate is proportional to the number of photons within the cavity, the total population inversion $N$, and the probability per unit time $B$ that a given photon will interact with a given inverted site. The interaction probability $B$ is the product of the probability that a photon will pass within the gain cross section $\sigma$ of a given inverted site as it traverses the laser cavity and the number of times per second the cavity is traversed. Mathematically, this reduces to $B = \sigma/A \times c/l = \sigma c/V$, where $A$ is the cross-sectional area of the laser mode, $l$ is the cavity length, and $V$ is the volume of the lasing mode. The stimulated emission rate is, therefore, $\dot{q}_{\mathrm{stim}} = qN\sigma c/V$.

The escape of photons from the laser cavity and their absorption within the cavity are characterized by the cold-cavity lifetime $\tau_{\mathrm{c}}$ (cavity lifetime in the absence of any inversion) and corresponds to $\dot{q}_{\mathrm{dec}} = q/\tau_{\mathrm{c}}$. The total rate of change of the number of photons within the laser cavity is thus

$$\dot{q} = \dot{q}_{\mathrm{stim}} - \dot{q}_{\mathrm{dec}} = \frac{qN\sigma c}{V} - \frac{q}{\tau_{\mathrm{c}}}. \tag{1}$$

We will now derive the rate equations for the population inversion of both a four-level and a three-level laser. The energy-level diagrams of a four- and a three-level laser are shown in Figs. 2 and 4. In both cases, the pump excites the active medium from the ground level (with population $N_{\mathrm{g}}$) to level e ($N_{\mathrm{e}}$). It is then assumed that the excited state quickly decays to the upper laser level, level u ($N_{\mathrm{u}}$), so that $N_{\mathrm{e}} \approx 0$. Lasing occurs between levels u and l ($N_{\mathrm{l}}$). The difference between a four- and a three-level laser is that for the three-level laser the lower laser level is also the ground state.

To derive the rate equation for the population inversion $N = N_{\mathrm{u}} - N_{\mathrm{l}}$, we start by considering the population of the upper laser level $N_{\mathrm{u}}$. The population of the upper level is affected by pumping $P_{\mathrm{r}}$, stimulated emission $\dot{N}_{\mathrm{u,stim}}$, and spontaneous decay $\dot{N}_{\mathrm{u,spon}}$. For most pumping schemes, the pump rate is proportional to the number of ions in the ground state and can be written as $P_{\mathrm{r}} = W_{\mathrm{p}} N_{\mathrm{g}}$. The stimulated-emission process decreases the population of the upper laser level by one for every photon created, so that $\dot{N}_{\mathrm{u,stim}} = -\dot{q}_{\mathrm{stim}}$. Spontaneous decay is characterized by the spontaneous lifetime $\tau$, corresponding to $\dot{N}_{\mathrm{u,spon}} = -N_{\mathrm{u}}/\tau$. The rate equation for the upper-level population is thus

$$\begin{aligned}\dot{N}_{\mathrm{u}} &= P_{\mathrm{r}} + \dot{N}_{\mathrm{u,stim}} + \dot{N}_{\mathrm{u,spon}} \\ &= W_{\mathrm{p}} N_{\mathrm{g}} - \frac{qN\sigma c}{V} - \frac{N_{\mathrm{u}}}{\tau}. \end{aligned} \tag{2}$$

In an ideal four-level laser there is a very rapid decay of the lower laser level to the ground state, so that $N_{\mathrm{l}}$ and $\dot{N}_{\mathrm{l}}$ are approximately equal to zero. Since the total number of active ions $N_t$ is constant, $N_t \approx N_g + N_u$. The rate equations for a four-level laser are, therefore,

$$\dot{q} = \left(\frac{N\sigma c}{V} - \frac{1}{\tau_c}\right) q, \quad (3)$$

$$\dot{N} = W_p(N_t - N) - \frac{qN\sigma c}{V} - \frac{N}{\tau}. \quad (4)$$

Since, in an ideal three-level laser, the lower laser level is the ground state, $N_t \approx N_l + N_u$ and $\dot{N}_l \approx -\dot{N}_u$. As a result, the rate equations for a three-level system reduce to

$$\dot{q} = \left(\frac{N\sigma c}{V} - \frac{1}{\tau_c}\right) q, \quad (5)$$

$$\dot{N} = W_p(N_t - N) - \frac{2qN\sigma c}{V} - \frac{(N_t + N)}{\tau}. \quad (6)$$

The photon rate equations (Eqs. 3 and 5) for the four-level and three-level lasers are the same. However, the rate equations for the population inversion are slightly different. In particular, the stimulated emission term for a three-level laser is twice that of a four-level laser.

### 5.2
### Buildup from Noise

In the photon rate equation derived in Sect. 5.1, the term corresponding to spontaneous emission was left out. Laser action is initiated by spontaneous emission, or noise. As a result, these rate equations cannot account for the onset of lasing, as is seen by setting $q = 0$ at time $t = 0$. When spontaneous emission is properly taken into account, the photon rate equation becomes

$$\dot{q} = (qN + N_u)\frac{\sigma c}{V} - \frac{q}{\tau_c}. \quad (7)$$

The net effect is as if there were initially one photon in the cavity. This one photon of noise stimulates optical transitions and initiates lasing.

### 5.3
### Threshold

The threshold inversion required for lasing is derived by requiring that the photon rate equation have a nontrivial solution in steady state. In steady state $\dot{q} = 0$, resulting in the condition $qN\sigma c/V - q/\tau_c = 0$. Physically, this states that the number of photons leaving the cavity must be balanced by the number of photons generated through the stimulated emission process. The threshold inversion is, therefore,

$$N_{thresh} = \frac{V}{\sigma c \tau_c}. \quad (8)$$

The pump rate $W_p$ required to reach threshold is obtained by setting $\dot{N} = 0$, $\dot{q} = 0$, and $N = N_{thresh}$. For a four-level laser

$$W_{p,thresh} = \frac{N_{thresh}}{(N_t - N_{thresh})}\tau \approx \frac{V}{N_t \sigma c \tau_c \tau}, \quad (9)$$

where we have assumed that $N_{thresh} \ll N_t$. For a three-level laser

$$W_{p,thresh} = \frac{(N_t + N_{thresh})}{(N_t - N_{thresh})}\tau \approx \frac{1}{\tau}. \quad (10)$$

For the same value of $\tau$, the threshold pump rate for a four-level laser is smaller than the threshold pump rate for a three-level system by a factor of $V/N_t\sigma c\tau_c$, which is usually quite large. This is the basis of the superior performance of a four-level system over a three-level system in cw operation.

### 5.4
### Gain Saturation

The gain $g$ of an active medium is defined as the fractional change in optical intensity per unit length as a light beam passes through. From the discussion in Sect. 5.1, it follows that $g = N\sigma/V$. In the presence of a strong optical field, the population inversion is reduced and the gain is saturated. The rate equation for the population inversion (of a three- or four-level gain medium) in steady state can be rewritten in the form

$$g = \frac{g_0}{(1 + I/I_s)}, \qquad (11)$$

where $g_0$ is the unsaturated gain (the gain in the absence of an optical field), $I = h\nu qc/V$ is the optical intensity ($h\nu$ is the energy of one photon), and $I_s$ is known as the saturation intensity.

The steady-state photon rate equation ($\dot{q} = 0$) predicts that above threshold ($q \neq 0$) the inversion density and the gain of a laser are clamped at their threshold values; the round-trip gain of the cavity is equal to the round-trip loss. With increased pumping, the gain remains fixed while the photon number and the output of the laser increase.

## 5.5 Laser Efficiency

The efficiency of a laser is often discussed in terms of the slope efficiency $\eta_s$. Slope efficiency is defined as the ratio of the change in output power to the change in pump power of a laser once it has reached threshold, and is determined by four factors: the pump efficiency $\eta_p$, the area efficiency $\eta_a$, the intrinsic efficiency $\eta_i$, and the output coupling efficiency $\eta_o$. Mathematically,

$$\eta_s = \eta_p \eta_a \eta_i \eta_o. \qquad (12)$$

The pump efficiency $\eta_p$ is the ratio of the energy absorbed by the gain medium to the energy of the pump source. In an optically pumped laser, part of the incident optical energy may be reflected by the gain medium and part may be transmitted. Both of these effects decrease the pump efficiency. The area efficiency $\eta_a$ is a measure of how well the pumped volume is used by the oscillating mode. If the cross section of the pumped volume is much larger than the cross section of the lasing mode, only a small portion of the pumped volume contributes to the gain of the system, and the area efficiency will be low. The intrinsic efficiency $\eta_i$ is simply the ratio of the energy of the photon created during lasing to the energy required to create one excitation. In an optically pumped system, the intrinsic efficiency is the ratio of the energy of a photon at the oscillating frequency to the energy of an absorbed pump photon. Finally, the output coupling efficiency $\eta_o$ is the ratio of the output coupling to the total round-trip loss of the laser cavity.

The total efficiency $\eta$ of a laser (power out divided by power in) is dependent on the slope efficiency and the laser threshold. Using the rate equation model, the slope efficiency is constant and the total efficiency is given by

$$\eta = \eta_s \left(1 - \frac{P_{\text{thresh}}}{P}\right), \qquad (13)$$

where $P$ is the total pump power and $P_{\text{thresh}}$ is the pump power required to reach threshold.

## 5.6 Multimode Operation

For most lasers, the frequency spacing between adjacent longitudinal modes is much less than the gain bandwidth. As a result, lasers tend to oscillate at several frequencies simultaneously.

Although the above statement is true, the reasons are subtler than they may initially seem. In the early days of lasers, it was believed that lasers with homogeneously broadened gain spectra should operate in a single longitudinal mode. The reasoning behind this can be understood from the rate equations. If we assume a uniform optical intensity within the laser cavity, the steady-state solution to the photon rate equation fixes the inversion density at its threshold value. The first cavity mode to lase (the one with the highest net gain) clamps the inversion density and no other mode can reach threshold. The flaw in this reasoning lies in the assumption of uniform optical intensity.

Experimentally, lasers with both homogeneously and inhomogeneously broadened gain media tend to oscillate in several longitudinal modes as a result of spatial and spectral hole burning.

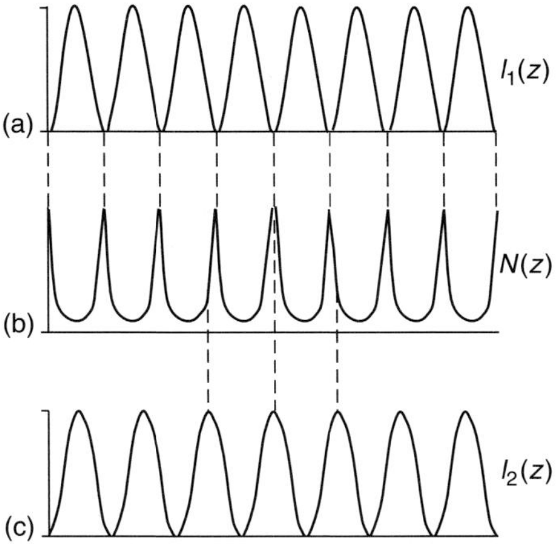

**Fig. 8** Illustration of spatial hole burning, showing (a) the intensity profile of the first longitudinal mode to lase, $I_1(z)$; (b) the population inversion in the presence of the first oscillating mode, $N(z)$; and (c) the intensity profile of the second longitudinal mode to lase, $I_2(z)$

### 5.6.1 Spatial Hole Burning

In standing-wave laser cavities, the coherent superposition of the optical fields traveling in two directions within the cavity results in a sinusoidal intensity distribution. At positions where the intensity distribution is at its maximum, there is strong gain saturation and the population inversion is depleted. However, at nulls in the optical field the oscillating mode is unable to deplete the inversion. As a result, the inversion density is no longer uniform, but has "holes" at the positions corresponding to the peaks in the optical intensity. This phenomenon is known as spatial hole burning. The gain at the nulls in the optical field will continue to increase as the gain medium is pumped harder. Since other cavity modes have a different spatial profile than the first mode, and can use the population inversion at these positions, this will lead to the onset of multimode operation. These ideas are illustrated in Fig. 8.

### 5.6.2 Spectral Hole Burning

Homogeneous gain broadening occurs when each excited state of the gain medium has exactly the same energy distribution. This is often the case in crystalline solid-state gain media at room temperature. In materials such as glasses, each excited ion sees a slightly different environment, resulting in a slightly different energy spectrum. The ensemble effect is inhomogeneous gain broadening. Another example of an inhomogeneous system is a gas, where each atom or molecule is moving at a slightly different speed and has its energy spectrum Doppler shifted by a different amount.

In an inhomogeneous system, only those excitations with gain at the lasing frequency are able to participate in the stimulated emission process. As a result, only those excitations become depleted, producing a gain spectrum that has a dip at the lasing frequency. This is known as spectral hole burning, and is illustrated in

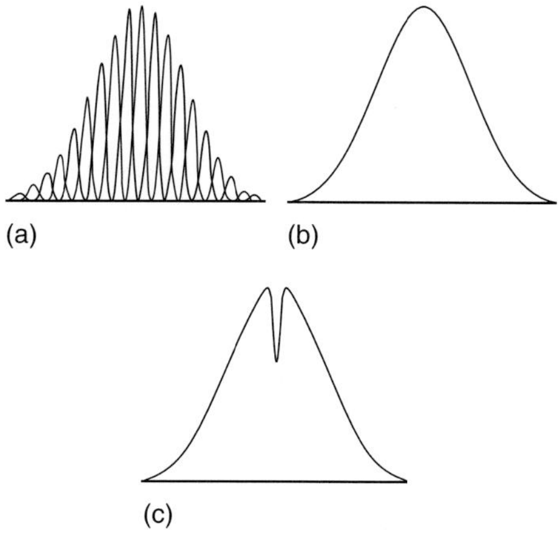


**Fig. 9** Illustration of spectral hole burning, showing (a) several closely spaced homogeneously broadened spectra; (b) the inhomogeneously broadened spectrum resulting from the sum of several homogeneously broadened spectra; and (c) the inhomogeneously broadened spectrum with a spectral hole burnt in the center as a result of the saturation of one of the homogeneously broadened components

Fig. 9. Excitations that cannot contribute to the lasing process for the first mode can contribute to the onset of lasing for other modes, resulting in multimode oscillation.

### 5.6.3 Single-frequency Operation

There are many techniques for obtaining single-frequency operation from a laser. Several of these involve introducing an element into the cavity such that the cavity sees a frequency-dependent loss, thereby decreasing the net gain bandwidth and selecting an individual longitudinal cavity mode. Examples of such elements are a prism, a grating, a Fabry-Pérot etalon, and the combination of a birefringent filter and a polarizing element. In some cases, a cavity may require more than one device in order to obtain enough frequency selectivity to ensure single-frequency operation.

Single-mode operation has also been obtained by reducing the length of the cavity so that the longitudinal mode spacing is comparable to, or less than, the gain bandwidth. This is most easily done with gas lasers, which have a narrow gain bandwidth, but has also been achieved in very short solid-state lasers.

Alternatively, single-mode operation can be achieved in a laser with a homogeneously broadened gain medium by reducing or eliminating the effects of spatial hole burning. A unidirectional ring cavity has a uniform optical intensity within the cavity, rather than the sinusoidal intensity distribution of a standing-wave cavity. Spatial hole burning is therefore eliminated, and such a laser may operate at a single frequency well above threshold. A variation of the same idea involves placing a quarter-wave plate on either side of the gain medium. As a result, the optical fields traveling in opposite directions in the gain medium are orthogonally polarized and do not interact coherently. The optical intensity within the gain medium is, therefore, uniform and there is no spatial hole burning.

In some gain media, there is a large amount of energy diffusion. Energy diffusion moves some of the excited states away from the peaks in the population inversion, toward the minima. This smoothes out the population-inversion profile, reducing the effects of spatial hole burning. One important example of such a gain medium is a semiconductor.

Finally, the effects of spatial hole burning are reduced if the gain medium is located very close to a cavity mirror. At a mirror, the phase of the optical spatial intensity distribution is pinned at zero for all of the cavity modes. The phase difference between longitudinal modes increases gradually as one moves toward the center of the cavity. Quite close to the mirror, the peaks and nulls of the optical intensities for each of the modes occur in approximately the same place, and that portion of the gain that is not depleted by one mode will not be in a good position to contribute to any other.

## 6
## Types of Pulsed Operation

### 6.1
### Long-pulse Operation

Long-pulse or quasi-cw operation refers to a pulsed laser with a pulse duration long enough for all relevant parameters within the system to come to their steady state value. Although the behavior of the system is cw-like at the end of the pulse, it will, in general, be quite different at the beginning of the pulse.

Let us consider a laser with a step-function pump source. The pump may quickly create a population inversion. It will take some time, however, for a lasing mode to buildup from spontaneous emission. During this time, the inversion density may greatly exceed threshold. The large inversion density eventually results in a large optical intensity, well in excess of the cw value. This optical intensity, in turn, drives the inversion density below threshold, substantially reducing the laser intensity. The entire process then starts again. For a single-mode laser, this often leads to regular spiking at the beginning of the pulse. The process is damped, however, and with time the intensity of the spikes decreases. Spiking eventually gives way to damped oscillations (known as *relaxation oscillations*) in the optical intensity and finally cw-like behavior.

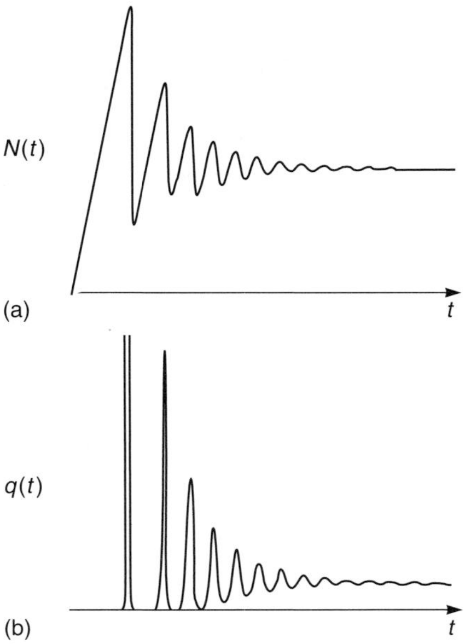


**Fig. 10** Computer solutions to the rate equations for a laser with a step-function pump source, showing (a) the population inversion $N(t)$ and (b) the photon number $q(t)$

Figure 10 shows the computer solutions to the rate equations for a laser under the conditions just described. In this computer simulation, the spiking was heavily damped and cw-like behavior was quickly obtained. In a multimode laser the interaction between modes often leads to mode hopping, mode beating, and very irregular spiking, which may never damp out.

#### 6.1.1 Relaxation Oscillations

Relaxation oscillations occur whenever the population inversion of a laser is disturbed from its steady-state value. It is a result of the coupling between the population inversion and the photon density within the laser cavity, as described above. From the rate equations, it can be shown that for a single-mode laser a small disturbance in the inversion density results in damped oscillations with an oscillation frequency

$$\omega = \left[ \frac{(N/N_{\text{thresh}} - 1)}{\tau_c \tau} \right]^{1/2} \tag{14}$$

and a damping constant

$$t_0 = \frac{2\tau N_{\text{thresh}}}{N}. \tag{15}$$

Note that if $1/t_0 > \omega$ the oscillations are overdamped and spiking will not occur. Although this condition is not satisfied in solid-state lasers, it is common in gas lasers.

### 6.2
### Q-switched Operation

For many applications it is desirable to obtain short, high-peak-power pulses from a laser. This can be achieved by creating a large population inversion and then quickly decreasing the cavity loss so that the inversion density is well in excess of its new threshold value. The large inversion density allows an intracavity optical field to rapidly develop. This field then depletes the population inversion and turns itself off. The cavity loss is subsequently increased to prevent the development of a second pulse. This technique is known as *Q* switching, since the quality factor, or *Q*, of the optical cavity is changed.

*Q* switching relies on the fact that the lifetime of the population inversion is much longer than the output pulse. The gain medium is therefore able to store energy that can be quickly released in the form of a short output pulse.

The length of the *Q*-switched output pulse is dependent on several factors. In order to obtain a rapid buildup of the optical pulse, it is desirable to have a large gain cross-section and a large population inversion when the *Q* of the cavity is switched. The rapid buildup of the output pulse also relies on the presence of a strong intracavity optical field, which argues in favor of a high cavity *Q*. In order to obtain a short output pulse, however, the intracavity intensity must also decay quickly after the peak of the pulse. Since the *Q* of the cavity is constant for the duration of the pulse (it is difficult to change the *Q* of the cavity significantly during the duration of a short output pulse, although it can be changed in the relatively long pulse buildup time preceding the pulse), this would argue for a low cavity *Q*. Solutions to the rate equations show that the minimum-width output pulse is obtained when the total round-trip loss of the cavity during the output pulse is adjusted so that the initial inversion density is about three times the threshold value.

Methods for *Q* switching a laser include the use of electro-optic shutters, acousto-optic *Q* switches, and mechanical devices. Passive *Q* switching can be obtained through the use of an intracavity saturable absorber with a long recovery time.

### 6.3
### Gain-switched Operation

Gain switching is another way to obtain short, high-peak-power pulses from a laser. The idea is to rapidly increase the pump power so that the population inversion of the laser is well in excess of the threshold value by the time the first output spike develops. The optical pulse then drives the population inversion below its threshold value. The pump power is subsequently reduced so that the inversion remains below threshold and only a single pulse is obtained.

The factors that are important for short gain-switched pulses are the same as for short *Q*-switched pulses.

### 6.4 Cavity-dumped Operation

Cavity dumping allows the energy in a laser cavity to be output in a time comparable to the cavity round-trip time. The concept is to rapidly (within a cavity round-trip time) introduce a large output coupling (nearly 100%) into a cavity that previously had no output coupling. Methods for cavity dumping include the use of an electro-optic Pockels cell and polarizing beam splitter and acousto-optic devices.

### 6.5 Mode-locked Operation

Mode locking refers to the situation when the phases of several cavity modes are fixed (or locked) with respect to each other such that the electric fields add coherently and constructively for a short period of time. This allows the generation of a train of high-intensity, ultrashort pulses. To understand this, consider the case of $2n + 1$ equally spaced longitudinal modes oscillating with the same amplitude $E_0$. Assume that the phases $\phi_m$ of the modes are locked according to $\phi_m - \phi_{m-1} = \phi$, where $\phi$ is a constant. The total electric field is the sum of all of these modes:

$$E(t) = \sum_{m=-n}^{n} E_0 \exp[i(\omega_0 - m\Delta\omega)t + m\phi] = A(t') \exp(i\omega_0 t), \tag{16}$$

where $\omega_0$ is the frequency of the center mode, $\Delta\omega$ is the frequency difference between two adjacent modes, $t' = t + \phi/\Delta\omega$, and

$$A(t') = E_0 \frac{\sin[(2n+1)\Delta\omega t'/2]}{\sin[\Delta\omega t'/2]}. \tag{17}$$

Equation (16) shows that $E(t)$ can be represented in terms of a wave with a carrier frequency $\omega_0$ whose amplitude $A(t')$ is time-dependent. The intensity of this wave is given by $A^2(t')$, which consists of a train of pulses whose peak intensity is $A^2(0) = (2n+1)^2 E_0^2$, pulse width is $\Delta\tau_p = 2\pi/(2n+1)\Delta\omega$, and separation between pulses is $\tau_p = 2\pi/\Delta\omega$. Since the total oscillating bandwidth $\Delta\nu_{osc}$ is given by $(2n+1)\Delta\omega/2\pi$, the pulse width can also by written as $\Delta\tau_p = 1/\Delta\nu_{osc}$. Using the relationship $\Delta\omega = \pi c/l$, where $l$ is the length of the laser cavity, the time between pulses is $\tau_p = 2l/c$. In words, a mode-locked laser produces a train of output pulses whose pulse width is given by the inverse of the oscillating bandwidth and whose separation between pulses is equal to the round-trip time of the laser cavity. Figure 11 shows $A^2(t')$ for $n = 5$.

In the above example, the phases of all of the modes were locked so that the output pulse had its minimum possible duration. Such a pulse is referred to as a *transform-limited pulse*, since its temporal profile is the Fourier transform of its spectral profile. This need not be the case – it is possible to obtain longer pulses, but not shorter pulses. Also, it is worth noting that, unlike the other pulsed schemes described in this section, a mode-locked laser is a

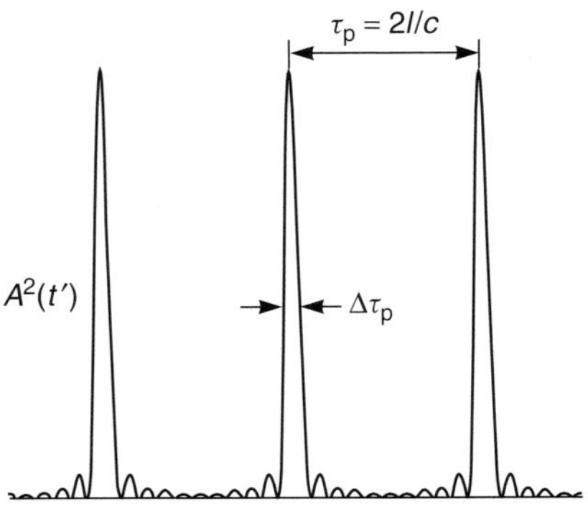


**Fig. 11** Train of mode-locked pulses, made up of 11 modes of equal intensity

cw device and there is phase coherence between pulses.

The above discussion tells us what mode locking is, but sheds little light on why or how it occurs. In general, mode locking will occur if the net gain for a mode-locked train of pulses is greater than the net gain of any other combination of cavity modes.

### 6.5.1 Active Mode Locking

Mode locking can be obtained actively or passively. Active mode locking can be broken into two categories, "AM mode locking" (produced using an amplitude modulator within the laser) and "FM mode locking" (produced using a frequency or phase modulator within the laser). In AM mode locking, the loss of some element in the laser cavity is modulated at the round-trip cavity frequency. Light circulating in the cavity will see less loss, and therefore more net gain, when it is incident at the loss element during the time of minimum loss. This encourages short-pulsed operation and mode locking can be induced. The same result occurs if the gain of the cavity is modulated. Gain modulation through modulation of the pump source is known as *synchronous pumping*.

In FM mode locking, the optical length of the laser cavity (length or refractive index) is modulated at the round-trip cavity frequency. For simplicity, let us consider the case where one of the mirrors is moved sinusoidally along the direction of the cavity axis. Light incident on the mirror during its motion will be Doppler shifted. As a result, it will not reproduce itself after one round trip, and will not produce a coherent oscillating mode. Light incident on the mirror at its turning points (maximum or minimum cavity length) sees a stationary mirror and will not experience a Doppler shift. The net result is that mode-locked pulses will tend to form. Modulation of the refractive index at some point in the cavity has the same effect.

### 6.5.2 Passive Mode Locking

Passive mode locking can occur when a laser cavity contains a nonlinear optical element, such as a saturable absorber. In this case, the more intense the light incident on the saturable absorber, the less the total absorption. The total loss of the cavity is therefore minimized by putting all the energy into short pulses. This is essentially self-induced AM mode locking. A similar effect is obtained by putting a Kerr lens and an aperture within the cavity. Other techniques include the use of interferometric elements containing nonlinear media.

Passive mode locking must be initiated by the presence of a pulse within the cavity. If the optical intensity within the cavity is uniform in time, there is no loss or gain element that is modulated at the round-trip cavity frequency in order to induce mode locking. Noise, however, is capable of introducing a small amplitude modulation on the optical field. In some lasers, this small modulation is sufficient to start the mode-locking process. Such lasers are referred to as self-starting. In other systems, a pulse (or AM modulation) must be intentionally introduced into the cavity to start the mode-locking process. Once started, however, mode locking can persist for a long time.

The introduction of an appropriate nonlinear optical element into a laser cavity is not sufficient to guarantee mode locking. In order for passive mode locking to work, the relative phases of all of the longitudinal modes must remain constant. One effect that can destroy the phase relationship, and therefore prevent mode locking, is dispersion. Passively mode-locked lasers must be dispersion free if mode locking is to occur. To accomplish this, prisms are often introduced into the laser cavity to compensate the dispersion of other intracavity elements, such as the gain medium. In a properly compensated cavity, non-mode-locked operation can be unstable.

The other effect that can destroy the phase relationship between the spectral components in a mode-locked pulse train is spontaneous emission, or noise. The phase of the noise is unrelated to the phase of the oscillating mode. The net phase will be shifted when the two are combined.

## 7 Control of Laser Output

### 7.1 Frequency Tuning

Frequency tuning of a laser can occur in one of two ways. If the longitudinal mode spacing of the laser cavity is much less than the gain bandwidth, the cavity is capable of supporting several modes, each at a different frequency. A single frequency is then selected through the insertion of an element into the cavity such that the cavity sees a frequency-dependent loss, as discussed in Sect. 5.6.3. In most of the examples listed in that section, a small repositioning of the frequency-selective element would result in a new longitudinal mode (and hence a new operating frequency) being selected. The frequency-selective element is used to select one of the several cavity modes, and discrete tuning is obtained. Fast tuning can be obtained through the use of electro-optic frequency-dependent components.

The other way a laser can be tuned is to change the frequency of a given cavity mode, by changing the optical length of the cavity. Since the cavity length can be changed continuously, this leads to continuous tuning. This type of tuning is often limited by the free spectral range of the cavity. Once the cavity modes are shifted by a full free spectral range, an adjacent cavity mode is positioned at the frequency where the initial mode started. For the same reasons that the initial mode was originally favored, the adjacent mode is now favored, and the laser will have a tendency to mode-hop back to the original frequency.

### 7.2 Amplitude Modulation

The output power of a laser can be controlled by changing the pump power, the output coupling, or the intracavity loss. This type of amplitude modulation is usually limited to frequencies below the frequency of the relaxation oscillations. The relaxation frequency characterizes the response time of the cavity. Near the relaxation frequency there is resonant enhancement of the modulation response; above the relaxation frequency the response rolls off.

Methods used for direct amplitude modulation of a laser may have the side effect of introducing frequency modulation as well. For example, changing the pump power affects the thermal load on the gain medium, and therefore the temperature. This, in turn, affects the refractive index, changing the optical length of the cavity and the oscillating frequency. For amplitude-modulation applications where frequency stability is critical, it is often better to modulate the laser power external to the cavity.

## 8 Oscillator–Amplifier Systems

As the required output power of the laser increases, the need for amplification stages becomes more apparent. Master oscillators (lasers) that operate at low powers can be easily controlled to produce a desired output. Successive amplification stages can be designed to provide increasingly higher output powers by increasing the pumping power and the amplifier aperture. Limiting the gain of each stage can eliminate parasitic oscillation. Isolators are used to prevent feedback from subsequent amplification stages. The design of each stage can also be optimized for heat removal and low optical distortion. Beam cleanup between stages can be used to suppress unwanted spatial frequencies and, thereby, maintain single-transverse-mode operation. Amplifiers can also be paralleled by splitting the master oscillator beam; if a coherent output is desired, the outputs of the amplifiers must be phased.

There are two basic types of amplifiers, regenerative and traveling wave. A regenerative amplifier provides a feedback loop and can oscillate without the laser (master oscillator) input. Control of the regenerative amplifier is achieved by injection seeding it with the master oscillator signal. In a traveling-wave configuration, the amplifier simply boosts the signal injected by the master oscillator. Regenerative amplifiers are, by nature, multi-pass. Traveling-wave amplifiers can be designed so that the radiation travels through the amplifying medium more than once without retracing its path.

Laboratory-scale oscillator–amplifier systems can be designed to produce petawatt ($10^{15}$ W) pulses. Very short (femtosecond) pulses cannot be directly amplified to very high powers without damage to the amplifier material. Instead, a dispersive element can be used to chirp and stretch the pulse in time. This lowers the peak power so that the pulse can be passed through a broadband amplifier, such as Ti:$Al_2O_3$, and later through another dispersive element that un-chirps and compresses the amplified pulse.

## 9 Issues in Laser Design

The design of a laser is dependent on many interdependent factors, including the requirements placed on the output beam (wavelength, spectral purity, tunability, divergence, polarization, power, and power stability), the operating environment (temperature, humidity, vibration, acceleration, and externally applied forces), and practical considerations (size, cost, and available power). There is an increasingly large number of gain media, cavity designs, and pump configurations that have been employed in lasers, and several texts have been written on the subject of laser design. No one design is well suited for all applications; every laser is optimized for operation at one point in the multidimensional parameter space outlined above.

A very important issue in the design of many lasers is the extraction of heat from the gain medium. In the process of pumping the gain medium, heat is generated. As the temperature of the gain medium changes, so too does its physical length and refractive index. Each of these contributes to a change in the optical length and resonant frequencies of the laser cavity. Nonuniform heating results in thermal lensing and internal stress. Thermal lensing changes the confocal parameters of the laser cavity and can destabilize an otherwise stable cavity (or vice versa). Internal stress leads to stress birefringence and, eventually, stress fracture.

Other issues that must be considered in high-power lasers are nonlinear optical effects and optical damage. The electrical field within the optical beam of a high-power laser can be large enough to damage optical components. This is particularly important in high-power pulsed lasers. At optical intensities below the optical damage level, deleterious nonlinear optical interactions can still degrade the performance of the laser, and even destroy the device. One example is stimulated Brillouin (acoustic wave) scattering in fiber lasers. In this case, nonlinear interactions create acoustic waves that can blow off the ends of the fiber.

## 10 Frequency Conversion and Nonlinear Control of Laser Radiation

Nonlinear optical techniques can be used to extend the frequency coverage of lasers as well as to modify other characteristics of laser radiation. Frequency conversion is a very important adjunct, converting the output of practical lasers to regions where primary laser sources may not exist or may not be very practical. Harmonic generation, frequency mixing, optical parametric oscillation, and stimulated Raman scattering have been used for frequency conversion. Nonlinear processes have also been used to produce mode locking (see Sect. 6.5.2), to improve transverse beam quality (i.e. to produce output closer to the diffraction limit), and to dampen relaxation oscillations in pulsed lasers. The article NONLINEAR OPTICS describes the nonlinear processes in detail; here, we will only discuss their general significance for lasers.

Finally, it should be mentioned that laser oscillators and amplifiers are nonlinear optical devices in that there is a reduction in optical gain caused by partial depletion of the population inversion by the laser radiation. This partial depletion, or saturation, stabilizes the laser output amplitude in cw lasers and plays a vital role in determining the operating characteristics of pulsed lasers, as we have seen in Sect. 6.

The most frequently used frequency-conversion techniques are second-harmonic and sum-frequency generation. Typically, the output of infrared lasers in the 1-μm region, such as Nd:YAG or Nd:glass, is frequency doubled into the green. Depending on the application, shorter wavelengths may be obtained (for example, by summing the green radiation with the infrared). There are a variety of uses for this short-wavelength output, including pumping short-wavelength lasers and optical parametric oscillators (OPOs). For second-harmonic generation to the green, conversion efficiencies as high as 75% have been reported and average powers of greater than 50 W have been obtained. Materials such as $KTiOPO_4$ (KTP), periodically poled $LiNbO_3$ (PPLN), $LiB_3O_5$ (LBO), and $\beta$-$BaB_2O_4$ (BBO) are used.

Difference-frequency generation and OPOs have been used as tunable sources of radiation for various spectroscopic applications, primarily in the infrared. Continuous-output difference-frequency generation, while producing very low average power with low efficiency, has been used for ultrahigh-resolution molecular spectroscopy in the mid-infrared. OPOs are useful sources of tunable pulsed output with high peak power and high efficiency.

Stimulated Raman scattering has been used to generate large pulse energies with essentially unity quantum efficiency at a variety of wavelengths. However, because energy is deposited in the Raman process, moving Raman media are often required to obtain high-average-power conversion. For high energies and powers, gases such as $H_2$ have been used. With a very high finesse cavity, a diode-laser-pumped hydrogen Raman laser has been operated continuously. Fiber Raman lasers have produced tens of watts of continuous power, while fiber Raman amplifiers have been used to increase the useful extent of the fiber-communications spectrum.

The optical Kerr effect, which is the change in refractive index proportional to the optical intensity, can cause short pulses, typically 0.1 to 10 ps in duration, to acquire a frequency sweep, or chirp. By sending the chirped pulse through a frequency-dispersive delay line, which might be a grating or prism pair, even shorter pulses, typically 5 to 50 fs, can result.

Nonlinear phase conjugation can reverse the phase distortion acquired from the active medium in a laser (for example, an optically imperfect laser crystal or a laser material that has large thermal gradients due to high-average-power operation). On a second pass through the distorting medium, the reversed phase distortion is canceled. Most often, backward stimulated Brillouin scattering is used to reverse the phase; the small acoustic frequency shift produces an optical wave that remains within the gain bandwidth of most lasers and has linear propagation characteristics that match the input to the Brillouin cell. Other techniques used take advantage of the photorefractive effect or absorptively induced nonlinearities.

Harmonic generation has been used to moderate the amplitude of the relaxation oscillations that often occur in long-pulse lasers. Relaxation oscillations (see Sect. 6.1.1) involve the flow of energy back and forth between the population inversion and the radiation field. Conversion to the second harmonic can clip the high initial peaks in the laser power, an action that reduces the amount of depletion in the population inversion, thereby damping the oscillations.

## Glossary

**Active Medium:** Material that amplifies radiation.

**Active Mode Locking:** Mode locking using a modulator.

**AM Mode Locking:** Active mode locking using amplitude modulation.

**Amplitude Modulation:** Temporal control of the amplitude of an electromagnetic field.

**Axial Mode:** Radiation pattern satisfying a resonance condition defined by the number of waves along the principal axis of a cavity; same as longitudinal mode.

**Bandwidth:** Width, in frequency or wavelength, of a radiative transition or the output of a laser; same as linewidth.

**Bandwidth Limited:** Having as short a duration as allowed by the bandwidth of the radiation.

**Brightness:** Measure of the spatial or spectral quality of a laser beam, see spatial brightness and spectral brightness.

**Cavity:** Region of space where electromagnetic radiation is confined, usually consisting of two or more reflecting surfaces.

**Cavity Dumping:** Using an active optical element to allow radiation to rapidly leave a cavity.

**Cavity Mode:** Radiation pattern satisfying the boundary conditions of a cavity.

**Cavity $Q$:** $2\pi$ times the damping time of the cavity divided by the period of the wave.

**Coherent:** Having a definite phase relationship between any two points in the radiation field.

**CW Operation:** Continuous-wave operation; steady-state operation.

**Diffraction Limited:** Having the minimum divergence allowed by diffraction for a given focal spot size.

**Diode Laser:** Semiconductor junction laser, where population inversion is created by injection of electrons and holes into the junction region; same as injection laser.

**Energy Gap:** Energy range where there are no allowed states in a pure material.

**Fabry-Pérot:** Cavity consisting of two parallel (but not necessarily flat) reflecting surfaces.

**Feedback:** Transfer of some of a system's optical output back to its input.

**FM Mode Locking:** Active mode locking using frequency modulation.

**Franck-Condon Principle:** Transitions between electronic states occur without changing nuclear coordinates.

**Free-running Mode:** Laser operation without active control.

**Frequency Conversion:** Nonlinear conversion of the frequency of laser radiation, includes harmonic generation, sum and difference frequency generation, parametric generation, and stimulated scattering.

**Frequency Tuning:** Changing the frequency of a laser.

**Gain Cross Section:** Parameter of an active medium that, when multiplied by the population inversion density, gives the gain per unit length.

**Gain Peak:** Frequency or wavelength where the gain of a laser transition is greatest.

**Gain Switching:** Increasing the gain of a laser cavity sufficiently fast that the population inversion is temporarily out of equilibrium with the optical field in the cavity resulting in pulsed laser output.

**Harmonic Generation:** Nonlinear generation of radiation at a multiple of the input frequency.

**Homogeneously Broadened:** Having a radiative transition where all contributors to the transition have the same environment and characteristics.

**Inhomogeneously Broadened:** Having a radiative transition where all contributors to the transition do not have the same environment or characteristics.

**Injection Current:** Current that injects electrons and holes into the junction region of a semiconductor diode.

**Injection Laser:** Semiconductor junction laser, where population inversion is created by injection of electrons and holes into the junction region; same as diode laser.

**Junction:** Interface region between *n*-type and *p*-type semiconductor materials.

**Laser Dynamics:** Temporal behavior of a laser.

**Laser Transition:** Change of material quantum state that occurs during the laser emission process.

**Linewidth:** Width, in frequency or wavelength, of a radiative transition or the output of a laser; same as bandwidth.

**Longitudinal Mode:** Radiation pattern satisfying a resonance condition defined by the number of waves along the principal axis of a cavity; same as axial mode.

**Long-pulse Operation:** Operation of a pulsed laser with a pulse long enough for the population inversion in the active medium to reach equilibrium with the optical field in the cavity.

**Mode Locking:** Forcing a phase relation between the axial (longitudinal) modes of a laser.

**Multimode Operation:** Laser operation with multiple cavity modes oscillating at the same time.

**Nonlinear Optics:** Optical processes that have a nonlinear dependence on the amplitude of the optical field.

***n*-type:** Semiconductor material where conduction-band electrons are the primary current carriers.

**Optical Kerr Effect:** Quadratic electro-optical effect induced by an optical field.

**Optical Pumping:** Producing a population inversion by pumping an active medium with an optical field.

**Output:** Optical radiation from a laser.

**Parasitic Oscillation:** Laser oscillation arising from spurious reflections in a laser oscillator or amplifier.

**Passive Mode Locking:** Mode locking using a nonlinear element in the laser cavity such as a saturable absorber or a Kerr lens.

**Phase Conjugation:** Reversal of the phase of an electromagnetic field.

***p-n* Junction:** Interface region between *n*-type and *p*-type semiconductor materials.

**Population Inversion:** State of a material in which a level with an energy greater than another level also has a larger occupation probability (population).

***p*-type:** Semiconductor material where valence-band holes are the primary current carriers.

**Pulsed Operation:** Laser operation where the radiation is produced in pulses.

**Pumping:** Producing a population inversion in an active medium.

**Q Switching:** Reducing the loss in a laser cavity sufficiently fast that the population inversion is temporarily out of equilibrium with the optical field in the cavity resulting in pulsed laser output.

**Quantum Well:** Semiconductor structure that spatially confines carriers and produces a new energy spectrum.

**Radiative Transition:** Transition that can produce electromagnetic radiation.

**Rate Equations:** Equations characterizing the number of photons in a laser cavity and the populations of the various energy levels of the active material system.

**Regenerative Amplifier:** Amplifier in which radiation is fed back through the active medium so that an input signal is not required for significant extraction of optical power, also called an oscillator or laser oscillator.

**Relaxation Oscillation:** Oscillatory condition that occurs when energy flows back and forth between an active material and a radiation field as a system approaches steady state.

**Saturation:** Reduction of laser gain (saturable-absorber loss) that occurs when the difference between upper and lower laser-level (saturable-absorber-level) populations approaches zero.

**Single-mode Operation:** Laser operation with only a single cavity mode oscillating; same as single-frequency operation.

**Single-frequency Operation:** Laser operation with only a single cavity mode oscillating; same as single-mode operation.

**Slope Efficiency:** Change in output power of a laser per change in pump power.

**Spatial Brightness:** Power per beam area at the focal point per beam solid angle.

**Spatial Hole Burning:** Spatial variation in the saturated inversion density in a standing-wave laser cavity.

**Spectral Brightness:** Power per unit frequency.

**Spectral Hole Burning:** Variation of population inversion in frequency due to saturation of a transition in an inhomogeneously broadened material.

**Spontaneous Emission:** Emission that occurs absent excitation by a radiation field.

**Spontaneous Lifetime:** Exponential time constant associated with spontaneous radiative decay.

**Stable Resonator:** Optical cavity with mirrors configured to generate modes that converge on the cavity axis.

**Stimulated Emission:** Emission induced by a radiation field.

**Stimulated Scattering:** Scattering induced by an output (scattered) field.

**Stokes Shift:** Shift of output radiation toward longer wavelength (lower photon energy), the energy difference is transferred to the material system responsible for the wavelength shift.

**Stored Energy:** Energy stored in an active medium that can be extracted optically.

**Threshold:** Onset of laser operation that occurs when the pump power is sufficient to achieve an amplifier gain equal to the cavity loss.

**Transition:** Change in the quantum state of a material system.

**Transition Frequency:** Energy difference between the two quantum states involved in a transition divided by Planck's constant.

**Transition Strength:** Magnitude of the interaction between the radiation field and the material system, proportional to the square of the transition dipole.

**Transverse Mode:** Radiation pattern transverse to the principal axis of a cavity.

**Traveling-wave Amplifier:** Amplifier in which radiation travels along a non-repeating path.

**Tunable Laser:** Laser whose output wavelength can be varied.

**Unstable Resonator:** Optical cavity with mirrors configured to generate modes that do not converge on the cavity axis.

## Further Reading

Popular discussions of lasers and their applications and history are given in:

Hecht, J. (1992), *Laser Pioneers*, (revised ed.), Boston, MA: Academic Press.

Hecht, J., Teresi, D. (1998), *Laser: Light of a Million Uses*. Mineola, New York: Dover Publications.

A general discussion of laser science and technology, accessible to anyone with a background of electromagnetic theory, basic quantum mechanics, and calculus, can be found in any of the following:

Eastham, D. A. (1986), *Atomic Physics of Lasers*. London: Taylor & Francis.

Kaminow, I. P., Siegman, A. E. (1973), *Laser Devices and Applications*. New York: IEEE.

Shimoda, K. (1991), *Introduction to Laser Physics*, (2nd ed.), New York: Springer.

Siegman, A. E. (1986), *Lasers*. Mill Valley, CA: University Science.

Silfvast, W. T. (2004), *Laser Fundamentals*, (2nd ed.), Cambridge: Cambridge University Press.

Svelto, O. (1998), *Principles of Lasers*, (4th ed.), New York: Plenum.

Verdeyen, J. T. (1995), *Laser Electronics*, (3rd ed.), Englewood Cliffs, NJ: Prentice Hall.

Yariv, A. (1989), *Quantum Electronics*, (3rd ed.), New York: Wiley.

Detailed theoretical discussion of lasers is contained in:

Haken, H. (1986), *Laser Theory*, (reprint ed.), New York: Springer.

Sargent, M. III, Scully, M. O., Lamb, W. E. Jr. (1974), *Laser Physics*. Reading, MA: Addison-Wesley.

The underlying physics of laser resonators and its application to the design of many types of lasers in use today is the topic of:

Hall, D. R., Jackson, P. E. (Eds.) (1989), *Physics and Technology of Laser Resonators*. Bristol, UK: Adam Hilgar.

Specific types of lasers are discussed in:

Brown, D. C. (1981), *High-Peak-Power Nd: Glass Laser Systems*. New York: Springer.

Casey, H. C. Jr, Panish, M. B. (1978), *Heterostructure Lasers*, Parts A and B, New York: Academic Press.

Coldren, L., Corzine, S. W. (1995), *Diode Lasers and Photonic Integrated Circuits*. New York: Wiley.

Duarte, F. J. (Ed.) (1995), *Tunable Lasers Handbook*. San Diego, CA: Academic Press.

Duarte, F. J., Hillman, L. W. (Eds.) (1990), *Dye Laser Principles*. Boston, MA: Academic Press.

Garrett, C. G. B. (1967), *Gas Lasers*. New York: McGraw-Hill.

Kaminskii, A. A. (1990), *Laser Crystals*, (2nd ed.), New York: Springer.

Kapon, E. (Ed.) (1999), *Semiconductor Lasers I: Fundamentals*. San Diego, CA: Academic Press.

Koechner, W. (1999), *Solid-State Laser Engineering*, (5th ed.), New York: Springer.

Koechner, W., Bass, M., Roth, H. (2003), *Solid State Lasers: A Graduate Text*. New York: Springer.

Mollenauer, L. F., White, J. C., Pollock, C. R. (Eds.) (1992), *Tunable Lasers*, (2nd ed.), New York: Springer.

Schäfer, F. P. (Ed.) (1990), *Dye Lasers*, (3rd ed.), New York: Springer.

Willett, C. S. (1974), *Introduction to Gas Lasers: Population Inversion Mechanisms*. Oxford: Pergamon Press.

A concise source of data in tabular and graphical form is provided for workers in the areas of laser research and development in:

Weber, M. J. (Ed.) (1982), *Handbook of Laser Science and Technology*, Vols. I and II; Weber, M. J. (Ed.) (1986), *Handbook of Laser Science and Technology*, Vols. III and IV; Weber, M. J. (Ed.) (1987), *Handbook of Laser Science and Technology*, Vol. V; Weber, M. J. (Ed.) (1991), *Handbook of Laser Science and Technology*, Suppl. 1; Weber, M. J. (Ed.) (1995), *Handbook of Laser Science and Technology*, Suppl. 2, Boca Raton, FL: CRC Press.

Weber, M. J. (Ed.) (2003), *Handbook of Lasers*. Boca Raton, FL: CRC Press.

# Lasers, Dye

*Francisco J. Duarte and David R. Foster*
*Research Laboratories, Eastman Kodak Company, Rochester, New York, USA*
*Phone: (1-585) 722 2528; Fax: (1-585) 477 7781; e-mail: francisco.duarte@kodak.com*

**Abstract**
Dye lasers are inherently tunable, and powerful, sources of narrow-linewidth radiation. This means that dye lasers can change the emission color continuously while emitting radiation of extremely high spectral purity. In addition, these tunable lasers can be configured to yield extremely short pulses in the femtosecond regime. As such, dye lasers have found a wide variety of applications in fundamental physics, spectroscopy, medicine, and the military. In this article, basic aspects of dye-laser operation are examined and explained. These include a discussion of the excitation process, via classical rate equations, the origin of tunability, and the theory of dispersion that explains the achievement of very narrow laser linewidths. Although most of the article deals with traditional liquid dye lasers, a section is devoted to newly developed solid-state dye lasers. Important applications, such as laser isotope separation and laser cooling are also discussed.

**Keywords**
lasers; tunable; tunable lasers; dye lasers; solid-state dye lasers; multiple prism; dispersion theory; femtosecond lasers.

## 1 Introduction

The study of dye lasers is motivated by their unparalleled versatility and intrinsic uniqueness as physical systems. For instance, dye lasers have been demonstrated to oscillate in the three states of matter, that is, in the gaseous, liquid, and solid states. However, it is largely in the liquid phase that this class of laser has shown its formidable utility and pervasiveness.

Basically, dye lasers are optically pumped, tunable, coherent light sources that can be divided into two large classes: pulsed dye lasers and continuous-wave (cw) dye lasers. Pulsed dye lasers are further categorized into the subclasses of laser-pumped pulsed dye lasers and flashlamp-pumped dye lasers. Femtosecond-pulsed dye lasers evolved from the technology of cw dye lasers. This latter class of dye lasers utilizes laser excitation.

The laser-active medium consists of a solution of organic dye molecules in a liquid such as ethyl alcohol, methyl alcohol, or water [1]. The most notable attributes of laser dye media are their broad wavelength tunability, wide spectral coverage, and practical simplicity. Hundreds of dyes have been reported to lase, and many of the more commonly used molecules for dye lasers belong to the oligophenylene, coumarin, xanthene, merocyanine, and cyanine classes. For a comprehensive listing of laser dye molecular species, the reader should refer to Maeda [2], Drexhage [3], and Steppel [4]. As illustrated in Fig. 1, the spectral coverage provided by these dye classes extends from ~0.3 to ~1.1 μm. Further regions of the spectrum can be reached via other infrared dyes and by using dye lasers in conjunction with nonlinear techniques such as frequency doubling and Raman shifting.

Laser dyes are large organic molecules having molecular weights between 175 and 1000 amu that have strong absorption bands in the ultraviolet and visible spectral regions, and when excited by absorbed light, they exhibit intense fluorescence bands at longer wavelengths than their absorption. This red shift in wavelength maxima between absorption and fluorescence bands is called the *Stokes shift*.

One of the best-known dye molecules is rhodamine 6G (a xanthene dye), depicted with its absorption and fluorescence spectra in Fig. 2. Rhodamine 6G is the most fully characterized laser dye and is often used as a standard to determine the relative performance of other laser dyes. This molecule has been reported to lase in the 550- to 650-nm spectral range with a variety of solvents, concentrations, and excitation sources.

The broad structure of the absorption spectrum shown in Fig. 2 corresponds to absorption transitions from the ground

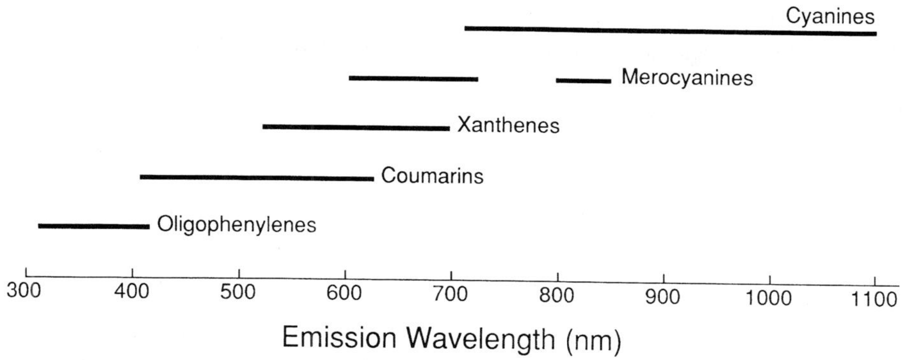


**Fig. 1** Approximate spectral emission range of various dye classes. Emission from the cyanines extends beyond 1100 nm

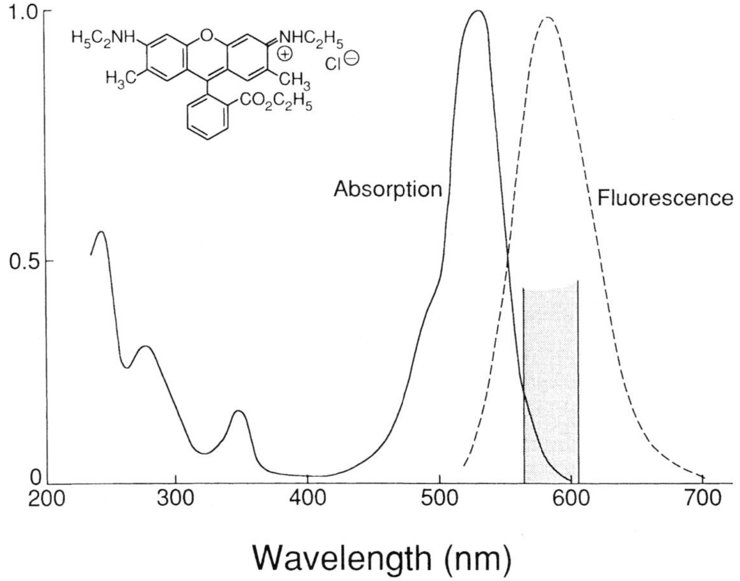


**Fig. 2** Rhodamine 6G absorption and fluorescence spectra. The main absorption maximum corresponds to the $S_0 \rightarrow S_1$ transition. The shaded region is the approximate tuning range achieved with copper vapor laser (CVL) excitation. The molecular weight of this molecule is 479

electronic state ($S_0$) to the first excited electronic state ($S_1$) and higher states. Each electronic state contains a multitude of vibrational and rotational levels. Because of line-broadening processes, the rotational levels are not resolved, thus giving rise to a continuum. This property explains the continuous tunability associated with dye lasers. In addition to spectrally broad emission, the dye gain medium exhibits homogeneous broadening that enables all the excited molecules to contribute efficiently to emission in a narrow frequency band defined by intracavity dispersive elements (see Sects. 2 and 6).

The versatility of dye lasers originates in their ability to emit tunable coherent radiation. Indeed, to a large extent, the words *dye laser* have acquired synonymous status with *tunability*. Moreover, dye lasers offer tremendous operational flexibility:

1. covering the spectrum from the near ultraviolet to the near infrared,
2. emitting from femtosecond pulses to the CW regime,
3. yielding radiation from large single-pulse energies to high-average powers.

Specifically, dye lasers have yielded up to 800 J in the single-pulse regime and average powers in *excess* of 2.5 kW in the high–pulse-repetition-frequency (prf) domain. Indeed, dye lasers are unrivaled in the generation of high–pulse-energy and high–average-power tunable coherent radiation in the visible. This unparalleled performance is inherently linked to the heat-dissipation ability characteristic of liquid-phase dye molecular media. High-average powers are essential to numerous applications including laser isotope separation (LIS, see also the article LASER ISOTOPE SEPARATION), in industry, and in astronomy.

In general, the operational flexibility of the dye laser has yielded numerous and varied applications ranging from cancer photodynamic therapy (PDT) to laser radar. The plethora of applications includes fields such as physics, chemistry, and biology. A perspective on applications is provided in Sect. 9.

The benefits of the dye laser to science extend beyond the area of applications, since dye-laser development has yielded significant contributions in the fundamental areas of amplified spontaneous emission (ASE), femtosecond laser physics, dynamical instabilities, and oscillator physics. Here, it should be emphasized that ASE refers to strong *spontaneous* emission that occurs *prior* to laser emission (see the Glossary for further information).

The dye laser has a rich literature. Since its discovery in 1966 (Sorokin and Lankard, [5]; Schäfer et al., [6]) until today, there have been hundreds of papers published on dye-laser development. A perspective on the dye-laser literature and its history can be obtained from the books suggested in Further Reading.

Although we focus here mainly on liquid-phase dye lasers, alternative solid-state dye lasers using polymers (see, for example, Pacheco et al., [7]) and silicate matrices are also available. Indeed, recent developments have produced very attractive organically modified silicate matrices that offer significant potential for long-lived, solid-state dye lasers for low-power applications. The oscillator and system configurations discussed here apply to liquid and/or solid-state lasers.

In this article, a basic and brief survey of the elements important to dye lasers is provided. In particular, attention is given to developments in the area of intracavity physics and resonator design. This focus is justified by the numerous contributions of a *fundamental nature* made by dye lasers, which are now being applied to lasers in general. Details on the science of the molecular media and other aspects of dye-laser physics are provided in [11].

## 2
## Energy Levels and Rate Equations

Laser action is intrinsically a quantum-mechanical phenomenon. However, in certain circumstances the basic excitation

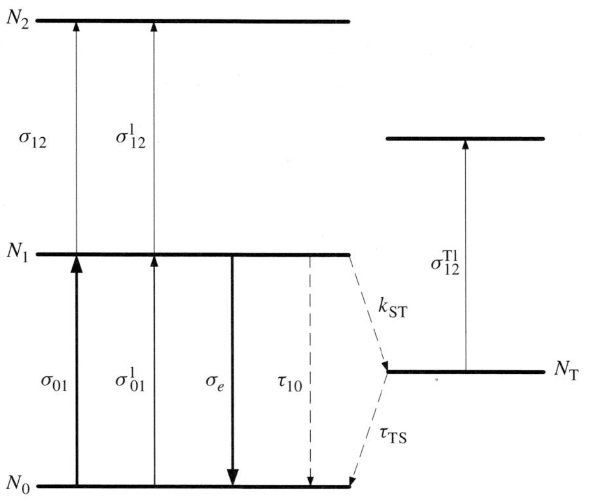


**Fig. 3** Energy-level diagram of a laser dye molecule under optical excitation

mechanism can be treated classically using rate equations. That is the case of laser dye molecules. As mentioned previously, dye molecules can populate several distinct energy levels known as electronic states. In turn, as illustrated in the energy diagram of Fig. 3, each electronic state contains a multitude of overlapping vibrational and rotational levels. It is precisely this characteristic that gives rise to spectral tunability. This is because $E = h\nu$, where $\nu$ is the frequency. Thus, a $\Delta E$ implies a $\Delta\nu$, which in turn means a change in the wavelength domain or $\Delta\lambda$. Hence, overlapping energy levels make possible the existence of continuous absorption and emission spectral bands, and hence the tunability of the dye laser.

The literature on rate equations for dye lasers is fairly extensive. For further details, the reader might refer to Ganiel et al. [8] and Teschke et al. [9]. In reference to the energy-level diagram of Fig. 3, a set of rate equations considering the vibrational manifolds at each electronic state was written by Duarte [10]. Replacing the vibronic manifolds by single levels, and neglecting the second excited state $N_2$, the set of equations simplifies to

$$N = N_0 + N_1 + N_T \tag{1}$$

$$\frac{\partial N_1}{\partial t} \approx N_0\sigma_{01} I_p(t) + (N_0\sigma^1_{01} - N_1\sigma_e - N_1\sigma^1_{12}) I_l(x, t, \lambda) - N_1(k_{ST} + \tau^{-1}_{10}) \tag{2}$$

$$\frac{\partial N_T}{\partial t} = N_1 k_{ST} - N_T\tau^{-1}_{TS} \tag{3}$$

$$\left(\frac{1}{c}\right)\frac{\partial I_p(t)}{\partial t} = -(N_0\sigma_{01} + N_1\sigma_{12}) I_p(t) \tag{4}$$

$$\left(\frac{1}{c}\right)\frac{\partial I_l(x, t, \lambda)}{\partial t} + \frac{\partial I_l(x, t, \lambda)}{\partial x} = (N_1\sigma_e - N_0\sigma^1_{01} - N_1\sigma^1_{12} - N_T\sigma^{T1}_{12}) I_l(x, t, \lambda) \tag{5}$$

Where, $N_0$ is the ground state, $N_1$ is the first excited state, and $N_T$ represents the triplet level. This set of equations is similar to the set of equations considered by Teschke et al. [9]. Here, $I_p(t)$ is the intensity of the pump laser beam, $I_l(x,t,\lambda)$ is the laser emission from the gain medium, $\sigma_{01}$ is the cross-section for absorption of the pump intensity by the ground level $N_0$, $\sigma_{12}$ is the cross-section for absorption of the pump intensity by the first excited state $N_0$, $\sigma_e$ is the cross-section for laser emission from the first excited state, and so on. Duarte [10] provides all the cross sections and excitation rates incorporated in these equations. For example, for rhodamine 6 G, $\sigma_{01} \approx 1.7 \times 10^{-16}$ cm$^2$ (at $\lambda \approx 510$ nm) and $\sigma_e \approx 1.3 \times 10^{-16}$ cm$^2$ (at $\lambda \approx 590$ nm). Using available excitation cross sections and rates these equations can be solved numerically for the case of pulsed excitation. For long pulse, or CW excitation, Eqs. (2) to (5) reduce to

$$N_0\sigma_{01}I_p + (N_0\sigma_{01}^1 - N_1\sigma_e - N_1\sigma_{12}^1) I_l(x,\lambda) = N_1(k_{ST} + \tau_{10}^{-1}) \quad (6)$$

$$N_1 k_{ST} = N_T \tau_{TS}^{-1} \quad (7)$$

$$N_0\sigma_{01} = -N_1\sigma_{12} \quad (8)$$

$$\frac{\partial I_l(x,\lambda)}{\partial x} = (N_1\sigma_e - N_0\sigma_{01}^1 - N_1\sigma_{12}^1 - N_T\sigma_{12}^{Tl}) I_l(x,\lambda) \quad (9)$$

From these equations some interesting features of CW dye lasers become apparent. For example, as indicated by Dienes and Yankelevich [11], from Eq. (6) just below threshold, that is, $I_l(x,\lambda) = 0$,

$$I_p = \sigma_{01}^{-1}(k_{ST} + \tau_{10}^{-1})(N_1/N_0) \quad (10)$$

which means that to approach population inversion, using rhodamine 6G under visible laser excitation, pump intensities exceeding $10^{22}$ photons cm$^{-2}$ s$^{-1}$ are necessary. A problem unique to long-pulse and CW dye lasers is intersystem crossing from $N_1$ into $N_T$. From Eq. (9) the differential gain can be written as

$$g = N_1\sigma_e\phi_T - N_0\sigma_{01}^1 \quad (11)$$

where

$$\phi_T = 1 - (\sigma_{12}^1/\sigma_e) - (k_{ST}\tau_{TS}\sigma_{12}^{Tl}/\sigma_e) \quad (12)$$

Values for the intersystem crossing rate are in the $0.4 \times 10^7$ s$^{-1}$ $\le k_{ST} \le 2.0 \times 10^7$ s$^{-1}$ range, while the lifetime of the triplet state is in the $0.5 \times 10^{-7}$ s $\le \tau_{ST} \le 2.5 \times 10^{-7}$ s range. Equation (11) implies that a necessary, but not sufficient, condition for lasing is to have a positive value for $\phi_T$. Since the ratio $(\sigma_{12}^1/\sigma_e) \approx 0.07$, or less, it can be neglected thus leaving the triplet state term as a dominant factor in Eq. (12). In order to ensure a positive value for $\phi_T$, researchers decrease the lifetime of the triplet level $\tau_{TS}$ via the introduction of a *triplet level quencher* such as $O_2$. Note that the values for $k_{ST}$ and $\tau_{ST}$ imply that for short-pulsed excitation, in the nanosecond regime, the effect of the triplet level can be ignored.

## 3
## Laser-pumped Pulsed Dye Lasers

Optical pumping from pulsed lasers was the first avenue of excitation demonstrated in dye lasers [5, 12]. At present there is a variety of solid-state and gas-pulsed lasers that are applied to the excitation of dye lasers. These lasers offer different excitation wavelengths and energetic, temporal, and prf characteristics. Pulsed lasers suitable for dye-laser pumping include KrF, XeCl, and XeF. These lasers emit via the $B^2\Sigma_{1/2}^+ - X^2\Sigma_{1/2}^+$ transition at $\lambda \sim$ 248, 308, and 351 nm, respectively. A simple and inexpensive source of ultraviolet radiation is the nitrogen laser that emits at $\lambda \sim$ 337 nm via the $C^3\Pi_u - B^3\Pi_g$ transition.

The copper vapor laser (CVL) emits mainly via the $^2P_{3/2} - {}^2D_{5/2}$ transition at

$\lambda \sim 510$ nm and also via the secondary $^2P_{1/2}-^2D_{3/2}$ transition at $\lambda \sim 578$ nm. The CVL has the advantage of high efficiency, in the 1 to 3% range, and operates at prfs up to 30 kHz. A widely applied solid-state source is the Nd:YAG laser (YAG denotes yttrium aluminum garnet) that emits via the $^4F_{3/2}-^4I_{11/2}$ transition at $\lambda \sim$ 1064 nm. The second, third, and fourth harmonics of this emission yield excitation at 532, 355, and 266 nm, respectively. In general, laser-pumped pulsed dye-laser emission has been shown to cover the spectrum from ~315 to 1100 nm.

The variety of excitation wavelengths available makes possible the design of very efficient dye-laser systems in which excitation is accomplished by populating directly the first excited electronic state via the $S_0 \rightarrow S_1$ absorption transition. An important example of this type of excitation is the efficient CVL pumping of rhodamine dye lasers reported to be as high as 50% when using the transition at 510.5 nm [13]. Furthermore, this dye under CVL excitation exhibits remarkably long lifetimes [14].

The nitrogen laser is a relatively low-prf excitation source of general utility. Its ultraviolet $C^3\Pi_u-B^3\Pi_g$ transition is used to excite a large number of dyes whose emissions span the electromagnetic spectrum from ~360 to 1000 nm. Excimer lasers, such as KrF and XeCl, can be used to excite dyes further into the ultraviolet. In addition, excimer lasers provide considerably higher pulsed energies and higher prfs. Using XeCl laser excitation, Bos [15, 16] demonstrated lasing in the 335 to 912-nm range in an eight-cell dye-laser system. Emission at shorter wavelengths, originating at ~315 nm can be accomplished using KrF laser excitation.

Narrow-linewidth dispersive oscillator configurations are described in Fig. 4. Oscillators are important since they form the first stage of any laser system. It is the oscillator that provides the vital wavelength and linewidth information that is subsequently augmented by the amplification stages of

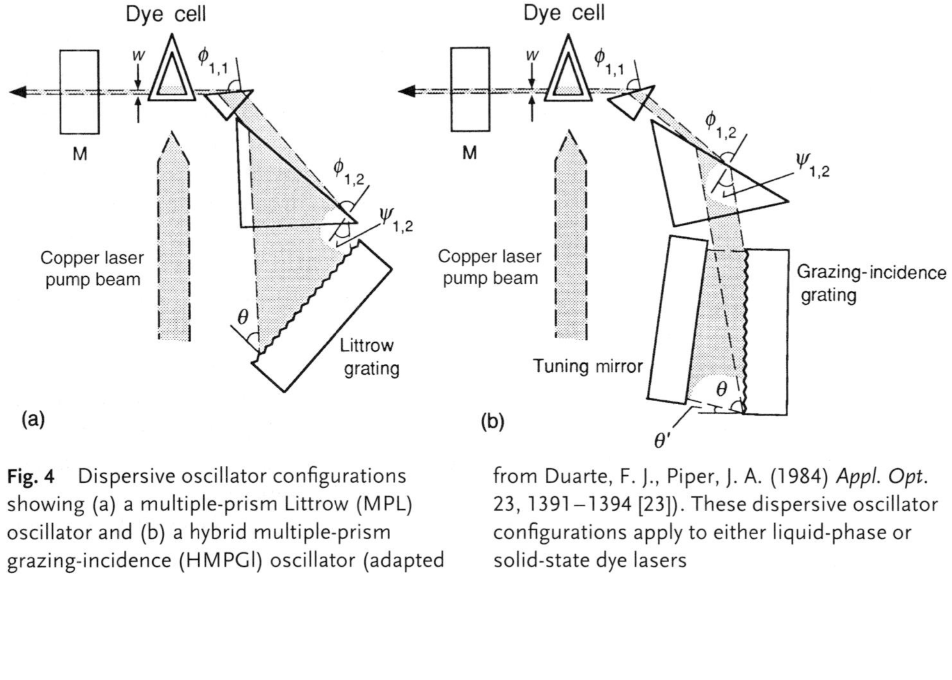


**Fig. 4** Dispersive oscillator configurations showing (a) a multiple-prism Littrow (MPL) oscillator and (b) a hybrid multiple-prism grazing-incidence (HMPGI) oscillator (adapted from Duarte, F. J., Piper, J. A. (1984) *Appl. Opt.* 23, 1391–1394 [23]). These dispersive oscillator configurations apply to either liquid-phase or solid-state dye lasers

the system. These master oscillators are usually composed of an output coupler mirror, a high–optical-quality quartz dye cell of trapezoidal geometry, a multiple-prism beam expander, and a grating either in Littrow configuration, where the angle of incidence equals the angle of diffraction (Fig. 4a), or near-grazing-incidence configuration (Fig. 4b). As illustrated in Fig. 4, the function of the beam expander is to magnify the narrow beam ($w$) emerging from the gain region in order to illuminate the grating completely. As explained in Sect. 8, the overall beam magnification factor is the product of the individual beam-expansion coefficients provided by each prism.

Typical cavity dimensions are 1 cm for the length of the dye cell and about 10 cm for the overall cavity. In the case of low-prf operation, the dye may be excited under static conditions. On the other hand, high-prf operation in the kilohertz range requires rapid flow of the dye in a direction perpendicular to the plane of propagation for heat-dissipation purposes. Linear speeds of dye flow are typically in the range of 5 to 10 m $s^{-1}$. These high velocities are achieved using high-pressure pumps and a very narrow cross-section at the excitation area that may be typically 10 mm in length by less than 1 mm in width. A typical excitation volume defined by the dimensions of the focused pump laser beam is 10 mm in length, with a cross-section diameter less than 200 μm.

In research prototypes optical components, such as mirrors and gratings, are usually mounted on high-resolution kinematic mounts that are in turn positioned on magnetic bases deployed over optical tables. The gratings can be rotated manually, using high-resolution Vernier micrometers, or automatically, using mechanical or piezoelectric drivers controlled electronically. Using these methods linear submicron resolutions can be achieved. The multiple-prism expanders are deployed over adjustable commercial optical surfaces or special platforms designed to minimize demands on intracavity space. This approach allows spatial flexibility that results in very compact cavities. In this regard, the oscillators shown in Fig. 4 can have a cavity length of 10 cm or less. Commercial cavities are often deployed along a linear Invar structure, and components are attached using traditional mechanical methods. These configurations tend to be permanent and thus provide fewer alternatives than their research counterparts.

Dispersive oscillators incorporating a multiple-prism beam expander and the grating in a Littrow configuration are classified as multiple-prism Littrow (MPL) oscillators [14]. Alternatively, if a multiple-prism beam expander is deployed with a grating in a near-grazing-incidence configuration, then the cavity is classified as a hybrid multiple-prism grazing-incidence (HMPGI) oscillator.

Intracavity prismatic preexpansion in conjunction with gratings deployed in a near-grazing-incidence configuration was introduced by Duarte and Piper [17]. The term *prismatic preexpansion* refers to the beam expansion necessary to illuminate completely a grating deployed in a near-grazing-incidence configuration as illustrated in Fig. 4(b). At present, HMPGI oscillators are widely used in research and commercial cavity designs.

MPL oscillators offer higher conversion efficiencies than HMPGI oscillators given the relatively higher angle of incidence utilized to deploy the grating in the latter configuration. On the other hand, an advantage of HMPGI cavities is their intrinsic higher dispersion that yields narrower linewidths than equivalent MPL

**Tab. 1** Performance of dispersive dye-laser oscillators[a]

| *Cavity configuration* | *Laser linewidth*[b] | *Efficiency*[c] *(%)* |
|---|---|---|
| Grazing incidence | $100\ \text{MHz} \leq \Delta\nu \leq 2.5\ \text{GHz}$ | 2–4 |
| MPL | $60\ \text{MHz}^{d} \leq \Delta\nu \leq 1.6\ \text{GHz}$ | 5–15 |
| HMPGI | $138\ \text{MHz}^{e} \leq \Delta\nu \leq 1.15\ \text{GHz}$ | 4–10 |

[a] All values refer to closed cavity configurations.
[b] Conversion to wavelength units can be done using $\Delta\lambda = \lambda^2(\Delta\nu/c)$. For example, $\Delta\nu = 100$ MHz becomes $\Delta\lambda = 0.000112$ nm at $\lambda = 580$ nm.
[c] Efficiency values refer to laser-pumped oscillators.
[d] Oscillator incorporated an intracavity etalon [21].
[e] Measured in a long-pulsed flashlamp-pumped HMPGI oscillator [22].

configurations. Alternative cavities include the pure grazing-incidence design [18] and the telescopic Littrow resonator [19]. For a description of distributed-feedback dye lasers see Schäfer [20].

As listed in Table 1, MPL and HMPGI oscillators produce very narrow linewidths and nearly diffraction-limited beam quality at relatively high conversion efficiencies. In this regard, linewidth values in the sub-GHz range and characteristic of single longitudinal mode oscillation are readily achieved. Oscillator cavities incorporating multiple-prism expanders offer energetic advantages over pure grazing-incidence cavities and are more compact, more rugged, and easier to align than telescopic resonators. In telescopic resonators, the beam-expansion function is performed by a conventional transmission or reflection telescope such as a Galilean or a Cassegrain telescope, respectively.

In addition to configurational and energetic differences, HMPGI oscillators have dispersive characteristics intrinsically distinct from those of the pure grazing-incidence cavity.

Wavelength tuning is inherent to dispersive oscillators, since both the prismatic beam expander and the grating provide angular dispersion. However, as outlined in Sect. 8 the prismatic beam expanders can be designed to yield zero dispersion, thus making wavelength tuning entirely dependent on the grating equation

$$m\lambda = d(\sin\theta + \sin\theta'), \qquad (13)$$

where $\lambda$ is the emission wavelength, $m$ is the diffraction order, $d$ is the groove spacing, $\theta$ is the angle of incidence, and $\theta'$ is the angle of diffraction (see Fig. 4b).

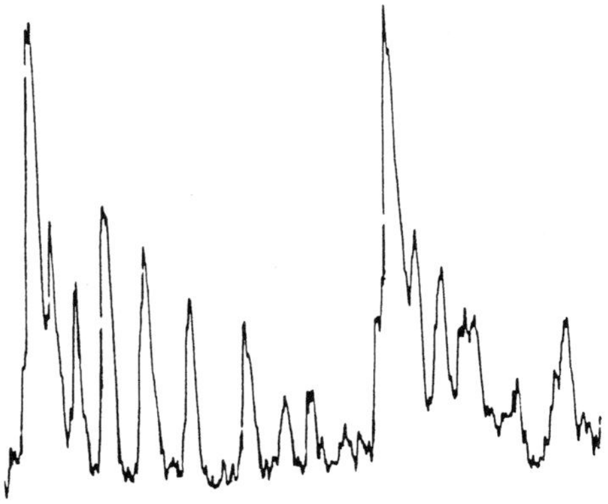

**Fig. 5** Fluorescence spectrum from $D_2CO$ due to excitation near 364 nm using a MPL oscillator ($\Delta\nu \approx 2$ GHz). The tuning wavelength range covered by the laser was 0.227 nm (~514 GHz). Tuning was achieved by rotation of the grating using a piezoelectric driver (adapted from Duarte, F. J. (**1988**), *Lasers Optron.* **7**(2), 41–45 [24])

For the special case of a grating deployed in the Littrow configuration where $\theta = \theta'$, the equation becomes

$$m\lambda = 2d \sin\theta. \qquad (14)$$

Thus, effective wavelength tuning can be achieved by angular displacement of the grating. Using grating tuning in a MPL oscillator similar to that described in Fig. 4(a) the fluorescence spectrum from molecules such as $D_2CO$ has been recorded as illustrated in Fig. 5. MPL and HMPGI oscillators can provide tuning ranges of up to 50 nm using a single dye.

Other means of tuning involve the use of Fabry–Pérot étalons and pressure-tuning techniques. Pressure tuning can be done by mounting the oscillator within a pressure chamber and then using gases such as $O_2$ and $N_2$. The method works because the index of refraction of the gas surrounding the cavity is a function of the change in pressure. In this manner, frequency can be changed almost linearly with pressure.

Dispersive oscillator configurations described here, and first developed for dye lasers, have found widespread applications in gas, solid-state, and semiconductor lasers.

Schematics of pulsed dye-laser systems are outlined in Fig. 6. As illustrated in the figure, a typical system can integrate an oscillator and several stages

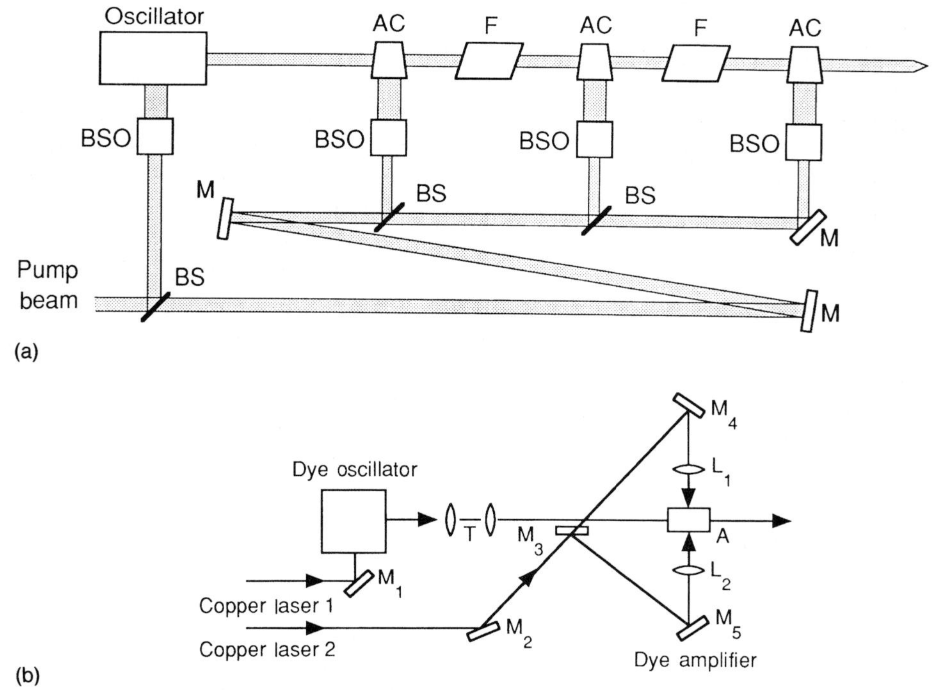


**Fig. 6** (a) Transverse excitation of multistage oscillator-amplifier dye-laser system. AC, amplifier cell; BS, beam splitter; BSO, beam-shaping optics; F, spectral filter; M, mirror. (b) Oscillator-amplifier system using collinear two-sided excitation of the amplifier (adapted from Hargrove, R. S., Kan, T. (**1980**), *IEEE J. Quantum Electron.* **QE-16**, 1108–1113 [26], © 1980 IEEE)

of amplification. Usually, research-class dye lasers incorporate one or two stages of amplification. However, larger high-power systems are composed of a master oscillator and numerous stages of amplification.

Amplification in laser-pumped dye lasers is accomplished in a single pass. In most research systems, the excitation laser beam is divided using beam splitters (BSs) (see Fig. 6) with a small portion of the total energy, usually $\leq 10\%$, used to excite the oscillator. Thus, most of the energy is applied to pump the amplifiers. For a three-stage amplification system Bos [15, 16] reports conversion efficiencies of $\sim$55% and an overall gain factor approaching 229. The efficiency of the oscillator was 6% and the efficiency of the amplifiers was 60%. This system delivered 165 mJ at a laser linewidth of $\Delta\nu \approx 320$ MHz.

Excitation of amplifiers demands careful design criteria to reduce the level of ASE. Since ASE occurs in the buildup time prior to the threshold of narrow-linewidth laser emission, an effective avenue to reduce the overall ASE of the system is to delay excitation of the amplifiers, using optical delay lines, until the narrow-linewidth emission from the oscillator arrives (see Fig. 6a). Typical delay times are within the 10-ns range. An additional technique to reduce ASE consists in the deployment of spatial and spectral filters between amplification stages as demonstrated by Wallenstein and Hänsch [25].

A useful measure of ASE can be given in terms of spectral energy density [14]:

$$\frac{\rho_{\mathrm{ASE}}}{\rho_{\mathrm{l}}} = \left((\Delta\Lambda)^{-1}\int_{\Lambda_1}^{\Lambda_2} W(\Lambda)\,\mathrm{d}\Lambda\right) \times \left((\Delta\lambda)^{-1}\int_{\lambda_1}^{\lambda_2} E(\lambda)\,\mathrm{d}\lambda\right)^{-1} \quad (15)$$

Here $W(\Lambda)$ is the ASE energy function and $\Delta\Lambda$ is the full width of the dye broadband spectral gain region. For narrow-linewidth lasing, $E(\lambda)$ approaches a Gaussian distribution and $\Delta\lambda$ is in the sub-pm range. This definition takes into account the spectral brightness of the laser. For CVL-pumped dispersive dye-laser oscillators $\rho_{\mathrm{ASE}}/\rho_{\mathrm{l}} \approx 5 \times 10^{-7}$ [14].

In large-scale dye-laser systems the amplifier cells are usually pumped using a collinear two-sided transverse excitation geometry [26] (see Fig. 6b). In systems using excitation from several pump lasers, the requirements for synchronous electronic triggering can be rather demanding.

Pulsed dye lasers have demonstrated large pulsed energies and high-average powers. In the area of high-average power, Tallman and Tennant [27] reported on a near-ultraviolet oscillator–amplifier system yielding up to 75 W of average power using XeCl laser excitation.

The large-scale CVL-pumped pulsed dye-laser system at Lawrence Livermore National Laboratory has demonstrated reliable operation at average powers exceeding 2500 W at a prf of 13.2 kHz [28]. This system comprises several master-oscillator–power-amplifier (MOPA) chains operating at 50 to 60% photon conversion efficiencies. Each power amplifier chain is composed of three to four amplifiers. The master oscillators are of the MPL class and incorporate an intracavity étalon. A partial view of a MOPA chain is shown in Fig. 7.

## 4 Flashlamp-pumped Dye Lasers

An alternative source of pulsed tunable coherent radiation from the near ultraviolet to the near infrared is the

**Fig. 7** Partial view near the first amplifier of a dye-laser MOPA chain at the Lawrence Livermore National Laboratory large-scale dye-laser system. Photograph courtesy of I. L. Bass, LLNL

flashlamp-pumped dye laser. A detailed review on this subject is given by Everett [29].

Flashlamp-pumped dye lasers offer unique output characteristics that have found many applications. Single-pulse energies vary from a few millijoules, for compact high-performance oscillators, to several hundred joules for large devices. Pulse lengths range from the nanosecond regime to tens of microseconds. Furthermore, average powers in excess of 1 kW have been demonstrated and overall efficiencies can approach 2%.

These lasers can be excited using coaxial or linear flashlamps. The coaxial devices (Fig. 8a) provide uniform, inherently concentric, excitation of the gain volume and are built using a concentric arrangement with an outer reflector surrounding the flashlamp that illuminates the dye region through the cooling fluid. Coaxial flashlamp-pumped dye lasers have yielded up to 400 J per pulse and a 0.8% efficiency [30].

The emission from flashlamps employed in dye-laser excitation can be characterized using the blackbody radiation model. A significant portion of this emission is in the 200 to 300-nm region. This ultraviolet radiation induces photodegradation of the dye solution via photochemical processes that cause the loss of dye molecules and the creation of photoproducts that absorb at the emission wavelength.

In addition to its thermal mission, the cooling fluid is used to absorb some of the ultraviolet radiation emitted by the lamp. Additives such as caffeine, ethanol, and $CuSO_4$ may be used as

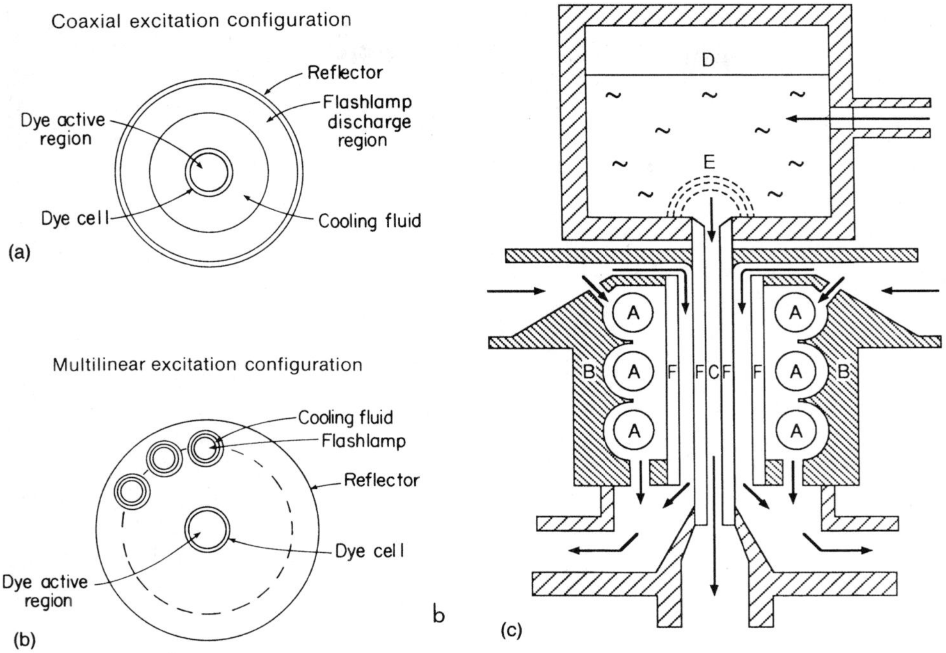


**Fig. 8** (a) Cross-section of a coaxial flashlamp dye laser and (b) cross-section of a device using multiple-lamp linear excitation (adapted from Duarte, [1990] [14]). (c) Cross-section of a transverse-flow waveguide flashlamp-pumped dye laser. Laser emission propagates normal to the page. *A*, flashlamp; *B*, flashlamp reflector; *C*, dye active region; *D*, expansion channel; *E*, filters; *F*, glass plates. The arrows indicate the flow direction of the dye and cooling fluid. (From Mazzinghi, P., Burlamacchi, P., Matera, M., Ranea-Sandoval, H. F., Salimbeni, R., Vanni, U. (**1981**), *IEEE J. Quantum Electron.* **QE-17**, 2245–2249 [32], © 1981 IEEE)

ultraviolet filters. Other filtering measures include the use of Pyrex tubing in the dye cell. These measures apply to both excitation methods.

A technique used to improve the conversion efficiency and lifetime of the dye solution is to employ spectral transfer dyes around the flashlamps to convert the ultraviolet radiation to longer-wavelength emission compatible with efficient excitation of the laser dye. Everett et al. [31] report a 75% efficiency improvement using the dye converter stilbene 420 and coumarin 504 as laser dye.

Linear flashlamp-pumped dye lasers often employ two or more flashlamps deployed symmetrically around the dye cell. Excitation with two flashlamps may involve the use of elliptical reflectors to illuminate the active region more efficiently. Some high-energy devices have used as many as 18 linear lamps configured to yield a nearly concentric excitation flux to the dye cell (Fig. 8b).

Both methods of excitation employ relatively large dye volumes with active lengths in the 10 to 100-cm range and diameters from a few millimeters to up to 6 cm. A common feature is the use of relatively low dye concentrations that may range from $10^{-5}$ M to a few times $10^{-4}$ M.

A further successful configurational alternative is the flashlamp-pumped waveguide dye laser. These devices utilize rapid dye flow that is *transverse* to the plane of propagation. Mazzinghi et al. [32] describe a planar waveguide dye-laser active volume with a 150-mm length, 55-mm height, and 0.5-mm width, excited by three flashlamps on each side (see Fig. 8c). These authors report an average power of 200 W at a prf of 50 Hz. The transverse-flow device described in Fig. 8(c) was designed to use either cooling fluid or a spectral transfer dye solution.

In the area of high-average power, Morton and Draggoo [33] reported on a transverse-flow linear device delivering 1.2 kW at a prf of 850 Hz. Recently, Klimek et al. [34] achieved ~1.4 kW at 1.8% efficiency. The pulse energy was 140 J and the prf 10 Hz. This performance was obtained with the excitation provided by 16 linear lamps, eight on each side, in a transverse-flow device using rhodamine 590 and a spectral transfer dye. In the high–average-power transverse-flow devices, the dye flows at speeds in the range 5 to 10 m $s^{-1}$.

Flashlamp-pumped dispersive dye-laser oscillators have yielded linewidths in the region $138 \leq \Delta\nu \leq 370$ MHz, at 3 to 5 mJ per pulse, in a nearly diffraction-limited beam [22]. Characteristic of these optical oscillators is their extremely low ASE level given by a figure of $\rho_{ASE}/\rho_l \approx 10^{-9}$. This performance is achieved by the use of low dye concentrations ($\sim 6 \times 10^{15}$ $cm^{-3}$) and the judicious application of oscillator physics to the cavity design. The MPL and HMPGI oscillators utilized in these experiments are shown schematically in Fig. 9. A temporal emission pulse from a

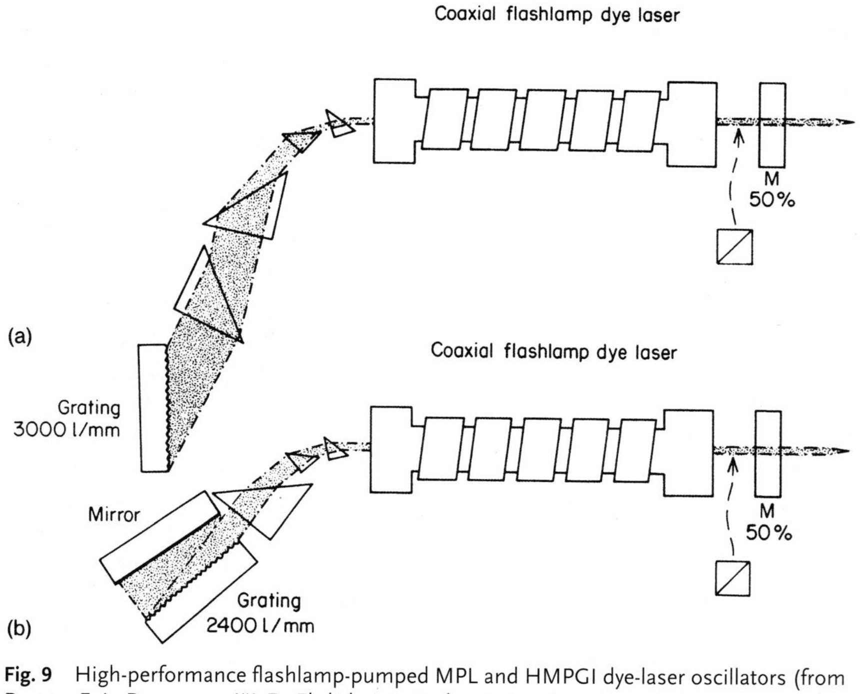


**Fig. 9** High-performance flashlamp-pumped MPL and HMPGI dye-laser oscillators (from Duarte, F. J., Davenport, W. E., Ehrlich, J. J., Taylor, T. S. (**1991**), *Opt. Commun.* **84**, 310–316 [22])

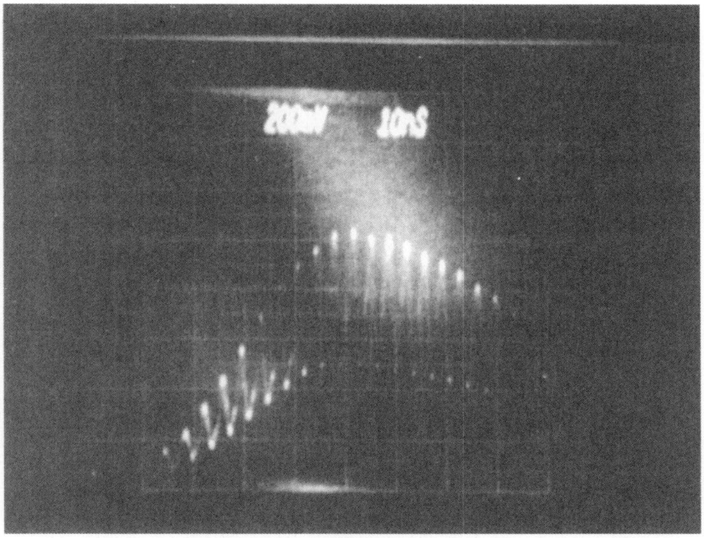


**Fig. 10** Temporal pulse demonstrating double–longitudinal-mode oscillation ($\Delta\nu \approx 300$ MHz) characteristic of flashlamp-pumped dispersive dye-laser oscillators. The time scale is 10 ns div

dispersive oscillator is shown in Fig. 10. The intensity modulation is due to mode beating between two longitudinal modes separated by $\Delta\nu \approx 300$ MHz. The magnitude of the modulation shown in this case is characteristic of mode beating between a very strong longitudinal mode and a weak secondary longitudinal mode. This oscillatory behavior is due to dispersive frequency selectivity exclusively and does not result from mechanisms associated with mode locking (see Sect. 6).

These high-performance dispersive oscillators were engineered in an all-Superinvar structure and tested for mechanical and thermal stability using vehicular displacement on a rugged terrain with excellent results [22]. A partial photograph of the ruggedized dispersive oscillator is shown in Fig. 11. Additional features of this laser are a stainless steel dye reservoir and a special dye and water circulation network consisting of Teflon protected by an outer stainless steel sleeve. In this system, a single dye solution can be used for a period in *excess of one year*. A specially designed thermal control network allows the matching of the dye temperature to the cooling-water temperature to within $0.01\,^{\circ}$C. This thermal control greatly reduces radial thermal gradients in the active medium, thus neutralizing an important source of active-medium inhomogeneities. The Superinvar laser chassis is attached to four pneumatic supports, each using an air pressure of $1.72 \times 10^5$ N m$^{-2}$. The engineering specifications given here far exceed those of typical commercial designs and respond to requirements of long-term stable operation with narrow linewidth and diffraction-limited beam quality.

Signal amplification in flashlamp-pumped dye-laser systems is often accomplished using forced oscillators rather than single-pass amplifiers. A

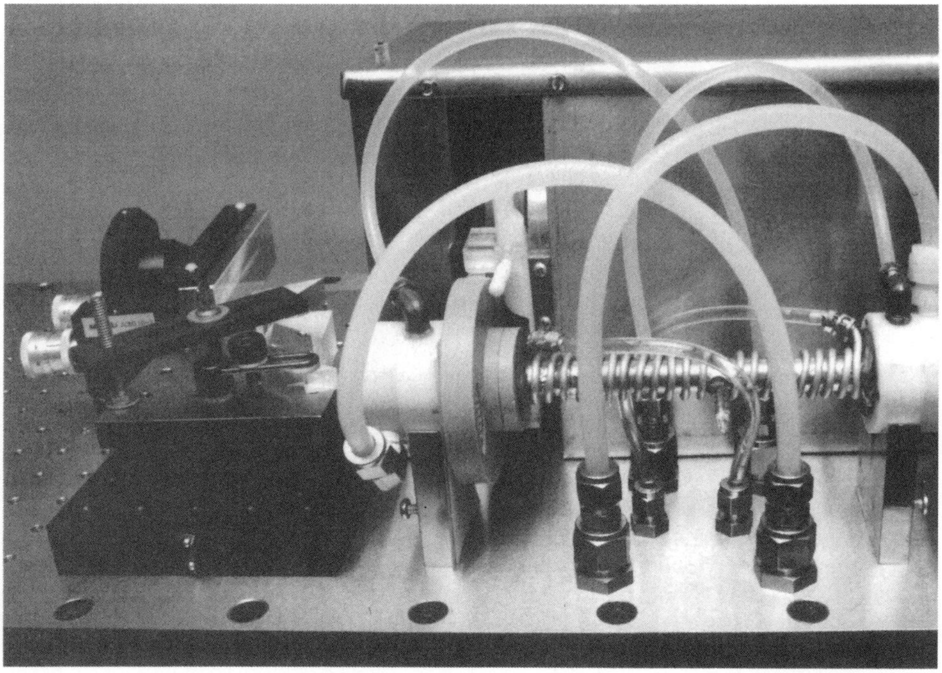

**Fig. 11** Partial perspective of a ruggedized dispersive dye-laser oscillator. The material of the cavity structure is Superinvar. This oscillator has been operated with a single rhodamine 590 dye solution, at $1 \times 10^{-5}$ *M* in methanol, for *more than one year.* (From Duarte, F. J., Davenport, W. E., Ehrlich, J. J., Taylor, T. S. (**1991**), *Opt. Commun.* **84**, 310–316 [22])

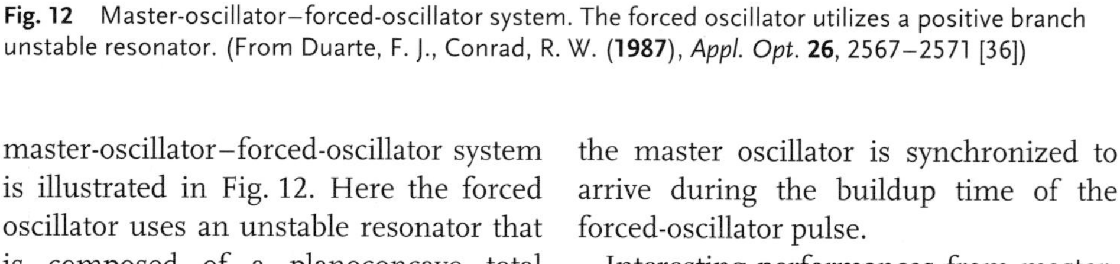


**Fig. 12** Master-oscillator–forced-oscillator system. The forced oscillator utilizes a positive branch unstable resonator. (From Duarte, F. J., Conrad, R. W. (**1987**), *Appl. Opt.* **26**, 2567–2571 [36])

master-oscillator–forced-oscillator system is illustrated in Fig. 12. Here the forced oscillator uses an unstable resonator that is composed of a planoconcave total reflector with a concentric injection hole and a convex–concave meniscus with a high-reflectance circular coating at its center. In these systems, the pulse from the master oscillator is synchronized to arrive during the buildup time of the forced-oscillator pulse.

Interesting performances from master-oscillator–forced-oscillator systems include that reported by Maeda et al. [35], who quote an output energy of 4 J at $\Delta\nu \approx$ 4 GHz using a multiple-étalon assembly at the oscillator. The energy gain was ~267. Using a multiple-prism grating master oscillator in conjunction with a forced oscillator utilizing an unstable resonator (Fig. 12), Duarte and Conrad [36] reported an output energy of 600 mJ at $\Delta\nu \leq 375$ MHz and an energy gain of ~60. These figures correspond to high-fidelity amplification where the interferometric characteristics of the master oscillator are preserved and the ASE level is very low. Further experiments using dispersive oscillators and a forced oscillator with an optimized unstable resonator have demonstrated high-fidelity energy amplification with gain factors in excess of 450.

## 5 Continuous-wave Dye Lasers

Continuous-wave dye lasers using $Kr^+$ or $Ar^+$ laser excitation span the electromagnetic spectrum from ~370 to ~1000 nm. Frequency-doubling techniques have been used to extend the output to the ultraviolet region from ~260 to ~390 nm. Given their unique stable single-frequency tunable operation capabilities, these lasers have made important contributions to physics, spectroscopy, chemistry, and related fields. An excellent and detailed review on the subject is provided by Hollberg [37].

The original pumping scheme of cw dye lasers, reported by Peterson et al. [38], is shown in Fig. 13. The pump beam from an $Ar^+$ laser is focused via a lens onto the dye gain region that is contained within the resonator. The dichroic mirrors transmit the blue-green excitation beam and reflect in the red region of the spectrum. An important feature of this design was the flow of the dye solution at a speed of about 2 m $s^{-1}$. This was necessary to remove the excess heat and to neutralize the influence of the triplet state. With this device Peterson et al. [38] reported an output of 30 mW with about 1 W of pump power delivered in a $TM_{00}$ beam profile focused to a $1/e^2$ diameter of 2 mm. A refinement in this type of longitudinally excited linear configuration was the introduction of external mirrors and the integration of intracavity dispersive elements for wavelength tuning [39].

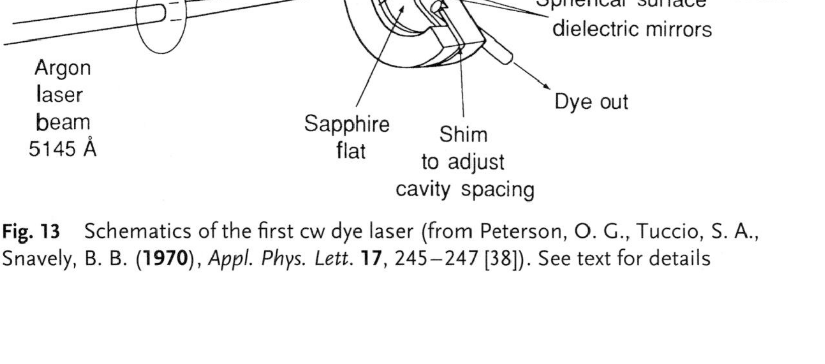


**Fig. 13** Schematics of the first cw dye laser (from Peterson, O. G., Tuccio, S. A., Snavely, B. B. (**1970**), *Appl. Phys. Lett.* **17**, 245–247 [38]). See text for details

In subsequent cw dye-laser designs the dye cell was replaced by a fluid jet created by the passage of dye through a nozzle [40]. This technique coupled with the use of high-viscosity solvents, such as ethylene glycol, produces high-optical-quality surfaces at the jet and flow speeds in the 10-m $s^{-1}$ range.

In addition to the original cavity design concepts introduced by Peterson et al. [38,] other cavity configurations of current use are the three-mirror folded resonator [41] and a four-mirror unidirectional ring resonator [37]. Both these cavities are depicted in Fig. 14. Excitation radiation is usually delivered semilongitudinally at a small angle, on the propagation plane, relative to the optical axis defined by $M_1$ and $M_2$. Dispersive and frequency-selective elements (FSE) are positioned between $M_2$ and $M_3$ and between $M_3$ and $M_4$ for the folded resonator and the ring resonator respectively. In addition, ring cavities incorporate what is called a *unidirectional device* (UDD), positioned next to the tuning elements, which controls the direction of the traveling wave. This optical diode selectively rotates the polarization of one of the propagating directions of the laser emission. Selective polarization rotation is accomplished by exploiting the Faraday effect on an appropriate optical material (see MAGNETO-OPTICAL DEVICES). The optical diode, or UDD, is integrated by a crystal

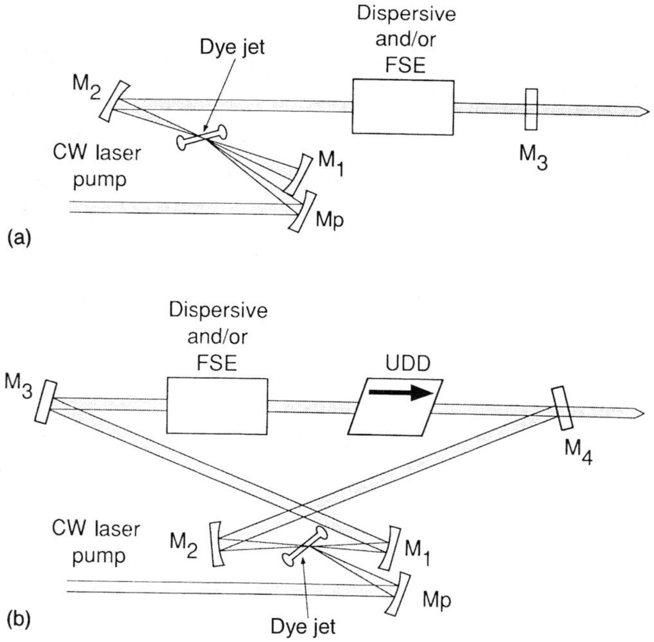


**Fig. 14** Typical cw dye-laser cavity configurations. (a) The linear three-mirror folded cavity; (b) the eight-shaped ring dye-laser cavity constituted by $M_1$, $M_2$, $M_3$, and $M_4$. Here, FSE are frequency-selective elements, and UDD represents a unidirectional device (see text). (Adapted from Hollberg, L. (1990), in F. J. Duarte, L. W. Hillman (Eds.), *Dye Laser Principles*. New York: Academic Press, pp. 185–238 [37])

or glass in a magnetic field that is applied parallel to the direction of propagation and a quartz plate that induces a reciprocal polarization rotation. The Faraday effect causes a rotation of the radiation of both propagating directions. Subsequent transmission by one of the propagating beams through the quartz plate neutralizes the initial rotation. On the other hand, the counterpropagating beam suffers losses induced by transmission through intracavity Brewster elements (see Hollberg, [37], for further details).

Cavity design analysis, determination of stability conditions, and evaluation of astigmatism can be established using propagation matrices as discussed by Diels [42]. These matrices can be used to characterize the propagation of Gaussian beams. Each optical component and propagation space is represented by a particular matrix so that an overall matrix for the resonator can be derived. This allows the prediction of important parameters such as beam expansion and beam divergence.

Frequency-selective elements used in cw dye lasers include gratings, prisms, étalons, and birefringent crystals. In addition, a number of interferometers have been used to control frequency characteristics. Usually, three optical elements are used to attain single-frequency operation. In a typical cavity the frequency-selective process may proceed with the use of a birefringent filter or a prism to provide the necessary bandwidth to discriminate in favor of a mode of a thin étalon, which, in turn, discriminates in favor of a mode of the second étalon with a considerably smaller free spectral range (FSR). The second étalon is designed with a FSR and finesse sufficient to select a single longitudinal mode of the cavity. In some designs, single-frequency emission is achieved by replacing the second étalon by an intracavity interferometer such as a Mach–Zehnder interferometer.

Linear standing-wave cw dye-laser cavities exhibit the effect of *spatial hole burning* that causes the laser to oscillate in more than one longitudinal mode (Hollberg, [37]. This problem can be avoided in ring-cavity configurations (Fig. 14b) where the oscillation is in the form of a traveling wave. Furthermore, ring cavities are known to provide, with sufficient pump power, greater single-mode power. On the other hand, ring cavities involve increased optical complexity and exhibit higher pump thresholds.

As outlined earlier, a very important contribution of cw dye lasers has been the generation of very stable single-frequency oscillation. This is a topic that involves considerable technical detail such as cavity structure, materials, thermal stabilization, control of dye-flow turbulence, active servomechanisms, and a series of frequency-locking techniques. For an in-depth discussion of this topic, including postlaser stabilization methods, the reader should refer to Hollberg [37]. In this regard, Hough et al. [43] demonstrated linewidths less than 750 Hz in an experiment that identified environmental noise as a principal limitation to stabilization. Postlaser stabilization using electro-optics modulators [44] can stabilize the frequency at the hertz range.

In the area of power performance, Johnston et al. [45] report 5.6 W of stabilized output power in a single longitudinal mode at 593 nm. This dye laser had a ring-cavity configuration and was excited by an $Ar^+$ laser yielding 24 W. On the other hand, Baving et al. [46] report a 43-W multimode output using rhodamine 6G in a simple folded cavity and 200 W of pump power from an $Ar^+$ laser.

## 6 Femtosecond Dye Lasers

The dye laser is central to the development of femtosecond laser physics and technology. Consequently, its impact in the science of femtosecond phenomena is of gigantic proportions. Here, we present an overview of the more important elements of femtosecond dye lasers. An extensive and detailed review of the subject is given by Diels [42].

The development of ultrashort-pulse, or ultrafast, dye lasers is intimately related to the phenomenon of *mode locking* (see also ULTRASHORT-PULSE PHENOMENA). Broadband lasers can oscillate in a number of transverse modes and a corresponding multitude of unrelated longitudinal modes whose relative phases vary randomly as a function of time. Using an intracavity aperture a single transverse mode can be selected, thus restricting oscillation to a sequence of equally spaced longitudinal modes in the frequency domain. Forcing these modes to have the same relative phase results in mode locking. This yields a train of pulses in the temporal domain separated by a period inverse to the mode spacing $\delta\nu = c/2L$, where $c$ is the speed of light and $L$ is the cavity length. The sequence of temporal pulses corresponds to a single intracavity pulse contributing to the output on every return pass. The number of longitudinal modes allowed to oscillate is given approximately by the gain bandwidth divided by $\delta\nu$. The shortest possible pulse length is determined by the reciprocal of the dye gain bandwidth, approximately.

Mode locking in dye lasers was first demonstrated in simple flashlamp-pumped dye-laser cavities incorporating an intracavity *saturable absorber* [47]. In this work, the broadband flashlamp-pumped rhodamine 6G dye laser emitted a relatively smooth temporal pulse with a few hundred-nanosecond pulse duration. Insertion of the saturable absorber 3, 3′-diethyloxadicarbocyanine iodide (DODCI) caused the temporal pulse to change into a pulse train of subnanosecond pulses.

Picosecond and subpicosecond emission was achieved using simple linear cw dye-laser cavities incorporating intracavity saturable absorbers. An important development in establishing shorter pulses was the concept of colliding-pulse mode locking (CPM) introduced by Ruddock and Bradley [48] and applied by Fork et al. [49] to a ring cavity. Here, a *collision* between two counterpropagating pulses is established at the saturable absorber. This creates a transient grating due to interference at the absorber, which has the effect of shortening the pulse and improving its synchronization and stability characteristics. Using the CPM technique, Fork et al. [49] obtained pulses as short as 90 fs.

An additional development of importance for the femtosecond laser was the introduction of prismatic intracavity negative group-velocity dispersion (GVD) [50]. Pulse broadening by GVD occurs at the dye gain and dye absorber regions since the pulse transit time is affected by the dispersion of the dye refractive index. The deployment of intracavity prisms makes possible the achievement of intracavity compensation of the GVD introduced at the gain and absorber regions. This is because the GVD also depends on the second derivative of the optical path length ($d^2P/d\lambda^2$) through the intracavity prismatic array (see Fig. 15). Furthermore, the value of $d^2P/d\lambda^2$ is a function of the prismatic dispersion and its derivative (see Sect. 8) and can be made negative by adjusting the separation distance of

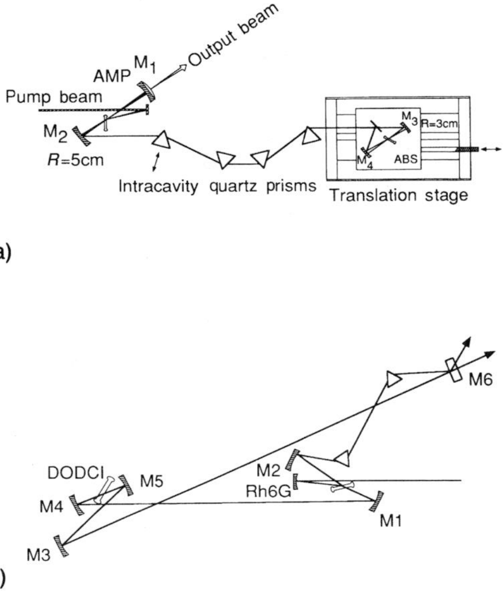


**Fig. 15** (a) Linear femtosecond dye-laser cavity terminated by an antiresonant ring that serves as a broadband reflector (adapted from Jamasbi, N., Diels, J. C., Sarger, L. (**1988**), *J. Mod. Opt.* **35**, 1891–1906 [64]) and (b) ring femtosecond dye-laser cavity (adapted from Diels, J. C., Dietel, W., Fontaine, J. J., Rudolph, W., Wilhelmi, B. (**1985**), *J. Opt. Soc. Am. B* **2**, 680–686 [65]). In both arrangements the saturable absorber is deployed to exploit the CPM effect

the prisms. Thus, the intracavity prismatic array is utilized to yield negative GVD. In other words, prismatic GVD compensation is used to remove temporal broadening (or *chirp*) introduced at the dye regions.

Examples of linear and ring femtosecond dye lasers are given in Fig. 15. The basic elements of these lasers are the cavities established for the gain region between $M_1$ and $M_2$, the saturable absorber in its counterpropagating arrangement, and the intracavity prisms that provide dispersion compensation. Again, a propagation analysis of these cavities can be done using ray-transfer matrices [42]. The use of GVD compensation techniques has yielded pulses shorter than 20 fs.

In addition to the passive mode locking considered so far some workers use *hybrid mode locking,* which occurs when synchronous pumping is used in conjunction with passive mode locking. Also, negative dispersion can be generated using pairs of gratings.

The shortest pulses reported to date correspond to 6 fs and were achieved by Fork et al. [51] using extra-cavity pulse compression. In this experiment, the authors applied a CVL to amplify, at a prf of 8 kHz, the pulses from a dye laser incorporating CPM and prismatic GVD compensation. The amplified 50-fs pulses were then propagated via grating pairs and a four-prism sequence. The optimum pulse duration was obtained by adjusting the prism and grating spacings while monitoring the compressed pulse using interferometric autocorrelation.

Measurement of femtosecond pulses involves the use of special techniques such as intensity and interferometric autocorrelations, self-convolution, asymmetric correlation, and single-shot detection. Diels [42] provides a detailed description of these techniques. In terms of physics, femtosecond dye lasers have led to the study of various nonlinear phenomena such as the Schrödinger equation with nonlinear terms, and solitons.

## 7 Solid-state Dye Lasers

Solid-state dye lasers were introduced shortly after the discovery of liquid dye lasers [52, 53]. However, active-medium inhomogeneities prevented significant activity in the field until the 1990s when improved, and highly homogeneous, forms of dye-doped polymeric matrices were reported [54]. This gain media was a modified form of polymethyl methacrylate, called *MPMMA*, doped with a laser dye in the process of polymerization. Subsequently, alternative matrices, such as, methal methacrylate copolymerized with 2-hydroxyethyl methacrylate (HEMA:MMA) also became available [55]. In the case of HEMA:MMA, the laser dye molecules can be either diffused into the matrix or they can be introduced via copolymerization. One of the most important characteristics of these gain media is their optical homogeneity, which enable the emission of clean, and well-defined laser beams with $TEM_{00}$ profiles.

For rhodamine 11 B dye-doped MPMMA at a concentration of 0.5 mM, Maslyukov et al. [54] reported efficiencies better than 60% for broadband emission under excitation at $\lambda = 532$ nm. For the same matrix doped with rhodamine 6 G, Duarte et al. [56] reported broadband efficiencies of ~49%, for excitation at $\lambda \approx 532$ nm. A detailed review of laser dye polymeric matrices has been given by Costela et al. [55].

For most scientific and analytical applications, tunable lasers need a laser linewidth below 3 GHz (or 0.0035 nm at $\lambda = 590$ nm). Thus, the performance and characteristics of narrow-linewidth dispersive solid-state dye-laser oscillators is important. The first laser oscillator of this class, incorporating a multiple-prism grating architecture, was introduced by Duarte [57]. It yielded a laser linewidth of $\Delta\nu = 1.12$ GHz (equivalent to 0.0013 nm at $\lambda = 590$ nm) at a conversion efficiency of ~9%. The reported tunability was 563 nm $\leq \lambda \leq$ 610 nm. An optimized multiple-prism grating tunable laser oscillator yielding single longitudinal mode emission at a linewidth of $\Delta\nu = 350$ MHz (equivalent to 0.0004 nm at $\lambda = 590$ nm) was reported by Duarte [58] and is illustrated in Fig. 16. Other characteristics of this optimized dispersive solid-state oscillator are a tuning range of 550 nm $\leq \lambda \leq$ 603 nm, nearly Gaussian temporal pulses

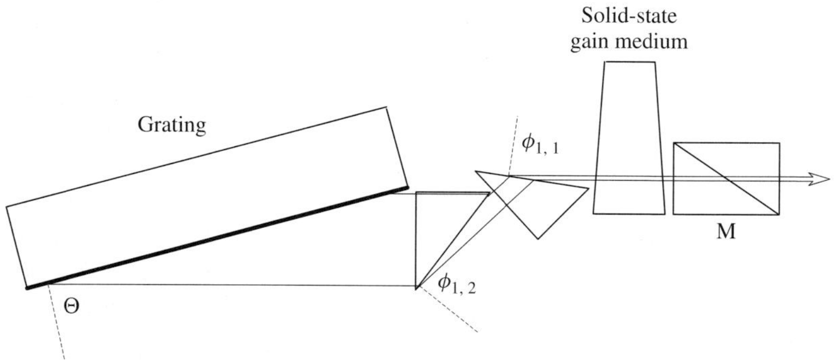


**Fig. 16** Solid-state MPL grating dye-laser oscillator designed to yield single longitudinal mode emission tunable in the visible. Optical elements integrating the cavity are shown to scale (after Duarte, F. J. (**1999**), *Appl. Opt.* **38**, 6347–6349 [58])

3 ns in duration (at full-width half maximum) and a conversion efficiency of ~5%. The peak power per pulse is ~33 kW.

In addition to multiple-prism grating solid-state state dye lasers, distributed-feedback architectures have also been demonstrated with polymeric [59] and sol-gel silica dye-doped matrices [60]. For these lasers, reported linewidths are in the $0.01\ \text{nm} \leq \Delta\lambda \leq 0.06\ \text{nm}$ range. In addition to dye-doped polymers, crystalline dye materials are also under investigation as solid-state tunable gain media [61, 62].

## 8 Linewidth Narrowing and Intracavity Dispersion

A very important subject in the physics of linewidth narrowing in high-gain pulsed lasers, and pulse compression in femtosecond lasers, is intracavity dispersion.

The return-pass dispersive linewidth in a multiple-prism grating oscillator is given by the expression

$$\Delta\lambda \approx \Delta\theta(\partial_\lambda\theta)^{-1}, \qquad (16)$$

where $\partial_\lambda\theta$ is the intracavity dispersion, $\Delta\theta$ is the beam divergence, and $\partial_\lambda = \partial/\partial\lambda$. Although this expression was first adopted from classical optics, it has been shown recently that it can be derived using Dirac's notation [63].

The cumulative single-pass dispersion at the $m$th prism (see Fig. 17) is given by Duarte [14]:

$$\partial_\lambda\phi_{2,m} = \mathcal{H}_{2,m}\partial_\lambda n_m + (k_{1,m}k_{2,m})^{-1} \times (\mathcal{H}_{1,m}\partial_\lambda n_m \pm \partial_\lambda\phi_{2,(m-1)}), \qquad (17)$$

where $k_{1,m} = (\cos\psi_{1,m}/\cos\phi_{1,m})$, $k_{2,m} = (\cos\phi_{2,m}/\cos\psi_{2,m})$, $\mathcal{H}_{1,m} = (\tan\phi_{1,m}/n_m)$, $\mathcal{H}_{2,m} = (\tan\phi_{2,m}/n_m)$, and $\partial_\lambda n_m = \partial n_m/\partial\lambda$ is characteristic of the prism material. Here, $\phi_{1,m}$ is the angle of incidence at the $m$th prism and $\phi_{2,m}$ is the exit angle. Thus, $\psi_{1,m}$ and $\psi_{2,m}$ are the corresponding angles of refraction. Equation (5) is a recursive function that evaluates $\partial_\lambda\phi_{2,m}$ in ascending values of $m$.

The *generalized* return-pass prismatic dispersion is given by Duarte [14]:

$$(\partial_\lambda\Phi)_P = 2M_1M_2\sum_{m=1}^{r}(\pm1)\mathcal{H}_{1,m}\left(\prod_{j=m}^{r}k_{1,j}\prod_{j=m}^{r}k_{2,j}\right)^{-1}\partial_\lambda n_m + 2\sum_{m=1}^{r}(\pm1)\mathcal{H}_{2,m}\left(\prod_{j=1}^{m}k_{1,j}\prod_{j=1}^{m}k_{2,j}\right)\partial_\lambda n_m, \qquad (18)$$

where

$$M_1 = \prod_{m=1}^{r}k_{1,m} \qquad (19a)$$

and

$$M_2 = \prod_{m=1}^{r}k_{2,m} \qquad (19b)$$

are the overall beam-expansion coefficients.

For identical right-angled prisms designed for orthogonal beam exit at the Brewster angle of incidence, Eq. (18) reduces to the elegant expression

$$(\partial_\lambda\Phi)_P = 2\sum_{m=1}^{r}(\pm1)n^{m-1}\partial_\lambda n \qquad (20)$$

and the overall beam expansion is given by $M = n^r$.

The overall return-pass intracavity dispersion is given by

$$\partial_\lambda\theta = [M_1M_2(\partial_\lambda\theta)_G + (\partial_\lambda\Phi)_P], \qquad (21)$$

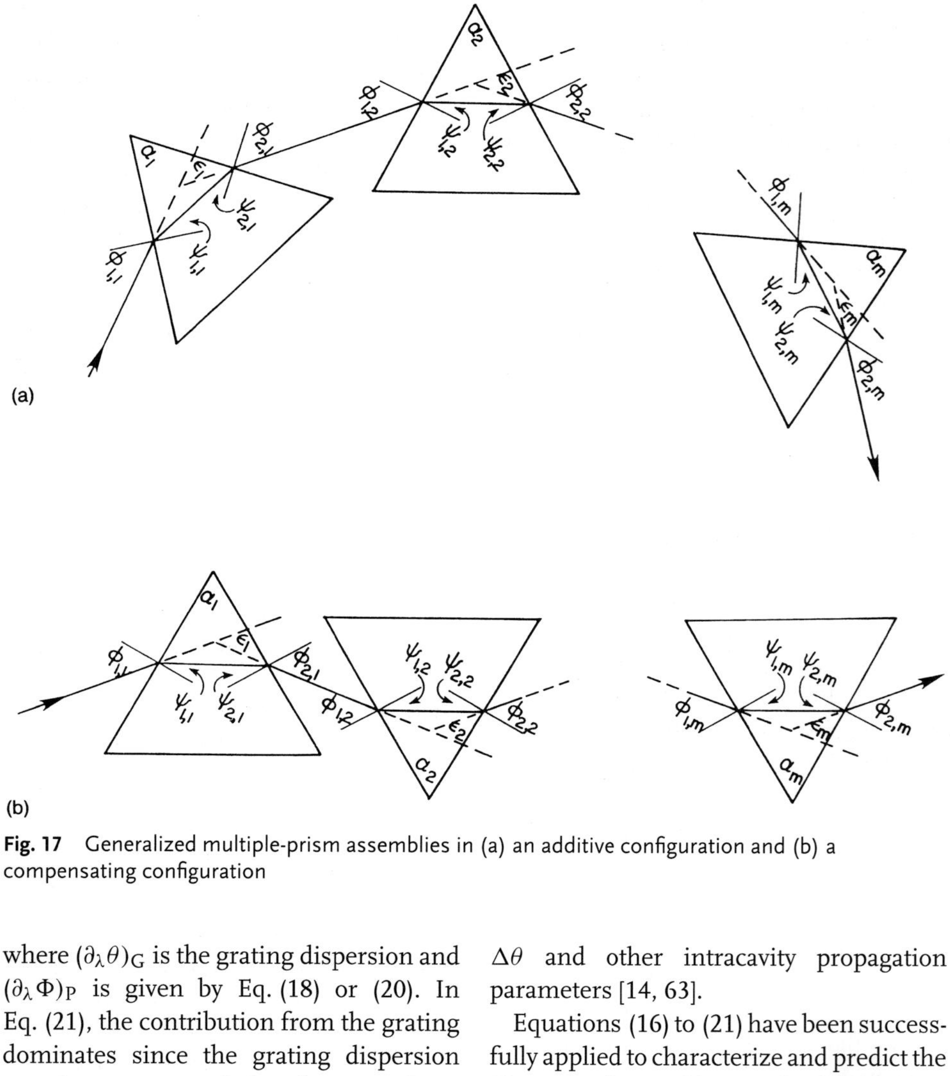

**Fig. 17** Generalized multiple-prism assemblies in (a) an additive configuration and (b) a compensating configuration

where $(\partial_\lambda\theta)_G$ is the grating dispersion and $(\partial_\lambda\Phi)_P$ is given by Eq. (18) or (20). In Eq. (21), the contribution from the grating dominates since the grating dispersion can be augmented by factors of up to 100 or more by intracavity beam expansion. Further, using Eq. (18) the prismatic contribution to dispersion can be made equal to zero at a given wavelength by design. This reduces beam deviations due to thermal effects and makes the tuning characteristics of the oscillator depend exclusively on the grating. In addition to dispersion equations, ray-transfer matrices are utilized to evaluate $\Delta\theta$ and other intracavity propagation parameters [14, 63].

Equations (16) to (21) have been successfully applied to characterize and predict the dispersive linewidth in pulsed high-gain, tunable, multiple-prism grating oscillators. Agreement with measured linewidths is particularly good in high-gain oscillators in which temporal and spatial dimensions allow stimulated emission to occur following a few intracavity passes. For the long-pulse regime, characteristic of flashlamp-pumped dye lasers, the dispersive linewidth value establishes the upper limit of the measured laser linewidth, which is determined by the interaction of dispersive and dynamic phenomena.

These equations transcend dye lasers and apply to pulsed lasers in general, irrespective of laser medium or emission wavelength. In this regard, the dispersive design concepts first developed for dye lasers have now been applied to gas, solid-state, and semiconductor lasers. Furthermore, the techniques of beam shaping using multiple-prism expanders have found widespread applications in electro-optics.

As mentioned previously, negative dispersion has a very important function in pulse compression, compensating for positive intracavity dispersion. Two parameters of importance in evaluating the negative dispersion effect on pulse compression are $\partial\phi_{2,m}/\partial n$ and $\partial^2\phi_{2,m}/\partial n^2$. This follows since the second derivative of the beam path length through the prismatic array is a function of these parameters. The first derivative of interest is obtained using the identity $\partial\phi_{2,m}/\partial n = \partial_\lambda\phi_{2,m}(\partial_\lambda n)^{-1}$, and the second derivative is obtained by direct differentiation [14]. For a single prism at Brewster's angle of incidence $\partial\phi_{2,1}/\partial n = 2$ and $\partial^2\phi_{2,1}/\partial n^2 = 4n - 2/n^3$.

## 9 Applications

Dye lasers have made significant contributions in a wide variety of scientific and technological fields including physics, spectroscopy, LIS, medicine, and lidar, and in industry.

Contributions to areas of laser physics include the study of dynamic instabilities, nonlinear phenomena, oscillator physics, and femtosecond laser development.

In the area of dynamic instabilities, an interesting result is the hysteresis effect observed in a cw dye laser by Hillman et al. [66]. These authors discovered that as the pump power was increased, thresholds were reached where the output power and spectrum of their simple ring dye laser changed discontinuously. Although dynamic instabilities in homogeneously broadened lasers have been identified with strong-field interactions in the gain medium, further studies are needed for a complete understanding of the phenomena. In addition, dynamic instabilities can also originate from optical inhomogeneities in the active medium due to flow turbulence and thermal gradients. In this regard, linewidth instabilities in narrow-linewidth dispersive oscillators were found to decrease substantially as radial thermal gradients and flow turbulence were neutralized [22].

The versatility of the dye laser created a renaissance in the field of spectroscopy. In this regard, dye-laser characteristics that are important to spectroscopy are tunability, extremely narrow-linewidth emission, low ASE levels, very stable oscillation in the cw regime, and the ability to emit very short pulses (down to the femtosecond regime). Spectroscopy is the study of the wavelength or frequency dependence of any optical process in which a substance absorbs or emits energy by the interaction with radiation. The tunable narrow-linewidth emission available from dye lasers allows selective excitation of atomic and molecular transitions with a high degree of specificity. The areas of applications are numerous and diverse including absorption, excitation, ionization, and saturation spectroscopy. Both cw and pulsed dye lasers have found useful roles in laser-induced fluorescence spectroscopy. On the other hand, the attainment of extremely short pulse widths has generated an explosion in the number of time-resolved and ultrafast spectroscopic measurements (see also SPECTROSCOPY, ULTRAFAST) that have affected every branch of science and technology. For detailed descriptions on the application of dye lasers to spectroscopy the reader should refer to Klick [67] and Stuke [68].

In addition to spectroscopy, femtosecond dye lasers have been applied to physical measurement devices such as the femtosecond-pulse laser gyroscope developed by Diels [69].

A very active area of research that has benefited significantly from cw dye lasers is *laser cooling* (see also LASER COOLING AND TRAPPING OF NEUTRAL ATOMS). Using counter propagating tunable laser beams, physicists have built magneto-optical traps that have been used to cool populations of atoms to extremely low temperatures. The initial experiments conducted using mainly cw ring dye lasers yielded temperatures below one mK [70]. Improvement in cooling techniques eventually made possible the attainment of Bose–Einstein condensation (BEC) where temperatures achieved are in the nK regime [71]. In BEC experiments, however, the illumination sources are tunable external cavity semiconductor lasers that are capable of delivering laser linewidths in the kHz regime. These external cavity semiconductor lasers often use diffraction grating tuning techniques originally developed for dye lasers [72].

A significant further application of dye lasers is LIS (see LASER ISOTOPE SEPARATION). In atomic-vapor LIS, a metallic element such as uranium is heated and vaporized by an electron beam. The resulting vapor is then illuminated by a high-prf CVL-pumped dye-laser system (see Sect. 3) that selectively excites the $^{235}$U isotope. Briefly, this is accomplished by exploiting the isotopic shift ($\delta\nu_i \approx$ few gigahertz) and the ability of narrow-linewidth dye lasers ($\Delta\nu \leq 1$ GHz) to tune into resonance with specific transitions. Thus, selective $^{235}$U excitation can be achieved in a multistep excitation process by tuning the dye laser(s) to specific wavelengths, in the visible spectral region, compatible with a transition sequence leading to photoionization. For instance, a three-step excitation sequence may proceed by exciting the atom from the ground state to an even-parity state from where it is further excited to a higher-energy odd-parity state, and subsequently to an even-parity state near the ionization potential (see Fig. 18). Then, the atom at the Rydberg state can be ionized using an infrared laser, or via collisional ionization or autoionization routes. The ion can then be extracted using electromagnetic methods. A possible alternative to the three-step excitation described here is a two-step excitation sequence involving near-ultraviolet dye lasers.

Interest in the atomic-vapor LIS process is derived from the considerable economic advantages that laser enrichment of uranium is reported to have over traditional methods such as gas diffusion or centrifuge techniques. For a

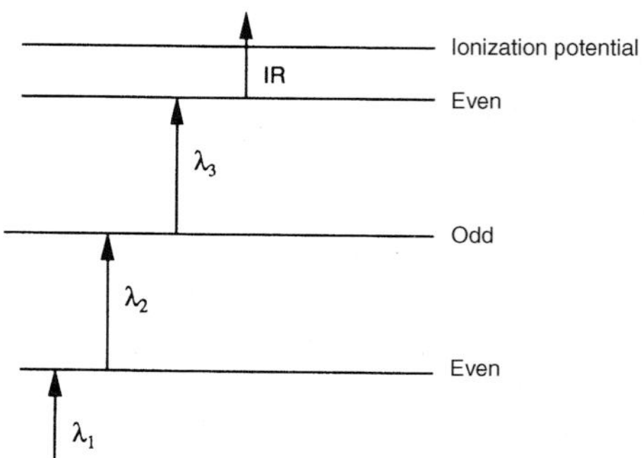


**Fig. 18** Energy-level diagram illustrating a three-step excitation sequence

detailed discussion of this topic and photoionization spectroscopy see Paisner and Solarz [73)].

High-prf CVL-pumped dye lasers can also be applied to material processing such as cutting and welding. In this area the high-prf characteristics of the laser yields fast processing velocities, and the good beam quality allows high spatial precision [74]. For a detailed review on industrial applications of dye lasers see Klick [67].

In addition, CVL-pumped dye lasers are suitable to applications in astronomy as a *guide star* [28] (see also ASTRONOMICAL TELESCOPES AND INSTRUMENTATION). Here, the laser is tuned to the well-known 589-nm line of the atomic sodium layer that is observed 90 to 100 km above the earth's surface. Light from the guide star, created by the laser illumination, is collected by a Cassegrain telescope and can then by used to characterize turbulence-induced distortions. Use of this concept in conjunction with adaptive optics can lead to the correction of atmospheric distortions and thus enhance the quality of astronomical images.

An important area of research is laser radar and *lidar* (see the article LIDAR (LASER RADAR)). Lidar is a technique that uses laser pulses for detection and ranging of atmospheric molecular species. The measurement consists in the emission of a laser pulse and the detection of the scattered return signal. Thus, tunable dye lasers are applied to detect and map concentrations of molecular elements in the atmosphere. For instance, Kölsch et al. [75] have used a mobile ground excimer-laser–pumped dye laser to map emission of nitric oxide in several European cities. Grant [76] provides a review on differential absorption lidar (dial) measurements of water vapor profiles and considers relevant requirements on laser characteristics.

Scientists at NASA use a pair of airborne tunable dye lasers to conduct tropospheric and stratospheric ozone and aerosol measurements [77]. These airborne experiments have led to the measurement and mapping of $O_3$ depletion in the stratosphere. This depletion has been found to be particularly pronounced over Antarctica and the Arctic. In the stratospheric experiments, two frequency-doubled dye lasers emit at 301.5 and 311 nm for the on- and off-line dial wavelengths of $O_3$. In addition, part of the fundamental dye-laser emission at 603 nm is used in conjunction with a portion of 1064-nm radiation from the pump laser for aerosol measurements. All wavelengths are transmitted simultaneously. These experiments are conducted to determine the dynamics and chemistry that control the distribution of $O_3$ in the atmosphere.

Applications including remote sensing, LIS, and spectroscopy demand tunable narrow-linewidth emission with significantly reduced levels of ASE. In this regard, the oscillator configurations described in Sect. 3 is particularly useful.

Finally, we consider medical applications of dye lasers. One of the first dye-laser applications in medicine was cancer PDT pioneered by Dougherty et al. [78]. In this technique, the patient is given internally 2 to 3 mg $kg^{-1}$ of a dye such as HpD (hematoporphyrin derivative). Following a given time (48–72 h) the photosensitizer is retained by the cancer tumor at an optimum concentration differential relative to healthy tissue. At that point, laser radiation is delivered to the tumor, at energy densities in the 100- to 500-J $cm^{-2}$ range, either directly or via an optical-fiber delivery system depending on the location of the tumor. A photosensitizer such as HpD absorbs mostly in the 400- to 450-nm region of the spectrum; however, since the body tissue transmits mainly in the red the dye lasers are operated at ~630 nm for cytotoxicity. Absorption of the laser light by the photosensitizer leads to a photochemical reaction that produces a reactive form of $O_2$ that causes loss of cellular structure and subsequently tumor destruction. The exact mechanism by which the oxygen singlet interacts with the cells continues to be subject to study.

In certain cases, a particular photosensitizer may be applied locally in conjunction with an appropriate biological vehicle. Also, given the characteristics of the absorption spectrum of HpD, blue–green lasers can be used for diagnostics using fluorescence techniques. Other photosensitizers are dihematoporphyrin ether (DHE), rhodamine dyes, chlorophyll derivatives, and phthalocyanines. Types of cancer reported to have been treated with PDT include cancers of the bladder, brain, breast, lungs, skin, stomach, and vagina.

It is interesting to note that although PDT does not require narrow-linewidth laser emission, the use of systems including dispersive narrow-linewidth oscillators (as described in Sects. 3 and 4) leads to high–beam-quality emission that can be advantageous in the transmission via optical-fiber delivery systems. Also, a laser system such as the CVL-pumped dye laser can be used for the dual purpose of diagnostics and treatment.

An additional, and very successful, medical application of the dye laser is *lithotripsy*, which involves the use of long-pulse flashlamp-pumped dye lasers, emitting at wavelengths such as 504 and 577 nm, to fragment ureteral calculi [79]. The average success rate for this procedure can be about 80%. Further medical applications of the flashlamp-pumped dye laser are in dermatology, the treatment of vascular lesions, and thrombolysis. For a detailed account of dye lasers in medicine the reader should consult Goldman [80]. See also LASERS, MEDICAL USE OF.

## Glossary

**Amplified Spontaneous Emission (ASE):** Spontaneous emission that is amplified in the gain medium. ASE exhibits some spectral narrowing and directionality. However, compared to coherent laser emission, ASE has very broadband spectral characteristics and much higher divergences. Considerable effort is devoted in the design of oscillators and amplifiers to minimize the ASE background.

**Brewster Angle:** Angle of incidence at which the reflection losses are a minimum. It is defined by $\tan\phi_{1,m} = n$, where $n$ is the refractive index of optical material.

**Chirp:** A temporal frequency sweep of a pulse. Typically a linear chirp will have the time-dependent frequency $\omega(t) = \omega_0 + bt$.

**Closed Cavity:** A cavity where the laser emission exits via the output coupler mirror and not via reflection losses of dispersive elements.

**Diffraction-limited Beam:** A Gaussian laser beam whose divergence is given by $\lambda/\pi w$, where $w$ is the beam waist.

**Dispersive Oscillator:** A narrow-linewidth laser oscillator utilizing intracavity dispersive elements.

**Fabry–Perot Etalon:** An interferometer, composed of two highly polished and parallel reflective surfaces, that is usually deployed intracavity for linewidth narrowing and fine frequency tuning.

**Forced Oscillator:** A multipass oscillator usually incorporating unstable resonator optics and no intracavity dispersive elements. Its function is to amplify the injected signal from a master oscillator.

**Free Spectral Range (FSR):** FSR of a cavity or Fabry–Perot etalon is defined by $FSR = c/2nd$ where $c$ is the speed of light, $n$ the refractive index of the medium between the reflective surfaces, and $d$ the distance between these surfaces.

**Frequency Doubling:** The use of a nonlinear medium such as a crystal to generate a second harmonic of the laser emission.

**Homogeneous Broadening:** A broadening mechanism associated with collisions and excited-state lifetimes. The broadening of the gain medium is said to be homogeneous when it can be characterized by the broadening of an individual atom or molecule in the collection. In homogeneously broadened gain media, all excited atoms or molecules can emit at the same single longitudinal mode, within the frequency gain profile.

**Intracavity Dispersion:** The dispersion created by the use of intracavity elements such as gratings and / or prisms.

**Lidar:** Light detection and ranging.

**Littrow Configuration:** Deployment of a diffraction grating so that the angle of incidence is equal to the angle of diffraction.

**Master Oscillator:** A narrow-linewidth oscillator usually of the dispersive class emitting a nearly diffraction-limited beam and a very low level of ASE.

**Mode Locking:** The condition in which all the modes in the cavity have the same relative phase. This can be achieved with the use of an intracavity saturable absorber. The output of a mode-locked laser is a sequence, or train, of short pulses separated by $\Delta t = 2L/c$, where $L$ is the cavity length.

**Photon Conversion Efficiency:** Efficiency quoted in optically pumped lasers and defined as the emission laser energy over the pump laser energy.

**Raman Shifting:** A nonlinear optics method utilized to generate coherently a series of Stokes and anti-Stokes frequencies. Relative to the laser fundamental, Stokes radiation is of longer wavelength and anti-Stokes radiation is of shorter wavelength. A common method to generate tunable Raman-shifted radiation is to use a dye laser to excite a gas like hydrogen contained at high pressure in an optical cell.

**Ray-transfer Matrices:** $2 \times 2$ and $4 \times 4$ matrices whose components are optical quantities such as beam expansion, optical path length, and dispersion. These matrices are used to characterize Gaussian beam propagation.

**Saturable Absorber:** A dye such as DODCI that is used intracavity. The absorption spectrum of such a dye overlaps with the emission spectrum of the gain dye. The absorption of saturable absorbers decreases as the incident intracavity emission intensity increases. This feature discriminates in favor of a high-intensity pulse and in detriment of lower-intensity pulses.

**Solid-state Dye Laser:** A dye laser using an active medium in which the dye is contained in a polymeric or a silicate matrix.

**Solitons:** Eigensolutions of the nonlinear Schrödinger equation $i(\mathrm{d}E/\mathrm{d}z) + \mathrm{d}^2E/\mathrm{d}t^2 + E^2E = 0$, where $E$ is proportional to the optical field.

**Spatial Hole Burning:** Refers to the spatial modulation of the gain saturation in an atomic or molecular medium. This modulation results from the interference of two propagating fields. In a linear cw dye-laser cavity, the interference establishes a standing-wave field at the gain region. This spatial standing wave does not use the available gain at the nodes of the field, which is then available for a secondary mode. Thus, more than one longitudinal mode can lase simultaneously even when the broadening is homogeneous.

**Synchronous Pumping:** This occurs when the gain of the dye laser is modulated by a mode-locked pump laser. The cavity mode spacing of the dye laser is adjusted to be synchronous with the mode-locking frequency of the pump laser.

**Unstable Resonator:** A resonator configuration often used in wide-aperture high-power pulsed lasers. Typically, the mirrors in these resonators are curved and form an intracavity reflective telescope such as a Cassegrain telescope.

## References


[1] Jones II, G., (1990), in F. J. Duarte, L. W. Hillman (Eds.), *Dye Laser Principles.* New York: Academic Press, pp. 287–343.

[2] Maeda, M. (1984), *Laser Dyes.* New York: Academic Press.

[3] Drexhage, K. H. (1990), in F. P. Schäfer (Ed.), *Dye Lasers.* Berlin: Springer, pp. 155–200.

[4] Steppel, R. N. (1991), in M. J. Weber (Ed.), *Handbook of Laser Science and Technology, Supplement 1: Lasers.* Boca Raton, FL: CRC Press, pp. 219–317.

[5] Sorokin, P. P., Lankard, J. R. (**1966**), *IBM J. Res. Dev.* **10**, 162–163.

[6] Schäfer, F. P., Schmidt, W., Volze, J. (**1966**), *Appl. Phys. Lett.* **9**, 306–309.

[7] Pacheco, D. P., Aldag, H. R., Itzkan, I., Rostler, P. S. (1988), in F. J. Duarte (Ed.), *Proceedings of the International Conference on Lasers '87.* McLean, VA: STS, pp. 330–337.

[8] Ganiel, U., Hardy, A., Neumann, G., Treves, D. (**1975**), *IEEE J. Quantum Electron.* **QE-11**, 881–892.

[9] Teschke, O., Dienes, A., Whinnery, J. R. (**1976**), *IEEE J. Quantum Electron.* **QE-12**, 383–395.

[10] Duarte, F. J. (1995), in F. J. Duarte (Ed.), *Tunable Lasers Handbook.* New York: Academic Press, pp. 167–218.

[11] Dienes, A., Yankelevich, D. R. (1998), in G. L. Trigg (Ed.), *Encyclopedia of Applied Physics,* Vol. 22. New York: Wiley-VCH, pp. 299–334.

[12] Spaeth, M. L., Bortfeld, D. P. (**1966**), *Appl. Phys. Lett.* **9**, 179–181.

[13] Webb, C. E. (1991), in F. J. Duarte (Ed.), *High Power Dye Lasers.* Berlin: Springer, pp. 143–182.

[14] Duarte, F. J. (1990), in F. J. Duarte, L. W. Hillman (Eds.), *Dye Laser Principles.* New York: Academic Press, pp. 133–183; 239–285.

[15] Bos, F. (**1981a**), *Appl. Opt.* **20**, 3553–3556.

[16] Bos, F. (**1981b**), *Appl. Opt.* **20**, 1886–1890.

[17] Duarte, F. J., Piper, J. A. (**1981**), *Appl. Opt.* **20**, 2113–2116.

[18] Littman, M. G., Metcalf, H. J. (**1978**), *Appl. Opt.* **17**, 2224–2227.

[19] Hänsch, T. W. (**1972**), *Appl. Opt.* **11**, 895–898.

[20] Schäfer, F. P. (1990), in F. P. Schäfer (Ed.), *Dye Lasers.* Berlin: Springer, pp. 1–89.

[21] Bernhardt, A. F., Rasmussen, P. (**1981**), *Appl. Phys. B* **26**, 141–146.

[22] Duarte, F. J., Davenport, W. E., Ehrlich, J. J., Taylor, T. S. (**1991**), *Opt. Commun.* **84**, 310–316.

[23] Duarte, F. J., Piper, J. A. (**1984**), *Appl. Opt.* **23**, 1391–1394.

[24] Duarte, F. J. (**1988**), *Lasers Optron.* **7**(2), 41–45.
[25] Wallenstein, R., Hänsch, T. W. (**1975**), *Opt. Commun.* **14**, 353–357.
[26] Hargrove, R. S., Kan, T. (**1980**), *IEEE J. Quantum Electron.* **QE-16**, 1108–1113.
[27] Tallman, C., Tennant, R. (1991), in F. J. Duarte (Ed.), *High Power Dye Lasers.* Berlin: Springer, pp. 93–142.
[28] Bass, I. L., Bonanno, R. E., Hackel, R. P., Hammond, P. R. (**1992**), *Appl. Opt.* **31**, 6993–7006.
[29] Everett, P. N. (1991), in F. J. Duarte (Ed.), *High Power Dye Lasers.* Berlin: Springer, pp. 183–245.
[30] Baltakov, F. N., Barikhin, B. A., Sukanov, L. V. (**1974**), *JETP Lett.* **19**, 174–175.
[31] Everett, P. N., Aldag, H. R., Ehrlich, J. J., Janes, G. S., Klimek, D. E., Landers, F. M., Pacheco, D. P. (**1986**), *Appl. Opt.* **25**, 2142–2147.
[32] Mazzinghi, P., Burlamacchi, P., Matera, M., Ranea-Sandoval, H. F., Salimbeni, R., Vanni, U. (**1981**), *IEEE J. Quantum Electron.* **QE-17**, 2245–2249.
[33] Morton, R. G., Draggoo, V. G. (**1981**), *IEEE J. Quantum Electron.* **QE-17**(12), 222.
[34] Klimek, D. E., Aldag, H. R., Russell, J. (1992), *Technical Digest, Conference on Lasers and Electro-Optics.* Washington, DC: Optical Society of America, p. 332.
[35] Maeda, M., Uchino, O., Okada, T., Miyazoe, Y. (**1975**), *Jpn. J. Appl. Phys.* **14**, 1975–1980.
[36] Duarte, F. J., Conrad, R. W. (**1987**), *Appl. Opt.* **26**, 2567–2571.
[37] Hollberg, L. (1990), in F. J. Duarte, L. W. Hillman (Eds.), *Dye Laser Principles.* New York: Academic Press, pp. 185–238.
[38] Peterson, O. G., Tuccio, S. A., Snavely, B. B. (**1970**), *Appl. Phys. Lett.* **17**, 245–247.
[39] Tuccio, S. A., Strome, F. C. (**1972**), *Appl. Opt.* **11**, 64–73.
[40] Runge, P. K., Rosenberg, R. (**1972**), *IEEE J. Quantum Electron.* **QE-8**, 910–911.
[41] Peterson, O. G. (1979), in C. L. Tang (Ed.), *Methods of Experimental Physics,* Vol. 15. New York: Academic. pp. 251–359.
[42] Diels, J. C. (1990), in F. J. Duarte, L. W. Hillman (Eds.), *Dye Laser Principles.* New York: Academic Press, pp. 41–132.
[43] Hough, J., Hils, D., Rayman, M. D., Ma, L. S., Hollberg, L., Hall, J. L. (**1984**), *Appl. Phys. B* **33**, 179–185.
[44] Hall, J. L., Hänsch, T. W. (**1984**), *Opt. Lett.* **9**, 502–504.
[45] Johnston, T. F., Brady, R. H., Proffitt, W. (**1982**), *Appl. Opt.* **21**, 2307–2316.
[46] Baving, H. J., Muuss, H., Skolaut, W. (**1982**), *Appl. Phys. B* **29**, 19–21.
[47] Schmidt, W., Schäfer, F. P. (**1968**), *Phys. Lett.* **26A**, 558–559.
[48] Ruddock, I. S., Bradley, D. J. (**1976**), *Appl. Phys. Lett.* **29**, 296–297.
[49] Fork, R. L., Greene, B. I., Shank, C. V. (**1981**), *Appl. Phys. Lett.* **38**, 671–672.
[50] Dietel, W., Fontaine, J. J., Diels, J. C. (**1983**), *Opt. Lett.* **8**, 4–6.
[51] Fork, R. L., Brito Cruz, C. H., Becker, P. C., Shank, C. V. (**1987**), *Opt. Lett.* **12**, 483–485.
[52] Soffer, B. H., McFarland, B. B. (**1967**), *Appl. Phys. Lett.* **10**, 266–267.
[53] Peterson, O. G., Snavely, B. B. (**1968**), *Appl. Phys. Lett.* **12**, 238–240.
[54] Maslyukov, A., Sokolov, S., Kaivola, M., Nyholm, K., Popov, S. (**1995**), *Appl. Opt.* **34**, 1516–1518.
[55] Costela, A., Garcia-Moreno, I., Figuera, J. M., Amat-Guerri, F., Sastre, R. (**1998**), *Laser Chem.* **18**, 63–84.
[56] Duarte, F. J., Costela, A., Garcia-Moreno, I., Sastre, R., Ehrlich, J. J., Taylor, T. S. (**1997**), *Opt. Quantum Electron.* **29**, 461–472.
[57] Duarte, F. J. (**1994**), *Appl. Opt.* **33**, 3857–3860.
[58] Duarte, F. J. (**1999**), *Appl. Opt.* **38**, 6347–6349.
[59] Wadsworth, W. J., McKinnie, I. T., Woolhouse, A. D., Haskell, T. G. (**1999**), *Appl. Phys. B* **69**, 163–165.
[60] Zhu, X.-L., Lam, S.-K., Lo, D. (**2000**), *Appl. Opt.* **39**, 3104–3107.
[61] Rifani, M., Yin, Y.-Y., Elliot, D. S., Jay, M. J., Jang, S.-H., Kelley, M. P., Bastin, L., Kahr, B. (**1995**), *J. Am. Chem. Soc.* **117**, 7572–7573.
[62] Braun, I., Ihlein, G., Laeri, F., Nöckel, J. U., Schulz-Ekloff, G., Schüth, F., Vietze, U., Weiß, Ö., Wöhrle, D. (**2000**), *Appl. Phys. B* **70**, 335–343.
[63] Duarte, F. J. (1991), in F. J. Duarte (Ed.), *High Power Dye Lasers.* Berlin: Springer, pp. 7–43.
[64] Jamasbi, N., Diels, J. C., Sarger, L. (**1988**), *J. Mod. Opt.* **35**, 1891–1906.
[65] Diels, J. C., Dietel, W., Fontaine, J. J., Rudolph, W., Wilhelmi, B. (**1985**), *J. Opt. Soc. Am. B* **2**, 680–686.

[66] Hillman, L. W., Krasinski, J., Boyd, R. W., Stroud, C. R. (**1984**), *Phys. Rev. Lett.* **52**, 1605–1608.
[67] Klick, D. (1990), in F. J. Duarte, L. W. Hillman (Eds.), *Dye Laser Principles*. New York: Academic Press, pp. 345–412.
[68] Stuke, M. (Ed.) (1992), *Dye Lasers: 25 Years*. Berlin: Springer.
[69] Diels, J. C. (1991), *Technical Digest, International Conference on Lasers '91*, San Diego, CA, Paper TA.2.
[70] Raab, E. L., Prentiss, M., Cable, A., Chu, S., Pritchard. D. E. (**1987**), *Phys. Rev. Lett.* **59**, 2631–2634.
[71] Donley, E. A., Anderson, B. P., Wieman, C. E. (**2001**), *Opt. Photon. News* **12**(10), 34–37.
[72] Fox, R. W., Hollberg, L., Zibrov, A. S. (1997), in F. B. Dunning, R. G. Hulet (Eds.), *Atomic, Molecular, and Optical Physics*. New York: Academic Press, pp. 77–102.
[73] Paisner, J. A., Solarz, R. W., (1987), in L. J. Radziemski, R. W. Solarz, J. A. Paisner (Eds.), *Laser Spectroscopy and its Applications*. New York: Marcel Dekker, pp. 175–260.
[74] Hargrove, R. S. (1991), *Technical Digest, International Conference on Lasers '91*, San Diego, CA, Paper THA.2.
[75] Kölsch, H. J., Rairoux, P., Wolf, J. P., Woste, L. (**1992**), *Appl. Phys. B* **54**, 89–94.
[76] Grant, W. B. (**1991**), *Opt. Eng.* **30**, 40–48.
[77] Browell, E. V. (**1991**), *Opt. Photon. News* **2**(10), 8–11.
[78] Dougherty, T. J., Kaufman, J. E., Goldfarb, A., Weishaupt, K. R., Boyle, D., Mittleman, A. (**1978**), *Cancer Res.* **38**, 2628–2635.
[79] Dretler, S. P. (**1987**), *J. Endourol.* **1**, 9–14.
[80] Goldman, L. (1990), in F. J. Duarte, L. W. Hillman (Eds.), *Dye Laser Principles*. New York: Academic Press, pp. 419–432.

## Further Reading

Duarte, F. J. (Ed.) (1991), *High Power Dye Lasers*. Berlin: Springer.
Duarte, F. J. (Ed.) (1992), *Selected Papers on Dye Lasers, SPIE Milestone Series*, Vol. MS 45. Bellingham, WA: SPIE.
Duarte, F. J., Hillman, L. W. (Eds.) (1990), *Dye Laser Principles*. New York: Academic Press.
Maeda, M. (1984), *Laser Dyes*. New York: Academic Press.
Schäfer, F. P. (Ed.) (1990), *Dye Lasers*. Berlin: Springer.
Stuke, M. (Ed.) (1992), *Dye Lasers: 25 Years*. Berlin: Springer.
Weber, M. J. (2001), *Handbook of Lasers*. Boca Raton, FL: CRC Press.

# Lasers, Extreme UV and Soft X-ray

*Joseph Nilsen*
*Lawrence Livermore National Laboratory, Livermore, CA, 94551-0808*
*Phone: (925)422-4766; Fax: (925)422-5102; e-mail: jnilsen@llnl.gov*

**Abstract**

Two decades ago, very large sources such as the building size ICF lasers were used as the energy sources to drive the first X-ray lasers. Tremendous progress has been made over the last two decades to produce tabletop, high-repetition rate X-ray lasers that can be used as tools for applications. With the use of high-reflectivity multilayer mirrors, many of the standard optical techniques such as interferometry can be extended to the soft X-ray regime between wavelengths of 10 and 50 nm. The dominant X-ray laser schemes are the monopole collisional excitation lasers either driven by chirped pulse amplification (CPA) laser systems or capillary discharge. The CPA systems drive lasing in neon-like or nickel-like ions, typically in the 10- to 30-nm range, while the capillary system is currently limited to driving neon-like ions at a longer wavelength near 50 nm. Most researchers use nickel-like ion lasers near 14 nm because they are well matched to the Mo : Si multilayer mirrors, which have peak reflectivity near 13 nm, and are an essential component of applications such as X-ray laser interferometry.

**Keywords**

neon-like; nickel-like; hydrogen-like; collisional excitation; recombination.

## 1 Introduction

Since the advent of the laser in the early 1960s, there has been interest in extending lasers to much shorter wavelength regimes. Most researchers realized that achieving shorter wavelengths requires creating highly ionized plasmas in order to utilize the higher-energy atomic transitions. Given the lack of good reflectors in this regime, it was realized that X-ray lasers would need to work in a single-pass amplified spontaneous emission mode with no external cavity. The high aspect ratio of a long cylindrical plasma would determine the mode structure, coherence, and divergence of the single-pass laser. Producing highly ionized plasmas require high-intensity sources and most researchers use a high-intensity optical laser focused to a line to transversely pump an X-ray laser target. The solid density target is heated and it expands to low enough density to allow lasing operation.

Reaching the shortest wavelength regimes, <1 nm, requires lasing on transitions between states with different principal quantum numbers, the $\Delta n = 1$ transitions, in high $Z$ plasmas. In the early 1970s, there were many ambitious proposals, an example being the proposal [1] to lase at 0.5 nm on the 3d → 2p transition in hydrogen-like krypton and the tremendous challenges required to pump that scheme. To achieve these conditions, one starts with neutral krypton ($Z = 36$) and ionizes all 36 electrons. This recombines to the hydrogen-like state, which has one bound electron, and lases. It was suggested that X-ray lasers would have far-reaching impacts on biology, chemistry, and crystallography and could someday be used to resolve the structure of molecules.

A more modest approach suggested in the 1970s to achieve X-ray lasers was to extrapolate from successful optical lasers, such as the argon-ion laser, that lases on the 4p $\rightarrow$ 4s transitions in single-ionized argon. An early proposal of this type was to lase on the 3p $\rightarrow$ 3s lines in neon-like ions [2–4] and that is one of the two dominant laser schemes today. These schemes have a metastable upper laser state that lases on a transition between levels with the same principal quantum number. The lower laser level decays by rapid radiative decay to the ground state.

## 1.1 First Demonstrations of X-ray Lasers

Shrouded in secrecy, researchers at Lawrence Livermore National Laboratory (LLNL) succeeded in demonstrating the first X-ray laser in the early 1980s using a nuclear explosion to drive the X-ray laser [5] as part of their effort to develop a missile defense system. This was followed in 1984 at LLNL with the first demonstration of the laboratory X-ray laser using the large Inertial confinement fusion (ICF) laser Novette to drive a neon-like selenium laser [6, 7]. In the same year, the effort at Princeton Plasma Physics Laboratory (PPPL) succeeded in demonstrating a hydrogen-like carbon laser [8].

### 1.1.1 Neon-like Selenium Collisional Excitation Laser

The neon-like selenium ($Z = 34$) laser is based on having a combination of collisional excitation from the neon-like ground state and recombination from the flourine-like states populate the 3p singly excited levels of neon-like selenium. Obtaining the neon-like state, which has 10 electrons, requires ionizing 24 electrons from the neutral selenium target. The 3p excited levels are metastable for decay to the ground state and are the upper laser levels that decay to the 3s lower laser levels. The 3s levels decay by rapid radiative decay to the neon-like ground state. The Novette laser used two 0.53-μm, 0.5-kJ, 450-ps pulses focused to a 200-μm wide by 1.1-cm long line to illuminate a thin foil target (75 nm of Se deposited on one side of 150-nm thick Formvar) from both sides at an intensity of $5 \times 10^{13}$ W cm$^{-2}$. This expanded the selenium to the density required for lasing, ionized it to the neon-like sequence, and heated it sufficiently to achieve gain on the lasing transitions. Typical plasma conditions for lasing are a mass density of 3 mg cm$^{-3}$, an electron density of $5 \times 10^{20}$ cm$^{-3}$, and an electron temperature of 1 keV. While the 20.6- and 20.9-nm lines lased well, a big surprise was the absence of lasing on the 3p $^1S_0 \rightarrow 3s^1P_1$ line at 18.2 nm that was predicted to be the dominant laser line [4, 6, 7].

For the next decade, this was the dominant X-ray laser scheme and worked in many other neon-like ions from copper ($Z = 29$) to silver ($Z = 47$) [9–11]. For many years, the neon-like germanium ($Z = 32$) laser with lines at 23.2 and 23.6 nm was the principal laser used at many laboratories around the world. At LLNL, the Nova ICF laser drove the neon-like yttrium ($Z = 39$) laser into saturation, producing an output of 1 mJ in a single pulse at 15.5 nm [12].

### 1.1.2 Hydrogen-like Carbon Recombination Laser

Hydrogen-like carbon lases by ionizing carbon to the bare nucleus and then recombining into the 3d single-excited upper laser level. The 3d state lases to the 2p lower laser level, which then decays rapidly to the 1s ground state. This was demonstrated at PPPL using a 1-kJ, 75-ns pulse from a $CO_2$ laser to heat a carbon plasma that was confined in a magnetic field.[8] The advantage of the recombination laser is that it has higher quantum efficiency owing to lasing on a $\Delta n = 1$ transition but is highly sensitive to radiation trapping whereby the population of the lower laser level increases and quenches the gain. The gain is low and researchers were frustrated in their attempts to increase the gain–length product and produce a saturated laser.

## 2
## Development of the Prepulse Technique

Over the two decades since the first X-ray lasers were demonstrated, the collisional excitation scheme has become the dominant laser used by most researchers. While substantial research has been devoted to trying other schemes, none has been very successful. With the recombination laser, researchers have achieved lasing in other ionization stages, such as the lithium-like stage, but the recombination lasers have all been plagued with low-gain–length product, typically less than 6, and have never reached saturated output, which requires gain–length product near 15 [13].

### 2.1
### Neon-like Collisional Excitation Lasers

Tremendous progress has been made in understanding and improving the collisional excitation lasers. The first leap forward was the development of the prepulse technique in 1992 [14]. This technique used a small prepulse to heat a solid target and create the large-scale-length plasma that was partially ionized. This was followed by the main heating pulse that would heat the plasma to lasing conditions. Using this technique, the $3p\,^1S_0 \rightarrow 3s\,^1P_1$ line suddenly became the dominant laser line for neon-like systems, as had always been predicted [4, 6, 7], and virtually every neon-like ion from silicon ($Z = 14$) to Se ($Z = 34$) has been made to lase [15–17]. Typical experiments illuminated a 2- to 3-cm-long target with a few 100 J of energy from a 0.5-ns pulse. This lower-energy requirement made X-ray lasers accessible to a larger number of laser facilities. This technique eliminated the need for thin-foil targets and allowed the use of simple, inexpensive, solid slab targets. The slab targets could be used many times by just repositioning the slab. The use of the prepulse enabled the creation of a plasma with small enough gradients so that the $3p\,^1S_0 \rightarrow 3s\,^1P_1$ X-ray laser line was no longer refracted out of the plasma before it could be amplified sufficiently [17–19]. This line has the highest gain but lases in a higher-density part of the plasma than the other neon-like lines, which makes it much more sensitive to refraction. The $3p\,^1S_0$ upper laser level is driven by a monopole collisional excitation of the neon-like ground state and is a simple three-level system that is now much better understood. The pair of $J = 2 \rightarrow 1$ laser lines that originally lased in neon-like selenium and other neon-like ions is no longer used.

### 2.2
### Nickel-like Collisional Excitation Lasers

Soon after the success of neon-like selenium, nickel-like europium ($Z = 63$)

was made to lase on the nickel-like 4d $^1S_0 \rightarrow 4p\ ^1P_1$ transition at 7.1 nm with the 4d $^1S_0$ upper laser state populated by monopole collisional excitation from the nickel-like ground state [20]. Nickel-like systems have the advantage of producing shorter wavelength lines than neon-like systems by a factor of 2 to 3 for similar excitation energies. The early nickel-like lasers were plagued with very low gain and never achieved saturated output but did reach wavelengths as short as 3.5 nm [21]. A modification of the prepulse technique where a series of 100-ps pulses was used to illuminate the target finally enabled the nickel-like lasers to perform well and reach saturated output [22–24].

## 3
## Tabletop Laser-driven X-ray Lasers

The newest variant of the prepulse technique uses a combination of a nanosecond laser prepulse to illuminate a solid target to create the preplasma and a picosecond laser pulse to heat the plasma to lasing conditions. This allows the use of much smaller lasers to drive the neon-like and nickel-like X-ray lasers. These experiments use chirped pulse amplification (CPA) [25, 26] to produce the picosecond duration drive pulse with typical energies less than 10 J. The nanosecond prepulse is created by using part of the uncompressed beam from the CPA system. Typical experiments use less than 10 J in each beam. These systems are compact, tabletop, higher-repetition rate systems that occupy a few standard optical benches and are largely replacing the enormous ICF lasers used in the past. Given the short pulse duration, these experiments use a traveling wave geometry to transversely illuminate the X-ray laser target and synchronize the pumping with the lasing process at the speed of light. Figure 1 shows the line-focused CPA pulse transversely illuminating the plasma created by the prepulse heating the solid slab target. The laser focus overfills the target to ensure uniform illumination of the plasma. Given the short pulse duration, the wavefront of the CPA pulse is tilted to create a traveling wave geometry that synchronizes the heating of the plasma with the X-ray laser propagation and amplification along the length of the plasma. This causes the X-ray lasing to preferentially emit from one end of the laser.

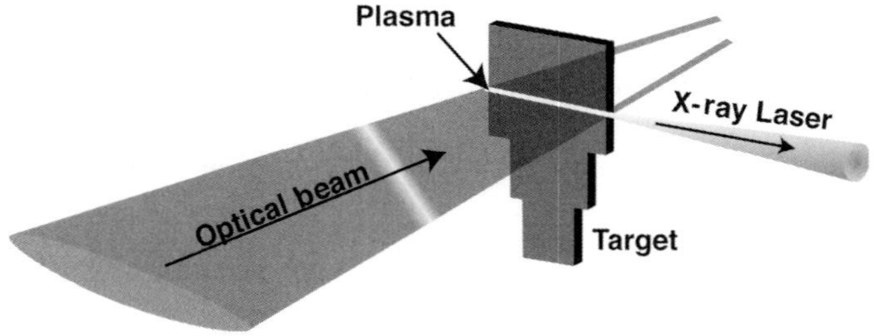


**Fig. 1** Schematic of the line-focused CPA pulse transversely illuminating the plasma created by the prepulse heating the solid slab target. The wavefront of the CPA pulse is tilted to create a traveling wave geometry that synchronizes the heating of the plasma with the X-ray laser propagation and amplification along the length of the plasma

### 3.1
### Neon-like Titanium Laser

The first experiments [27] of this type in 1995 used a 4-J, 1.5-ns pulse focused to a 30-μm by 0.5-cm line to illuminate a solid titanium ($Z = 22$) target. This created the preplasma that was then heated by a similarly line-focused 7-J, 0.7-ps pulse. The titanium lased on two neon-like transitions, the 3p $^1S_0 \rightarrow$ 3s $^1P_1$ transition at 32.6 nm and the 3d $^1P_1 \rightarrow$ 3p $^1P_1$ transition at 30.1 nm. Figure 2 shows the energy-level diagram for neon-like titanium. LS coupling notation is used where the state is written as $^{2S+1}L_J$ where S is the electron spin, L is the orbital angular momentum and J is the total angular momentum of the electrons. Lasing is driven by two different mechanisms[28.] The 3p $^1S_0$ upper laser state is populated by monopole collisional excitation of the neon-like ground state, while the 3d $^1P_1$ upper state is populated primarily by photopumping on the strong 2p $^1S_0 \rightarrow$ 3d $^1P_1$ transition. In subsequent experiments [29], the titanium laser achieved saturated output on both these lines using 1-cm long targets with 100-μm wide line focus. Small signal gain coefficients of 46 and 35 cm$^{-1}$ were measured for the 32.6- and 30.1-nm lines with total gain–length products of 16.7 and 16.9, respectively. Using up to 30 J from the Vulcan CPA Nd:glass laser at Rutherford Appleton Laboratory, lasing near saturation has also been achieved in neon-like germanium ($Z = 32$) on the 3p $^1S_0 \rightarrow$ 3s $^1P_1$ transition at 19.6 nm [30].

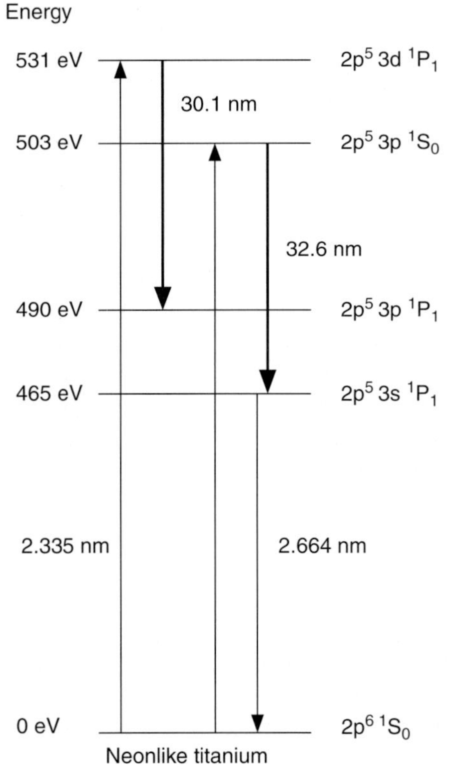


**Fig. 2** Energy-level diagram for lasing at 32.6 and 30.1 nm in neon-like titanium. The 3p $^1S_0$ upper laser state is populated by monopole collisional excitation of the neon-like ground state while the 3d $^1P_1$ upper state is populated primarily by photopumping on the strong 2p $^1S_0 \rightarrow$ 3d $^1P_1$ transition

### 3.2
### Nickel-like Palladium Laser

After the success of the neon-like titanium laser, attention was quickly turned to using the CPA lasers to drive nickel-like ions. Nickel-like palladium ($Z = 46$) lased in 1997 on the 4d $^1S_0 \rightarrow$ 4p $^1P_1$ transition at 14.7 nm [31]. Figure 3 shows the energy-level diagram for this laser. The 4d $^1S_0$ upper laser state is populated by monopole collisional excitation of the nickel-like ground state. The 4p $^1P_1$ lower state decays by rapid radiative decay to the nickel-like ground state. Optimizing the drive conditions, palladium reached saturated output using a total of 7 J from

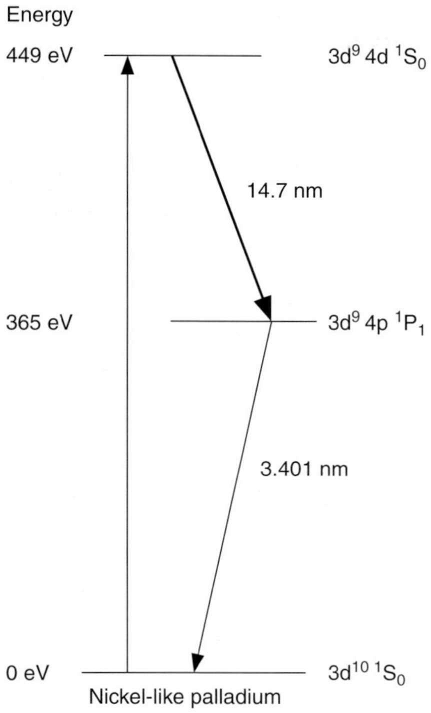


**Fig. 3** Energy-level diagram for lasing at 14.7 nm in nickel-like palladium. The 4d $^1S_0$ upper laser state is populated by monopole collisional excitation of the nickel-like ground state

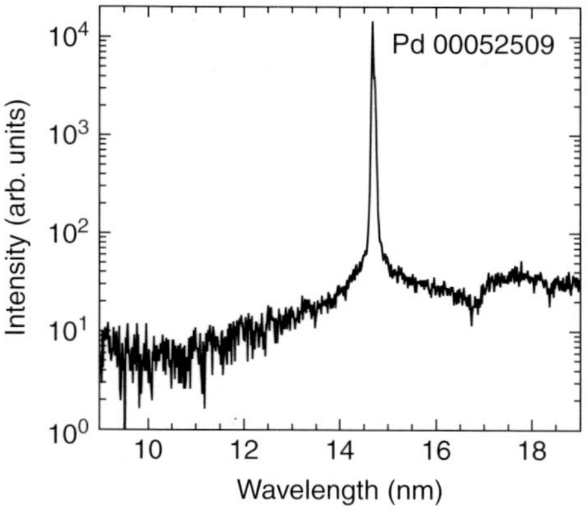


**Fig. 4** Spectrum of the palladium plasma taken from the output end of the X-ray laser. The spectrum shows saturated output on the 4d $^1S_0 \rightarrow$ 4p $^1P_1$ transition at 14.7 nm that dominates the spectrum. A 0.6-J, 600-ps long pulse illuminated the 1-cm long target 700 ps before the 5-J, 1.5-ps short pulse

a 1.05-µm Nd : glass laser to drive a 0.9-cm long target. Typical conditions use a 1-J, 600-ps pulse followed 700 ps later by a 5-J, 1-ps pulse to illuminate a solid palladium target with a 80-µm by 1.1-cm line focus [32]. A typical spectrum is shown in Fig. 4. Small signal gain coefficients as large as 61 $cm^{-1}$ were measured with a total gain–length product of 18.1. Typical output is 12 µJ per pulse with divergence of 2 to 3 mrad. The CPA laser system used in this experiment has a repetition rate of one shot every four minutes. Experiments with silver have measured an X-ray laser pulse duration of 2 ps [33]. Many nickel-like ions from yttrium ($Z = 39$) to samarium ($Z = 62$) have lased [33–36] using CPA systems with saturated output observed for as short as 7.3 nm. Nickel-like palladium at 14.7 nm and silver, which lases at 13.9 nm, have become the lasers of choice for many applications developments. In the case of silver, strong lasing has also been observed on a second line, the self-photopumped 4f $^1P_1 \rightarrow$ 4d $^1P_1$ transition at 16.0 nm [37, 38].

## 3.3 Longitudinally Pumped Lasers

To improve the efficiency of the X-ray lasers, researchers have tried to longitudinally pump an X-ray laser by focusing a spot on the end of the plasma and having the optical drive laser copropagate with the X-ray laser. This offers the potential for more efficient energy coupling with the lasing plasma and naturally creates a traveling wave configuration. Recent experiments focused a 10-Hz,

100-mJ, 100-fs Ti:sapphire laser pulse into a 0.2-cm-long molybdenum capillary with 0.45 cm inner diameter and observed lasing on the nickel-like molybdenum line at 18.9 nm [39]. Similar pumping configurations without a capillary have been proposed and are being tested [40].

### 3.4 Gas Puff Targets

Another approach to improving laser efficiency is to use gas puff targets that are already at the final lasing density instead of expending laser energy to do the hydrodynamic work to expand the solid target. The gas puff targets have the advantage of less debris for a high-repetition system. Experiments using argon gas puffs have observed strong lasing on the neon-like argon line at 46.9 nm when driven by CPA systems [41, 42].

## 4 Tabletop Capillary Discharge Lasers

An alternative to driving an X-ray laser plasma with a high-intensity laser is to use an electric discharge to heat and ionize the plasma. Tremendous effort was put into using large plasma machines such as SATURN at Sandia National Laboratory to drive an X-ray laser. The difficulty with these $Z$ pinch experiments was that the plasma had large axial inhomogeneities that distorted the plasma column and prevented amplification. In 1994, there was a tremendous breakthrough in this area when the group at Colorado State University was able to achieve lasing in neon-like argon at 46.9 nm using an argon-filled capillary driven by an electric discharge [43].

### 4.1 Neon-like Argon Capillary Discharge Laser

This laser uses a plastic or aluminum oxide capillary filled with argon gas that is heated using the rapid discharge of a Marx generator through the capillary channel. A typical capillary has an inside diameter of 0.3 cm, an argon pressure of 460 mtorr, lengths up to 35 cm driven by peak currents of 26 kA with a rise time of 40 ns. Lasing occurs on the 3p $^1S_0 \rightarrow$ 3s $^1P_1$ transition at 46.9 nm in neon-like argon with the upper laser state driven by monopole collisional excitation from the neon-like ground state. Typical electron temperature in the plasma is 60 to 80 eV with the electron density in the range 0.3 to $1.0 \times 10^{19}$ cm$^{-3}$. This laser has achieved saturated laser operation with outputs of 1 mJ per pulse and has been operated at 4-Hz repetition rate [44]. The lasing duration is 1 ns. Average power of 3.5 mW has been produced with a single capillary lasting for 5000 shots. The output has been shown to be almost completely spatially coherent with a divergence of 5 mrad [45]. Figure 5 shows a photograph of this tabletop X-ray laser. Lasing has also been achieved in neon-like chlorine at 52.9 nm and neon-like sulfur at 60.8 nm. To reach shorter wavelengths, researchers are trying to build a higher-current, more

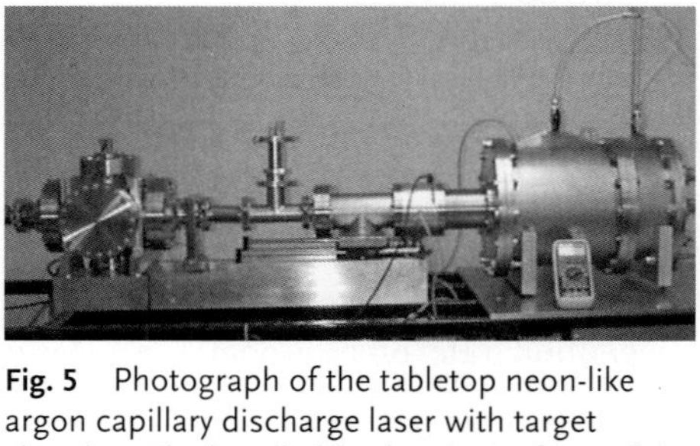

**Fig. 5** Photograph of the tabletop neon-like argon capillary discharge laser with target chamber. The handheld voltmeter in front of the laser shows the small size of the laser

rapid discharge system to drive the nickel-like cadmium laser at 13.2 nm.

# 5 Alternative X-ray Laser Schemes

## 5.1 Optical Field Ionization

Optical field ionization (OFI) uses a very high-intensity laser to generate electric fields that are comparable to the binding energy of the electrons. Typically, these involve laser intensities greater than $10^{17}$ W cm$^{-2}$. The bound electrons are ionized by the tunneling process. The use of linearly polarized light keeps the electron temperature low, which is required for recombination laser schemes, while circularly polarized light generates hot electrons, which is ideal for collisional excitation schemes. The laser pulses are subpicosecond and the scheme uses low-energy lasers with less than 1 J, which are focused to a spot and used to longitudinally pump a short plasma.

### 5.1.1 Hydrogen-like Lithium Recombination Laser

Lasing has been observed on the hydrogen-like lithium 2p $\rightarrow$ 1s transition at 13.5 nm. Using a linearly polarized KrF laser with a 50 mJ, 0.5-ps pulse focused at $10^{17}$ W cm$^{-2}$ into a lithium plasma, a gain of 20 cm$^{-1}$ was measured for a 0.2-cm-long plasma during the recombination phase [46].

### 5.1.2 Palladium-like Xenon and Nickel-like Krypton Collisional Excitation Laser

Using a 330-mJ, 35-fs, 10-Hz Ti : sapphire laser focused to $3 \times 10^{17}$ W cm$^{-2}$ in a xenon gas cell filled to 15 torr, saturated output with a gain–length product of 15 was observed on the 5d $^1S_0 \rightarrow$ 5p $^1P_1$ transition at 41.8 nm in palladium-like xenon ($Z = 54$) [47]. Using similar drive conditions, lasing has recently been observed on the 4d $^1S_0 \rightarrow$ 4p $^1P_1$ transition at 32.8 nm in nickel-like krypton.

## 5.2 Resonant Photopumping Schemes

The idea to create an inversion by using radiation from a bright line in one ion to pump a resonant transition in another ion and populate an excited state that could lase has been investigated for many decades. A classic scheme proposed in the 1970s was to use the He-$\alpha$ line of Na to resonantly photopump the He-$\gamma$ line of Ne. Lasing would then be expected on several 4f $\rightarrow$ 3d lines in helium-like neon at wavelengths near 23 nm [48]. Many experiments were tried using pulsed power machines but gain and lasing have never been observed even though measurements of the line emission suggests that gain does exist in the plasma [49–51]. Many photopumped schemes have been suggested and even tested experimentally but the results have been disappointing. Several challenges need to be overcome including finding an adequate resonance between a strong pump line and a potential lasing material. The bright pump line usually requires a higher density plasma than the lasing plasma. This presents the challenge of efficiently coupling together two plasmas with different temperatures and densities.

## 5.3 Inner-shell Excitation

To reach shorter wavelength more efficiently, researchers are trying to lase on inner-shell transitions, such as the K$\alpha$ line [52], in single-ionized carbon [53]. The idea, suggested in the 1970s, is to use a short pulse of X rays to selectively ionize the 1s electron and lase on the 2p → 1s transition before autoionization destroys the inversion [54]. Since the autoionization rates are typically $10^{14}$ sec$^{-1}$, it requires a pump with duration of a few femtoseconds. While femtosecond lasers have been available for many years, only now are powerful lasers becoming available with this short-pulse duration [55]. Ideally, this scheme would have high quantum efficiency and carbon would lase at 4.5 nm. There are still many challenges to overcome, which include converting the femtosecond laser into a femtosecond X-ray source.

Another approach under investigation is to use a femtosecond laser to collisionally ionize the inner-shell electrons directly. Lasing has been proposed in single-ionized titanium by ionizing a 2p electron and lasing on the 3s → 2p transition at 3.1 nm [56].

### 5.4 X-ray Free Electron Laser

Perhaps, the best candidate for a new X-ray laser scheme is the extension of the free-electron laser technology into the X-ray regime. Recent experiments have observed lasing at 100 nm using the TESLA Test Facility at the Deutsches Elektronen-Synchrotron in Hamburg. This and other facilities are using self-amplified spontaneous emission (SASE) to obtain high gain in a single pass. This concept offers the potential to produce subpicosecond laser pulses that could be tuned continuously down to 0.1-nm wavelength [57].

## 6 Conclusion

Tremendous progress has been made over the last two decades to produce a tabletop, high repetition rate X-ray laser that can be used as a tool for applications. With the use of high-reflectivity multilayer mirrors,

**Tab. 1** Wavelengths of some of the prominent X-ray lasers being used or studied today

| *Material* | *Ion stage* | *Transition* | *Wavelength (nm)* |
|---|---|---|---|
| Argon | Neon-like | $3p\,^1S_0 \rightarrow 3s\,^1P_1$ | 46.9 |
| Xenon | Palladium-like | $5d\,^1S_0 \rightarrow 5p\,^1P_1$ | 41.8 |
| Krypton | Nickel-like | $4d\,^1S_0 \rightarrow 4p\,^1P_1$ | 32.8 |
| Titanium | Neon-like | $3p\,^1S_0 \rightarrow 3s\,^1P_1$ | 32.6 |
| Titanium | Neon-like | $3d\,^1P_1 \rightarrow 3p\,^1P_1$ | 30.1 |
| Germanium | Neon-like | $3p\,^1S_0 \rightarrow 3s\,^1P_1$ | 19.6 |
| Molybdenum | Nickel-like | $4d\,^1S_0 \rightarrow 4p\,^1P_1$ | 18.9 |
| Palladium | Nickel-like | $4d\,^1S_0 \rightarrow 4p\,^1P_1$ | 14.7 |
| Silver | Nickel-like | $4d\,^1S_0 \rightarrow 4p\,^1P_1$ | 13.9 |
| Lithium | Hydrogen-like | $2p \rightarrow 1s$ | 13.5 |
| Cadmium | Nickel-like | $4d\,^1S_0 \rightarrow 4p\,^1P_1$ | 13.2 |
| Tin | Nickel-like | $4d\,^1S_0 \rightarrow 4p\,^1P_1$ | 11.9 |
| Xenon | Nickel-like | $4d\,^1S_0 \rightarrow 4p\,^1P_1$ | 10.0 |
| Samarium | Nickel-like | $4d\,^1S_0 \rightarrow 4p\,^1P_1$ | 7.3 |

many of the standard optical techniques such as interferometry can be extended to the soft X-ray regime between 10 and 50 nm. Examples of many of these applications can be found in the series of X-ray laser conference proceedings listed under Further Reading. The dominant laser scheme is the monopole collisional excitation laser either driven by CPA laser systems or capillary discharge. The CPA systems drive lasing in both the neon-like and nickel-like ions, while the capillary system is currently limited to driving neon-like ions at longer wavelength. Most researchers use nickel-like silver or palladium lasers near 14 nm because they are well matched to the Mo : Si multilayer mirrors that have peak reflectivity near 13 nm.

Table 1 gives a short list of some of the X-ray lasers being used today. An extensive list of laser wavelengths can be found in the handbook by Weber and Refs. [11, 34, and 38].

The efficiency of pumping X-ray lasers is very low, typically $10^{-6}$. The low efficiency offers the potential for tremendous improvement in the future. One area that looks very promising is to use lower-energy CPA systems to longitudinally pump a low-density gas target so that most of the drive energy is deposited in the lasing plasma. This would both improve the efficiency and allow high-repetition rate operation (>1 Hz).

## Acknowledgment

The author would like to thank R. C. Elton, E. E. Fill, J. Dunn, R. F. Smith, and J. J. Rocca for their kind assistance and advice in preparing this article. Work performed under the auspices of the US Department of Energy by the University of California Lawrence Livermore National Laboratory under Contract No. W-7405-Eng-48.

## Glossary

**Collisional Excitation:** Process by which free electrons in the plasma excite bound electrons to a higher-energy bound state by transferring part of their energy.

**Gain–length Product:** The integral of the gain over the length of the gain medium including the effect of saturation on the gain. Typically, gain-length products of about 15 are needed for the X-ray laser to reach saturation intensity.

**Hydrogen-like:** An ionic stage where an atom has been ionized such that there is one bound electron. The energy-level structure is very similar to that of neutral hydrogen.

**ICF:** Inertial confinement fusion. The largest optical lasers such as Nova at LLNL were built to demonstrate ICF.

**Metastable:** An excited bound state in the ion that cannot decay to another bound state with a different principal quantum number by direct dipole-allowed radiative transition.

**Neon-like:** An ionic stage where an atom has been ionized such that there are 10 bound electrons. The energy-level structure is very similar to that of neutral neon and the ground state has filled K and L shells.

**Nickel-like:** An ionic stage where an atom has been ionized such that there are 28 bound electrons. The ground state has filled K, L, and M shells.

**Palladium-like:** An ionic stage where an atom has been ionized such that there are 46 bound electrons. The ground state has filled K, L, M shells, and full 4s, 4p, and 4d subshells.

**Radiation Trapping:** A process by which the radiation emission from a strong resonance line builds up sufficient intensity such that the population of the lower laser state increases and starts to quench the gain of the X-ray laser transition.

**Recombination:** Process by which free electrons in the plasma recombine with an ion and become bound electrons.

**Saturation Intensity:** The X-ray laser intensity at which the gain is reduced by a factor of 2 due to stimulated emission in the plasma modifying the populations of the laser states.

**Traveling Wave Geometry:** A situation by which the optical laser illuminates the target synchronously with the X-rays traveling in the amplification direction. For transverse illumination, this is typically done by changing the phase front of the optical laser using a grating or stepped mirror.

## References

[1] Chapline, G., Wood, L. (**1975**), *Phys. Today* 40–48.

[2] Elton, R. C. (**1975**), *Appl. Opt.* **14**, 97–101.

[3] Zherikhin, A. N., Koshelev, K. N., Letokhov, V. S. (**1976**), *Sov. J. Quantum Electron.* **6**, 82–84.

[4] Vinogradov, A. V., Sobel'man, I. I., Yukov, E. A. (**1977**), *Sov. J. Quantum Electron.* **7**, 32–35.

[5] Nilsen, J. (**2003**), *Quantum Electronics (Russia)* **33**, 1–2.

[6] Matthews, D. L., Hagelstein, P. L., Rosen, M. D., Eckart, M. J., Ceglio, N. M., Hazi, A. M., Medecki, H., MacGowan, B. J., Trebes, J. E., Whitten, B. L., Campbell, E. M., Hatcher, C. W., Hawryluk, A. M., Kauffman, R. L., Pleasance, L. D., Rambach, G., Scofield, J. H., Stone, G., Weaver, T. A. (**1985**), *Phys. Rev. Lett.* **54**, 110–113.

[7] Rosen, M. D., Hagelstein, P. L., Matthews, D. L., Campbell, E. M., Hazi, A. U., Whitten, B. L., MacGowan, B., Turner, R. E., Lee, R. W., Charatis, G., Busch, Gar. E., Shepard, C. L., Rockett, P. D. (**1985**), *Phys. Rev. Lett.* **54**, 106–109.

[8] Suckewer, S., Skinner, C. H., Milchberg, H., Keane, C., Voorhees, D. (**1985**), *Phys. Rev. Lett.* **55**, 1753–1756.

[9] Lee, T. N., McLean, E. A., Elton, R. C. (**1987**), *Phys. Rev. Lett.* **59**, 1185–1188.

[10] Fields, D. J., Walling, R. S., Shimkaveg, G. M., MacGowan, B. J., Da Silva, L. B., Scofield, J. H., Osterheld, A. L., Phillips, T. W., Rosen, M. D., Matthews, D. L., Goldstein, W. H., Stewart, R. E. (**1992**), *Phys. Rev. A* **46**, 1606–1609.

[11] Nilsen, J., Scofield, J. H. (**1994**), *Phys. Scr.* **49**, 588–591.

[12] Da Silva, L. B., MacGowan, B. J., Mrowka, S., Koch, J. A., London, R. A., Matthews, D. L., Underwood, J. H. (**1993**), *Opt. Lett.* **18**, 1174–1176.

[13] Zhang, J., Key, M. H., Norreys, P. A., Tallents, G. J., Behjat, A., Danson, C., Demir, A., Dwivedi, L., Holden, M., Holden, P. B., Lewis, C. L. S., MacPhee, A. G., Neely, D., Pert, G. J., Ramsden, S. A., Rose, S. J., Shao, Y. F., Thomas, O., Walsh, F., You, Y. L. (**1995**), *Phys. Rev. Lett.* **74**, 1335–1338.

[14] Nilsen, J., MacGowan, B. J., Da Silva, L. B., Moreno, J. C. (**1993**), *Phys. Rev. A* **48**, 4682–4685.

[15] Fill, E. E., Li, Y. L., Schlögl, D., Steingruber, J., Nilsen, J. (**1995**), *Opt. Lett.* **20**, 374–376.

[16] Li, Y. L., Pretzler, G., Lu, P. X., Fill, E. E., Nilsen, J. (**1997**), *Phys. Plasmas* **4**, 479–489.

[17] Nilsen, J., Moreno, J. C. (**1995**), *Phys. Rev. Lett.* **74**, 3376–3379.

[18] Li, Y. L., Pretzler, G., Fill, E. E., Nilsen, J. (1996), *J. Opt. Soc. Am. B* **13**, 742–750.
[19] Moreno, J. C., Nilsen, J., Li, Y. L., Lu, P., Fill, E. E. (**1996**), *Opt. Lett.* **21**, 866–868.
[20] MacGowan, B. J., Maxon, S., Hagelstein, P. L., Keane, C. J., London, R. A., Matthews, D. L., Rosen, M. D., Scofield, J. H., Whelan, D. A. (**1987**), *Phys. Rev. Lett.* **59**, 2157–2160.
[21] MacGowan, B. J., DaSilva, L. B., Fields, D. J., Keane, C. J., Koch, J. A., London, R. A., Matthews, D. L., Maxon, S., Mrowka, S., Osterheld, A. L., Scofield, J. H., Shimkaveg, G., Trebes, J. E., Walling, R. S. (**1992**), *Phys. Fluids B* **4**, 2326–2337.
[22] Zhang, J., MacPhee, A. G., Nilsen, J., Lin, J., Barbee, Jr, T. W., Danson, C., Key, M. H., Lewis, C. L. S., Neely, D., O'Rourke, R. M. N., Pert, G. J., Smith, R., Tallents, G. J., Wark, J. S., Wolfrum, E. (**1997**), *Phys. Rev. Lett.* **78**, 3856–3859.
[23] Tommasini, R., Löwenthal, F., Balmer, J. E. (**1999**), *Phys. Rev. A* **59**, 1577–1581.
[24] Sebban, S., Daido, H., Sakaya, N., Kato, Y., Murai, K., Tang, H., Gu, Y., Huang, G., Wang, S., Klisnick, A., Zeitoun, Ph., Koike, F., Takenaka, H. (**2000**), *Phys. Rev. A* **61**, 043810.
[25] Kalachnikov, M., Nickles, P. V., Schnürer, M , Will, I., Sandner, W. (**1997**), *Opt. Commun.* **133**, 216–220.
[26] Sullivan, A., Bonlie, J., Price, D. F., White, W. E. (**1996**), *Opt. Lett.* **21**, 603–605.
[27] Nickles, P. V., Shlyaptsev, V. N., Kalachnikov, M., Schnürer, M., Will, I., Sandner, W. (**1997**), *Phys. Rev. Lett.* **78**, 2748–2751.
[28] Nilsen, J., Li, Y. L., Dunn, J. (**2000**), *J. Opt. Soc. Am. B* **17**, 1084–1092.
[29] Kalachnikov, M. P., Nickles, P. V., Schnürer, M., Sandner, W., Shlyaptsev, V. N., Danson, C., Neely, D., Wolfrum, E., Zhang, J., Behjat, A., Demir, A., Tallents, G. J., Warwick, P. J., Lewis, C. L. S. (**1998**), *Phys. Rev. A* **57**, 4778–4783.
[30] Warwick, P. J., Lewis, C. L. S., Kalachnikov, M. P., Nickles, P. V., Schnurer, M., Behjat, A., Demir, A., Tallents, G. J., Neely, D., Wolfrum, E., Zhang, J., Pert, G. J. (**1998**), *JOSA B* **15**, 1808–1814.
[31] Dunn, J., Osterheld, A. L., Shepherd, R., White, W. E., Shlyaptsev, V. N., Stewart, R. E. (**1998**), *Phys. Rev. Lett.* **80**, 2825–2828.
[32] Dunn, J., Li, Y., Osterheld, A. L., Nilsen, J., Hunter, J. R., Shlyaptsev, V. N. (**2000**), *Phys. Rev. Lett.* **84**, 4834–4837.
[33] Klisnick, A., Kuba, J., Ros, D., Smith, R., Jamelot, G., Chenais-Popovics, C., Keenan, R., Topping, S. J., Lewis, C. L. S., Strati, F., Tallents, G. J., Neely, D., Clarke, R., Collier, J., MacPhee, A. G., Bortolotto, F., Nickles, P. V., Janulewicz, K. A. (**2002**), *Phys. Rev. A* **65**, 033810.
[34] Li, Y., Nilsen, J., Dunn, J., Osterheld, A. L., Ryabtsev, A., Sergey, C. (**1998**), *Phys. Rev. A* **58**, R2668–R2671.
[35] Dunn, J., Nilsen, J., Osterheld, A. L., Li, Y. L. Shlyaptsev, V. N. (**1999**), *Opt. Lett.* **24**, 101–103.
[36] King, R. E., Pert, G. J., McCabe, S. P., Simms, P. A., MacPhee, A. G., Lewis, C. L. S., Keenan, R., O'Rourke, R. M. N., Tallents, G. J., Pestehe, S. J., Strati, F., Neely, D., Allott, R. (**2001**), *Phys. Rev. A* **64**, 053810.
[37] Kuba, J., Klisnick, A., Ros, D., Fourcade, P., Jamelot, G., Miquel, J. L., Blanchot, N., Wyart, J. F. (**2000**), *Phys. Rev. A* **62**, 043808.
[38] Nilsen, J., Dunn, J., Osterheld, A. L., Li, Y. L. (**1999**), *Phys. Rev. A* **60**, R2677–R2680.
[39] Li, R., Xu, Z. Z. (**2001**), *J. Phys. IV (France)* **11**(Pr 2), 27–34.
[40] Li, R., Ozaki, T., Kanai, T., Kuroda, H. (**1998**), *Phys. Rev. E* **57**, 7093–7102.
[41] Fiedorowicz, H., Bartnik, A., Dunn, J., Smith, R. F., Hunter, J., Nilsen, J., Osterheld, A. L., Shlyaptsev, V. N. (**2001**), *Opt. Lett.* **26**, 1403–1405.
[42] Lu, P., Kawachi, T., Suzuki, M., Sukegawa, K., Namba, S., Tanaka, M., Hasegawa, N., Nagashima, K., Daido, H., Arisawa, T., Kato, Y., Fiedorowicz, H. (**2002**), *Jpn. J. Appl. Phys.* **41**, L133–L135.
[43] Rocca, J. J., Shlyaptsev, V., Tomasel, F. G., Cortázar, O. D., Hartshorn, D., Chilla, J. L. A. (**1994**), *Phys. Rev. Lett.* **73**, 2192–2195.
[44] Macchietto, C. D., Benware, B. R., Rocca, J. J. (1999), *Opt. Lett.* **24**, 1115–1117.
[45] Liu, Y., Seminario, M., Tomasel, F. G., Chang, C., Rocca, J. J., Attwood, D. T. (**2001**), *Phys. Rev. A.* **63**, 033802.
[46] Nagata, Y., Midorikawa, K., Kubodera, S., Obara, M., Tashiro, H., Toyoda, K. (**1993**), *Phys. Rev. Lett.* **71**, 3774–3777.

[47] Sebban, S., Haroutunian, R., Balcou, Ph., Grillon, G., Rousse, A., Kazamias, S., Marin, T., Rousseau, J. P., Notebaert, L., Pittman, M., Chambaret, J. P., Antonetti, A., Hulin, D., Ros, D., Klisnick, A., Carillon, A., Jaeglé, P., Jamelot, G., Wyart, J. F. (**2001**), *Phys. Rev. Lett.* **86**, 3004–3007.
[48] Nilsen, J., Scofield, J. H., Chandler, E. A. (**1992**), *Appl. Opt.* **31**, 4950.
[49] Porter, J. L., Spielman, R. B., Matzen, M. K., McGuire, E. J., Ruggles, L. E., Vargas, M. F., Apruzese, J. P., Clark, R. W., Davis, J. (**1992**), *Phys. Rev. Lett.* **68**, 796–799.
[50] Apruzese, J. P., Davis, J. (**1985**), *Phys. Rev. A* **31**, 2976–2983.
[51] Nilsen, J., Chandler, E. (**1991**), *Phys. Rev. A* **44**, 4591–4598.
[52] Elton, R. C. (**1975**), *Appl. Opt.* **14**, 2243–2249.
[53] Moon, S. J., Eder, D. C. (**1998**), *Phys. Rev. A* **57**, 1391–1394.
[54] Axelrod, T. S. (**1976**), *Phys. Rev. A* **13**, 376–382.
[55] Kato, Y., Nagashima, A., Nagashima, K., Kado, M., Kawachi, T., Tanaka, M., Hasegawa, N., Sukegawa, G., Nanba, S., Lu, P., Sasaki, A. (**2001**), *J. Phys. IV (France)* **11**(Pr 2), 3–10.
[56] Kim, D., Tóth, C., Barty, C. P. J. (**1999**), *Phys. Rev. A* **59**, 4129–4132.
[57] Feldhaus, J. (**2001**), *J. Phys. IV (France)* **11** (Pr 2), 237–244.

## Further Reading

Elton, R. C. (1990), *X-ray Lasers*. San Diego, CA: Academic Press.

Weber, M. J. (Ed.) (2001), *Handbook of Lasers*. Boca Raton: CRC Press.

The following set of conference proceedings are from a series of international conferences held every 2 years since 1990.

Tallents, G. J. (Ed.) (1991), *IOP Conference Series 116 – X-ray Lasers 1990*. Bristol: IOP Publishing Ltd.

Fill, E. E. (Ed.) (1992), *IOP Conference Series 125 – X-ray Lasers 1992*. Bristol: IOP Publishing Ltd.

Eder, D. C., Matthews, D. L. (Eds.) (1994), *AIP Conference Proceedings 332 – X-ray Lasers 1994*. New York: American Institute of Physics.

Svanberg, S., Wahlström, C.-G. (Eds.) (1996), *IOP Conference Series 151 – X-ray Lasers 1996*. Bristol: IOP Publishing Ltd.

Kato, Y., Takuma, H., Daido, H. (Eds.) (1999), *IOP Conference Series 159 – X-ray Lasers 1998*. Bristol: IOP Publishing Ltd.

Jamelot, G., Moller, C., Klisnick, A. (Eds.) (2001), X-ray Lasers 2000 *J. Phys. IV (France)* **11** (Pr 2).

Rocca, J. J. (Ed.) (2002), *AIP Conference Proceedings 641 – X-ray Lasers 2002*, New York: American Institute of Physics.

# Lasers, Free-electron

*Claudio Pellegrini and Sven Reiche*
*University of California, Los Angeles, CA 90095-1547, USA*
*Phone: +1 310 206 4540; Fax: +1 310 825 8432;*
*e-mail: pellegrini@physics.ucla.edu; reiche@stout.physics.ucla.edu*

**Abstract**
Free-electron lasers are radiation sources, based on the coherent emission of synchrotron radiation of relativistic electrons within an undulator or wiggler. The resonant radiation wavelength depends on the electron beam energy and can be tuned over the entire spectrum from micrometer to X-ray radiation. The emission level of free-electron lasers is several orders of magnitude larger than the emission level of spontaneous synchrotron radiation, because the interaction between the electron beam and the radiation field modulates the beam current with the periodicity of the resonant radiation wavelength. The high brightness and the spectral range of this kind of radiation source allows studying physical and chemical processes on a femtosecond scale with angstrom resolution.

**Keywords**
free-electron laser; undulator; microbunching; SASE FEL; FEL oscillator; FEL amplifier; FEL parameter; gain length.

## 1 Introduction

A free-electron laser (FEL) [1] transforms the kinetic energy of a relativistic electron beam produced by a particle accelerator like a microtron, storage ring, or radio frequency (RF) linear accelerator into electromagnetic (EM) radiation. The transformation occurs when the beam goes through an alternating magnetic field, produced by a magnet called an *undulator* [2], that forces the electrons to move in an oscillatory trajectory about the axis of the system, as shown in Fig. 1. An electromagnetic wave propagates together with the electron beam along the undulator axis, and interacts with the electrons.

The undulator magnet is a periodic structure in which the field alternates between positive and negative values and has zero average value. It produces an electron trajectory having a transverse velocity component perpendicular to the axis and parallel to the electric field of the wave, thus allowing an energy exchange between the two to take place. One can either transfer energy from the beam to the wave, in which case the device is an FEL, or from the wave to the beam, in which case it is an inverse FEL (IFEL) [3]. In the second case, the system is acting like a particle accelerator and can be used to accelerate the electron beam to higher energies.

The energy transfer can take place only if a condition of synchronism between the wave and the beam oscillations is satisfied (Sect. 2.2). This condition gives a relationship between the radiation wavelength $\lambda$, the electron beam velocity $\beta_z$ along the undulator axis (measured in units of the light velocity $c$), and the undulator period $\lambda_0$:

$$\lambda = \frac{\lambda_0(1-\beta_z)}{\beta_z} \tag{1}$$

For relativistic electrons, this condition can also be written in an approximate, but more convenient, form using the beam energy $\gamma$ (measured in the rest energy units $E = mc^2\gamma$):

$$\lambda = \left(\frac{\lambda_0}{2\gamma^2}\right)(1+a_w^2). \tag{2}$$

The dimensionless quantity

$$a_w = \frac{eB_0\lambda_0}{2\pi mc} \tag{3}$$

is the undulator vector potential normalized to the electron rest mass $mc^2$; $e$ is the electron charge and $B_0$ is the undulator peak magnetic field. (Here, and in the rest of the paper, we use MKS units.) The quantity $a_w$ is called the *undulator parameter*,

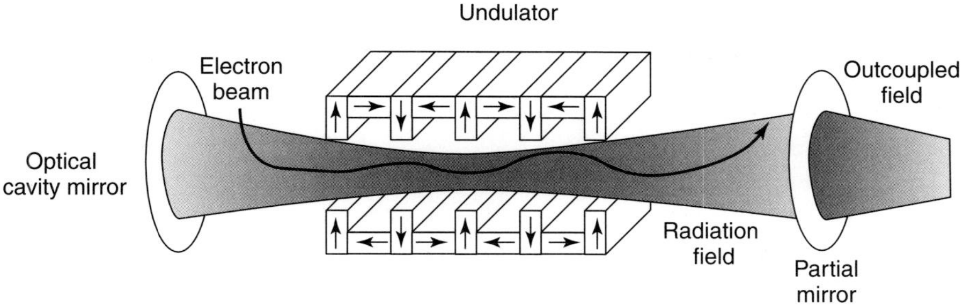


**Fig. 1** Schematic representation of an FEL oscillator, showing its main component: the electron beam, undulator magnet, and optical cavity. The undulator shown is a permanent-magnet planar undulator. The arrows indicate the magnetic field direction. In an FEL amplifier of SASE FEL setup the optical cavity is omitted. The FEL amplifier is seeded by an external radiation field

which is the normalized vector potential of the undulator field. The undulator can be of two types: helical and planar (Sect. 2). Equation (2) is valid for a helical undulator. In the case of a planar undulator, Eq. (2) is still valid if we replace the undulator parameter with its rms value $a_w/\sqrt{2}$.

Because of the dependence of the radiation wavelength on the undulator period, magnetic field, and electron-beam energy – quantities that can be easily and continuously changed – the FEL is a tunable device that can be operated over a very large frequency range. At present, the range extends from the microwave to the UV [4]. A new FEL is now under construction in the United States [5] to reach the X-ray region, about 0.1 nm. A similar program is being developed in Germany [6], and other FEL to cover the intermediate region between 0.1 nm and the UV [7] are also being considered by several countries.

The efficiency of the energy transfer from the beam kinetic energy to the EM wave is between 0.1 to a few percent for most FELs, but it can be quite large, up to about 40%, for specially designed systems. The beam energy not transferred to the EM wave remains in the beam and can be easily taken out of the system, to be disposed of, or recovered elsewhere. This fact suggests that high- to average-power FELs can be designed without the problem, common in atomic and molecular lasers, of heating the lasing medium.

The time structure of the laser beam mirrors that of the electron beam. Depending on the accelerator used, one can design systems that are continuous-wave (cw) or with pulses as short as picoseconds or subpicoseconds.

Tunability, high efficiency, and time structure make the FEL a very attractive source of coherent EM power. In some wavelength regions, like the X-ray, the FEL is unique. Its applications range from purely scientific research in physics, chemistry, and biology to military, medical, and industrial applications.

FELs originate in the work carried out in the 1950s and 1960s on the generation of coherent EM radiation from electron beams in the microwave region [2, 8]. As scientists tried to push power sources to shorter and shorter wavelengths, it became apparent that the efficiency of the microwave tubes, and the power they produced, dropped rapidly in the millimeter region. It was then realized that this problem could be overcome by using an undulator magnet to modify the beam trajectory [1], making it possible for the beam to interact with a wave, away from any metallic boundary. Two pioneering experiments at the Stanford University [9, 10] proved that the FEL is a useful source of coherent radiation.

The current disadvantage of FELs is the greater complexity and cost associated with the use of a particle accelerator. For this reason, the use and development of FELs are mainly oriented to the following:

1. portions of the EM spectrum, like the far infrared (FIR), or the soft and hard X-ray region, where atomic or molecular lasers are not available or are limited in power and tunability;
2. large-average power, high-efficiency system.

An order-of-magnitude comparison of the peak brightness obtained or expected

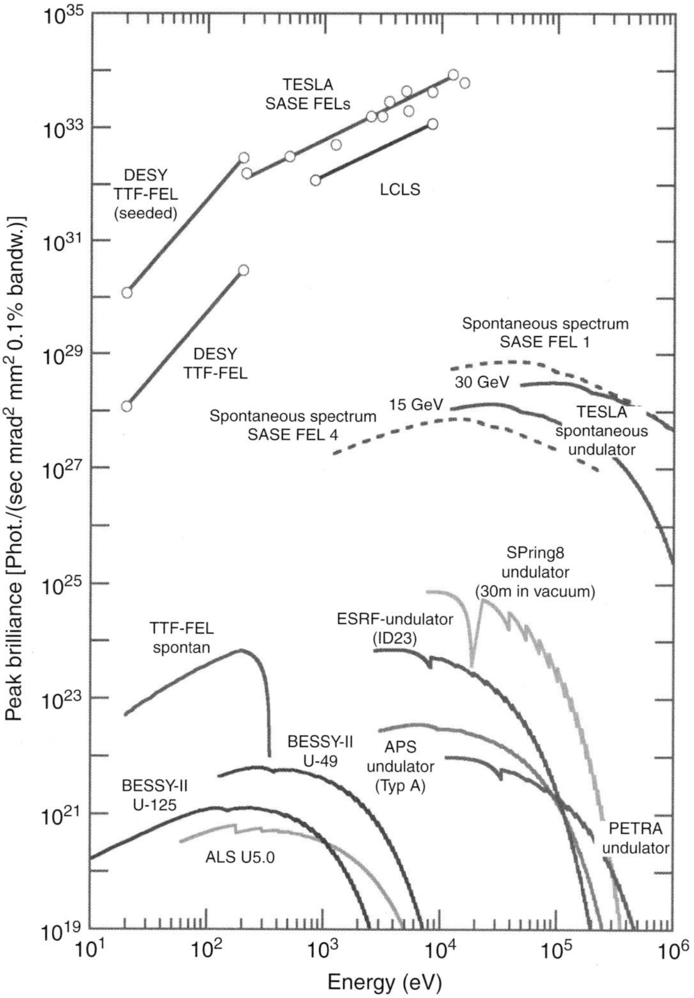


**Fig. 2** Peak brightness of free-electron lasers in the VUV and X-ray regime, compared to 3rd generation light sources [11]

for the FEL, compared to other sources of EM radiation, is given in Fig. 2, pointing out again the interest for FELs in the soft-to-hard X-ray regions.

In Sect. 2 of this article, we discuss the general physical principles of FELs and the properties and characteristics of their main components, that is, the undulator magnets, electron beams, and optical systems. We also describe the main properties of the spontaneous radiation from an undulator, the process leading to the amplification of radiation, the high-gain and small-signal-gain regimes, diffraction effects, saturation effects, the limits on the system efficiency, the dependence of gain on beam parameters.

In Sect. 3, we give a short review of the experimental results obtained to date for various FELs. The future development of FELs is described in Sect. 4.

## 2 Physical and Technical Principles

An FEL has three fundamental components: an electron beam of given energy and intensity and the associated accelerator used to produce it; the undulator magnet; and the EM wave and the associated optical components controlling its propagation. A schematic representation of an FEL is given in Fig. 1.

The accelerators used to provide the electron beam are of many types: electrostatic, induction line, radio-frequency (rf) linear accelerator, pulsed diode, or storage rings. Some of their basic characteristics, their energy ranges, and the FEL wavelengths for which they are more commonly used are given in Table 1.

Undulator magnets are of two main types: helical or planar. In the first case, the magnetic field vector rotates around the axis as a function of axial distance; in the second case, its direction is fixed, and its amplitude oscillates along the axis. These magnets can be, and have been, built using a wide variety of technologies: pulsed or DC electromagnets, permanent magnets, and superconducting magnets. The field amplitude can vary from a fraction of a tesla to over 1 T, and the period from 1 cm to many centimeters. A great effort is under way to develop undulators with periods in the millimeter range; these would allow a reduction of the beam energy for a given radiation wavelength, thus reducing the cost and complexity of the FEL. A helical undulator was used by Madey and coworkers [9] in the first FEL. This undulator was a superconducting magnet, built with two helical windings with current flowing in opposing directions.

The FEL can be operated as an oscillator, using an optical cavity to confine the radiation in the undulator region, as shown in

**Tab. 1** Particle accelerators for FELs

| | *Energy* | *Peak current* | *Pulse length* | *FEL wavelength* |
|---|---|---|---|---|
| Electrostatic | 1–10 MeV | 1–5 A | 1–20 μs | mm to 0.1 mm |
| Induction linear accelerator | 1–50 MeV | 1–10 kA | 10–100 ns | cm to μm |
| Storage ring | 0.1–10 GeV | 1–1000 A | 30 ps–1 ns | 1 μm to nm |
| RF linear accelerator | 0.01–25 GeV | 100–5000 A | 0.1–30 ps | 100 μm to 0.1 nm |

Fig. 1, or it can be used as an amplifier. Oscillator–amplifier combinations [master oscillator, power amplifier (MOPA)], as well as systems in which one amplifies the electron spontaneous radiation emitted in traversing the undulator [self-amplified spontaneous radiation (SASE) [12]], have been used.

The performance characteristics of some of the existing FELs are discussed in detail in Sect. 3.

### 2.1
### Undulator Spontaneous Radiation

For this discussion, we use a reference frame with $z$ along the undulator axis and $x$ and $y$ perpendicular to $z$. For a helical undulator, the field near the axis $z$ is given by

$$\vec{B} = B_0 \left[ \vec{e}_x \cos\left(2\pi \frac{z}{\lambda_0}\right) + \vec{e}_y \sin\left(2\pi \frac{z}{\lambda_0}\right) \right], \qquad (4)$$

where $\vec{e}_x$, $\vec{e}_y$, and $\vec{e}_z$ are unit vectors along the $x$, $y$, and $z$ axes. The electron velocity in this field, in units of $c$, is

$$\vec{\beta} = \frac{a_w}{\gamma} \left[ \vec{e}_x \sin\left(2\pi \frac{z}{\lambda_0}\right) + \vec{e}_y \cos\left(2\pi \frac{z}{\lambda_0}\right) \right] + \vec{e}_z \beta_z. \qquad (5)$$

The amplitude of the transverse velocity is $a_w/\gamma$. This quantity also represents the maximum angle between the electron trajectory and the undulator axis. For relativistic beams, this quantity is much less than 1, and the longitudinal velocity is near 1. Notice that in the case of a helical undulator, the axial component of the velocity, $\beta_z$, remains constant. For a planar undulator, only one of the two transverse components in Eqs. (4) and (5) is nonzero. In this case, the longitudinal velocity is not constant; in fact, it can be written as the sum of a constant term plus one oscillating at twice the undulator period. The electron trajectory in a helical undulator is a helix of radius $a_w \lambda_0 / 2\pi\gamma\beta_z$ and period equal to the undulator period. For a planar undulator, it is a sinusoid in the plane perpendicular to the magnetic field.

When traversing the undulator, the electrons are subject to an acceleration and radiate EM waves. This spontaneous emission is fundamental to the operation of an FEL. One relativistic electron traversing an undulator magnet emits radiation in a narrow cone, with an angular aperture of order $1/\gamma$ [13]. The radiation spectrum is a superposition of the synchrotron radiation emitted when the electron trajectory is bent in the undulator field, a narrow line introduced by the periodic nature of the motion in the same field, and its harmonics. The wavelength of the fundamental line is of the order of $\lambda_0/2\gamma^2$, which can be interpreted as the undulator period $\lambda_0$ doubly Doppler shifted by the electron motion.

The aperture $1/\gamma$ of the synchrotron radiation cone can be compared with the angle between the electron trajectory and the undulator axis, $a_w/\gamma$. If $a_w$ is smaller than 1, the emitted radiation is contained within the synchrotron radiation cone, and the emission is predominantly in a single line. If $a_w$ is larger than 1, the synchrotron radiation cone sweeps an angle larger than its aperture. In this case, the spectrum is rich in harmonics and approaches the synchrotron radiation spectrum when $a_w$ is very large. When $a_w \gg 1$, the magnet is usually referred to as a wiggler, reserving the name undulator for the case $a_w$ less than or on the order of one. In this article, we use the generic term "undulator" to describe a magnet with a periodic transverse field described by Eq. (4).

The fundamental wavelength seen by an observer looking at an angle $\theta$ to the undulator axis is $\lambda = (\lambda_0/2\gamma^2)(1 + a_w^2 + \gamma^2\theta^2)$, and is the wavelength at which there is positive interference between the radiation emitted in two undulator periods. This is also, when $\theta = 0$, the wavelength corresponding to the "synchronism condition" Eq. (1).

In a planar undulator, the magnetic field has only one component, say along $y$. Then the velocity has components in the $x$ and $z$ directions. The wavelength of the spontaneous radiation is the same as that for a helical undulator if the vector potential $a_w$ is replaced by its rms value, $a_w/\sqrt{2}$.

In a helical undulator, the radiation on axis contains only the fundamental harmonic given by Eq. (1). In a planar undulator, the longitudinal velocity is modulated twice at the undulator period. This leads to a richer spectrum: all the odd harmonics of Eq. (1) appear on axis. For both type of undulators, even and odd harmonics appear off axis.

The linewidth of the radiation is determined by the number $N_w$ of undulator periods; the total length of the pulse emitted by one electron is that of a wave train with a number of periods equal to $N_w$, the number of undulator periods. The corresponding linewidth of the fundamental is

$$\frac{\Delta\omega}{\omega} = \frac{1}{2N_w}. \tag{6}$$

When more than one electron is producing the radiation, the pulse length and angular distribution can be affected by the electron-beam characteristic. The pulse length is increased by the electron bunch length when this is larger than $N_w\lambda$. The radiation linewidth is increased if the beam energy spread is larger than $1/2N_w$, the single electron width. Similarly, the effective source radius and angular distribution can be increased by the electron beam. This effect can reduce the FEL gain, as we will discuss in the next section.

## 2.2 The FEL Amplification

Undulators are widely used in storage rings as brilliant sources of synchrotron radiation. In this case, the superposition of spontaneous radiation of many electrons has no phase correlation, so that the total intensity is proportional to the electron number $N_e$. In an FEL, as in an atomic laser, there is a phase correlation between emitting electrons. This correlation is obtained by modulating the longitudinal beam density on the scale of the radiation wavelength, a process called *bunching*. When electrons are bunched together at a distance that is short compared to the wavelength of the radiation, they emit in phase, thus increasing the intensity in the emitted line and simultaneously reducing its width.

The formation of bunching is caused by the interaction between the electrons and the radiation field. It can be summarized by three actions, fundamental to any FEL process:

1. the modulation of the electron energy due to the interaction with the radiation field,
2. the change in the longitudinal position of the electrons owing to the path length difference of the trajectories within the undulator of electrons with different energies,
3. the emission of radiation by the electrons and, thus, growth in the radiation field amplitude.

Each point will be explained in detail below. The FEL amplification couples these basic processes and exploits the inherent instability of this system. A stronger modulation in the electron density increases the emission level of the radiation, which, in return, enhances the energy modulation within the electron bunch. With stronger energy modulation, the formation of the bunching (modulation in the electron positions) becomes even faster. The FEL amplification is stopped, when the electron density modulation has reached a maximum. At this point, all electrons have the same phase of emission and the radiation is fully coherent. The free-electron laser has reached saturation.

The FEL process is either initiated by a seeding radiation field or by the inherent fluctuation in the electron position at the undulator entrance. The first classifies the free-electron laser as an FEL amplifier, while the latter is called a *self-amplifying spontaneous radiation* (SASE) FEL, where the initial radiation level is produced by the spontaneous radiation, as described in Sect. 2.1.

In the following, we describe the basic FEL process. We start with a simplified model, where transverse effects such as the diffraction of the radiation field due to its finite size is ignored. The discussion of these three-dimensional effects is postponed till Sect. 2.5. Let

$$\vec{E} = E_0\left[\vec{e}_x \cos\left(2\pi\frac{z}{\lambda} - \omega t + \Psi\right) + \vec{e}_y \sin\left(2\pi\frac{z}{\lambda} - \omega t + \Psi\right)\right] \quad (7)$$

be the electric field of a plane wave propagating along the undulator axis. The initial phase of the radiation field at $t$, $z = 0$ is $\Psi$. This field is transverse to the undulator axis, and thus has components parallel to the transverse electron velocity, as given by Eq. (5). An energy transfer between the EM field and the electron beam can take place. Using Eqs. (5) and (7), one finds

$$mc^2\frac{\mathrm{d}\gamma}{\mathrm{d}t} = eE_0\frac{a_w}{\gamma}\sin(\Phi + \Psi), \quad (8)$$

where the phase $\Phi$ is

$$\Phi = 2\pi\left(\frac{1}{\lambda} + \frac{1}{\lambda_0}\right)z - \omega t + \Phi_0. \quad (9)$$

This quantity is called the *ponderomotive force phase* [14] and plays a key role in FEL physics. The ponderomotive force phase has terms changing at the "fast" EM field frequency and terms changing with the "slow" undulator period. The time-average value of the electron energy change (8), averaged over a few radiation periods, is 0 unless we satisfy a "synchronism condition" $\mathrm{d}\Phi/\mathrm{d}t = 0$. This condition can be written as $\lambda = \lambda_0(1 - \hat{\beta}_z)/\hat{\beta}_z$ with $\hat{\beta}_z$ as the "resonant" velocity for a given undulator period and radiation field wavelength. In the case of relativistic particles, using the relationship $\hat{\beta}_z = (1 - 1/\gamma^2 - \beta_x^2 - \beta_y^2)^{1/2}$, and assuming $\gamma \gg 1$, $\beta_z, \beta_y \ll 1$, this condition can be approximately written as $\lambda = \lambda_0(1 + a_w^2)/2\gamma^2$. These are Eqs. (1) and (2), given in the introduction. If the synchronism condition is satisfied, the ponderomotive force phase is constant, and the electron beam, oscillating at the slow undulator frequency, can exchange energy with the fast oscillating EM wave.

If the longitudinal velocity $\beta_z$ differs from the resonant velocity $\hat{\beta}_z$, the electron slips in the ponderomotive force phase as

$$\frac{\mathrm{d}\Phi}{\mathrm{d}t} = \frac{2\pi c}{\lambda}\left(\frac{\beta_z}{\hat{\beta}_z} - 1\right). \quad (10)$$

If the electron moves with the resonant velocity, its ponderomotive force phase is constant according to the synchronism condition. Particles with higher energy move forward with respect to the initial phase $\Phi_0$, while particles with lower energy fall back.

The distribution of the initial position of the electrons within the bunch is determined by the electron source and the acceleration process. If the electron beam is produced by an accelerator using RF cavities to accelerate the beam, the distribution is characterized by two scale lengths. One is the RF wavelength; the other is the radiation wavelength. In an FEL, the first is much larger than the second. The wavelengths used in RF cavities vary from a minimum of about 10 cm for an RF linear accelerator to infinity for an electrostatic accelerator. At the scale length defined by the RF system, the electrons are distributed longitudinally into bunches separated by one RF wavelength. The length of these bunches is a fraction of the wavelength of the accelerating RF and varies from about 1 mm for the case of a 3-GHz ($\lambda_{rf} = 10$ cm) RF linear accelerator to a continuous beam for the electrostatic case.

At the scale length of the radiation wavelength, the electron distribution is in good approximation uniform; there is no correlation between the longitudinal electron positions at the scale of the radiation wavelength $\lambda$ except for a residual, but small bunching owing to the finite number of electrons per radiation wavelength. It represents the spontaneous radiation level at the given wavelength $\lambda$, which is amplified in an SASE FEL. Section 2.6 discusses in detail the properties of this device.

If the electrons have the resonant velocity or are close to it, the interaction with the radiation field results in a sinusoidal modulation of the electron energy. Electrons, lying within the phase interval $[\Psi, \ \Psi + \pi]$, gain energy and become faster ($\mathrm{d}\Phi/\mathrm{d}t > 0$), while electrons in the interval $[\Psi + \pi, \Psi + 2\pi]$ are slowed down ($\mathrm{d}\Phi/\mathrm{d}t < 0$). As a result, all electrons tend to drift toward the phase $\Psi + \pi$ and become bunched with the periodicity of the radiation field. The electrons emit in phase and the radiation is coherent.

The radiation emitted by the electrons adds up to the interacting EM wave. The amplitude $E_0$ and the phase $\Psi$ of the EM field change with the ongoing interaction between radiation field and electron beam. This is apparent because the electrons tend to bunch at a phase different to the radiation field. The radiation phase is slowly adapting to the new phase defined by the bunching phase of the electrons. For a self-consistent description of the FEL interaction, the Maxwell equation for the EM wave has to be solved as well. If the fast oscillation of the EM wave $\exp[i(2\pi/\lambda)z - i\omega t]$ is split from the amplitude and phase information, the complex amplitude $E \equiv E_0 \exp[i\Psi]$ changes slower than the "slow", transverse oscillation of the electrons (Eq. 5). The change is given by [15]:

$$\left(\frac{\partial}{c\partial t} + \frac{\partial}{\partial z}\right) E = i\frac{\mu_0}{2} a_w \sum_j \frac{\mathrm{e}^{-i\Phi_j}}{\gamma_j} \qquad (11)$$

with $\mu_0$ as the magnetic permeability and the sum over all electrons. Equations (8), (10), and (11) are the mathematical representation of the three basic processes, listed at the beginning of this chapter. Note that the right-hand side of Eq. (11) is determined by the degree of bunching in the electron density and is at maximum when all electron phases $\Phi_j$ are identical. It is also clear from the same equation that if more electrons are contributing to the FEL process, the growth in the radiation field is stronger. The strength of this coupling scales with the FEL parameter [16]

$$\rho = \left( \frac{a_w}{4\gamma} \frac{\omega_p}{\omega_0} \right)^{\frac{2}{3}}, \qquad (12)$$

where $\omega_p = (4\pi r_e c^2 n_0/\gamma)^{1/2}$ is the beam plasma frequency, $n_0$ is the electron-beam density, and $\omega_0 = 2\pi c/\lambda_0$. Typical values of $\rho$ vary from $10^{-2}$ for a millimeter FEL to $10^{-3}$ or less for a visible or UV FEL.

The solutions for Eqs. (9), (10), and (11) are of type $E \propto \exp[i\Lambda 2\omega_0 \rho t]$, where the growth rate $\Lambda$ depends on the electron distribution in energy and the deviation of the mean energy from the resonance energy $\gamma_R = \sqrt{(\lambda_0/2\lambda) \cdot (1 + a_w^2)}$. For the simplest case of a mono-energetic beam with $\langle\gamma\rangle = \gamma_R$, $\Lambda$ is the solution of the dispersion equation $\Lambda^3 = -1$ [17]. Although energy deviation or finite energy spread adds additional terms, the resulting dispersion equation is always cubic for this one-dimensional model.

Of particular interest for the FEL process is the solution of the dispersion equation, which has a negative imaginary part. It corresponds to an exponentially growing mode. The maximum growth rate in this one-dimensional model is $-\Re e(i\Lambda) = \sqrt{3}/2$. Inserted into the ansatz for the radiation field amplitude $E$, the e-folding length of the amplification is

$$L_g = \frac{\lambda_0}{2\pi\sqrt{3}\rho}, \qquad (13)$$

which is also referred to as the gain length [18]. Besides the exponentially growing mode, the other modes are either oscillating or decaying. In the beginning of the free-electron laser, all modes are of the same magnitude and it requires a few gain lengths before the exponentially growing mode dominates. This retardation in the amplification of the seeding field is called *lethargy regime* [16].

The growth in the radiation field amplitude stops at saturation, when the modulation in the electron distribution has reached a maximum. At this point, the bunch is broken up into multiple microbunches. The spacing of the microbunches is the radiation wavelength. Because the wavelength is tunable with the electron beam energy, the free-electron laser is also a device to modulate the current of a long pulse with free control on the periodicity. This property is useful for some experiments on plasma-beam interaction [19].

### 2.3
### The Small-signal Gain Regime

The first FEL experiments had an enhancement of a few percent of the seeding radiation field. In this low-gain regime, the undulator length is shorter than the gain length. All modes – growing, decaying, and oscillating – have approximately the same amplitude and the radiation power at the end of undulator as a result of the interference of these modes. If the wavelength of the seeding radiation field fulfills the synchronism condition, the interference is completely destructive and the gain – the relative change in the radiation field intensity – is zero. The growth rate of each mode depends differently on the deviation from the synchronism condition (Eq. 1) and a slight deviation $\Delta\omega$ results in an enhancement of the EM field.

The small-signal gain was first evaluated by Madey [1]. Although his original theoretical discussion of the FEL used a quantum-mechanical description, it was soon realized that under most conditions, a classical description is sufficient to describe the system, since the most relevant quantities, such as gain, Eq. (14), are independent of the Planck constant.

When the initial EM field amplitude is small – and the relative change in the intensity is also small – the gain is called the *small-signal gain*. For an initially monoenergetic electron beam, the small-signal gain, calculated by Madey, is

$$G_s = 4\sqrt{2}\pi \frac{\sqrt{\lambda\lambda_0}}{w_0^2} \frac{a_w^2}{\sqrt{(1+a_w^2)^3}} \frac{I_p}{I_A} N_w^3 \times \frac{d}{d(x/2)} \frac{\sin^2(x/2)}{(x/2)^2}, \qquad (14)$$

where $x = 2\pi N_w \Delta\omega/\omega$, $N_w$ is the number of undulator periods, $I_p$ is the beam current, $I_A = ec/r_e$ is the Alfvén current (about 17 kA), $r_e$ is the classical electron radius ($r_e \approx 2.8 \times 10^{-15}$ m), and $\pi w_0^2$ is the transverse cross section of the EM wave (assumed to be larger than that of the electron beam).

The gain curve in Eq. (14) is proportional to the derivative of the spectrum of the spontaneous radiation. The gain is zero for $x = 0$, at synchronism, and is maximum for $x \simeq 2.6$. The linewidth is on the order of $1/N_w$, the inverse of the number of periods in the undulator.

Since the EM radiation frequency and the electron energy are related by the synchronism condition, Eq. (2), we can also redefine $x$, for a fixed frequency, as $x = 4\pi N_w \Delta\gamma/\gamma$. The gain linewidth then corresponds to an energy change $\Delta\gamma/\gamma$ on the order of $1/2N_w$. When the radiation field increases, the beam kinetic energy decreases; and when the change is about $1/2N_w$, the gain becomes 0. Hence the maximum FEL efficiency, defined as ratio of the laser intensity to the initial beam kinetic energy, is on the order of $1/2N_w$.

### 2.4 High-gain Regime and Electron Beam Requirement

The FEL saturates when the electron bunch has achieved maximum modulation and all electrons radiate coherently. Early free-electron lasers could not reach saturation with a single pass through an undulator. The gain can be accumulated when the undulator is enclosed by an optical cavity and the radiation field is reflected back to seed the FEL for succeeding electron bunches. On the other hand, optical cavities restrict the FEL to radiation wavelength where the losses of the cavity are smaller than the gain per single pass. To avoid this restriction, the undulator length has to be several gain lengths long to obtain a large gain and to reach saturation [18].

To reduce the overall undulator length, the gain length (Eq. 13) must be made as short as possible. The undulator period length is typically limited to a few centimeters and found by optimization of the gain length. Shorter values would reduce the undulator parameter and, thus, the coupling of the electron beam to the radiation field, while longer values would increase the resonant wavelength (Eq. 2). The energy has to be increased to keep the radiation wavelength constant, which reduces the $\rho$-parameter.

The gain length depends on the electron beam parameters as well. The gain length is reduced for a higher electron density, although there are limits on this, given by the transverse velocity spread in the electron beam. We discuss this and other three-dimensional effects in the next section.

The gain length of Eq. (13) is based on a beam with no energy spread. Energy spread prevents bunching of all electrons at the same ponderomotive force phase, because the electrons have different longitudinal velocities and the bunching is smeared out. The bandwidth in energy, which contributes to the FEL performance, is given by the $\rho$-parameter. A larger value means that the energy exchange between radiation field and electron beam is stronger. The undulator length, over which the synchronism condition must be fulfilled, is shorter. This allows a larger tolerance in the spread of the longitudinal velocity and thus, in energy. To keep the increase in the gain length reasonable, the initial rms energy spread $\sigma_\gamma$ must be

$$\frac{\sigma_\gamma}{\gamma_R} \ll \rho. \tag{15}$$

Following the same argument, the deviation of the mean energy from the resonant energy is limited to

$$\left|\frac{\langle\gamma\rangle - \gamma_R}{\gamma_R}\right| \ll \rho. \tag{16}$$

Note that in the high-gain regime, the gain is the largest if the beam energy is on resonance, as shown in Fig. 3. This is opposite to the small-signal gain regime with no gain at $\langle\gamma\rangle = \gamma_R$.

Only electrons that lie within the energy acceptance bandwidth of the FEL around $\gamma_R$ contribute to the FEL interaction. The relative width is approximately $\rho$. During the FEL amplification, energy is transferred from the electron beam to the radiation field. The amplification stops when most of the electrons are dropped out of the bandwidth. After that the interaction with the radiation field is negligible because the synchronism condition is no longer satisfied. Thus, the efficiency of the FEL, which is the relative amount of energy transferred from the electron beam to the radiation field, is $\rho$. The saturation power of the radiation field [20] is given by

$$P_{sat} = \rho P_b, \tag{17}$$

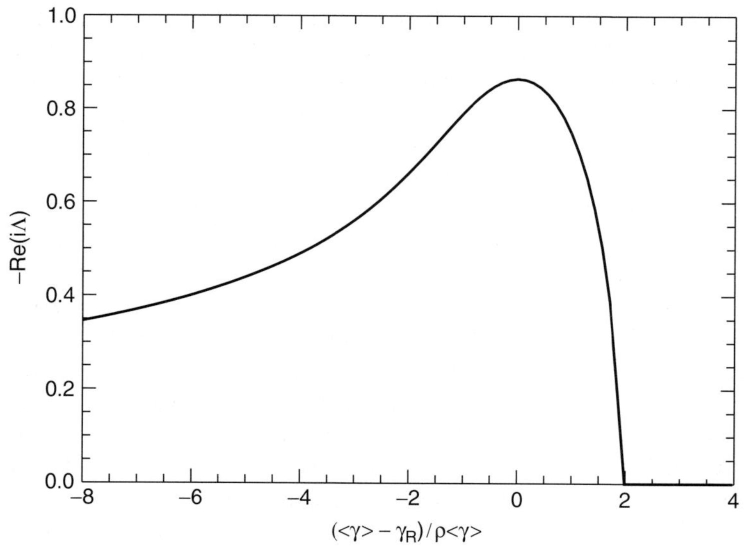


**Fig. 3** Growth rate of the FEL amplification as a function of the detuning–the deviation of the mean energy of the electron beam from the resonant energy

where $P_b = \langle\gamma\rangle I_p mc^2/e$ is the electron beam power and $I_p$ is the peak current. Because of this scaling, it is of interest to maximize the $\rho$-parameter not only for reducing the needed undulator length for saturation but also for increasing the output power level of the FEL.

Because the FEL amplification is an exponential process, the major change in the electron energy occurs close to saturation, when the electrons are pushed out of the FEL bandwidth. Although the FEL amplification stops at saturation, further energy can be extracted from the electron beam. For this, the synchronism condition has to be extended beyond saturation. The radiation wavelength is fixed, so the undulator field or period has to be reduced to compensate for the beam energy, which has been lost due to the FEL interaction. This process to gradually reduce the undulator field or period is called *tapering*. Although the energy transfer during tapering is slow compared to the exponential regime of the FEL, the process has no fundamental limits beside the given length of the undulator. Experiments have shown that a tapered system can enhance the efficiency up to 40% [21].

## 2.5 Three-dimensional Effects

The basic FEL process, as described in Sect. 2.2, remains the same when three-dimensional effects are included in the discussion, although the one-dimensional FEL parameter $\rho$ (Eq. 12) might not be valid anymore. However, an effective three-dimensional FEL parameter can be found, which satisfies Eqs. (13), and (15) to (17).

Two major effects contribute to the extended, three-dimensional model of the FEL process, which are

1. the spread in the transverse velocity of the electron beam and
2. the diffraction of the radiation field.

A measure for the transverse spread in the electron beam is the normalized emittance $\varepsilon_n$ [22], which is the area covered by the electron beam distribution in transverse phase space. It is an invariant in linear beam optics. If the beam is more focused, the spread in the transverse momenta is increased. Focusing along the undulator is necessary to prevent the growth of the transverse electron beam size, reducing the electron density and decreasing the gain. Electrons that drift away from the axis are deflected back to the undulator axis. As a result, electrons perform an additional oscillation even slower than the "slow" oscillation due to the periodic magnetic field of the undulator (Eq. 5). Because part of the electron energy goes into this so-called betatron oscillation, the electron is slower in the longitudinal direction than an electron without a betatron oscillation. The spread in the betatron oscillation, which scales with the normalized emittance, has a similar impact as an energy spread. Thus, the requirement for the transverse emittance is

$$\varepsilon_n \ll \frac{4\lambda\beta\langle\gamma\rangle}{\lambda_0}\rho, \qquad (18)$$

where the beta-function $\beta$ [22] is a measure for the beam size of the electron beam. Stronger focusing would increase the electron density ($\beta$ becomes smaller) and consequently, the $\rho$ parameter. But at the same time, the impact of the emittance effect is enhanced, reducing the amount of electrons that can stay in phase with the radiation field. It requires the optimization of the focusing strength to obtain the shortest possible gain length. A rough estimate for the optimum $\beta$-function is $\beta \approx L_g \approx \lambda_0/4\pi\rho$. Inserting this into Eq. (18) shows that the beam emittance $\varepsilon = \varepsilon_n/\gamma$ has to be smaller than the photon beam "emittance" $\lambda/4\pi$. If this condition is fulfilled, the electron beam does not diverge faster than the radiation field and all electrons stay within the radiation field.

The diffraction of the radiation field transports the phase information of the radiation field and the bunching of the electron beam in the transverse direction. This is essential to achieve transverse coherence of the radiation field, in particular, if the FEL starts from the spontaneous radiation (SASE FEL). SASE FELs do not guarantee the transverse coherence as a seeding laser signal of an FEL amplifier would do. The betatron oscillation, which yields a change in the transverse position of the electron, contributes to the growth of transverse coherence as well.

As a degrading effect, the radiation field escapes from the electron beam in the transverse direction. The field intensity at the location of the electron beam is reduced and the FEL amplification is inhibited. The compensation for field losses due to diffraction is the FEL amplification itself. After the lethargy regime, the FEL achieves equilibrium between diffraction and amplification. The transverse radiation profile becomes constant and the amplitude grows exponentially. This "quasi-focusing" of the radiation beam is called *gain guiding* [23] and the constant profile of the radiation field is an eigenmode of the FEL amplification.

Similar to the eigenmodes of the free-space propagation of a radiation field (e.g., Gauss–Hermite modes), there are an infinite number of FEL eigenmodes [24]. Each couples differently to the electron beam and thus has a different growth rate or gain length. That mode, which has the largest growth rate, dominates after several gain lengths and the radiation field becomes transversely coherent.

At saturation, gain guiding vanishes and the electron beams radiate into multiple modes. Typically, the fundamental FEL-eigenmode is similar to that of free-space propagation and the resulting reduction in transverse coherence at saturation is negligible.

The characteristic measures for diffraction and FEL amplification are the Rayleigh length $z_R$ and gain length $L_g$. To calculate $z_R$, we approximate the radiation size at its waist by the transverse electron beam size $\Sigma$ as the effective size of the radiation source, resulting in $z_R = k\Sigma/2$. For $z_R \ll L_g$, the FEL amplification is diffraction-limited with a gain length significantly larger than that in the one-dimensional model (Eq. 13). In the opposite limit ($z_R \gg L_g$), the one-dimensional model becomes valid.

### 2.6
### Longitudinal Effects, Starting from Noise

The radiation field propagates faster than the electron beam and advances one radiation wavelength per undulator period. Thus, information cannot propagate further than the slippage length $N_w\lambda$, where $N_w$ is the total number of undulator periods. If the bunch is longer than the slippage length, parts of the bunch amplify the radiation independently. While an FEL amplifier is seeded by a longitudinal coherent signal, this is not the case for an FEL, starting from the spontaneous emission (SASE FEL). SASE FELs with electron bunches longer than the slippage length generate longitudinally incoherent signals. Further processing, succeeding the FEL, is required for fully longitudinal coherence (e.g., it can be achieved with a monochromator with a bandwidth smaller than the Fourier limit of the electron bunch length).

An FEL amplifier has a coherent seeding signal and the electron beam can be more or less in resonance with it. The output power level depends on the deviation of the mean energy of the electron beam from the resonant energy $\gamma_R$. An SASE FEL, on the other hand, uses the broadband signal of the spontaneous emission to start the FEL amplification. Independent of the beam energy, the resonance condition is always fulfilled, but the SASE FEL amplifies all frequency components within the acceptance bandwidth of the FEL as well. The relative width is $\Delta\omega/\omega = \rho$ and is typically much larger than the observed width of an FEL amplifier.

The initial emission level of the spontaneous radiation depends on the fluctuation in the electron density. Because there are only a finite number of electrons per radiation wavelength and the initial ponderomotive force phase is random, electrons can cluster together at random. The radiation from parts of the bunch with a larger clustering will have larger intensity than that of other parts. The variation in these clusters and, thus, the fluctuation in the beam current, depend on the total number of electrons. Because of the random nature of the electron positions, more electrons result in a smoother distribution. The chances for large clusters are reduced.

The effective power level, emitted because of spontaneous radiation and then further amplified by the FEL interaction, is called the *shot noise power* [25]. An FEL amplifier, seeded with a power signal smaller than the shot noise power level does not operate as an FEL amplifier but as an SASE FEL instead. Typical values are a few watts for free-electron lasers in the IR regime to a few kilowatts for FELs in the VUV and X-ray regime.

When the electron bunch length is longer than the slippage length, the radiation profile contains many spikes. The spikes are also present in the frequency spectrum. The origin is the random fluctuation in the beam density. The shot-to-shot fluctuation in the radiation pulse energy follows a Gamma distribution [26]. The only free parameter of this distribution, $M$, can be interpreted as the number of the spikes in the radiation pulse. The relative width of the distribution is $1/\sqrt{M}$. The length of the spikes is approximately $(\lambda/\lambda_0)L_g$ [27]. A shorter pulse would result in a larger fluctuation of the radiation energy. The fluctuation of the instantaneous power is given by a negative exponential distribution.

From the optical point of view, the electron beam acts as a dielectric, dispersive medium for the radiation field. The field amplitude $E$ evolves as $\exp[i\Lambda 2\omega_0 \rho t]$. The imaginary part of $\Lambda$ results in the growth of the radiation field and thus, gain guiding, while the real part adds a phase advance to the fast oscillating wave $\exp[ikz - i\omega t]$. The effective phase velocity differs from $c$, the speed of light. The electron beam is dielectric and the radiation field is focused due to the same principle of fiber optic cables. In addition, $\Lambda$ depends on the deviation of the radiation frequency from the frequency that fulfills the synchronism condition in Eq. (1). This dispersion reduces the group velocity below the speed of light [28] and spikes advance less than one radiation wavelength per undulator period. Amplification stops at saturation and the "dielectric" electron beam becomes nondispersive.

## 2.7 Storage Ring–based FEL Oscillators

Free-electron lasers are mainly based upon linear accelerator and storage rings. Electron beams, provided by linear accelerators, have the beam quality needed to reach saturation in a single pass. Because this setup exhibits a strong amplification, the single-pass FEL has properties that are a disadvantage in comparison to quantum lasers or storage ring–based FEL oscillators. Radiation of a single-pass FEL suffers from the inherent fluctuation of the beam parameters such as energy and position jitter. In addition, an SASE FEL does not necessarily provide longitudinal coherence and has a spectral bandwidth, which is large compared to standard lasers. The average power is limited by the repetition rate of a linear accelerator, small compared to that of a storage ring. The single-pass high-gain FEL is the only solution for a high-brightness radiation source for wavelength regions, where an FEL oscillator system will not work owing to the losses in the optical cavity.

The limitations for storage ring–based FEL are given by the instabilities that occur in a storage ring. These instabilities yield an upper limit on the electron beam current of typically a few hundred amperes [29]. In addition, the energy spread and normalized emittance are large compared to those in a linear accelerator [30].

The FEL oscillator accumulates the single-pass gain of typically a few percent. As long as the amplitude of the EM field is small, the gain per pass is constant (small-signal gain) and the field grows exponentially with the number of passes. The oscillator reaches saturation when the FEL interaction is strong enough to induce complete bunching within a single pass. Because the electron emission is fully coherent at saturation, the field cannot be further amplified. The oscillator reaches an equilibrium between the losses of the optical cavity and the reduced gain is closed to saturation.

When the FEL oscillator has reached saturation, the power level within the optical cavity is small compared to that of a single-pass FEL because the effective $\rho$ parameter and beam power are lower (Eq. 17). In addition, only a fraction can be coupled out of the cavity for succeeding experiment because the out-coupling contributes to the net loss of the optical cavity. Higher, out-coupled radiation power requires a higher gain to compensate for the losses. In the equilibrium state, the FEL oscillator's temporal and frequency stability, longitudinal coherence, and repetition rate are better than those of a linear accelerator–based FEL. However, the official quality of linear accelerator–based FELs can be improved using optical instruments, like monochromators.

The radiation of an FEL oscillator is typically longitudinal coherent, because the phase information of the radiation field is distributed over the entire electron bunch mainly due to two reasons. The first reason is the same as for a single-pass FEL, because the radiation field propagates faster than the electron beam. The second reason depends on the design of the optical cavity. If the cavity is "detuned", the time of the radiation pulse to pass the cavity is different from the arrival time of the next electron bunch. The radiation field has a longitudinal offset with respect to the electron bunch. This offset allows a faster buildup of coherence.

The FEL oscillator can start from the spontaneous radiation. If the gain is larger than a few percent and the detuning is small, the random nature of the spontaneous emission phase can manifest in the occurrence of superradiant spikes [31]. The peak power can be considerably larger than the power level of the fully coherent signal, while the spike length is shorter than the electron bunch length. The nature of these superradiant spikes is the same as those occurring in the temporal radiation profile of an SASE FEL.

## 3 Present Status

The current status of the FEL research program around the world can be classified into two groups with different objectives. The single-pass free-electron laser is aiming toward shorter wavelength, while storage ring–based FELs are developed to overcome the limitation of the losses of the optical cavities. Free-electron lasers, which have finished their research program, are often converted to user facilities, where researchers can use the unique properties of the FEL radiation for research in their field.

### 3.1 Single-pass Free-electron Lasers

Single-pass free-electron lasers are frequently based on linear accelerators. Because the FEL does not rely on any optical cavity to accumulate the gain, there is no restriction on the wavelength. A beam energy of about 10 MeV corresponds to a resonant wavelength of about 10 microns, while a 10- to 25-GeV beam can drive an FEL in the VUV and X-ray region. The energy of the electron beam is determined by the length of the linear accelerator.

The requirements on the beam quality (Sect. 2.4 & 2.5) are more stringent for FELs operating at shorter wavelength, and the first single-pass FELs started at wavelengths in the millimeter or infrared regime [32]. The major motivation of the ongoing FEL research is to shorten the wavelength down to the angstrom level, where no alternative radiation sources with the same radiation properties exist. The total number of single-pass FEL is numerous and Table 2 lists a selection, namely, those with the shortest wavelength. The record for the shortest wavelength so far lies at around 80 nm [33]. The majority of future

**Tab. 2** Linear accelerator–based free-electron lasers

| *FEL* | *UCLA* | *VISA* | *LEUTL* | *TTF-FEL I* |
|---|---|---|---|---|
| Energy [MeV] | 18 | 71 | 217/255 | 250 |
| Energy spread [%] | 0.25 | 0.1 | 0.4/0.1 | 0.06 |
| Norm. emittance [mm·mrad] | 5.3 | 2.3 | 8.5/7.1 | 6 |
| Peak current [A] | 170 | 250 | 630/180 | 1300 |
| Wavelength [nm] | 12 000 | 840 | 530/385 | 95 |
| Peak power [MW] | 0.0001 | 15 | 500 | 1000 |
| Pulse length [ps, RMS] | 5.5 | 0.3 | 0.2/0.7 | 0.02 |
| Gain length [cm] | 25 | 18.7 | 97/76 | 67 |

FELs will operate at shorter wavelengths (Sect. 4).

The FEL performance is limited by the electron beam quality, the transverse emittance, the energy spread, and the beam current. The linear accelerators of all the latest single-pass free-electron lasers are based on RF photoelectron guns [34]. A conventional laser in the visible or UV illuminates a cathode, embedded into an RF cavity. The photo effect generates the electrons and the RF field accelerates them to minimize the beam "blowup" due to the electric field of each electron (space charge field). The value of the transverse emittance arises mainly from this space charge field at the cathode region [35]. At higher energy, the electron beam is relativistic and is less sensitive to space charge effects.

Because a more intense pulse of the photo gun laser produces a higher electron density, the space charge field is stronger. As a consequence, there is a limit on the achievable emittance and current from the RF photoelectron gun [36]. Increasing the current would increase the emittance as well. Typical values are an emittance of around 2 mm·mrad at a current of about 50 A. The total charge of the electron bunch is on the order of 1 nC. Single-pass free-electron lasers typically require beam currents of 100 A and above to keep the saturation length within reasonable limits. This is achieved by including bunch compressors in the beam line of the linear accelerator.

A bunch compressor consists of bending magnets, which deflect the beam off the axis of the linear accelerator and then bend it back to the axis. Because the bend angle depends on the electron energy, electrons with a lower energy have a stronger distortion and, thus, a longer path length. The electron bunch is given a correlated energy variation along its length by the RF structure, so that electrons at the head of the bunch have lower energy than those in the tail. The path-length difference in the bunch compressor reduces the bunch length and increases the beam current. The beam energy correlation is obtained when the RF field is still building up ("off-crest" injection). Electrons entering first see a lower accelerating gradient than those entering later. At the cavity exit, the bunch has an energy correlation along bunch, needed to compress the bunch in the bunch compressor.

### 3.2 Free-electron Laser Oscillators

Storage rings impose a limit on the achievable electron beam quality. Because bending magnets are required to transport the electron bunches back to the undulator entrance, electrons emit spontaneous radiation. When an electron looses energy owing to the spontaneous radiation, its orbit shifts with respect to the design orbit. The energy losses are not the same for every electron due to quantum fluctuations of the emitted photons. This results in a limitation of the achievable transverse emittance [37]. To increase the performance of a storage ring–based FEL oscillator, this limit has to be reduced, which often requires a special lattice of bending and focusing magnets. Bends with more, but shorter, bending magnets alternating with quadrupole magnets, perform better than those with less magnets. This approach is not only used for storage ring–based FEL but also for most third-generation light sources [38].

Since electron bunches are kept in a storage ring for minutes or hours, even small effects can be accumulated and can degrade the beam quality. One effect is intrabeam scattering, in which, due to Coulomb collision electrons are gradually kicked out of the orbit and lost in the storage ring aperture [39]. Another effect is the "communication" between bunches by means of wakefields [40]. Each storage ring has RF cavities to compensate energy losses, which occur because of spontaneous radiation and the FEL interaction. The modes of the cavities are excited whenever an electron bunch passes the cavities. These higher modes are not damped quickly enough, so that the next bunch interacts with them. The resulting energy modulation is not constant over the entire bunch and dispersion – energy dependence of the orbit – tears the bunch apart, spoiling the beam quality or even dumping part of the beam in the vacuum chamber wall.

Besides an advanced lattice design of the storage ring to minimize the effects described above, storage ring limitation can be diminished by improving the optical cavity of the FEL oscillator. The current research on oscillators is focused to improve the reflecting mirrors in two areas such as

1. the extension of the wavelength range in the VUV range by novel coating or material,
2. increasing average power by improved power handling.

The shortest wavelength an FEL oscillator is operating is around 100 nm [41] with significant power emitted at higher harmonics as well. The FEL oscillator at Jefferson Lab achieved an average power in the IR region above 2 kW [42]. In slight contrast to the single-pass FEL, which are mostly FEL experiments to extend the wavelength range, storage ring–based FELs mostly serve as a user facility, where the radiation wavelength and power is tuned to the needs of experiments from all fields of science (see Table 3).

The limitations in beam current and bunch length can be overcome in an Energy Recovery Linear accelerator (ERL) [43], where the electron bunches are only kept for a few turns in the storage ring. The beam energy is extracted back to the accelerating field prior to the ejection and dumping of the bunch. Since the peak current in an ERL is higher than that in storage ring, the brightness and the saturation power is increased.

**Tab. 3** Free-electron laser oscillators

| *FEL* | *Duke*[a] | *Elletra*[a] | *Clio*[b] | *Felix*[c] | *Jefferson Lab*[d] |
|---|---|---|---|---|---|
| Energy [MeV] | 200–1200 | 1000 | 15–50 | 15–50 | 48 |
| Avg. current [mA] | 115 | 25 | 2–40 | 2.5–100 | 4.8 |
| Bunch length [ps] | <50 | 6.3 | 0.5–6 | 0.8 | 0.4 |
| Rep. rate [MHz] | 13.95 | 4.63 | 62 | 25/1000 | 74.8 |
| Wavelength [μm] | 0.19–0.4 | 0.19–0.20 | 3–90 | 4.5–250 | 3–6.2 |
| Avg. power [W] | 0.025 | 0.01 | 3 | 25–5000 | 2000 |

[a] Storage ring.
[b] 10 μs electron bunch train from thermionic gun at 25 Hz.
[c] 10 μs electron bunch train from thermionic gun at 10 Hz.
[d] Energy recovery linear accelerator.

## 4 Future Development

All recent single-pass free-electron laser experiments have not shown any fundamental limits that would prevent a resonant wavelength at the angstrom level. The typical beam energy would be around 10 to 25 GeV. The required undulator length would be around 100 m. Several projects have been proposed or are under construction to have single-pass FELs operating in the wavelength regime between 1 and 1000 Å [5–7]. Table 4 lists a selection of them. All these FELs are planned to be user facilities. Most of them have several beam lines and undulators to serve multiple users simultaneously. To reduce the limitation on the average brightness, most of the linear accelerators are based on superconducting technology. It allows to lengthen the RF pulse and fill the beam line with multiple bunches per RF pulse, while operating on a lower power level of the RF system.

A combination of a storage ring with a linear accelerator is the Energy Recovery Linear accelerator (ERL) [43]. After the electron bunch is accelerated and passes the undulator, it is transported back to the beginning of the beam line. It propagates to the accelerating cavities a second time, but with a 180° difference in the RF phase. The residual kinetic energy of the bunch is transferred back to the RF field, and thus is available for succeeding bunches. It requires superconducting cavities; otherwise the stored energy will be lost by thermal losses before the next bunch enters the beam line. Using this setup, the repetition rate of the electron bunches are not limited by the RF system and FELs can be operated in cw mode, similar to storage ring but with the advantage that there are no limitations on the beam current or emittance.

The research in FEL physics is shifting from the proof of principle toward enhancement of the radiation properties, in particular, in the X-ray regime. SASE FELs have an intrinsic instability in the output energy of a few percent up to 100%. In addition, the operation of a linear accelerator is not as stable as that of a storage ring. The system fluctuation adds up to the intrinsic fluctuation of the SASE pulse. Another disadvantage is the reduced longitudinal coherence. Because of the short cooperation length of FELs in the VUV and X-ray regime, part of the bunch amplifies the spontaneous radiation independently. The longitudinal coherence can be achieved by a pulse stretcher, namely, monochromator, with a bandwidth slightly smaller than the Fourier limit of the electron bunch length. All spikes are stretched to the bunch length and add up coherently. Each spike in the SASE pulse has 100% fluctuation, so has the resulting coherent pulse after the monochromator. Stability of the output power is achieved if the monochromator is not placed after the SASE FEL but between, splitting the undulator in two parts [44]. The second undulator is seeded by the output signal of the monochromator and operates as an FEL amplifier

**Tab. 4** VUV & X-ray free-electron lasers, planned or under construction (*)

| *FEL* | *Wavelength [Å]* | *Location* |
|---|---|---|
| TESLA* | 1–50 | Germany |
| SCSS | 1–10 | Japan |
| LCLS* | 1.5–15 | USA |
| MIT Bates FEL | 5–1000 | USA |
| FERMI | 12–100/400 | Italy |
| BESSY | 12–1500 | Germany |
| SPARX | 15–140 | Italy |
| TTF-FEL II* | 60–1200 | Germany |
| 4GLS | 120–4000 | UK |
| NSLS DUV-FEL | 200–2000 | USA |

instead of an SASE FEL. Because the saturation power of an FEL amplifier does not depend on the input signal, the output power level in the deep saturation regime is stable despite the 100% fluctuation of the input signal.

Another focus is on short radiation pulses. Standard methods are pulse compression and slicing, similar to conventional lasers. It requires an energy correlation along the electron bunch ("chirp"), driving an SASE FEL. Because of the square dependence of the resonant wavelength on the energy (Eq. 1), the radiation frequency chirp is twice as large. A monochromator selects only that part of the pulse in which frequency components fall within the bandwidth. The monochromator bandwidth has to be optimized in order to avoid a pulse stretching for a too-small bandwidth or large chirps for a too-wide bandwidth. For X-ray FELs, a monochromator bandwidth of $10^{-4}$, which is roughly a tenth of the intrinsic FEL bandwidth, and a frequency chirp of around 1%, results in a 10-fs slice [45].

An alternative method for short bunches is to condition the driving electron bunch. A stronger compression spoils the emittance and energy spread by wakefields and coherent synchrotron radiation in the linear accelerator and bunch compressor, so that the electron pulse length is limited to roughly 200 fs FWHM for the shortest gain length. On top of that, the beam can be conditioned by various methods (RF cavities, wakefields, external laser pulses) to keep the beam quality good over only a short subsection of the bunch. Although the entire bunch emits spontaneous radiation, only the preserved part of the bunch amplifies it and reaches saturation [46].

A free-electron laser in the X-ray regime requires about 100 m of undulator and a linear accelerator of about 1 km, using the present technology. It is desirable to reduce the size of the FEL facility or make it even portable and robust. The FEL can benefit from results of other branches in the accelerator physics. Acceleration based on plasmas are promising higher acceleration gradient and lower emittances [47], thus shortening both the undulator and the linear accelerator. The FEL physics is unchanged, when the undulator is replaced by a device that also enforces a transverse oscillation of the electron bunch. A possible solution would be a high-intensity optical laser pulse [48]. Because the period length is several orders of magnitude smaller than that for an undulator, the FEL interaction requires less energy to obtain the same wavelength. Other proposals are based on plasma columns [49], transverse wakefields [50] or crystal lattices [51].

## Glossary

**3rd Generation Light Sources:** Spontaneous emission from dipoles and undulator with radiation pulse lengths in the picosecond regime and recirculating electrons with an energy between 3 and 10 GeV.

**$\beta$-function:** Characteristic function of the focusing property of a given beam line. Depending on the initial values of the $\beta$-function and its derivative, it describes the evolution of the beam size along the beam line.

**Bending Magnet:** see *dipole magnet*.

**Bunch Compressor:** A series of dipole magnets to deflect the electron beam from the orbit and then back onto the orbit. The resulting orbit and thus the path length of this orbit bump depends on the electron energy. With a correlation between longitudinal position and energy within an electron bunch, the path-length difference yields a reduction in the bunch length and an increase in the beam current.

**Bunching:** Density modulation of the electron beam on a scale shorter than the bunch length. Also called *microbunching*.

**Coherent Synchrotron Radiation:** Strong enhancement of the synchrotron radiation for wavelength longer than the bunch length. The radiation can interact back with the electron bunch and because of its large amplitude alter the electron energy and, thus, the orbit of the electron bunch.

**Dipole Magnet:** Magnet with a constant, transverse magnet field to deflect electron and thus, cause a bend in the orbit of the electron bunch.

**Emittance:** Phase space area, covered by the electron bunch distribution.

**Energy Recovery Linear Accelerator:** Circular accelerator with continuous injection of electron bunches. After acceleration, each electron bunch passes each RF cavity again but with a 180° phase difference. The deceleration transfers back the kinetic energy of the electron bunch to the RF field, which can be used to accelerate the succeeding bunch.

**Linac:** Linear accelerator where the electron passes the beam line only once. In contrast to the storage ring, the linear accelerator consists mainly of accelerating structures, such as RF cavities. Quadrupole magnets keep the transverse beam size within the acceptable limits.

**Microbunching:** See *bunching*.

**Photoelectron Gun:** RF cavity with an embedded photo cathode. A UV laser pulse illuminates the cathode and generates free electrons due to the photo effect. The RF field accelerates these electrons to relativistic energies within the length of the RF gun.

**Plasma Frequency:** Measure for the strength of the beam size growth due to the repulsive electrostatic field between each electron.

**Ponderomotive Phase:** Phase of the electron–radiation interaction due to the beat wave of the transverse oscillation of the electrons and the oscillation of the EM wave.

**Quadrupole Magnet:** Magnet with a linear growing transverse magnetic field with respect to the orbit of the electron bunch. The field configuration yields a focusing in one transverse plane but a defocusing in the perpendicular transverse plane. A net focusing in both planes is achieved by a series of quadrupoles with alternating polarity.

**Radiofrequency Cavity:** Cavity, which resonantly holds an oscillating EM field at well-defined frequencies. The fundamental mode has a longitudinal electric field component, which is used to accelerate an electron beam, passing the cavity.

**Shot noise Power:** Power of the spontaneous radiation, which falls within the bandwidth of the free-electron laser and which is amplified in the SASE FEL configuration.

**Storage Ring:** Circular accelerator to keep injected electron bunches on a closed orbit for many turns. RF cavities compensate the energy loss due to the spontaneous radiation within the dipole magnets and other transverse deflecting magnets (e.g., undulators)

**Superconducting Magnet:** Electromagnet, based solely on superconducting windings to overcome saturation effects of any yoke material.

**Undulator:** Device with transverse, alternating magnetic field on the main axis in planar and helical configuration. In a planar undulator, the transverse magnetic field is oscillating within only one transverse plane, while the magnetic field is constant but rotates along the longitudinal axis for a helical undulator.

**Undulator Parameter:** Normalized amplitude of the vector potential of the undulator field. It is proportional to the product of peak undulator field and undulator period and can be regarded as the total deflection strength of the undulator field per half period.

**Undulator Taper:** Variation of the undulator field and/or period along the longitudinal axis of the undulator.

**Wakefield:** Radiation, caused by the interaction of the electric static field of an electron with the vacuum chamber. The radiation can interact with succeeding electrons, changing the electron energy or transverse momentum.

**Wiggler:** See *undulator*.

## References


[1] Madey, J. M. J. (**1971**), *J. Appl. Phys.* **42**, 1906.
[2] Motz, H. (**1951**), *J. Appl. Phys.* **22**, 527.
[3] Palmer, R. B. (**1972**), *J. Appl. Phys.* **43**, 3014.
[4] Freund, H. P., Granatstein, V. L. (**1999**), *Nucl. Inst. & Methods* **A429**, 33; Colson, W. B. (**1999**), *Nucl. Inst. & Methods* **A429**, 37; online at http://sbfel3.ucsb.edu/www/vl_fel.html
[5] Linac Coherent Light Source (LCLS), SLAC-R-521, UC-414 (1998).
[6] TESLA – Technical Design Report, DESY 2001-011, ECFA 2001-209, TESLA Report 2001-23, TESLA-FEL 2001-05, Deutsches Elektronen Synchrotron, Hamburg, Germany (2001).
[7] Various proposals are listed in the Proc. of the 24th Free-Electron Laser Conference, Argonne, USA (2002).
[8] Phillips, R. M. (**1960**), *IRE Trans. Electron Devices* **7**, 231.
[9] Elias, R. L. et al. (**1976**), *Phys. Rev. Lett.* **36**, 717.
[10] Deacon, D. A. G. et al. (**1977**), *Phys. Rev. Lett.* **38**, 892.
[11] Litvinenko, V. (1999), *Storage Ring-based Light Sources*, Proc. of the 17th Advanced Beam Dynamics Workshop on Future Light Sources, Argonne, USA.
[12] Derbenev, Ya. S., Kondradenko, A. M., Saldin, E. L. (**1982**), *Nucl. Inst. & Methods* **193**, 415.
[13] Nodvick, J., Saxon, D. *Phys. Rev.* (**1954**), **96**, 180.
[14] Dawson, J. M. (**1989**), *Sci. Am.* **260**, 54.
[15] Born, M., Wolf, E. (1980), *Principles of Optics*. Pergamon Press, London.
[16] Bonifacio, R., Pellegrini, C., Narducci, L. M. (**1984**), *Opt. Commun.* **50**, 373.
[17] Kim, K.-J. (**1986**), *Nucl. Inst. & Methods* **A250**, 396.
[18] Kondradenko, A. M., Saldin, E. L. (**1980**), *Part. Accel.* **10**, 207.
[19] Ogata, A. (1992), in J. Wurtele (Ed.), *Advanced Accelerator Concepts*. Port Jefferson, New York, p. 420, AIP Conf. Proc. 279, New York, 1993.
[20] Bonifacio, B., Pellegrini, C., Narducci, N. (1984), in J. M. J. Madey, C. Pellegrini (Eds.), *Free Electron Generation of Extreme Ultraviolet Coherent Radiation*. New York: American Institute of Physics, p. 236.
[21] Orzechowski, T. J. et al. (**1985**), *Phys. Rev. Lett.* **54**, 889.
[22] Edwards, D. A., Syphers, M. J. (1992), *Introduction to the Physics of High Energy Accelerators*. John Wiley & Sons, New York.

[23] Krinsky, S., Yu, L. H. (**1987**), *Phys. Rev.* **A35**, 3406.
[24] Moore, G. T. (**1984**), *Opt. Commun.* **52**, 46.
[25] Kim, K.-J. (**1986**), *Phys. Rev. Lett.* **57**, 1871.
[26] Goodman, J. (1985), *Statistical Optics.* New York: John Wiley & Sons.
[27] Bonifacio, R. et al. (**1994**), *Phys. Rev. Lett.* **73**, 70.
[28] Saldin, E. L., Schneidmiller, E. A., Yurkov, M. V. (1997), TESLA-FEL 1997-02, Deutsches Elektronen Synchrotron, Hamburg, Germany.
[29] Chao, A. W. (1993), *Physics of Collective Instabilities in High Energy Accelerators.* New York: John Wiley & Sons.
[30] Renieri, A. (**1979**), *Nuovo Cimento* **53B**, 160.
[31] Jaroszynski, D. A. et al. (**1997**), *Nucl. Inst. & Methods* **A393**, 332.
[32] Hogan, M. (**1998**), *Phys. Rev. Lett.* **81**, 4867.
[33] Ayvazyan, V. et al. (**2002**), *Phys. Rev. Lett.* **88**, 104802.
[34] Palmer, D. T. et al. (1997), Proc. of the Part. Accel. Conf., Vancouver, p. 2843.
[35] Carlsten, B. E. (**1989**), *Nucl. Inst. & Methods* **A285**, 313; Serafini, L., Rosenzweig, J. B. (**1997**), *Phys. Rev.* **E55**, 7565.
[36] Rosenzweig, J. B., Colby, E. (1995), *Advanced Accelerator Concepts.* AIP Conf. Proc. 335, New York, p. 724.
[37] Sands, M. (**1955**), *Phys. Rev.* **97**, 470.
[38] Litvienko, V. (1999), *Storage ring-based light sources,* Proc. of the 17th Advanced Beam Dynamics Workshop on Future Light Sources, Argonne, USA.
[39] Bruck, H. Accélérateurs circulaires de particules, Paris.
[40] Krinsky, S., Wang, J. M. (**1985**), *Part. Accel.* **17**, 109.
[41] Walker, R. P. et al. (**2001**), *Nucl. Inst. & Methods* **A475**, 20.
[42] Neil, G. R. et al. (**2000**), *Phys. Rev. Lett.* **84**, 662.
[43] Tigner, M. (**1965**), *Nuovo Cimento* **37**, 1228.
[44] Saldin, E. L., Schneidmilller, E. A., Yurkov, M. V. (**2000**), *Nucl. Inst. & Methods* **A445**, 178.
[45] Schroeder, C. B. et al. (**2002**), *J. Opt. Soc.* **19**, 1782.
[46] Reiche, S., Emma, P. (2002), *Proc. of the 24th Free-Electron Conference,* Argonne, USA.
[47] Suk, H., Barov, N., Rosenzweig, J. B., Esarey, E. (**2001**), *Phys. Rev. Lett.* **86**, 1011.
[48] Warren, B. (1969), *X-Ray Diffraction.* Addison-Wesley.
[49] Joshi, C. et al. (**2002**), *Phys. Plasmas* **9**, 1845.
[50] Zhang, Y., Derbenev, Y. (2002), *Proc. of the 24th Free-Electron Conference,* Argonne, USA.
[51] Avakian, R. O. et al. (**2001**), *Nucl. Inst. & Methods* **B173**, 112.

## Further Reading

A complete description of the FEL theory is given in the books *Thy Physics of Free Electron Lasers* by E. L. Saldin, E. A. Schneidmilller, M. V. Yurkov (Springer, Berlin, 2000) or *Free-Electron Lasers* by C. A. Brau (Academic Press, Boston, 1990). A brief overview on the high gain FEL can be found in the paper "Introduction to the Physics of the FEL" by J. B. Murphy and C. Pellegrini (Proc. of the South Padre Island Conference, Springer (1986) 163).

The book *An Introduction to the Physics of High Energy Accelerators* by D. A. Edwards and M. J. Syphers (John Wiley and Sons, New York, 1993) covers the physics, needed to generate and accelerate an electron beam, driving the free-electron laser. A review of undulator design is given by J. Spencer and H. Winick in *Synchrotron Radiation Research* (Eds. H. Winick and S. Doniach, Plenum Press, New York, ch. 21, 1980).

The radiation process of relativistic particles are covered in great detail by J. D. Jackson's *Classical Electrodynamics* (John Wiley and Sons, New York, 1975) while the description of the random nature of the SASE FEL follows the treatment of *Statistical Optics* by J. Goodman (Wiley and Sons, New York, 1985).

# Lasers, Gas

*J. Gary Eden*
*University of Illinois, Everitt Laboratory, Urbana Illinois, USA*

*Jürgen Eichler*
*Technische Fachhochschule and Technische Universität Berlin, Germany*
*Phone: 0049-30-4504 3917; Fax: 0049-30-4504 3959; e-mail: eichler@tfh-berlin.de*

**Abstract**
This article reviews the characteristics of gas lasers that are chiefly responsible for their effectiveness as both scientific and industrial tool. The basic principles common to operation of all gas lasers and a few of their general properties are discussed. The lasers are categorized according to the pumping mechanisms: electron-impact excitation (copper and gold laser, Ar- and Kr-ion lasers), excitation transfer (He–Ne laser, He–Cd laser, $CO_2$ laser), reactive collision lasers (excimer laser (ArF, KrF, XeCl, XeF, $F_2$), chemical lasers (HF, COIL)), and other lasers in the far infrared, ultraviolet, and soft X-ray region. Selected applications of pulsed and continuous gas lasers are presented in material processing, marking and laser evaporation, microlithography, photochemical processing, stereolithography, and medicine.

**Keywords**
gas laser; pumping mechanisms; He–Ne laser; ion laser; excimer laser; $CO_2$ laser; laser application; material processing; laser medicine.

## 1 Introduction

Although the existence of stimulated emission was described by Einstein ("On the quantum theory of radiation") in [1], and Tolman [2] envisioned the potential of a population inversion to amplify an optical wave by stimulated emission, it was not until the 1950s that Schawlow and Townes [3] articulated the basic features of the optical maser or laser – features such as the open resonator that would later prove to be essential to its success.

In December of 1960, Ali Javan, William R. Bennett, and Donald R. Herriott observed laser oscillation from an electric discharge in a mixture of helium and neon gases in Bell Labs, New Jersey [4]. Following closely on the heels of Theodore Maiman's demonstration of the ruby laser in the spring of that year, this laser not only was the first gas laser but it also operated continuously. More than 10 000 different transitions in atoms and molecules in the gas phase have lased in the intervening four decades, generating radiation at wavelengths ranging from the far-infrared ($\lambda > 100$ μm) to the soft X-ray region (>1 nm). From among these have emerged several lasers that are workhorses in research as well as industrial or medical applications. Initially only a laboratory curiosity, gas lasers have been engineered to the point where they have an impact on our lives in a variety of areas. Their applications range from supermarket checkout (bar-code) scanners to material processing to ophthalmic surgery and laser light shows. The utility of gas lasers is reflected by the prominent position they have held and continue to occupy in the laser industry. Sales in 2003 totaled approximately $750 million – or roughly one half of the entire nondiode laser market – with the remainder divided among solid-state and dye (liquid) lasers ($900 million and $7 million respectively).

This article briefly reviews the characteristics of gas lasers that are chiefly responsible for their effectiveness as both scientific and industrial tools. Since the breadth of this subject precludes an exhaustive treatment, attention is focused on those gas lasers that are widespread and have proven to be useful for research and yet are of commercial value. The basic principles common to the operation of all gas lasers and a few of their general properties are discussed in Sects. 2 and 3 respectively. Gas lasers can be broadly categorized according to the methods by which one drives (or "pumps") the laser. Section 4 describes specific parameters of the most important laser systems in each class. Section 5 concludes, presenting selected applications of pulsed and continuous gas lasers in materials processing, medicine photochemical synthesis, and electronics.

## 2 Principles of Operation

It would be well to begin by considering several of the principles that undergird the operation of all lasers – gas, solid, or liquid. The laser is an optical oscillator – that is, a device that generates coherent radiation at optical frequencies – and, viewed in its most basic form, is quite similar conceptually to a radio-frequency (rf) oscillator. Although their operating frequencies differ by several orders of magnitude (rf, $10^6-10^9$ Hz; laser, $10^{13}-10^{16}$ Hz), both are based on identifying a medium or device that is capable of amplifying radiation of the desired frequency. As depicted in Fig. 1, the existence of amplification relies

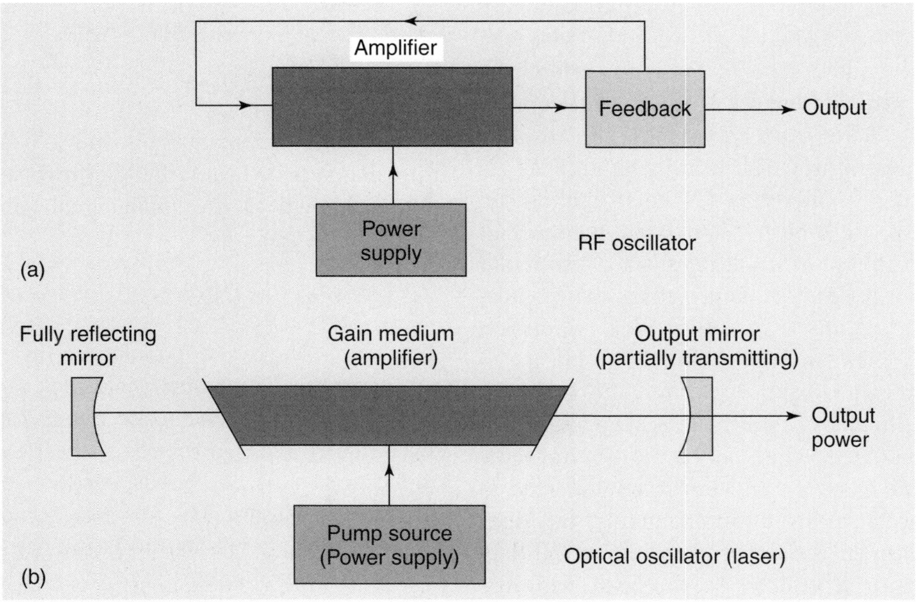


**Fig. 1** Generalized diagrams for (a) radio-frequency and (b) optical (laser) oscillators. Despite their widely disparate emission frequencies, the principles underlying the operation of both are similar

on the ability to couple power in some manner to the amplifying medium. The process or system by which this transfer of energy occurs is known as the *pump*, and several specific examples will be presented in Sect. 4. Once amplification at the desired frequency (or wavelength) is realized, oscillation is generally achieved by introducing feedback – redirecting a portion of the output power back into the amplifier. At optical frequencies, this is conveniently accomplished with mirrors.

For amplification to occur in the optical region, it is necessary that a *population inversion* be established between two energy states of the atom or molecule in question. Normally – that is, in the absence of an external source of excitation – the population of a particular excited state will be smaller than that for every state lying beneath it in energy. Said another way, at thermal equilibrium, the populations of excited states diminish monotonically with increasing energy in a manner prescribed by the Boltzmann distribution function: $\exp(-\Delta E/kT)$, where $\Delta E$ is the energy of the excited state in question, referenced to ground, and $kT$ is the thermal energy ($k =$ Boltzmann constant, $T =$ absolute temperature). The factor $\exp(-\Delta E/kT)$ is much smaller than 1. If this situation can be partially reversed such that the number of atoms or molecules in a particular energy level is *larger* than the population of any state below it, a population inversion and the potential for optical amplification exist.

*Stimulated emission* is the radiation generated when an atom or molecule undergoes a transition from one state to another lying lower in energy, but does so under the influence of an optical field whose frequency $f$ is related with the energy difference between the two states involved in the transition. To explore this further, let us consider the ramifications of having a population inversion between two states at energies $E_2$ and $E_1$ (where $E_2 > E_1$). If an optical field of photon energy

$$hf = E_2 - E_1 \tag{1}$$

($h = 6.63 \times 10^{-34}$ Js is Planck's constant) illuminates this two-state system, then transitions from state 2 to state 1 will be induced (stimulated) by the field. Since a photon is produced in the process, this additional radiation is emitted coherently – that is, in phase with the driving optical field. Of course, absorptive transitions (which have the effect of attenuating the field) will also be taking place from state 1 to state 2, but since we started with a population inversion, stimulated emission dominates and net optical gain results. Consequently, the intensity of the optical field grows as it propagates through the inverted medium, and amplification has been realized. In the limit of weak optical fields, the intensity $I$ of the radiation increases exponentially with distance $z$:

$$I = I_0 e^{g_0 z}, \tag{2}$$

where $I_0$ is the intensity of the optical field immediately prior to entering the amplifier and $g_0$, known as the small-signal gain coefficient, is given by

$$g_0 = \sigma(N_2 - N_1). \tag{3}$$

Here, $N_2 - N_1$ is the population inversion (expressed in atoms/molecules per centimeter cube) between the upper and lower-energy states involved in the transition, and $\sigma$ is the cross section for stimulated emission. The wavelength $\lambda$ of the radiation is given by the frequency $f$ and the velocity of light $c$:

$$f = \frac{c}{\lambda}. \tag{4}$$

Since the small-signal gain coefficient $g_0$ is proportional to the magnitude of the population inversion, much of the effort in developing new gas lasers or improving existing ones has centered on different approaches to populating specific excited states of atoms and molecules. The key is to put the species of interest into a given excited state while simultaneously minimizing the rate at which lower-energy levels are populated. Several of the processes that have most often proven to be effective in pumping gas lasers are the subject of Sect. 4.

Finally, it is clear that the energy difference between the two states involved in the optical transition determines the wavelength $\lambda$ of the laser. When attempting to develop lasers in a particular region of the spectrum, therefore, we must look for groups of atomic or molecular states whose energy differences lie in the proper range. As illustrated qualitatively in Fig. 2, this requires transitions between *electronic* states of atoms and molecules if laser wavelengths below $\sim 1\,\mu m$ are desired. The generation of longer wavelengths (infrared–submillimeter) usually entails *vibrational and rotational* transitions in molecules. It is also necessary that quantum-mechanical selection rules allow for transitions between the desired states.

## 3 Characteristics of Gas Lasers

Although gas lasers produce a wide array of wavelengths and power levels, operate continuously or pulsed, and offer beams of differing cross-sectional profiles, they share several characteristics that distinguish them as a family. An inherent property of the gas-laser medium is its tolerance of excessive currents and voltages. That is, the gas or gas mixture is usually "self-healing" and is capable of withstanding the deposition of large amounts of power (known as specific power loading and expressed in W $cm^{-3}$). Gas lasers are also comparatively straightforward to scale in volume – that is, to construct lasers having a large aperture or length. The ability to optimize quickly the composition of the gas mixture in a laser is another advantage in that it allows one to test rapidly a range of gas combinations and pressures.

For many laser applications, the high optical quality beams and pointing stability available from gas lasers represent their greatest assets. Because of the homogeneity of gaseous media and their typically low index of refraction, numerous gas lasers are noted for routinely producing beams in the fundamental spatial mode ($TEM_{00}$), having divergences that are close to the diffraction limit. The red He–Ne gas laser

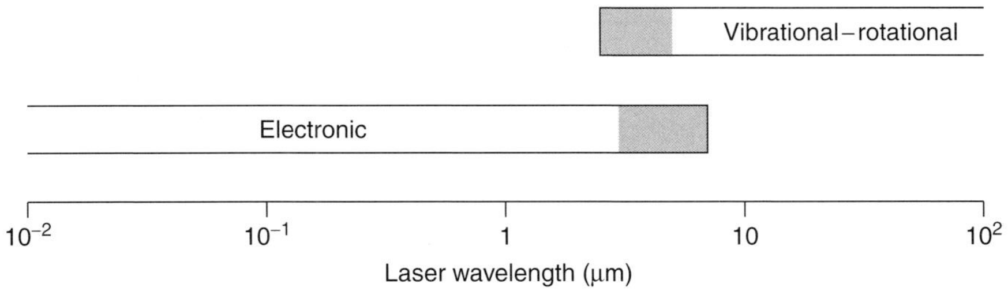


**Fig. 2** Shorter-wavelength lasers involve transitions between electronic states of atoms and molecules, whereas wavelengths beyond a few microns (μm) are usually the result of vibrational–rotational transitions

($\lambda = 632.8$ nm), for example, typically produces a beam having a divergence angle of 1 mrad. Consequently, for 10 mW of output power, the spectral brightness of this laser – defined as intensity (W cm$^{-2}$) per steradian (sr) – is $\sim 10^5$ W cm$^{-2}$ sr$^{-1}$, a value that is three orders of magnitude larger than that for the sun.

If we assume the laser beam to have a Gaussian cross-sectional intensity profile and to be operating in the $TEM_{00}$ mode, then the *diffraction-limited spot diameter w* for a focused beam is given by

$$w \simeq \frac{2\lambda f}{\pi d} \approx 0.64\lambda\frac{f}{d}, \qquad (5)$$

where $\lambda$ is the laser wavelength and $f$ and $d$ are the focal length and beam diameter at the focusing lens respectively. To a first approximation, then, the smallest spot size obtainable is on the order of the wavelength $\lambda$ (assuming, of course, that $f/d$ – the inverse of the focusing system's numerical aperture – is approximately unity), and this limit is accessible *only* when the beam being focused is diverging at the diffraction limit. As a result, visible and ultraviolet gas lasers are frequently applied to those industrial tasks (such as microelectronic circuit inspection) for which spatial resolution and, hence, the focusability of the optical source are of primary concern.

Entire families of gas lasers are based on the fact that producing large number densities of transient small molecules can often be accomplished in the gas phase but may be difficult or impossible to do in a solid or liquid. If lasers employing vapors (such as those of volatile metals – including Hg and Cs, metal halide salts, the halogens, etc.) as the active medium are included, the gas-laser scientist has an enormous range of gas and vapor combinations from which to choose and an almost endless list of electronically excited species (that do not normally exist in nature) to explore.

Finally, it should be mentioned that, for wavelengths below roughly 200 nm (i.e., in the ultraviolet, $hf = 5$ eV), virtually all of the available laser sources are gas systems. It is important to remember that the power required to obtain oscillation scales nonlinearly with the optical frequency $f$. Consequently, demonstrating lasers that operate in the short-wavelength portion of the visible, the ultraviolet, or beyond demands the deposition of a minimum of tens of kilowatts of power per centimeter cube to maintain the necessary population inversion. Such power levels frequently have catastrophic consequences for most solids and liquids, particularly for those crystals and glasses having poor thermal conductivity. As we pass the ultraviolet in moving to shorter wavelengths, the photon energies rise above those available from transitions involving outer-shell electrons in atoms and molecules. Exciting transitions involving inner-shell electrons becomes necessary, and, while this has been accomplished with gases and highly ionized plasmas, this constraint imposes increasing difficulties for condensed media as the laser wavelength decreases.

The two major drawbacks of gas lasers are their size and power consumption. Although gas lasers as small as several centimeters in length have been demonstrated, most manufactured lasers are considerably larger (tens of centimeters), and pulsed, high-power commercial devices are typically 50–100 cm (or more) in length. Also, because of their low electrical-to-optical conversion efficiencies, large-frame argon-ion lasers, for example, require water cooling and more than 10 kW of electrical power to produce 20 W in the visible. As a result, gas lasers are best suited for those applications in which beam quality, peak or average power, or short wavelength (rather than size or weight) are the driving considerations.

## 4
## Laser Excitation Processes and Specific Gas Lasers

The manner in which power is delivered to the upper laser level has far-reaching consequences for a gas laser, particularly insofar as its overall efficiency and engineering design are concerned. While a variety of atomic and molecular optical and collisional processes have been shown to be capable of pumping gas lasers, a handful appear repeatedly and are described here. Several of the technologically most important gas lasers that exemplify each of the pumping mechanisms are also discussed.

### 4.1
### Electron-impact Excitation

One of the simplest approaches to populating excited states of an atom or molecule is by electron-impact excitation. As illustrated in Fig. 3, if a fast electron has the energy necessary to excite an atom or molecule to a particular state, it is possible to transfer this energy to the heavier particle in a collision. If $X$ represents the species of interest as the laser medium, this process can be expressed as

$$e + \mathrm{X} \longrightarrow \mathrm{X}^* + e, \qquad (6)$$

where the asterisk denotes an excited state and, of course, energy conservation requires that, if the electron's initial energy was $E$, it will emerge from the collision with energy $E - E_2$, where $E_2$ is the excitation energy of the $X^*$ state. The high-energy electrons required for the process of Eq. (6) are usually produced in a low-pressure discharge. It is also clear from Fig. 3 that it is essential for both

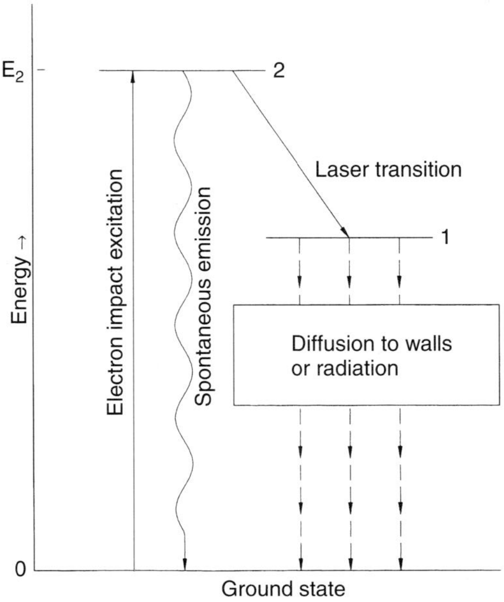


**Fig. 3** Partial energy-level diagram for an atom or molecule illustrating electron-impact excitation of the upper laser level (2). The upper laser level can also decay by spontaneous emission. State 1 (lower laser level) may be deactivated by one of several processes, including spontaneous emission or collisions with the discharge cell wall (after Willett [5])

of the energy levels participating in the laser transition to be chosen carefully so that the electron-impact excitation process populates primarily the upper laser level.

### 4.1.1 Copper-vapor Laser

An immediate example of a system that is pumped by direct electron impact is the copper-vapor laser that emits at two wavelengths: 510.6 nm in the blue–green and 578.2 nm in the yellow. Figure 4 shows the states and processes underlying its operation. Collisions between high-energy (about 4 eV) electrons and ground-state copper atoms preferentially populate the $4p^2P$ excited states that subsequently emit to one of the $^2D$ levels. Because of quantum-mechanical selection rules, the $^2D$ states are unable to radiate to the ground state and are said to be "metastable". Consequently, excited atoms in the $^2D$ states can be removed only by slow processes, including diffusion of the atoms to the walls of the laser cell. As a result, population of the $^2D$ state steadily builds, the inversion is quickly lost, and lasing stops. If the pump source is turned off, the $^2D$ number density will eventually decay, and the laser may once more be operated briefly. For these reasons, the copper-vapor laser is able to operate only in a pulsed mode, emitting pulses that are roughly 3 ns in duration. By operating the laser at a pulse repetition frequency (PRF) of several kilohertz, however, large average powers can be readily generated. Commercial devices producing up to 100 W in a beam 6 cm in diameter and operating at a PRF of 10 kHz are currently available. If gold is substituted for copper as the lasant metal, similar considerations hold except that oscillation now occurs in the red at 628 nm. The copper (or gold) vapor laser is an excellent example of the principle that the operational specifications of gas lasers (as with all lasers) are intimately tied to the details of the atomic and molecular processes occurring in the laser medium. Making *a priori* generalizations concerning the output power, repetition rate, or any other significant parameter of a group of lasers is usually not possible, and the physics of the medium in question

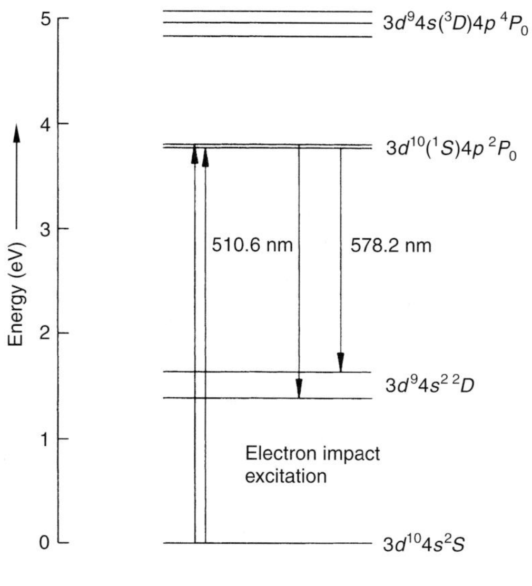


**Fig. 4** Partial energy-level diagram for atomic copper showing the states relevant to the 510.6- and 578.2-nm laser lines. This laser only operates pulsed because the $^2D$ lower laser level is metastable to radiative decay to ground – hence, the transition "self-terminates" (after Walter [6])

must be closely examined for the utility of the laser to be determined.

### 4.1.2 Rare-gas Ion Lasers

One of the major gas lasers today, from both technological and scientific standpoints, is the rare-gas ion laser. Discovered by William Bridges [7], the argon-ion laser is capable of continuously producing tens of watts of power (up to 30 W in commercially available systems) at several discrete wavelengths in the blue and green. Lasing occurs on several transitions of the singly or doubly charged atomic ion ($Ar^+$ or $Ar^{2+}$ respectively) that is produced from the ground state by electron impact (Table 1a). Because the states that produce the laser lines lie at about 35 eV or more above the atomic ground state (Fig. 5), the efficiency of the laser is low. Exciting the upper laser levels requires highly energetic electrons that are produced in an intense, continuous discharge in low-pressure argon (approximately 0.2 Torr). Consider the visible $Ar^+$ lines at 514.5, 501.7, 496.5, 488.0, 476.5, 472.7, 465.8,

**Tab. 1a** Power of a 20-W argon-ion laser at different laser lines

| *Visible* | | *UV-optic* | |
|---|---|---|---|
| ***Wavelength [nm]*** | ***Power [W]*** | ***Wavelength [nm]*** | ***Power [W]*** |
| 528.7 | 2 | 385.1–351.1 | 3 |
| 514.5 | 10 | 363.8–333.6 | 5 |
| 501.7 | 2 | 335.8–300.3 | 2 |
| 496.5 | 3 | 305.5–275.4 | 0.6 |
| 488.0 | 10 | | |
| 476.5 | 3 | | |
| 472.7 | 1 | | |
| 465.8 | 1 | | |
| 457.9 | 1 | | |
| 454.5 | 1 | | |
| All lines | 20 | | |

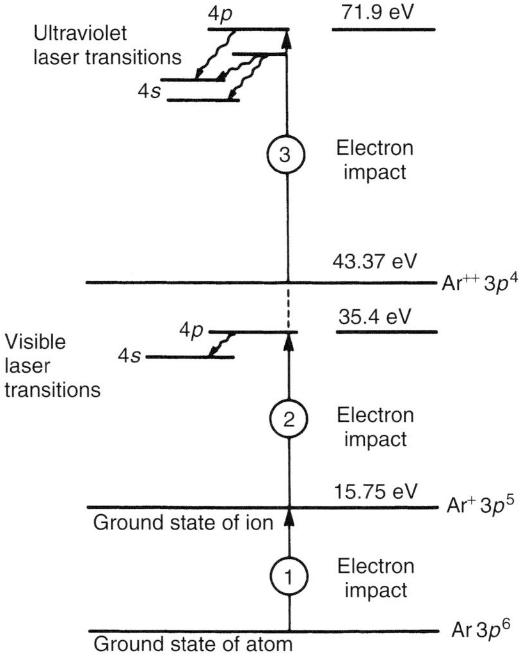


**Fig. 5** Energy levels of argon that are involved in the argon-ion visible and ultraviolet lasers. The visible laser transitions range in wavelength from 454.5 nm (blue) to 514.5 nm (green) and are excited by Steps 1 and 2. The ultraviolet transitions require higher plasma-tube currents because of the additional electron-impact Step 3 (diagram courtesy of Spectra-Physics, Inc.)

457.9, and 454.5 nm, for example. The upper states for these transitions ($Ar^+$ $4p^2S$, $^2P$, $^2D$) lie almost 20 eV above the Ar-*ion* and 35.4 eV above the Air-*atom* ground state. The maximum theoretical efficiency for this laser is <8% since a total of 35 eV of energy must be invested in order to obtain 2.5–2.7-eV photons. In practice, the overall efficiencies of rare-gas ion lasers are considerably smaller (<0.1%), since pumping of the upper laser levels occurs in two steps but both require electrons having energies that are well beyond the average electron energy in the discharge. Said another way, the energies of the electrons in a plasma are described by a distribution, and the electrons required for exciting the $Ar^+$ laser must be in the high-energy "tail" of that distribution. Lower-energy electrons and other unwanted excitations reduce the efficiency.

The output power of an argon-ion laser, $P_0$, can be expressed as

$$P_0 \propto \frac{g_0 L}{T + a}, \tag{7}$$

where $g_0$ is again the small-signal gain coefficient, $L$ is the length of the discharge (i.e., active length), $T$ is the transmission of the output coupling mirror, and $a$ represents all losses in the optical cavity other than the output coupler's transmission. In the $Ar^+$ laser, $g_0$ is approximately linear in the *average* electron energy which, in turn, is proportional to the discharge current density $J$. Consequently, realizing optimal efficiencies from this laser requires large density current – typically 800–1000 A $cm^{-2}$ for a plasma-tube bore diameter of 3 mm. Because of the low electrical efficiencies for the laser, large-frame ion lasers yielding 4–30 W of output power require water cooling of the plasma tube and 5–35 kW of electrical input power. A magnetic field of ~1000 G, which serves to increase the ion density for a given discharge current, is often imposed on the plasma of the higher-power ion lasers.

Figure 6 gives a cutaway view of the structure of an $Ar^+$-laser plasma tube. The high discharge currents that are characteristic of the $Ar^+$ laser place stringent demands on the design and construction of the discharge vessel and, in particular, the structures that confine the plasma. Erosion of the walls of the tube must be minimized since an increase in the bore diameter directly translates into a lower value for $J$. By using refractory

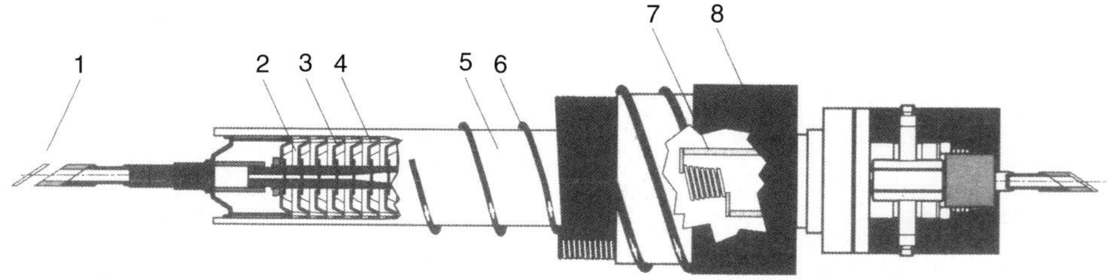


**Fig. 6** Key components of the plasma tube for a large-frame (water-cooled) argon-ion laser: (1) quartz windows at Brewster's angle; (2) tungsten disks for confining the plasma and defining the gain region; (3) copper support disks for heat removal; (4) copper-to-ceramic braze; (5) ceramic cylindrical enclosure; (6) water-cooling line; (7) tungsten cathode; and (8) external magnet for plasma confinement and power enhancement (courtesy of Spectra-Physics, Inc.)

**Tab. 1b**
Wavelength of a krypton-ion laser

| *Wavelength [nm]* |
|---|
| 799.3 |
| 752.5 |
| 676.4 |
| 647.1 |
| 568.2 |
| 530.9 |
| 520.8 |
| 482.5 |
| 468.0 |

materials such as beryllium oxide (BeO) and graphite, manufacturers are now able to specify tube lifetimes beyond several 1000 h. Similar comments could be made for the krypton-ion laser, which emits a dozen strong lines ranging in wavelength from 337.5 to 799.3 nm (Table 1b).

Smaller, air-cooled versions are also commercially available that produce typically 50–100 mW of visible radiation from a much more compact package. A photograph of the head (plasma tube and cooling assembly) for a small (<50 mW), air-cooled ion laser is shown in Fig. 7. Since the plasma-tube bore is only 1 mm, however, the plasma current densities (>1000 A $cm^{-2}$) can actually exceed those of their larger, water-cooled counterparts. Intermediate power levels (0.3–1 W) have been generated by adopting a hybrid approach to cooling the discharge tube. The tube itself is cooled by a closed-loop water system that is interfaced to a water–air heat exchanger. $Ar^+$ lasers are frequently used in laser printing, compact disc manufacturing, medicine, and light shows.

A development by laser manufacturers with plasma tubes capable of operating at still higher current densities (>1500 A $cm^{-2}$) is the demonstration of new laser lines in the deep ultraviolet (200–300 nm) from argon, krypton, and xenon. In krypton, for example, 2.5 W of total power can now be obtained from several lines oscillating in the 242- to 266-nm interval; similarly, a group of argon lines in the 275- to 306-nm range has produced 6 W.

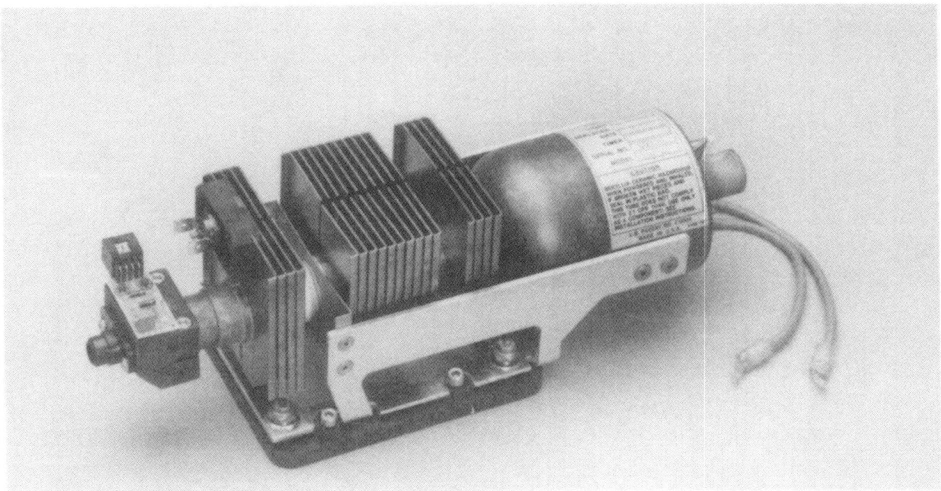

**Fig. 7** Photograph of an air-cooled argon-ion laser head (courtesy of S. Jarrett, Spectra-Physics, Inc.)

## 4.2 Excitation Transfer

In many instances, it is impractical or inefficient to populate directly a particular state that is of interest as the upper level for a laser transition. If, however, the energy level in question lies close in energy to a state of another atom (or molecule) that can be efficiently populated, then it is frequently possible to transfer the required energy from one atom to another through a collision. The rate at which energy can be transferred from the initial to the final state is strongly dependent upon the difference in energy (denoted $\Delta E = E_A - E_B$) between the two states. In most cases, this process is increasingly efficient as $\Delta E$ decreases, and when $\Delta E$ is comparable to or less than $kT$ (thermal energy), the collision is said to be *resonant*. This pumping process, known as *excitation transfer*, is illustrated qualitatively in Fig. 8 and is at the heart of three of the best-known gas lasers: He–Ne, He–Cd, and $CO_2$.

### 4.2.1 He–Ne Laser

The first gas laser, reported by Javan et al. [4], emitted continuously on the 1.15-μm transition of atomic neon in the infrared. Shortly thereafter, oscillation on the red transition at 632.8 nm was also reported [8]. The critical feature was the use of helium as a vehicle to transfer energy to specific energy levels of neon. As depicted in Fig. 9, the lowest two excited states of He ($2^3S$ and $2^1S$) lie within 50 meV (or roughly twice thermal energy at 300 K) at the 2 s and 3 s levels of Ne. The He ($2^1S$) state, for example, rapidly transfers its energy to the 3*s* levels of Ne by the reaction

$$\mathrm{He}^*(2^1S) + \mathrm{Ne} + 48\ \mathrm{meV} \longrightarrow \mathrm{He\ (ground\ state)} + \mathrm{Ne}^*(3s), \quad (8)$$

where the asterisk denotes an excited state. A similar relation can be written for the transfer of energy from the lowest-energy He excited state ($2^3S$) to one of the 2*s* levels of Ne. The He–Ne laser, then, relies on populating the $2^1S$ and $2^3S$ states of He

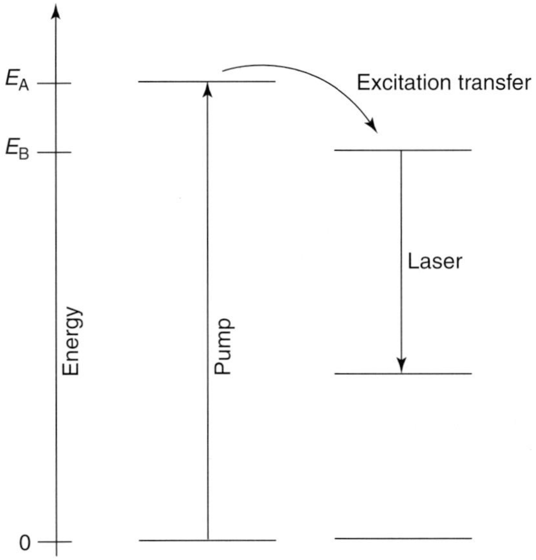


**Fig. 8** Excitation transfer from atom (or molecule) *A* to *B*. Energy can generally be delivered to species *A* by a variety of processes, including electron-impact excitation (described earlier) or optical pumping

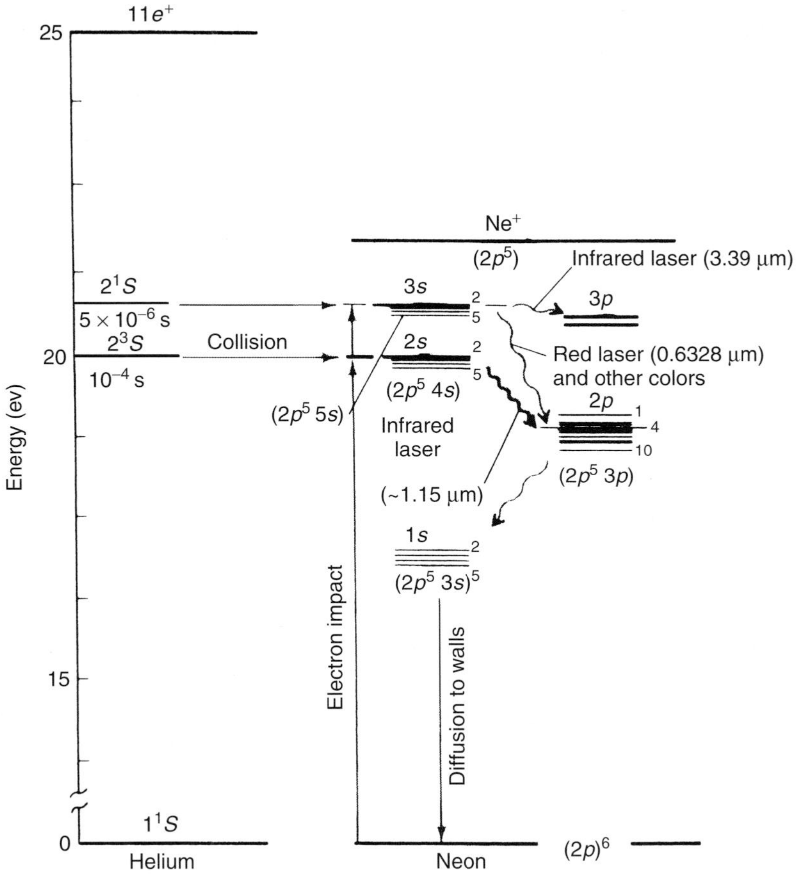


**Fig. 9** Energy-level diagram for atomic He and Ne indicating excitation transfer from the $2^3S$ and $2^1S$ states of helium to energy levels of neon (after Bennett [9] and Yariv [10])

in a discharge and transferring that energy to Ne excited states. Lasing subsequently occurs on several $3s \rightarrow 2p$, $3p$ and $2s \rightarrow 2p$ transitions, the strongest of which are the 1.15-μm and 632.8-nm lines mentioned earlier but also the $3s \rightarrow 3p$ transition at 3.39 μm. The lower laser levels are depleted by spontaneous emission and collision of the atoms with the tube wall after diffusion. However, direct electron-impact excitation of the Ne levels also occurs in the He–Ne gas mixture. Since a fraction of the energy delivered to the Ne excited states will reside in the lower laser levels, a reduction in the steady state population inversion that is attainable results, which, in turn, limits the efficiency and output power available from the laser.

The key component of the He–Ne laser is a discharge tube having an active length of 10 to 100 cm (commercial devices tend to the short end of this range). The bore of the discharge $d$ (in millimeters) is related to the optimal gas fill pressure $p$ (in torr) by

$$pd \approx 3.6 \text{ torr mm.} \quad (9)$$

So, for a discharge diameter of 4 mm, the He–Ne fill pressure should be approximately 0.9 torr. To encourage the preferential excitation of He over Ne excited states in the discharge, the He : Ne gas ratio is typically between 5 : 1 and 20 : 1. The dependence of the output power of a He–Ne laser having a 3.0-mm bore and an active length of 55 cm on the total gas pressure is shown in Fig. 10. Under these conditions, the optimal He : Ne ratio is (5–7) : 1. Also, the discharge current is generally between 5 and 50 mA, which translates into current densities of 30 to 300 mA $cm^{-2}$ for a 4-mm-bore tube.

Because of the low population inversions that are inherent with continuous (cw) He–Ne lasers, the small-signal gain coefficient $g_0$ of the red line is on the order of 0.1 $m^{-1}$. Power outputs as large as 50 to 100 mW are available from commercial lasers but, for low-power applications, models producing 0.5 mW in the $TEM_{00}$ mode are also offered. Other $3s \to 2p$ transitions also lase in the visible, and several manufacturers produce a He–Ne laser having a prism (or selective mirror) in the optical cavity that enables one to select a particular (green, yellow, orange, red) transition (543, 594, 612, or 633 nm) (Table 2). In many applications the red He–Ne laser is now replaced by more compact and efficient diode laser. But these have still lower spatial beam quality and spectral stability.

**Tab. 2** Color, wavelength λ, power, and small-signal gain coefficient $g_0$ of the He–Ne laser

| *Color* | *λ [nm]* | *Power [mW]* | *$g_0$ [% $m^{-1}$]* |
|---|---|---|---|
| Infrared | 3391 | >10 | 10 000 |
| Infrared | 1523 | 1 | |
| Infrared | 1153 | 1 | |
| Red | 640 | | |
| Red | 635 | | |
| Red | 633 | >10 | 10 |
| Red | 629 | | |
| Orange | 612 | 1 | 1.7 |
| Orange | 604 | | |
| Yellow | 594 | 1 | 0.5 |
| Green | 543 | 1 | 0.5 |

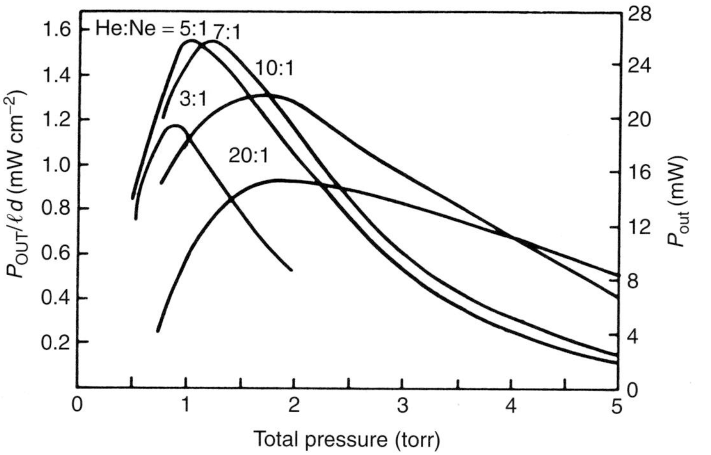


**Fig. 10** Output power as a function of total pressure for a He–Ne laser having a discharge tube diameter (*d*) of 3.0 mm, and a plasma length of 55 cm (after Field, Jr [11])

### 4.2.2 He–Cd Laser

Excitation transfer from helium is also the process underlying the success of the helium–cadmium (He–Cd) laser [12, 13]. In this instance, however, the He ($2^1S, 2^3S$) states have sufficient energy to ionize cadmium atoms, thereby producing excited states of the metal ion. The reaction, written as

$$\mathrm{He}^*(2^1S, 2^3S) + \mathrm{Cd}\ (\text{ground}) \longrightarrow (\mathrm{Cd}^+)^* + \mathrm{He} + e, \qquad (10)$$

is known as *Penning ionization* and produces an inversion on the $4d^9 5s^2\,D_{5/2} \rightarrow 4d^{10}5p^2\,P_{3/2}$ and $5s^2\,D_{3/2} \rightarrow 5p^2\,P_{1/2}$ transitions of the ion at 441.6 and 325.0 nm respectively. Typically, the Cd metal vapor and helium pressures in the discharge are roughly 2 mtorr and 3 to 4 torr respectively, and to maximize excitation of the upper laser levels by electron-impact collisions requires that the current in the plasma be ~100 mA.

The He–Cd laser is a low-power laser that is usually operated continuously (cw). Commercially available models produce up to 70 mW (cw) in the blue and 25 mW in the ultraviolet (40 and 10 mW, respectively, in the $\mathrm{TEM}_{00}$ mode). The lifetime of the laser tube (5000–6000 h) is limited by several factors, including the deposition of cadmium metal on the optics. The short-wavelength emission of the He–Cd laser, combined with its low beam divergence (0.6–2.0 mrad) and near-diffraction-limited spot size, make it well suited for applications such as flow cytometry (cell detection and counting) or circuit inspection that require high photon energies and excellent beam quality.

A particular discharge version of the helium–cadmium laser, known as a *hollow-cathode laser,* will, in addition to the familiar blue visible transition, provide red and green beams as well. Figure 11 is a schematic diagram of a commercially available version of such a laser. Rather than having a longitudinal discharge (that is, the discharge axis coinciding with the optical axis), the electrode separation is much smaller because of the transverse geometry of the discharge, and one consequence is that the plasma generates electron energies and densities higher than those typical of the longitudinal geometry. Output powers (all lines) are generally as high as 50 mW. Because of their ability to produce white light (i.e., red, blue, and green beams simultaneously), hollow-cathode He–Cd lasers are currently being applied to optical image processing systems.

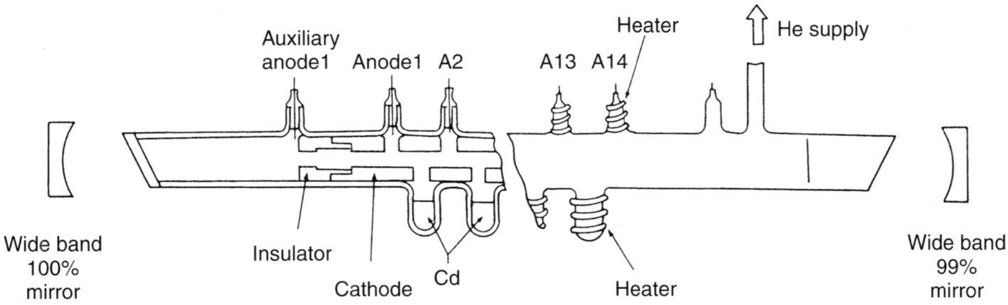


**Fig. 11** Schematic diagram of a commercial hollow-cathode helium–cadmium laser that is capable of supplying the three primary colors (red: 635.5, 636.0 nm; green: 533.7, 537.8 nm; blue: 441.6 nm) in a single output beam

### 4.2.3 $CO_2$ Laser

Most gas lasers (and particularly those operating in the visible or at shorter wavelengths) have overall efficiencies that are typically well under 1%. In fact, few lasers of any type are as efficient as the $CO_2$ gas laser, which produces power at 10.6 μm in the mid-infrared with an efficiency (wall plug) of 20%. The success of this laser is attested to by its prevalence in a range of industrial applications but particularly those – such as cutting and welding metals, fabric, and wood – requiring as much as one or more kilowatts of average power.

The unusually high efficiency of the $CO_2$ laser is partially attributable to a bit of serendipity that was first recognized by Patel [14]. To appreciate this, we must first consider the energy states of the $CO_2$ molecule. This triatomic molecule is able to vibrate in the three ways portrayed in Fig. 12. The three "modes" are known (from top to bottom) as the symmetric stretch, bending, and asymmetric stretch modes. Each mode has associated with it a set of energy levels. For the symmetric stretch, they are denoted ($n$00), where $n$ is an integer. The bending-mode states are given as (0$n$0) and those for the asymmetric stretch as (00$n$). The key to the $CO_2$ laser's efficiency and enormous available output powers is that – as shown in Fig. 13 – the first excited asymmetric stretch state (001) is almost exactly in resonance with the lowest vibrational level of nitrogen ($N_2$). Therefore, if we attempt to populate $CO_2$ (001) directly by electron collisions, little power will actually be delivered to this particular state. Rather, most of the pump energy will go into the low-lying (010) and (020) states. If, however, one takes the tack exploited so successfully for the He–Ne and He–Cd lasers and electrically excites nitrogen gas having a small amount (typically 10–20%) of $CO_2$ added, then power is delivered efficiently into the first vibrational state ($\nu = 1$) of $N_2$ since no intervening levels (i.e., between ground ($\nu = 0$) and $\nu = 1$) exist. Subsequently, energy is rapidly transferred to the $CO_2$ (001) state by $CO_2 - N_2(\nu = 1)$ collisions, a population inversion is created, and lasing ensues on the (001) → (100) and (001) → (020) transitions of $CO_2$ at 10.6 and 9.6 μm respectively. This pumping process, known as *vibrational–vibrational* ($\nu$–$\nu$) energy transfer, is instrumental to the efficiency and, therefore, the utility of this laser.

Another reason for the remarkable efficiency of the $CO_2$ laser is the small energy ($\sim\frac{1}{4}$ eV) required to reach the

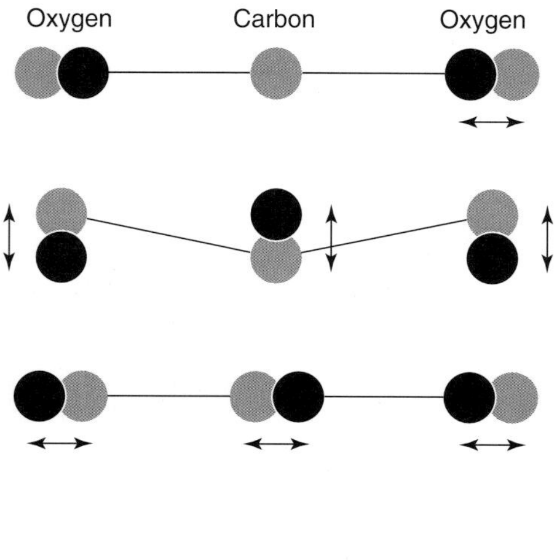


**Fig. 12** $CO_2$ is a linear triatomic molecule that has three stretching and bending modes: (top) symmetric stretch, (middle) bending, and (bottom) asymmetric stretch (after Yariv, A. [10])

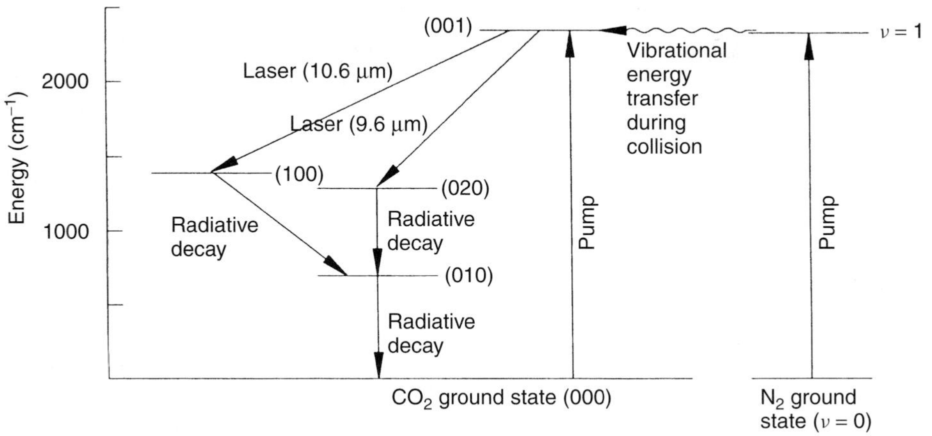


**Fig. 13** Several of the energy states of the $CO_2$ and $N_2$ molecules showing the near coincidence between the first excited vibrational state ($\nu = 1$) of $N_2$ and the lowest excited asymmetric stretch (001) state of $CO_2$ (after Yariv [10])

upper laser level – an energy possessed by a significant fraction of the discharge electrons. Contrast this situation with the He–Ne laser, which requires almost 40 times as much energy for excitation of the upper laser level. Of course, placing such stringent demands on the electron energy relegates the He–Ne laser to the high-energy tail of the electron energy distribution and severely limits the overall efficiency of the device. For a more detailed discussion of the mechanisms critical to the operation of the $CO_2$ laser, the reader is referred to Verdeyen [15].

The gas mixture for a typical $CO_2$ laser consists of He, $N_2$, and $CO_2$. Adding helium serves several purposes, among which is the improvement of heat removal from the laser discharge. When pumped by a longitudinal discharge that is 50 to 100 cm in length, the laser will produce tens of watts of power continuously. Obtaining much higher output powers (i.e., in the multikilowatt range) requires a transverse discharge such as that illustrated schematically in Fig. 14. To minimize the "residence time" of gas in the discharge, the direction of gas

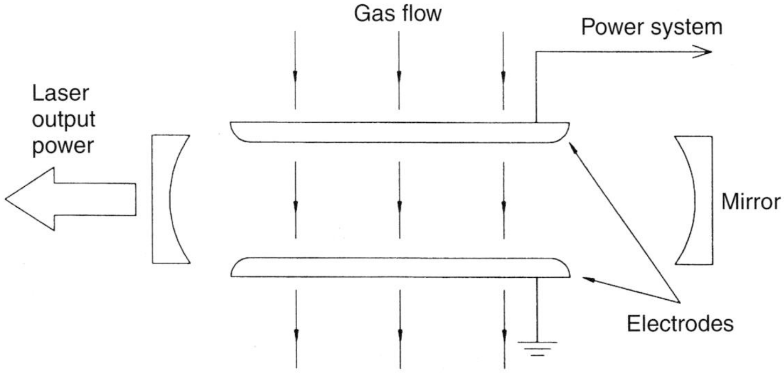


**Fig. 14** Basic features of a transverse discharge-pumped $CO_2$ laser

**Tab. 3** Key parameter and application of different $CO_2$ lasers

| ***Type (applications)*** | ***Power [CW]*** | ***Pulse energy peak power*** | ***Pulse duration/frequency*** |
|---|---|---|---|
| Sealed-off laser (Ma, Me) | 50 W $m^{-1}$ | | |
| Wave-guide laser (Mo, S) | 50 W | | |
| Slow axial flow (Ma, Me) | kW | | |
| –pulsed (Ma) | | 1 J $l^{-1}$; kW $m^{-1}$ | >μs/100 Hz |
| –Q-switch (Ma) | | | 100 ns/1 kHz |
| Fast flow (Ma) | 100 kW | | |
| TEA-laser (Ma) | | 10 J $l^{-1}$; GW | 100 ns/1 kHz |
| Gas dynamic laser (Ma, F) | 100 kW | 10 J | |
| High-pressure laser (S, U) | see TEA and Wave-guide laser | | |

Note: Ma: Material processing; Me: Medicine; Mo: Monomode laser; S: Spectroscopy; F: Laser fusion; U: Ultrashort pulses.

flow and the discharge are at right angles to the laser axis. Table 3 shows the key parameters of several $CO_2$ laser operating continuously (cw) or pulsed.

A number of other molecular lasers emitting in the infrared (such as CO, 4.9–6.66 μm) depend on vibrational–vibrational energy transfer as the dominant means for pumping the laser. A detailed account of the historical background and principal physical mechanisms for a few of these systems as well as the $CO_2$ laser is given by Garscadden et al. [16].

### 4.3 Reactive Collisions: Excimer Lasers

Several of the most important lasers – particularly in the visible and ultraviolet regions of the spectrum – rely on the transient production of small molecules that do not normally exist in the gas phase but are readily formed by collisions. If $A$ is an atom and $BX$ is a molecule (where $B$ and $X$ may consist of more than one atom each), this process is represented as

$$A^* + BX \longrightarrow (AX)^* + B. \quad (11)$$

Clearly, a new species (i.e., one that was not present in the original gas mixture) is formed that is derived from the reactants. Note that $A$ must itself be electronically excited before the reaction can proceed. A collision analogous to (11), where all of the reactants are in their ground states may well produce no new species.

The family of lasers that best exemplifies this pumping mechanism is the excimer lasers. Excimer (a contraction for "excited dimer") is the term given to a group of molecules (usually diatomic) that are not stable in the ground state but yet have chemically stable excited states. Figure 15 is a generalized energy-level diagram for a diatomic excimer molecule. Since the ground state is dissociative (unstable), obtaining a population inversion between the excited and ground states is simplified because a molecule making a transition from $(AX)^*$ to $AX$ finds itself on a repulsive potential curve and quickly dissociates (as $AX$ moves down the potential hill, $A$ and $X$ separate). The excimer laser $A$ is a rare-gas atom and $X$ a halogen, for example, $(AX)^* \rightarrow [KrF]^*$.

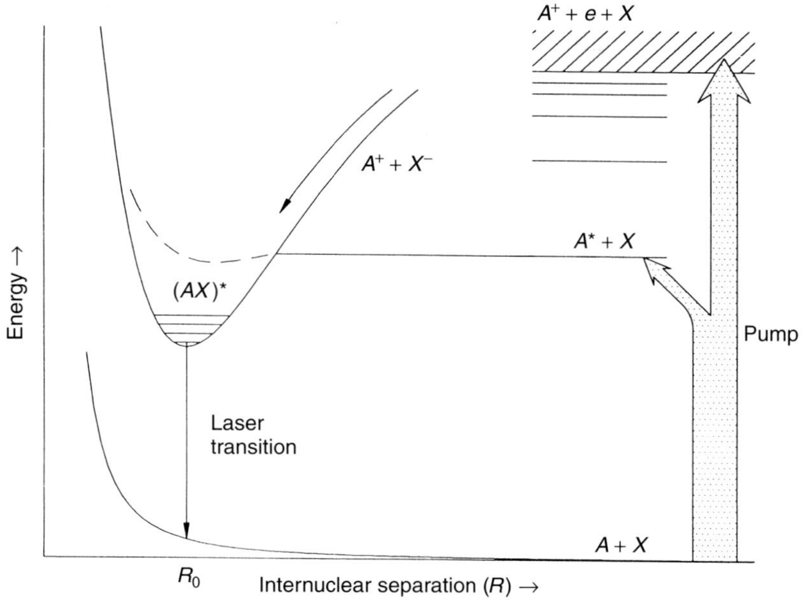


**Fig. 15** Generalized energy-level diagram for the excimer molecule $AX$ showing (1) its dissociative (unstable) ground state; (2) a stable, electronically excited state, $(AX)^*$; and (3) two collisional approaches to forming $(AX)^*$ from $A^* + X$ or $A^+ + X^-$ collisions ($A$ = rare gas, $X$ = halogen)

Consequently, the number density of ground-state molecules ($AX$) is quite small. If a method can be found for populating the $(AX)^*$ state, not only should lasing be obtainable but, because of the nature of the ground state, it will also be possible to tune the laser's output continuously over a modest spectral range.

Electronically excited excimer molecules can usually be produced by two types of collisions. The first involves only electrically neutral reactants and was given earlier in Eq. (11). As $A^*$ and $BX$ approach one another, $A^*$ ejects one of its outermost electrons, and, in so doing, becomes the cation $A^+$. The ejected, or "harpoon," electron transforms $BX$ into the anion $BX^-$. As $A^+$ and $B^-$ are drawn closer under the influence of the Coulombic force, the internal strain in $BX^-$ increases until $B$ is expelled and the $A^+$ and $X^-$ ions form the stable complex $AX$. It is easy to see why Herschbach [17] has applied the picturesque term *harpoon collision* to reactions of this type. A more direct approach is to begin with $A^+$ and $X^-$ and form $(AX)^*$ by the *ion–ion recombination process*:

$$A^+ + X^- + M \longrightarrow (AX)^* + M, \quad (12)$$

where $M$ is any third body and is required to conserve momentum. In summary, pumping the excimer lasers usually entails delivering energy to an atomic species (excited neutral $A^*$ or ion $A^+$) that is subsequently incorporated (by a collision) into an electronically excited molecule. That is, the sole purpose of the pump is to generate $A^*$ or $A^+$ (generally by electron-impact excitation of ground-state $A$ atoms), thereby initiating production of the $(AX)^*$ molecule by the harpoon or ion–ion recombination process (reaction (11) or (12) respectively).

From among those groups of excimer molecules that have lased, the rare-gas–halide lasers have emerged as the clear leader because of their efficiencies and commercial impact. Although they have been in existence only since 1975, the rare-gas–halide lasers have transformed the ultraviolet in much the same way as the $CO_2$ laser has dominated the infrared region.

It is well known that the chemistry of ground-state rare-gas atoms is limited since their outer electron shells are full. However, if the outermost electron is placed in an excited state, the atom's reactivity increases markedly, and research of the past 20 years has revealed at least half a dozen families of rare-gas–containing excimer molecules that can be produced quite readily. In collisions with halogen-containing molecules, excited rare-gas atoms produce excimers with efficiencies approaching 100%. The KrF molecule, for example, is produced by the harpoon reaction

$$\mathrm{Kr}^* + \mathrm{F}_2 \longrightarrow (\mathrm{KrF})^* + \mathrm{F} \qquad (13)$$

where the asterisk associated with the krypton atom denotes the lowest excited states ($5s^3P_{1,2}$) of the atom lasing occurs in the ultraviolet at 248 nm. A comparison of the fluorescence (spontaneous emission) and laser spectra for KrF is given in Fig. 16.

Early rare-gas–halide lasers were pumped by relativistic electron beams, but today the majority are driven by pulsed, transverse discharges. Several of the subsystems for a typical discharge-pumped excimer laser are illustrated in Figs. 17 and 18. Figure 17 is a diagram of the gas-flow loop for a commercially available laser. The gas mixture consists of two or more rare gases and a halogen-containing molecule known as the "donor" or "precursor." One of the rare gases is necessary as a component of the lasing molecule itself (such as Ar in ArF, Xe in XeCl, etc.), whereas the second (or third) is included for the purpose of stabilizing the discharge (by improving its heat-transport properties) or for increasing the efficiency with which rare-gas excited atoms or ions are generated. For the XeCl laser, for example,

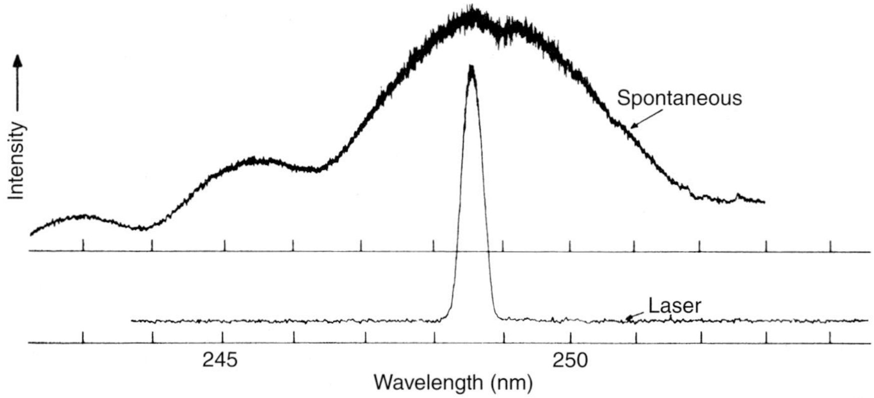


**Fig. 16** Spontaneous-emission (top) and laser spectra for the KrF molecule. For the sake of comparison, the laser emission intensity has been attenuated (after Ewing and Brau [18])

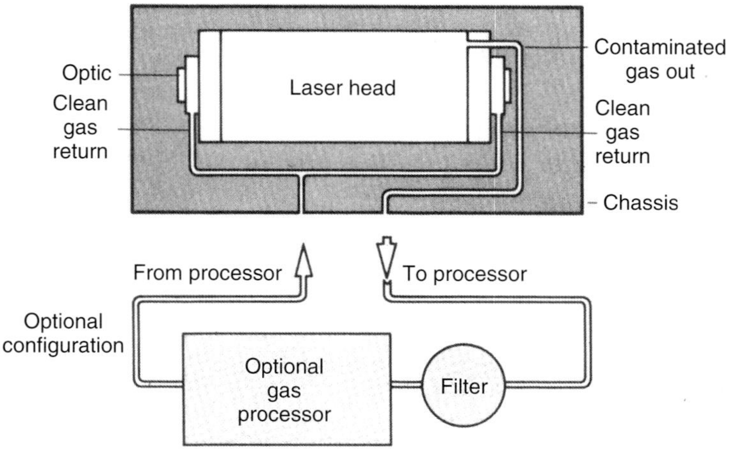


**Fig. 17** Gas recirculation system for a rare-gas–halide excimer laser (courtesy of Questek, Inc.)

a typical gas mixture is 3- to 4-torr HCl, 60-torr Xe, and 1800-torr He (where 1 atm $\approx$ 760 torr). For long gas "lifetimes," the mixture is recirculated and periodically replenished.

Continuous purification of the laser gas is useful because the reaction of the halogen donor molecule (or halogen atoms themselves) with the walls of the laser chamber frequently produces volatile molecules that contaminate the gas mixture and either interfere with the chemical reaction chain that pumps the laser or directly absorb the laser radiation. Most of these contaminants, such as $CF_4$ and other "halocarbons" (including Freons), are readily removed from the gas-flow stream by a cryogenic trap. Another factor responsible for reducing the effective lifetime of the gas mixture is the production of aerosols by the pulsed discharge. Since particulates degrade the beam quality (by scattering) and are capable of damaging the laser's optics, efforts are made to remove them with a filter Recently, sealed-off excimer laser became available.

Producing stable and uniform discharges in gas mixtures containing a halogen poses formidable engineering obstacles because of the electron affinity of the halogen atoms. As a consequence, it has proven necessary to produce a uniform background of electrons in the gas immediately prior to firing the main discharge. The subsystem responsible for doing this is known as the *preionizer* (see Fig. 18). It often consists simply of an array of sparks (produced by discharging a capacitor across two metal electrodes separated by a few millimeters) that emits ultraviolet and vacuum-ultraviolet ($\lambda =$ 100–200 nm) photons having sufficient energy to photoionize one constituent of the gas mixture, resulting in an electron density of at least $10^9$ $cm^{-3}$. This apparently mundane step is absolutely necessary for the uniform delivery of electrical energy into the gas and hence the proper performance of the laser. Without it, the main discharge is severely nonuniform and rapidly collapses into an arc.

Table 4 lists the critical characteristics of the excimer and $F_2$ lasers. Although the

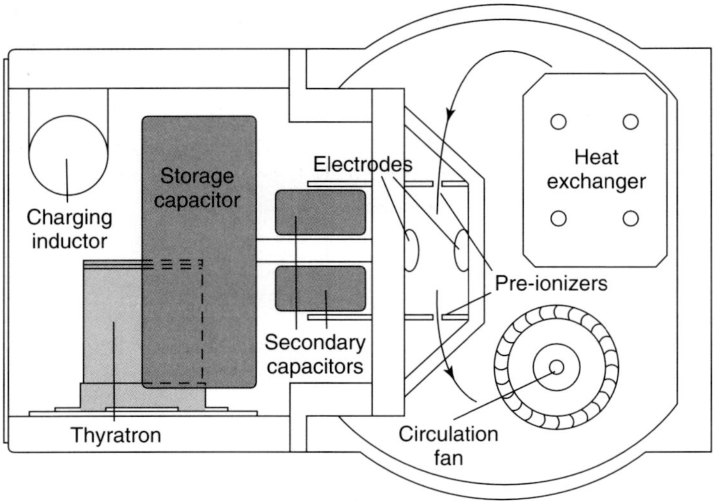


**Fig. 18** Diagram of a transverse discharge system for pumping a rare-gas–halide excimer laser. The gas mixture consists of two or more rare gases and a halogen donor such as HCl or $F_2$, and the total gas pressure is typically 1 to 2 atm (absolute). The preionizer is necessary for discharge stability. Output beams are usually rectangular in cross section with an area of 1 to 5 $cm^2$ (courtesy of Questek, Inc.)

lowest-lying excited states of the diatomic molecule (and the upper laser level, in particular) are stable, their spontaneous-emission lifetimes are typically less than 20 ns, which, in turn, demands that specific pumping powers exceeding (and generally much higher than) several tens of kilowatts per centimeter cube must be delivered to the gas mixture in order for reasonable optical gains to be realized. Such an enormous amount of power cannot be conveniently supplied on a continuous basis (nor long sustained) and, for this reason alone, most excimer lasers emit pulses of less than 30 ns in duration. Longer pulses have been obtained by pumping small volumes (a few to tens of centimeter cube) with microwave pulses or larger volumes (liters) with electron beams.

Note, too, from Table 4 that the wavelengths available from the rare-gas halides range from 351 nm (for xenon fluoride, XeF) in the near-ultraviolet to below

**Tab. 4** Key parameter for excimer and $F_2$ lasers

| *Laser* | | *$F_2$* | *ArF* | *KrF* | *XeCl* | *XeF* |
|---|---|---|---|---|---|---|
| Wavelength $\lambda$ | [nm] | 158 | 193 | 248 | 308 | 351 |
| Pulse energy | [mJ] | 60 | 650 | 1200 | 600 | 400 |
| Pulse duration | [ns] | | | 10 to 30 | | |
| Pulse frequency | [Hz] | 50 | | 250 | | |
| Mean power | [W] | 10 | 10 | 5 | 10 | 10 |
| Gas life time | [$10^6$ pulses] | 5 | 100 | 250 | 100 | 70 |

193 nm ($hf = 6.4$ eV), which lies at the threshold of the vacuum ultraviolet. As far as the average power available from these lasers is concerned, more than 100 W are now available from "off-the-shelf" devices, while the peak (instantaneous) output powers are generally 10 to 50 MW in 15- to 25-ns pulses. To achieve these results requires population inversions of $10^{14}$ to $10^{15}$ cm$^{-3}$, which manifest themselves in a small-signal gain coefficient in the vicinity of 10 to 20% cm$^{-1}$. The wide span in photon energies, when combined with the reasonable overall efficiencies (1–2%) and high peak powers of discharge devices, has prompted a virtual revolution in the applications of ultraviolet lasers to medicine, materials processing, and areas such as remote sensing. Several of these are discussed specifically in Sect. 5.

Another recently developed laser that is of increasing interest is the molecular fluorine ($F_2$) laser that operates at 157 nm (photon energy of 7.8 eV). Similar to the rare-gas–halide lasers in several respects – gas chemistry, design, and operation of the pulsed-power driver – this laser provides megawatts of peak power in a spectral region (vacuum ultraviolet) in which few other efficient lasers exist. It is well suited for discharge pumping, and recent devices have produced pulse energies >100 mJ in 99.85% He –0.15% $F_2$ gas mixtures at a total pressure of several atmospheres. However, efficient operation of this laser requires specific power loadings of the active medium in the ∼4 to 23 MW cm$^{-3}$ range, which is roughly an order of magnitude larger than that required for the rare-gas–halide excimer lasers. This is a reflection of the large ionization potentials for both of the constituents of the $F_2$ laser-gas mixture (He, $F_2$). As indicated in Table 4, the $F_2$ laser has been developed to such an extent that commercially available, discharge-pumped systems now offer watts of average output power at pulse repetition rates exceeding 50 Hz. Applications of this laser that appear to be particularly promising include microlithography and photochemistry.

## 4.4
## Chemical Lasers: HF, Coil

Another, albeit less common, manner of producing population inversions in atoms or molecules is through highly exothermic chemical reactions involving reactants that are in their ground electronic states. Of course, for such an approach to be effective, it is necessary that the reaction culminate in the production of excited species rather than the generation of heat. The best-known example of a chemically pumped laser is the hydrogen fluoride system, which is pumped by the reaction of $H_2$ and $F_2$:

$$H_2 + F_2 \rightarrow 2HF \qquad (14)$$

Usually, it is necessary to initiate the chemical reaction with an electrical spark, and the reaction leaves the HF molecule in a vibrationally excited level of its ground electronic state. Lasing occurs at 2.6 to 3 μm in the infrared from optical transitions occurring between inverted vibrational states of the molecule. The HF laser is capable of producing enormous pulse energies (>1 kJ) and continuous-wave powers (1 kW) from a commercial system. Another example of a chemically pumped laser is the iodine atomic laser, which has produced more than 10 kW of power continuously at 1.3 μm. Since oxygen is also important for laser excitation, the system is called Coil (= chemical oxygen–iodine laser).

The drawbacks associated with this pumping technique are that

1. it is usually difficult to recycle the reactants and
2. the exothermicities available from chemical reactions severely complicate efforts to obtain lasing at wavelengths shorter than the near-infrared ($\lambda \sim 1\ \mu m$).

## 4.5
## Other Lasers in the Far-infrared, UV, and Soft X-ray Regions

No discussion of gas lasers would be complete without briefly mentioning other systems that are not currently as prominent as those discussed earlier but were nevertheless instrumental as scientific tools or in developing gas-laser science and technology. The longest-wavelength lasers rely on obtaining a population inversion between rotational states of polyatomic molecules such as HCN and $H_2O$ to produce stimulated emission at a wide range of discrete wavelengths in the far-infrared ($\lambda = 10-100\ \mu m$) or submillimeter ($100-1000\ \mu m$) spectral regions. Most are driven by a shorter-wavelength laser (such as $CO_2$ at $10.6\ \mu m$) and have served primarily in laboratory roles. Applications can be found in molecular spectroscopy and other scientific experiments.

At the other wavelength extreme lie the $N_2$ and $H_2$ lasers in the ultraviolet and vacuum-ultraviolet regions respectively. Both were discovered in the 1960s, and for more than a decade, the hydrogen molecular laser (116 and $\sim$160 nm) was the shortest-wavelength laser in existence. Over the last several years, however, that distinction has been claimed by multiply charged atomic ions (such as $Se^{24+}$) that lase in the soft X-ray region when a thin metal foil of the desired material disintegrates under the influence of an intense pulse ($10^{12}-10^{14}$ W cm$^{-2}$) from a Nd:YAG (YAG denotes yttrium aluminum garnet) laser (1.06 or $0.532\ \mu m$).

The $N_2$ laser, which lases predominantly at 337 nm, served for a number of years as the optical pump for dye (liquid) lasers but in the mid 1970s was quickly displaced by the excimer and Nd:YAG (frequency-doubled or -tripled) lasers. However, the $N_2$ laser is still an interesting light source in the $\mu$s-region for fluorescence spectroscopy. Other gas (and vapor) lasers that have been studied extensively but have had a less pervasive impact are the pulsed rare-gas atomic lasers, various metal vapors, and the rare-gas diatomic excimers $Ar_2$, $Kr_2$, and $Xe_2$.

# 5
# Applications

Since its discovery, the laser has fired the imaginations of many, and it was viewed early on by some as a practical device with almost limitless industrial potential. Although applications were initially slow to develop, lasers were the sources of many technical innovations (see LASERS, INDUSTRIAL USE OF).

Scientific applications of the laser, including the study of atoms and molecules by laser spectroscopy, were among the earliest to be explored, and the ability of laser techniques to elucidate the structure of matter on the atomic scale must be considered to be among the most notable scientific achievements of the past 30 years. Other areas in which the laser has provided new scientific opportunities include environmental sensing (and probing of the earth's upper atmosphere, in particular) and the separation of atomic isotopes from one another in a process that relies on the narrow spectral bandwidth of the laser (see LASER ISOTOPE SEPARATION).

Military applications of gas lasers have been well publicized, but laser radar (Lidar) is now often used for pollution and air turbulence detection (see also LIDAR (LASER RADAR)). The generally low efficiencies of gas lasers preclude their use as a weapon, but, because of excellent beam quality (near-diffraction-limited divergence), they are well suited for target identification, imaging, and tracking.

Dramatic progress in medicine, chemistry, and materials processing, among other fields, has been made possible by the advent of the gas laser, but electronics may well, in retrospect, be the area most profoundly impacted by laser technology. Inroads made by lasers into diverse areas of electronic device and system manufacturing have recently expanded significantly and in the last decade, in particular. Because of the attractive spectral and spatial properties of laser radiation, it is fair to say that lasers are well into the process of revolutionizing electronics and computer technology along two lines. The first is as a tool in processing electronic materials by trimming, drilling, and marking, for example. Future applications are expected in large-scale microlithography for semiconductor device fabrication and the deposition of films by photoablation or photochemical processes. Because of space limitations, this section reviews only a few of the emerging and established commercial applications of gas lasers in electronics manufacturing and photochemical synthesis.

### 5.1 Material Processing

Research on laser material processing is today concentrating on diode lasers as well as diode-pumped solid-state systems but industrial applications are still dominated by lamp-pumped solid-state lasers and in an even greater extent by $CO_2$ lasers. This is due to the fact that in most applications cost efficiency is more important than using the most advanced technology. Apart from that, commercially available $CO_2$ systems for material processing from 500 W to 20 kW are established and very reliable (Fig. 19).

*Laser cutting* of sheet metal is performed by $CO_2$ and other lasers (Fig. 20). The material is melted in the focus of the laser beam and disturbing material may be removed by a gas jet. If oxygen is added, additional heat is produced by the oxidation process and the cutting velocity is raised. The laser beam can follow complicated cutting curves in three dimensions. Special advantages of laser cutting arise for hard and coated materials, plastics, ceramics, glass- and rubber-containing metallic wires.

*Laser welding* uses different processes. Welding by heat conduction is applied for thin metallic sheets. Two metals are joined and melted at the welding seam without evaporating metal. In the second process, the temperature is higher and boiling takes place. The vapor pressure pushes the liquid material in the melting seam giving an intense heat transfer.

*Laser surface processing* improves the properties of the materials by a thermal treatment. The main applications are hardening, coating, and alloying. An example is the hardening of gear wheels, valves, and other motor parts.

Gas lasers have become very widespread and are replacing conventional processing equipment for cutting and welding. For example, the Trumpf Group, Germany, sells about 1000 $CO_2$ lasers annually for macro material processing, which is about

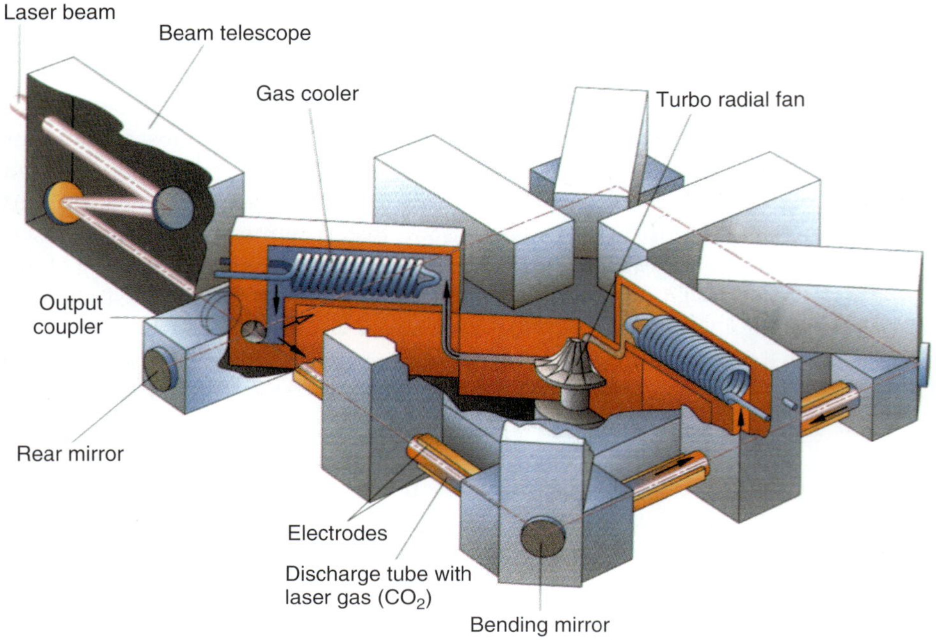


**Fig. 19** Multikilowatt $CO_2$ laser for material processing. The $CO_2$ gas in the transversal excited discharge tubes (length about 20 cm) flows with about the velocity of sound. The hot gas is cooled using a turbo radial fan and a gas cooler and leads back to the discharge tubes. A two-storied version of this laser has about 15 kW. (Courtesy of Trumpf Laser Company)

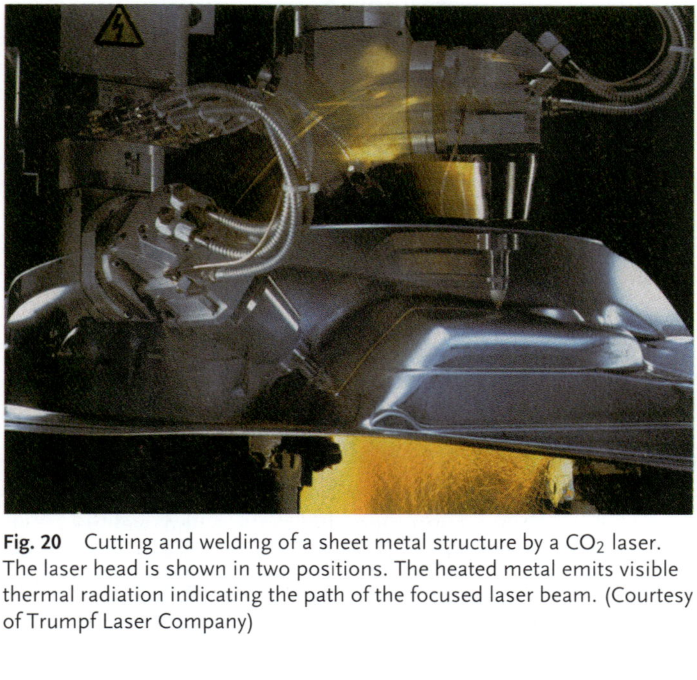

**Fig. 20** Cutting and welding of a sheet metal structure by a $CO_2$ laser. The laser head is shown in two positions. The heated metal emits visible thermal radiation indicating the path of the focused laser beam. (Courtesy of Trumpf Laser Company)

30% of the worldwide market in 2003. The market share for $CO_2$ and solid-state lasers is nearly equal.

In 2002, the market for laser material processing accounted for a worldwide total volume of about \$4 billion if the value of the laser system is included, \$3 billion for macro, and \$1 billion for micro processing. The laser sources account for \$1.5 billion. The average annual growth rate for the last 10 years has been more than 10%. For the next 10 years, a similar growth is expected.

## 5.2 Marking and Laser Evaporation

Several of the first applications to be pursued by the electronics industry – drilling, trimming, and marking of materials (see LASERS, INDUSTRIAL USE OF) – exploit the capability of coherent infrared, visible, or ultraviolet radiation to be focused to beam diameters below 100 μm. Trimming of thick-film resistors, for example, typically requires a beam spot size of ~50 μm, which is defined by the numerical aperture of the beam-delivery optics.

Marking of materials is another attractive industrial application of lasers since mechanical contact with the surface to be marked is no longer required. Rather, the laser can be located well away from the workpiece or electronic component to be identified, and the laser beam is delivered to the target via an optical transmission and focusing system. Materials that are normally difficult to mark (such as ceramics and glass) with conventional approaches are quickly handled by laser systems, which are typically built around $CO_2$ (infrared, $\lambda = 10.6\ \mu m$) or the excimer (ultraviolet, $193 \leq \lambda \leq 351$ nm) lasers. Figure 21 is a diagram of a commercially available $CO_2$-laser–based marking system capable of making up to 200 000 marks per hour on paper, polyvinyl chloride (PVC), coated metals, plastic, and glass. Since the interaction of radiation with materials is a strong function of wavelength, excimer-laser marking systems yield superior results when working with certain materials including several plastics and metals. Marking glass with a $CO_2$ laser, for example, produces extensive microcracking in the bulk because of dissipation of heat into the material surrounding the marked region. An ArF excimer laser ($\lambda = 193$ nm), however, produces clean cuts with steep sidewalls, as the physical mechanism responsible for the removal of material is primarily photochemical rather than thermal in nature. Wire cladding is also readily marked with an excimer laser. A photograph of an EFTE (Tefzel)-clad wire that has been marked with characters 0.8 mm in height by an XeCl excimer laser ($\lambda = 308$ nm) is shown in Fig. 22. Ultraviolet laser powers up to 75 W (250 mJ per pulse at 300 Hz) are commercially available from excimer marking systems. These industrial-duty systems are also widely employed in hermetic seam welding of electronic packages, spot welding of wire to lead frames, and integrated-circuit (IC) mask design and repair.

All of the applications discussed above depend on the rapid and localized heating of a surface that accompanies the absorption of laser radiation. This approach has also been successfully applied to depositing thin films and is known as laser evaporation or ablation. When intense laser pulses ($\gtrsim 1\ J$ pulse energy, 10–30 ns pulse widths) are focused onto one of a wide array of materials, the "target" is heated to the point that material is ablated from the surface. A second surface located nearby will be coated with the

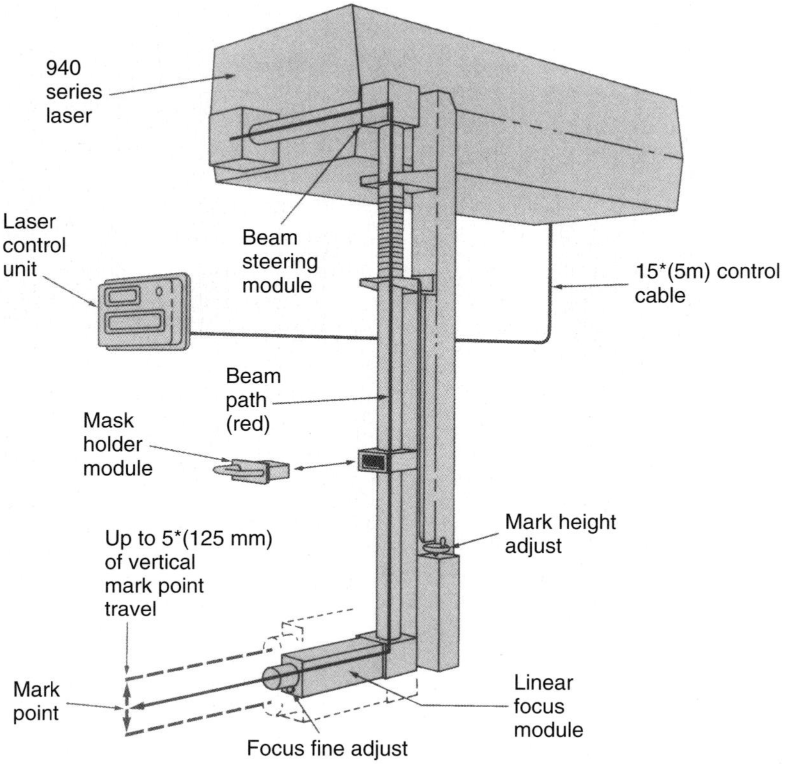


**Fig. 21** Diagram of a $CO_2$ laser marking system, illustrating details of the beam-delivery configuration (courtesy of Lumonics, Inc.)

ejected material and studies have shown that the chemical stoichiometry of the target is, surprisingly, preserved in the deposited films. One area for which laser evaporation has provided the solution to a difficult problem is in the deposition of high-temperature superconducting films. Several research groups have deposited Y-Ba-Cu oxide films on a variety of substrates by evaporating targets with $\sim 2\ \mathrm{J\,cm^{-2}}$ ultraviolet (UV) pulses from a krypton fluoride (KrF, $\lambda = 248$ nm) or xenon chloride (XeCl, $\lambda = 308$ nm) laser. The unique ability of laser evaporation to preserve the stoichiometry of even complex target materials has also permitted modulated structures composed of alternating epitaxial layers of II–VI or III–V compound semiconductor materials (such as gallium nitride) to be deposited. This feature, as well as the simplicity of the laser evaporation technique and the high deposition rates available (>5 Å/laser pulse), suggests that its importance to electronic device fabrication will grow in the next few years. Also, laser crystals like Nd:YAG and others are produced in thin-film structures by laser deposition.

### 5.3 Microlithography

Another process critical to the electronics industry that is being markedly affected by maturing laser technology

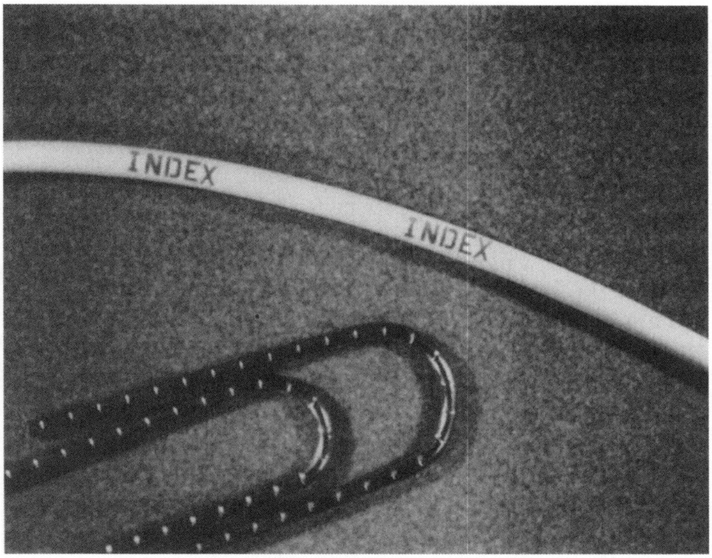


**Fig. 22** Photograph of an ETFE (Tefzel)-clad wire that has been marked by a pulsed XeCl ($\lambda = 308$ nm) laser. The characters are 0.8 mm in height, and similar results are obtained for Teflon cladding (photo courtesy of Lumonics, Inc.)

is microlithography. Traditionally, the exposure of a photoresist in IC production has been accomplished with the *g* or *i* lines emitted by a mercury vapor lamp (436 or 365 nm respectively). Shorter optical source wavelengths are desirable, however, in order to permit smaller features to be patterned onto the chip, thereby increasing the packing density. For this reason, considerable effort is being devoted worldwide to engineering development of the fluorine and rare-gas–halide excimer lasers, which offer six major, discrete wavelengths between 157 and 353 nm. Several members of this laser family, when operated untuned or "free-running," emit a beam having a spectral bandwidth of ~3 THz (1 part in 500) or, in terms of wavelength, ~400 pm). This breadth is a liability in microlithographic applications as it results in chromatic aberrations, which, in turn, degrade the system's spatial resolution. Consequently, firms in several countries have introduced frequency-stabilized, "injection-locked," excimer lasers having bandwidths of $50 \pm 15$ GHz ($3 \pm 1$ pm) that are capable (in conjunction with the proper photoresist and optical system) of patterning features smaller than 0.25 μm. Both the ArF (193 nm) and $F_2$ (158 nm) lasers are particularly promising sources for sub-0.1-μm lithography, and, it appears that Hg lamps are being displaced in production by these lasers.

### 5.4 Photochemical Processing

Virtually all of the applications discussed above draw on the spatial properties of laser radiation to induce thermal processes at a surface. One emerging area that holds great promise for microelectronics device fabrication – photochemical processing – is driven by the spectral purity of a laser. The basic concept underlying the efforts in this area is that the strengths of the chemical bonds in many small molecules are less than 6 eV, an energy that corresponds to a deep UV photon. Furthermore, if a polyatomic molecule absorbs a photon of the proper wavelength, one or more bonds within the molecule are ruptured, freeing an atom or molecular radical (see PHOTOCHEMISTRY). Exploiting such *photodissociation* processes for depositing or etching thin films has several advantages over conventional technology. A major one is that large densities of a desired atom can be produced without the need for high processing temperatures. Second, the distribution of products issuing from a photodissociation reaction changes with the photon energy (wavelength). Consequently, photochemical techniques offer considerably more control over the process than is normally available with pyrolytic (thermal) approaches.

More than 20 elemental and compound films have been deposited (or etched) by the photodissociation of polyatomic molecules. One well-studied example is aluminum. The alkyl trimethylaluminum [$Al(CH_3)_3$] is absorbing for wavelengths below ~255 nm and, for $230 \lesssim \lambda \lesssim 255$ nm, a single photon has sufficient energy to free an aluminum atom:

$$Al(CH_3)_3 + hf(\geq 4.9\ \text{eV}) \longrightarrow Al(CH_3)_2 + CH_3 \longrightarrow Al + C_2H_6 + CH_3. \quad (15)$$

Photodissociating trimethylaluminum with a KrF excimer laser has been found to be capable of depositing Al films more than 0.5 μm in thickness at deposition rates up to 1000 Å $min^{-1}$. The resistivity of films deposited from both $Al(CH_3)_3$ and dimethylaluminum-hydride (DMAH) compares favorably with bulk resistivity values and shows the films to be suitable as Ohmic contacts and interconnects.

Similarly, silicon thin films have been deposited on a variety of surfaces by photodissociating the silanes $SiH_4$, $Si_2H_6$, or $Si_3H_8$ with an excimer laser (typically ArF, 193 nm). Deposition rates generally range from 100 to 300 Å $min^{-1}$, and the film properties are comparable to those for films deposited by plasma processes. With care given to surface preparation, gas-mixture flow, and substrate temperature, epitaxial (crystalline) films have also been grown at temperatures 200 to 400 °C lower than those normally required by existing (thermal) deposition processes. Compound films such as $SiO_2$ and $Si_3N_4$ have been deposited by irradiating mixtures of a silane (typically $Si_2H_6$) and either $N_2O$ or $NH_3$ respectively, with UV photons from a laser or lamp. The high quality of the photochemically deposited $SiO_2$ films (as evidenced by pinhole densities, breakdown voltages, etc.), in particular, has led to their incorporation into commercial dynamic random access memories (DRAMs).

A similar approach has succeeded in etching semiconductor materials such as Si and GaAs except that, in this case, photodissociation is employed to free a chemically reactive species, which results in the removal of surface atoms. Photodissociating $CH_3Br$ in the UV, for example, produces free bromine atoms that will effectively etch both metals and semiconductors. Although these processes are still in the early developmental stages, several specific reactions have already been implemented in IC production and mask repair.

## 5.5
## Stereolithography

Of the industrial applications that have been envisioned and realized for UV lasers, many of the newest and most exciting involve materials processing. Another immediate example is stereolithography, in which plastic parts and models are produced by photopolymerization with a computer-controlled UV laser. Images generated by computer-aided design (CAD) systems or obtained from X-ray or magnetic resonance imaging data, for example, can now be "printed" in three dimensions, thereby allowing accurate models to be made rapidly and inexpensively. Commercial systems utilize the focused beam from a He–Cd (325 nm) or $Ar^+$-ion laser to "write" objects in photopolymer contained in vats as large as $50 \times 50 \times 60\ cm^3$ ($20 \times 20 \times 24\ in.^3$). Writing speeds up to $254\ cm\ s^{-1}$ (100 in.) $s^{-1}$ are available, which dramatically reduce the prototype production time. Automotive parts, such as an exhaust manifold prototype, which routinely took 16 weeks to fabricate, are now produced in less than 24 h. Further advances in both resolution and speed are forthcoming as several photopolymer manufacturers are developing new families of resins offering a range of mechanical properties. Other industrial materials processes that are benefiting from the properties of UV lasers include cleaning of plastic parts and the removal of polymers from electronic circuit boards.

## 5.6
## Lasers in Medicine

Lasers are used widely in medicine for diagnostics and therapy (see LASERS, MEDICAL USE OF). The diagnostic methods are applications of spectroscopy, interferometry, Doppler velocimetry, and others. There are many advantages in using lasers in surgery. They produce sterile incisions and result in reduced bleeding and operation time especially when treating highly vascular tissue. The laser beam can be coupled with different endoscopes, operation microscopes, and the slit lamp. For a wavelength less than 3 μm, flexible optical fibers are available, whereas for a longer wavelength, rotating mirror beam transport systems or flexible hollow fibers can be used. The most important gas lasers in medicine are the $CO_2$ laser, the excimer laser, the Ar laser, and, in few cases, the metal vapor (Au) laser.

The most widespread medical laser application is the retinal photocoagulation in the case of different indications, for example, diabetes. Until now, the Ar laser was used in combination with the slit lamp producing coagulation spots with a diameter of about 0.1 mm or more. The laser radiation produces a temperature rise of up to 60 °C and the tissue coagulates. In the future, the complicated and expensive Ar laser may be replaced by the frequency-doubled Nd:YAG laser or the diode laser.

Cornea surgery with the ArF excimer laser is another important application of gas lasers in ophthalmology. The photons with an energy of 6.4 eV are highly absorbed by the cells, resulting in disintegration of the molecules. The penetration depth for the 193 nm UV radiation of the laser is less than 1 μm. Thus, every laser pulse removes with high precision a small amount of cell material from the cornea. By this process of photoablation, the cornea can be reshaped, thereby minimizing or entirely eliminating the need of eyeglasses.

The $CO_2$ laser is important in various areas of surgery, for example, dermatology,

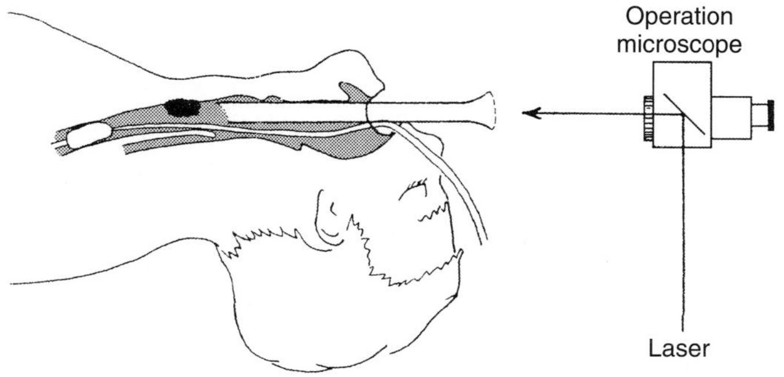


**Fig. 23** Laser microsurgery of the larynx

ENT, neurosurgery, gynecology, and others (Fig. 23). In the continuous-wave mode, the laser is used for cutting, evaporation, and coagulation with high precision. In the pulsed mode, this lasers produces surface ablation in dermatology and cosmetics.

## Glossary

**Efficiency:** A measure of the ability of the laser to convert pump power into optical radiation. For lasers pumped electrically, efficiency is often defined as the optical power emerging from the laser divided by the electrical power delivered to the gain medium. If, on the other hand, the efficiency is computed by dividing the output power of the laser by the AC power consumed by the power supply, this is said to be the "wall plug" efficiency.

**Electron-impact Excitation:** The process of exciting an atom or molecule (i.e., promoting it into an excited state) through a collision with an electron having the necessary energy.

**Excimer:** (1) A molecule that does not normally exist (that is, its ground state is unstable) but, which has one or more excited states that are relatively stable; (2) a name given to a specific family of lasers – the rare-gas–halide lasers – that emit at a number of wavelengths in the 193- to 353-nm region in the ultraviolet.

**Excitation Transfer:** A common method for pumping lasers that entails delivering energy to the upper laser level from an excited state of another species (atom or molecule) that lies nearby in energy; this is usually done by a collision.

**Infrared:** Region of the electromagnetic spectrum with a wavelength from 780 nm to 1 mm.

**Photodissociation:** Rupturing of one or more chemical bonds within a molecule by the absorption of a photon – often employed as a means of producing specific products or fragments from a precursor or "parent" molecule.

**Population Inversion:** A condition required for stimulated emission in which the number density of atoms or molecules in a particular excited state exceeds that for a lower-lying level.

**PRF (Pulse Repetition Frequency):** The number of laser pulses per second, expressed in Hertz.

**Pump:** The process or device by which the upper laser level is populated.

**Stereolithography:** The process of producing three-dimensional models of objects by photopolymerization of a liquid with a computer-controlled ultraviolet laser.

**Stimulated Emission:** Radiation generated when an atom or molecule falls from an excited state to one lying lower in energy and does so under the influence of an optical field. For this to occur, the wavelength of the field is dictated by the energy difference between the two states involved in the transition.

**Ultraviolet:** Region of the electromagnetic spectrum in which the photon wavelength is roughly between 100 and 380 nm.

**Vacuum Ultraviolet:** 100 to 200 nm portion of the spectrum.

**Visible:** Generally taken to be the region of the electromagnetic spectrum lying between 380 nm (violet) and 780 nm (red).

## References


[1] Einstein, A. (**1917**), *Phys. Z.* **18**, 121–128.

[2] Tolman, R. C. (**1924**), *Phys. Rev.* **23**, 693–709.

[3] Schawlow, A. L., Townes, C. H. (**1958**), *Phys. Rev.* **112**, 1940–1949.

[4] Javan, A., Bennett, Jr, W. R., Herriott, D. R. (**1961**), *Phys. Rev. Lett.* **6**, 106–110.

[5] Willett, C. S. (1971), Neutral gas lasers, in R. J. Pressley (Ed.), *Handbook of Lasers*. Cleveland: CRC Press, pp. 183–241.

[6] Walter, W. T., Solimene, N., Piltch, M., Gould, G. (**1966**), *IEEE J. Quantum Electron.* **QE-2**, 474–479.

[7] Bridges, W. B. (**1964**), *Appl. Phys. Lett.* **4**, 128–130.

[8] White, A. D., Rigden, J. D. (**1962**), *Proc. IRE* **50**, 1697.

[9] Bennett, W. R. (**1962**), *Appl. Opt. Suppl.* **1**, Optical Masers, 24–61.

[10] Yariv, A. (1985), *Introduction to Optical Electronics*. New York: Holt, Rinehart, & Winston.

[11] Field, Jr, R. L. (**1967**), *Rev. Sci. Instrum.* **38**, 1720–1722.

[12] Fowles, G. R., Hopkins, B. D. (**1967**), *IEEE J. Quantum Electron.* **QE-3**, 419.

[13] Silfvast, W. T., Fowles, G. R., Hopkins, B. D. (**1966**), *Appl. Phys. Lett.* **8**, 318–319.

[14] Patel, C. K. N. (**1964**), *Phys. Rev. Lett.* **13**, 617–619.

[15] Verdeyen, J. T. (1989), *Laser Electronics*. New York: Prentice Hall, pp. 334–341, 434–444.

[16] Garscadden, A., Kushner, M. J., Eden, J. G. (**1991**), *IEEE Trans. Plasma Sci.* **19**, 1013–1031.

[17] Herschbach, D. R. (1966), Reactive scattering in molecular beams, in J. Ross (Ed.), *Advances in Chemical Physics*, Vol. 10. New York: Interscience, pp. 319–393.

[18] Ewing, J. J., Brau, C. A. (**1975**), *Appl. Phys. Lett.* **27**, 350–352.

[19] Patel, C. K. N. (**1968**), *Sci. Am.* **219**(2), 22–33.


## Further Reading


Berlin, H., Müller, G. (2003), *Applied Lasers Medicine*. Berlin, Heidelberg: Springer.

Carruth, J. A. S., Mckienzie, A. L. (1986), *Medical Lasers – Science and Clinical Practice*. Bristol: Adam Hilger.

Cherrington, B. E. (1979), *Gaseous Electronics and Gas Lasers*. Oxford: Pergamon Press.

Eichler, H. J., Eichler, J. (1995), *Laser-HiTech mit Licht*. Berlin, Heidelberg: Springer.

Eichler, J., Eichler, H. (2003), *Laser*. Berlin, Heidelberg: Springer.

Garscadden, A., Kushner, M. J., Eden, J. G. (**1991**), *IEEE Trans. Plasma Sci.* **19**, 1013–1031. An in-depth treatment of the underlying physical mechanisms occurring in a number of gas lasers pumped by electrical discharges.

Hecht, J. (1992), *The Laser Guidebook*. Toronto: McGraw-Hill.

Jain, K. (1990), *Excimer Laser Lithography*. Bellingham, WA: SPIE Optical Engineering. An excellent discussion of the use of excimer lasers in microlithography.

Menzel, R. (2001), *Photonics*. Berlin, Heidelberg: Springer.

Rhodes, Ch. K. (Ed.) (1984), *Excimer Lasers*, Topics in Applied Physics, Vol. 30. Springer, Berlin, Heidelberg).

Rosenwaks, S. (1987), *Gas Flow and Chemical Lasers*, Springer Proceedings in Physics, Vol. 15. Berlin, Heidelberg: Springer.

Steen, W., Wutuins, V. (2004), *Laser Material Processing*. Berlin, Heidelberg: Springer.

Svanberg, S. (1991), *Atomic and Molecular Spectroscopy – Basic Aspects and Practical Applications*. Berlin: Springer. Discusses a wide array of spectroscopic applications of gas lasers.

Svelto, O., Hanna, D. C. (1989), *Principles of Lasers*, (3rd ed.), New York, London: Plenum Press.

Weber, M. J. (Ed.) (1982), *Handbook of Laser Science and Technology*, Vol. 2, Gas Lasers, Boca Raton, FL: CRC Press, and more recent supplements. Contains a summary of most of the atoms and molecules that have produced stimulated emission to date and their operating wavelengths.

Winburn, D. C. (1987), *Lasers*, What Every Engineer Should Know About, Vol. 22. New York: Marcel Dekker.

Willett, C. S. (1974), *Introduction to Gas Lasers: Population Inversion Mechanisms*. Oxford: Pergamon Press.

For a periodic update on laser products and applications, see the monthly publications *Laser Focus World* and photonics.

# Lasers, Industrial Use of

*David M. Roessler*
*General Motor Research and Development Center, Warren, Michigan, USA*
*e-mail: david.m.roessler@gm.com*

*Wiliam M. Steen*
*Emeritus Professor, The University of Liverpool, Liverpool UK; Distinguished Research Fellow, Material Science and Metallurgy Department, The University of Cambridge, UK*
*e-mail: W.Steen@btinternet.com*

**Abstract**
Lasers are capable of generating large and controllable quantities of optical energy and delivering it at intensities sufficient to produce permanent changes in materials. This new form of industrial energy has led to a wide range of laser-based manufacturing processes, such as cutting, welding, surface treatment, bending, cleaning, rapid prototyping, direct casting, and so on. Although these form the focus of this section, it must be noted that there are many other industrial applications that are not covered here, ranging from communications and compact disc players to medicine and metrology. Although laser-materials processing has developed and grown rapidly in the last few decades, it is still far from being a mature technology. New lasers continue to be developed as do new applications.

**Keywords**
laser-material processing; welding; cutting; drilling; surface treatment; cleaning; bending; rapid prototyping.

## 1 Introduction

The early history of the laser is well documented [1–3]. The first man-made laser, producing pulses of red light from a synthetic ruby rod, was demonstrated in 1960 by Theodore Maiman of the Hughes Research Laboratories. During the next few years, other materials were also found to lase but it was ruby that was employed in 1965 for the first significant industrial machining application of lasers. The automotive industry began using lasers for materials processing in 1969 and there are now few industries that do not utilize lasers in some way. Currently, about 60% of the value of nondiode commercial lasers sold is used for material processing, whereas only 3% of diode lasers are used this way. In terms of the number of units sold annually approximately 10% of all commercial lasers are used for materials processing, with over a third of these being used in the electrical and electronics industries.

This review excludes the many lasers used for nonprocessing applications, ranging from inspection and measurement to communication and entertainment. The average power of such lasers is usually lower than that of those used in machining and there is no long-term change resulting from the interaction of the laser beam. Also excluded are applications for medical or military purposes. Several books exist that discuss a wide range of both machining and nonmachining uses of lasers [4, 5] (Steen, 1998).

The first section of this review introduces the lasers most used in materials processing and notes the relevant properties of laser beams. A very brief account is given of the physics underlying laser processing: useful references to its extensive literature can be found in the works recommended for further reading. Each application is then discussed separately, first in terms of the process itself and then by industrial examples. Understanding the former and expanding the latter are still vigorous pursuits.

## 2 Basic Principles

The laser generates optical energy. The invention of reliable and powerful lasers has thus provided a new form of industrial energy. Indeed, the laser is currently capable of delivering the most intense form of energy available to industry today. When optical energy is absorbed by a material it may cause heating, melting, evaporation, plasma formation or even atomic fusion depending on the quantity and rate of delivery. On the other hand, the direct interaction of photons in a chemical process opens up a new area in photochemistry (see PHOTOCHEMISTRY). The prime role of the laser in materials processing is to produce permanent changes in a material. The laser may provide nearly all, or only a small part, of the total energy required, the remainder being supplied by mechanical, electrical or chemical forces.

## 2.1 Lasers and Beam Properties

### 2.1.1 Primary Industrial Lasers

Many materials, solid, liquid, and gas, have been made to lase but practical requirements greatly restrict the number used industrially. The principal commercial lasers are shown in Table 1.

**Tab. 1** Principal lasers employed for materials processing (Laser Focus World, Jan and Feb 2002)

| *Type of laser* | *% Material processing laser market* | |
|---|---|---|
| | *Value* | *Number of Units* |
| $CO_2$ laser – flowing gas | 36 | 12 |
| $CO_2$ laser – sealed | 6 | 29 |
| Solid-state laser, lamp pumped | 33 | 22 |
| Solid-state laser, diode pumped | 4 | 9 |
| Excimer | 18 | 2 |
| Diode >10 W | 1 | 0.001 |

Kumar Patel discovered laser action in carbon dioxide gas in 1964. The first commercial $CO_2$ laser appeared in 1966, with an output power of 100 W. Since then, continuous output powers of 100 kW have been obtained and 25 kW commercial lasers are already being used for manufacturing operations. Laser emission is in the 9- to 11-μm wavelength region, the 10.6-μm line usually having highest gain. The laser is excited by an electrical discharge in the gas and has an overall ("wall plug") power efficiency (i.e., optical energy output/electrical energy input) exceeding 10% (see also LASERS, GAS).

Solid-state lasers commonly used for processing applications include those made from YAG (yttrium aluminum garnet, $Y_3Al_5O_{12}$) or glass, doped with a few percent of neodymium (Nd). Nd:YAG lasers are usually excited by xenon or krypton flash lamps and emit near-infrared radiation of about 1-μm wavelength. The wall plug efficiency is only about 1% for YAG, and 0.1% for glass lasers. YAG is much more expensive than glass but has better thermal conductivity, allowing higher average output powers and even continuous operation. Glass can be made with much bigger dimensions but is more difficult to cool: this limits the firing rate to about one pulse per second, though each pulse may contain up to 100 J.

Nd:YAG lasers with high average output powers often suffer from poor beam quality, which affects how finely they can be focussed. This can be improved by means such as aperturing the beam (but thereby losing power) or changing the lasant geometry (from a rod to a slab, for example). A major development in recent years has been the use of diode-laser arrays, rather than flash lamps, to pump Nd:YAG lasers. Diode lasers both last much longer than flash lamps (e.g., 20 000 h or so compared to approximately 1000 h) and have a spectral emission that is much more efficiently matched to the absorption lines of the neodymium, thus reducing the heating effects that are partially responsible for the poor beam quality (the other significant factor is the cavity design).

The advent of multikilowatt arrays of diodes has also increased their use for direct materials processing. The wavelength of the radiation is typically below 1 μm (≈0.8 μm), a feature that can improve the ability of some materials to absorb the incident energy. A current disadvantage is that the beam quality is usually poorer than that of other materials processing lasers. The capital cost of diode-laser arrays, initially very high, is now low enough for them to be significant competitors with more traditional lasers in some applications, particularly those requiring only modest power densities such as soldering.

As shown in Table 1, over 72% of all industrial machining lasers are based upon either carbon dioxide gas or neodymium-doped solids because of their ruggedness and reliability. The remainder includes both gas (excimer, argon, copper vapor, etc.) and solid-state (ruby, alexandrite, diode, etc.) lasers.

Laser action in excimers was first reported in 1975. Emission wavelengths range from 193.3 nm (ArF) to 354.0 nm (XeF). Average output powers are typically 10 to 100 W. Excimer lasers represent a difficult technology but their short wavelength gives them unique advantages in certain applications demanding high precision or minimal heating, such as microlithography.

Other short-wavelength sources (ultraviolet and visible range) can be achieved by frequency doubling, tripling or quadrupling Nd:YAG radiation; as well as certain gas lasers, which usually have low output powers (a few milliwatts for He–Cd, a few watts for argon-ion lasers and copper vapor lasers). They are used where high-photon energies or short wavelengths are more critical than beam power, as in processing applications such as printing, lithography, or stereolithography.

The ruby laser, based on an aluminum oxide rod doped with about 0.05% $Cr^{3+}$ by weight, has a dominant emission line at 694.3 nm. Its low efficiency limits operation to a few pulses per second, though pulse energies of several hundred joules can be obtained. While still used for drilling (and in some nonmachining applications), it has largely been supplanted by newer lasers.

Industrial lasers, like other tools, must be operated safely. In the US, the Center for Devices and Radiologic Health (CDRH) requires lasers and laser systems to be certified and classified according to the nature of the hazards they represent, while the American National Standards Institute (ANSI), British and European Standards (BS and IEC) advises on their safe operation [6].

Class I lasers are sufficiently low powered to be considered nonhazardous and require no control or warning labels. The most powerful lasers, class IV, common in processing applications, can cause biological damage to the eyes and skin, whether by direct or diffuse radiation. The potential hazards in laser processing include not only the laser beam itself but also the associated power supply (a source both of potential electric shocks and of X rays), waste products (particulates, toxic vapors), fire and noise. In practice, beam-related accidents are less frequent than those from the other hazards, the most common cause of fatalities being electric shock.

### 2.1.2 Beam Properties

The high spatial and temporal coherence of laser light distinguishes it from other light sources. It is also polarized: the degree of polarisation may be constant or may fluctuate. For many processing applications, the most important characteristics are its high radiance and low beam divergence. Not only can a laser beam be efficiently propagated and steered over long distances if necessary but it can also be focused to a very small spot, producing a high irradiance (intensity or power density).

Even moderate focusing can give power densities well above those available from other light sources. A 1-kW $CO_2$ laser beam, focused to a spot of 0.1-mm radius, has an intensity of $3 \times 10^6$ W cm$^{-2}$: this is enough to melt and vaporize most solids and is several orders more intense than an electric arc.

The properties of laser beams are associated with the various longitudinal and transverse standing waves that occur in the laser cavity. The transverse electromagnetic (TEM) modes have characteristic spatial intensity distributions of light in a plane perpendicular to the direction of propagation. The simplest is the fundamental or lowest-order mode, $TEM_{00}$ (also loosely termed the *Gaussian mode*), for which the radial distribution of the intensity is given by:

$$I(r) = I_0 \exp\left(\frac{-2r^2}{w^2}\right) = I_0 \exp\left(\frac{-r^2}{a^2}\right). \qquad (1)$$

The intensity $I(r)$ decreases from its on-axis maximum value, $I_0$, to a value of $I_0/e$ at $r = a$, and $I_0/e^2$ at $r = w = a\sqrt{2}$. The total beam power, $P$, is $\pi a^2 I_0$, where a circle of radius $a$ would contain 63.2% of this power while one of radius $w$ would contain 86.5%. Higher order or mixed modes have more complex spatial intensity distributions but, for practical purposes, the beam size can still be defined by the area containing a given fraction of the total power.

The spreading of a laser beam as it propagates out of the resonator cavity depends on the resonator design and on the modes that are present. It is characterized by the half-angle divergence, $\theta$ (the semivertex angle of the conical envelope of the spreading beam), and increases with mode order. It may be only 0.2 mrad in He–Ne lasers, from 1 to 10 mrad for $CO_2$ lasers and generally near 10 mrad for high-powered YAG lasers.

Focusing of laser beams can be accomplished by conventional optics. For a $TEM_{00}$ mode, the radius, $w_0$, of a diffraction-limited focal spot is given by:

$$w_0 \cong \frac{\lambda f}{\pi w_l}, \qquad (2)$$

where $w_1$ is the radius of the beam at the focusing optic (of focal length $f$). With good optics the minimum spot size can be made comparable to the wavelength of the light. In practice, the spot size will often be larger than this but, for both low and higher-order mode beams, is linearly related to the half-angle beam divergence $\theta$ by:

$$w_0 \cong f\theta, \qquad (3)$$

The intensity decreases with distance from the focal plane because the beam radius increases from its minimum value of $w_0$. The working depth of focus, $\delta$, is the length on either side of the focal plane for which the intensity can adequately sustain the process of interest. The "depth of focus" is defined as the region in which the focussed diameter changes $\pm 5\%$. For a Gaussian beam, $\delta$ is about $w_0^2/\lambda$. The best position of the focal plane is not necessarily the top of the workpiece: in some applications the two may be displaced by over 1 mm.

The polarisation of the beam describes the alignment of the electric vector of the electromagnetic wave stream. If a laser beam is not perpendicular to the work surface, the amount of absorption will generally vary with the state of polarization and the angle of incidence. The effect is significant in cutting processes and circularly polarizing optics is usually incorporated to minimise such variation (see below).

## 2.2 Principles of Laser Processing

### 2.2.1 Laser–material Interaction

Laser radiation, which is a form of optical energy, is an electromagnetic wave stream, an oscillating electric and magnetic field. This radiation can therefore exert a force on small electrically charged particles such as electrons. If they move they will reradiate, leading to reflection, but if they are held within the lattice structure (phonon structure) they will exert a force on the lattice leading to vibrations. This vibration is initially perceived as heating. As the incident beam power is increased, these vibrations increase in intensity, the lattice breaks down and melting or even vaporization can occur. If the beam intensity is increased further, the collisions in the vapor remove electrons from atoms and partial ionization occurs, forming a plasma. The plasma has free electrons and is more able to absorb the beam; the plasma temperature may rise to near 8000 K, as in keyhole welding, giving an intense blue–white light in argon-shrouded processes. If the plasma becomes detached [7] it may interfere with the process. Table 2 shows the power intensities usually required for these different events to occur together with the associated areas of material processing.

Only the laser light that is absorbed can contribute to changes in the material. The fraction of the total incident laser energy that is transferred to the workpiece is termed the *coupling coefficient*, $\gamma$. It depends on numerous parameters, including the angle of incidence and polarization of the radiation, the optical properties of the material (which vary with wavelength, temperature, and surface condition) and the presence of any intervening matter. The instantaneous value of $\gamma$ may change considerably during the course of the irradiation. In deep-penetration welding, for

**Tab. 2** Relation between laser power intensity, subsequent thermal process and corresponding materials processes

| *Power intensity [W cm$^{-2}$]* | *Stage of heating* | *Associated material processes* |
|---|---|---|
| $<10^5$ | Thermal conduction heating | Transformation hardening, bending |
| $10^5$ | Melting | Welding, surface melting, alloying, cladding, marking, glazing. |
| $10^6$ | Vaporization | Cutting, drilling, lithography |
| $10^7$ | Plasma production | Cleaning, shock hardening |
| $10^9$ | Atomic fusion | Nuclear power production |

example, the initial value of $\gamma$ may be much less than 0.1 before melting occurs but will be close to unity once substantial vaporization has occurred.

Figure 1 summarizes some general features of laser-surface interaction.

Few laser-material processing interactions are sufficiently simple to be described by models that yield analytic expressions for the temperature increase. Such models usually ignore the temperature dependence of the thermal and optical properties of the irradiated material and have limited value once melting or vaporization occur. There are useful collations of many of the more interesting analytic solutions [8–11] but most detailed analyses of laser processing rely heavily on numerical techniques. Numerous summaries of laser-processing models are available [12–16].

### 2.2.2 Processing Regimes

In general, energy densities of at least 0.1 J $cm^{-2}$ are needed to produce permanent changes in materials. Significantly higher-energy densities ($10^2$–$10^5$ J $cm^{-2}$) are needed for many common applications such as cutting or welding. The energy density alone does not define what process occurs: that depends on the rate at which the energy is supplied. Figure 2 shows how different processes are controlled by the incident intensity and the interaction time. It is schematic: the limits for each regime will depend strongly on the material of interest.

Low-power densities for relatively long interaction times allow significant thermal penetration prior to surface melting. This regime is useful for solid-state changes. Higher intensities and shorter interaction times steepen the temperature gradient within the material. Once the surface melts, processes such as welding can occur. For many materials, intensities near $10^7$ W $cm^{-2}$ and interaction times of 1 ms produce surface vaporization before there is significant thermal penetration. In such cases, ablation rather than melting will occur.

While the interaction depends strongly on the laser power density and interaction time, it is also affected by the wavelength

**Fig. 1** Laser-surface interaction. Schematic of the primary features and phases involved in many laser machining processes. The relative contributions of melting, vaporization, and plasma formation are determined by the intensity and duration of the irradiation

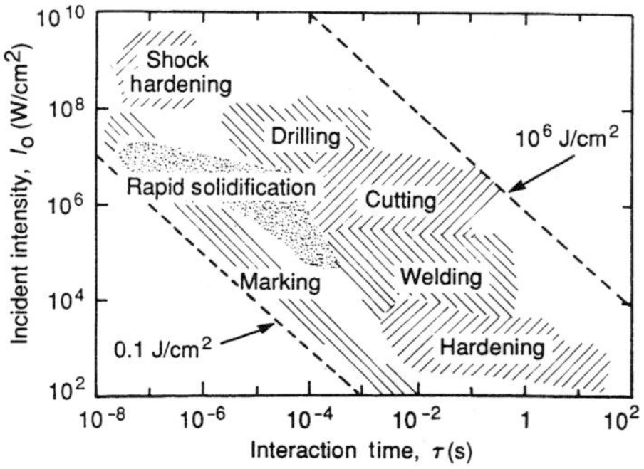


**Fig. 2** Roles of incident power density (intensity) and interaction time in determining what laser processes occur. Short interaction times limit thermal penetration: high-power densities increase vaporization. Energy densities used for laser-processing range from a few hundred mJ $cm^{-2}$ to many thousands of J $cm^{-2}$. For a given energy density, the *rate* at which the energy is delivered controls the particular process that occurs

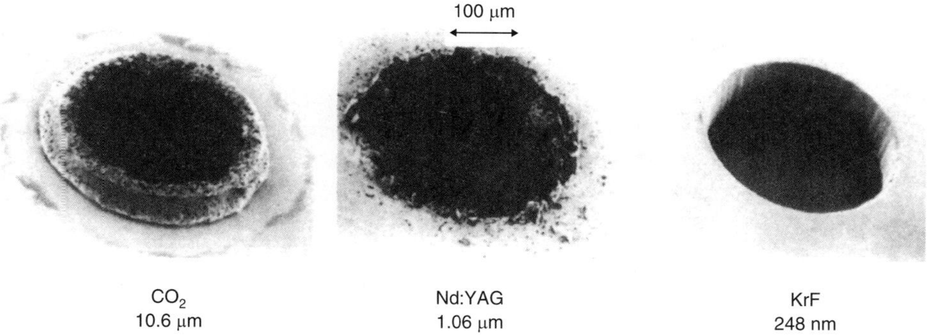


**Fig. 3** Scanning electron micrographs of 300-μm diameter holes made in a 75-μm-thick polyimide sheet with three different lasers. The amount of thermal damage depends strongly on the type of laser and is least evident for the excimer (KrF). (Adapted from work of, and used by permission of, T. Znotins, Lumonics, Inc.)

of the radiation. Figure 3 shows scanning electron micrographs of 300-μm diameter holes made in 75 μm-thick polyimide sheet with three different lasers: a long-pulsed $CO_2$; a *Q*-switched, short-pulsed, Nd:YAG; and a KrF excimer laser. The laser-material interaction is different for each case, varying partly with the wavelength-dependent coupling coefficient and partly with the nature of subsequent heat transfer and material removal. Thermal effects are least with the excimer laser [17] possibly due to direct bond breaking with the high-energy photons of ultraviolet light (≈4.9 eV, characteristic of bonds in organic material), as well as the shorter interaction time.

### 2.2.3 **Beam Optics and Delivery**

Key elements include mirrors, lenses, beam splitters, circular polarizers and fiber-optic guides. In all cases, absorption should be low to avoid thermal distortion or damage by the beam. Air or water cooling may be used to limit temperature increases and a slight overpressure of gas is often used to prevent contamination of the optics.

Lenses are sometimes antireflection coated to increase their transmittance; mirrors may have dielectric coatings to enhance their reflectance and damage threshold. Quartz is commonly used as the lens material for visible and near-IR lasers: GaAs or ZnSe is used for medium- or high-power $CO_2$ lasers.

Mirror substrates should have high thermal conductivity, $k$, low coefficient of thermal expansion, $X$, and low absorptivity, $A$. Copper is frequently used, with or without coatings. However, while bare copper can withstand irradiances of $10^5$ W cm$^{-2}$, it is soft and oxidizes readily. A molybdenum surface improves the durability but lowers the reflectance. Silicon has a low coefficient of thermal expansion and the highest figure of merit ($k/AX$).

Diffractive optic elements (DOE) are commonly used for beam shaping. They are essentially holographic mirrors or lenses etched to several layers. They will transform a beam of known mode into almost any shape required, from a simple series of points to a complex pattern. Adaptive optics are also used for transmissions of power over long distances. They consist of flexible mirrors whose shape is adjusted by pistons (e.g., piezo-electric transducers) and can, in real time, correct for mirage effects and atmospheric fluctuations.

Circular polarizers are fitted to most cutting lasers to eliminate the directional effects of a polarised beam. They consist of two mirrors. The first mirror is inclined at an angle of 45° to the plane of polarisation thereby splitting the polarisation into two orthogonal vectors. The second mirror has a 1/4-wave film capable of retarding the phase of the vector in the plane of incidence by a half-wavelength. The beam from the second mirror recombines the two polarisation vectors with the result that the plane of polarisation changes by 90° every wave, thus appearing to rotate.

Processing applications commonly entail relative motion between the beam and the workpiece. There are three possibilities: the beam is fixed and the part is moved; the beam is moved and the part is fixed; or both the beam and the part are moved. In low-power systems, the laser head itself, not just the beam, may be moved. Other refinements include the possibility of a single laser servicing several workstations, via beam splitting or time-sharing. A further general classification occurs depending upon the laser wavelength and the availability of suitably transmissive or reflective optics for beam manipulation.

Beam manipulation via mirrors is possible for all laser wavelengths, and considerable flexibility is afforded by the use of articulated arms. The disadvantages increase with the number of mirrors. They include: the cumulative loss of power when many successive reflections are involved; the need for added cooling (increasing mass and complexity) in high-powered laser systems to avoid distortion and damage to the optics; difficulties in alignment of the beam within the system; and difficulties in maintaining pointing accuracy and reproducibility over long path lengths (e.g., mirage effects).

Beam manipulation via flexible fibers requires the availability of low-loss optical light guides that have mechanical properties compatible with the bending and durability requirements. Currently, no materials adequate for high-power $CO_2$ lasers have been developed. However, quartz fibers have greatly increased the versatility of high-powered Nd:YAG lasers. Such fibers can transmit average powers exceeding 1 kW and enable the output from a single laser to be shared in power or time between several workstations or the outputs from several lasers to be combined.

Metallized waveguides, hollow channels with highly reflecting internal walls, have been used primarily with lower powered lasers. They can be flexible and may transmit over 90% of the initial power per meter of length.

Costs can be reduced, and performance enhanced, by integration of the laser with the beam delivery system. Several laser-robot devices are now available that give at least five axes of motion for the laser beam but where only four or five internal mirrors are required. Telescopic and articulated arms give flexibility but, where reach is extensive, positioning accuracy may suffer.

#### 2.2.4 General Advantages and Limitations

The laser provides unique processing opportunities that can be summarized as follows:

1. Delivers precisely controlled localized irradiance to any point in "line of sight".
2. No mechanical contact needed.

3. Can transmit power through transparent materials to enclosed or restricted targets (e.g., radioactive areas).
4. Requires no special atmosphere except process shrouding (whereas electron-beam processes require at least a partial vacuum).
5. A chemically clean form of energy.
6. Fast manipulation of the low inertia beam is possible.
7. Fixturing is simplified because of the lack of vibration or mechanical stress from the tool. Some fixturing or clamps may be required to ensure part fit-up or to contain the slight thermal distortion.
8. Total heat input can be regulated by process speed, power, focus or pulse shape (width and repetition rate).
9. Thermal damage or distortion can be very small, allowing treatment of sensitive components.
10. Can treat most materials regardless of hardness or other properties.
11. Easily automated with good access for sensors and other process monitors.
12. Little after-treatment is usually required.
13. High processing speeds, hence high productivity.
14. Equipment reliability (unscheduled downtime is typically less than 1%).

There can be disadvantages, some of which stem simply from unfamiliarity with the technology and how best to use it. However, the decision whether to use lasers is usually based upon economic considerations. Capital costs for laser installations are high: they include both the laser and the associated apparatus (beam delivery system, gas handling, chiller, parts handling, waste disposal, and safety features). The laser itself may account for well under a half of the total cost of some systems. In general, Nd:YAG lasers are more expensive (per output watt), and excimer lasers are much more expensive, than comparably powered $CO_2$ lasers.

Operating costs are moderate compared with alternative processes. Major items are the high electrical power consumption, flash lamps (for solid-state lasers), and the amount and nature of any assist gases used. Until more detailed data for a particular application are obtained, a very crude but useful starting assumption is that the cost of operating the laser over its useful life will probably be comparable to the initial capital cost of the system.

The laser is an expensive source of energy. The business case for its use can therefore be greatly enhanced by taking advantage of its high productivity and by ensuring that there is a high use factor (i.e., that the laser beam is being used for a significant fraction of the total process cycle time).

## 3 Major Applications

This section discusses some of the common industrial applications of lasers by process. Useful compilations of materials processing data have now been published [18, 19].

### 3.1 Surface Treatment

In many applications, the surface properties of a component (optical, electrical, chemical, mechanical, and so on, including cleanliness) are critical. Much industrial interest has focused on improving resistance to impact, wear, corrosion, and oxidation. Materials that intrinsically possess satisfactory properties are often expensive and may be difficult to machine by virtue of the very attributes desired (hardness, for example).

The surface properties of some common materials can be improved by laser irradiation and made comparable to those of more expensive ones. This modification may require only a thermally induced change in the material or it may involve addition of other materials to the surface. Only a thin surface region is affected and the process can be precisely controlled [20].

### 3.1.1 **Transformation Hardening**

It has long been known that the mechanical properties, such as hardness, of some materials may be significantly altered by heat treatment. The transformation hardening of ferrous alloys takes advantage of certain metallurgical changes that can be thermally induced while such metals remain solid.

Lasers have been used for heat treatment since the early 1960s and offer several benefits over traditional furnace, flame or induction hardening. Distortion is low because only the area of interest need be heated; the case depth (hardened thickness) can be tightly controlled; and the heat produced by optical absorption at the surface rapidly diffuses into the underlying metal, allowing self-quenching without the need for water or gas jets.

For such applications, the cross-sectional area of the incident beam may be as large as several square centimeters. The beam shape can be modified by rastering, defocussing or the use of DOE. The requirement is to create a uniform temperature, typically just below the melting point, at the surface of the area to be hardened for long enough for the transformation temperature to be reached at the required depth of hardening (usually $<1$ mm) at an economic covering rate. The depth of hardening is related to the parameter $P/\sqrt{(D/V)}$ – where $P =$ absorbed power (W), $D =$ spot size (m) and $V =$ traverse speed ($ms^{-1}$) [21]. Typical values of these parameters are $P = 100$ to 1000 W, $D = 0.5$ to 2 cm and $V = 5$ to 150 mm $s^{-1}$. The absorbed power, $P$, is usually enhanced by the use of an absorbing coating such as graphite or special high-temperature paint, or by surface sand blasting, oxidation or corrosion. Hardening inside holes can be achieved by using a polarised beam incident at the Brewster angle (approximately 80° for metals) [22]. The transformation of ferrous alloys to the hard martensite phase also results in an expansion of the structure (up to 4%). This leaves the hardened zone under compression and is useful for increasing fatigue resistance, as in the treatment of crankshafts.

Figure 4 shows the depth profile of a steel alloy that has been hardened by a 2-kW $CO_2$ laser moved across the surface

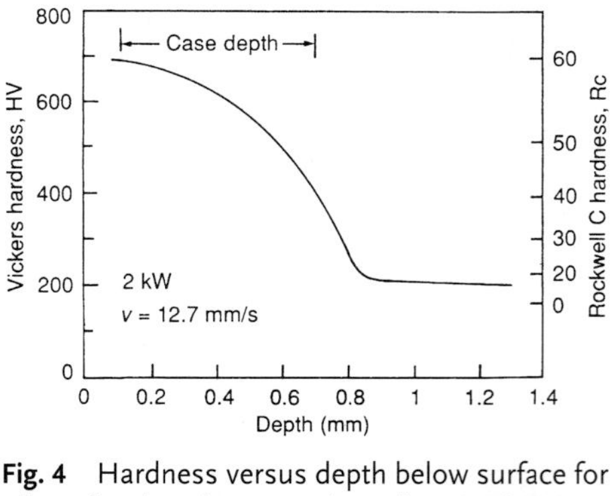


**Fig. 4** Hardness versus depth below surface for a laser-hardened iron–carbon alloy. A 2-kW $CO_2$ laser was scanned at 12.7 mm $s^{-1}$ across the surface, the subsequent heating and cooling rates being greatest near the surface. The temperature of material at least 0.9 mm below the surface did not reach that needed for austenitization to begin. (Adapted from work of J. R. Bradley, General Motors R&D Center)

at 12.7 mm $s^{-1}$. The alloy contained about 0.4% C and, initially, was mainly pearlitic, a lamellar mixture of ferrite ($\alpha$-Fe) and carbide ($Fe_3C$, cementite) phases. After the laser treatment, the surface zone is fully martensitic with a Vickers hardness approaching 700 (Rockwell C hardness of 60). The hardness remains above 600 HV (55 Rc) for a case depth of about 0.5 mm. The underlying layer, corresponding to incomplete dissolution of cementite ($Fe_3C$) and diffusion of carbon, shows decreasing hardness with depth. At depths greater than 0.9 mm, temperatures did not increase enough for austenite to form and the material retains its original structure and softness.

Industrial applications are numerous, including the edges of many cutting tools, gear teeth, crankshafts, diesel engine cylinder liners, valves, piston ring grooves and cylinder bores. The first intensive use was the hardening of the cast iron housing of automobile power steering gears, begun in 1973 to contend with higher front-end loads on vehicles. A 1-kW continuous wave (CW) $CO_2$ laser puts a stripe pattern in the housing, selectively hardening only those regions known to experience high stress. Heat input is thus minimized, less than 0.1% of the total housing mass being significantly affected. Distortion is very low and there is little scrap. The rate of wear was decreased by 90% at a cost that was only 20% of that of alternative techniques, such as furnace hardening or nitriding. Furthermore, castings can be ordered on the soft side of their hardness specifications to improve machinability.

### 3.1.2 Surface Melting

With adequate heating the surface melts, allowing homogenization of material. At high laser intensities and short interaction times, the cooling rates following irradiation are very high ($\approx 10^6$ $Ks^{-1}$), far exceeding those found in many conventional techniques, and the uniformity becomes frozen in. Metastable glassy or amorphous phases can be formed in certain alloys that have wear, corrosion resistance, and anomalous magnetic properties that are markedly different from those of the bulk material [23].

The technique is relatively new and little exploited, although interest is growing for applications such as automotive camshafts and tool steels, and in the modification of coatings applied by conventional plasma and flame-spray. While preprocess and postprocess heating can reduce surface stress cracking and warping (from the rapid heating and cooling), there is generally more distortion than in transformation hardening. The surface melting of grouting in tiles using diode lasers is a novel way to ease cleaning.

### 3.1.3 Alloying

Major surface modifications are possible when chemical additions are made. The desired material may be added to the laser-generated melt pool in diverse ways: as a powder, reactive gas, wire, sheet or coating prior to the laser scan. Strong convection currents in the melt pool, due mainly to the steep thermal gradients, ensure thorough mixing. The alloyed material forms a dense, fine-grained, homogeneous layer extending up to 1 mm or so below the surface [24]. Industrial applications commonly employ kilowatt-class $CO_2$ lasers and interaction times of 10 to 50 ms.

This technique has been used for numerous engine components (valves, piston rings) and diverse tools, such as saw blades. Inexpensive steels can be surface-alloyed with elements such as nickel or chromium to become as corrosion-resistant as stainless steel. Variants of the process include "layer glazing", particle injection (by metal carbides, for example) and patterning of ceramic tiles by melting painted glazes.

### 3.1.4 Cladding (Hardfacing)

Like alloying, laser cladding requires the addition of a new material to the workpiece surface: unlike alloying, not all the additive is mixed into the substrate to form an alloy. Some remain undiluted and constitute the new surface, which therefore has the same properties as those of the added material. The laser provides enough heat to produce a metallurgical bond between the cladding material and the base metal. The heat input must be carefully controlled: extensive melting of the base metal will result in undesirable dilution of the cladding material and possible loss of the properties required. A good bond demands some dilution but this is typically less than 10%, which may be held in the interface area.

Laser cladding has been investigated since 1974 and has advantages over alternatives such as plasma spraying and arc welding [25]. Many of the applications are similar to those for alloying, ranging from automotive valve seats to oil-drilling tools. A popular example is the use of a cobalt-based alloy to clad wear pads on jet engine turbine blades. These are made of a cast nickel-based superalloy that is optimized for creep strength. The high ambient temperatures (1600 K) and continuous blade vibration shorten blade edge life unless wear-resistance measures are used. Manual TIG (tungsten inert gas) processes to improve wear-resistance were time consuming, often diluted the nickel in the blade and had poor reproducibility. A 2-kW $CO_2$ laser avoids these problems and is ten times as fast, greatly reducing costs. Consumption of the expensive cladding material, which is deposited via an inert gas stream into the melt pool, is only half of that in TIG. The small heat-affected zone (HAZ) reduces the likelihood of heat cracking.

Cladding is also being used as a form of "reverse machining", in which worn parts such as dies or vehicle components can be rebuilt. The aerospace industry uses this technique extensively to repair turbine blades. Cladding has also been developed for low-volume manufacture by weld-build processes (see Sect. 3.2).

### 3.1.5 Other Surface Treatments

The appearance and properties of surfaces are critically affected by their topography and roughness. The ability of the laser to alter surface roughness, either in degree or nature, gives it a role in changing these characteristics.

Applications range from improving friction grip, paint flow over surfaces, optical and electrical properties, to preparing the surface for subsequent adhesive bonding. Lasers have been used for purposes as diverse as preconditioning brake linings to improving the brightness of painted metals. The latter, for example, depends both on the painting procedure and on the texture of the bare metal surface underlying the primer. A pulsed laser can be used to pattern the work roll surface at the steel mill, thereby controlling the finish of the steel sheet being produced.

Diverse surfaces, from metals and semiconductors to glasses and plastics, can be cleaned by laser irradiation. The process avoids chemical solvents and the waste disposal and safety hazards they pose (see Sect. 4.4).

### 3.1.6 Advantages and Limitations of Laser-surface Treatments

The common advantages of the laser for surface treatments are: chemical cleanliness; controlled thermal penetration and therefore distortion; controlled thermal profile and therefore shape and location of the heat-affected region; less after-machining, if any; remote noncontact processing; and relative ease of automation. The particular advantages of the laser for transformation hardening are, in addition, the self-quenching nature of the process and line-of-sight application. There is usually little scrap and the technique is particularly useful for applications in which dimensions must be tightly controlled, such as engine cylinders.

For alloying and cladding, additional advantages include: material cost savings by the use of inexpensive materials, rather than specialized alloys, for the bulk; and near-net-shape finish (thus reducing the need for subsequent machining.

## 3.2 Welding

Laser welding has been employed with many materials, including the joining of dissimilar metals. It is best suited for shallow penetrations (less than 10 mm) although much greater penetrations are possible. Speeds range from a few mm $s^{-1}$ to well over 1 m $s^{-1}$.

### 3.2.1 Process

If a laser beam of adequate power is scanned across the surface of a metal, material melts, and the melt front propagates some distance below the surface. At high speeds the heat penetrates by conduction and convection through the melt pool – "conduction-limited welding". Surface melting is achieved in less than a microsecond. At lower speeds or higher powers the surface may boil, leading to a pit that helps to trap the beam by multiple reflections enhancing the depth of the boiling front – "keyhole welding". The rate of penetration of the boiling front is of the order of 1 m $s^{-1}$. Figure 5 shows cross sections of a 6.5-mm-thick AISI 4140 steel plate whose surface was traversed by a 5-kW $CO_2$ CW laser beam at speeds of 21 to 85 mm $s^{-1}$. The etched microstructures correspond to material that melted and then resolidified (the fusion zone), and to metal that was modified by the heat but did not melt (the HAZ). The structure of a keyhole penetration is shown in Fig. 5(a). It is similar to that of an electron-beam weld. A conduction-limited weld is shown in Fig. 5(c). It has a semicircular cross section with an aspect ratio of approximately 1.5 : 1 and is similar to arc or TIG welds. For a focused CW $CO_2$ laser, the transition from conduction to keyhole welding occurs at powers near 1 kW and power densities of about $10^6$ W $cm^{-2}$.

Numerous models of laser welding have appeared since the early 1970s [26–28] and have become increasingly sophisticated as they have tried to include more of the complex phenomena in the melt pool and plasma [29–33]. There are now excellent reviews of such work [8, 12, 34].

In the keyhole regime, the irradiated surface begins to evaporate during the propagation of the melt front. The vapor produces a cavity below the beam, forcing molten material aside and allowing the incident beam to penetrate deeper. The liquid flows around the keyhole, collapses back into the hole and resolidifies while the latter moves forward. The cavity is dynamically stable for a restricted range of scanning speeds. If the speed is too low, the amount of molten material is too great to be supported by the vapor pressure

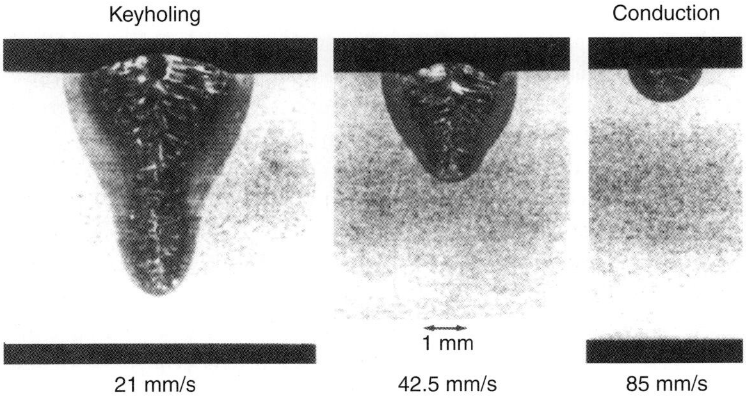


**Fig. 5** Cross sections of an AISI 4140 steel plate traversed by a cw 5-kW $CO_2$ laser at speeds of 21 to 85 mm $s^{-1}$. The semicircular profile of the fusion zone becomes elongated as the speed is decreased and as keyhole welding takes over from conduction welding. The increased penetration occurs because a cavity is produced by vaporized material below the laser beam

and no cavity can be sustained. If too fast, inadequate vaporization occurs and again no cavity is formed.

The keyhole regime has considerable practical value. The penetration is much deeper than in the conduction mode, giving aspect ratios as high as 10 : 1. The HAZ is narrowed to less than a quarter of the weld width, the welding efficiency increases and the near-parallel-sided fusion zone reduces any tendency to angular distortion. Simple energy balance considerations predict that penetration depth should increase linearly with power and inversely with speed. Empirical data for $CO_2$ laser powers up to 20 kW and for a wide range of speeds are in general agreement but give a slightly nonlinear dependence on power. Figure 6 shows the approximately inverse relationship between penetration depth and welding speed, at a given beam power, for a cw $CO_2$ laser beam scanned across a thick 304 stainless steel plate. The large bandwidth for each curve reflects the dependence upon other processing variables (focal spot size, assist gas parameters, and so on) and shows the stability of the process.

The reflectivity and thermal capacity (the product of mass density and specific heat) of metals may determine the

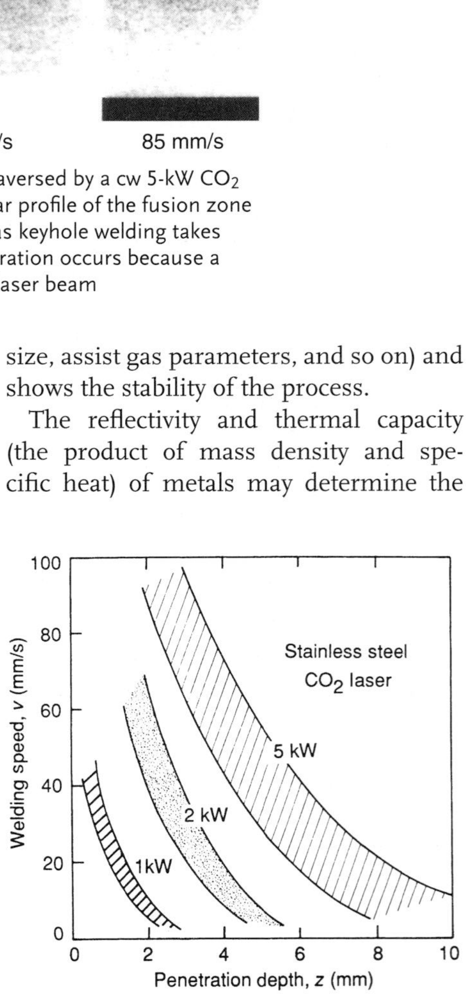


**Fig. 6** Welding speed versus penetration depth of the melt zone for $CO_2$ laser scans across a 304 stainless steel plate. The large bandwidth associated with each laser power reflects the dependence upon laser beam parameters, including mode and focusing, and upon gas-assist parameters

optimal welding procedure. Aluminum, with its high reflectivity of 95 to 98% for $CO_2$ radiation requires either a shorter wavelength or laser powers greater than 1.5 kW to initiate a keyhole. However, once formed, the keyhole will absorb nearly all the incident power. The workpiece is often tilted to avoid back reflections that could damage the beam delivery system or even the laser itself. Copper presents several challenges: its high reflectivity yields initially low coupling efficiency; its high thermal conductivity produces wide melt pools; and there is a wide temperature span between melting and boiling (1083 to 2595 °C). Welding efficiency can therefore be improved by the use of short, high-power pulses.

Some of the metallurgical properties of the weld, such as tensile strength, toughness, and hardness, may exceed those of the parent metal. In some cases, particularly for dissimilar metals or where the mechanical fit-up of the parts is poor, filler metals may be added to improve the metallurgy. Benefits range from reduction of porosity to elimination of centerline cracking.

Various gases and flow geometries, from coaxial to cross flow, are utilized in laser welding. Purposes include protection of the focusing optics from contamination (especially by debris and spatter from the workpiece), protection of the weld bead from contamination, reduction in porosity and control of the plasma or weld plume.

Inert gases are often used to blow the plasma out of the path of the incident laser beam. The direction must be controlled to avoid disruption of the weld pool. Helium reduces oxidation and plasma effects, but is expensive. Argon has a much higher molecular weight and gives excellent coverage against oxidation. However, it ionizes more readily and may produce an intense plasma. Results with $CO_2$ lasers show that the type of assist or shield gas used greatly affects the weld penetration depth. Penetration is least with argon, moderate with nitrogen or carbon dioxide and greatest with helium. The plasmas produced by Nd:YAG radiation are considerably weaker than those encountered in $CO_2$ laser welding because they are more transparent to the shorter wavelength. Penetration is virtually unaffected by the shroud gas for the 1.06-μm radiation from a Nd:YAG laser because the plume is mainly hot dust rather than partially ionized gas (the temperature is less than 2000 °C compared to nearer 8000 °C for $CO_2$ welding) [35].

The laser provides considerable flexibility in joint configurations, allowing significant materials and cost savings if the component can be designed to take advantage of the beam's unique properties. These offer one-sided access to the workpiece, precise positioning and localized heat input, and good access for in-process monitoring and control. No mechanical contact is needed and the material can be isolated if desired, the laser beam being introduced via a window or other optical means.

The low heat input results in narrow HAZs, little distortion, and negligible thermal damage to surrounding areas, especially an advantage in the welding of fragile or heat-sensitive components. Heat inputs can be reduced further by the use of a pulsed laser rather than continuous mode. The laser avoids some of the limitations imposed by electron beams (*e*-beams), the competing keyhole welding process: there are few X-ray hazards or potential problems from strong electric and magnetic fields, and the welding does not require a low pressure ambient.

For many metals, weld quality is consistent and good. Laser welding is fast, has an attractive weld bead and is easily automated. Uptime is high (90% is routine and 95% is sometimes stipulated at the time of purchase) and operating and maintenance costs can be kept below those of alternative welding techniques.

Not all metals are readily laser welded. The very rapid temperature changes induced by laser irradiation can lead to very hard, brittle, weld beads in some metals and can also produce hot and cold cracking – especially in high carbon steels. There may be unacceptably high porosity (for example, in the lap welding of zinc-coated steels), embrittlement and cracking, or the metallurgical properties of the weld may be inadequate in general. Filler metals may not always solve such problems. Useful data have been collated on the laser weldability of many materials [18, 19].

The advent of Nd:YAG lasers and diode lasers with average powers of several kilowatts means that high-powered $CO_2$ lasers are no longer the only option for some deep welding applications, although they continue to dominate for continuous welding at high speed. However, lasers are generally inefficient at making very deep welds. Many industrial processing lasers can make shallow penetration welds (less than a few mm) but depths above 1 cm usually require lasers with very high powers and for fit-up reasons require a large spot size, which reduces the advantage of the high beam quality of $CO_2$ lasers. Even though 100-kW-level $CO_2$ lasers have achieved weld depths of several centimeters at speeds exceeding 1 m $min^{-1}$, their high capital costs usually make them less attractive than electron beams.

In recent years, there have been some interesting developments in the use of hybrid techniques in which the materials processing is no longer accomplished by a single laser beam. Variations can be as simple as splitting the original beam to produce two adjacent spots on the workpiece or as complex as combining the laser beam with an electric arc. Benefits range from extension of the melt zone to decreased costs. The former may, *inter alia*, reduce weld porosity: the latter reduces the energy required from an expensive laser by supplementing it with energy from a much less expensive source. Most of these variations can also provide some degree of compensation for less-than-ideal joint configurations.

Poor mechanical fit-up in the components to be welded may necessitate expensive fixturing of the workpiece. Careful attention must be given to the nature and degree of any clamping that may be needed. Further, although the welding process can be automated, the laser requires special maintenance and operating skills, as well as the necessary safety precautions.

### 3.2.2 Applications

The laser has proved very versatile in a wide variety of welding applications. These range from low-power uses in small electronic components to the multikilowatt powers for ship building. Keyhole welding is not always preferred to conduction welding. For example, the latter is used in applications such as the hermetic sealing of electronic packages or where surface distortion must be negligible. Not all applications are metallic. The laser has proved useful in the packaging industry because of its ability to heat seal or weld a variety of thermoplastics. Also, clothing material can now be laser welded instead of stitched.

If speed is critical, as in the hermetic sealing of some components, the laser must be either CW or possess a high pulse repetition rate, limiting the choice to lasers such as Nd:YAG or $CO_2$. However, low pulse rates are sometimes adequate for applications such as the welding of wires to terminals in circuits. In such cases, a less costly Nd:glass laser can be used, despite a repetition rate of only about 1 pulse/s. In one application, a laser with a single operator replaces a dozen employees who had manually soldered the terminals: the laser welds are much faster and far more uniform. There is considerable interest in the use of DOE to allow higher powered lasers to weld many spots simultaneously for soldering applications or to make pulse welds of specified linear shape in a single pulse.

The laser is especially useful when existing electric and magnetic properties of the workpiece must remain undisturbed. Thus, a laser is preferred to TIG for the welding of stator cores since the smaller fusion zone reduces eddy current losses.

Thin walled or other delicate components can be laser welded because of the very low heat input, precise spot positioning and absence of mechanical contact. Sometimes the ability of the laser to make hermetic seals is critical, as in the welding of battery cases for heart pacemakers. Another example is the use of a 50-W $CO_2$ laser to make a hermetic weld between the Kovar header and the Inconel body of a microdetonator can. In this application, the heating effects must be tightly controlled to prevent damage to the electrode and to avoid premature firing of the charge. A further example is the use of lasers by jewelers to fix claws holding gemstones in place without having to remove the gemstone and with considerable saving in after-cleaning. Heat effects can be further reduced, if necessary, by use of a pulsed laser. It is sometimes possible to minimize distortion in wires or terminals if the pieces being welded together can be simultaneously irradiated (by a split laser beam, for example) rather than if all of the heat is generated in only one part.

Lasers are widely employed to perform welding operations in microcircuits, the tightly focused beam allowing the welding of very small components, such as connecting wires on computer chips, spaced closely together. Hand soldering of wires in some circuits has now been superseded by automated, pulsed laser welding, resulting in both higher quality welds and higher processing speeds while reducing labor needs. In some electronic applications, no prestripping of the insulation is needed.

The laser butt welding of aluminum window spacers (common in double-glazed windows) has become the standard production technique, outclassing alternatives in terms of quality and productivity. For this purpose, over 100 $CO_2$ lasers were employed by the 1980s.

The automotive industry employs laser welding extensively. Low-power applications include the welding of components in electronic circuit boards used by the auto manufacturers. High-power (greater than 5 kW) $CO_2$ lasers are used extensively for welding transmission components, in which there is good fit and where the low distortion and precise positioning of the weld are key benefits. Further, a single laser can serve multiple workstations, the beam being shuttled between them. The transition from *e*-beam welding to laser welding was motivated partly by poor uptimes with the former in North America. This trend toward laser welding was somewhat slower in Japan.

The laser welding of automotive body sheet metal is currently of great interest because of the savings in weight and material costs that can be realized through joint redesign and the more efficient utilization of sheet metal. Linear laser-welded joints, as opposed to resistance spot welds, offer higher strength and crash resistance as well as being a cosmetic improvement. Further advantages stem from increased productivity and reductions in floor space and workers. However, many of the potential savings are offset by difficulties associated with poor fit-up and the use of coated materials. Recently, "Tailored Blanks" have swept the automotive industry. A tailored blank is a laser-welded assembly of flat sheets of metal (usually steel) having only the properties (corrosion protection, strength, etc.) needed at a particular location. This composite sheet is then stamped into the final desired form, whether it be a door panel, bonnet, or even the complete side of a vehicle. This can greatly reduce the scrap generated in stamping operations, the use of expensive, coated, or heavy alloys only where really needed, and eliminate the need for reinforcements to be attached. These blanks can be joined together by other means (mash seam welding, for example) but the laser has now taken over because of the speed and quality of the laser welds.

A recent high-power application is the welding of ship plate and the assembly of lightweight sandwich sections for hold covers etc. The prime advantage for shipbuilding is the reduced distortion, saving many hours of straightening.

The ability of the laser beam to be directed to normally inaccessible areas makes it particularly useful in hazardous work, such as the on-site repair of nuclear fuel rods. In such cases, a Nd:YAG laser beam can be directed to the site by a long fiber.

## 3.3 Drilling (Piercing)

Drilling was probably the first major industrial application of laser processing: a ruby laser was used for the initial piercing of diamond dies used to extrude wire.

### 3.3.1 Process

The laser can be used for drilling by single or multiple pulses in one location ("percussion drilling") or by cutting a hole ("trepanning"). Material can be removed as solid, liquid, and/or vapor. Several analyses of drilling have been given [36–38] and useful references can be found in the books suggested for further reading. A simple model indicates that, for a fixed irradiance, $H$, (W m$^{-2}$), the temperature increases roughly with the square root of the duration of the irradiation. In terms of the total energy density, $E$, (J/m$^2$), delivered in a pulse of length $\tau$:

$$\Delta T \propto H\sqrt{\tau} \propto \frac{E}{\sqrt{\tau}}. \qquad (4)$$

Thus for a given amount of laser energy, higher surface temperatures are achieved with short pulse lengths. Long pulses allow greater time for thermal diffusion into the bulk: this may create more melt than can be efficiently ejected by the escaping vapor. A short powerful pulse creates more vapor and hence uses more energy to remove a given volume. There is thus an optimal pulse length. A given amount of material can be removed by either a few long (several ms) pulses or many short (less than 0.5 ms) pulses. The latter may take more time but give less taper and distortion, resulting in generally better hole quality. Lasers are now available with long pulses (longer than nanoseconds, $10^{-9}$s), ultrashort pulses (less than picoseconds, $10^{-12}$s), or any length between. Long-pulse drilling can be modelled as implied above. However, as pulse length is reduced, the absorption is eventually by electrons, which do not have enough time to equilibrate with the lattice temperature. This leads to nonthermal melting processes because of the repulsion between charged atoms. The effect on drilling is to create holes with only a small HAZ and to allow very fine lithography [39]. If peak intensities are too high, or if pulse lengths are too long, plasma formation may disrupt the process. Drilling efficiency can be improved if the time between pulses is made long enough for the plume to dissipate [4, 40].

Assist gases are often used to accelerate removal of loose or molten material and to flush out the hole. If the gas is inert, it supplies only mechanical energy. If reactive gases such as oxygen are used, the heated material may burn vigorously in an exothermic reaction. This energy supplements that from the laser and from mechanical processes. Sometimes the formation of oxides at the erosion front reduces cutting efficiency: this happens in the oxygen-assisted drilling (and cutting) of steels with high nickel or chromium content. There is generally poorer control of hole dimensions when reactive-assist gases are used.

The lack of mechanical contact avoids problems of distortion, contamination, or tool wear. The flexibility of the laser allows holes of diverse geometry, not merely circular, to be made. Holes can be made in curved surfaces, at large angles of incidence to surfaces, in hard materials and in regions that are difficult to reach by other means. HAZs are small and the process is both fast and reproducible.

The laser outclasses other methods for drilling holes only a few microns in diameter. Laser-drilled holes range from a few μm to over 1 mm in diameter. For wider holes, prohibitively high laser energies are required and the hole quality is degraded by the amount of material vaporized. Such holes are usually trepanned, the laser beam being brought off-axis through a rotating lens to cut out a circular slug of material (see also laser punch blanking in Sect. 2.4.1).

Maximum hole depth is typically up to 25 mm deep. They can be deeper but the quality usually suffers. High aspect ratios (depth-to-width) are possible: values may be 10 : 1 in metals but even higher in ceramics. Limitations are imposed by taper, especially in high aspect ratio holes, though this can be reduced by the use of long focal length optics or by continual lowering of the focal plane. Precision holes may be possible only in thin stock. In some materials, wall roughness may be undesirable. Dross on the exit side can be removed by shot-blasting or light grinding: recast on the entrance side is less easily removed but, if unacceptable, can be reduced by proper choice of processing parameters.

### 3.3.2 Applications

Lasers can drill or pierce holes in any material, whether hard like diamond or soft like cloth, brittle like ceramic or flexible like rubber. One of the earliest demonstrations of the ability to make holes was the piercing of razor blades. The punching of holes in diamonds used for drawing wires was the first industrial machining application of lasers. The laser-produced hole requires subsequent mechanical polishing but the process is still very much faster than mechanical drilling. It has now become the standard method but generally employs a Nd:YAG, rather than a ruby, laser.

Among numerous other industrial drilling applications are the holes in rubies used in watch timepiece movements, mounting holes in ceramic substrates for circuits, lubrication holes in transmission components, bleed holes in hydraulic brakes, holes in plastic medical catheters, liquid feed holes in polyethylene irrigation tubing, via holes in electronic circuits and the holes in the plastic upholstery covering the walls of passenger aircraft cabins (for sound deadening).

For ceramic substrates, rapidly repeated short pulses ($CO_2$ or Nd:YAG) are used to lessen both thermal stress cracking at the walls and the amount of resolidified material. Microcracks can be further reduced by a preheat process. The laser is faster and cheaper than diamond-tipped drill bits, and there is no wear or breakage. The latter can be reduced if the mechanical drilling is done in the "green" state, before the ceramic is fired to harden it. However, there is subsequent shrinkage and loss of precision, problems avoided in laser drilling of the fired material.

The turbine blades of aircraft engines contain numerous small air bleed holes. These allow cool air from the intake to mix with the hot combustion gases and to keep the turbine blades themselves at a desired temperature. Mechanical methods are precluded because of the difficulties in drilling 1-mm diameter or smaller holes at various angles through the curved surfaces of very tough, heat-resistant alloys. Pulsed Nd:YAG lasers can drill such components at about 1 hole/s.

New aerospace applications include the drilling of millions of holes in aircraft wings for purposes such as improving air flow characteristics and lift, and to allow a very thin sheet of oil to be sprayed out to combat ice build up.

The laser is widely used to make the holes in the plastic nozzles of aerosol cans. The submillimeter-diameter holes are difficult to make with conventional techniques and it is time consuming to make the modifications required when hole sizes are changed. The laser has the necessary flexibility, and can drill differently sized holes in the same component if necessary by simple changes in the beam parameters. High-precision holes are drilled by excimer lasers for ink jet printers. The hole shape can be adjusted for taper, circularity, and even fluting.

Baby bottle nipples contain small-sized holes, typically less than 0.5 mm in diameter. The soft rubber is easily deformed and traditional techniques give clogged or nonuniform holes, usually with a residual flap. These problems have been avoided since the mid-1960s by use of a pulsed $CO_2$ laser system that can drill several nipples per second, each nipple containing one hole for milk and two for air.

An intriguing application is the laser drilling of holes in time-release medicinal capsules. These holes permit the slow, controlled release of the contents once the capsule is ingested. The laser is uniquely suited to making the very precise holes needed in the curved surfaces of the capsules with negligible thermal or mechanical damage.

## 3.4 Cutting

Cutting with laser radiation is the most common form of material processing today. It differs from laser drilling in that the fully penetrating hole is traversed, creating a cut. The label "trimming" is sometimes used for cutting applications specifically involving the removal of flash and burrs from curved parts.

### 3.4.1 Process

The laser can cut in several ways: these are primarily by vaporization; fusion and blowing; fusion in a reactive gas and blowing; chemical bond breaking; and thermal stress cracking. It is also used to scribe a series of perforation holes allowing subsequent mechanical breaking, as with alumina substrates for electronic applications. The laser may provide only part, or all, of the energy required for these processes, the remainder being contributed by assist gases or by exothermic reactions at the cutting surface [13, 41]. From these processes there are two main classes: sublimation cutting, in which material is removed primarily by vaporization; and fusion cutting, in which molten and other material is flushed out by assist gases. The main features of a laser cutting head are shown in Fig. 7.

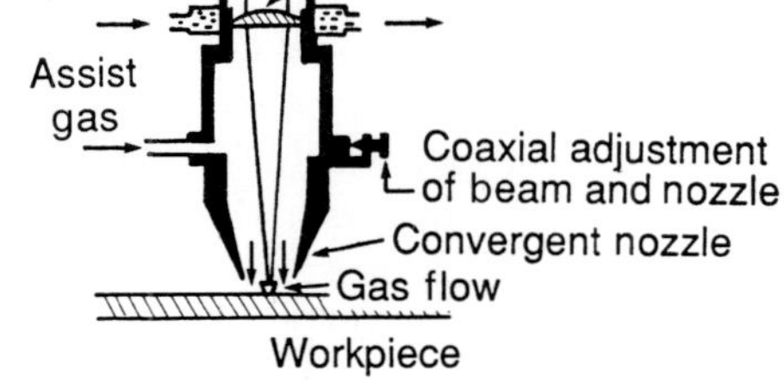


**Fig. 7** Primary features of a laser cutting head, including focusing optics and assist gas provision. Accessories to monitor or control height of the head above the workpiece are not shown

Sublimation cutting is suitable for materials such as plastics, which have low vaporization temperatures and small latent heats of vaporization. Fusion cutting includes the use of either inert gases, to provide the mechanical force to flush out material, or reactive gases, which additionally promote exothermic or other chemical reactions at the surface. Assist gases are commonly directed into the kerf by a jet that is coaxial with the laser beam (Fig. 7). They also serve to cool the cut edges, reduce back-spatter, and protect the focusing optics.

Inert gases reduce unwanted oxidation and are imperative if burning is to be avoided in flammable materials. Some steels form high melting point oxides (e.g., stainless steel, forming chromic oxide dross): these may increase the formation of dross. Some oxides are very hard and, if left on the cut edges, may damage other materials that they may contact. Cleaner and smoother edges can be obtained with an inert assist gas such as argon or nitrogen. In cutting metals with inert gas, it is usual to use very high pressures (e.g., 14 bar) and suitably tough optical components together with supersonic nozzle designs.

Reactive-assist gases can dramatically increase cutting rates. In cutting steel, for example, with oxygen assist there is an operating window. If the speed is too low, sideways burning may occur giving a rough striated edge; at too high a speed, the dross does not clear the kerf, leading once again to poor edge quality and variable heat-affected zone. The oxygen pressure must also be controlled within a certain band of the order of 1 to 2 bar. If it is very high, sideways burning will again result. One notable variation is to use the laser to initiate "self-burning." as in oxygen cutting. This process allows thick section cutting, for example, 80 mm or so, with only a low-powered laser of 900 W acting as the "match". The cut quality in this laser/oxygen process is similar to oxygen cutting but with squarer top edges [42].

In all styles of laser fusion cutting there is a slight ripple, or striation pattern on the cut edge. With reactive-assist gas the striations are more marked. The cause is either a pulsation in the sideways burning reaction when the laser is moving slower than the oxygen burning front, or periodic beam coupling as the beam travels faster than the cut rate for the power absorption on the sloping cut front, or pulsation in the droplet formation at the exit of the cut (Fig. 8). In some materials, these striations can adversely affect fatigue strength although it can usually be restored by a light abrasive sanding.

In many applications, the characteristically narrow cut width (kerf) provided by lasers is an important benefit. The beam quality and mode structure, which affect focusing, are therefore significant. High-power densities are required and are best provided by low-order-mode beams.

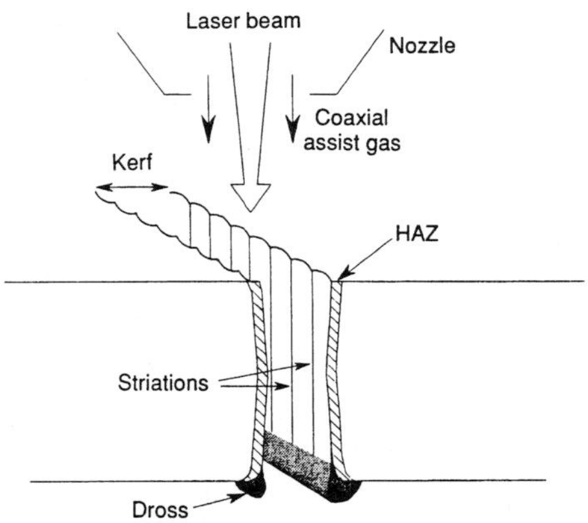


**Fig. 8** Schematic of laser cutting, defining the kerf (width of cut), showing dross on the underside, and formation of striations on the walls

The laser beam is incident on the steeply inclined cut front; this is usually at an angle of approximately 14°. At such grazing angles the plane of polarization of the laser radiation is significant. If the plane is perpendicular to the plane of incidence, then the absorption is low since the beam is readily reflected; if the plane of polarization is in the plane of incidence then the absorptivity may be doubled. When cutting with a polarised beam at 45° to the plane of incidence cutting will proceed faster to one side than the other, leading to angled penetration and the failure of circular pieces to drop out after cutting. The problem is often remedied by use of a circular polarizer in the beam, ensuring uniform cutting independent of direction.

Focal control is critical for good cutting. If the workpiece is not flat, some type of automatic height adjustment may be needed to ensure continuous control of the focal plane position with respect to the workpiece surface. This may be provided by mechanical means, such as rollers supporting the focusing head, or via optical and electrical sensors monitoring the distance to the workpiece (not shown in Fig. 7).

Laser cutting involves complex interactions between the laser beam, the workpiece, and the various gases present. Any material, from cloth and wood to metals and ceramics, can be laser cut although the best values for the laser and gas-assist parameters will vary widely. The cutting speed, $v$, should increase with incident laser power, $P$, but decrease with increased material thickness, t. The specific energy ($P/vt$ in $J/mm^2$) is approximately constant for a given material and optical arrangement. Figure 9 indicates typical ranges of cutting speeds as a function of laser power normalized by the material thickness. Both

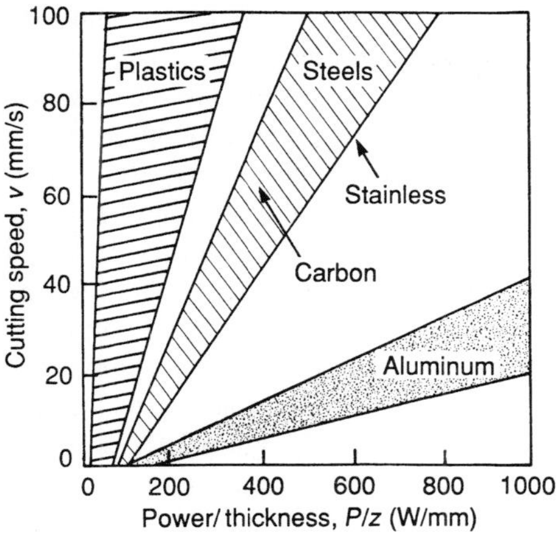


**Fig. 9** Cutting speed for diverse materials as a function of power per unit thickness of workpiece. The width of the bands indicates variations due to materials and laser and gas 'assist parameters. In practice, cutting speeds used in practice will be dictated by requirements on edge quality

$CO_2$ and Nd:YAG lasers have been used to cut metals well over 10 mm in thickness. The speed will vary both with material and processing parameters: the conditions that produce the highest speeds may not give acceptable cut quality.

The cutting of metals is usually slower than that of nonmetals because the laser coupling efficiency is often poorer and because the heat generated diffuses rapidly away from the interaction zone. Good beam quality and high peak power densities are needed for the cutting of highly reflective and thermally conductive metals such as copper and aluminum. The green radiation from a copper vapor laser (see LASERS, GAS) is sometimes used for cutting or drilling copper.

The benefits offered by the laser for cutting purposes overlap those for drilling. These include the diversity of materials that can be treated, total absence of tool wear, the narrow HAZ and localized material damage, the lack of distortion, low noise level, and the precise control possible even in complex geometries. Limitations include the high capital cost and, if large amounts of oxygen or nitrogen assist gas are needed, significant operating costs. Cutting thick samples may be slow and edge quality poor. It is difficult to expel molten debris in deep kerfs: when left, it absorbs the incident beam, lowers the cutting efficiency and degrades edge quality. Waterjet cutting is less versatile than laser cutting but is often preferred for thick materials or where thermal effects must be minimized. There is growing interest in feeding the laser beam down the water jet.

Significant cost savings can be achieved in time and material for short-run or prototype work, where the volume does not justify hard tooling or where rapid changes in cutting geometry are desired. The recent introduction of linear motor drives, swift computing protocol and piercing on the fly, has allowed the laser to compete favorably with punch blanking equipment for speed and to surpass by far such equipment for flexibility of punch shape. The narrow kerf (comparable in width to the focused beam diameter) is much smaller than that achieved in mechanical cutting. Together with the noncontact nature of the process and narrow HAZ, this allows close nesting of parts and efficient area utilization of material. Even the cutouts may be useful and there is generally less debris and waste. Multiaxis laser cutters can handle three-dimensional parts, whereas most conventional cutting systems are restricted to cutting flat workpieces.

As in other laser processes, the same laser may be used for diverse materials and different machining patterns, even intricate shapes, by simple variation of the beam and table parameters or some modifications to the gas controls. The process is easily automated and, at least on thin materials, fast.

### 3.4.2 Applications

Cutting applications constitute the main use of industrial machining lasers and range from superalloy aircraft engine parts to wooden toys and electronic lead frames. They vary widely with the particular advantage afforded by the laser over traditional techniques. The ease with which a laser can cut a material does not depend on its mechanical properties, a feature especially useful in the cutting of tough composites.

The cutting of sheet metal with lasers became significant in the early 1970s. Contour or profile cutting with a $CO_2$ laser remains very popular, with good economic returns. Thin-walled metal tubing is particularly suited to laser, rather than mechanical, cutting. The laser produces neither burrs nor part distortion.

The aircraft industry uses titanium extensively, despite its high cost, because it is both strong and lightweight. Lasers give much narrower kerfs and HAZs than possible with techniques such as oxy-acetylene or plasma arc cutting. Extensive machining had previously been necessary to remove heat-affected material from the edges, whereas little treatment is needed for laser-cut material – a chemical etch suffices for millimeter-thick stock. The materials savings offset the capital cost of the laser. Kilowatt-class $CO_2$ lasers give adequate speed in reactive gas cutting because titanium oxidizes so readily.

The ship building industry is finding that the cutting of 10- to 20-mm-thick sheets results in far less distortion and subsequent flattening than competing processes. The laser can also be used to mark the cut pieces in the same program [72].

Paper cartons and boxes are stamped by flat or cylindrical plywood dieboards containing metal-cutting knives or creasing blades. These are positioned in narrow slots in the board that must be cut very precisely according to the stamping pattern. A $CO_2$ laser, operating at several hundred watts, can give taper-free cuts with the necessary accuracy and reproducibility in an automated process that is almost ten times as fast as the previous manual technique. This process, introduced in 1971, was one of the first industrial applications of lasers for cutting.

The speed and flexibility with which laser cutters can be programmed with different patterns has led to their widespread use in prototype and low-volume manufacture, including the manufacture of dies. They are also valuable for "late option" cutting: the laser can be used to customize a product at a late stage in its manufacture. In the automotive industry, for example, options such as roof rack fixtures, sun roofs, left-hand versus right-hand drive features, air vents, and so on, can be inserted with a laser as required. This is much less expensive than having special tooling for each option.

Trimming tools are routinely used in automotive manufacture to remove excess material from deep-drawn parts. Establishing the necessary contour to the required precision can be time consuming and expensive. The final contour can be established very quickly with a laser cutter, because of the ease with which it can be programmed to accommodate modifications in the contour. The correct coordinates are then used to build the trim tool.

The lack of mechanical contact and distortion or deformation is of value in the cutting of both hard (metal pipes) and soft materials (plastic, rubber). Rubber seals or neoprene diving suits, of diverse geometries can be cut by a laser without any of the mechanical distortion that would be produced by a knife.

Speed (well over 1 m $s^{-1}$), precision, close nesting, and the self-sealing of edges, are some of the benefits provided by the $CO_2$ laser cutting of fabrics, whether men's suits or automotive components such as mats, seat belts or airbags. Different patterns can be easily accommodated by reprogramming of the laser beam path. There is no thermal damage or charring and the fraying characteristic of mechanical cuts is eliminated.

Charring may occur in some plastics, composites, and wood, but can be removed in a later process if not reduced to acceptable levels with appropriate assist gases. Wood must invariably be cut with a gas-assist and cutting speeds are strongly affected by the moisture content of the wood. In general, the laser cutting of wood is more costly than mechanical methods.

Glass has a low thermal diffusivity and may crack when laser heated. The problem can be reduced in some glasses (soda lime and borosilicate) by the use of pulsed lasers or by a preheating process. Low-power lasers are used for cutting by controlled fracture: higher powers and assist gases are used when cutting proceeds via a significant melting phase. Quartz tubes used in the lighting industry are now cut by a laser instead of a diamond saw. There is far less dust and waste. Further, the laser not only gives a much narrower kerf, it also fire-polishes the edges.

## 3.5 Marking

By the 1990s, the number of lasers being used for various marking purposes passed 10 000, comparable to the number used for welding and second only to the number used for cutting. The laser is superseding many traditional marking methods, such as silk screening, acid etching, labeling, stamping, and mechanical engraving. The potential speed of application and the permanence of the mark characterize the laser process.

### 3.5.1 Process

Marking, by local optical modification of a surface, can be achieved in numerous ways, from color change and brightness contrast, to the deep scoring and ablation of surfaces. Contrast can be enhanced by deliberately coloring components. Use of absorbing paints and dyes also facilitates the marking of materials normally transparent to the laser wavelength. The laser can modify chemical or physical properties of the surface and the corresponding range of operating parameters is very broad, as indicated by Fig. 2.

Two techniques are common: mask marking (or stencilling) and beam deflection [43–45]. In the former, one or two laser pulses are fired through a mask to create the whole of the desired pattern at once. It is particularly suited to standard or repeated patterns, such as decorative engravings or identification marking of electric cables. A focusing lens may be placed between the mask and the workpiece to reduce the image size. A TEA (transversely excited atmospheric) $CO_2$ laser is commonly used, each pulse delivering about 1 J of energy in about 1 μs. Penetration depths are shallow (1–25 μm) but can be increased by repetition. A recent variant is to use a liquid crystal display as the mask. This allows greater flexibility in the mark.

The second technique uses computer-controlled galvanometer mirrors to steer the laser beam across the workpiece, tracing out the pattern as it moves. This method is appropriate when the pattern varies rapidly, as in the sequential numbering of parts. Nd:YAG lasers with average powers up to 100 W are popular. They are often *Q*-switched, to give peak powers 1000 times greater than the average value. Penetration depths are usually several times deeper than those from mask marking. Character sizes range from 0.1 to 100 mm, with line thicknesses up to 3 mm. The trace speed is on the order of 3 m $s^{-1}$, which usually allows 70 characters/s. A dot matrix, rather than a continuous, mode may be used for elongated characters. Rapid dithering of the beam can be used to broaden lines.

Both $CO_2$ and Nd:YAG lasers are in common use, the choice depending on the absorption of the workpiece, the contrast required, the marking technique and so on. Excimer lasers are particularly useful if thermal effects must be minimized or where optical absorption of longer wavelength radiation is poor.

Durable marks of high quality and readability can be made on a wide range of materials. The high speed allows even rapidly moving objects to be marked on the fly. The process is easily automated and computer-controlled. It is very flexible: patterns can be very complex and changes can be easily accommodated, whereas changing metal stamps is time consuming.

The finely localized and controlled heat input allows very small items to be marked as well as those susceptible to thermal damage. The lack of mechanical pressure allows delicate components to be marked (a problem with mechanical stamping) and there is no chemical contamination (a problem with inks). Selected areas difficult to access by other means can be marked by lasers, provided there is line-of-sight. Operating costs are low compared to those of the consumables in ink jet marking or adhesive labeling. The marks are more durable than those in either process though the ease of marking and the legibility depend strongly on the material.

The primary disadvantage, as in most laser applications, is that the capital investment cost is high compared to that of some of the alternative methods (ink jet, stamping, or etching, for example).

### 3.5.2 Applications

Laser marking applications span a wide range of industries, from transportation and electronics to cosmetics and food, and from vehicle identification numbers (VINs) on automobiles to labels on beverage containers. Purposes include batch/date coding, inventory control, quality assurance, safety information, government requirements, security (theft protection), personal identification, logos, and decoration.

The VIN plate in all General Motors' vehicles made since 1990 is laser-marked. The aluminum plate is first blackened by anodization to give good final contrast. A 100-W Nd:YAG laser then inscribes a 17-character 3-of-9 bar code and duplicate alphanumeric data, detailing serial numbers, emission standards, and manufacture. Automotive marking is rapidly expanding, both in off-line and on-line applications. One of the first industrial applications of excimer lasers was the marking of the heater symbol on the heatable, outside mirrors offered on some automobiles. It is done by laser evaporation of the metallic surface. The ArF excimer laser used gives precise control and causes negligible damage to the glass, unlike the former sandblasting technique. The quality is higher and the amount of scrap is greatly reduced. The marking of triplex glass can be made by laser marking the central plastic film between the two sheets of glass, creating a truly secure mark.

Serial numbers are engraved on hand guns by a Q-switched Nd:YAG laser. The assembled gun cannot withstand the mechanical pressure of steel stamping, and the laser marks are both durable and difficult to duplicate. Precious stones can be marked with identification numbers as a security measure. The numbers are barely discernible to the naked eye but cannot be removed without recutting (and loss in value) of the gem. Hallmarks, used for quality control on gold and silver items manufactured in the United Kingdom, are made both by stamping and by lasers: many jewelers now prefer the laser.

Aircraft cables can be marked on the fly with the 9 ns pulses from an excimer laser and a mask marking technique.

Fragile semiconductor components, such as capacitors or silicon substrates, can be marked with manufacturing data and material properties. Excimer lasers are often used for silicon wafers. Nd:YAG lasers both hermetically weld and mark the titanium and stainless steel cases of heart pacemakers. The very low heat input is localized and does not affect the calibration. Similarly, a pulsed TEA $CO_2$ laser is used to mark the sensitive thermal sensors used in automotive emission controls. The laser has a higher capital cost, but lower running costs, than alternative marking techniques. Further, the process is fully automated, giving permanent marks of high quality. Metal stamping would destroy the sensor calibration, labels would become detached in the fuel atmosphere near the engine, wax dots would melt and electrochemical etching would be slow and difficult to automate.

One recent variation of the process is to have a tightly focussed Nd:YAG beam whose focal point lies within a glass block. At the focal point the intensity is just sufficient to generate a small crack of a few microns in size, but large enough to scatter light. Accurate movement of the focal spot and subsequent crack pattern within the glass block allows 3D objects to be "printed" within the slab. The applications for this technique are currently in the art and trophy market.

## 4
## Miscellaneous Applications

### 4.1
### Electronic Circuits

The precisely positioned, finely focused, input of the laser, together with its low heat input, has opened up a variety of applications in the electronics industry. A few of the more common ones are included here.

#### 4.1.1 Annealing

Ion implantation is a popular means of creating desired dopant profiles in semiconductors. However, the process damages the crystal structure, often rendering a 100-nm thick region amorphous. Thermal annealing is then used to provide the needed electrical properties. This can be done with lasers or simply hot wires.

Laser annealing has found diverse uses, from improving the conductivity of polycrystalline silicon, to modifying the magnetic properties of garnet films and the superconducting properties of other metallic films [46].

#### 4.1.2 Trimming

The electronics industry employs lasers to modify the electrical values of components such as resistors in thin and thick film circuits. These consist of material such as tantalum nitride ($Ta_2N$) on an insulating substrate such as alumina ($Al_2O_3$). The resistors must have very precise values but the fabrication process commonly used (screen printing) may give variations of up to 15%.

Lasers were first used to counter this problem in 1967, by trimming the resistance to the required value. It is now standard practice to print all conductor paths wider than required (thus having resistances that are too low) and then use a laser to perform final adjustments by trimming off excess material.

*Q*-switched Nd:YAG lasers with average power of only a few watts are frequently used for resistor trimming. The process can be monitored while the component is active in the circuit, and the resistance values desired can be achieved with an accuracy of better than 0.1% within a few seconds. The low heat input prevents thermal damage to adjacent components. The precise positioning of the laser beam allows trimming to be done at a late stage, even when circuits are closely nested together. Older techniques, such as air-blasting the edges of the resistor sheet with an abrasive, were practical only if components were not too densely packed.

The resistance also depends on the chemistry of the material, such as its surface oxidation. Currently, there is increased interest in the ability of the laser to produce such chemical changes.

#### 4.1.3 **Scribing**

Please refer to Sect. 3.4.

#### 4.1.4 **Soldering**

In this variant of welding, the laser supplies the heat needed to melt solder that is already positioned on the components of interest. Where wires are insulated with plastics, a $CO_2$ laser can automatically burn off or melt away this coating prior to the soldering process. Power levels used in soldering, whether with $CO_2$ or Nd:YAG lasers, are typically in the 10- to 100-W range.

Electronic circuits contain fragile parts and a low-power laser (10–20 W) restricts the heating effects to the areas needed. Traditional techniques (including infrared and vapor-phase soldering) raise the temperature of the component to that of the solder and would produce thermal damage or distortion. A $CO_2$ laser beam can be readily focused to spot sizes less than those of the solder pads on electronic components and can also be directed to areas normally difficult to access. This is increasingly important as circuits continue to be miniaturized. Components must be accurately positioned and precautions may be necessary to avoid damage from stray reflections of the laser beam.

Laser soldering systems are expensive and give low production rates compared to wave soldering or continuous vapor-phase soldering. However, the laser offers unusually high reliability and the ability to access difficult locations and intricate geometries. Quality is often superior to that in other techniques: the rapid heating and cooling rates restrict the time available for undesirable intermetallic layers to form.

#### 4.1.5 **Repairs**

Defects during manufacture can often be remedied by a laser, thereby reducing waste and increasing productivity. The laser may be needed to repair broken connections or to trim off surplus material and burrs.

For example, a pulsed laser has been used to repair defective terminal connections in an electronic tube. The glass wall is transparent to the beam and the tubes can thus be salvaged rather than scrapped.

Redundancy is often built into modern integrated circuits. A pulsed Nd:YAG laser can easily vaporize defective components, without damaging adjacent parts or the substrate, and redirect channels.

Metallic masks can be repaired by selectively heating appropriate parts of the pattern [47]. For example, various defects can occur in the fabrication of IC photomasks used for microlithography, such as too much or too little chromium deposited onto the glass. Excess material can be removed by a *Q*-switched, frequency-doubled Nd:YAG laser, operating at a wavelength of 0.532 μm, enabling smaller regions to be repaired than what is possible at 1.06 μm. Thin areas or pinholes can be repaired by vapor deposition of chromium with a CW argon-ion laser or a frequency-quadrupled Nd:YAG laser.

#### 4.1.6 Deposition

Materials can be deposited, often as thin films, by the laser-initiated pyrolytic or photolytic decomposition of gaseous media. Many variants of chemical vapor deposition (CVD) are possible, including use of the laser to ablate a solid source material or to heat a substrate. Novel materials can be produced because the laser can heat substrates to temperatures higher than usual and because nonequilibrium phases can occur during the rapid heating and cooling cycle. There are many potential applications but many of the phenomena are only poorly understood [46, 48].

Primary application is to integrated circuits because of the high spatial resolution and control (although the electronics industry is not the only user of laser-assisted CVD). For example, thin gold films can be deposited over transparent defects in photomasks. Direct maskless writing applications are proliferating. Excimer lasers are of increasing interest in the fabrication of films, from high-temperature superconductors to polymers for waveguides. Whatever laser is used for film deposition, however, the localization of the effect makes the process inefficient for large area devices.

#### 4.1.7 Oxidation and Nitridation

The electrical characteristics of many metal and semiconductor films can be drastically changed by oxides or other surface layers. For silicon systems, the oxide passivates the surface and controls interfacial diffusion. Similarly, nitrides often provide good insulating properties and, further, may have excellent microhardness. Such films may be grown by heating a wafer in an appropriate atmosphere.

Conventional furnace treatment requires raising the temperature of the whole wafer for a prolonged period. Lasers provide very precise localization of heating not only for selected areas but also for controlled depths. While silicon treatments have been widely explored, many other systems are also under investigation and several types of laser are being used [48].

### 4.2 Prototype or Low-volume Manufacture

A wide range of layer manufacturing processes has been developed since the early 1980s. They are usually based on designing a 3D object within a CAD package in a computer, mathematically slicing the design into layers, and then converting each layer into machine code to drive a multiaxis fabricator to create one layer at a time. The fabricator may use a manipulator such as an X-Y stage moving the object or the optics or galvanometer mirrors moving the beam, as in marking. The end product is a 3D object built directly from a computer design without the need for ancillary tooling, such as moulds.

There are over 25 different rapid build methods with the majority being based on the use of a laser [49]. The four main processes are: liquid polymerization (stereolithography); cutting solid layers from sheet (Laminated Object Manufacturing, LOM); sintering powders (Selective Laser Sintering, SLS); or powder fusion (Laser Direct Casting, LDC or Laser Weld Build, LWB).

Stereolithography was the first process to become commercial in 1988. It is sometimes called a *desktop manufacturing process*. In this process an ultraviolet laser beam from a He–Cd laser ($\approx$10 mW, 325 nm wavelength) or argon-ion lasers of several watts is flickered over the surface of a pool of photo-monomer, which it rapidly causes to polymerize only where it passes. This solidification of the monomer syrup is performed in layers starting on a platform, which is carefully lowered between layers. The resulting prototype models are cured in an ultraviolet chamber or heated in an oven and are then ready for further use. They are used as either demonstration models for design decisions, models for preassembly of complex machinery, models as teaching aids or as formers for making rubber moulds for casting further copies or moulds for electroforming. The photo-polymers used are based on epoxy, vinyl ether, or acrylic monomers mixed with a photo-initiator. There are variations on this process based on more powerful lasers being capable of printing whole layers at a time by imaging a mask, or, for greater flexibility, the mask may be a liquid crystal mask.

Laminated object manufacture, LOM, is a process whereby each layer is cut in turn out of a sheet material and mounted on top of the previous layer. In one variation a sticky paper roll is moved over the object and the laser, of just the correct power (usually $CO_2$ of about 25 W, for example) cuts that layer alone, which is stuck to the one beneath. The laser also makes cuts to the side to aid the final removal of the waste material. In another variation, the layers may be metallic and can be laser welded or brazed one layer to the next. The application being considered for the tough metallic models is for tooling, the paper models being mainly design aids. The first commercial process came out in 1991.

Selective laser sintering, also first commercially available in 1991, uses equipment similar to that in stereolithography but, instead of the photo-monomer pool, it has a powder bed of plastic or metal particles. The laser, which is usually a $CO_2$ or Nd:YAG of approximately 50 to 100 W, is passed over the bed, layer by layer. The platform holding the powder is precisely lowered and the powder bed carefully levelled between each layer. The process forms a sinter of plastic or metal particles. The metal particles used are often coated with a lower melting point material to aid accuracy of the finished object. The accuracy can be $\pm$0.05 to $\pm$0.25 mm. This porous structure can then be heated in a vacuum and filled with a lower melting material, for example, solder in a bronze casting. These castings are used for tooling, design aids, and various complex final components.

Laser weld build or LDC is the latest variant in which each layer is laid down by laser cladding (Sect. 3.1.4) from a coaxial powder feed. The metallic object is built layer by layer with considerable accuracy. The levelling between each run can be done aerodynamically during cladding. The microstructure of the casting is unlike standard castings in that the grain growth is unidirectional toward the growth direction. This makes for a structure that is tougher than a standard casting. It is also possible to change the material during cladding: this allows the article to be constructed from whatever material is needed with the possibility of different materials on the inside and the outside. The article made could thus be the part required, not just a model. Hence this is a form of low-volume manufacture suitable for short runs of less than 100 pieces. Applications for this process, which was first commercialized in the late 1990s, have included for the manufacture of extrusion dies with conformal cooling channels, the repair of heavily worn parts and the alteration of press tools [50].

### 4.3 Forming

Laser forming, or bending, of materials is fast becoming a standard manufacturing tool for adjusting, aligning, low-volume manufacturing or rapid prototyping in the aerospace, automotive, and electronic industries. The process creates a thermal stress that causes plastic flow and, hence, on cooling, leaves a permanent, precision bend with no spring-back or tool contact.

Flame bending or "line heating" with an oxyacetylene torch has been used for straightening and flattening for a long time in the heavy engineering industries, such as shipbuilding. The laser differs through the precision with which the heat can be located. The laser acts by three distinct mechanisms:

**The Thermal Gradient Method (TGM)** The laser is passed relatively swiftly over the sheet to be bent enough to heat the surface, but not enough to melt it or for the heat to penetrate the sheet. This creates a thermal gradient through the sheet thickness. The top layer is under compression and plastically deforms. On cooling it will be shorter than before and hence the sheet will bend towards the laser. The bend rate is approximately 2 to 3°/pass. The process for larger angles is thus slow but can be very precise.

**Buckling** If the laser passes more slowly, or the sheet is thinner, then there will be no significant thermal gradient across the sheet. The whole local sheet thickness expands on heating and will bend upwards or downwards dependent on some initial tendency. This buckle can then be led across the sheet along the line required for bending. The bend rate is still only a few degrees/pass.

**Upsetting** If the piece to be treated is firmly held so that buckling cannot occur, then the whole local thickness of the sheet will plastically deform under the compression stress caused by heating. On cooling it will then be slightly shorter than before, with a thickening where it had been heated.

In general, the thermal gradient method (TGM) or upsetting are methods for shortening the material in the top layer or the section. Thus, it is possible to calculate how to organize the pass sequences to create a wide variety of shapes without any tooling except a laser and beam manipulator. The process was first reported by Namba [51] and has been extensively modelled and developed by Vollertsen [52]. There is now global interest in some of the more subtle detail [53, 54].

The applications in industry are in the adjustment and alignment of electronic devices [55] and the straightening and aligning of car body components [56]. Applications that are being studied include bending of ship plate [57]; tube forming [58]; dish and spherical cap shapes [59]; rapid prototyping [60]; and space applications [61].

The alignment of electronic devices, such as miniaturized hard disc drives, is crucial but difficult to achieve with mechanical adjustment or extremely accurate manufacturing stages. The use of a laser pulse to bend the support mechanism can be arranged to give an X-Y-Z movement of a few microns simply by varying the location of the pulse [55], thereby saving considerable cost and manufacturing time.

A lightweight bending tool with no need for opposing forces to achieve a bend is ideal for applications in space. The vocabulary of shapes possible with the laser has been studied by Silve [54] with implications for bending awkward tubes structures, such as exhaust systems and artwork. A virtual tooling system for manufacturing large bent sheets with precision, once the bend sequence and some form of feedback control has been established, is of interest to low-volume manufacturers such as Aerospace.

## 4.4 Cleaning

Short sharp pulses of optical energy can remove rust, oil, paint, oxide scale, grease chewing gum and ultrafine dust [62]. The process mechanism appears to be based on: (1) photo-thermal events, such as evaporation of the surface contaminant; (2) photochemical-bond breaking of the linkage of the contaminant to the surface; and (3) photomechanical events, such as blast wave from the surface plasma, vibration of the substrate surface due to thermal expansion shaking the contaminant free, or direct photon pressure. Probably each route has some part to play. All these mechanisms are enhanced by the use of short pulses ($\approx$1–10 ns) and short wavelengths (e.g., green or ultraviolet radiation). The overall process also requires a careful balance of power and spot size to achieve a quality clean surface without significant surface damage.

The applications to date have been in the area of art conservation, one of the pioneering papers being that by Asmus [63]. One research group (FORTH, Crete) developed an in-process control system based on the plasma spectrum so that paint layers could be selectively removed from a painting [64]. A 2-kW $CO_2$ TEA laser, with 9 J pulses at 350 Hz, can strip paint from aircraft bodies at 8 to 20 $m^2\,h^{-1}$, with a reduction in dangerous waste from chemical cleaners by a factor of 10 or so [65]. A handheld laser cleaner based on a Nd:YAG laser working at 25 kHz and 100 W with optical fiber delivery has been developed by the Fraunhofer group. The equipment could be used for cleaning buildings or artwork. There is some question as to whether the use of an optical fiber may "blunt" the sharpness of a single laser pulse by time smearing. For smaller scale operation, the cleaning of electronic circuits has been shown to be possible [66]. Microscopic particles can account for 50% of yield losses, hence the need for ultraclean rooms; perhaps the laser offers an alternative.

The advantages of this noncontact cleaning process are the lack of any abrasion or waste chemicals, the low level of noise, the detailed control possible with in-process sensing and the totally noncontaminating nature of optical energy. It is an ideal tool for cleaning the inside or outside of a building, cleaning statues, coins or paintings, paint stripping aircraft or bridges and the localized cleaning of electronic substrates. It also sterilizes as it cleans.

## 4.5 Other Processes

Many other laser-processing applications exist, as in the following examples. Some, like 3D and laser-assisted machining, are still under development.

### 4.5.1 Ablation

In many laser processes, much of the material is removed by vaporization. Where, as is often the case with excimer laser irradiation, the process is not dominated by thermal energy, it is sometimes termed *photoablation*. This typically requires pulsed lasers with high peak intensities ($10^6$ W cm$^{-2}$ or greater) but produces very clean boundaries with negligible heat-affected zones.

The laser offers even more precise depth control than does chemical etching but produces no chemical contamination. It can very selectively strip off only the layer desired from a multiply layered part, such as a coating from a glass substrate or the insulation from a coaxial cable. Mechanical stripping techniques might easily damage underlying material. Provided adjacent layers differ enough in their optical and thermal properties, the laser can remove one layer while not affecting the next. For coaxial cables, the radiation from a $CO_2$ laser is efficiently absorbed by the insulating layers but highly reflected by the metal wire.

The electronics industry uses lasers to etch photoresists and in a host of other selective material removal applications. The excimer laser is especially useful in applications requiring high spatial resolution and no damage to adjacent material.

Laser ablation is used for the dynamic balancing of gyroscopes and other rotating components. A sensor detects any imbalance as the part rotates and triggers the firing of a Nd:YAG laser pulse when the excess material lies beneath the beam. The laser has reduced the time required to balance a gyroscope from several hours to less than a minute. It is used in a similar manner for the tuning of quartz oscillators for watches and electronic timers.

Greetings cards often exhibit filigree patterns that result from selective removal of paper by a high-power $CO_2$ laser directed through a photo-etched template. The beam is focused onto the paper several meters from the template. The paper issues from a modified printing press and moves at high speed, together with the template, across the path of the laser beam. The process is not as fast as die cutting but leaves no frayed or curved edges and can give high resolution. Close examination of the edges of the pattern occasionally reveals faint charring, indicating that the process was indeed laser ablation rather than mechanical cutting. The same technique is used to engrave wood to a depth of approximately 3 mm.

### 4.5.2 Shock hardening

Until recently shock hardening by laser was a laboratory curiosity [67]. Today it is an industrial process. Developed at the Lawrence Livermore Laboratories [68] the process is used to generate a compressive layer (up to 60% of the ultimate tensile strength) on the surface of a work hardening material to improve fatigue resistance. The method is to fire short (10 – 100 ns), powerful ($10^{10}$ W), high-energy (100 J) pulses, which on striking the surface to be hardened generate a plasma, creating a recoil pressure on the order of several GPa. The beam has a spot size of approximately 100 $mm^2$ and the plasma generated is inertially confined by firing the laser beam through water onto a painted surface, which enhances the absorption. The resulting high pressure over a large area, gives a four or five times increase in the depth of the surface compressive layer created, compared to mechanical shot peening. This is the reason the laser shock hardening process gives better fatigue properties than does conventional shot peening. The larger treated area/pulse also explains why the surface treated is almost unaltered by the process. It is currently used for shock hardening aerospace components such as turbine blades, leading edges of fan blades, engine discs, and rotors. It is expected to find applications in the automotive industry, power industry, and other industries where it is preferable to fire photons rather than solid shot at the manufactured part.

#### 4.5.3 Grain Refining

Cast alloys often contain undesirable inclusions, whether chemical impurities or pores. Those near the surface can degrade properties such as fatigue strength and resistance to corrosion or wear. Polycrystalline semiconductors, such as CVD silicon, often exhibit low electrical conductivity because of electron scattering at grain boundaries. Laser scanning can result in the refining of grains or growth in their size and promote the desired properties.

High-power, CW Nd:YAG lasers have been used to refine the magnetic domain structure in steel used for electrical transformer cores. The high magnetic permeability steel contains primarily iron with about 3% silicon and has large grains. In operation, motion of domain walls in the applied alternating electric field results in significant energy losses. The laser thermally shocks the microstructure and thereby controls the magnetic domain wall boundaries. Electrical energy losses have been reduced by up to 10%.

#### 4.5.4 Curing

Ultraviolet light is used to cure paints and polymers in a number of industries. The process is slow, lasting days in some cases, but can be accelerated by means such as ovens. A short-wavelength laser, such as the excimer, can stimulate a fast chemical reaction in the polymer molecules, leading to shorter curing times.

Generally, the applications best suited to the laser are those that require only small areas to be treated, such as microprinting and microelectronics.

#### 4.5.5 3D Laser Machining

Two or more laser beams can be directed onto a workpiece so that each creates an independent groove. When the grooves intersect, their geometry determines how much material falls out. Unlike cuts, the grooves do not fully penetrate the workpiece and debris can be flushed out only via the front face [15].

#### 4.5.6 Laser-assisted Machining

In most processes, the laser is used as a replacement tool. It can be used to bend metals without mechanical contact, for example. In some new applications, however, lasers are used to supplement rather than supplant traditional tools [69]. The laser heats the workpiece just ahead of the tool, making it easier to machine. This is of interest in the cutting of superalloys, ceramics, and other hard substances. The required tool force is reduced and so, consequently, is tool wear.

In "chip management", the laser is used to break up the long continuous spirals of thin material produced in turning operations. This simplifies the management of materials such as titanium alloys or nickel-based superalloys that are difficult to cut.

These applications of the laser, whether before, during or immediately after, conventional machining, are still being developed and are less well-known than welding or marking, for example.

#### 4.5.7 Laser Scabbling

This is a process for roughening stone or concrete by removing the surface layer. The laser beam heats the surface of concrete, preferably over a substantial area (e.g., 3 kW over 10 $cm^2$), creating a stress field that causes violent cracking of the surface and ejection of chunks of material to a depth of a centimeter or so. It is currently under study as a technique for decommissioning nuclear power stations or cleaning contaminated water tanks and concrete structures that are superficially dirty but not strong enough to withstand mechanical scabbling, as with road construction. One of the attractions of the process is the need to have large spot sizes to create sufficient thermal stress. This means that the focussing or rather collimating optics can be far removed from the process. It can also be a fairly quick process removing material at a typical rate of 20 $cm^3\ s^{-1}$.

The process was originally conceived by Li et al [70] and has been studied by Johnston et al. [71] and theoretically modeled by Dowden [8].

## 5 New Developments

There is little sign that the pace of laser-processing developments is lessening, either in the laser sources themselves or in the applications for which they are used. Industrial lasers of all types (excimer, solid-state, $CO_2$, diode) continue to be improved. Greater attention is being given to refinements such as pulse shaping and beam spatial control. There is interest both in increasing output powers and in producing more compact and reliable systems, together with ultrashort pulses (femtoseconds, $10^{-15}$s) and very short wavelengths (X rays). (See also ULTRASHORT-PULSE PHENOMENA and LASERS, EXTREME UV AND SOFT X-RAY).

Excimer lasers, because of their short-wavelength radiation, are expected to increase their role significantly in applications ranging from marking to microelectronics. Their potential is being enlarged by intensive efforts to increase their output powers: several hundred watts have already been obtained from XeCl (308 nm). The major impediments to their use will continue to be their high capital cost and maintenance requirements (compared to more common lasers). They are also being challenged in the ultraviolet by frequency-quadrupled Nd:YAG or Nd:YLF (yttrium lithium fluoride) lasers.

The power available from diode arrays continues to increase and, by 2000, portable systems capable of yielding multikilowatt outputs were already commercially available. Although there are numerous applications in which their relatively poor beam quality is not critical, efforts to improve beam focusability continue. The use of diodes coupled to fiber delivery (as in endoscopy) is becoming extensively used for low-power medical applications. It is probable that the laser diode may, in the next decade or so, replace the incandescent bulb as a source of domestic, traffic or street lighting.

Gas lasers being further developed include kW-level iodine lasers (which emit at a wavelength of 1.315 μm) and multi-kW carbon monoxide lasers ($\lambda = 5{-}7$ μm). Even $CO_2$ laser technology continues to be improved: alternative types of excitation (RF and microwave) are being used; output powers are increasing even while the size of the laser head is decreasing; and the control of beam power and quality is becoming more sophisticated. Very rugged, near-maintenance-free, compact, sealed lasers offering near-kW powers are becoming available.

While standard beam delivery systems, both mirror and fiber, are becoming more efficient and versatile; new optical systems are being developed. There is a growing interest in DOE for beam shaping. These devices are basically holographic lenses or mirrors and are capable of shaping a beam of known mode into almost any shape required. One application being studied is that of multispot soldering.

There have been remarkable developments with Stimulated Brillouin Scattering (SBS) phase conjugators. These devices reflect the laser beam but, unlike the situation with normal mirrors, the sign of the wave front is reversed so that, after travelling back through the distorting optics that formed it, the distortion may be cancelled.

Interest continues in real-time, *in situ* monitoring. The simplest type of process monitoring is to ensure that the operating parameters (laser power, mode, gas pressures, etc.) are maintained within specific ranges. It is usually more difficult to conduct real-time *in situ* monitoring of the workpiece itself but several diagnostic techniques, notably optical, electrical, thermal, and acoustic, are being assessed. They include visual appearance of the irradiated spot, the brightness and spectral distribution of the plasma, plasma space charge, temperatures at the workpiece, noise in various frequency ranges, and so on. While this information helps and can be used in closed-loop control systems, it seldom distinguishes adequately between acceptable and unacceptable conditions and postprocess evaluations may still be needed.

Despite the successes already seen since the laser was first used for materials processing, the field is far from mature. Our understanding of the interaction of laser beams with materials is still limited and we can anticipate significant improvements and refinements with time. We can also expect better appreciation of the need to design products for laser processing in order to maximize the benefits that the laser can bring. When this happens it will be clear that the laser has made possible an entirely new range of manufacturing techniques such as additive machining, noncontact bending, marking on the fly, cleaning, mouldless casting, and other processes described here. The market for laser applications continues to expand.

## 6
## Summary

The ability of the laser to produce carefully controlled thermal effects in highly localized areas has opened up numerous applications. In some cases, lasers are used as direct replacement tools for existing technologies. In others, their use may allow redesign of a whole manufacturing system, with savings in materials and time by the elimination of several operations. Progress continues both in the design of industrial lasers and in the range and sophistication of processing operations.

The high capital costs of lasers limited their rapid adoption for applications in which traditional methods exist. However, technological developments and increased competition, from diode lasers, for example, continue to reduce the costs of owning a laser. Hence, the use of lasers for industrial applications is steadily increasing: this review has included only the better known.

A laser is not always the best tool for an application, even though its use is technically feasible. However, production floor experience has demonstrated that lasers can be reliable processing tools, that they can provide unique advantages in many applications and that they can be economically viable. The laser is proving itself to be an effective tool, not an exotic toy.

## Glossary

**Assist Gas:** A gas or mixture of gases, other than those used for the laser operation itself, that are employed near the workpiece to improve processing performance.

**Beam Diameter:** The effective width of a laser beam, often defined on the basis of an area that encompasses a given fraction of the total beam power.

**Beam Divergence $\theta$:** A measure of the angular spreading of a laser beam as it propagates.

**$CO_2$ Laser:** Gas laser in which carbon dioxide is the medium responsible for the 10-mm-wavelength radiation emitted.

**Continuous Wave (CW):** Continuous rather than pulsed emission from a laser.

**Coupling Coefficient $\gamma$:** The fraction of incident laser energy that is ultimately transferred to the workpiece and hence effective for processing.

**Heat-affected Zone (HAZ):** Region adjacent to that processed that is thermally affected by the laser irradiation.

**Irradiance:** The optical flux per unit area falling on a surface (often, but ambiguously, termed *intensity*).

**Kerf:** Width of the material removed by a cutting process.

**Keyhole:** Elongated hole produced in deep-penetration welding, when a dynamically stable cavity is formed amid molten material.

**Mode:** A standing-wave pattern generated by the radiation field in a resonator cavity.

**Nd Laser:** Solid-state laser, comprising a glass or YAG (yttrium aluminum garnet) rod or slab that contains a few percent of neodymium ions ($Nd^{3+}$) .

**Polarized Beam:** A laser beam in which the electric vector is aligned in one plane.

**Q Switch:** A device that controls laser cavity oscillation until a large amount of energy is available for release.

**Recast:** Material that, having been melted, is not removed entirely but resolidifies as a layer on the walls of a hole or cut.

**Trepanning:** Removal of material by the cutting out of a circular slug, usually by an off-axis motion of the laser beam with respect to the workpiece.

## References


[1] Bromberg, J. L. (1991), *The Laser in America*. Cambridge, MA: MIT Press.
[2] Hecht, J. (1985, 1992), *Laser Pioneers*. London, UK, San Diego, USA: Academic Press.
[3] Bertolotti, M. (1983), *Masers and Lasers: an historical approach*. Bristol, UK: Adam Hilger.
[4] Ready, J. F. (1978), *Industrial Applications of Lasers*. New York: Academic Press.
[5] Duley, W. W. (1983), *Laser Processing and Analysis of Materials*. New York: Plenum.
[6] Rockwell, R. James Jr, (1990), in D. Belforte, M. Levitt (Eds.), *The Industrial Laser Annual Handbook*. Tulsa, OK: PennWell Books, pp. 131–148.
[7] Ready, J. F. (1971), *Effects of High-Power Laser Radiation*. New York: Academic Press.
[8] Dowden, J. M. (2001), *The Mathematics of Thermal Modeling*. New York: Chapman & Hall.
[9] Shkarofsky, I. P. (**1975**), *RCA Rev.* **36**, 336–368.
[10] Duley, W. W. (1976), *$CO_2$ Lasers: Effects and Applications*. New York: Academic Press.
[11] Modest, M. F., Abakians, H. (**1986**), *Trans. ASME J. Heat Transfer* **108**, 597–601.
[12] Mazumder, J. (1983), in M. Bass (Ed.), *Laser Materials Processing*. Amsterdam: North Holland, Chap. 3.
[13] Steen, W. M., Kamalu, J. N. (1983), in M. Bass (Ed.), *Laser Materials Processing*. Amsterdam: North Holland, Chap. 2.
[14] Von Allmen, M. (1987), *Laser-Beam Interactions with Materials*. Berlin: Springer.
[15] Chryssolouris, G. (1991), *Laser Machining: Theory and Practice*. New York: Springer.
[16] Ion, J. C. (1992), in D. Belforte, M. Levitt (Eds.), *The Industrial Laser Handbook 1992–1993 Edition*. New York: Springer, pp. 39–47.
[17] Znotins, T. A., Poulin, D., Reid, J. (**1987**), *Laser Focus* **23**(5), 54–70.
[18] Belforte, D., Levitt, M. (Eds.) (1986–1990), *The Industrial Laser Annual Handbook*. Tulsa, OK: PennWell Books.
[19] Ready, J. F., Farson, D. F. (Eds.) (2001), *LIA Handbook of Laser Materials Processing*. Orlando, FL: Laser Institute of America.
[20] Steen, W. M. (**1985**), *Metals Mater.* **1**, 730–736.
[21] Steen, W. M., Courtney, C. (**1979**), *Metals Tech.* **6**(12), 456.
[22] Dausinger, F., Beck, M., Rudlaff, T., Wahl, T. (1988), in H. Hugel (Ed.), *Proc. LIM 5*. Bedford, UK: IFS Publ., pp. 177–186.
[23] Breinan, E. M., Kear, B. H. (1983), in M. Bass (Ed.), *Laser Materials Processing*. Amsterdam: North Holland, Chap. 5.
[24] Draper, C. W., Poate, J. M. (**1985**), *Int. Metals Rev.* **30**, 85–108.
[25] Li, L. J., Mazumder, J. (1985), in K. Mukherjee, J. Mazumder (Eds.), *Laser Processing of Materials*. Warrendale, PA: Metallurgical Society of AIME, pp. 35–50.
[26] Swift-Hook, D. T., Gick, A. E. F. (**1973**), *Welding J.* **52**, 492S–499S.
[27] Klemens, P. G. (**1976**), *J. Appl. Phys.* **47**, 2165–2174.
[28] Cline, H. E., Anthony, T. R. (**1977**), *J. Appl. Phys.* **48**, 3895–3900.
[29] Mazumder, J., Steen, W. M. (**1980**), *J. Appl. Phys.* **51**, 941–947.
[30] Dowden, J., Postacioglu, N., Davis, M., Kapadia, P. (**1987**), *J. Phys. D* **20**, 36–44.
[31] Mazumder, J. (**1991**), *Opt. Eng.* **30**, 1208–1219.
[32] Kapadia, P., Ducharme, R., Dowden, J. (1992), *Proceedings of ICALEO'91 (International Congress on Applications of Lasers and Electro-Optics)*, Vol. 74. Orlando, FL: Laser Institute of America, pp. 53–63.
[33] Ki, H., Mohanty, P. S., Mazumder, J. (**2002**), *Metall. Mater. Trans. A* **33A**, 1–10.
[34] Schnöcker, D. (1992), in D. Belforte, M. Levitt (Eds.), *The Industrial Laser Handbook 1992–1993 Edition*. New York: Springer, pp. 67–73.
[35] Greses, J., Barlow, C. Y., Steen, W. M., Hilton, P. A. (2001), *Proceedings of ICALEO 2001 (International Congress on Applications of Lasers and Electro-Optics)*. Jacksonville, FL: Laser Institute of America, Paper 808.
[36] Paek, U.-C., Gagliano, F. P. (**1972**), *IEEE J. Quantum Electron.* **QE-8**, 112–119.
[37] Von Allmen, M. (**1976**), *J. Appl. Phys.* **47**(12), 5460–5463.
[38] Minardi, A., Bishop, P. J. (**1989**), *Heat Transfer Manu. Mater. Process. HTD* **113**, 39–44.
[39] Chan, C. L., Latham, W. P., Chen, J. K. (2001), *Proceedings of ICALEO 2001 (International Congress on Applications of Lasers, and Electro-Optics)*. Jacksonville, FL: Laser Institute of America, Paper M704.

[40] Rykalin, N., Uglov, A., Kokora, A. (1979), *Laser Machining and Welding*. Oxford: Pergamon Press.

[41] Schuocker, D. (1987), in D. Belforte, M. Levitt (Eds.), *The Industrial Laser Annual Handbook*. Tulsa, OK: PennWell Books, pp. 65–79.

[42] O'Neill, W., Gabzdyl, J. (**2000**), *J. Opt. Lasers Eng.* **34**, 355–367.

[43] Willis, J. B. (1983), in M. F. Kimmitt (Ed.), *Proceedings of the 1st International Conference on Lasers in Manufacturing (LIM)*. Kempston, Bedford, UK: IFS and North Holland, pp. 53–62.

[44] Klimt, B. H. (**1988**), *Lasers Optron.* **7**, 61–67.

[45] Kaplan, A. F. H. (1998), in D. Schnöcker (Ed.), *Marking and Scribing. Handbook of Eurolaser Academy*, London: Chapman & Hall, Chap. 7, pp. 455–472.

[46] Lugomer, S. (1990), *Laser Technology: Laser Driven Processes.* Englewood Cliffs, NJ: Prentice Hall.

[47] Tison, J. K. (**1988**), *Laser Focus/Electro-Opt.* **24**(5), 88–94.

[48] Boyd, I. W. (1987), *Laser Processing of Thin Films and Microstructures*. Berlin: Springer.

[49] Pera, L., Marinsek, G. (1992), *Proc. NATO ASI Laser Applications for Mechanical Industry*, Erice, Sicily, Vol. 238, pp. 293–303.

[50] Mazumder, J., Dutta, D., Kikuchi, N., Gosh, A. (**2000**), *J. Opt. Lasers Eng.* **34**, 397–414.

[51] Namba, Y. (1987), *Proc. Laser Advanced Materials Processing (LAMP) Conf.*, Osaka, Japan, pp. 601–606.

[52] Vollertsen, F. (1994), in M. Geiger, F. Vollertsen (Eds.), *Proc LANE'94 (Laser Assisted Net Shape Engineering)*, Vol. 1. Germany: Meisenbach Bamberg, pp. 345–360.

[53] Magee, J., Watkins, K. G., Steen, W. M., Calder, N. J., Sidhu, J., Kirby, J. (**1998**), *J. Laser Appl.* **10**(4), 149–155.

[54] Silve, S. (2000), Laser Forming and Creative Metal Work, Ph.D. Thesis, Brunel University, UK.

[55] Hoving, W. (1997), *Proc. European Symposium on Laser and Optics Manufacturing*, Vol. 3097. Munich: SPIE, pp. 284–295.

[56] Geiger, M., Vollertsen, F., Deinzer, G. (1993), SAE Paper 930279.

[57] Skully, K. (**1987**), *J. Ship Prod.* **3**(4), 237–246.

[58] Frackiewicz, H., Trampczynski, W., Przetakiewicz, W. (1992), *Proc. ISATA 25th (Int. Symp. on Automotive Technology and Automation)*, pp. 373–380.

[59] Hennige, T. (1997), in M. Geiger, F. Vollertsen (Eds.), *Proc. LANE'97 (Laser Assisted Net Shape Engineering)*, Vol. 2. Germany: Meisenbach Bamberg, pp. 409–420.

[60] Vollertsen, F. (1995), in M. Geiger (Ed.), *Proc. 12th Int. Congress LASER95*. Germany: Meisenbach Bamberg, pp. 151–162.

[61] Namba, Y. (1986), in C. P. Wang (Ed.), *Proc. Laser'85*. McClean: STS Press, pp. 403–407.

[62] Watkins, K. (2000), *High Power Lasers in Manufacturing Technology*, Vol. 3888. Bellingham, WA: SPIE, pp. 69–79.

[63] Asmus, J. F. (**1978**), *Technol. Conservation* **3**(3), 14–189.

[64] Tornari, V., Zafiropulos, V., Bonaru, A., Vainos, N. A., Fotakis C. (**2000**), *J. Opt. Lasers Eng.* **34**, 309–326.

[65] Ploner, L. (**1995**), *Laser Focus World* **26**, 27.

[66] Walz, J. Y. (**1997**), *Photon. Spectra*, 116–121.

[67] Clauer, A. H., Fairand, B. P. (1979), in E. A. Metzbower (Ed.), *Applications of Lasers in Materials Processing.* Metals Park, OH: American Society for Metals, pp. 291–315.

[68] Hammersley, G., Hackel, L. A., Harris, F. (**2000**), *J. Opt. Lasers Eng. (JOEL)* **34**, 327–337.

[69] Copley, S., Bass, M., Jau, B., Wallace, R. (1983), in M. Bass (Ed.), *Laser Materials Processing.* Amsterdam: North Holland, Chap. 6.

[70] Li. L., Moderm. P. J., Steen, W. M. (1994), *Proc. RECOD'94 4$^{th}$ Conf on Nuclear Fuel Reprocessing and Waste Management* Vol. III, pp. 427–440.

[71] Johnston, E. P., Shannon, G., Steen, W. M., Jones, D. R., Spencer, J. T. (1999), *Proc. ICALEO'98 (International Conference on the Applications of Lasers and Electro-Optics)*, Vol. 85. Orlando, FL: Laser Institute of America, pp. A210–A218.

[72] Perryman, I. (1997), *The Industrial Laser User*. Oxon, UK: AILU, Abingdon, May (1997), p. 17.


## Further Reading


Bass, M. (1983), *Laser Materials Processing*. Amsterdam: North Holland. A collection of reviews, with practical examples, of laser manufacturing processes.

Belforte, D., Levitt, M. (Eds.) (1986–1990), *The Industrial Laser Annual Handbook*, Tulsa, OK: PennWell Books, and (1992–1993), *The Industrial Laser Handbook*. New York: Springer. A useful resource on many aspects of laser processing.

Charschan, S. S. (Ed.) (1972), *Lasers in Industry*. Orlando, FL: Laser Institute of America. An early, but extensive and still useful, treatment of a wide variety of laser applications, including machining.

Chryssolouris, G. (1991), *Laser Machining: Theory and Practice*. New York: Springer. A graduate-level text, succinctly covering much material.

Dawes, C. (1992), *Laser Welding*. Cambridge, UK: Abington Publishing. An especially practical guide.

Duley, W. W. (1976), *$CO_2$ Lasers: Effects and Applications*. New York: Academic Press. A comprehensive account, with emphasis on the physics involved.

Duley, W. W. (1999), *Laser Welding*. John Wiley, Toronto, Canada. A detailed and perceptive review of the literature on laser welding.

Lugomer, S. (1990), *Laser Technology: Laser Driven Processes*. Englewood Cliffs, NJ: Prentice Hall. A graduate-level review of diverse laser material processing, emphasizing the science involved.

Luxon, J. T., Parker, D. E., Plotkowski, P. D. (1987), *Lasers in Manufacturing: An Introduction to the Technology*. Kempston, Bedford, UK: IFS and Springer. A good nonspecialist introduction to both low and high-power applications.

Ready, J. F. (1978), *Industrial Applications of Lasers*, New York: Academic. About 20% of this wide-ranging book is devoted to machining applications.

Ready, J. F., Farson, D. F. (Eds.) (2001), *LIA Handbook of Laser Materials Processing*. Orlando, FL: Laser Institute of America.

Schouker, D. (Ed.) (1998), *The Handbook of the Eurolaser Academy*, Vols. 1 and 2. London: Chapman & Hall.

Steen, W. M. (1998), *Laser Material Processing*. London: Springer. A tutorial treatment that includes both principles and numerous practical applications.

# Lasers, Medical Use of

*Irving Itzkan*
*Massachusetts Institute of Technology, G.R. Harrison Spectroscopy Laboratory, Cambridge Massachusetts, USA*

*Joseph A. Izatt*
*Massachusetts Institute of Technology, Research Laboratory of Electronics, Department of Electrical Engineering and Computer Science, Cambridge Massachusetts, USA*

*Holger Lubatschowski*
*Laser Zentrum Hannover, Hannover, Germany*
*Phone: +49 511 2788 279; Fax: +49 511 2788 100; e-mail: hl@lzh.de*

**Keywords**
laser–tissue interaction; laser surgery; laser tissue welding; refractive surgery; photoablation; photodynamic therapy (PDT); optical tweezer; laser scanning confocal microscopy; optical tomography.

## 1
## Introduction

Lasers are now used routinely in medicine to cut, shape, remove, heat, and examine biological tissues. Largely in the last decade, lasers have made the transformation out of the research laboratory and into the medical clinic, garnering an impressive penetration into many medical specialties (e.g., Fig. 1). In ophthalmology, a classic example for medical laser applications, the so-called Laser in Situ Keratomileusis (LASIK) procedure, which is a method to correct refractive errors of the eye has been performed more than 2 million times worldwide in 2001. The total revenues in the medical laser market worldwide are estimated to about 2.2 billion dollars in 2003.

The growth of interest in recent decades in the applications of light, optics, and spectroscopy in medicine is related to several key technological advances. The principal advance is the development of lasers to the state where they can be used in medical environments. The special qualities of lasers that make the light they generate especially attractive for medical applications include high spatial coherence, tunability, and generation of

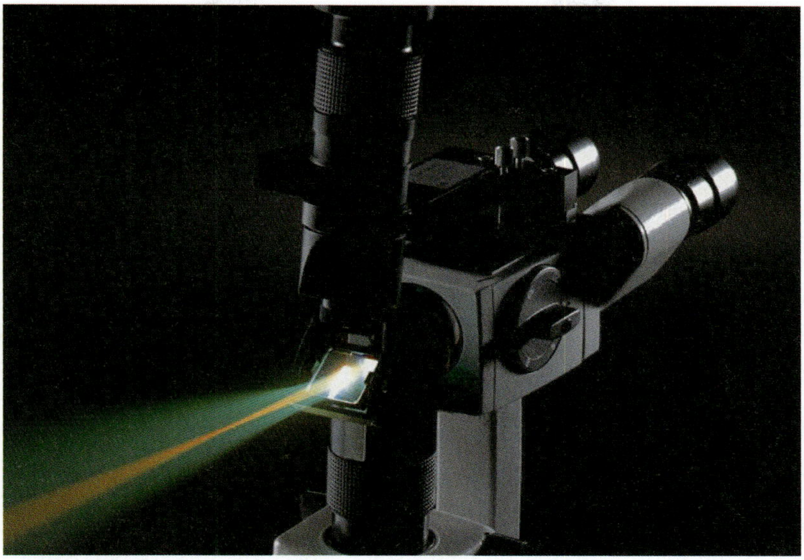

**Fig. 1** Medical laser system integrated in an operation microscope (slit lamp) for the treatment of a variety of retinal deseases (courtesy of Carl Zeiss Meditec)

high peak and average power. Spatial coherence is important for beam handling, shaping, and focusing, and for efficient coupling into optical fibers. Tunability provides the capability of selecting optimal wavelengths for specific applications, matching the absorption and scattering properties to the problem at hand. High power accompanying the spatial coherence allows for the precise delivery of high-energy densities to selected tissue sites.

The second major technical advance that has been critical to laser–tissue interactions in research and applications is the development of optical fibers, which has been driven primarily by the communications industry. The spatial coherence of laser light enables it to be coupled efficiently into low-loss optical fibers, which can deliver the light to remote sites in the body, minimizing trauma to adjacent tissues. Where normal body openings exist, treatment can be performed through endoscopes. Otherwise, the fiber can be passed through a catheter, a laparoscope, or even a hypodermic needle in a minimally invasive manner.

The development of intensified detector arrays, particularly optical multichannel analyzers with high quantum efficiencies and large dynamic ranges, has brought real-time spectroscopic analysis of tissue to the medical clinic and the operating room.

Finally, the development of high-speed, large-capacity computers for analysis and real-time control of medical instrumentation has permitted the use of elaborate algorithms for therapy, imaging, and diagnostics, and has amplified the impact of the other technical advances already mentioned.

Current research in laser applications in medicine may be roughly divided into four areas: biomedical optics, therapeutic applications, diagnostic applications, and biomedical instrumentation. Biomedical optics is the study of light transport in tissue and the optical properties of tissue. This includes theory, such as analytical and numerical modeling of radiation transport, and experiment, such as techniques for measuring tissue optical properties. The primary difficulty in modeling light propagation in tissue is that, at many wavelengths, scattering dominates absorption in most biological tissues [1]. While many approximate or numerical models have been suggested, no universally acceptable analytical models have been developed.

Recent developments in biomedical optics include the study of time-resolved photon migration in tissue. Using fast electronics and operating in either the time or frequency domain, this technique reduces the light-scattering problem, since the transit time of photons (and, hence, their approximate path length) is measured. Another advantage is that it is a particularly direct way to obtain absorption coefficients of tissues. Applications of time-resolved radiation transport have included *in situ* optical ranging in biological tissue, and initial steps toward optical tomography (see OPTICAL TOMOGRAPHY and [2–6]).

Lasers are being used in a wide range of therapeutic applications. The most widely known and commercially successful is laser cutting or ablation of tissue, the "laser scalpel." Although laser cutting is a capital-intensive technology, its advantages are still compelling under the proper circumstances. Since it requires no physical contact between the light-delivery system and the target tissue, the possibility of infection is greatly reduced. In contrast to a sharp scalpel blade, which cuts by exerting pressure in excess of the local tissue tensile strength, laser ablation exerts only minimal pressure in the form of acoustic transients on the tissue generated by the thermoelastic expansion of the tissue and by the recoil momentum of the expelled ablation products. Unfortunately, owing to the lack of physical contact with the tissue, the tactile feedback to the surgeon is also absent. This may lead to complications if the surgeon cannot get out of the way of sensible structures beneath the tissue surface.

Avoiding the traction caused by the laser scalpel is particularly important in nerve and brain tissue surgery. The use of the laser for excision of brain tumors decreases the likelihood of initiating convulsions, and requires a smaller orifice through the skull and intervening brain tissue. With the use of optical fibers inserted through endoscopes, catheters, and needles, surgery is possible in regions of the body that are inaccessible to the scalpel blade without the collateral damage that results from conventional surgery.

Other applications of lasers in medical therapy involve tissue heating only. Nonablative laser treatments have been developed in dermatology, for example, which involve coagulation of subcutaneous blood vessels or destruction of pigment cells for treatment of skin disorders [7]. Lasers are used as distributed heat sources for hyperthermia treatment [8]. Laser welding of tissue is used in vascular surgery and wound repair [9]. The activation with laser light of phototoxic exogenous chromophores, primarily hematoporphyrin derivatives (HPDs), continues to receive attention as a treatment for cancer and certain immune disorders [10–13].

Lasers are also used for medical diagnosis and monitoring. Laser-induced fluorescence in tissues can serve as a probe for architectural and compositional changes signifying the onset and development of the disease. Work in this area has demonstrated the successful identification and discrimination of the various stages of atherosclerotic plaque, and neoplastic conversion in urinary bladder and gastrointestinal tissues. Related techniques using infrared wavelengths, which yield molecular level information, hold out the promise of gaining functional as well as morphological information.

Finally, several novel applications of lasers in biomedical instrumentation deserve comment. Laser-induced fluorescence microscopes are commonly used in pathology research, contributing to the study of disease (see MICROSCOPY). Systems have been developed for rapid cell sorting by laser-induced fluorescence-based flow cytometry [14]. A novel technique for suspending transparent microscopic objects at the focus of a laser beam, or "optical tweezers," is capable of holding and manipulating individual, living biological cells [15]. This technique has been used to study the mechanics of the biochemical motors contained in cells.

## 2
## Physics of laser–tissue Interaction

### 2.1
### laser–tissue Interaction Regimes

The nature of the laser–tissue interaction process may be divided into several regimes, determined primarily by the intensity of the laser beam and its interaction time with the tissue (Fig. 2). The physical basis of most surgical applications of lasers is tissue cutting and removal at relatively high laser-beam intensities with exposure times of milliseconds to seconds, resulting in the rapid deposition of heat and subsequent vaporization or decomposition. These are the processes that are characteristic of thermal or thermoacoustic ablation. For nanosecond pulses at relatively high photon energies (ultraviolet wavelengths), photons can directly break specific chemical bonds, resulting in strong absorption and particularly clean cuts. Picosecond and shorter pulses can induce nonlinear absorption in biological

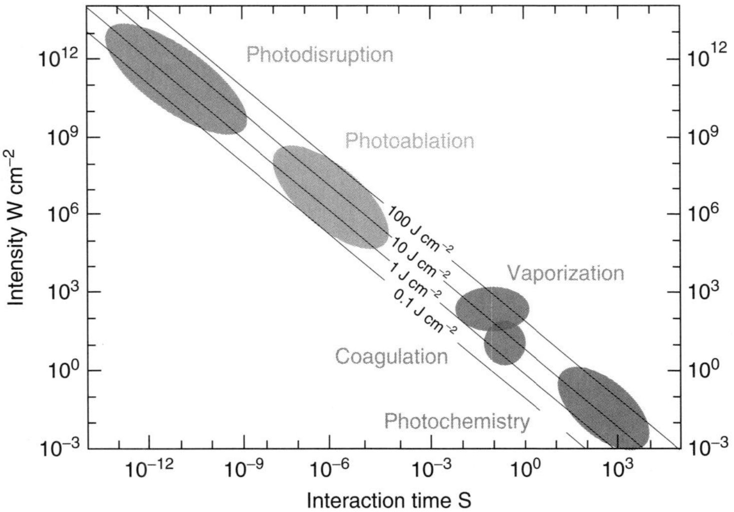


**Fig. 2** laser–tissue interaction map, indicating primary laser–tissue interaction regimes (after Boulnois, J.-L. [16])

tissue, providing the capability for localized absorption of laser energy in otherwise transparent tissue such as the eye. Finally, some interactions have been observed at very low laser-intensity levels delivered over minutes or hours. These interactions are strongly coupled to the host tissue response to photochemistry, and remain as yet poorly understood.

Following a review of the physical properties of biological tissue that are of concern for laser–tissue interactions, these mechanisms are discussed in detail.

## 2.2 Tissue Physical Properties

### 2.2.1 Tissue Optical Properties

The ability to deliver therapeutic laser light to collect diagnostic light from a specific location within the body depends upon an accurate understanding of how light is distributed and absorbed in biological tissues. The distribution of laser light in tissue is, in general, a very complicated function of its wavelength-dependent intrinsic optical properties. Light propagation in tissue can be considered as a special case of radiation transport in random media [17]. Conventional problems in electromagnetics usually deal with deterministic wave propagation, such as is encountered in antenna, waveguide, and diffraction theory. In contrast, problems of wave propagation in random media may involve unpredictable variations in wave amplitude and phase. Optically, a tissue is a continuous medium with a spatially varying complex index of refraction. For modeling purposes, it is usefully approximated as a distribution of discrete absorbers and scatterers, each of which has intrinsic scattering properties. Under the assumption of noninteracting particles, the bulk absorption ($\mu_a$) and scattering ($\mu_s$) coefficients (units of $cm^{-1}$), are connected to the cross sections for absorption ($\sigma_a$) and scattering ($\sigma_s$) of individual particles by their number density $n$ (particles/$mm^3$):

$$\mu_a = n\sigma_a, \quad (1)$$

$$\mu_s = n\sigma_s. \quad (2)$$

Light is absorbed in tissue through the action of a variety of chromophores, including water, various proteins, hemoglobin in blood, and tissue pigments such as melanin. A graph of the wavelength dependence of absorption for several prominent tissue chromophores appears in Fig. 3. The low absorption of almost all the tissue constituents in the 600–1300-nm region has resulted in that wavelength range being referred to as the "diagnostic window" because of the resulting accessibility of deep tissues to light in that range. Light is very strongly scattered in tissue from index gradients within organs and cells, as well as from discrete structures such as collagen fibrils and red blood cells. A significant characteristic of tissue optics is that in almost all the tissues except for portions of the eye, the scattering coefficient is much larger than the absorption coefficient at most wavelengths. A summary of the relevant parameters for radiative transport calculations in biological tissues and typical values appear in Table 1.

Several methods are available for modeling the interaction of light with large collections of absorbers and scatterers. Multiple-scattering theory starts with a description of the interaction of waves with individual particles consistent with Maxwell's equations, and then introduces the interaction effects of many particles with statistical averaging. This method is very difficult and has not been applied to light propagation in tissues. Another

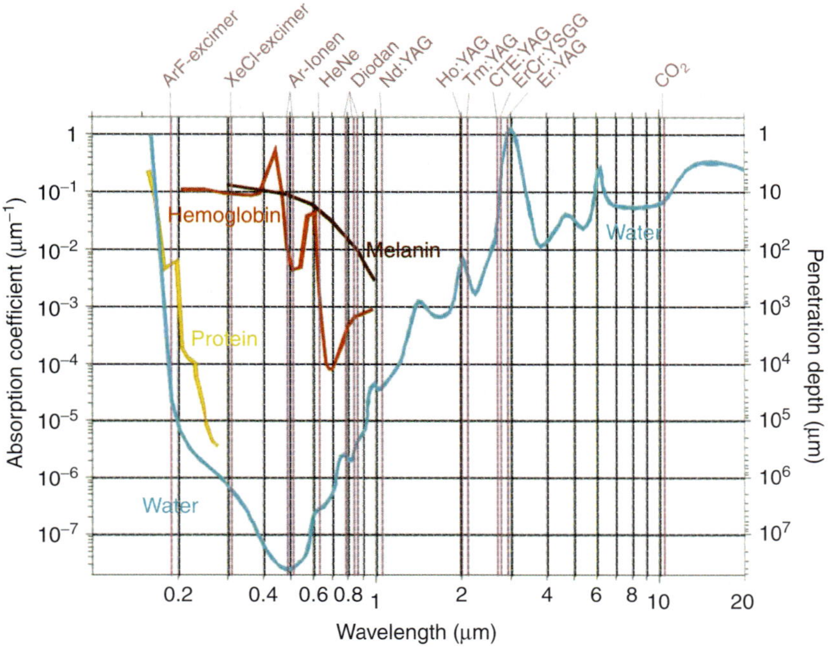


**Fig. 3** Absorption spectrum of several important tissue chromophores, together with the wavelength positions of the most widely used medical lasers. (after Boulnois, J.-L. [16]. The $H_2O$ spectrum is adapted from Hale, G. M., Querry, M. R. [18])

method for describing light distribution within a large collection of absorbers and scatterers is transport theory. Transport theory describes the propagation of intensities on the basis of the phenomenological observations of transport characteristics. It does not include diffraction and interference between fields; in this sense it is less rigorous than multiple-scattering theory.

**Transport Theory** Transport theory describes light propagation in a random collection of scatterers, and absorption by the stationary radiative transport equation

$$\mathbf{s} \cdot \nabla L(\mathbf{r}, \mathbf{s}) = -(\mu_a + \mu_s) L(\mathbf{r}, \mathbf{s}) + \mu_s \times \int_{4\pi} p(\mathbf{s}, \mathbf{s}') L(\mathbf{r}, \mathbf{s}') \, d\omega'. \quad (3)$$

where $L(\mathbf{r}, \mathbf{s})$ is the radiance (W mm$^{-2}$-sr) of light at position $\mathbf{r}$ traveling in the direction of the unit vector $\mathbf{s}$, and $\omega'$ is the differential solid angle in the direction of $\mathbf{s}'$. The first term on the right represents the decrease in the radiance resulting from absorption and scattering out of the direction $\mathbf{s}$, while the second term describes the increase in the radiance resulting from scattering from other directions $\mathbf{s}'$ into the direction $\mathbf{s}$. Absorption and scattering are described by the coefficients $\mu_a$ and $\mu_s$ respectively, and the phase function by $p(\mathbf{s}, \mathbf{s}')$. The phase function $p(\mathbf{s}, \mathbf{s}')$ represents the normalized probability that light scattered out of the direction $\mathbf{s}'$ is scattered into the direction $\mathbf{s}$. Often, the effect of the phase function can be adequately characterized

**Tab. 1** Summary of tissue physical properties

| ***Parameter*** | ***Symbol*** | ***Optical properties*** | |
|---|---|---|---|
| | | ***Typical values in therapeutic window ($\lambda \sim$ 600–1300 nm)*** | ***Approximate value in breast ($\lambda$ = 635 nm) [19]*** |
| Absorption coefficient | $\mu_a$ | 0.1–10 $cm^{-1}$ | 0.2 $cm^{-1}$ |
| Scattering coefficient | $\mu_s$ | 10–1000 $cm^{-1}$ | 200 $cm^{-1}$ |
| Total attenuation coefficient | $\mu_t = \mu_a + \mu_s$ | 10–1000 $cm^{-1}$ | 200 $cm^{-1}$ |
| Albedo | $\mu_s/(\mu_a + \mu_s)$ | 0.5–0.99 | 0.99 |
| Anisotropy | $g$ | 0.7–0.95 | 0.95 |
| Reduced scattering coefficient | $\mu_s' = \mu_s(1 - g)$ | | 10 $cm^{-1}$ |
| Forward attenuation coefficient | $\mu_{forward} = \mu_s' + \mu_a$ | | 10 $cm^{-1}$ |
| Forward-scattering mean free path | $(\mu_s' + \mu_a)^{-1}$ | | 1 mm |
| Diffusion coefficient | $D = (3[\mu_a + \mu_s(1 - g)])^{-1}$ | | 300 μm |
| Effective attenuation coefficient | $\mu_{eff} = \sqrt{\mu_a/D}$ | | 2.5 $cm^{-1}$ |
| ***Parameter*** | ***Symbol*** | ***Thermal properties*** | |
| Specific heat/unit volume | $\rho C$ | $3.96 \times 10^{-3}$ (J/°C-$mm^3$) | |
| Heat diffusivity | $\kappa$ | 0.106 $mm^2$ $s^{-1}$ | |
| | | ***Mechanical properties*** | |
| ***Parameter*** | ***Symbol*** | ***Soft tissue*** | ***Bone*** |
| Speed of sound | $c$ | 1500 m $s^{-1}$ | 2800 m $s^{-1}$ |
| Grüneisen coefficient | $\Gamma$ | 0.3 | 1–2 |
| Maximum elongation | | 25–100% | 2.5% |
| Maximum stress | $\sigma$ | 600 bars | 500–2500 bars |

by the average value of the cosine of the angle between **s** and **s**′, denoted by $g$.

$$g = \int_{4\pi} p(\mathbf{s}, \mathbf{s}')(\mathbf{s} \cdot \mathbf{s}')\, d\omega'. \qquad (4)$$

The three quantities $g$, $\mu_a$, and $\mu_s$ represent the fundamental tissue optical properties used in most treatments of laser light distribution in tissues.

The exact form of the phase function $p(\mathbf{s}, \mathbf{s}')$ is often unknown. In many problems in tissue optics, the Heyney–Greenstein phase function, first used for scattering problems in astrophysics, has been found to be useful.

$$p(\mathbf{s}, \mathbf{s}') = \frac{1 - g^2}{[1 + g^2 - 2g(\mathbf{s} \cdot \mathbf{s}')]^{3/2}}. \qquad (5)$$

When defined in this way, the parameter $g$ in the Heyney–Greenstein phase function has the same meaning as in the defining equation, Eq. (4).

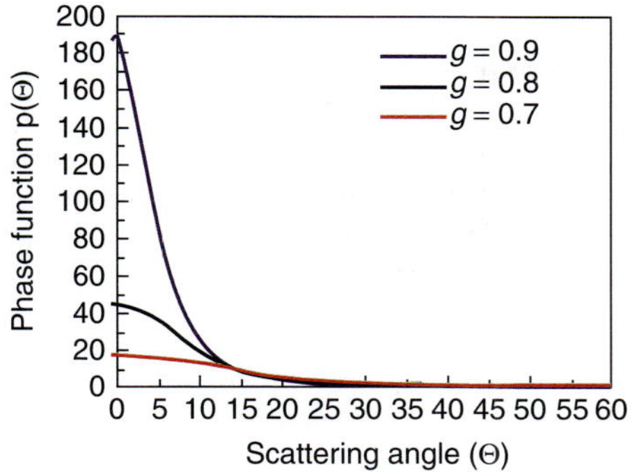


**Fig. 4** Heyney–Greenstein phase function as a function of scattering angle for different values of g

If $g$ is equal to zero, the phase function has the constant value of 1, which describes homogeneous scattering. A positive value of $g$ between 0 and 1 describes forward scattering and a negative value of $g$ means backward scattering. Biological tissue usually has values between 0.7 and 0.99. In Fig. 4, the Heyney–Greenstein phase function is plotted for three different values of g as an example.

Heat deposition in tissue occurs by light absorption from all directions. The integral of the radiance over all solid angles is the fluence rate $F$ ($\mathrm{W\,cm^{-2}}$, also called the *space irradiance*). The rate of energy deposition in the tissue per unit volume, the absorbed power density $U$ ($\mathrm{W\,cm^{-3}}$), is given by the product of the local absorption coefficient and the fluence rate,

$$U = \mu_\mathrm{a} F. \qquad (6)$$

Analytical solutions for the radiative transport equation are very difficult to obtain. Two approaches are generally followed to obtain approximate solutions. The first is the numerical simulation, in which individual photon trajectories through a model tissue are calculated with Monte–Carlo methods, and the resulting fluence rate is summed over a large number of trajectories. A second method that gives useful, although approximate, analytical results is the diffusion approximation.

**Diffusion Approximation** One approach to solving the radiative transport equation is the diffusion approximation. Here the radiance in Eq. (3) is separated into scattered and unscattered components:

$$L(\mathbf{r}, \mathbf{s}) = L_\mathrm{c}(\mathbf{r}, \mathbf{s}) + L_\mathrm{d}(\mathbf{r}, \mathbf{s}). \qquad (7)$$

The unscattered component is attenuated exponentially according to Beer's law. The scattered component then approximately satisfies the diffusion-like equation

$$-D\nabla^2\Phi(\mathbf{r}) + \mu_a\Phi(\mathbf{r}) = S(\mathbf{r}), \qquad (8)$$

where

$$\Phi(\mathbf{r}) = \int_{4\pi} L_\mathrm{d}(\mathbf{r}, \mathbf{s}')\,\mathrm{d}\omega' \qquad (9)$$

is the diffuse flux,

$$D = \frac{1}{3}[\mu_\mathrm{a} + \mu_\mathrm{s}(1 - g)] \qquad (10)$$

is the diffusion coefficient, and $S(\mathbf{r})$ is a source term.

Solutions to the diffusion equation have been obtained for the fluence rate, total reflectance, and transmittance through a tissue slab, including time dependence [20, 21]. The diffusion approximation is accurate for optically dense and highly scattering media, far from tissue boundaries or internal sources of light. A particularly useful result of the diffusion approximation is that the diffuse flux attenuates exponentially according to an effective attenuation coefficient.

$$\mu_{\text{eff}} = \sqrt{\frac{\mu_a}{D}} = \sqrt{3\mu_a[\mu_a + \mu_s(1-g)]}. \tag{11}$$

A problem with the diffusion approximation is that the assumptions are poorest for forward scattering, that is, where $g$ is near unity. Unfortunately, biological tissue is predominantly forward scattering.

### 2.2.2 Tissue Thermal Properties

Absorption of visible and near-ultraviolet light in tissue occurs by exciting electronic transitions in tissue molecules, primarily proteins and lipids. Near- and mid-infrared light are absorbed via vibrational-rotational excitations in biomolecules, as well as in water. The lifetimes of both the electronic and vibrational-rotational states typically are in the range of picoseconds. As soon as the initial excited-state distributions thermalize, the energy that is deposited by the laser pulse may be considered as heat, and its dissipation is described by thermal diffusion.

The thermal properties of tissue relevant to heating (without phase change) and heat diffusion are the density ($\rho$), specific heat ($C$), and thermal conductivity ($\sigma$). For thermal diffusion calculations, the important quantity is the thermal diffusivity obtained from these quantities via

$$\rho C = \text{specific heat per unit volume}, \tag{12}$$

and

$$\kappa = \frac{\sigma}{\rho C} = \text{thermal diffusivity}. \tag{13}$$

Estimates for $\rho C$ and $\kappa$ in biological media are listed in Table 1. These values are very close to the thermal properties of pure water, which comprises ~60–90% of all the biological tissues except bone, teeth, hair, and pathological calcifications.

The attenuation depth $\mu_{\text{eff}}^{-1}$ is inversely proportional to the concentration of the absorbing chromophores times their molar extinction coefficient. Figure 3 shows the spectral variation of the molar extinction coefficients of the principal tissue chromophores and provides a qualitative sense of why the body is most translucent in the range of 600 nm to 1.3 μm. Measured attenuation depths in most tissues of clinical interest range from fractions of a micron in the deep-ultraviolet and mid-infrared regions up to several millimeters in the near-infrared. In many surgical applications, the laser-spot diameters on tissue, on the other hand, are rarely less than a few hundred microns. For these situations of clinical interest, it is thus sufficient to consider the limit in which the illuminated region of the tissue corresponds to a relatively thin disk. Most of the heat loss from the disk will occur out of the bottom, and so a semi-infinite one-dimensional slab geometry may be used to describe heat diffusion.

The time-dependent equation of heat conduction in this geometry reduces to

$$\frac{\partial T(z,t)}{\partial t} = \kappa \frac{\partial^2 T(z,t)}{\partial z^2}. \tag{14}$$

where $T(z,t)$ is the temperature, a function of both distance into the tissue ($z$) and time ($t$) after the (assumed short) laser pulse is over. A solution to this equation for an exponential initial temperature distribution

$$T(z', t'=0) = T_0 e^{-z'} \tag{15}$$

is

$$T(z', t') = \frac{1}{2} T_0 \left\{ e^{(t'-z')} \left[ 1 + \text{erf}\left( \frac{z'}{2\sqrt{t'}} - \sqrt{t'} \right) \right] + e^{(t'+z')} \right.$$

$$\times \left[ \text{erfc} \left( \frac{z'}{2\sqrt{t'}} + \sqrt{t'} \right) \right] \Bigg\}, \quad (16)$$

where

$$z' = \mu_{\text{eff}} z \quad \text{and} \quad t' = \mu_{\text{eff}}^2 \kappa t \quad (17)$$

are dimensionless intermediate variables.

It is common to refer to

$$\tau = \frac{1}{4} \kappa \mu_{\text{eff}}^2 \quad (18)$$

as the so-called "thermal relaxation time," which is an approximate characteristic time for thermal diffusion in tissue. This value is used for determining how short laser-pulse durations must be in order to ablate without causing thermal damage (so-called "thermal confinement") in peripheral tissue. For example, if the laser has an optical penetration depth of only 1 μm, the pulse duration should be shorter than 1 μs according to the thermal relaxation time. Otherwise, if the pulse is longer, heat would dissipate into the tissue deeper than 1 μm during the irradiation time. At 1 ms the thermal diffusion length is 23 μm and at 1 s it is 0.72 mm.

### 2.2.3 **Tissue Structural Properties**

Human tissue can be conveniently divided into two classes based on its response to laser ablation. Soft tissue includes muscle, fat, skin, blood vessels and the various internal organs. Hard tissue includes bone, teeth, and calcified plaque, which is often the result of a disease process. Soft tissue consists primarily of proteins, with a large amount of water and many trace constituents, such as electrolytes and complex organic molecules, which frequently determine the optical-absorption coefficient. Hard tissue is a composite material consisting of calcium phosphate salts embedded in a soft tissue matrix.

An important point applicable to most biological tissues is that the optical scattering coefficient is usually much larger than the absorption coefficient. Depending on the wavelength, it can be ten to one hundred times larger; that is, the biological tissue is a translucent material. In modeling the effects of lasers on tissue, it is often appropriate to assume that the material has the optical properties of the particular type of tissue, the thermodynamic properties of water, and the mechanical properties of structural proteins.

Almost all calcified tissues of the body are composite materials, in which microcrystallites of a calcium salt are embedded in a soft connective tissue matrix. The matrix provides tensile strength, while the salt provides compressive strength. In compact bone, the inorganic fraction is mostly made up of hydroxyapatite salts, densely deposited in a highly organized collagen matrix, with a small percentage of residual water (Table 2). Individual salt crystals in bone are needle-shaped, with diameters on the order of nanometers; however, their deposition into the tissue matrix organizes them into granules with dimensions on the order of a few microns. Cardiovascular calcified deposits are not organized, but consist mostly of mixed apatite salts embedded in a soft-component matrix of lipids and cross-linked proteins. The hardest calcified tissues have little or no soft component and are very brittle. Tooth enamel consists of almost all hydroxyapatites, with a small amount of keratin as the soft component. Urinary calculi (kidney stones) have no soft component at all, and so do not constitute a composite substance. For this reason, they are very brittle and can be broken up by ultrasound. Laser ablation of these stones proceeds by depositing energy in a surface plasma rather than in the stone, and using

**Tab. 2** The composite nature of calcified tissues mp = melting point; bp = boiling point

| *Tissue* | *Hard component* | *Soft component* |
|---|---|---|
| Cardiovascular calcified deposits | 75% mixed apatite salts (mp ~1500 °C), granules 1 to 10 μm in size | 20% lipids, cross-linked proteins (bp ~300 °C)<br>5% water (bp 100 °C) |
| Bone | 75% hydroxyapatite (mp ~1500 °C), granules 1 to 10 μm in size | 20% collagen (bp ~300 °C)<br>5% water (bp ~100 °C) |
| Tooth enamel | Hydroxyapatite | Keratin |
| Urinary calculi | Calcium oxalate | None |

the shock wave generated in the plasma to disintegrate the stone.

## 2.3 Nonablative Tissue Effects

Significant tissue alterations may be induced by the cumulative effects of relatively slight laser heating. A moderate rise in temperature on the molecular level effects the breakage of hydrogen and other van der Waals bonds that stabilize the conformation of the molecules. Such breakage can lead to denaturation of proteins and enzymes, resulting in the inactivation of their function. It has been shown that denaturation reactions are fairly well described by a rate-process formalism from first-order chemical kinetics, the Arrhenius damage integral:

$$\Omega = P\int_0^t \exp\frac{-E_a}{RT}\,dt'. \qquad (19)$$

Here $\Omega$ is a measure of the cumulative damage, $P$ is a pre-exponential constant, $E_a$ is an activation energy for the reaction, $R$ is the universal gas constant, $T$ is the temperature, and $t$ is the time of exposure. For denaturation reactions, $P$ and $E_a$ are determined experimentally, with typical values in the range of $P \sim 10^{70}\,\mathrm{s}^{-1}$ and $E_a \sim 4 \times 10^5\,\mathrm{J\,mol}^{-1}$ for liver tissue. The cumulative damage is thus a highly nonlinear function of the tissue temperature and heating history.

## 2.4 Laser Tissue Ablation

A review of the experimental evidence shows, for a wide range of parameters, that the energy density in tissue during ablation is an order of magnitude less than that required for vaporization of tissue water alone. Consideration of the thermodynamics of water shows that, under the conditions of inertial confinement, when there is not enough time for material to move, these energy densities can lead to the generation of tremendous pressure. This pressure exceeds a threshold that depends on the structural properties of the tissue and leads to ablation [22].

Since soft biological tissue contains up to 70% water, the extremely well-known thermodynamic properties of water can be used to model its thermal behavior. These properties, including the density, pressure, internal energy, entropy, enthalpy, and temperature, have been tabulated over a wide range of values. Figure 5 shows the most important parameters required for an understanding of the ablation process. On this plot of pressure versus internal energy, contours of constant temperature and density are displayed, and the boundary of the phase-change region is marked with a bold line.

The value of 2500 J cm$^{-3}$ required to vaporize pure water applies to a constant-pressure process. This process is illustrated in Fig. 5 by the horizontal path A–C. This path assumes that the conditions of the heating are such that the pressure remains constant. However, if the heating is rapid compared to the time of pressure relaxation, the pressure will increase rapidly, and the system will traverse a different thermodynamic path. In the extreme case of instantaneous heating, the volume (or density) of the system remains constant. To illustrate the difference between heating at constant pressure and heating at constant volume, consider paths A–B and A–D in Fig. 5. In path A–B, 300 J cm$^{-3}$ are slowly added to the water. The pressure remains constant at 1 bar while the temperature increases to 92 °C, and the density decreases to 0.964 g cm$^{-3}$. In path A–D, 300 J cm$^{-3}$ are added to the water faster than the pressure can relax. While the density remains constant at 1.0 g cm$^{-3}$ and the temperature increases to 96 °C, the pressure rises to 900 bars.

A concept of inertially confined ablation may be defined on the basis of the generation of pressure that occurs while heating a substance at constant volume. In pulsed laser ablation of tissue, the energy

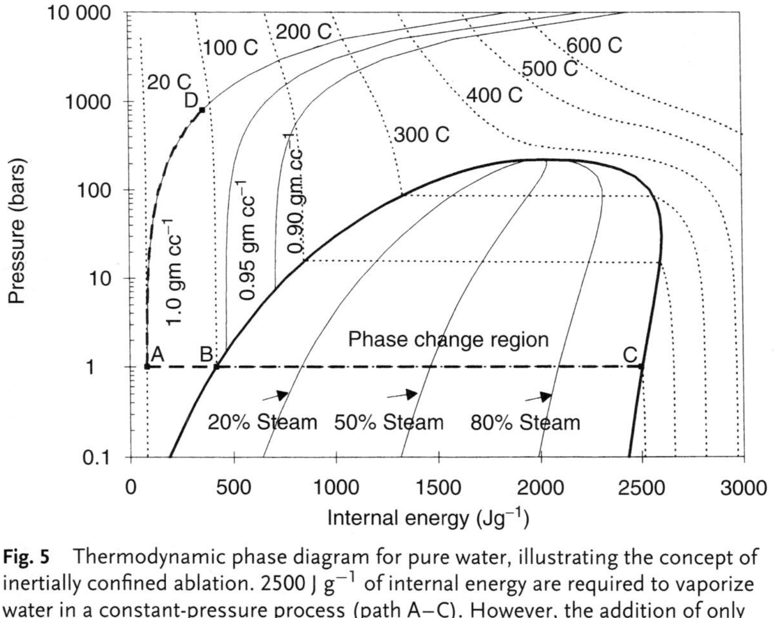


**Fig. 5** Thermodynamic phase diagram for pure water, illustrating the concept of inertially confined ablation. 2500 J g$^{-1}$ of internal energy are required to vaporize water in a constant-pressure process (path A–C). However, the addition of only 300 J g$^{-1}$ of internal energy to water at STP in a constant volume process increases the pressure to 900 bars (path A–D). This pressure rise, when achieved in tissue water by the rapid deposition of energy by a pulsed laser, is sufficient to exceed the tensile strength of tissue and lead to tissue fracture and removal (reprinted with permission from Itzkan et al. [23])

can be supplied so rapidly that the material is inertially confined during the length of the laser pulse. For an instantaneous pressure generation the so-called stress-confinement has to be fulfilled, which means that the laser-pulse duration has to be short compared to the time the stress transient needs to propagate with the speed of sound through the optical penetration depth:

$$\tau_L \cdot \mu_a \cdot c_0 \ll 1$$

This high pressure, which corresponds to large compressive stresses in tissue, leads to dislocations between the layers of tissue and the nucleation and growth of microcracks, which can be modeled by solving the kinetic equation for the total volume of cracks. Ablation occurs when the laser-induced pressure exceeds a threshold value (which is dependent on the structural properties of the tissue) and proceeds through a disassembly of the tissue at the microcracks and subsequent acceleration away from the crater as a result of the stress gradient. Although a tremendous pressure is generated, the small temperature increase is about the same as that predicted by the assumption of heating at constant pressure (the lines of constant temperature on the left portion of Fig. 5 are nearly vertical).

Ablation of hard tissue can be described in terms of the inertial confinement model for soft tissue discussed above. Under these conditions, inertial confinement causes the same large pressures and pressure gradients as in soft tissue. The soft-component bonds fail in tension, and the material is accelerated, ejected, and leaves behind a crater in the tissue. The salt crystals remain intact; however, the soft component rapidly expands and exerts a force on the salt granules through momentum transfer and viscous drag. At sufficiently high irradiance, the salt granules become dislodged and are accelerated out of the crater. Experimentally, ablation of calcified tissue is achieved at all wavelengths (ultraviolet, visible, and infrared) at irradiances above $10^6$ W cm$^{-2}$. This is consistent with the calculations that indicate that vapor velocities of a few hundred meters per second are needed to drag hydroxyapatite particles 1–10 μm in size out of the crater. Gas-dynamic calculations show that these vapor velocities are obtained with incident laser irradiances of about $10^6$ W cm$^{-2}$.

## 2.5 Nonlinear laser–tissue Interactions

The sharpness of a laser scalpel cut and the efficiency of tissue ablation are both limited by the depth over which laser light is absorbed in the target tissue and the time while the laser energy is deposited, without dissipation of heat out of the interaction zone. The absorption depth is a function of the laser wavelength being used and the optical properties of the tissue. In some applications, where a sufficiently short absorption depth cannot be obtained through the choice of an optimum wavelength or the use of a dye additive, it is possible to use the high intensities available from short-pulse laser sources to induce nonlinear absorption. The most successful application of nonlinear absorption to achieve a desired clinical end point is in intraocular surgery, since the eye has negligible absorption for visible and near-infrared wavelengths.

The process through which nonlinear absorption is used in laser medicine is termed laser-induced optical breakdown, and it involves the interplay of several complex physical mechanisms. When a short laser pulse is brought to a tight focus within a target tissue, nonlinear absorption can occur through the processes of multiphoton absorption and/or avalanche ionization. For sufficiently high pulse intensity, the resulting dielectric breakdown of the medium results in a microexplosion accompanied by plasma formation and, if the process takes place in a liquid medium, by cavitation. The rapidly expanding plasma generates strong acoustic transients, including shock waves that propagate outward from the breakdown site. When the pulse is over, the extinguished plasma leaves behind a cavity, which may continue to expand and, depending upon the tissue mechanical properties, may contract and reexpand repeatedly. These explosive actions lead to the desired therapeutic cutting effect but also may produce undesirable collateral damage. Studies conducted using pulses from nanoseconds to femtoseconds have shown that, in general, the volume of the collateral damage varies as the pulse energy, while the threshold of the laser breakdown process is more or less intensity dependent (Table 3). Thus, shorter pulses are capable of initiating a breakdown resulting in less collateral damage and finer cuts. This rule applies until the pulses are so short (femtoseconds) that nonlinear propagation effects such as dispersive pulse broadening and/or self-focusing begin to inhibit the delivery of pulses to the target tissue.

Laser-induced breakdown using nanosecond pulses is used in intraocular surgery for the cutting of transparent structures within the eye, such as piercing the lens' posterior capsule if it opacifies in the months or years following cataract surgery. Femtosecond photodisruption is used where collateral damage is an important consideration, such as refractive corneal surgery to correct myopia or in keratoplasty, where parts of the corneal tissue has to be cut out and replaced by donor tissue [24].

## 2.6 Laser Micromanipulation

### 2.6.1 Optical Tweezer

A novel use of lasers in medicine is to use the pressure exerted by light itself to exert controlled forces on cells or microscopic

**Tab. 3** Typical parameters and tissue effects for photodisruption

| | *ns* | *ps* | *fs* |
|---|---|---|---|
| Intensity ($10^{12}$ W cm$^{-2}$) | 0.05[a] | 0.5 . . . 1[a] | 5 . . . 10[a,f] |
| Fluence (J cm$^{-2}$) | 10 . . . 100[b] | 2 . . . 10[b,c] | 1 . . . 3[b,c,f] |
| Pulse energy (μJ) | 100 . . . 10 000[a] | 1 . . . 5[a] | 0.5 . . . 3[a,f] |
| Amplitude of the accoustic transient (1 mm distance) (bar) | 100 . . . 500[e] | 10 . . . 100[e] | 1 . . . 5[f] |
| Diameter of the cavitation bubble (μm) | 1000 . . . 2000[d] | 200 . . . 500[d] | <30[f] |

[a] Vogel et al. (**1998**), *SPIE 3255*, **34**.
[b] Niemz et al. (**1993**), *Laser and Light in Ophthalmol.* **5**(3), 149.
[c] Loesel et al. (**1997**), *SPIE 2923*, **118**.
[d] Vogel et al. (**1993**), *SPIE 1877*, **312**.
[e] Vogel et al. (**1996**) *J. Acoust.Soc. Am.* **100**(1), 148.
[f] Results by one of the authors (H.L.).

particles without physical contact. A laser beam of a few watts intensity focused by a microscope (Fig. 6) to a few microns spot size exerts sufficient light pressure to nudge or grasp objects of up to hundreds of microns in dimension. The principle of operation of such an optical "tractor beam" can be understood with simple ray optics, as illustrated in Fig. 7. Light propagating through a transparent particle positioned at the focus of a laser beam is refracted in a pattern of directions such that the forces

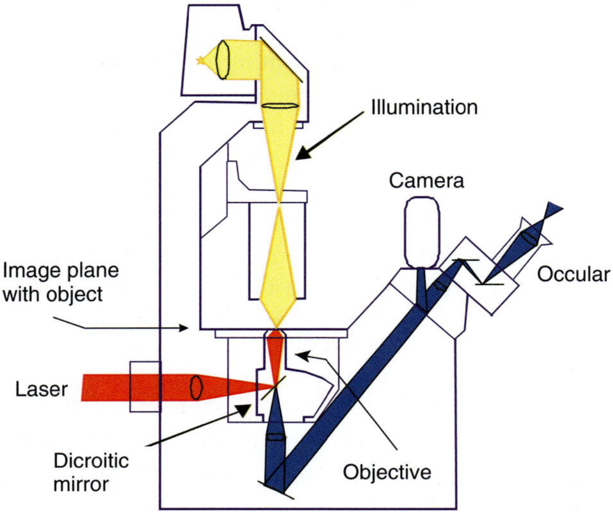


**Fig. 6** Typical setup of an optical tweezer with its beam path for trapping and imaging the objects

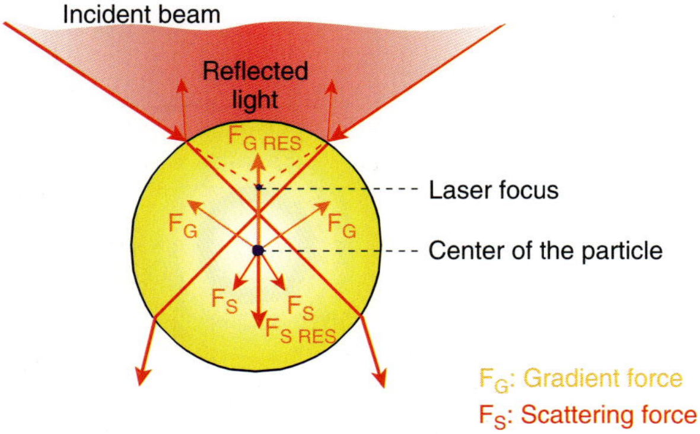


**Fig. 7** Schematic of an optical trap, illustrating how a spherical particle is captured near the focus of a strongly focused laser beam. The particle is held in balance between the force of light pressure due to backscattering and reflection of the beam, and the sum of reaction forces resulting from refraction at the particle boundaries (gradient force)

exerted by the light on the particle tend to keep the particle near the focus. The particle must be relatively transparent at the laser wavelength used so that it will not be destroyed or killed in the process. This ingenious device was conceived by Ashkin et al. [15].

Laser traps have been used for holding and manipulating single biological cells and cell components such as chromosomes. Laser traps are used in research to measure the strength of intracellular biochemical motors, and may be useful in practice to select and position individual cells for operations such as selection of viable sperm cells in fertility enhancement, and in genetic engineering.

#### 2.6.2 Laser Microdissection

Microdissection is a method for separating single cells or a cell cluster from specific regions of microscopic tissue sections. The cells of interest, such as precancerous cells or invading groups of cancer cells, are usually surrounded by heterogeneous tissue elements. Here, microdissection is essential to apply molecular analysis methods to study evolving disease lesions inside the tissue.

In a laser microdissection setup a pulsed UV laser, typically an $N_2$ laser is coupled into a microscope and focused through the objective to a spot size in the range of 1 to 10 μm in diameter. The principle of laser cutting is based on the ablative photodecomposition process without the heating of adjacent material.

Recently, this laser microdissection has been used to isolate Alzheimer's plaques from brain, hepatocytes for subsequent Hepatitis B analysis, and single fetal cells from a cervical mucous smear. Moreover, this technique is used in conjunction with fluorescence or immunohistochemical staining to select specific cells. For example, in tumor research it has been used for the recovery of expressing tumour cells from breast cancer.

In combination with an optical tweezer, microdissection enables cutting cytoplasmic fragments or fusion of cells in culture and it can be used to perform noncontact laser microinjection for the transfer of substances and genetic material into living cells without vectors or mechanical stress.

### 2.7 Sub-diffraction-limit Microscopy and Ablation

The resolution of a conventional optical microscope depends upon the quality of the lenses and/or mirrors of which it is constructed, but is ultimately limited by diffraction to approximately the wavelength of light. The same limit applies to the minimum spot size achievable when focusing a laser beam for ablation. In many applications, however, it is desirable to have the benefits of optical techniques, such as the noninvasive pump/probe analysis of spectroscopic and polarization properties, available at nanometer resolution. It was realized early in this century that the diffraction limit could be overcome if a sample were illuminated through an aperture much smaller than the wavelength of light, with the sample positioned very close to the aperture, that is, in the near field. In this case, the resolution depends on the size of the aperture rather than on the light wavelength. It was not until the scanning tunneling microscope was developed in 1982, that the technology for creating such small apertures could be matched with the technology for positioning them precisely enough with respect to the sample to create the scanning near-field optical microscope (SNOM). An optical waveguide with a nanometer-scale aperture can be made by stretching out a heat-softened pipette tip to a very long length, after which the inside is coated with a reflective metal (to enhance the pipette's intrinsic waveguide properties). The tip is then held in very close proximity to a sample with feedback techniques developed in scanning tunneling microscopy. This tip can then be used to obtain nanometer-resolution optical images by raster-pattern scanning or, alternatively, can be used to deliver laser energy to vaporize specific cellular components with the same spatial resolution.

The ablative technique has been used recently to make a hole in the outer coating of a mouse ovum, providing just enough space for the passage of a sperm cell, thus rendering fertile a group of ova that was previously infertile because of a defect in the outer coating.

### 2.8 Multi Photon Microscopy

In normal fluorescence microscopy, a molecule is excited by absorption of a single photon, typically in the visible spectrum, and then releases another photon of slightly lower energy for its fluorescence emission. At high photon densities (high fluence rates), more than one photon can be absorbed in a few femtoseconds to attain the excited state.

This effect can be used to get a new imaging technique in microscopy. Unlike confocal microscopy, the site of two-photon excitation is limited by the focal point of the microscope. As a consequence, a detection pinhole is no longer necessary, thus not limiting the number of photons being detected. Another advantage is that the two-photon microscope, using infrared light, can improve the depth of focus, allowing the study of events far below the surface of complex tissues and organs. Moreover, photodamage caused by light absorption is restricted to only a small focal point. In normal microscopy, fluorophores above and below the focal plane continue to absorb the light and potentially enhance tissue damage.

Recently, Multi-photon fluorescence microscopy has emerged as a new optical sectioning technique. The technique allows direct visualization of optical sections within thick, fluorescently labeled tissue, and enables the use in the study of living cells and tissues (see also the section on Laser-scanning microscopy in the article on MICROSCOPY).

## 3 Therapeutic Applications

### 3.1 Ophthalmology

The use of lasers in ophthalmology is more mature than in any other medical specialty, and with good reason. Because of the transparent nature of the eye to optical wavelengths from 400 to 1200 nm, lasers can deliver the energy required to probe, heat, and cut ocular tissue in a far less invasive manner than any other surgical technique. Figure 8 is a diagram of the essential parts of the eye. Lasers are currently in clinical use in a variety of procedures addressing diseases and defects in many different parts of the eye. The therapeutic uses of lasers in ophthalmology can be divided into three classes: photocoagulation, photodisruption, and photoablation.

Laser wavelengths that are transmitted freely through the optical pathway of the eye, but absorbed in the surrounding tissues, can be used at relatively low intensities to coagulate structures and thus alter or disable their function. Diabetic

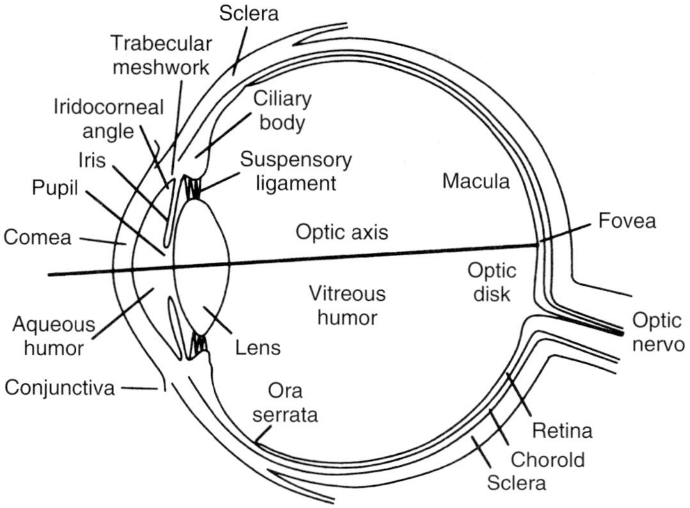


**Fig. 8** Schematic of the anatomy of the human eye, illustrating the primary structures relevant to ophthalmic laser therapy and diagnostics

retinopathy, which is characterized by retinal hemorrhages, microaneurisms, and retinal neovascularization, has become a leading cause of blindness in the United States. If caught early, the progress of the disease can be halted by the technique of panretinal photocoagulation, in which short bursts of cw argon laser light are used to create an array of thermal damage spots in the portion of the retina surrounding the macula. The macula is the central portion of the retina, which contains most of the color receptors and is the region of highest visual acuity. The laser light, which is absorbed in the pigment epithelium of the retina, creates a form of scar tissue, which reduces the rate of severe visual loss in diabetic patients by blocking neovascularization. Thus, the macula is saved at the expense of the loss of some peripheral vision.

Glaucoma is also a leading cause of blindness in the western world. It is characterized by elevated intraocular pressure, which leads to visual field loss. Under normal conditions, aqueous humor, the fluid that fills the space between the lens and the cornea, is continually produced in the ciliary body, an organ located next to the suspensory ligaments. This fluid enters the space between the lens and the iris, flows through the pupil into the space between the iris and the cornea, and is drained to the venous system through a porous network located at the iridocorneal angle called the trabecular meshwork. Glaucoma can occur when the aqueous outflow is reduced by a contact between the iris and the lens, which leads to bulging of the iris and blocking of the anterior angle (angle-closure glaucoma), or by an excessive resistance to aqueous flow in the trabecular meshwork (open-angle glaucoma).

In argon laser trabeculoplasty, laser light is used to coagulate the trabecular meshwork, allowing increased aqueous outflow in open-angle glaucoma.

Another approach to open the meshwork by means of photoablation is called *trabeculotomy*. For this, 308 nm radiation of the XeCl excimer laser is delivered through an optical fiber. Because of the higher absorption of the UV radiation, the trabecular tissue is ablated without significant thermal damage. The edges of the openings are very smooth, thus only minimal wound healing reactions are induced and the openings will not close again. Only a few openings (6–8) are necessary to reduce the resistance of the outflow and consequently decrease the intraocular pressure.

Pulsed laser light can also be used to cut transparent structures of the eye through the process of photodisruption. In photodisruption, high-intensity laser light is strongly focused at the desired incision location, leading to the creation of a plasma. The plasma, consisting of ionized material at very high temperature, creates a local damage zone whose volume scales with the laser-pulse energy. Cataract is a disease in which the lens of the eye becomes cloudy, usually caused by aging, and exacerbated by excess exposure to ultraviolet or infrared light. The treatment is to surgically remove the lens and replace it with a plastic implant. During surgery, it is customary to leave the posterior capsule (the sac which encloses the lens) in place, to serve as a barrier against infection, to retain the vitreous humor, and to provide a bed for the implant. In approximately 30% of the cases, the posterior capsule, which is initially clear, will become cloudy within the first year after surgery. Posterior capsulotomy is an outpatient procedure during which this opaque capsule is opened using a Q-switched Nd:YAG laser. It replaces the older, invasive technique in which the posterior capsule was simply torn open with a needle. Posterior capsulotomy is the most commonly performed photodisruption surgery. Photodisruption with a Q-switched Nd:YAG is also used in a procedure known as iridectomy to puncture holes in the iris for the treatment of closed-angle glaucoma.

In 1983, a new kind of refractive surgery has been proposed by Steve Trockel and R. Srinivasan. They used an UV-Excimer laser, emitting at 193-nm wavelength in order to reshape the surface of the human cornea by means of photoablation. In the early nineties, the first clinically controlled studies were performed and today it is almost a routine surgical procedure, shared by two different methods: Photo refractive Keratectomy (PRK) and Laser in Situ Keratomileusis (LASIK).

In PRK the corneal surface of a myopic eye will be flattened by ablation of corneal tissue in the center, in order to reduce its refractive power (Fig. 9). In myopic eyes, the refractive power of the cornea and the lens is too strong in relation to the length of the eyeball. According to the amount of the refractive error, some 10 μm to 100 μm of the superficial layer of the cornea will be removed. Before the laser treatment starts, a 70-μm-thick superficial cell layer, the so-called epithelium of the cornea is removed mechanically. The epithelium is a thin multilayer of cells with a rapid turnover. The cell layer heal up within 24 to 48 h after removal, in contrast to the underlying corneal stroma, which shows almost no regression after ablation.

The postoperative recovery of the epithelium in combination with the tear film has a smoothing effect. Thus little irregularities in the submicrometer scale, which will occur during laser processing of the tissue, will cause no problems for the visual acuity of the patient's eye.

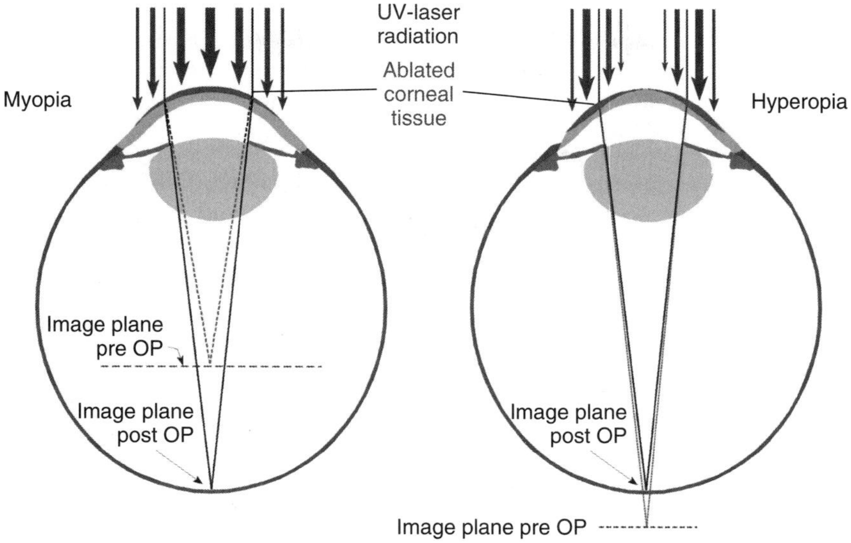


**Fig. 9** Principle of Photo Refractive Keratectomy (PRK). On myopic eyes (a), where the refractive power of the cornea and the lens is too strong in relation to the length of the eyeball, the corneal surface will be flattened by ablation of corneal tissue in the center, in order to reduce its refractive power. On the other hand, the cornea of a hyperopic eye (b), where the refractive power is too low, the surface will be steepened by removing most of the tissue at the edge of the cornea

At low and medium myopia (up to −6 dpt.), PRK stands for a safe and established procedure. At higher corrections of the refractive error, however, complications such as scar formation and regression can occur. This is in the first place caused by a stronger induction of wound healing induced by the removal of a higher amount of tissue. These difficulties can be avoided by a procedure, which was proposed by Pallikaris in the early nineties and is called LASIK. Here, not the superficial layer of the corneal stroma but the inner part of the cornea will be processed with the ArF-excimer laser. For this, a layer with a thickness of approximately 160 μm is separated partially by a knife, the so-called micro keratom, and opened. Afterwards, corneal tissue will be ablated by the laser, for it is necessary to correct the refractive error. Finally, the top lamella will be replaced. This flap joins to the new surface and fits without any suture due to adhesion forces. By means of LASIK, myopic corrections well over −6 dpt. can be performed and also hyperopia and astigmatism of more than 4 dpt. can be corrected.

## 3.2 Surgery

### 3.2.1 Urology

Unlike bone, many of the various types of urinary calculi (kidney stones) are brittle and are thus susceptible to fragmentation by ultrasound. They can also be fragmented by the pressure wave generated by laser-induced plasmas created on their surface. The smaller fragments can then be passed out through the urinary tract. The laser procedure is useful in removing stones lodged in the portion of the ureter (the tube which connects the kidney to the bladder) that is shielded by the pelvic bone and is therefore inaccessible to external ultrasound. This is accomplished by inserting an optical fiber via the bladder into the ureter until it is in contact with the stone, and then applying pulsed dye laser pulses until the stone is fragmented.

Bladder tumors on the inner wall of the bladder can also be destroyed with a laser by ablation and coagulation of the tissue and by Photodynamic Therapy (PDT) (Sect. 3.7). This procedure has the advantage of leaving the bladder wall intact and watertight, a major advantage for any form of bladder treatment.

Hypertrophy of the prostate gland, a common problem among older men, causes difficulty in urination. This condition is usually treated surgically, often through the urethra. Laser treatment of prostate hypertrophy is currently being actively studied. It has the major advantage of minimizing blood loss, since avoiding the need for blood transfusion has become a high priority in medical practice.

#### 3.2.2 **Gynecology**

Cervical dysplasia is a common precancerous condition of the surface of the cervix. Surgical treatments may create a problem for women wishing to bear children by compromising the cervix's ability to retain a fetus to full term. Laser removal of the diseased tissue can preserve the mechanical strength of the cervix, retaining its ability to perform its function during pregnancy. Here too, decreased blood loss is a major advantage.

Endometriosis is also a common condition affecting women in the middle-age group leading to infertility, pain, and dysmenorrhoea. Lasers are quite effective in the treatment of endometriosis because of their ability to coagulate and evaporate the endometriotic implants very precisely.

#### 3.2.3 **Otolaryngology**

A frequent cause of hearing loss is the calcification of, or damage to, the tiny bones of the middle ear. Today, this problem can often be corrected by surgically removing one or more of these bones and replacing them with a prosthesis. Because of the confined space and the small amount of tissue to be removed, endoscopic delivery of laser energy is an ideal way to perform this operation.

One of the earliest uses of laser surgery was in the larynx, to remove growths on the vocal cords and the larynx. Because of the oxygen being supplied to the breathing patient, some cases of airway fires occurred. Great care is now taken to avoid such accidents in all laser surgeries performed near the airways by the use of special endotracheal tubes.

### 3.3 Dermatology

Anderson and Parrish [25] invented the term *selective photothermolysis* to describe the process in which specific cell types are targeted for destruction by a laser while leaving intact the adjacent cells of differing types. The technique involves selecting the laser wavelength to coincide with the absorption spectrum of a particular chromophore present in the targeted cells, but absent in the adjacent cells. The concentration of laser energy deposited in the target cells causes their temperature to rise to the point where cell death is induced by photocoagulation or hyperthermia. An additional requirement is selecting the laser-pulse duration such that the diffusion of heat from the targeted cells does not injure the adjacent cells.

A successful application of selective photothermolysis has been the treatment of vascular lesions. By targeting the hemoglobin present in the unwanted blood vessels, these vessels are selectively destroyed (the absorption spectrum of hemoglobin is shown in Fig. 3). Port wine stain (PWS), for example, is a congenital birthmark resulting from an assembly of abnormal blood vessels located in the dermis, often appearing on the face or neck. Normal blood vessels in the skin are usually about 90% empty (except when flushed as a result of exercise, cold, spicy food, or embarrassment). In contrast, PWS vessels are normally filled. By irradiating the area containing PWS vessels, the abnormal vessels are selectively destroyed while sparing the remainder of the dermis. The thermal diffusivity of biological tissue (see Table 1) leads to an approximate thermal diffusion time of about 1 μs for abnormal vessels, which are typically 40 to 100 μm in diameter. Many different lasers have been used to treat PWS. Considerable success has been achieved with a pulsed dye laser with a 300-μs pulsewidth, tuned to a wavelength near the 577-nm absorption peak of oxyhemoglobin [7]. Usually, multiple treatments are required to obtain a satisfactory result.

Other successful applications of selective photothermolysis include pigmented lesions, where the pigmentation can be natural, due to melanin (see spectrum, Fig. 3), or artificial, as in tattoos. The basal layer, which is the boundary between the dermis and epidermis, contains cells that continually manufacture new epidermal skin cells. These new cells dehydrate as they migrate to the surface, where the outermost layer of dead cells is continually being sloughed off. In the basal layer, there is also a sprinkling of melanocyte cells, whose function is to generate the pigment melanin and inject it into the newly forming skin cells. Benign pigmented lesions containing an excess of melanin are successfully treated with lasers, which selectively destroy the melanocytes. On suspicion of malignant lesions a biopsy is advised with subsequent conventional complete excision.

Tattoos consist of artificial pigments that have been deposited in the skin. The pigments are insoluble, and the granules are too large to be removed by the normal body processes, such as phagocytosis. By using short-pulse high-intensity lasers, the pigments can be broken into smaller fragments that are then cleared by the body, causing minimal damage to the surrounding tissue. Lasers that have been successfully used to remove tattoos include alexandrite, 1-μs 510-nm pulsed dye, Q-switched ruby, and Q-switched Nd:YAG (both the fundamental and the second harmonic). The choice of laser depends upon the pigment being attacked.

Permanent removal of unwanted hair has recently emerged as a viable modality. The basic principle of epilation with laser light is the destruction of the hair follicle by selectively targeting the melanin in the areas of hair sheath and hair shaft. An irradiation time of 3–50 ms is presumed to get a selective heating of the follicle structure. Patients with light blond hair cannot be treated due to the absence of the target chromophores melanin.

By ablating the outer layer of epidermis with a $CO_2$ (10.6 μm) or an Er:YAG laser (2.94 μm), a rejuvenation of the skin by stimulating its regenerative potential can be achieved. Sun damage and age-related changes like small wrinkles can be improved by this kind of skin resurfacing. However, it should be noted, that in some cases complications such as dyspigmentation, infections, and scar formation have been reported [26, 27].

## 3.4 Cardiology

Although the subject of intense study for the past decades, the utility of lasers in cardiology is not yet clearly established. Initially, it was conjectured that a fiber optic passed through a catheter to the site of a coronary artery blockage could be used to deliver laser energy to ablate and remove the diseased tissue causing the blockage, thus successfully recanalizing the artery. This procedure would avoid the need for costly and painful bypass operations involving open-heart surgery. Balloon angioplasty, in which a deflated balloon is inserted into the partially blocked opening and then inflated to force an enlargement of the opening, has achieved partial success toward this goal, but is only about 50% successful. This is principally because restenosis can occur as early as during the first six months after treatment. Restenosis is usually not a recurrence of the original disease, but is caused by the growth of scar tissue called intimal hyperplasia, which closes down the balloon-inflated opening. The balloon is also contraindicated for use in many of the more difficult lesions.

Initially, the argon-ion laser was used for laser angioplasty. Although this laser could ablate soft plaque, it was unable to penetrate calcified plaque, and would occasionally perforate vessel walls. The use of pulsed ultraviolet excimer lasers solved the problem of ablating the calcified tissue, and using various forms of guidance, such as guidewires, reduced the danger of perforation. However, the restenosis rate for laser angioplasty has proven to be no better than that of the balloon. There are a limited number of special lesions for which the laser may have a useful and unique role, such as enlarging the opening of a lesion so that a balloon may be inserted, or recanalizing partially occluded bypass grafts. Laser systems are also being studied to determine their utility in opening partially occluded vessels in the legs.

In mammals, the oxygenated blood supply to the heart muscle is via the coronary arteries, and it is the blockage of these vessels that is responsible for most heart diseases. In reptiles, the heart muscle is supplied with oxygenated blood directly from the heart chamber. A repair method which mimics the reptilian heart has been pioneered by Mihroseini [28]. In a procedure, called Transmyocardial Laser Revascularization (TMLR) a $CO_2$ laser is used to drill a hole directly through the heart muscle into the left ventricle. Oxygenated blood then perfuses into the heart muscle directly via this new passage. The outside of the hole is sealed by blocking the opening until a clot forms.

Acute heart attacks are very often the result of a thrombus (blood clot) flowing downstream and blocking the opening of an already partially occluded coronary artery. Emergency treatment involves dissolving this clot quickly, before serious damage can occur to oxygen-starved tissues. There are several effective drugs available for this purpose; however, in many cases, these drugs are contraindicated or unsuccessful. The clot is very soft compared to artery wall or plaque and is easily ablated using laser energies well below levels, which could damage adjacent vessel walls. At these levels, the energy can be delivered using a fluid-filled catheter, in which the fluid guides the light in much the same manner as an optical fiber.

## 3.5 Orthopedics

Initial attempts to use lasers in arthroscopic surgery and, in particular, surgery involving the knee began with the use of the $CO_2$ laser. This laser had two serious drawbacks. First, this wavelength is strongly absorbed in water, and, therefore, the knee had to be flushed with $CO_2$ gas to provide a dry field. This added complexity to the procedure and occasionally led to embolisms. Second, acceptable optical fibers at this wavelength are not yet available, thus requiring the use of articulated arms or hand-held units in which the surgeon manipulates the entire laser. Both were cumbersome approaches.

More recently, surgeons have investigated using the Ho:YAG laser, taking advantage of the fact that its 2.1-μm light can propagate through conventional fibers. Although it can be used in a wet field, light at this laser wavelength is partially absorbed by water, and, thus, the distance from the fiber tip to the tissue must be precisely controlled. A substantial amount of the energy is dissipated in the saline field. The Ho:YAG laser is used for meniscal-cartilage smoothening or capsular shrinkage of the shoulder joint.

## 3.6 Dentistry

In recent years, the laser has been suggested for use in preventive and reconstructive dentistry. Among the potential applications that have been considered in the literature are removal of caries, sealing fissures in enamel, inhibition of demineralization, etching of enamel, preparation of retention pin holes, and cleaning and preparation of root canals. The principal laser studied, the $CO_2$ laser, has two serious drawbacks. First, although useful for cutting and removing soft tissues, such as during gum surgery or frenectomies, the $CO_2$ laser does not ablate hard tissues without depositing considerable heat. This can raise the temperature of the pulp, causing pulpal damage. The maximum permissible temperature rise for pulp is 5.5 °C, after which irreversible damage occurs. Second, until today there exists no practicable fiber delivery to transport the radiation to the target tissue.

Since ultraviolet (193 nm) and 2.9-μm lasers have proven effective with other hard tissues, (see Sect. 2.4), these sources have been considered for ablation of tooth enamel. First experiments with the ArF-excimer laser have shown, that only little thermal effects were induced due to the short-pulse duration (15 ns) and low-penetration depth ($<1$ μm) of the uv radiation. However, the ablation rate, that is, the ablated volume per unit time, is by far too low for clinical applications.

The wavelength of the Er:YAG laser at 2.94 μm matches the highest absorption peak of water and shows good absorbance in hydroxyapatite, the mineralized component of dental hard substance, resulting in a better ablation efficiency than the excimer laser. However, although the treatment of dental tissue with the Er:YAG laser is almost free of pain, it could not get a breakthrough on the medical market. Conventional mechanical devices are still much more effective and clearly cheaper in acquisition and maintenance.

## 3.7 Photodynamic Therapy (PDT)

The two principal methods now used to destroy malignant cells in the treatment of cancer *in situ* are radiation and chemotherapy. In both cases, the treatments also damage healthy cells. This damage is minimized by careful aiming during radiation treatment and by precise dose control in chemotherapy; however, the fine line between destroying all of the tumor but not the healthy tissue is a difficult one to maintain. Oncologists have long wished for a "magic bullet" that would selectively destroy only tumor cells and not normal cells. Photodynamic therapy (PDT), offers the prospect of such a magic bullet.

Photodynamic therapy relies on the fact that a great many organic compounds possess the property of phototoxicity; that is, in the absence of light they are benign, but, when irradiated with certain specific wavelengths, they become toxic. Some well-known drugs, such as tetracycline, possess this property, and users are cautioned to avoid exposure to sunlight. Another essential property of a good PDT drug, often called a sensitizer dye, is that it is preferentially absorbed by tumor tissue, or else more rapidly cleared from normal tissue and therefore selectively retained in the tumor.

Figure 10 is a schematic of photodynamic therapy of cancer. Prior to irradiation, the patient is injected with the phototoxic dye. To date, the preponderance of clinical trials have used a commercial dye, Photofrin®, which is a form of hematoporphyrin derivative (HPD). This dye depends on selective clearing, requiring 2 to 3 days for the dye concentration in the normal tissue surrounding the tumor to be reduced to a safe value. Phototoxic dyes are chosen to have their excitation wavelength in the wavelength region in which tissue is translucent, from approximately 600 to 1200 nm (Fig. 3). The excitation wavelength for HPD is 630 nm.

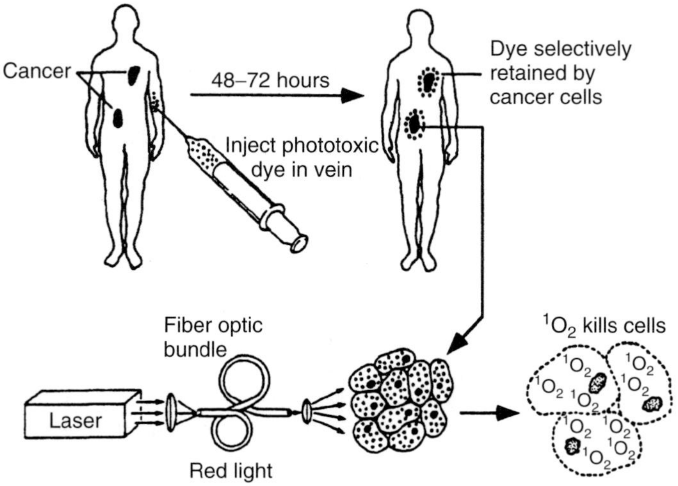


**Fig. 10** Schematic of photodynamic therapy of cancer. The patient is first injected with a phototoxic dye. After dye localization in tumor tissue has occurred, the drug is activated by illumination with laser light delivered through a catheter, endoscope, or laparoscope. Tumor tissue containing the drug is selectively destroyed by singlet oxygen generated by the activated drug (after Berns, M. W. [10])

Although incoherent light can be used for surface tumors, the light source for PDT is usually a laser, so that the light may be efficiently coupled into optical fibers. This enables the light to be directed to the tumor through endoscopes, laparoscopes, or even hypodermic needles.

The mechanism for cell destruction is thought to be via the generation of singlet oxygen ($^1O_2$), a metastable excited state of molecular oxygen, which is known to be toxic to cells. The incoming photon excites the dye to a singlet state. Good phototoxic dyes have a high probability for intersystem crossing to the triplet manifold, and a large population of metastable triplets is formed. When one of these triplets collides with a ground-state oxygen molecule, excited-state singlet oxygen is created, which can then destroy membranes or vital proteins.

Photodynamic therapy has been in clinical use for more than 10 years, mostly for palliative tumor treatment. At present, a number of dyes are under active study, although none has yet proven to be the sought-after magic bullet. There are several major classes of principal interest, including the phthalocyanines (of which the nonsulfonated zinc analog is about to undergo clinical trials), the benzoporphyrin derivatives, various cationic dyes, which include benzothiazines and rhodamines, and the chlorins, which are chlorophyll derivatives. Results to date show great promise. The most likely result will be that specific dyes prove to be effective for specific types of tumors.

One of the most successful applications of PDT was found in ophthalmology in treating some forms of age-related macular degeneration (AMD), which is the most frequent cause of legal blindness in the elderly in the industrial world. Mostly the loss of vision originates from subretinal neovascularization, creating a macular edema with consecutive decrease in vision. In a later stage, the neovascularization may bleed, which leads to further decrease of visual acuity, close to blindness. To treat AMD, a photosensitizer (Verteporphin) is injected intravenously to accumulate in neovascular tissue. The irradiation of the dye with light in its maximum of absorption (690 nm) then leads to cytotoxicity and destroys the neovascular tissue.

### 3.8 Biostimulation

An area of great controversy has been variously termed biostimulation or low-light-level laser therapy. There have been a large number of reports in the literature of the observation of enhanced wound healing, accelerated tissue growth, and a measurable decrease in pain induced by the application of doses of low-intensity laser light. Many of these reports have come from reputable and careful workers. Unfortunately, much of this work proves to be irreproducible, and there are a host of reports in which no effect is observed.

There is some physiological basis for the possibility of such effects. Even though the light intensities used are too low to cause measurable temperature changes, significant changes in the rate of cell growth *in vitro* have been observed by Karu [29]. These changes are highly dependent on the wavelength, fluence, and cell type. Itzkan and Bourgelais [30] have postulated a mechanism involving the photodissociation of oxyhemoglobin, in which physiologically significant levels of additional $O_2$ could be liberated. Nevertheless, until this field can be placed on a sounder scientific basis, biostimulation will remain controversial.

### 3.9 Laser Tissue Welding

The first attempts at laser welding were used to perform blood vessel anastomoses, that is, reconnecting the severed ends of two blood vessels. This was accomplished by heating the ends of the vessels with a laser, causing the tissue proteins to become sticky, and adhere on cooling. A successful result requires that the joined vessels remain patent (have an adequate internal opening), withstand the blood pressure, and not leak. Initially, "stay" sutures were used to keep the ends of the vessels together during the procedure. This vitiated some of the advantage over conventional suturing of avoiding the scar tissue caused by the sutures. The results were no better than those obtained by a good vascular surgeon.

Recent improvements include the use of protein materials, such as fibrin, to act as "solders," providing more strength and better sealing, and the use of soluble mandrels to hold the ends in apposition while the laser heat is being applied. The mandrel then dissolves, leaving the vessel patent. It also ensures that tissues of the same type align with one another, which speeds healing.

These techniques are especially useful in reconnecting nerves and bowel, where the absence of leakage is more important. Blood vessels have the ability to repair small leaks by forming clots. On the other hand, a micron-sized hole in a repaired nerve sheath will allow the extremely fine axons to grow out of the sheath, impairing the result, while leakage in a bowel can cause peritonitis.

### 3.10 Laser Energy-delivery Systems

Optical fibers capable of transmitting substantial amounts of energy are available for wavelengths from about 300 nm to 3 μm. When the emerging energy is to be used directly from the fiber, various focusing and diffusing tips may be used, the choice depending on the application. For example, the Nd:YAG laser can conveniently supply large amounts of average power, but its 1.06-μm wavelength penetrates too deeply for some applications. For surgery, it may be coupled to a shaped sapphire tip, which is heated by the laser and then serves as a combination of mechanical knife, heated tip, and laser light deliverer.

For external work, the practitioner may use a handpiece that provides a focus at a fixed point in space. The handpiece is usually optically coupled to the laser via a fiber for flexibility and convenience. For wavelengths longer than about 3 μm, especially the 10.6-μm of the $CO_2$ laser, the laser is usually optically coupled to the handpiece via an articulated arm, containing folding mirrors, similar in mechanical operation to the kind used for dentist's drills. Some optical fibers and hollow flexible waveguides are available for these IR wavelengths, but they are, as yet, not entirely satisfactory. Small lasers are sometimes mounted directly on an articulated arm. Handpieces have been designed incorporating a computer-controlled scanning function to provide more precise control of the delivered intensity and spatial uniformity of the dose. Surgical laser systems are usually fitted with a low-power aiming beam to permit precise placement of the high-power beam on the tissue.

### 3.11 Surgical Laser Safety

Safety issues that arise during laser surgery are eye protection, fire, and infection. Protocols requiring the use of safety glasses or goggles for all personnel when the laser is operating are mandatory in most medical facilities. Because a stray laser beam can easily ignite flammable material, all surgical drapes in use during laser surgery must be fire resistant and must be kept wet at all times when the laser is in use, particularly if oxygen is also being used. Special care is taken in or near the patient's airway. Also, live organisms and viable DNA have been found in the laser plume. Cases of practitioners developing warts in their nostrils after removing warts from patients have been reported. The laser plume must be considered potentially infectious, and appropriate ventilation and masking systems must be in use at all times.

## 4 Diagnostic Applications

One of the most active areas of research in laser medicine in recent years has been the application of optics and photonics technology for tissue imaging and diagnosis of disease. This area has included the conversion of several state-of-the-art technologies to medical applications, such as ultrafast optics, holography, confocal microscopy, and various forms of high-resolution spectroscopy. In comparison to other advanced medical diagnostics developed in recent decades, such as magnetic resonance imaging (MRI) and computed tomography (CT), optical techniques for imaging and diagnosis have the advantages of being simultaneously nonionizing, potentially inexpensive, and often noninvasive. In this section, highlights of recent advances in this active field are reviewed.

### 4.1 Laser-based Optical Imaging in Medicine and Biology

#### 4.1.1 Laser Scanning Confocal Microscopy

An important application of the laser in medicine is its use as a high-brightness light source for confocal microscopy. In the decades since the first demonstrations of the laser and of confocal microscopy, the combination of these two inventions has revolutionized high-resolution optical imaging in biology and medicine. In a confocal microscope, illumination light emanating from a point source is focused with a condenser lens to a minimal spot size in the sample, and only that light is collected that passes through a pinhole placed in an image plane of the point source (Fig. 11). Two-dimensional images are built up by scanning the incident beam or the sample in two dimensions, while synchronously recording the transmitted light-intensity level. The brightness of the image depends upon the intensity and the degree of spatial coherence of the illumination light, making the laser an ideal source. The confocal microscope provides a resolution superior to that of the conventional microscope in both lateral and axial directions. In the lateral direction, the overlap of the illuminating light-intensity distribution and the objective point-spread function is slightly smaller than the resolution that could be obtained by the objective lens alone. The primary benefit is in the axial direction, however; almost all of the light scattered from planes in the sample either above or below the object plane (the plane of dual focus of both the illumination light and the objective lens) does not pass through the

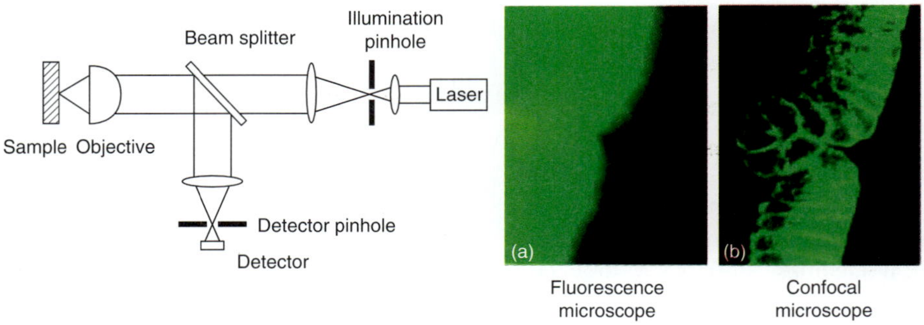


**Fig. 11** (a) Schematic setup of a confocal microscope and (b) comparison of a fluorescence and a confocal microscope image. (B. Alberts et al. (2002), *Molecular Biology of the Cell*, (4th ed.), London: Taylor & Francis, p. 559)

pinhole aperture, and is rejected from the image. By translating the focal plane axially through a relatively thick sample, one can thus optically "section" planes with micron resolution without having to slice the sample physically. This feature is particularly useful in living biological samples, permitting detailed examination of living organisms in their natural environments.

### 4.1.2 Laser Imaging for Ophthalmic Diagnosis

An application of laser scanning confocal microscopy that has made the leap into current medical practice is in imaging of the eye and retina. Confocal microscopic images of the living cornea can be obtained and reconstructed in three dimensions with sufficient resolution to discern cellular structure. The scanning laser ophthalmoscope is a confocal microscope optimized for imaging of the retina [31]. It provides substantial advantages over the conventional hand-held indirect ophthalmoscope for retinal viewing, in that the images are crisper and cleaner (because of confocal rejection of scattered light), and, at the same time, it is much more comfortable for the patient because of the more efficient usage of the illumination intensity.

High-resolution imaging of the living retina is necessary for diagnosis and treatment of almost all retinal diseases. These include emergency conditions, such as torn retinas and dislocations, as well as degenerative diseases, such as retinopathy. Of particular current research interest, is the imaging of the retina and optic disk for early diagnosis of glaucoma (see Sect. 3.1). The retinal nerve fiber layer (RNFL) is the uppermost layer of the retina, consisting of nerve axons carrying visual stimuli that converge from all parts of the retina toward the optic disk and out of the eye to the brain. In glaucoma, significant degeneration of the RNFL can occur before conventional tests of intraocular pressure and visual field can detect the presence of the disease. This layer is less than 50-μm thick in most regions of the eye. Unfortunately, this is too thin to be sectioned by confocal microscopy, since the axial resolution obtainable at the back of the eye is limited by the pupil and ocular aberrations to a few hundred microns.

Two new laser-based retinal imaging devices have recently been developed that have the capability to measure directly the thickness of the RNFL, in the hope that such a direct measurement will

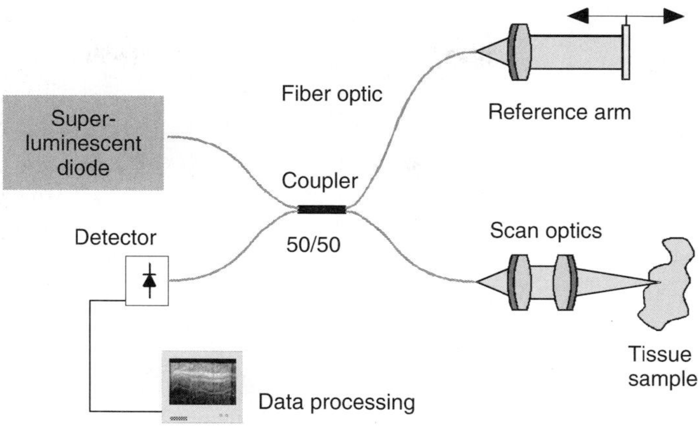


**Fig. 12** Schematic for optical coherence tomography setup

provide for an early diagnosis of glaucoma before vision is threatened. Scanning laser tomography is an enhanced form of scanning laser ophthalmology, in which the thickness of the RNFL is estimated by measuring the optical retardation of backscattered light [32]. This technique depends upon the fact that the RNFL is a birefringent structure, composed of thin, directionally oriented nerve fibers. The birefringence is measured with Fourier ellipsometry, in which the polarization states of the incident and scattered light are modulated and analyzed, respectively, in such a way as to provide a fast, accurate estimation of the layer's birefringence.

Optical coherence tomography (OCT) is a new scanning laser confocal imaging technique for micron-resolution imaging at arbitrary working distance, which can perform direct measurements of RNFL thickness (see OPTICAL TOMOGRAPHY and [33, 34]). Optical coherence tomography is based on low-coherence interferometry, in which light from the sample arm of a Michelson interferometer is directed onto the tissue to be imaged. The interferometer is illuminated with a broad-bandwidth light source, having a coherence length of only a few microns. Measurements of reflectivity as a function of depth into the tissue are obtained by scanning the reference arm of the interferometer while synchronously monitoring the signal at the interferometer output. Interference occurs only when the distance to a reflectance in the sample matches the reference arm length to within the coherence length of the light source (Fig. 12). Micron-resolution cross-sectional images of tissue microstructure are built up by transverse scanning of the sample arm beam. Optical coherence tomography is particularly useful for imaging structures in the back of the eye, since the axial resolution depends only on the source coherence length and is independent of the available pupil aperture and the optical quality of the eye. This technique has produced the highest-resolution images of the living retina to date (Fig. 13).

#### 4.1.3 **Optical Tomography**

The ultimate goal in optical imaging of tissue would be to develop a general-purpose tomographic imaging technology, which could obtain images of optical scattering or tissue spectroscopic properties from deep within the body. A practical goal for

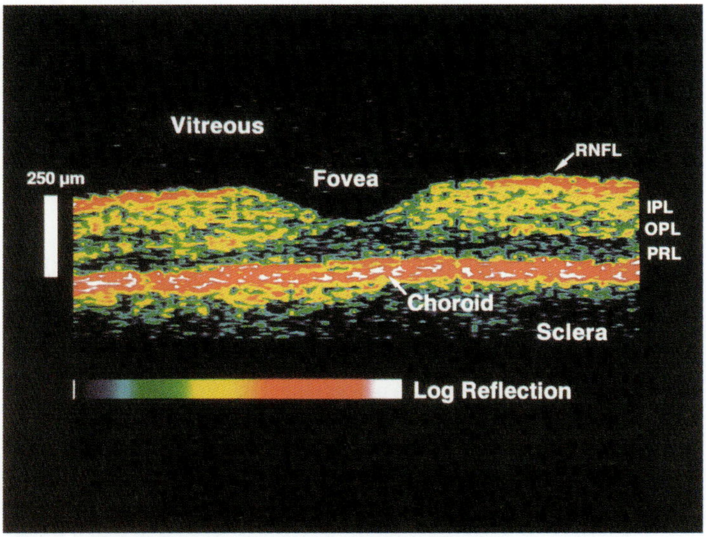


**Fig. 13** Optical coherence tomography image of the retina taken in a living human subject. The fovea (the region of highest visual acuity) is centrally located; lateral to the fovea, identifiable retinal layers include RNFL, retinal nerve fiber layer; IPL, inner plexiform layer; OPL, outer plexiform layer; PRL, photoreceptor layer. The false color scale represents the logarithm of the power reflectivity of the retina, ranging 35 dB from a minimum detectable reflectivity of approximately 1 part in $10^9$ (photo courtesy of Prof. J. G. Fujimoto, Massachusetts Institute of Technology, and Prof. C. A. Puliafito, Tufts University School of Medicine)

such investigations has been to develop a method for optical transillumination imaging of the breast with sufficient sensitivity and resolution to visualize the early stages of breast cancer, which could serve as an alternative to X-ray mammography. The primary difficulty in imaging deep into tissues is that almost all the tissues, except for the eye, are highly scattering, and image information is severely degraded after just a few scattering mean-free paths into the tissue, typically ~1 mm in breast tissue (see Table 1).

Several approaches are under investigation for resolving the problem of optical imaging through highly scattering media. One is to attempt to reconstruct an image of the interior of an illuminated diffusing medium by recording the position and direction of all of the photons that emerge from it, and then to trace them backward in time computationally to derive the positions of the internal absorbers and scatterers. This "inverse problem" is tremendously complex at best and has not yet been shown to be soluble. Another approach is to use time-resolved methods to detect only those photons that have not lost image-information in passing through the tissue. A pulse of light propagating through a tissue sample will be severely lengthened by multiple scattering inside the sample. The photons in the transmitted pulse can be classified into so-called ballistic, snake, or diffuse components, depending upon their arrival times (Fig. 14). Ballistic light is the unscattered component, which has taken a straight-line path

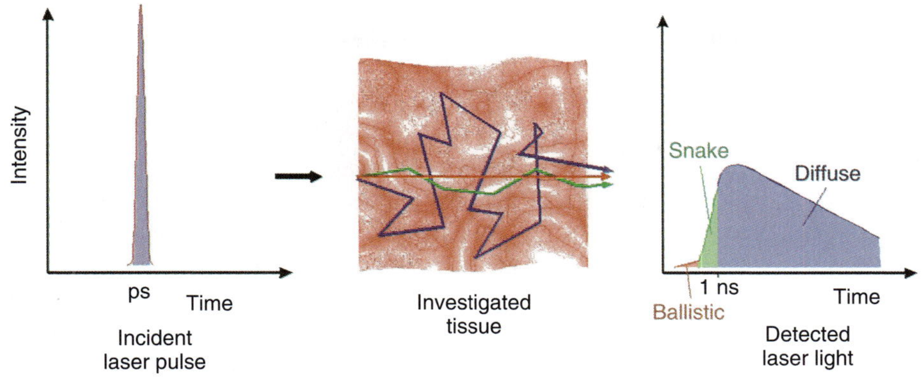


**Fig. 14** Schematic of time-resolved optical transillumination imaging in a turbid medium, illustrating the ballistic, snake, and diffuse components of a transmitted short pulse of light (after Wang, L. et al. [35])

through the sample. Snake light is that component that has been scattered but has remained close to the straight-line path, and diffuse light is the remainder. A shadowgram of hidden objects within the medium may be obtained by time-gating the earliest-arriving light, either the ballistic or snake light components.

Several techniques are available for distinguishing between the different components of light transmitted through a turbid medium. Conceptually simplest is time gating, in which ultrashort pulses are used to illuminate a tissue sample, and the transmitted light is classified according to its arrival time through the use of fast electronic detectors (such as streak cameras), or even faster nonlinear optical detection techniques (such as the ultrafast Kerr shutter). This approach has resulted in shadowgraph images of small animals [36] and in tissue phantoms up to a few centimeter thickness [35], as well as in first clinical trials [37]. Two additional techniques have been introduced for time-resolved imaging that do not use explicit time gating. The first is the frequency-domain imaging, in which the amplitude and phase characteristics of the transmitted portion of light from a radio-frequency-modulated continuous light source are used to distinguish photon arrival time [38]. The second is the coherence-domain imaging, in which the coherence length of a broad-bandwidth light source is used in place of a time gate in order to distinguish early-arriving photons. The longitudinal ranging capability of OCT (Sect. 4.1.2) is an example of the use of this technique in one dimension, but it is also possible to image in two or three dimensions using holography [39]. These latter techniques have the advantage that ultrashort-pulse sources and fast time gates are not required. More detailed information on time-resolved medical imaging techniques may be found in the list of suggestions for further reading.

A few theoretical results have been obtained that are useful for evaluating different transmitted light components for optical tomography. Since ballistic photons are unscattered in traversing the tissue, diffraction-limited imaging theoretically is possible. However, the ballistic component is severely attenuated in the scattering medium. The maximum tissue thickness for ballistic light imaging may be written as

$$L_{\max} = \frac{1}{\mu_t} \ln\left(\frac{1}{2}\frac{E}{h\nu}\right), \qquad (20)$$

where $\mu_t$ is the total attenuation coefficient, $h\nu$ is the photon energy, $E$ is the light energy deposited per resolution pixel, and a signal-to-noise ratio of 1 has been assumed as the detectivity limit. Tissue optical damage thresholds limit the maximum energy that can be deposited in the tissue in a reasonable image-acquisition time. This limits the maximum ballistic imaging thickness to approximately $\mu_t L_{\max} \sim 40$ mean-free paths, corresponding to $\precsim$4 mm of breast tissue [40].

The attenuation of snake light is experimentally observed to be about a factor of 10 less than ballistic light, so that imaging through several centimeters of tissue is possible. However, widening of the time gate to accept snake photons inevitably degrades the resolution of the resulting image. An expression for the full-width–half-maximum spatial image resolution $\Delta x$ as a function of the tissue thickness has been obtained from diffusion theory, which is independent of the tissue-scattering properties as long as the conditions for the diffusion approximation are satisfied:

$$\Delta x \sim 0.2d. \qquad (21)$$

Here $d$ is the tissue thickness, and a detectivity of 1 part in $10^{10}$ of the incident optical signal is assumed [41]. These two theoretical results demonstrate that this approach for imaging through thick tissue using time-resolved methods, that is, accepting snake photons, comes at a considerable cost in image resolution.

## 4.2 Spectroscopic Monitoring and Diagnosis

An exciting new application for lasers in medicine is the prospect of locating, identifying, and classifying diseased tissue inside the living patient in a minimally invasive manner. Among current clinical methods available to study the disease, pathology is the gold standard. Pathology requires biopsy samples, which can be difficult and painful to obtain and can involve considerable delay before the results are available, and requires a highly skilled pathologist to interpret the results. Even highly skilled pathologists often differ on interpretation and conclusions. Clinical techniques capable of evaluating the nature of diseased tissue *in situ*, in real time, and in a safe, valid, and reproducible manner are therefore of critical importance for diagnosis, for selecting and evaluating the effects of various therapies, and for epidemiological and clinical research. Other technologies that can provide some of this information include magnetic resonance imaging (MRI), X-ray computed tomography (CT), radionuclear imaging, and ultrasound. All of these techniques have limitations. It is therefore desirable to develop complementary techniques for probing and quantitatively monitoring the disease.

Using flexible optical fibers to excite the tissue and collect spectroscopic signals, optical systems provide access to many organs in the body through endoscopes, laparoscopes, catheters, and hypodermic needles, and high-performance diagnoses have been demonstrated. Optical spectroscopy using natural fluorescence emission from tissue has been demonstrated to yield valuable information about the state of the health of many tissues. For example, fluorescence spectroscopy can distinguish between oxygenated and ischemic tissue. The principal source of energy for cells is chemical processes localized in the mitochondria, which employ a cyclical process called the Krebs cycle, followed by oxidative phosphorylation. A simplified view of this process is that an energy-rich compound called adenosine triphosphate (ATP) is produced, and molecular oxygen and various carbohydrates are consumed. Two important routes for transporting electrons from the foodstuffs to the oxygen involve the equilibria of the reactions

$$\mathrm{FAD} + 2\mathrm{H}^+ + 2e^- \longleftrightarrow \mathrm{FADH_2}, \quad (22)$$

$$\mathrm{NAD}^+ + \mathrm{H}^+ + 2e^- \longleftrightarrow \mathrm{NADH}. \quad (23)$$

The compounds FAD and $FADH_2$ are the oxidized and reduced forms, respectively, of flavin adenine dinucleotide, and $NAD^+$ and NADH are the corresponding forms of nicotinamide adenine dinucleotide. The compound NADH has a fluorescence peak in the blue, FAD has a fluorescence peak in the green, and $FADH_2$ and $NAD^+$ have no appreciable fluorescence. Consequently, when there is insufficient $O_2$ to accept the electrons from NADH and $FADH_2$, the equilibrium shifts to the right, and the ratio of blue to green fluorescence increases. Conversely, insufficient nutrient shifts the equilibrium to the left, and the ratio of blue to green decreases. By this means, laser-induced fluorescence becomes a sensitive measure of the metabolic state of cellular activity.

It has been demonstrated that, by analyzing the shape of the fluorescence emission spectrum using a suitable algorithm, it is possible to distinguish between the various forms of arterial plaque (fibrous, fatty, calcified) and normal arterial wall. It has also been demonstrated that it is possible to distinguish between the malignant and normal tissue in many organs, including breast and colon. A comparison of fluorescence spectra obtained from normal and adenomatous colon tissue is shown in Fig. 15. The capacity to differentiate is clearly evident.

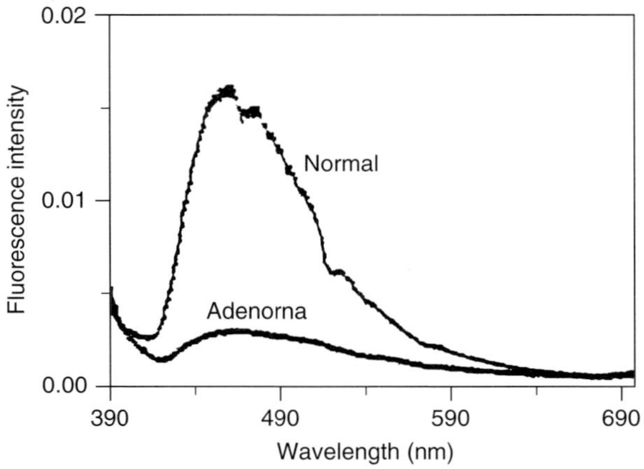


**Fig. 15** Laser-induced fluorescence spectra of normal and adenomatous colon tissue, illustrating the diagnostic capability of this technique. The excitation wavelength was 369.9 nm (reprinted with permission from Richards-Kortum [42])

Fluorescence techniques, however, are not capable of providing detailed biochemical information suitable for histochemical analysis of the tissue. For this purpose, the use of Raman spectroscopy as an *in situ* optical probe of disease holds the promise of providing information unobtainable by other *in situ* methods. The Raman spectrum contains information about the quantitative biochemical composition of the tissue. The signals are inherently weak, but a number of recent technological advances have combined to permit data to be obtained on clinically useful time scales. Figures 16 and 17 illustrate what can be accomplished. Atheromatous plaque is the fatty deposit that occurs within the blood vessels of a patient with coronary artery disease, and, as it progresses, it narrows the vessel lumen, resulting in a decrease in the blood supply to tissue downstream. Figure 16 compares the Raman spectrum obtained from such a deposit with that of pure cholesterol. The detailed similarity shows that this form of plaque contains a high concentration of cholesterol and its esters, and that such a spectrum can serve as a unique detector for these kinds of deposits. In the more advanced stages of this disease, calcified deposits form within the plaque. These deposits contain phosphates that have a unique Raman peak at 960-$cm^{-1}$. Figure 17 shows that the peak from a subsurface calcification can be clearly observed even when it is covered by a fibrous cap about 1.5-mm thick. The signal increases by about a factor of 8 when the cap is removed. Thus, calcified tissue can be detected even when hidden by intervening tissue.

The techniques described above rely on the naturally occurring, or endogenous, fluorophores present in the tissues. As was discussed in the section on Photodynamic Therapy (Sect. 3.7), fluorescent dyes exist that, when introduced into the body,

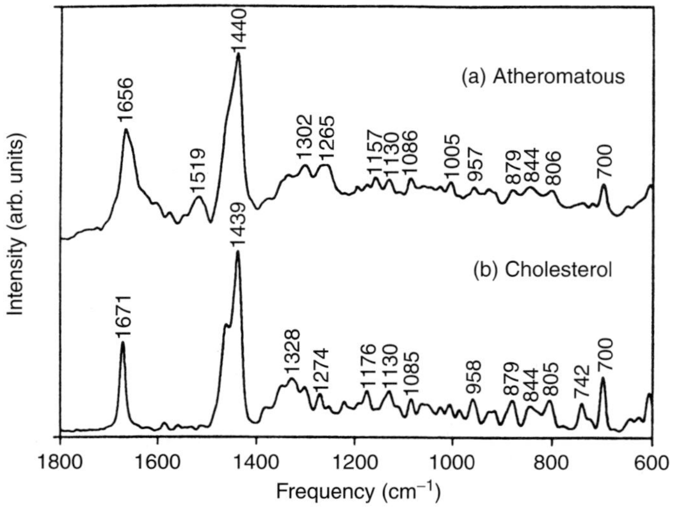


**Fig. 16** Near-infrared Fourier transform Raman spectra, illustrating the correspondence between (a) typical atheromatous plaque and (b) pure, anhydrous cholesterol. The well-characterized vibrational spectrum of cholesterol is clearly identified in the atheromatous data (reprinted with permission from Baraga [43])

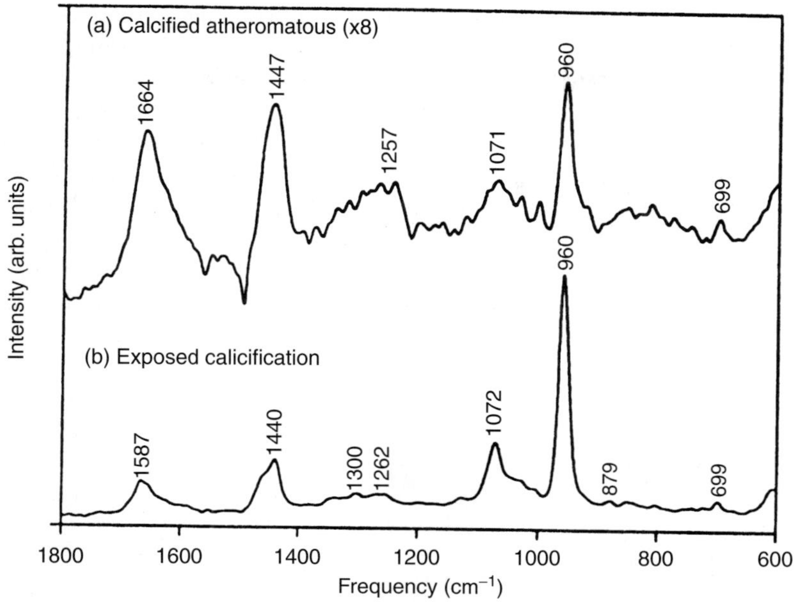


**Fig. 17** Near-infrared Fourier transform Raman spectrum of calcified atheromatous plaques: (a) obtained through an intact fibrous cap approximately 1.5-mm thick and (b) with the calcification exposed. The 960-$cm^{-1}$ peak of the $PO_3^-$ ion, which is unique to calcification, is clearly observed through the fibrous cap. Spectrum (a) is expanded by a factor of 8 relative to spectrum (b) (reprinted with permission from Baraga [43])

selectively deposit in specific tissues, such as tumors. The presence of these exogenous dyes can be detected by their characteristic fluorescent spectra, and the information can be used to localize the diseased tissue.

### 4.3
### Laser-based Biomedical Instrumentation

Lasers are playing an increasingly important role in the biomedical instrumentation that underlies much of modern-day medicine. Examples of laser-based biomedical instrumentation include pulsed laser oximetry, laser-induced fluorescence microscopy and cytometry, and laser scanning confocal microscopy.

The pulsed laser oximeter uses infrared laser light to monitor a patient's heart rate and blood oxygenation in clinics. This is especially important in newborn and premature infants. The operation of the oximeter depends on the fact that oxygenated and deoxygenated hemoglobin possess different absorption spectra (see Fig. 3). This information is used to determine the degree of oxygen saturation in the blood by monitoring the differential transmission of laser light between two different wavelengths through the patient's finger or earlobe.

The monochromaticity of laser light is useful in creating fluorescence images in microscopy samples. With a spectrally narrow laser illumination source, the laser light can be blocked from saturating the image with a line-blocking filter. Laser-induced fluorescence of endogenous (native) tissue chromophores

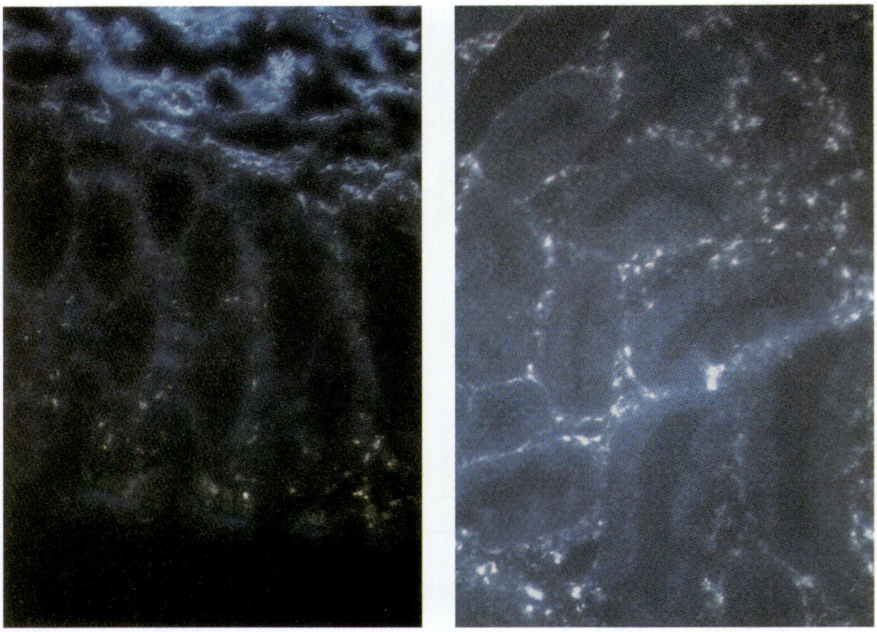

(a) (b)

**Fig. 18** Laser-induced fluorescence microphotographs revealing differentiation between (a) normal and (b) dysplastic colon crypt cells. The crypt cells line the crypts (the oval structures) and do not fluoresce in normal tissue, and therefore are not discerned. In abnormal tissue, the crypt cells emit a fluorescence which appears light blue. The bright white spots are eosinophylls in the spaces between the crypts (photo courtesy of Prof. M. Fitzmaurice, Case Western Reserve University School of Medicine)

provides information supplemental to pathological analysis for discrimination between normal and diseased excised tissues during microscopic analysis (Fig. 18). In the last few years, several commercial laser scanning confocal microscopes using fluorescence have been introduced to the market. These devices are capable of generating bright, clear images of optically thin sections of living cells and embryos. These instruments are of great interest in pathology, as well as in general biomedical research.

Cells tagged with exogenous fluorescent chromophores can be detected and manipulated in special flow cytometers [14]. In these instruments, the cells are injected into a stream of droplets and directed to fall through an excitation beam. When a tagged cell passes through the beam, the fluorescence signal is detected, and that droplet is deflected by a pair of electrostatic plates into a collecting cup.

## 5 Conclusion

In this article, we have endeavored to illustrate the increasing importance of the use of lasers in medicine. Lasers are now routinely used in many medical applications, both diagnostic and therapeutic. We have briefly discussed the physics of the various types of laser–tissue interaction processes. We have surveyed a number of the more interesting therapeutic laser procedures, and described highlights of recent work on laser-based imaging of tissue and spectral diagnosis of disease. These discussions have included both those procedures and diagnostics in actual clinical practice, as well as those which are still in the research phase or in clinical trials.

## References


[1] Cheong, W. F., Prahl, S. A., Welch, A. J. (**1990**), *IEEE J. Quantum Electron.* **QE-26**, 2166–2185.

[2] Wilson, B. C., Sevick, E. M., Patterson, M. S., Chance, B. (**1992**), *Proc. IEEE* **80**, 918–930.

[3] Muller, G. (Ed.) (1993), *Medical Optical Tomography: Functional Imaging and Monitoring*. Bellingham, WA: SPIE Optical Engineering Press.

[4] Morin, M., Mailloux, A., Painchaud, Y., Beaudry, P. (**1999**), *Appl. Opt.* **38**(16), 3681.

[5] Bartel, S., Hielscher, A. H. (**2000**), *Appl. Opt.* **39**(10), 1580.

[6] Hielscher, A. H., Hanli, L., Britton, C., Tittel, F. K., Jacques, S. L. (**1996**), *Appl. Opt.* **35**(4), 719–727.

[7] Tan, O. T., Sherwood, K., Gilchrest, B. A. (**1989**), *New Engl. J. Med.* **320**, 416–421.

[8] Svaasand, L. O., Boerslid, T., Oeveraasen, M. (**1985**), *Laser Surg. Med.* **5**, 589–602.

[9] Schober, R., Ulrich, F., Sander, T., Durselen, H., Hessel, S. (**1986**), *Science* **232**, 1421.

[10] Berns, M. W. (**1984**), *Laser Surg. Med.* **4** (1–4); Vol. 4, No. 1 is a special issue on Hematoporphyrin Derivative Photoradiation Therapy.

[11] Peng, Q., Warloe, T., Berg, K., Moan, J., Kongshaug, M., Giercksky, K. E., Nesland, J. M. (**1997**), *Cancer* **79**(12), 2282–2308.

[12] Oschsner, M. (**1997**), *J. Photochem. Photobiol., B: Biol.* **39**(1), 1–18.

[13] Moore, J. V., West, C. M. L., Whitehurst, C. (**1997**), *Phys. Biol. Med.* **42**(5), 913–935.

[14] Melamed, M. R., Lindmo, T., Mendelsohn, M. L. (Eds.) (1990), *Flow Cytometry and Sorting*. New York: Wiley-Liss.

[15] Ashkin, A., Dziedzic, J. M., Yamane, T. M. (**1987**), *Nature* **330**, 769–771.

[16] Boulnois, J.-L. (**1986**), *Laser Med. Sci.* **1**, 47–66.

[17] Ishimaru, A. (1978), *Wave Propagation and Scattering in Random Media*. New York: Academic Press.

[18] Hale, G. M., Querry, M. R. (**1973**), *Appl. Opt.* **12**, 555–565.

[19] Key, H., Davies, E. R., Jackson, P. C., Wells, P. N. T. (**1991**), *Phys. Med. Biol.* **36**, 579–590.

[20] Patterson, M. S., Chance, B., Wilson, B. C. (**1989**), *Appl. Opt.* **28**, 2331–2336.

[21] Wang, L.-H., Jacques, S. L. (**1993**), *J. Opt. Soc. Am. A* **10**, 1746–1752.

[22] Vogel, A., Venugopalan, V. (**2003**), *Chem. Rev.* **103**, 577–644.

[23] Itzkan, I., Albagli, D., Banish, B. J., Dark, M., von Rosenberg, C., Perelman, L. T., Janes, G. S., Feld, M. S. (1993), in: J. C. Miller, D. B. Geohegan (Eds.), *Laser Ablation: Mechanisms and Applications-II*, AIP Conference Proceedings No. 28, New York: American Institute of Physics, p. 491.

[24] Heisterkamp, A., Thanongsak, M., Kermani, O., Drommer, W., Welling, H., Ertmer, W., Lubatschowski, H. (**2003**), *Graefes Arch. Clin. Exp. Ophthalmol.* **241**(6), 511–517.

[25] Anderson, R. R., Parrish, J. A. (**1983**), *Science* **220**, 524–527.

[26] Grema, H., Greve, B., Raulin, C. (**2003**), *Lasers Surg. Med.* **32**(5), 405–412.

[27] Papadavid, E., Katsambas, A. (**2003**), *Int. J. Dermatol.* **42**(6), 480–487.

[28] Mihroseini, M., Shelgikar, S., Cayton, M. M. (**1988**), *Ann. Thorac. Surg.* **45**, 415–420.

[29] Karu, T. (**1987**), *IEEE J. Quantum Electron.* **QE-23**, 1703–1717.

[30] Itzkan, I., Bourgelais, D. B. C. (**1988**), *Laser Life Sci.* **2**, 249–255.

[31] Webb, R. H., Hughes, G. W., Delori, F. C. (**1987**), *Appl. Opt.* **26**, 1492–1499.

[32] Bille, J. F., Dreher, A. W., Zinser, G. (1990), in: B. R. Masters (Ed.), *Noninvasive Diagnostic Techniques in Ophthalmology*. New York: Springer-Verlag, p. 528.

[33] Huang, D., Swanson, E. A., Lin, C. P., Schuman, J. S., Stinson, W. G., Chang, W., Hee, M. R., Flotte, T., Gregory, K., Puliafito, C. A., Fujimoto, J. G. (**1991**), *Science* **254**, 1178–1181.

[34] Radhakrishnan, S., Rollins, A., Roth, J. E., Yazdanfar, S., Westphal, V., Bardenstein,

D. S., Izatt, J. A. (**2001**), *Arch. Ophthalmol.* **119**(8), 1179–1185.
[35] Wang, L., Ho, P. P., Liu, C., Zhang, G., Alfano, R. R. (**1991**), *Science* **253**, 769–771.
[36] Benaron, D. A., Stevenson, D. K. (**1993**), *Science* **259**, 1463–1466.
[37] Fujimoto, J. G., Patterson, M. S. (Eds.) (1998), *Advances in Optical Imaging and Photomigration*, OSA TOPS, Vol. 21, Washington, DC: Optical Society of America.
[38] Fantimi, S., Franceschimi, M. A., Fishkin, J. B. (**1994**), *Appl. Opt.* **32**, 5204–5212.
[39] Leith, E., Chen, C., Chen, H., Chen, Y., Dilworth, D., Lopez, J., Rudd, J., Sun, P.-C., Valdmanis, J., Vossler, G. (**1992**), *J. Opt. Soc. Am. A* **9**, 1148–1153.
[40] Hee, M. R., Izatt, J. A., Jacobson, J. M., Fujimoto, J. G., Swanson, E. A. (**1993**), *Opt. Lett.* **18**, 950–952.
[41] Moon, J. A., Mahon, R., Duncan, M. D., Reintjes, J. (**1993**), *Opt. Lett.* **18**, 1591–1593.
[42] Richards-Kortum, R. R. (1990), *Fluorescence Spectroscopy as a Technique for Diagnosis in Human Arterial, Urinary Bladder, and Gastrointestinal Disease*, Ph.D. Thesis, MIT Archives, p. 200.
[43] Baraga, J. J. (1992), *Situ Chemical Analysis of Biological Tissue: Vibrational Raman Spectroscopy of Human Atherosclerosis*, Ph.D. Thesis, MIT Archives, p. 80.

## Further Reading

### Major Journals Specific to Laser Medicine

Puliafito, C. A. (Ed.) *Lasers in Surgery and Medicine*. New York: Wiley-Liss, published 6 times per annum.
Tromberg, B. J. (Ed.) *J. Biomed. Opt.* SPIE, published 12 times per annum.
Wolbarsht, M. (Ed.) *Lasers in the Life Sciences*. New York: Harwood Academic Publishers, published 4 times per annum.

### Recent Journal Issues Devoted to Laser Medicine

Birngruber, R. (Ed.) (**1990**), *IEEE J. Quantum Electron.* **26**(12).
Fujimoto, J. G. (Ed.) (**1993**), *Opt. Photon. News* **4**(10).
Motamedi, M. (Ed.) (**1993**), *Appl. Opt.* **32**(4).
Welch, A. J. (Ed.) (**1989**), *IEEE Trans. Biomed. Eng.* **36**(12).

### Books

Berlien, H. P., Mueller, G. (Eds.) (2003), *Applied Laser Medicine*. Springer.
Chance, B. (Ed.) (1990), *Photon Migration in Tissues*. New York: Plenum Press.
Chance, B., Alfano, R. R. (Eds.) (1993), *Photon Migration and Imaging in Random Media and Tissues*, SPIE Proceedings Vol. 1888, Bellingham, WA: Society of Photo-Instrumentation Engineers.
Dixon, J. A. (1987), *Surgical Applications of Lasers*. Chicago: Year Book Medical Publishers.
Minet, O., Mueller, G., Beuthan, J. (Eds.) (1998), *Selected Papers on Optical Tomography, Fundamentals and Applications in Medicine*. Bellingham, WA: SPIE Press.
Müller, G. (Ed.) (1993), *Medical Optical Tomography: Functional Imaging and Monitoring*. Bellingham, WA: SPIE Optical Engineering Press.
Tuchin, V. V. (2000), *Tissue Optics: Light Scattering Methods and Instruments for Medical Diagnosis*. Bellingham, WA: SPIE Tutorial Texts in Optical Engineering, SPIE Press.
Welch, van G. (1995), *Optical-Thermal Response of Laser-irradiated Tissue*. New York: Plenum Press.

# Lasers, Semiconductor

*Nagaatsu Ogasawara*
*University of Electro-Communications, Department of Electronics Engineering, Tokyo, Japan*

*Ryoichi Ito*
*Meiji University, Department of Physics, Kawasaki, Japan*
*Phone/Fax: +81-44-934-7425; e-mail: ito-ryoichi@sam.hi-ho.ne.jp*

**Abstract**
Following a brief history of semiconductor lasers, this article describes their operation principles in relation to their device structures and materials. Their fundamental operation characteristics, including the static, spectral, and dynamic aspects, are then presented. Finally, the current state of the art in semiconductor laser technology is described by reviewing up-to-date devices incorporating new structures and new functions.

**Keywords**
semiconductor laser; laser diode; photonics; compact light source; heterostructure; fiber-optic telecommunications; optical disk.

## 1 Introduction

Semiconductor lasers have now grown to be key components in modern photonics technology, most notably as light sources in fiber-optic telecommunications systems and optical disk systems. They have also found an increasing number of applications ranging from instruments such as bar code scanners and computer printers to solid-state laser pumps and measuring and sensoring equipment for engineering use.

Semiconductor lasers have the following features that distinguish them from other lasers.

1. *Compactness*: The typical size of a laser chip is $300 \times 200 \times 100\,\mu m^3$. The small chip contains all the ingredients of a laser structure: a resonator, a waveguide, an active medium, and a *p-n* junction to pump the active medium.

2. *High efficiency*: Semiconductor lasers can be driven by low electrical power [(several tens of mA) × (1–2 V)]. The efficiency of converting electrical power into optical power is several tens of percent, which should be compared with that of 0.1% or less for gas and other lasers.

3. *Capability for high-speed direct modulation*: Light output can be modulated at frequencies of 10 GHz or more simply by modulating the pumping current.

4. *Wide emission spectrum*: The emission wavelength is determined by the band-gap energy of the active-layer material. The blue-near-infrared region is covered by III–V compounds. The infrared region (3–30 µm) is covered by IV–VI compounds.

5. *High reliability*: Semiconductor lasers have become highly reliable devices because of the remarkable improvement in crystal quality, although rapid degradation was a great problem at the early stage of their development.

6. *Sensitivity to temperature*: The threshold current and the lasing wavelength strongly depend on temperature. This is a disadvantage for most applications. On the other

hand, wavelength tunability is useful in spectroscopic research.

This article is intended to serve as a useful reference for those who wish to grasp the basic concepts of semiconductor lasers and utilize them in engineering and research activities. Following this introduction is a brief history of semiconductor lasers. Next, the operation principles of semiconductor lasers are described in relation to their structures and materials. The fourth section presents a general view of fundamental operation characteristics covering the static, spectral, and dynamic aspects. The fifth section gives the current state of the art in semiconductor laser technology by reviewing up-to-date devices incorporating new structures and new functions. Because of the limited space, citation of published papers has been kept to a minimum. Several books listed in the (Further Reading) section can be consulted both for further study and for access to the relevant published papers.

## 2
## History

The advent of semiconductor injection lasers dates back to 1962, only two years later than the first achievement of laser action in ruby, when stimulated emission from forward-biased GaAs diodes was demonstrated [1–3]. The stimulated emission resulted from the radiative recombination of electrons and holes in the direct-gap semiconductor GaAs. The electrons and holes were injected by forward bias into the depletion region in the vicinity of the *p-n* junction. Optical feedback was provided by polished facets perpendicular to the junction plane. The demonstration of GaAs lasers prompted the exploration of many other III–V and IV–VI compound semi-conductors. Unfortunately, these early homojunction devices, consisting of a single semiconductor, were not considered for serious applications since continuous operation at room temperature was not feasible because of their high threshold current densities for lasing (several tens of kA $cm^{-2}$).

A breakthrough was brought about in 1970 by employing a double heterostructure grown by liquid-phase epitaxy [4, 5]. In the double heterostructure, stimulated emission occurred only within a thin active layer of GaAs sandwiched between *p*- and *n*-doped AlGaAs layers that have a wider band gap. The threshold current density was dramatically reduced to several kA $cm^{-2}$ or less, and continuous operation at room temperature was eventually accomplished. The invention of the double heterostructure was obviously the most important step in the history of the development of semiconductor lasers toward practical utility. For the invention of heterostructure, Alferov and Kroemer were honored with a Nobel Prize in physics in 2000, while Alferov, Hayashi and Panish shared the Kyoto Prize in 2001.

There still remained, however, a troublesome problem of degradation; the reliability of the early double-heterostructure devices was very poor and many of them stopped their continuous operation within a few or several tens of minutes. Fortunately, intensive investigations of the degradation mechanism led to the identification of the causes of major failure: the recombination-enhanced growth of dislocations, and facet erosion [6 Part B, Chapter 8]. These studies have revealed that a long operating life is obtained by eliminating dislocation generation during material growth and device processing and by coating the mirror facets with dielectric thin films.

Another task undertaken in parallel with the improvement of reliability was the stabilization of the optical mode in the lateral transverse direction (parallel to the junction plane). It had been recognized by the late 1970s that the lateral-mode instability often observed at higher output powers was detrimental to most applications and that such an instability could be effectively suppressed by incorporating a built-in refractive-index profile in the lateral direction [7]. A large number of laser structures have been proposed and demonstrated to introduce lateral variations of the refractive index.

As a result of these research and development efforts during the first two decades, the maturity of AlGaAs lasers reached the stage of practical use in printers and disk players early in the 1980s. Meanwhile, the rapid reduction of transmission loss in optical fibers achieved during the 1970s stimulated intensive development efforts on GaInPAs lasers, which have emission wavelengths (1.3–1.55 μm) in the low-loss and dispersion-free region of silica fibers. During the first half of the 1980s, GaInPAs lasers emerged as indispensable light sources in telecommunications systems using silica fibers (see OPTICAL COMMUNICATIONS).

Progress has continued to date in both device structures and materials. In the following, let us look at some major topics.

The idea of utilizing periodic gratings incorporated in a laser medium as a means of optical feedback was proposed by Kogelnik and Shank in [8]. Distributed-feedback and distributed-Bragg-reflector semiconductor lasers utilizing this principle have since become the major device structures used in long-haul, high bit-rate telecommunications systems because of the stable single–longitudinal-mode oscillation they exhibit even under high-speed modulation.

Recent advances in growth technology of semiconductor ultrathin layers by molecular-beam epitaxy and metal-organic vapor-phase epitaxy have also created a new class of laser structures. Quantum-well lasers incorporating such ultrathin active layers ($\sim$20 nm or less) have been shown to display a variety of superior laser characteristics, such as low-threshold currents and wide modulation bandwidths, which are attributed to the size quantization of electrons in the ultrathin active layers [9]. Quantum-well structures as well as multiple quantum-well structures are now commonly employed in most semiconductor lasers.

In the visible spectrum, GaInP lasers were first commercialized as red-emitting lasers (0.63–0.69 μm) in 1988 by Sony, Toshiba, and NEC. They are now extensively used in DVD systems. Blue–green laser emission was achieved first in ZnCdSe lasers [10] and then in InGaN lasers [11, 12]. InGaN blue lasers are being employed in second-generation DVD players (see DATA STORAGE, OPTICAL).

## 3
## Structures, Materials, and Operation Principles

### 3.1
### Double Heterostructure

A semiconductor laser is a diode, as shown schematically in Fig. 1. When a forward current is passed through the *p-n* junction, electrons and holes injected into the active region from the *n* and *p* regions respectively, recombine to emit radiation. The wavelength $\lambda$ of the radiation is

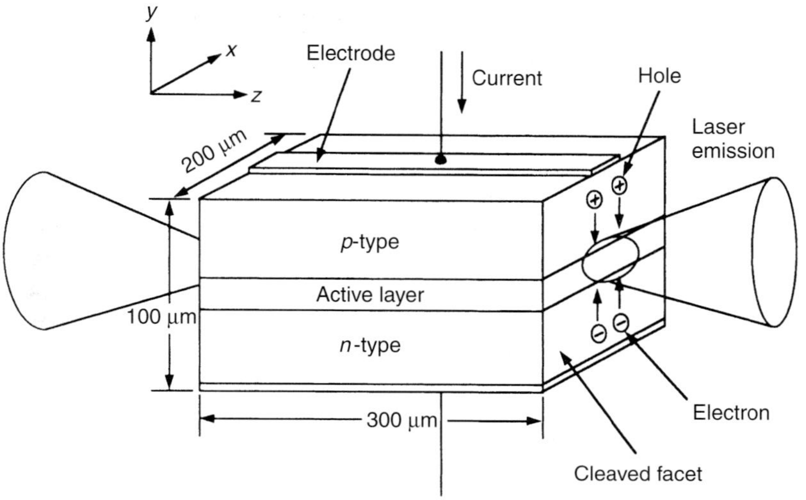


**Fig. 1** Schematic illustration of a double-heterostructure semiconductor laser

basically determined by the relation $\lambda \sim hc/E_g$, where $h$ is the Planck constant, $c$ is the light velocity, and $E_g$ is the band-gap energy of the active-region material.

For low injection currents, light is emitted through spontaneous emission. In order for lasing to take place, a sufficiently high concentration of carriers must be accumulated within the active region to induce population inversion. This is effectively accomplished by adopting a double-heterojunction (DH) structure, where a thin active layer, typically ~0.1 μm thick, but as thin as 10 nm in quantum-well lasers, is sandwiched between $n$- and $p$-type cladding layers, which have wider band gaps than the active layer. Electrons and holes injected into the active layer through the heterojunctions are confined within the thin active layer by the potential barriers at the heteroboundaries, as illustrated in Fig. 2(a). The DH structure forms an efficient optical waveguide as well, because of the refractive-index difference between the active and cladding layers shown in Figs. 2(b) and 2(c). Thus, the DH structure facilitates the interaction needed for laser action between the optical field and the injected carriers.

Let us use a GaAs(active)/$Al_xGa_{1-x}As$ (cladding) DH structure as an example. The dependence of the band-gap energy of $Al_xGa_{1-x}As$ on the AlAs mole fraction $x$ can be approximated for $x < 0.42$ (direct-gap region) by [6 Part A, p. 193].

$$E_g = 1.424 + 1.247x \text{ eV}. \tag{1}$$

The energy-gap difference between the active and cladding layers $\Delta E_g$ is divided into the band discontinuities $\Delta E_c$ in the conduction band and $\Delta E_v$ in the valence band according to [13].

$$\Delta E_c \sim 0.62 \Delta E_g \tag{2}$$

and

$$\Delta E_v \sim 0.38 \Delta E_g. \tag{3}$$

The efficiency of the carrier confinement depends strongly on the magnitude of the band discontinuity. The AlAs mole fraction $x$ of the cladding layers is usually chosen to be greater than that of the active

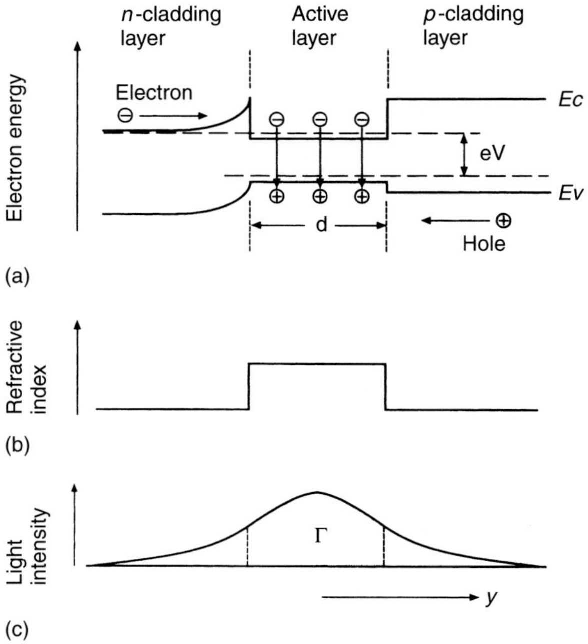


**Fig. 2** Diagram illustrating carrier confinement and waveguiding in a double heterostructure (a) Energy-band diagram at high forward bias; (b) refractive-index distribution; (c) light intensity distribution. $E_c$ and $E_v$: The edges of the conduction and valence bands. $V$: Applied voltage. $d$: Active-layer thickness (~0.1 μm). Γ: Confinement factor

layer by ~0.3. This gives rise to a $\Delta E_c$ approximately 10 times as large as the room-temperature thermal energy, which is sufficient to suppress electron diffusion over the heterobarriers. The hole leakage current is less important because of the smaller diffusion constant for holes.

The characteristics of a three-layer slab waveguide are conveniently described in terms of the normalized waveguide thickness $D$, defined as

$$D = \left(\frac{2\pi}{\lambda}\right) d\sqrt{\eta_a^2 - \eta_c^2}, \qquad (4)$$

where $\eta_a$ and $\eta_c$ are the refractive indices of the active and cladding layers respectively and $d$ is the active-layer thickness. For example, the condition that a waveguide supports only the lowest-order fundamental mode is expressed as

$$D < \pi. \qquad (5)$$

Meanwhile, the refractive index of $Al_xGa_{1-x}As$, $\eta(x)$, can be approximated for a light wavelength of ~0.9 μm by [6, Part A, p. 45]

$$\eta(x) = 3.590 - 0.710x + 0.091x^2. \qquad (6)$$

Thus, for a waveguide with a GaAs active layer surrounded by $Al_{0.3}Ga_{0.7}As$ layers and a wavelength $\lambda$ of 0.87 μm, the condition of Eq. (5) simply becomes

$$d < 0.36\ \mu m. \qquad (7)$$

Since the thickness $d$ is typically 0.1 to 0.2 μm, the single, fundamental-mode condition is almost always satisfied in practical devices.

The confinement factor $\Gamma$, defined as the fraction of the electromagnetic energy of the guided mode that exists within the active layer, is an important parameter representing the extent to which the waveguide mode is confined to the active layer. $\Gamma$ for a fundamental mode is approximately given by [14]

$$\Gamma \sim \frac{D^2}{2 + D^2}. \tag{8}$$

For a $GaAs/Al_{0.3}Ga_{0.7}As$ waveguide with $d = 0.1$ μm, $\Gamma \sim 0.27$.

3.2
## Fabry–Pérot Cavity

In addition to the optical gain, optical feedback is the other ingredient for laser oscillation. This is provided by a pair of mirror facets at both ends of the devices. These facets are normally formed simply by cleaving the crystal. Since the refractive index of major semiconductor laser materials is ~3.6, the reflectivity of the mirror is ~0.3. This is very low compared to other types of lasers but is still sufficient to provide optical feedback in semiconductor lasers.

3.3
## Lasing Threshold Condition

When a sufficient number of electrons and holes is accumulated to form an inverted population, the active region exhibits optical gain and can amplify light passing through it since stimulated emission overcomes interband absorption. The condition for self-sustained laser oscillation to occur is that the light makes a full round trip in the cavity without attenuation; that is, the optical gain should equal the losses both inside the cavity and through the partially reflecting end facets. Thus, the gain coefficient at threshold $g_{\mathrm{th}}$ is given by the relation

$$\Gamma g_{\mathrm{th}} = \underbrace{\Gamma\alpha_{\mathrm{a}} + (1 - \Gamma)\alpha_{\mathrm{c}} + \alpha_{\mathrm{s}}}_{\alpha_{\mathrm{i}}} + \frac{1}{L}\ln\frac{1}{R}. \tag{9}$$

Here, $\alpha_{\mathrm{a}}$ and $\alpha_{\mathrm{c}}$ denote the losses in the active and cladding layers respectively, due to free-carrier absorption. $\alpha_{\mathrm{s}}$ accounts for scattering loss due to heterointerfacial imperfections. The first three loss terms on the right-hand side combined are termed *internal loss* $\alpha_{\mathrm{i}}$ and add up to 10 to 20 $\mathrm{cm}^{-1}$ [6, Part A, p. 174–176]. The reflection loss $L^{-1}\ln R^{-1}$ due to output coupling (~40 $\mathrm{cm}^{-1}$ for $L \sim 300$ μm, $R \sim 0.3$) is normally the largest among the loss terms. Despite the fact that both the internal and the reflection losses are exceptionally large as compared to other lasers, lasing is brought about by virtue of the high optical gain in semiconductors.

It is not an easy task to analyze precisely the optical gain spectrum in semiconductor lasers. Fortunately, however, we can utilize a phenomenological linear relationship between the maximum gain $g$ (the peak value of the gain spectrum for a given carrier density) and the injected carrier density $n$,

$$g(n) = \frac{\partial g}{\partial n}(n - n_{\mathrm{t}}), \tag{10}$$

to a good approximation [15–17]. Here, $\partial g/\partial n$ is termed *differential gain*, and $n_{\mathrm{t}}$ denotes the carrier density required to achieve transparency where stimulated emission balances against interband absorption corresponding to the onset of population inversion. For GaAs lasers,

$$\frac{\partial g}{\partial n} \sim 3.5 \times 10^{-16}\ \mathrm{cm}^2 \qquad (11)$$

and

$$n_t \sim 1.5 \times 10^{18}\ \mathrm{cm}^{-3}. \qquad (12)$$

Substituting Eqs. (10) to (12) into Eq. (9) and assuming that $\Gamma = 0.27$, $\alpha_i = 10\ \mathrm{cm}^{-1}$, and $L^{-1} \ln R^{-1} = 40\ \mathrm{cm}^{-1}$, we get a threshold carrier density $n_{th}$ of $\sim 2 \times 10^{18}\ \mathrm{cm}^{-3}$.

The threshold current density $J_{th}$ is expressed as

$$J_{th} = \frac{edn_{th}}{\tau_s}, \qquad (13)$$

where $\tau_s$ is the carrier lifetime due to spontaneous emission. Assuming that $\tau_s = 2$–$4$ ns and $d = 0.1\ \mu$m, we obtain a threshold current density $J_{th}$ or $\sim 1\ \mathrm{kA\ cm}^{-2}$.

### 3.4 Stripe Geometry

Most modern semiconductor lasers adopt a stripe geometry, where current is injected only within a narrow region beneath a stripe contact several μm wide, in order to keep the threshold current low and to control the optical field distribution in the lateral direction. As compared with broad-area lasers, where the entire laser chip is excited, the threshold current of the stripe-geometry lasers is reduced roughly proportional to the area of the contact.

Figure 3(a) shows the simplest form of the stripe geometry [18]. Lasing occurs in a limited region of the active layer beneath the stripe contact where a high density of current flows. Such lasers are termed *gain-guided lasers* because the optical intensity distribution in the lateral direction ($x$ direction) is determined by the gain profile produced by carrier density distribution.

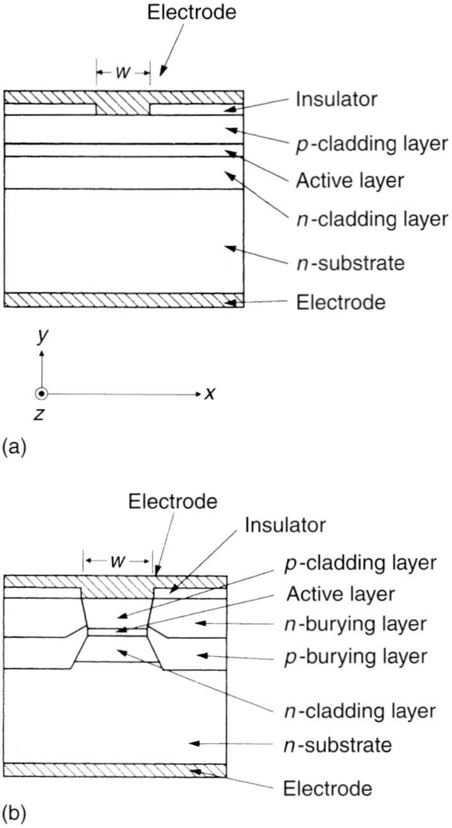


**Fig. 3** Two representative examples of stripe-geometry lasers (front view of the end facet) (a) Stripe-contact laser; (b) Buried heterostructure (BH) laser. *W*: Stripe width (2–10 μm)

Devices incorporating a built-in refractive-index variation in the lateral direction are termed *index-guided lasers.* Figure 3(b) shows the structure of a buried heterostructure (BH) laser [19] as a representative example of index-guided lasers. The active region is surrounded by materials with lower refractive indices in both the vertical ($y$) and lateral ($x$) transverse directions, thus forming a waveguide structure in both directions. In the BH laser, the width of current flow is delineated by the *p-n* junction formed by the burying layers, which is reverse biased when the active region is forward biased.

### 3.5
### Materials and Emission Wavelengths

Semiconductor-laser materials that have been studied range over III–V, IV–VI, and II–VI compounds as listed in Table 1.

#### 3.5.1 III–V Compounds

III–V compounds are the most important and popular laser materials [6, Part B, Chapter 5]. In particular, AlGaAs (active layer)/AlGaAs (cladding layer)/GaAs (substrate) lasers emitting at 0.7 to 0.9 μm and GaInP/AlGaInP/GaAs lasers emitting at 0.63 to 0.69 μm are extensively used in optical disk systems and laser printers. GaInPAs/InP/InP lasers emitting at 1.2 to 1.6 μm are used in fiber-optic communications systems. GaInN/AlGaN/sapphire lasers emitting at 0.38 to 0.45 μm are starting to be used in second-generation DVD players. All of them exhibit low-threshold, room-temperature continuous wave (cw) operation with high reliability.

The common features among them are as follows:

1. The active layer consists of direct-gap materials.
2. Binary compounds (GaAs, InP) are used as the substrate except for GaInN/AlGaN/sapphire lasers, for which GaN substrate will be used in the future.
3. The active and cladding layers have nearly the same lattice constant as the substrate.

**Tab. 1** Major semiconductor-laser materials and their emission wavelengths

| *Materials (active/cladding/substrate)* | *Emission wavelengths [μm]* |
|---|---|
| III–V compounds | |
| AlGaAs/AlGaAs/GaAs | 0.7–0.9 |
| GaInPAs/InP/InP | 1.2–1.6 |
| GaInP/AlGaInP/GaAs | 0.66–0.69 |
| GaInPAs/GaInP/GaAs or GaPAs | 0.65–0.9 |
| GaInPAs/AlGaAs/GaAs | 0.62–0.9 |
| AlGaAsSb/AlGaAsSb/GaSb | 1.1–1.7 |
| GaInAsSb/AlGaAsSb/GaSb or InAs | 2–4 |
| InPAsSb/InPAsSb/GaSb or InAs | 2–4 |
| GaInAs(strained)/AlGaAs/GaAs | 0.9–1.1 |
| GaInPAs(strained)/GaInPAs/InP | ~1.55 |
| GaInN/AlGaN/sapphire | 0.38–0.45 |
| IV–VI compounds | |
| PbSnTe/PbSnSeTe/PbTe | 6–30 |
| PbSSe/PbS/PbS | 4–7 |
| PbEuTe/PbEuTe/PbTe | 3–6 |
| II–VI compounds | |
| ZnCdSe/ZnSSe/GaAs | ~0.5 |

The use of direct-gap semiconductors, featuring efficient radiative recombination, as the active-region material (**1**) is essential in achieving laser action. Features **2** and **3** are related to the crystal quality of the hetero-interfaces. To minimize the generation of lattice defects that may impair the device reliability, the lattice parameter of the active and cladding layers should be matched to that of the high-quality binary substrate (the lattice-matching condition). The lattice parameter of $Al_xGa_{1-x}As$ is essentially independent of alloy concentration $x$. This is the only exception in ternary alloys, arising from the fortuity that GaAs and AlAs have virtually the same lattice parameter, the difference being as small as ~0.1%. In quaternary alloys like $Ga_xIn_{1-x}P_yAs_{1-y}$ and $Al_xGa_yIn_{1-x-y}P$, the band-gap energy can be tuned in a certain range for a fixed lattice parameter by choosing appropriate pairs of $x$ and $y$ values. The design of lattice-matched DH structures is facilitated by taking advantage of this degree of freedom in quaternary alloys. In some quantum-well structures, a certain degree of strain in the active material caused by inevitably or deliberately introduced slight lattice mismatching is used to provide better laser performance.

#### 3.5.2 IV–VI Compounds

PbSnTe and PbSSe lasers emitting in the infrared region (3–30 μm) [20] are used in high-resolution gas spectroscopy and in air pollution monitoring. These lasers operate only at cryogenic temperatures. However, an appreciable spectral tuning is feasible by simply changing temperature and injection current, which is extremely desirable for spectroscopy.

#### 3.5.3 II–VI Compounds

ZnCdSe/ZnSSe/ZnMgSSe/GaAs quantum-well lasers emitting in the blue–green spectral region emerged around 1991 as a result of success in overcoming the difficulty of *p*-type doping into these materials. [10, 21] Despite extensive efforts, however, these lasers have never attained device life times exceeding several hundred hours and hence have not been put into practical use.

## 4 Fundamental Characteristics

In this section, several aspects of fundamental characteristics are described. Numerical examples are given based on AlGaAs ($\lambda \sim 0.8$ μm) and GaInPAs ($\lambda \sim 1.3$ μm) lasers.

### 4.1 Light–Current Characteristics

Figure 4 shows an example of output power ($P$) versus DC injection current ($I$) characteristics. The ordinate is the light power emitted from one end facet, and essentially the same power is emitted from the other end facet.

The threshold current at room temperature is typically several tens of mA. The dependence of the threshold current $I_{th}$ on device temperature $T$ is phenomenologically expressed as

$$I_{th} \propto \exp\left(\frac{T}{T_0}\right), \qquad (14)$$

where $T_0$ is a constant referred to as the characteristic temperature and takes a value of 100 to 150 and 50 to 80 K for AlGaAs and GaInPAs lasers, respectively.

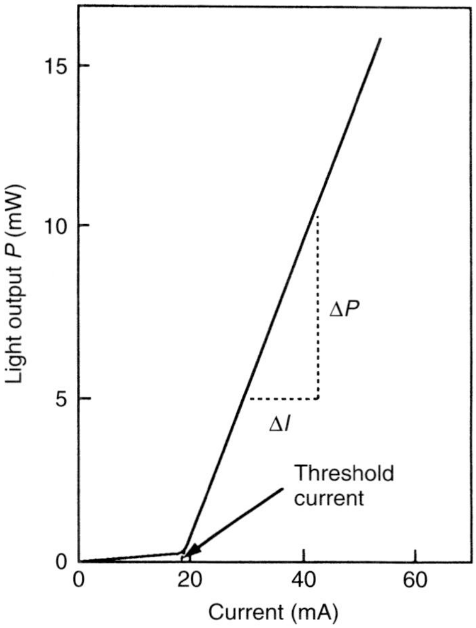


**Fig. 4** Light output versus injection current curve in an AlGaAs laser

The light power emitted by injection above threshold is expressed as

$$P = \hbar\omega_l \frac{I - I_{th}}{e} \eta_i \frac{\frac{1}{2}L^{-1}\ln(1/R)}{\alpha_i + L^{-1}\ln(1/R)}. \quad (15)$$

Here, $(I - I_{th})/e$ is the rate of excess carrier injection beyond the threshold. $\eta_i$ is the internal quantum efficiency representing the fraction of injected carriers that recombine radiatively and generate photons of energy $\hbar\omega_l$. The factor $1/2L^{-1}\ln R^{-1}/(\alpha_i + L^{-1}\ln R^{-1})$ represents the fraction of the generated photons that are coupled out of the cavity through one of the end facets. From Eq. (15), the differential efficiency $\eta_D$, defined as the slope of output power versus current curve above threshold, is written as

$$\eta_D = \frac{\Delta P}{\Delta I} = \frac{\hbar\omega_l}{e}\eta_i \frac{\frac{1}{2}L^{-1}\ln(1/R)}{\alpha_i + L^{-1}\ln(1/R)}. \quad (16)$$

The differential external quantum efficiency $\eta_{ext}$, defined as the ratio of the number of photons coupled out to the number of carriers injected, is given by

$$\eta_{ext} = \frac{\Delta P/\hbar\omega_l}{\Delta I/e} = \eta_i \frac{\frac{1}{2}L^{-1}\ln(1/R)}{\alpha_i + L^{-1}\ln(1/R)}. \quad (17)$$

Above threshold, the stimulated emission predominates over the nonradiative processes and $\eta_i \sim 1$. Typical values of $L = 300\,\mu\text{m}$, $R = 0.3$, and $\alpha_i = 20\,\text{cm}^{-1}$ give $\eta_{ext} \sim 0.3$. Since $\hbar\omega_l \sim E_g$, $\eta_D \sim (E_g/e)\eta_{ext}$. For GaAs lasers ($E_g = 1.424$ eV), we get $\eta_D \sim 0.4$ W/A.

The maximum rating for output power is typically 5 to 30 mW. Figure 5 shows schematically the three phenomena that determine the maximum power rating.

1. *Kink*: The nonlinearity in the curve of light versus current caused by lateral-mode instability is referred to as a kink. Since a kink is often associated with a beam-profile shift and the generation of intensity noise, lasers cannot be used

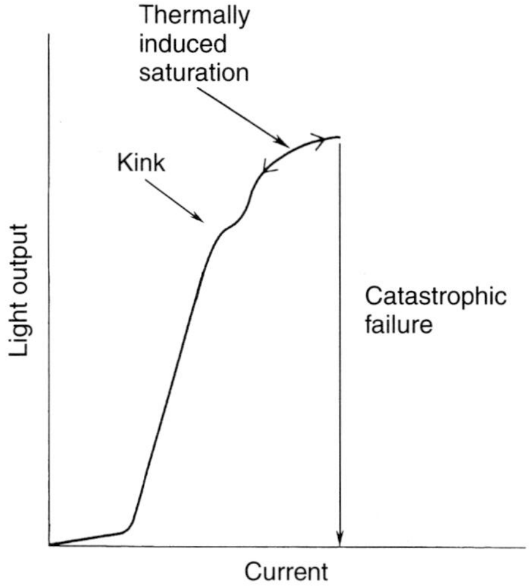


**Fig. 5** Schematic illustration of the three phenomena occurring at high injection levels

in the kink region for most applications. The lateral-mode instability arises from the optical power dependence of the refractive index. By incorporating a refractive-index profile in the lateral direction to form a waveguide, the power level for kink generation can be appreciably increased.

2. *Catastrophic optical damage*: For optical power densities higher than several MW cm$^{-2}$, the end facets of semiconductor lasers may be melted. This results in a sudden decrease of output power and failure of the device. The catastrophic damage is considered to be brought about by the enhanced light absorption associated with surface states at the end facets. The importance of this process seems to depend on materials and photon energies. In GaInPAs lasers ($\lambda \sim 1.3$ μm), the catastrophic optical damage practically does not occur.

3. *Thermally induced saturation*: Heating of the junction under high current injections raises the threshold current. This may lead to saturation of output power with an increase in injection current, especially in lasers with lower $T_0$ values.

4.2
### Beam Profile and Polarization

Typical radiation intensity patterns of laser diodes are shown in Figs. 6 and 7. Shown in Fig. 6 are the near-field patterns, that is, the spatial distributions of optical intensity on the end facet in the directions (a) perpendicular and (b) parallel to the junction plane. The far-field pattern (Fig. 7) is the angular distribution of radiant intensity measured at distances several mm or more away from the facet, which is mathematically a Fourier transform of the near-field pattern. The single-lobed radiation patterns show that the transverse mode of the device is single and fundamental. Transverse-mode-stabilized lasers incorporating appropriate waveguide structures exhibit such beam profiles stably up to reasonable output power levels.

Normally, the angular width or beam divergence (the full width at half maximum) perpendicular to the junction, $\theta_\perp$ (20°–60°), is larger than that parallel to the junction $\theta_\perp$ (10°–30°) since the near-field spot is an ellipsoid because of the large difference between the stripe width

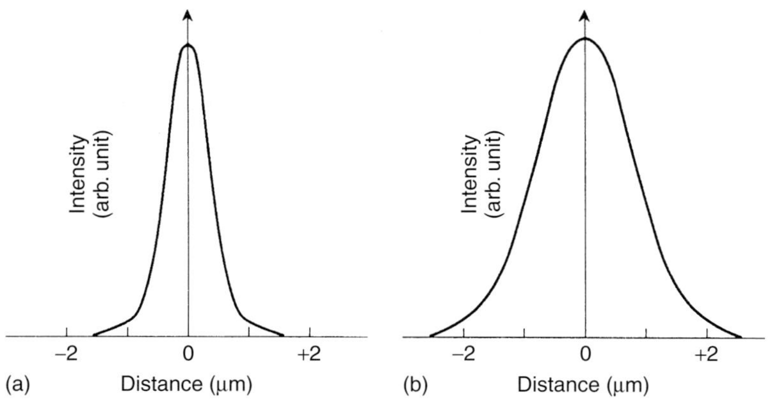


**Fig. 6** Near-field optical intensity patterns (a) perpendicular; (b) parallel to the junction plane

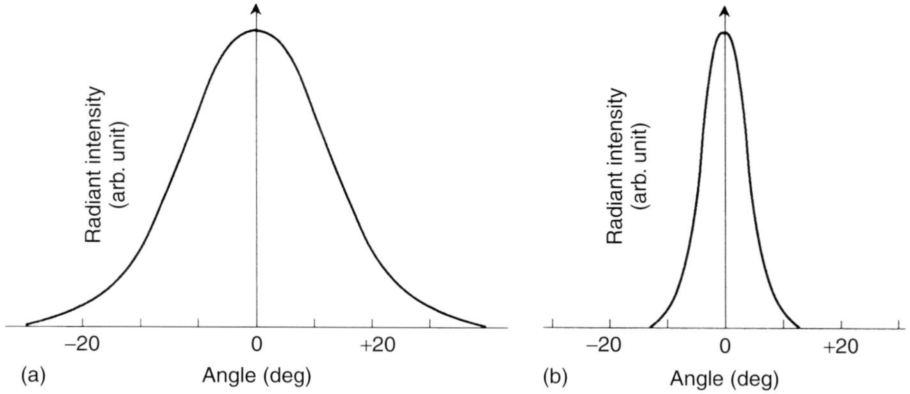


**Fig. 7** Far-field radiant intensity patterns (a) perpendicular; (b) parallel to the junction plane

(several μm) and the active-layer thickness (~0.1 μm). $\theta_{\perp}$ can be reduced by adopting thinner active layers; then, most of the optical energy penetrates into the cladding layers, increasing the spot size perpendicular to the junction.

The output beam from gain-guided lasers can be considerably astigmatic; the beam waist perpendicular to the junction plane is located essentially on the end facet, while the virtual beam waist along the junction plane is formed in the cavity at a position several tens of μm behind the end facet. This is brought about by the difference of the waveguiding mechanism between the two directions. In contrast, in index-guided lasers, the beam waist, both parallel and perpendicular to the junction plane, is located within several μm of the facet. The astigmatism should be taken into consideration when the beam is coupled into lenses.

The laser beam is linearly polarized along the junction plane. This is because, in a slab waveguide, the facet reflectivity for the TE mode is higher than that for the TM mode [22]. Since the gain needed to reach threshold depends on the facet reflectivity (see Eq. (9)), the TE mode is selected for oscillation in DH lasers because of the higher reflectivity.

## 4.3 Spectral Characteristics

Figure 8 shows the variations of lasing spectrum with output power in an index-guided AlGaAs laser. At lower powers, the laser oscillates in several longitudinal modes. Here, the longitudinal-mode wavelength $\lambda_N$ is determined by the relation

$$\left(\frac{1}{2}\frac{\lambda_N}{\eta}\right) N = \mathrm{L} \quad (N: \text{positive integers}), \tag{18}$$

where $\eta$ is the effective refractive index of the waveguide and $L$ is the cavity length. The mode spacing $\Delta\lambda$ is given by

$$\Delta\lambda = \frac{\lambda^2}{2\eta_{\mathrm{g}} L}, \tag{19}$$

where $\eta_{\mathrm{g}} = \eta - (\partial\eta/\partial\lambda)\lambda$ is the group index. $\Delta\lambda$ is typically 3 Å ($\lambda = 0.8$ μm)–8 Å ($\lambda = 1.3$ μm) for $L = 300$ μm.

As injection is increased, the laser power tends to concentrate in a single-longitudinal mode, while the power in the

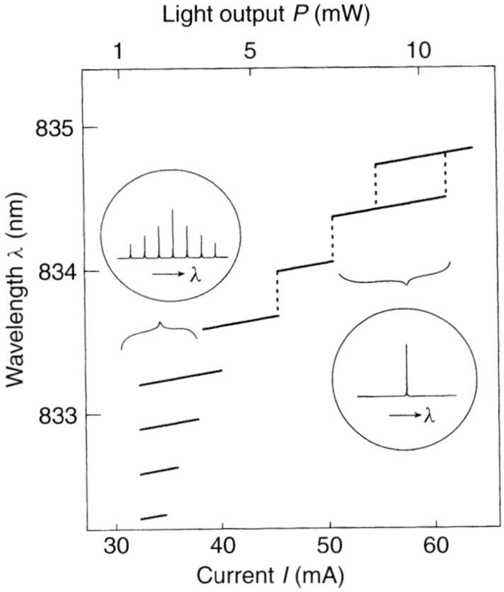


**Fig. 8** Lasing wavelength versus injection current characteristics in an AlGaAs laser

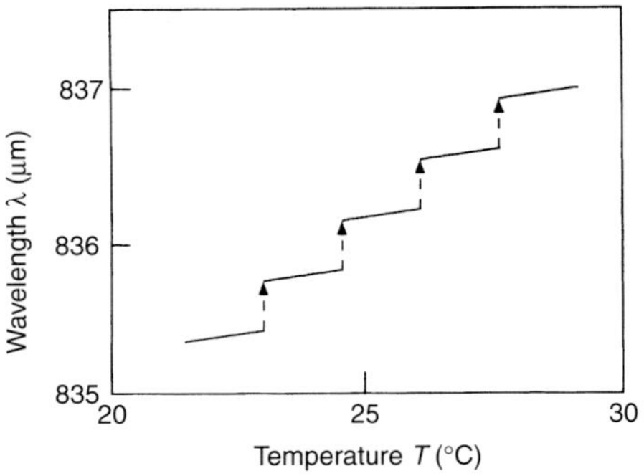


**Fig. 9** Temperature dependence of the lasing wavelength

remaining modes saturates. This is common among index-guided lasers. Gain-guided lasers, in contrast, tend to exhibit multiple–longitudinal-mode spectra even at higher powers.

It can be seen in Fig. 8 that the lasing wavelength shifts toward longer wavelengths as the output power is increased. This is induced by the temperature rise in the active region with the increase in injection current. Shown in Fig. 9 is an example of the wavelength shift due to a heat sink temperature change at a fixed injection current. Each longitudinal-mode shifts at a rate of 0.5 Å $K^{-1}$ ($\lambda \sim 0.8$ μm)–0.8 Å $K^{-1}$ ($\lambda \sim 1.3$ μm) because of the temperature dependence of the refractive index. In addition, the lasing wavelength jumps toward longer wavelengths, as temperature is raised, at a rate of 2 Å $K^{-1}$ ($\lambda \sim 0.8$ μm)–5 Å $K^{-1}$ ($\lambda \sim 1.3$ μm). This is caused by the temperature dependence of the band-gap energy.

At ranges of temperature and bias current where the oscillation mode jumps to a neighboring mode, it is often observed that the lasing is randomly switched between the two longitudinal modes. The random switching is associated with a large-intensity noise since the output power is different between modes by (0.1–1%). Some devices exhibit hysteresis in the lasing wavelength versus temperature and/or injection-current characteristics. These phenomena have been interpreted in terms of nonlinear mode coupling among longitudinal modes [17].

The spectral linewidth of a single-longitudinal mode is inversely proportional to the output power to a good approximation. The product of the linewidth $\Delta f$ and the output power $P$ is typically in the range of 1 to 100 MHz mW. This value is 10 to 50 times larger than $\Delta f_{ST}$ obtained from the well-known Schawlow–Townes formula. An analysis shows that the enhancement is expressed as [23]

$$\Delta f = \Delta f_{ST}(1 + \alpha^2), \qquad (20)$$

where $\alpha$ is the linewidth enhancement factor defined as

$$\alpha = -\frac{4\pi}{\lambda}\frac{(\partial\eta/dn)}{(\partial g/dn)} \qquad (21)$$

by the derivatives of the refractive index $\eta$ and the gain $g$ with respect to the carrier density $n$. $\alpha$ represents the magnitude of the amplitude-phase coupling, inherent in semiconductor lasers, originating from the strong dependence of the refractive index on the carrier density. The linewidth is determined not only by the direct phase fluctuation caused by spontaneous emission but there also exists an additional contribution from the amplitude fluctuation since it is coupled into the phase fluctuation through the carrier density fluctuation.

$\alpha$ takes a value of 2 to 6 depending on the active-region material, the injected carrier density, and the lasing photon energy [24], while in most gas and solid-state lasers, $\alpha$ can be virtually taken to be zero. This is because the gain spectrum in semiconductor lasers based on a band-to-band transition is asymmetric with respect to the gain-peak frequency (corresponding to the laser frequency) and, as a consequence, the associated refractive index in the vicinity of the lasing frequency varies appreciably with the injected carrier density. In contrast, in ordinary lasers where the lasing transition takes place between two discrete levels, the gain spectrum is symmetric and the associated refractive-index dispersion crosses zero at the lasing frequency. The nonzero value of this parameter affects a number of semiconductor laser characteristics including linewidth broadening, lateral-mode instability, and frequency chirping.

### 4.4
### Dynamic Characteristics

The capability of direct-current modulation is one of the important advantages of semiconductor lasers. The laser response to a stepwise current pulse is schematically shown in Fig. 10. There is a delay time $\tau_d$ for the turn-on of lasing because a finite time is required before the injection carriers are accumulated to form population inversion. $\tau_d$ is approximated by [25]

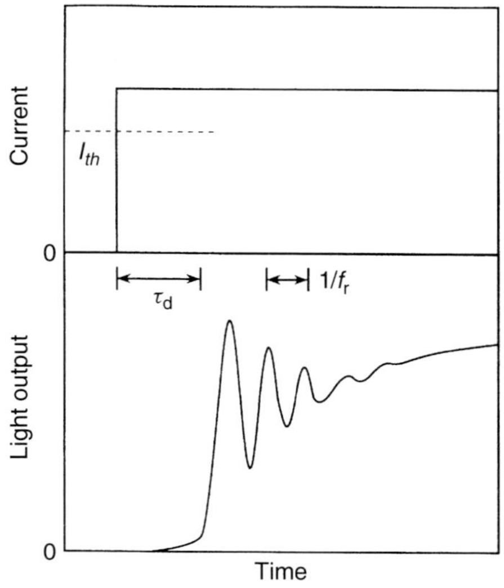


**Fig. 10** Laser response to a staircase injection current. $\tau_d$: Turn-on delay time. $f_r$: Relaxation oscillation frequency

$$\tau_d = \tau_s \sqrt{\frac{I_{th}}{I}} \tanh^{-1} \sqrt{\frac{I_{th}}{I}}, \qquad (22)$$

where $\tau_s$ is the carrier lifetime at threshold (~2 ns), $I_{th}$ is the threshold current, and $I$ is the current pulse height. $\tau_d$ can be reduced by prebiasing the laser. Then $\tau_d$ is expressed by the DC bias current $I_0$ ($<I_{th}$) as

$$\tau_d = \tau_s \sqrt{\frac{I_{th}}{I}} \left( \tanh^{-1} \sqrt{\frac{I_{th}}{I}} - \tanh^{-1} \sqrt{\frac{I_0}{I}} \right). \qquad (23)$$

After the onset of laser oscillation, relaxation oscillations are generated for a time interval of several nanoseconds before the steady state is attained. The relaxation oscillation occurs because of the resonant interaction between photons and carriers; the energy stored in the cavity is transferred back and forth between the photon and carrier subsystems. Therefore, the periodic oscillation of the laser output is accompanied by a modulation of carrier density. The temporal variation of gain spectrum caused by the carrier density modulation leads to a multiple–longitudinal-mode oscillation. Furthermore, the frequency of each individual mode is modulated because of the carrier-density–dependent refractive index, leading to a broadening of the time-averaged spectral linewidth. This phenomenon is referred to as *frequency chirping*. The magnitude of the chirping is governed by the linewidth enhancement factor $\alpha$.

When a laser is biased above threshold, the bandwidth of the dynamic response of the laser is essentially determined by the relaxation oscillation frequency $f_r$. Let $\Delta I(\omega)e^{i\omega t}$ be a small-amplitude sinusoidal current superimposed on a DC bias $I_0$ ($>I_{th}$). Then the photon density inside the active region is made up of a DC component $S_0$ and an AC component $\Delta S(\psi)e^{i\omega t}$. The magnitude of the modulated component $\Delta S(\omega)$ is analyzed to be [26]

$$\Delta S(\omega) = \frac{\Gamma \tau_p \omega_r^2}{-\omega^2 + i\omega\Omega + \omega_r^2}\frac{\Delta I(\omega)}{eV_a}. \quad (24)$$

Here, the angular relaxation oscillation frequency $\omega_r$ is given by

$$\omega_r = 2\pi f_r = \sqrt{\frac{S_0}{\tau_p}\frac{c}{\eta_g}\frac{\partial g}{\partial n}}. \quad (25)$$

$\Omega$ represents the damping of the relaxation oscillation and is given by

$$\Omega = \frac{1}{\tau_s} + S_0\frac{c}{n_g}\frac{\partial g}{\partial n}. \quad (26)$$

$\tau_p$ is the photon lifetime of the cavity (~2 ps) expressed as

$$\tau_p^{-1} = \frac{c}{\eta_g}\left[\alpha_i + \frac{1}{L}\ln\left(\frac{1}{R}\right)\right]. \quad (27)$$

$V_a$ denotes the active-region volume. The magnitude of the modulated component of output power $\Delta P(\omega)$ is proportional to $\Delta S(\omega)$; that is,

$$\Delta P(\omega) = \frac{1}{2}\frac{c}{\eta_g}\hbar\omega_l\frac{1}{L}\ln\left(\frac{1}{R}\right)\Delta S(\omega)\frac{V_a}{\Gamma}. \quad (28)$$

Figure 11 shows an example of relative modulation response $|\Delta P(\omega)/\Delta P(0)| = \omega_r^2/[(\omega^2 - \omega_r^2)^2 - \omega^2\Omega^2]^{1/2}$. A flat response at modulation frequencies less than $f_r$ is followed by a peak at $f_r$ and a sharp drop at modulation frequencies exceeding $f_r$. Therefore $f_r$ is the uppermost useful modulation frequency. $f_r$ is proportional to the square root of output power

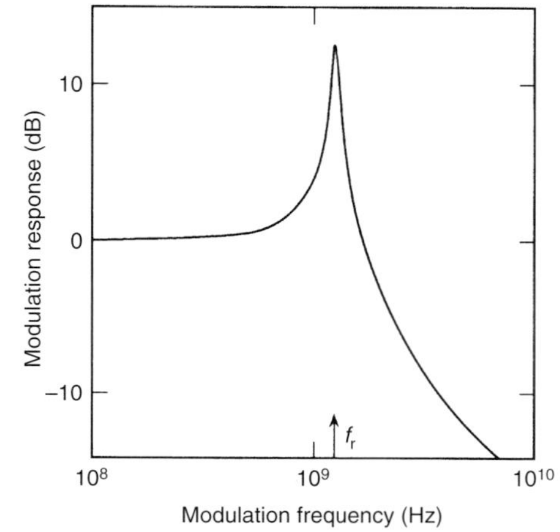


**Fig. 11** Calculated small-signal modulation response. $f_r$ is chosen to be 1.3 GHz

$P$ and typically takes a value of 3 to 5 GHz at $P = 10$ mW. In practice, however, the modulation bandwidth may be limited by the electrical parasitics associated with a specific device structure that leads to a rolloff in the modulation response. The cutoff frequency due to the parasitic rolloff varies 3 to 30 GHz depending on the device structure.

## 5
## New Structures and Functions

### 5.1
### Quantum-well Lasers

Quantum-well (QW) lasers have an active region composed of ultrathin layers (~20 nm thick or less) forming narrow potential wells for injected carriers. Illustrated in Fig. 12 are the two representative examples of the active-region structures in AlGaAs QW lasers. Lasers comprising multiple wells (Fig. 12a) are termed *multiple-quantum-well* (MQW) lasers [27, 28]. Shown in Fig. 12(b) is the active region of a GRIN (graded index)–SCH (separate confinement heterostructure)–SQW (single-quantum-well) laser [27, 28]. In the latter structure, the injected carriers are confined in the SQW, while the laser light is guided by the GRIN waveguide, thus ensuring an effective interaction between carriers and light without resort to multiple-well structure.

The quantization of electronic states in the well gives rise to a variety of superior laser characteristics over the conventional lasers with bulk active region. A major benefit is the reduction of the threshold current. Figure 13 helps us understand the low-threshold characteristics of QW lasers, where the energy distributions of the density of states and the injected carriers are compared between a bulk material and a QW structure. As the parabolic density of states in the bulk material changes into the staircase density of states in the QW structure, the energy distribution of the injected carriers narrows. The narrower energy distribution of the injected carriers leads to a narrower gain spectrum, with a higher peak gain value for a given carrier

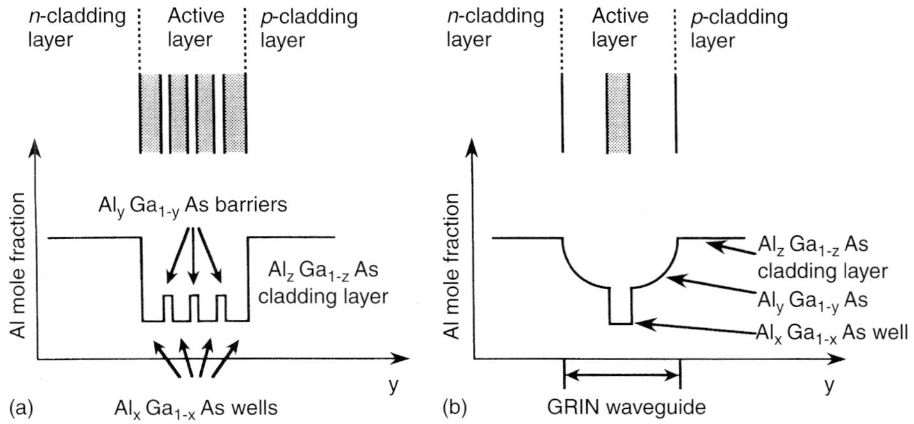


**Fig. 12** Two representative examples of active-region structures in AlGaAs quantum-well (QW) lasers (a) Multiple-quantum well (MQW) laser; (b) Graded index (GRIN)–separate confinement heterostructure (SCH)–single-quantum-well (SQW) laser

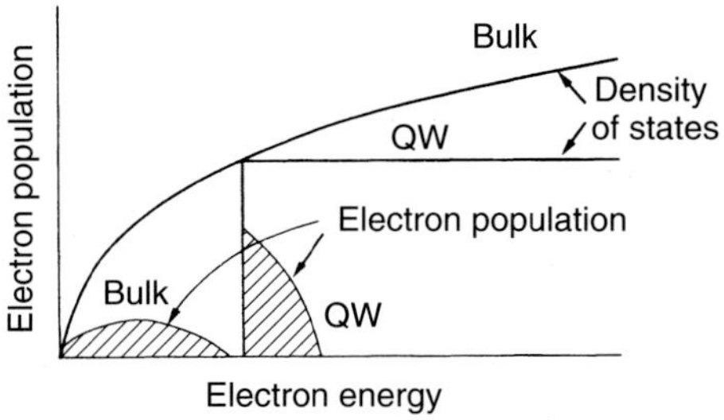


**Fig. 13** Comparison of energy distribution of injected carriers between bulk and quantum-well active regions

density. Because of the staircase density of states, however, optical gain in QW lasers saturates for higher injection levels. Since the saturated gain value depends on the number of QW layers, SQW is preferable for high-Q cavities while MQW structure should be employed for lossy cavities. A threshold current as low as 0.55 mA has been demonstrated in a high-reflectivity coated ($R \sim 0.8$) GRIN-SCH-SQW laser [29]. The narrowing of the gain spectrum in QW lasers as compared to bulk lasers is accompanied by an enhancement of $\partial g/\partial n$ (nearly doubled) and a reduction of $\alpha$ (see Eq. (21)). This leads to an enhancement of modulation bandwidths (see Eq. (25)) and a reduction of linewidths (see Eqs. (20) and (21)) under both DC and pulsed excitation [30].

The DH structure is generally formed by successive epitaxial growth of materials lattice-matched to the substrate. The lattice-matching condition imposes severe limitations on the combination of DH structure materials. However, the limitation is alleviated to a certain extent if the grown layer is thin enough to be elastically strained without any generation of dislocations. Applying this principle to the active region, we can obtain lasers whose emission wavelengths cannot be realized by lattice-matched systems. In QW lasers comprising a strained $Ga_xIn_{1-x}As$ well and AlGaAs cladding layers grown on a GaAs substrate, laser emission at 0.9 to 1.1μm can be realized by varying the composition $x$ and the well width. This wavelength region corresponds to the gap between AlGaAs and GaInPAs laser wavelengths and is fit for such applications as the excitation of Er-doped fibers and blue–green light generation by the second-harmonic generation technique.

Furthermore, it is predicted theoretically that the strain is beneficial in obtaining such good laser properties as low-threshold currents, high-relaxation oscillation frequencies, and low linewidth-enhancement factors [31, 32]. This is related to the strain-induced alleviation of the asymmetry in the density of states between the conduction and valence bands. The strain effects on high-speed modulation characteristics in GaInPAs lasers are now being studied experimentally.

## 5.2
## Distributed-feedback and Distributed Bragg-reflector Lasers

In place of mirror facets in Fabry–Pérot cavities, periodic gratings incorporated within laser waveguides can be utilized as a means of optical feedback. Integrated optical feedback from the periodic grating provides strong wavelength selectivity. Devices incorporating the grating in the pumped region are termed *distributed-feedback* (DFB) lasers (Fig. 14a), while those incorporating the grating in the passive region are termed *distributed Bragg-reflector* (DBR) lasers (Fig. 14b) [33]. By virtue of the strong wavelength selectivity, DFB and DBR lasers oscillate in a single-longitudinal mode even under high-speed modulation, in contrast to Fabry–Pérot lasers, which

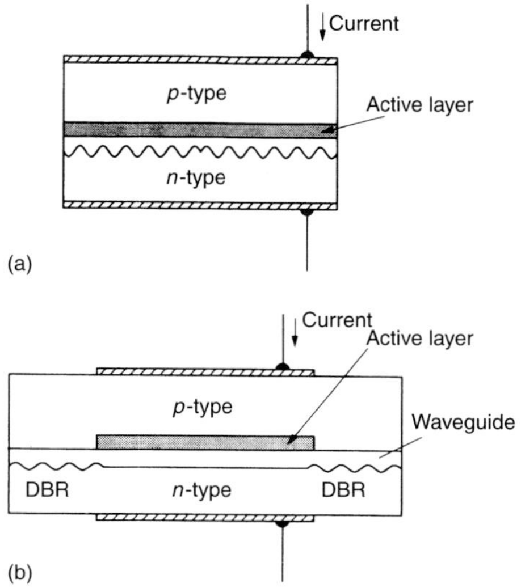


**Fig. 14** (a) Distributed-feedback (DFB); (b) distributed Bragg-reflector (DBR) lasers

exhibit multiple–longitudinal-mode oscillation when pulsed rapidly. This is an advantageous feature for optical data transmission using fibers because the spectral width determines the maximum bit rate transmitted in the presence of fiber-chromatic dispersion.

The carrier-density dependence of the refractive index can be exploited to provide wavelength tunability to the periodic gratings [34]. Illustrated in Fig. 15 is a tunable

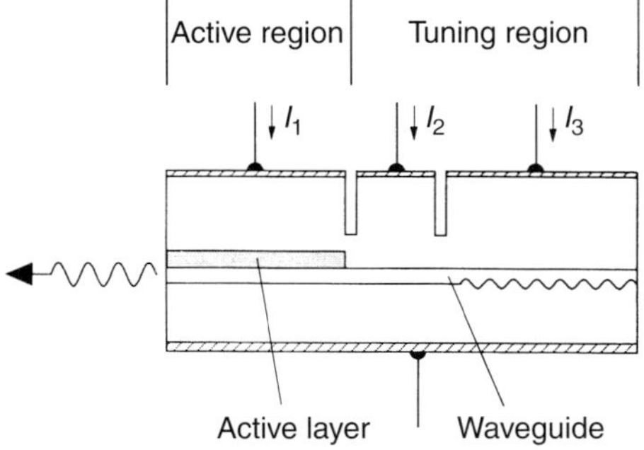


**Fig. 15** Wavelength-tunable distributed Bragg-reflector (DBR) laser

laser utilizing this principle, which can be tuned continuously over a wavelength range of several nanometers. Here, the Bragg wavelength of the DBR structure, at which the reflection loss is minimized, is tuned by the injection current $I_3$. The current $I_2$ is adjusted so that one of the resonant wavelengths of the cavity is brought into coincidence to the Bragg wavelength. Thus, by choosing appropriate pairs of $I_2$ and $I_3$, the oscillation wavelength can be varied continuously without mode jumping. Tunable lasers are expected to be used in future wavelength-division multiplexing communications systems.

## 5.3
## Semiconductor-laser Arrays

The major drawbacks of semiconductor lasers, the relatively low output powers, and the conspicuous beam divergence can be alleviated by integrating multiple laser stripes into an array as shown in Fig. 16. The stripes are closely spaced so that the radiation from neighboring stripes is coupled to form coherent modes of the entire array. The array modes, often referred to as supermodes, are phase-locked combinations of the individual stripe modes and are characterized by the phase relationship between the optical fields supported by adjacent stripes. If an array is properly designed so that

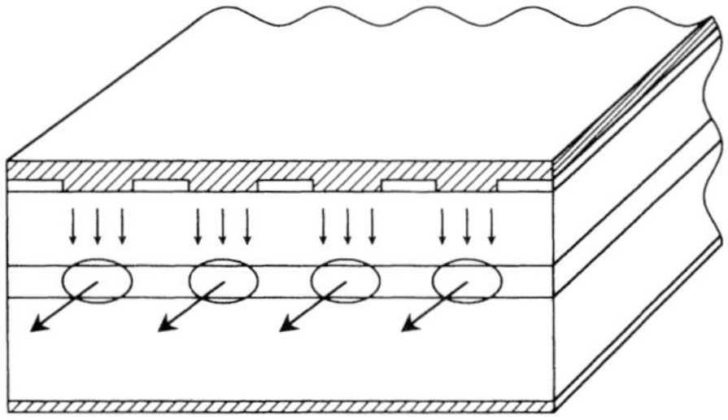

**Fig. 16** Semiconductor-laser array

one particular supermode, which is a uniphase superposition of the individual stripe modes (so called $0^\circ$ shift mode), is excited, a single-lobed, low-divergence beam is attained; the divergence angle $\theta_\parallel$ is approximated by the relation $\theta_\parallel \sim \lambda/Ns$, where s is the center-to-center separation of adjacent stripes and $N$ is the number of stripes.

A serious problem in the performance of laser arrays is their liability to multisupermode oscillation; the difference among the modal gains of the individual supermodes is so small that the lateral-mode behavior of an array is susceptible to spatial hole burning and temperature distribution in the lateral direction induced at high-excitation levels. The multisupermode oscillation is generally accompanied by a multilobed output beam, which greatly reduces the utility of arrays for most potential applications. The maximum output power for the single-lobed beam operation is typically 300 to 500 mW.

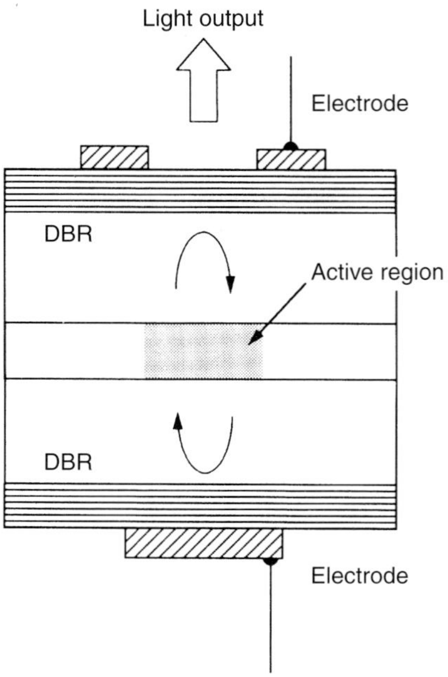


**Fig. 17** Vertical-cavity Surface-emitting Laser

## 5.4 Surface-emitting Lasers

Conventional semiconductor lasers utilizing cleaved end facets as a means of optical feedback are not fit for monolithic integration. For monolithic integration, it is desirable to have laser output normal to the wafer surface. Illustrated in Fig. 17 is the basic configuration of a surface-emitting laser incorporating two mirrors parallel to the surface to form a vertical cavity typically 5 to 10 μm long [35]. It is crucially important to increase the reflectivity of mirrors, since the reflection loss otherwise becomes very high because of the very short gain region. Bragg reflectors composed of multilayered semiconductors or dielectrics can be exploited to provide reflectivities exceeding 95%. Carrier injection is performed through a pair of electrodes in the upper and lower surfaces to excite a cylindrical active region typically of 10-μm diameter, 1 to 3 μm thick.

The surface-emitting laser has several advantages over the conventional edge-emitting laser, such as the capability of being integrated on a wafer to form a two-dimensional densely packed laser array, the feasibility of stable single–longitudinal-mode operation because of the large mode spacing due to the short cavity length, and the attainability of a narrow circular output beam.

## 6 Summary

We have reviewed the history and the state of the art of semiconductor lasers. We note here that their successful development is the result of contributions of scientists and engineers from a number of fields, including semiconductor physics, quantum electronics, crystal growth, processing technologies, and system applications, as is most notably represented by the application to the fiber-optic communications systems.

Ongoing research and development activities indicate that the same will be true in the future. There is growing interest in improving laser characteristics by miniaturizing device structures. The natural extension of the QW lasers will be quantum wire and box lasers, where we can take advantage of the two- and three-dimensional quantization of electronic systems. Researchers in quantum optics are interested in the physics of microcavity semiconductor lasers, which have cavity lengths comparable to light wavelength. Such a microcavity will drastically alter the nature of spontaneous emission and can lead to ultralow-threshold (~1 μA or less), low-noise, and high-speed lasers. In order to convert the microcavity lasers as well as the quantum wire and box lasers from a laboratory curiosity to a practical light source, further progress in growth and processing technology of semiconductor microstructures is indispensable. Integration of these superior devices will find new application fields such as optical interconnection in large-scale-integration circuits and two-dimensional information processing.

## Glossary

**Active Layer:** A layer, typically ~0.1 μm thick in which stimulated emission occurs.

**Cladding Layers:** Layers that sandwich an active layer made of materials that have wider band gaps than the active layer, $p$ and $n$ doped to facilitate carrier injection into the active layer.

**Distributed Bragg Reflector Laser:** A laser incorporating a periodic grating at both ends of the cavity as a means of optical feedback.

**Distributed-feedback Laser:** A laser incorporating a periodic grating in the vicinity of the active layer as a means of optical feedback.

**Double Heterojunction Structure:** An active-region structure in which an active layer of one material is sandwiched between two cladding layers of another material.

**Gain-guided Laser:** A laser in which the optical field distribution in the transverse direction is determined by the gain profile produced by carrier density distribution.

**Index-guided Laser:** A laser in which the optical field distribution in the transverse direction is determined by a built-in refractive-index profile.

**Linewidth-enhancement Factor:** The ratio between the carrier-induced variations of the real and imaginary parts of susceptibility $\chi(n)$, i.e.,

$$\alpha = \frac{\partial[\mathrm{Re}\chi(n)]/\partial n}{\partial[\mathrm{Im}\chi(n)]/\partial n}, \tag{29}$$

where $n$ is the carrier density. An equivalent expression is given in Eq. (21).

**Phase-locked Laser Array:** An array of stripe lasers spaced closely so that a phase-locked oscillation among the laser stripes is attained.

**Quantum-well Laser:** A laser that has ultra-thin active layers forming narrow potential wells for injected carriers and giving rise to a size quantization of the electronic states.

**Stripe-geometry Laser:** A laser in which the current is injected only within a narrow region beneath a stripe contact several μm wide.

**Surface-emitting Laser:** A laser that emits light normal to the wafer surface.

## References


[1] Hall, R. N., Fenner, G. E., Kingsley, J. D., Soltys, T. J., Carlson, R. O. (**1962**), *Phys. Rev. Lett.* **9**, 366–368.

[2] Nathan, M. I., Dumke, W. P., Burns, G., Dill Jr, F. H., Lasher, G. (**1962**), *Appl. Phys. Lett.* **1**, 62–64.

[3] Quist, T. M., Rediker, R. H., Keyes, R. J., Krag, W. E., Lax, B., McWhorter, A. L., Zeigler, H. J. (**1962**), *Appl. Phys. Lett.* **1**, 91–92.

[4] Alferov, Zh. I., Andreev, V. M., Garbuzov, D. Z., Zhilyaev, Y. V., Morozov, E. P., Portnoi, E. L., Trofim, V. G. (**1970**), *Fiz. Tekh. Poluprovodn.* **4**, 1826–1829. [English Translation: (**1971**), *Sov. Phys. Semicond.* **4**, 1573–1575.]

[5] Hayashi, I., Panish, M. B., Foy, P. W., Sumski, S. (**1970**), *Appl. Phys. Lett.* **17**, 109–111.

[6] Casey Jr, H. C., Panish, M. B. (1978), *Heterostructure Lasers.* Parts A and B, New York: Academic Press.

[7] Lang, R. (**1979**), *IEEE J. Quantum Electron.* **QE-15**, 718–726.

[8] Kogelnik, H., Shank, C. V. (**1971**), *Appl. Phys. Lett.* **18**, 152–154.

[9] Arakawa, Y., Yariv, A. (**1986**), *IEEE J. Quantum Electron* **QE-22**, 1887–1899.

[10] Haase, M. A., Qiu, J., Depuydt, J. M., Cheng, H. (**1991**), *Appl. Phys. Lett.* **59**, 1272–1274.

[11] Nakamura, S., Senoh, M., Nagahama, S., Iwasa, N., Yamada, T., Matsusita, T., Kiyoku, H., Sugimoto, Y. (**1996**), *Jpn. J. Appl. Phys.* **35**, L74–L76.

[12] Akasaki, I., Sota, S., Sakai, H., Tanaka, T., Koike, M., Amano, H. (**1996**), *Electron. Lett.* **32**, 1105–1106.

[13] Kroemer, H. (**1986**), *Surf. Sci.* **174**, 299–306.

[14] Botez, D. (**1978**), *IEEE J. Quantum Electron.* **QE-14**, 230–232.

[15] Yamamoto, Y., Saita, S., Mukai, T. (**1983**), *IEEE J. Quantum Electron.* **QE-19**, 47–58.

[16] Yamada, M. (**1983**), *IEEE J. Quantum Electron.* **QE-19**, 1365–1380.

[17] Ogasawara, N., Ito, R. (**1988**), *Jpn. J. Appl. Phys.* **27**, 607–626.

[18] Dyment, J. C. (**1967**), *Appl. Phys. Lett.* **10**, 84–86.

[19] Tsukada, T. (**1974**), *J. Appl. Phys.* **45**, 4899–4906.

[20] Horikoshi, Y. (1985), in W. T. Tsang (Ed.), *Semiconductors and Semimetals,* Vol. 22. Part C, Orlando, FL: Academic Press, Chap. 3.

[21] Ohkawa, K., Karasawa, T., Mitsuya, T. (**1991**), *Jpn. J. Appl. Phys.* **30**, L152–L155.

[22] Ikegami, T. (**1972**), *IEEE J. Quantum Electron.* **QE-8**, 470–476.

[23] Henry, C. H. (**1982**), *IEEE J. Quantum Electron.* **QE-18**, 259–264.

[24] Osinski, M., Buus, J. (**1987**), *IEEE J. Quantum Electron.* **QE-23**, 9–29.

[25] Chinone, N., Ito, R., Nakada, O. (**1974**), *IEEE J. Quantum Electron.* **QE-10**, 81–84.

[26] Yariv, A. (1989), *Quantum Electronics.* New York: Wiley, Chap. 11.

[27] Tsang, W. T. (**1981a**), *Appl. Phys. Lett.* **39**, 786–788.

[28] Tsang, W. T. (**1981b**), *Appl. Phys. Lett.* **39**, 134–137.

[29] Lau, K. Y., Derry, P. L., Yariv, A. (**1988**), *Appl. Phys. Lett.* **52**, 88–90.

[30] Uomi, K., Sasaki, S., Tsuchiya, T., Okai, M., Aoki, M., Chinone, N. (**1990**), *Electron. Lett.* **26**, 52–53.

[31] Suemune, I., Coldren, L. A., Yamanishi, M., Kan, Y. (**1988**), *Appl. Phys. Lett.* **53**, 1378–1380.

[32] Ohtoshi, T., Chinone, N. (**1989**), *IEEE Photon. Technol. Lett.* **1**, 117–119.

[33] Mroziewicz, B., Bugajski, M., Nakwaski, W. (1991), *Physics of Semiconductor Lasers.* Amsterdam: North Holland, Chap. 6.

[34] Yoshikuni, Y. (1991), in Y. Yamamoto (Ed.), *Coherence, Amplification, and Quantum Effects in Semiconductor Lasers.* New York: Wiley, Chap. 4.

[35] Iga, K., Koyama, F., Kinoshita, S. (**1988**), *IEEE J. Quantum Electron.* **QE-24**, 1845–1855.

## Further Reading

Agrawal, G. P., Dutta, N. K. (1986), *Long-Wavelength Semiconductor Lasers.* New York: Van Nostrand Reinhold.

Casey Jr, H. C., Panish, M. B. (1978), *Heterostructure Lasers,* Parts A and B, New York: Academic Press.

Kressel, H., Butler, J. K. (1977), *Semiconductor Lasers and Heterojunction LEDs.* New York: Academic Press.

Mroziewicz, B., Bugajski, M., Nakwaski, W. (1991), *Physics of Semiconductor Lasers.* Amsterdam: North Holland.

Nakamura, S., Chichibu, S. F. (2000), *Nitride Semiconductor Blue Lasers and Light Emitting Diodes.* London: Taylor & Francis.

Nakamura, S., Fasol, G., Pearton, S. J. (2000), *The Blue Laser Diode: The Complete Story.* New York: Springer.

Peterman, K. (1988), *Laser Diode Modulation and Noise.* Dordrecht: Kluwer Academic Publishers.

Thompson, G. H. B. (1980), *Physics of Semiconductor Laser Devices.* Chichester: Wiley.

Yamamoto, Y. (Ed.) (1991), *Coherence, Amplification, and Quantum Effects in Semiconductor Lasers.* New York: Wiley.

# Lasers, Solid-state

*Tso Yee Fan*
*Lincoln Laboratory, Massachusetts Institute of Technology, Lexington, Massachusetts, USA*
*Fax: 1 781 981 0602; e-mail: fan@LL.mit.edu*

**Abstract**
An overview of solid-state lasers, which are lasers using doped, dielectric solids as the gain media, is presented. The article includes information on the history, materials physics, technology, and application of these devices.

**Keywords**
solid-state lasers; fiber lasers; rare-earth-doped materials; transition-metal-doped materials.

## 1 Introduction

Solid-state lasers are a class of lasers that use an insulating dielectric solid, as opposed to a gas, liquid, semiconductor, or plasma, as the gain medium. These dielectric solids can be crystalline or amorphous, and optical gain is due to transitions between discrete electronic energy levels of dopants such as rare-earth ions or transition-metal ions with unfilled outer shells, or defect centers, also known as color centers. Solid-state lasers have been used to generate a wide range of wavelengths from the vacuum ultraviolet to the mid-infrared. These lasers are optically excited, from either broadband sources such as flashlamps or narrowband sources such as semiconductors or other lasers. Among the most common examples of solid-state lasers are ruby, which is sapphire doped with chromium: titanium-doped sapphire ($Ti:Al_2O_3$); neodymium-doped yttrium aluminum garnet (Nd:YAG); neodymium-doped glasses (Nd:glass); and erbium-doped glasses.

Solid-state lasers have a set of unique properties that set them apart from other lasers. Because they are solid, they are relatively easy to handle, are more inert, and are less susceptible to contamination compared with gas and liquid lasers. They are capable of generating high peak powers by Q-switching owing to relatively long metastable lifetimes, which allow higher energy storage compared with other classes of gain media; solid-state laser systems with $\sim 10^{15}$ W peak power have been demonstrated. They are also capable of high average power, up to ~1 kW, while maintaining nearly ideal output beams. Many lasers doped with transition-metal ions and color-center lasers are capable of broadband tunability, up to 20% of the center frequency, which means solid-state lasers are a powerful tool in optical spectroscopy. The broad gain bandwidth also allows short pulses to be generated using mode-locking and pulse-compression schemes; pulses of <5 fs in length have been generated. High Q-laser cavities can be built because solid-state laser materials have low optical loss. This leads to the ability to attain narrow linewidth laser operation; subhertz laser linewidths have been demonstrated (see LASER PHYSICS). Rare-earth-doped glasses can be drawn into optical fibers and can be used as amplifiers in fiber-optic communications systems.

However, solid-state lasers also have significant limitations. They are relatively inefficient; many commercial solid-state lasers have an overall power efficiency of ~1% or lower. There are also difficulties with the pump sources. Typically, lamps have lifetimes of only hundreds of hours when operated continuously before they need to be replaced. In many wavelength regions, solid-state lasers are either nonexistent or perform poorly. Finally, there are significant engineering difficulties in designing and building solid-state lasers with high average power, greater than a few hundred watts, due to heat dissipation in the gain medium. However, there is active research ongoing to overcome all these difficulties. Some of the limitations are beginning to be addressed by using semiconductor diode lasers as pump sources for solid-state lasers. Solid-state lasers pumped by semiconductor diode lasers have been demonstrated in the laboratory with nearly 20% overall power efficiency, and they have lifetimes of tens of thousands of hours. Active research is also underway to develop new materials for solid-state laser gain media that offer the possibility of increased performance and new wavelengths.

## 2
## History

Solid-state lasers have been in existence longer than any other type of laser by virtue of Theodore Maiman's demonstration of a ruby laser in July 1960 at the Hughes Research Laboratory [1]. In Maiman's experiment, ruby rods of about 10 mm diameter by 19 mm long were polished flat and parallel, and the ends were coated with evaporated silver that served as the resonator mirrors. The rod was then placed in the middle of a pulsed helical flashlamp, which was in turn inserted into a reflective aluminum cylinder as shown in Fig. 1. Laser oscillation was observed on the three-level laser transition at 694.3 nm.

Following Maiman's announcement, lasers in other solid-state materials were

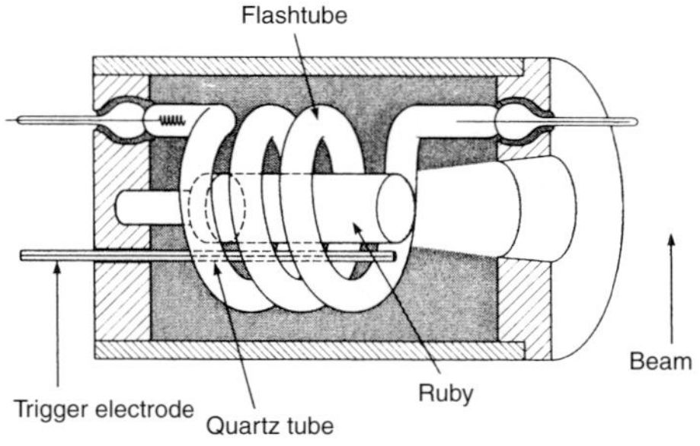


**Fig. 1** Schematic diagram of the ruby laser demonstrated by Maiman. The size of the device was about one inch in diameter and two inches in length (adapted from Maiman, T. H., Hoskins, R. H., D'Haenens, J., Asawa, C. K., Evtuhov, V. (**1961**), *Phys. Rev.* **123**, 1151–1157 [2])

also demonstrated. A low-threshold four-level laser was demonstrated in uranium-doped $CaF_2$ at 2.5 μm [3]. This laser was operated at liquid-helium temperature because the lower level of this transition in $U^{3+}$ is located only ~500 cm$^{-1}$ above the ground state; at room temperature the laser is three level, but at low temperature the lower laser level is depopulated. The first neodymium-doped laser was demonstrated in $CaWO_4$ in 1961; this allowed low-threshold room temperature operation owing to its four-level laser energy-level diagram at room temperature. Shortly thereafter, the first continuous-wave solid-state laser operation was achieved in Nd : $CaWO_4$ and ruby. What followed was an explosion in the number of host materials and activator ions used for solid-state lasers with rare-earth, actinide, and transition-metal ions as dopants. One of the most important solid-state laser materials, Nd-doped yttrium aluminum garnet ($Y_3Al_5O_{12}$ or YAG) came out of this early work [4]. The first broadly tunable solid-state lasers, $Ni^{2+}$ and $Co^{2+}$ in $MgF_2$, were also demonstrated in this period [5]. They operated at liquid nitrogen temperature and were tuned with intracavity prisms.

The first glass lasers were Nd-doped fibers of ~32 μm in diameter [6], and thus these glass lasers were also the first fiber lasers. The capability of Nd-doped glasses to be used in high energy per pulse laser systems was also demonstrated in a 100 J/pulse laser; subsequent engineering of high energy per pulse Nd : glass laser systems has lead to the development of a 100 kJ/pulse laser system, the Nova laser, for laser fusion experiments.

In the late 1960s and early 1970s, development of solid-state laser technology seemed to stagnate. The only practical solid-state lasers were ruby, Nd : YAG, and Nd : glass. While many other solid-state lasers had been demonstrated, these devices remained primarily laboratory curiosities, either due to cryogenic temperature requirements, high threshold, or low efficiency. However, three developments have helped revive solid-state laser technology. The first was the demonstration of the alexandrite ($Cr^{3+}$-doped $BeAl_2O_4$) laser in which efficient, broadband tunable performance (~700–820 nm) was obtained [7]. This set off a search for other materials doped with transition-metal ions in which broadband tunability with efficient laser operation could be achieved. The second development was improvements in the lifetime and efficiency of semiconductor diode lasers, in particular, AlGaAs diode lasers, that allowed them to be used as practical pump sources. Higher efficiency, longer lifetime, and more compact solid-state lasers have been built using these diode lasers. The concept of diode pumping was not new; the use of a semiconductor diode laser as a pump source was initially demonstrated in 1964 [8], but at the time, diode lasers worked only at cryogenic temperatures. The third development was the practical demonstration of erbium-doped fiber amplifiers (EDFAs) for use in optical communications systems. The primary gain band in Er-doped silica glass at 1.55 μm overlaps the minimum of optical loss in fused-silica optical fiber, and consequently the use of EDFAs in communications systems became widespread in the 1990s [9].

## 3
## Properties of Gain Media and Pump Sources

The output wavelengths of solid-state lasers and the appropriate pump sources are determined by the positions of the energy levels of the activator ion or of the color center in a host material. Solid-state lasers have been demonstrated using many different activator ions in many host materials; however, only a few of these activator ion–host material combinations have been found to be practical. The viability of any particular solid-state gain medium is dependent on many factors. For example, efficient, low-threshold operation is generally desired, which means that efficient pumping of the laser is needed. This requirement means that there must be optical absorption in wavelength regions where practical pump sources exist.

## 3.1
## Gain Media

Gain media for solid-state lasers can basically be broken into three categories, which are materials containing rare-earth ions, transition-metal ions, and color centers [10, 11]. Each of these classes of gain media has its own characteristics. Rare-earth-ion lasers have relatively narrow gain bandwidth and long (>50 μs) metastable-level lifetimes. In addition, the energies of the electronic states of a particular rare-earth ion change only slightly when doped into different host materials. Transition-metal-ion lasers can have relatively broad gain bandwidth and intermediate to long (>1 μs) metastable-level lifetimes, and the energy-level positions can be changed significantly by doping into different host materials. Finally, color-center lasers are characterized by relatively broad tunability and short metastable-level lifetimes (<1 μs). The energy-level positions of these defect centers are strongly host-dependent. Color-center lasers have largely fallen out of favor because of instability in many color centers over time.

Host materials are chosen on the basis of many factors. First, the host material interacts with the dopant ion, and this interaction determines the positions of the energy levels and the radiative and nonradiative transition strengths between energy levels. The host material must be able to incorporate the dopant ion or defect centers at sufficiently high concentrations, and good chemical stability of the dopant ion or defect center in the host is needed. Another factor is fabrication of the gain medium; it must be growable with high optical quality and sufficient size at a reasonable cost. For ease of handling, the material should be hard, have high strength, be chemically stable, and be non-hydroscopic. The thermo-optic properties of the host material are also important. Heat is dissipated in the gain medium owing to nonradiative relaxation; thus high thermal conductivity is desired to allow efficient heat extraction from the gain medium, and low thermal expansion coefficient and change of refractive index with temperature are needed to reduce thermo-optic aberrations. Improved materials continue to be an active area of research.

### 3.1.1 Gain Media Doped with Rare-earth Ions

The largest and most important class of solid-state lasers are materials doped with rare-earth ions. Most of the rare earths are stable in the trivalent state when doped into solids, which leaves an outer shell of 4f electrons. The energy-level diagrams of 4f multiplets of the trivalent rare-earth ions are shown in Fig. 2. Each multiplet contains several energy levels that are degenerate if the ion is in free space, and the energies of these multiplets are

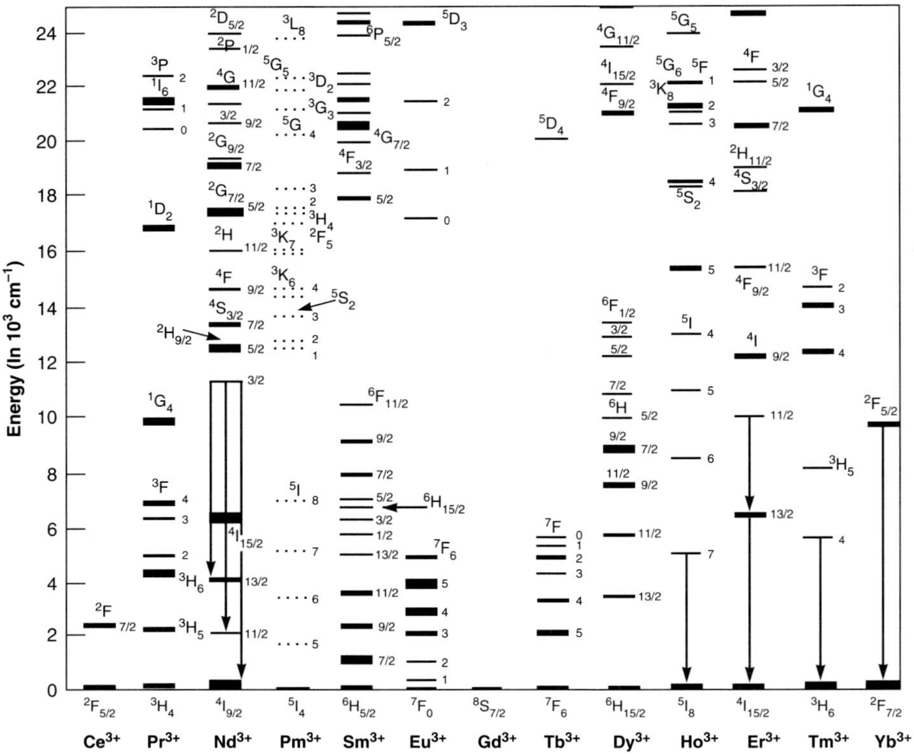

**Fig. 2** Energy levels of the trivalent rare-earth ions in $LaCl_3$, from [12]. The positions of the multiplets of the trivalent rare earths vary only slightly as the host material is changed. $Pm^{3+}$ is shown in dotted lines because it is radioactive. While many transitions have been used for solid-state lasers, only the most important are shown with the arrows. Only a limited number of these multiplets can be upper laser levels; only those with a gap in energy to the next lowest-lying multiplet of $\sim$3000 $cm^{-1}$ or more are metastable

determined primarily by the Coulomb interaction between the 4f shell electrons and spin-orbit coupling. The effect of the host lattice, also known as the crystal field, is small because the 4f shell is well shielded and splits the energy levels within a multiplet by only hundreds of wavenumbers.

Wavelengths that have been generated in the various rare-earth ions are shown in Fig. 3. The ability to use a particular transition is determined by its spectroscopic properties in the host material. First, the host material must be transparent at the wavelength of the transition and the pump wavelengths. Most of the hosts used for solid-state lasers are oxides or fluorides; thus, the wavelength range is limited by host transparency to about 120 nm to 6 μm. Second, the upper level must be metastable. Electric-dipole transitions between multiplets in rare-earth ions are only weakly allowed, which leads to long radiative lifetimes (>50 μs); however, there are also nonradiative relaxation mechanisms. The nonradiative relaxation rates are dependent on the host material and the energy gap between the multiplet

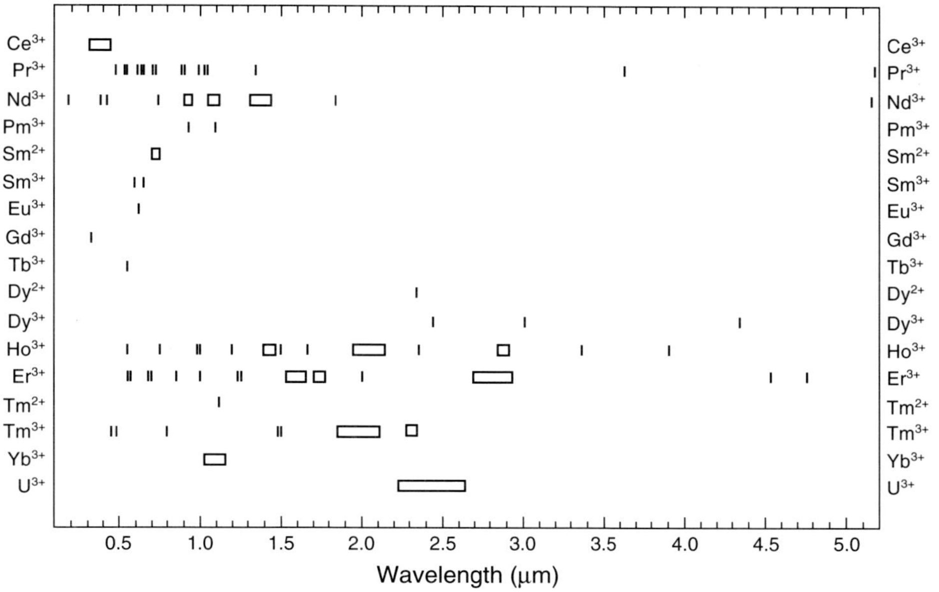


**Fig. 3** Wavelengths demonstrated in rare-earth- and actinide-ion solid-state lasers

in question and the next lower-lying multiplet. High nonradiative relaxation rates occur for small energy gaps and host materials with high characteristic phonon energies.

Only a few of these transitions are used in commercial devices, which are denoted in Fig. 2. These are the $^4F_{3/2}-^4F_{9/2}$ transition around 0.94 μm, $^4F_{3/2}-^4I_{11/2}$ transition around 1.06 μm, and the $^4F_{3/2}-^4I_{13/2}$ transition around 1.32 μm in $Nd^{3+}$; the $^4I_{13/2}-^4I_{15/2}$ transition around 1.5 μm and the $^4I_{11/2}-^4I_{13/2}$ transition around 2.8 μm in $Er^{3+}$; the $^5I_7-^5I_8$ transition around 2.1 μm in $Ho^{3+}$; and the $^2F_{5/2}-^2F_{7/2}$ transition around 1.03 μm in $Yb^{3+}$. These transitions are characterized by either low threshold and efficient operation or by output wavelengths that have important applications. The other laser transitions are for the most part limited by inefficient or currently impractical (for commercially viable products) pump schemes.

Many host materials have been used for rare-earth-ion dopants, but once again the number of host materials used in commercial products is limited. The most important set of host materials are the garnets such as YAG and $Gd_3Sc_2Ga_3O_{12}$ (gadolium scandium gallium garnet or GSGG). The garnets have good mechanical properties, high thermal conductivity, well-understood crystal growth, and substitutional sites for rare-earth ions. Some other important host materials are $LiYF_4$ (YLF), $YAlO_3$ (YALO or YAP), $YVO_4$, and glasses. There are many other host materials that have been demonstrated to accommodate rare earths, and a few of these are subjects of active research to see whether they can serve as practical laser gain media.

### Nd:YAG and Other Nd-doped Materials

The most widely used bulk solid-state laser material is Nd:YAG. YAG has a

**Tab. 1** Properties of Nd : YAG, Nd : glass, and Nd : YLF

| ***Property*** | ***Nd : YAG*** | ***Nd : glass (LG-760)*** | ***Nd : YLF*** |
|---|---|---|---|
| Peak stimulated-emission cross section ($10^{-19}$ cm$^2$) | 3 | 0.4 | 1.8 |
| $^4F_{3/2}$ lifetime (μs) | 230 | 350 | 480 |
| Thermal conductivity at 300 K (W cm$^{-1}$ K$^{-1}$) | 0.11 | 0.006 | 0.06 |
| Thermal expansion at 300 K ($10^{-6}$ K$^{-1}$) | 6 | 12.5 | 8 (*c* axis)<br>13 (*a* axis) |
| Refractive index *n* at 1.06 μm | 1.815 | 1.508 | 1.4704 ($n_e$)<br>1.4481 ($n_o$) |
| d*n*/d*T* ($10^{-6}$ K$^{-1}$) | 7.6 | −6.8 | −4.3 ($n_e$)<br>−2 ($n_o$) |
| Hardness (Knoop) (kg mm$^{-2}$) | 1350 | 300 | 300 |
| Crystal system | Cubic | Amorphous | Tetragonal |

combination of attractive properties for a laser host including high thermal conductivity, hardness, and strength. Some of the properties of Nd : YAG are listed in Table 1. The $Nd^{3+}$ ion substitutes for the $Y^{3+}$ in this material, and all sites are equivalent, which means the material is homogeneously broadened. The spectroscopic properties of the $Nd^{3+}$ ion, in particular, the transition at 1.064 μm, are favorable; the gain cross section is large $\sim 3 \times 10^{-19}$ cm$^2$, the upper-state lifetime is long, and low laser threshold can be obtained because this laser transition is four level. There are also absorption bands in the near infrared and visible portions of the spectrum so that the laser can be pumped efficiently by either AlGaAs diode lasers in the 0.81-μm wavelength range or lamps. The absorption spectrum is shown in Fig. 4 for nominal 1.1 at.% doping of Nd. After the Nd ion is excited in these absorption bands, the ion then relaxes nonradiatively to the $^4F_{3/2}$ manifold, which is a metastable level; the fluorescence spectrum from this state in the 1.06-μm and 1.32-μm bands is shown in

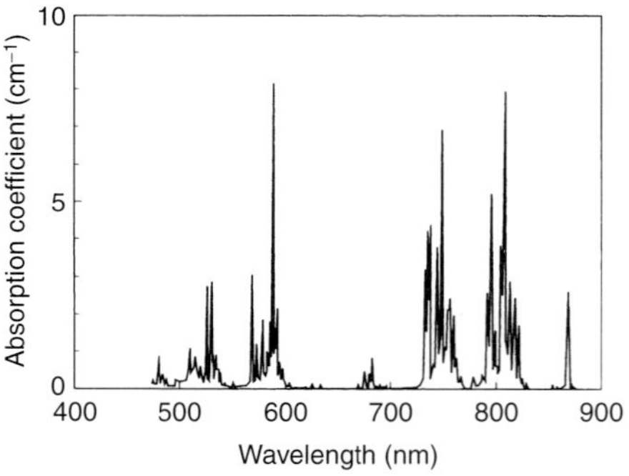


**Fig. 4** Absorption spectrum in nominally 1.1 at.% Nd-doped YAG

Fig. 5. The spectrum in Nd : YAG around 1.06 μm is composed of 12 lines (two of the lines overlap), which are transitions from 2 crystal-field splittings in the upper manifold to 6 splittings in the lower manifold. Nd : YAG lasers have been tuned to most of the fluorescence lines shown in the spectrum.

Nd : YAG crystal boules are grown by the Czochralski technique up to ~7 cm diameter and 20 cm long. The doping level of Nd is limited to ~3 at.% because of a mismatch in size of the $Nd^{3+}$

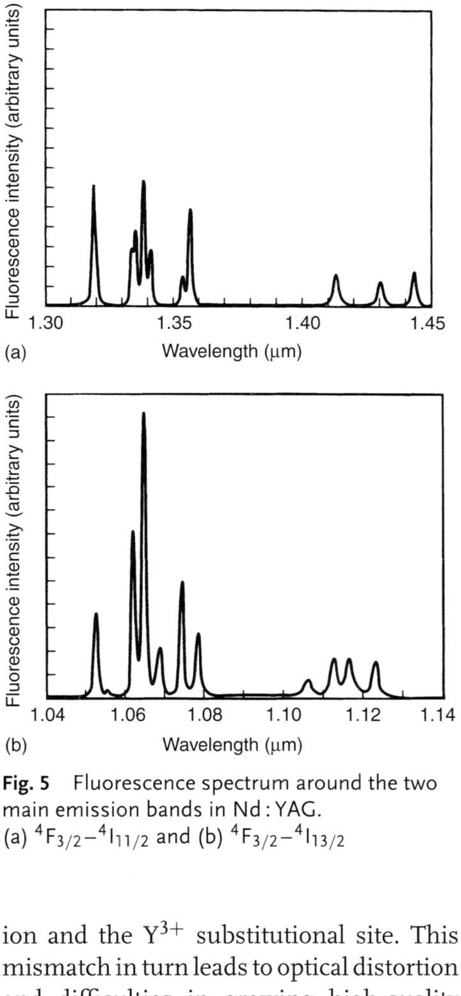


**Fig. 5** Fluorescence spectrum around the two main emission bands in Nd : YAG. (a) $^4F_{3/2}$–$^4I_{11/2}$ and (b) $^4F_{3/2}$–$^4I_{13/2}$

ion and the $Y^{3+}$ substitutional site. This mismatch in turn leads to optical distortion and difficulties in growing high-quality crystals, and in practice, Nd : YAG crystals are grown with 1.1 at.% doping or less to keep these effects within reasonable bounds. Also, as the concentration is increased beyond about 1 at.%, the upper-state lifetime begins to decrease because of interactions between neighboring $Nd^{3+}$ ions. This effect is known as *concentration quenching* and is another limiting factor in the concentration of $Nd^{3+}$ in YAG. Not all of the cross-sectional area of a boule is useful for fabrication of gain elements because Nd : YAG grows with facets at the liquid–solid interface. At the intersections of these facets and in the center of the boule, there is a higher incorporation of Nd and also stress that leads to optical distortion. Gain elements are cut from boules with the optical axis parallel to the growth axis, and rods are limited in size to ~15 mm diameter by faceting.

Other important $Nd^{3+}$-doped crystals include Nd : YLF and Nd : $YVO_4$, which are both inherently birefringent. Nd : YLF can be grown to good optical quality and has the advantages relative to Nd : YAG of much lower change in refractive index with temperature, a two-times longer upper-state lifetime, and a somewhat larger linewidth. This material has found applications in lasers where small thermally induced optical distortion is required, in continuously pumped Q-switched lasers because larger energy per pulse can be obtained because of the longer lifetime, and in mode-locked lasers where shorter pulses than Nd : YAG lasers can be generated. The main disadvantage relative to Nd : YAG is poorer thermo-mechanical properties such as higher thermal expansion, reduced hardness, lower thermal conductivity, and lower yield stress; consequently, Nd : YLF is not attractive for high average-power laser systems. Nd : $YVO_4$ has been of particular interest because its optical absorption band around 0.81 μm is relatively strong compared with Nd : YAG, and therefore it is favorable for pumping with diode lasers. It also has a higher gain cross section that enables shorter Q-switched pulses. $YVO_4$ is difficult to grow in large sizes, and some of its thermo-optic properties are inferior to YAG.

Nd-doped glasses are an important class of gain media. Glasses have several advantages as host materials. They are relatively inexpensive compared with single crystals, they can be fabricated in very large sizes, they can be optically polished to a high degree of accuracy, and their properties can be engineered to some extent by changing the glass composition. Major disadvantages relative to many single crystals are low thermal conductivity, hardness, and strength. Some of the properties of one of the more popular Nd-doped phosphate glasses are listed in Table 1.

The transitions in glasses are inhomogeneously broadened because of different local site symmetry of the neighboring ions around the activator ion. The fluorescence spectrum in the 1.06-μm wavelength region and the absorption spectrum in a Nd-doped phosphate glass are shown in Fig. 6. In contrast to the spectra of Nd : YAG, the Nd : glass spectrum is relatively featureless, and the increase in linewidth leads to reduced stimulated-emission cross section relative to most single crystals. In some applications this change is desirable; for example, in mode-locked lasers, the wider linewidth makes it possible to attain shorter pulses, and the reduced stimulated-emission cross section makes Nd : glass lasers attractive for high-energy, short-pulse length lasers because relatively high energy can be stored in the metastable level with reduced amplified spontaneous emission and parasitic oscillation. In general, glasses can be doped more heavily than crystals, and in practice, the upper bound on $Nd^{3+}$ doping is established by concentration quenching, which is weaker in glasses than in YAG. This effect limits the doping to about 8 wt.% $Nd_2O_3$ in phosphate glass.

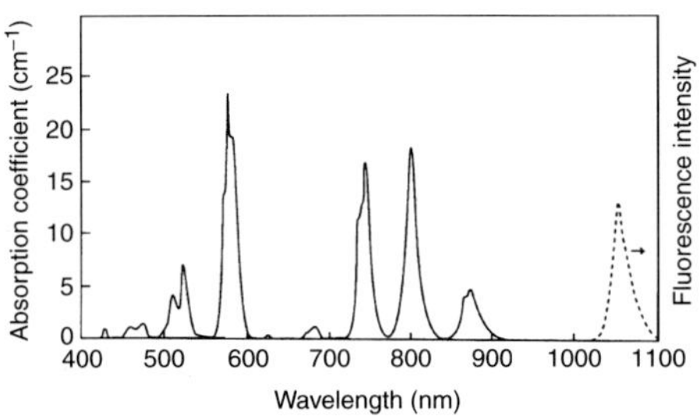


**Fig. 6** Absorption spectrum and fluorescence spectrum around 1.06 μm in Nd-doped phosphate glass. The absorption spectrum is for nominal 6 wt.% $Nd_2O_3$ doping

**Er-doped Glass** The $^4I_{13/2}$–$^4I_{15/2}$ transition in $Er^{3+}$ near the 1.5-μm wavelength range matches the window of minimum loss in fused-silica optical fiber. The ability to pump this transition using high-power diode lasers on either the $^4I_{15/2}$–$^4I_{9/2}$ transition near 980 nm or the $^4I_{15/2}$–$^4I_{13/2}$ transition near 1480 nm has enabled the development of EDFA with large gain (>30 dB) [13, 14].

Figure 7(a) shows the absorption spectrum of an EDFA in silica glass with the primary absorption bands, and Fig. 7(b) shows more details of the absorption and emission spectra near 1.5 μm. The broad, relatively featureless spectra are a result of large crystal-field splitting, a large homogeneous linewidth, and significant inhomogeneous broadening that occurs for erbium in silica glass, and these spectra change significantly as the glass composition changes. However, given the large number of crystal-field splittings and the large homogeneously broadened linewidth, the transitions can be treated as being essentially homogeneously broadened, at least at room temperature and above. The $^4I_{13/2}$ metastable-level lifetime is dependent on the glass composition, but in typical silica-based glass used in telecommunications fiber, this lifetime is on the order of 10 ms. Dopant concentrations are limited to typically $<10^{19}$ cm$^{-3}$ in silica-based glasses because of solubility limitations, upconversion effects, and concentration quenching of the metastable lifetime.

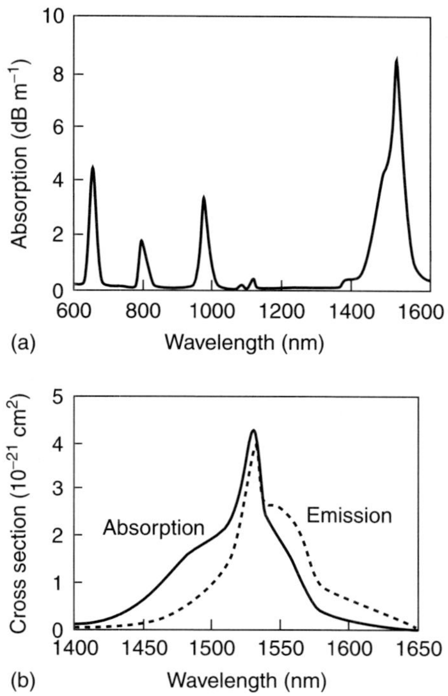


**Fig. 7** Spectra of Er-doped silica fiber (a) absorption spectrum and (b) absorption and fluorescence spectra around 1.5 μm

Bulk Er-doped glass lasers are also of interest for generating energetic pulses, particularly in applications that require a degree of eye safety since the 1.5-μm wavelength is in the relatively eye-safe portion of the spectrum. The glass compositions for these bulk lasers are generally based on phosphate glasses, and $Yb^{3+}$ is often a codopant. The role of the $Yb^{3+}$ is as a sensitizer to absorb pump radiation and then transfer the energy to the $Er^{3+}$ ion.

**Yb-doped Materials** Yb-doped crystals (primarily YAG) and Yb-doped glass optical fiber became of interest in the 1990s. The simple energy-level diagram of $Yb^{3+}$ implies that there should be no intrinsic concentration quenching nor other parasitic energy transfer effects. The simple energy-level structure also prevented the practical development of these lasers because of the lack of optical absorption bands. Only with the development of high-power diode-laser pump sources based on InGaAs in the 0.91 to 0.98-μm wavelength region did Yb-doped lasers become of interest.

The operating wavelength of these Yb-doped materials is similar to Nd-doped materials, and so it is useful to compare and contrast Yb- and Nd-doped lasers. In Yb-doped lasers, generally the difference between the pumping wavelength and operating wavelength is smaller. This difference in photon energies goes into heat dissipation in the gain medium, and therefore, Yb-doped lasers have the potential for better average-power handling than Nd-doped lasers. On the other hand, the lower laser level of the Yb gain transition is near the ground state, so at room temperature, the threshold is raised by the need to put significant population in upper level in order to reach population inversion. Thus, bulk Yb-doped lasers are typically less efficient than Nd-doped lasers.

In contrast, in Yb-doped glass optical fiber, the performance is superior to Nd-doped optical fiber. The need for significant excited-state population density is mitigated by the ability to confine and guide the pump radiation along the fiber. The lack of parasitic energy-transfer effects makes Yb-doped fibers work nearly ideally, whereas parasitic energy-transfer effects caused by the more complicated energy-level structure reduce the performance of Nd-doped optical fiber.

**Other Rare-earth-ion-doped Lasers** There are a large number of rare-earth ion–host combinations that have been demonstrated, and a few of these may become of some importance.

Er-doped lasers operating in the 2.8-μm wavelength range are becoming more important. The 2.8-μm wavelength is strongly absorbed by water and is of interest for use in medical applications. The most commonly used host materials for 2.8-μm $Er^{3+}$ lasers are the garnets, in particular, YAG and $Y_3Sc_2Al_3O_{12}$ (YSAG). These 2.8-μm $Er^{3+}$ materials are sometimes codoped with $Cr^{3+}$, which provides additional absorption bands for lamp pumping. In this case, the $Cr^{3+}$ dopant is known as a sensitizer because the energy absorbed by the sensitizer is transferred to the activator to increase the efficiency and pump density. Both $Tm^{3+}$- and $Ho^{3+}$-doped lasers have outputs near 2 μm, and these lasers are of interest for applications in which eye-safe wavelengths are required and in medical applications. These lasers have been demonstrated using YAG and YLF as host materials.

### 3.1.2 Gain Media Doped with Transition-metal Ions

Transition-metal ions are used as activators in solid-state lasers to provide broad tunability. Lasers have been demonstrated using the first-row transition-metal ions that have 3d electron outer shells with the laser transitions occurring between energy levels of the 3d configurations. The position of the energy levels in transition metals is affected by the host matrix more than in rare-earth ions, and an example of the dependance of the energy-level positions on the strength of the crystal field is shown in Fig. 8 for an outer electron shell of $d^3$ in octahedral symmetry. The crystal field is a measure of the strength of the interaction between the dopant ion and the host matrix. Varying the crystal field by using hosts with different nearest-neighbor distances, the energy-level positions, and thus the output wavelength can be changed. The linewidth of transitions can be either broadband or narrowband depending on the characteristics of the upper and lower electronic states.

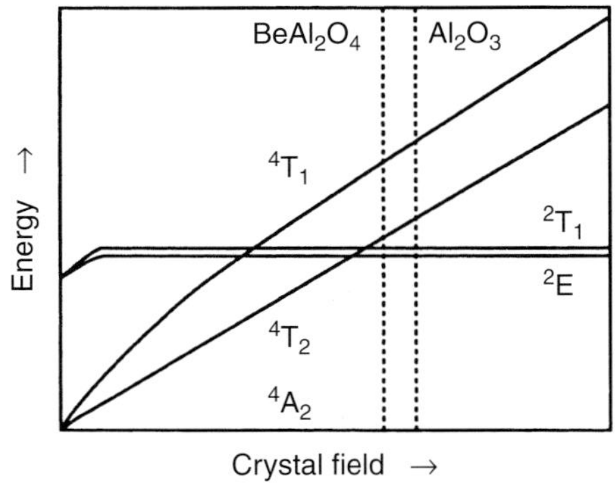


Fig. 8 Simplified crystal-field diagram for $d^3$ outer shell, which represents $Cr^{3+}$ showing only the energy levels important for laser action. The upper laser level is $^2E$ for high crystal fields, $^4T_2$ for low crystal fields and a combination of the two states for intermediate crystal fields. The vertical lines represent the crystal fields for ruby and alexandrite

The wavelength range over which laser oscillation has been demonstrated using various transition-metal ions is shown in Fig. 9. The utility of any given possible transition between two electronic

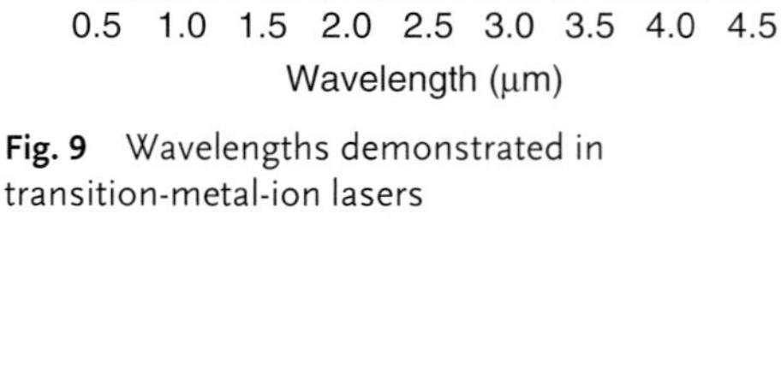


Fig. 9 Wavelengths demonstrated in transition-metal-ion lasers

states is determined by similar factors as in rare-earth ions. One of the additional complications is that the chemistry of transition-metal-ion dopants is more complex than rare earths since many transition-metal ions can exist in host materials in multiple valance states, whereas the rare earths are almost always trivalent. Also, nonradiative relaxation rates tend to be higher than that of rare earths. Finally, excited-state absorption (ESA) is a limiting factor in many solid-state lasers doped with transition-metal ions. ESA is the absorption of a photon by an ion in the upper metastable level, and this process reduces the efficiency of laser operation if the ESA occurs for either the laser or the pump wavelengths. The broadband spectra of transition-metal ions often leads to ESA at the laser wavelengths; this is a particular problem for $Ni^{2+}$-, $V^{2+}$-, and some $Cr^{3+}$-doped materials.

**Ruby, Alexandrite, and Other $Cr^{3+}$-doped Materials** $Cr^{3+}$-doped solid-state lasers are the most studied class of transition-metal-ion lasers. They have an outer shell of $3d^3$; thus the crystal-field diagram of Fig. 8 applies. $Cr^{3+}$ is of interest because its spectroscopic characteristics, in particular, its tunability, upper-state lifetime, and gain cross section, can be engineered to some degree by changing host materials. Figure 10 shows the fluorescence spectrum of ruby and alexandrite, and some of the properties of these materials are listed in Table 2. The change in spectroscopic properties with the host can be described by variations in crystal-field strength; ruby has high strength, and alexandrite has intermediate strength. In high crystal field, the $^2E$ energy level is the upper state; the transition between the $^2E$ state and $^4A_2$ lower state is characterized by narrow linewidth and long metastable-state lifetime. In intermediate crystal fields, $^2E$ and $^4T_2$ are simultaneously populated in the excited state leading to both narrowband and broadband features in the spectrum. For low crystal fields, the upper level is the $^4T_2$, and the transition between this state and the $^4A_2$ lower state is characterized by shorter lifetime and broadband gain.

Ruby lasers are useful in applications that require high energy per pulse because ruby can be efficiently pumped by lamps, it has a low-gain cross section that reduces amplified spontaneous emission and parasitics, and $Al_2O_3$ is an excellent host material due to its high thermal conductivity, strength, chemical stability, and ease of growth. This is a three-level laser; thus the threshold is high, but its broadband absorption and long metastable-level lifetime allow efficient utilization of the pump radiation from flashlamps. The detailed energy-level diagram of ruby is shown in Fig. 11. Alexandrite was the first broadly tunable $Cr^{3+}$ laser, and its discovery initiated efforts to understand the effect of host material on energy-level positions. Alexandrite lasers are useful as tunable lasers in the 700- to 820-nm wavelength range. Laser systems with high energy per pulse and high average power can be built using alexandrite because of its excellent thermo-mechanical properties. Many other $Cr^{3+}$-doped laser materials have been demonstrated, but ESA limits the efficiency and tuning range of many of these other host materials, particularly those with low crystal fields, and in some cases, crystal growth is difficult. Two host materials for $Cr^{3+}$ that are becoming commercially available are $LiSrAlF_6$ and $LiCaAlF_6$, which can both be grown to good optical quality with relative ease and have only weak ESA.

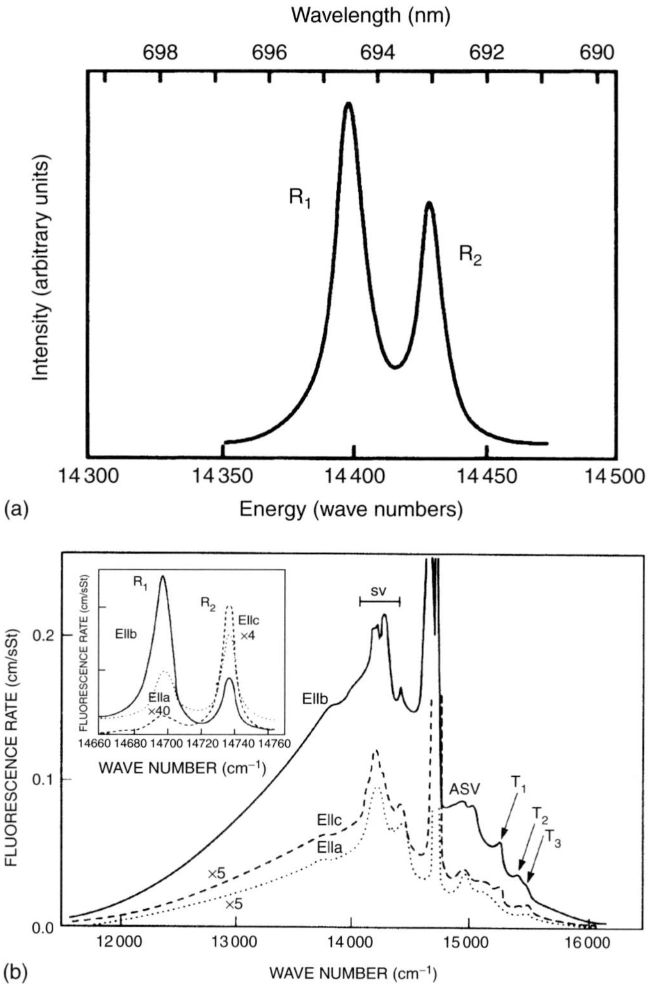


**Fig. 10** Fluorescence spectrum of ruby and alexandrite. (a) ruby and (b) alexandrite (from Walling, J. C., Peterson, O. G., Jenssen, H. P., Morris, R. C., O'Dell, E. W. (**1980**), *IEEE J. Quantum Electron.* **QE-16**, 1302–1315 [7], © 1980, IEEE)

**Ti : $Al_2O_3$ Lasers** The most broadly tunable solid-state laser is Ti : $Al_2O_3$ [15]. The fluorescence and absorption spectra of this material are shown in Fig. 12. Titanium is trivalent in this material, which leads to a single 3d electron in its outer shell and to a simple energy-level diagram with no chance for ESA to other 3d energy levels. Wide tunability of this laser has been demonstrated from 670 to 1100 nm, and thus, this laser has found wide acceptance as a replacement for near-infrared dye lasers in scientific applications. It is the workhorse for ultrashort-pulse laser systems by virtue of its large gain bandwidth (see ULTRASHORT-PULSE PHENOMENA). Other host materials with $Ti^{3+}$ have also been investigated, but

**Tab. 2** Properties of ruby and alexandrite

| ***Property*** | ***Ruby*** | ***Alexandrite*** |
|---|---|---|
| Stimulated-emission cross section at 300 K ($10^{-20}$ cm$^2$) | 2.5 | 0.7 (at 0.7 μm) |
| Upper-state lifetime (ms) | 3 | 0.2 |
| Thermal conductivity at 300 K (W cm$^{-1}$ K$^{-1}$) | 0.35 ($\parallel c$) | 0.23 |
| | 0.33 ($\perp c$) | |
| Thermal expansion at 300 K ($10^{-6}$ K$^{-1}$) | 5.3 ($\parallel c$) | 6 ($\parallel a$) |
| | 4.8 ($\perp c$) | 6 ($\parallel b$) |
| | | 7 ($\parallel c$) |
| Refractive index $n$ at 0.7 μm | 1.7635 ($n_o$) | 1.7381 ($n_a$) |
| | 1.7555 ($n_e$) | 1.7436 ($n_b$) |
| | | 1.7361 ($n_c$) |
| $dn/dT$ ($10^{-6}$ K$^{-1}$) | 13 ($n_o$) | 8 |
| | 17 ($n_e$) | |
| Crystal system | Trigonal | Orthorhombic |

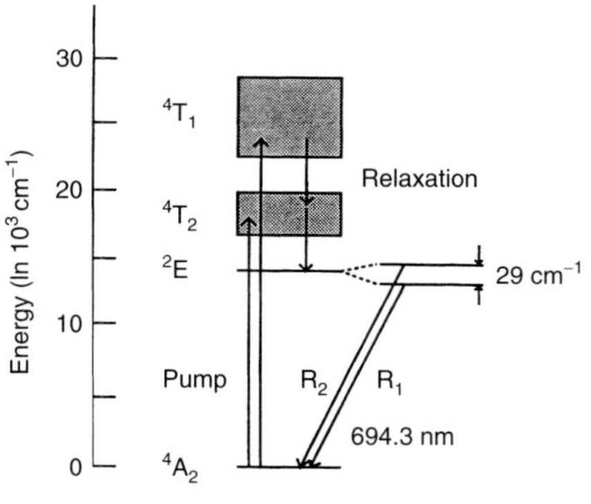


**Fig. 11** Energy-level diagram of ruby

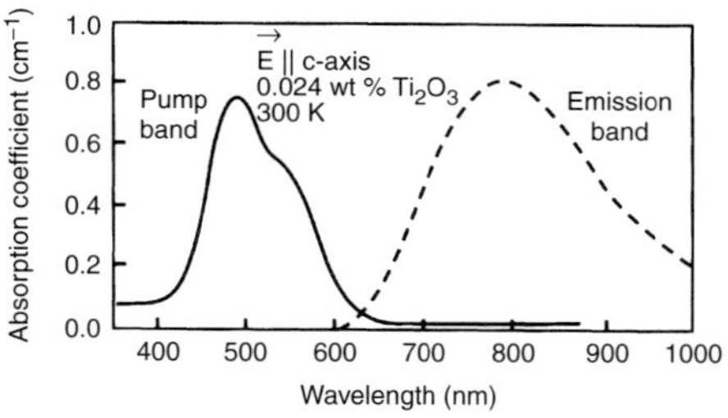


**Fig. 12** Absorption and fluorescence spectra in $Ti:Al_2O_3$

$Ti:Al_2O_3$ is the only material in which efficient laser operation using this ion has been demonstrated. $Ti:Al_2O_3$ lasers can be effectively pumped by either lamps or lasers such as Cu vapor, argon ion, or frequency-doubled Nd:YAG.

### Other Transition-metal-ion-doped Materials

Lasers using other transition-metal ions are subjects of research. $Cr^{4+}$ has been doped into hosts such as YAG and forsterite ($Mg_2SiO_4$) and shown to have outputs in the 1.2- to 1.5-μm wavelength range depending on the host material. The zinc chalcogenides and similar materials (ZnS, ZnSe, CdSe, etc.) have been shown to be useful hosts for divalent transition metals ($Fe^{2+}$, $Cr^{2+}$) that occupy a site of tetrahedral symmetry. The tetrahedral symmetry relative to the more typical octahedral symmetry more strongly breaks the parity of the d-shell electronic states, and consequently, the transitions have much stronger oscillator strength (shorter radiative lifetime, larger integrated cross sections).

3.2
## Pump Sources

Solid-state lasers are optically pumped by either a lamp or another laser. The desired characteristics of these pump sources are that they be efficient converters of electrical energy to optical energy, that the wavelength of the pump source spectrally match the pump bands of the gain medium, and that the energy or power density be high. The choice of a particular pump source for a particular gain medium is dependent on how the characteristics of the pump source match the requirements for efficient, low-threshold operation of the laser and the requirements for the application, for example, the cw or pulsed operation, the average output power, and the energy per pulse.

### 3.2.1 Lamps

Lamps can be used to pump both pulsed and cw lasers. For cw lasers, linear xenon or krypton arc lamps or tungsten–halogen lamps are used. Pulsed flashlamps are gas discharge tubes filled with xenon or krypton and come in linear or helical shapes. Helical lamps are useful for high energy per pulse laser systems due to their longer arc length per length of laser rod. Other types of lamps such as Na vapor discharge lamps can be used but are less common.

The output spectrum for a cw krypton arc lamp and a pulsed xenon flashlamp are shown in Fig. 13. The spectrum of a tungsten–halogen lamp closely approximates a blackbody emitter. The spectrum of a cw krypton arc lamp is dominated by line emission; these lamps are particularly useful in pumping Nd:YAG lasers because of the excellent spectral match between some of the strong lines from the lamp with

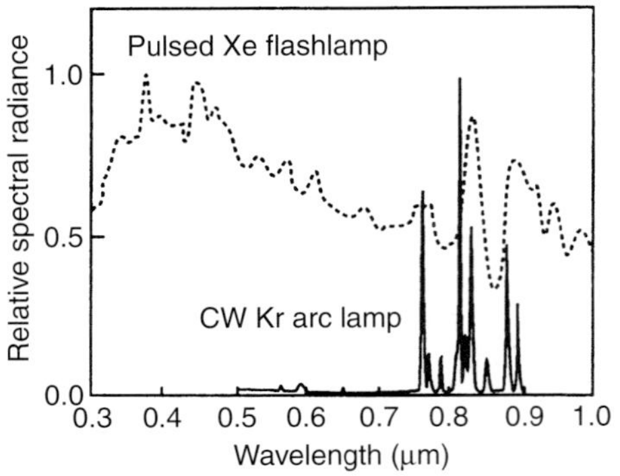


**Fig. 13** Spectral power density of the outputs for cw Kr and pulsed Xe arc lamps

absorption features in Nd:YAG. The operating temperatures and ion and electron densities are very different in pulsed flashlamps from the cw regime, which leads to a spectrum that is dominated by continuum radiation as opposed to line spectra, and the ratio of the two contributions is determined to a large extent by the current density in the flashlamp.

Lamps are available in diameters from ~2 to 20 mm and arc lengths of 25 mm to greater than 1 m and can be cooled with either air or liquids. The size of the lamp used is determined by the required spectral characteristics, size of the gain medium, energy per pulse and/or average power of lamp operation, and desired lamp lifetime. For efficient coupling of the lamp light into a gain medium, the size of the lamp should not be larger than the gain medium itself. In particular for pumping rods, imaging techniques that work efficiently only for small lamps are often used. The area of the lamp jacket must be sufficient to remove heat to avoid excessive temperature rise, and in the pulsed mode the lamp must be large enough to avoid explosion caused by shock waves that arise from heating of the gas. The lifetime of pulsed flashlamps is a strong function of the ratio of the operating energy per pulse relative to the explosion limit. For pulsed flashlamps operating at 20% of the explosion limit, the lamp lifetime is $\sim 10^7$ pulses. In cw arc lamps, the lifetime is limited by deposition of material on the lamp wall from the electrode, which reduces the light output; lamp lifetime is typically limited to a few hundred hours.

### 3.2.2 Laser Sources

Lasers are becoming increasingly common as pump sources for solid-state lasers because they have higher brightness than lamps and have directional output, which means that focusing of their beams can be used to obtain higher excitation densities than that available from lamps. Their output is also in only a narrow wavelength range; thus very efficient utilization of their output is achieved provided that the output wavelength is at an absorption feature. While it may not seem to make sense to use lasers to pump other lasers, laser-pumped lasers can provide wavelength shifts, peak-power enhancement, beam-quality improvement, and spectral-brightness improvements relative to the pump laser.

Semiconductor diode lasers (see LASERS, SEMICONDUCTOR) are the most common pump lasers and are used primarily for pumping $Nd^{3+}$ lasers. AlGaAs diode lasers can be fabricated to emit near 0.8 μm where there is a strong absorption in $Nd^{3+}$. Fabrication of large arrays of diode lasers for solid-state laser pumping has been demonstrated, and these arrays are capable of emitting $>1$ kW cm$^{-2}$ from areas of $>10$ cm$^2$ on a pulsed basis [16]. This technology is an area of very active research that promises to lead to efficient, all solid-state laser devices [17, 18]. As pump sources, the advantages relative to lamps are low voltage requirements and longer operational lifetimes, with lifetimes of tens of thousand of hours cw and $10^9$ shots pulsed. Diode lasers also reduce the thermal load for $Nd^{3+}$ lasers. In $Nd^{3+}$ lasers, the dominant contribution to heat generation per absorbed pump photon is nonradiative relaxation that generates heat nominally equal to the difference between the pump photon energy and the output photon energy. In pumping $Nd^{3+}$ lasers, the average photon energy for pulsed flashlamps is significantly larger than that for AlGaAs diodes operating at 0.81 μm; the difference in thermal loading for these two pump sources in Nd : YAG is about a factor of 3.

Other examples of laser-pumped lasers are Ti : $Al_2O_3$ and Er-doped fibers. Ti : $Al_2O_3$ requires pump sources in the blue-green part of the spectrum and can be pumped by Ar ion, Cu vapor, and frequency-doubled Nd : YAG lasers. Lasers are especially useful in pumping Ti : $Al_2O_3$ because of its low metastable-level lifetime and stimulated-emission cross section product, which requires a high pump rate to achieve laser threshold (see LASER PHYSICS). Er-doped fibers are used in fiber-optic communication, and here diode lasers operating around either 0.98 or 1.48 μm are the pump sources of choice because of their long lifetime, high efficiency, and the ability to couple their output efficiently into fibers.

## 4
## Laser Design

The laser engineer is confronted with many design choices including size and shape of the gain element, the method of coupling the pump radiation into the gain element, and the optical resonator design.

The choices are driven by the characteristics of the gain medium, such as the stimulated-emission cross section, upper-state lifetime, and thermal properties; the type of pump source, lamp, or laser; and the requirements of the application such as beam quality, average power, energy per pulse, and spectral output.

## 4.1
## Gain-element Geometries

Laser gain elements in solid-state lasers have been fabricated primarily in the form of rods, slabs, disks, and optical fibers. The choice of gain-element geometry is governed by the average power and/or energy per pulse requirements, raw material (i.e., crystal boule) shape and size, desired application, and cost.

### 4.1.1 Rods

The most common form of a bulk solid-state gain element is a cylindrical rod in which the optical axis of the laser resonator is collinear with the axis of the cylinder. In rods, the length of the cylinder is larger than the beam size of the laser. Rods are relatively inexpensive to fabricate, and the pump radiation from either linear or helical lamps can be coupled efficiently using simple pump cavities as shown in Fig. 14. Rod sizes range from ~2 mm diameter to as large as ~10 cm diameter and 2 m long. These larger-diameter rods are available only in glasses; crystalline rods are limited to ~20 mm diameter by crystal size.

A key issue in all geometries is the effect of heat dissipation on the optical properties of the gain element. Rods are typically cooled by flowing a coolant such as water across the barrel, which leads to a radial temperature gradient in the

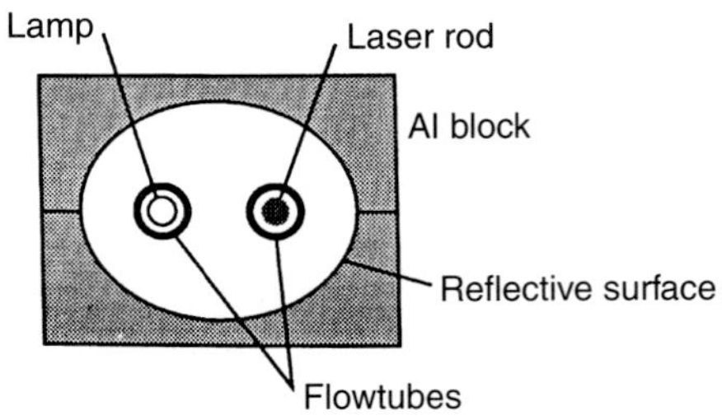


**Fig. 14** Cross-sectional schematic of a pump cavity for coupling of cylindrical lamps to a laser rod. The rod and lamp are located at the foci of an ellipse so that radiation from the lamp is imaged onto the rod

rod. This gradient leads to two separate effects: (1) a radially varying index of refraction due to change in refractive index with temperature and (2) radially varying stress due to differential thermal expansion.

To first order, the former effect acts as a lens with a focal length that decreases with increasing thermal load. This causes the cavity modes to vary in size with pump power, and over a sufficiently wide variation in thermal load, the cavity can change from being stable to unstable or vice versa (see LASER PHYSICS). To ameliorate this effect, laser cavities are often designed to accommodate thermal lensing at the nominal operating point of the laser.

The stress also causes a change in the refractive index and causes an isotropic gain element to become birefringent with the characteristic that the laser rod acts as a waveplate with variable retardation across its aperture and with principal axes whose directions also change across the rod aperture. If polarization elements are used inside the laser resonator, substantial loss will occur, although there are partial compensation techniques that can reduce this loss by using multiple laser rods and additional intracavity optical elements.

Stress-induced birefringence can be essentially eliminated by using materials that are inherently birefringent such as YLF (as long as laser beam propagation is not along the optical axis of the crystal) since the magnitude of the inherent birefringence is much greater than the stress-induced birefringence. At high-enough thermal load, stress can eventually lead to fracture. In YAG, the stress-fracture limit is $\sim$100 W of thermal load per linear centimeter of rod length, in YLF it is $\sim$25 W cm$^{-1}$, and in phosphate glass it is $\sim$10 W cm$^{-1}$. The stress-fracture limit is independent of the rod diameter, and thus, the average power limit of rod lasers can be increased only by lengthening the rod. Rod length is ultimately limited by crystal size and lamp characteristics. High-power rod lasers ($>$1 kW average) have been demonstrated using multiple rods.

#### 4.1.2 **Zig-zag Slabs and Disks**

Gain elements in the form of zig-zag slabs reduce the thermal lensing and birefringence problems. In these slabs, the laser beam is incident on the slab at Brewster's angle and then propagates down the slab by making total internal reflections off the two faces as shown in Fig. 15. The slab is pumped and cooled through the bounce faces while the other two faces are insulated. By cooling through only the pump faces, the thermal gradient is essentially one dimensional assuming that the pump density is uniform throughout the slab.

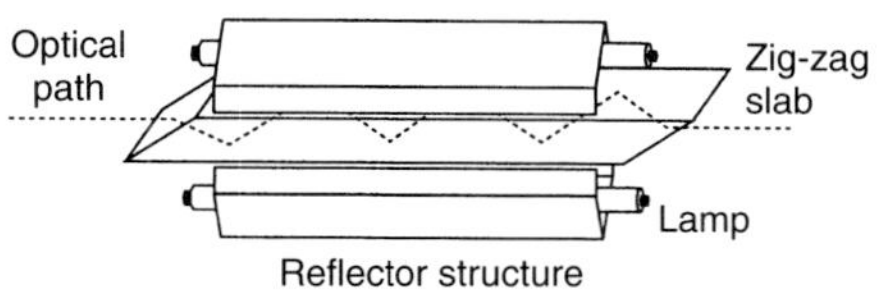


**Fig. 15** Schematic of a zig-zag slab laser head

By zig-zagging the beam across the slab, the refractive index variation due to the thermal gradients are averaged to zero because all parts of the beam sample all parts of the slab. Thermally induced depolarization of linearly polarized beams is also reduced in slabs relative to rods. While thermally induced birefringence still exists, the principal axes of the induced birefringence are aligned across the entire slab aperture. Thus, linearly polarized light with polarization along one of the principal directions will not suffer depolarization.

The stress-fracture limit in slabs scales differently from rods; it is proportional to the area of a pump face divided by the thickness of the slab. In Nd:YAG, the stress-fracture limit in zig-zag slabs is nominally 200 $(A/t)$ W cm$^{-1}$ where $A$ is in units of cm$^2$ and $t$ is in units of cm; the quantity $(A/t)$ is also known as the *aspect ratio*. While large aspect ratio slabs are desirable for high average power, there are practical limits to using large aspect ratio slabs. A slab cannot be made too thin, otherwise it will absorb pump light inefficiently, and large aspect ratio slabs are difficult to fabricate. Typical Nd:YAG slab sizes are 10 cm long by 2 cm wide by 0.6 cm thick. High-power slab lasers ($\sim$1 kW) have been demonstrated with nearly ideal output beams and with little change in beam size as the power is varied.

Slabs do have some significant drawbacks. They are more costly to fabricate than rods because of the larger area of optical surface. Another problem is coolant flow across the optical faces of the slab because these faces can be eroded by the flow, and deposits can become attached to the slab faces leading to optical loss and/or damage. Finally, careful engineering is required to realize the benefits of zig-zag slab lasers. Uniform deposition of the pump power is extremely important.

In lamp-pumped systems, reflector mirror structures for directing the pump radiation are often empirically designed to achieve uniform pumping. Also, there are significant nonideal effects that occur at the ends of the slab, which are difficult to model and predict. These difficulties have led to slow marketplace acceptance of slab laser technology.

Disks are used as amplifiers in laser systems with high peak power and energy per pulse, such as fusion lasers. They are typically inserted into the beam at Brewster's angle as shown in Fig. 16, and their large area, up to ~1 m diameter, ensures that the fluence is kept below the damage fluence. These disks are pumped by lamps through their faces and are also cooled through the faces. As in zig-zag slab lasers, uniform pumping is needed to reduce thermo-optic aberrations. One difficulty with operating these lasers at high average power is that the laser beam propagates through the coolant, which can be either gas or liquid. Significant distortion of the beam can result because of thermally induced refractive index changes in the coolant; thus, disk amplifier systems have been limited primarily to low pulse-repetition frequencies (<1 Hz).

### 4.1.3 **Optical Fiber** (see also FIBER OPTICS)

An optical fiber is a rod in the limit of small diameter and long length. The operation of fiber lasers and amplifiers is not limited by thermal effects because of their large surface-to-volume ratio and their long length; instead the power obtainable from a fiber laser or amplifier is limited by damage at the ends of the fiber owing to the small area and consequent high surface power density. For single-mode fiber devices, surface damage limits the average power to ~10 W. High gain can be achieved in fiber devices when laser pumping is employed, and the pump radiation is guided by the core because this arrangement gives high pump intensity over a long distance. In $Er^{3+}$-doped fiber amplifiers, unsaturated power gains in excess of 30 dB have been demonstrated (see AMPLIFIERS, OPTICAL).

The rare-earth-ion dopants are introduced into a fiber preform only in the core region before the fiber is drawn. Dopant concentration is limited to typically $<10^{19}$ cm$^{-3}$ because of limited solubility. Otherwise, the fiber is similar to undoped optical fiber. For typical EDFAs, the pump radiation is coupled into the single-mode core by use of standard fiber couplers.

In optical fibers for high-power lasers and amplifiers, a somewhat different fiber geometry called *double-clad fibers* is used. These fibers still have a single-mode, doped core that is surrounded by a slightly lower index inner cladding, but this inner cladding is (1) typically not round in cross section and (2) is surrounded by a much

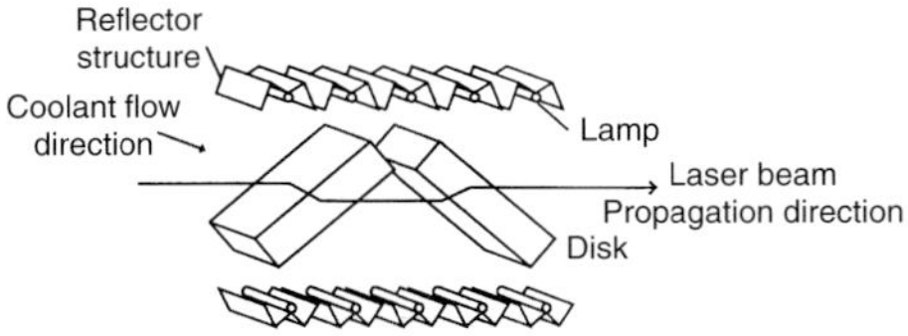


**Fig. 16** Schematic of a laser disk amplifier

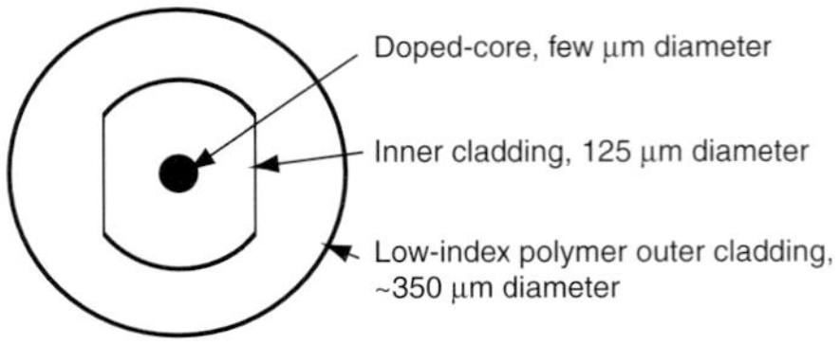


**Fig. 17** Schematic of double-clad rare-earth-doped fiber. The single-mode core is doped with the rare earth and is surrounded by a cladding that has a slightly lower refractive index. This inner cladding is surrounded by an outer cladding with much lower refractive index for guiding of the pump radiation

lower index polymer clad so that the index difference gives a numerical aperture (NA) on the order of 0.4 or more for the inner cladding. A cross section of this type of fiber is shown in Fig. 17. Pump radiation is coupled into the inner cladding, either using some sort of fiber coupler or end pumping. The large size of the inner cladding (compared with the single-mode core) and its large NA enables high-power diode-laser arrays to be coupled into the inner cladding. The pump radiation is absorbed as it passes through the core; the noncircular inner-clad cross section mixes the modes of the inner cladding to enable more efficient absorption of the pump radiation. Fiber lasers have been made with >100-W output power using this type of fiber.

### 4.2
### Pump Geometries

There are two basic pumping geometries, longitudinal, (or end-pumped) in which the pump radiation is directed nominally collinear to the optical axis of the laser, and transverse, (side-pumped) in which the pump radiation is directed nominally perpendicular to the optical axis of the laser. The choice is governed by the pump source characteristics, the gain-element geometry, and the characteristics of the gain medium. Almost all lamp-pumped solid-state lasers are side-pumped because efficient coupling of the pump radiation to rods is easily achieved since the diameter-to-length ratio of a lamp can be matched to the rod's diameter-to-length ratio. It is difficult to effectively couple lamps to rods in an end-pumped geometry because of the lack of directionality to the lamp output.

On the other hand, many laser-pumped lasers are configured in an end-pumped geometry to take advantage of the focusability of laser radiation; this geometry is preferred for efficient, low-power devices. The ability to focus laser radiation leads to high pumping rates and consequently low threshold and excellent matching to the $TEM_{00}$ mode of the solid-state laser. However, end pumping becomes more inconvenient for higher-power devices when large numbers of pump lasers are needed, and thermo-optic distortions become more of a problem. In such cases, side-pumping becomes more attractive. In particular, in high energy per pulse diode-pumped lasers, a large array containing many pump lasers is required, and focusing the output of such a pump source becomes difficult.

### 4.3
### Laser Resonators (see also LASER PHYSICS)

The vast majority of bulk solid-state lasers use simple two-mirror resonators. Both stable resonators and unstable resonators are used in commercial devices, the difference is that stable resonators are used in relatively low-gain lasers, while unstable laser resonators are used in high-gain lasers with large output apertures. Stable resonators are those in which Gaussian laser modes are confined in their transverse extent upon propagation around the laser resonator; resonators not satisfying this condition are unstable. In stable resonators, the output is extracted from the cavity using a partially reflective output mirror, while in unstable resonators, the output is extracted by diffraction around a cavity mirror.

The high-gain laser systems that use unstable resonators are almost exclusively Q-switched Nd:YAG lasers operating at relatively low repetition rates, <100 Hz, and high energy per pulse, >100 mJ; almost all other solid-state lasers use stable resonators. Unstable resonators are appropriate for systems with high gain because they allow efficient laser operation in the fundamental transverse mode of the resonator across gain-media apertures that are large. For lower-gain systems that require low-magnification unstable resonators, the mirror-alignment requirements become increasingly stringent. Efficient operation in the fundamental transverse mode can be obtained with stable resonators but only for devices in which the limiting cross-sectional aperture size is less than $\sim(\lambda L)^{1/2}$, where $\lambda$ is the laser wavelength and $L$ is the cavity length. For larger limiting apertures, the laser will run the multitransverse mode. In fact, one challenge in the engineering of solid-state lasers is the resonator design for low-gain lasers that yields laser output in the fundamental transverse mode with good extraction across a large aperture [19].

Laser resonators are often designed to take into account thermal lensing in laser rods. For example, many Nd:YAG rod lasers have resonators that are unstable or at the edge of the stability region with no input power, but as the input power is increased, thermal lensing of the rod is used to bring the laser cavity into stability or to its desired operating point. However, this technique is useful only over a limited output power range of the laser.

One trend in laser resonator design is to use graded-reflectivity mirrors. Traditionally, output mirrors for stable resonators have uniform reflectivity across their surface, and unstable resonator output mirrors have a region that is highly reflective over a portion of their area with the rest having no reflectivity. A typical unstable resonator for a rod laser has a highly reflective dot at its center. A graded-reflectivity mirror has a reflectivity that varies across its face with the highest reflectivity in the center of the mirror. These mirrors add an additional degree of freedom to resonator design. They have been used with stable and unstable resonators to achieve output in a single transverse mode with good beam quality and high efficiency.

More complicated resonators with a larger number of mirrors are also used in solid-state lasers. These resonators find applications in tunable lasers or single-frequency lasers. In tunable lasers, additional laser cavity elements such as birefringent tuning filters are often inserted in the cavity, and the gain element often has Brewster angle faces. These elements add astigmatism to the laser mode, which can be compensated by using a three- or four-mirror cavity. Single-frequency performance from solid-state lasers is often obtained by using a unidirectional ring resonator that requires additional mirrors to form a ring path.

## 5
## Laser Manufacture

The key components in solid-state lasers are the gain element, pump source, and resonator optics. High-quality components are required for solid-state lasers if high power, good efficiency, good beam quality, and minimal environmental sensitivity are desired.

### 5.1
### Gain-element Fabrication

Most of the important solid-state laser crystals are grown using the Czochralski technique, including Nd:YAG, Nd:YLF, ruby, and $Ti:Al_2O_3$. Provided that reasonable size crystals of the host material can be grown, there are additional issues that need to be addressed to determine the suitability for laser gain elements. These include the impact of dopants on the crystal growth process, facet formation, impurity incorporation, and adequate control of crystal growth conditions.

For example, in Nd:YAG, the $Nd^{3+}$ ion is significantly larger than the $Y^{3+}$ ion for which it substitutes (radius of 0.098 nm vs 0.090 nm). This leads to stress in the crystal, which in turn leads to refractive index distortion and limits the Nd concentration. The size difference also causes the distribution coefficient for neodymium to be only 0.18; in other words, the ratio of the dopant concentration that solidifies to that in the melt is only 0.18. This leads to an increase in Nd concentration in the melt as the boule is grown and a resulting increase in Nd concentration along the length of the boule. Because of differences in crystal growth rates along various crystallographic planes, the YAG solid–liquid interface is faceted. Additional dopant and stress accumulates at the intersection of the facets, leading to these parts of the boule being unusable as seen in Fig. 18. It is important that impurities be kept to a minimum because they can lead to formation of second phases or inclusions that optically scatter, act as damage sites, or cause optical absorption at the laser wavelength. Finally, good control of mixing of the melt and temperature needs to be maintained to achieve uniform growth rates. Poor control can lead to stress caused by nonuniform growth rates and striations, which are quasi-periodic composition and refractive index variations along the length of the boule.

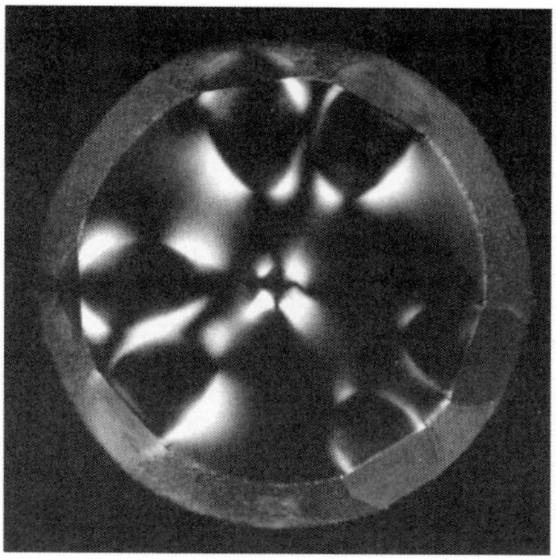

**Fig. 18** A piece of YAG viewed along the crystal growth axis undercrossed polarizers. The light regions indicate strain. The highly strained center region is known as the core. The strained regions of the crystal cannot be used because they are optically distorted and contain stress-induced birefringence

In glasses, there are fewer problems in gain-element fabrication. The ability to mix and cast large pieces of glass doped with rare-earth ions with the high uniformity, large size, and optical quality needed for laser gain elements is within the state of the art. Gain elements in Nd-doped phosphate glasses have been fabricated in sizes nearly as large as 1 m in diameter. Single-mode optical fibers doped with rare-earth ions have activator-ion concentration on the order of a few hundred parts per million. The doping is lower than that for most bulk gain elements because the silica glass used for most optical fibers has low solubility for rare earths. This low doping is typically not a problem in optical fiber devices because long fiber lengths can be used to achieve efficient absorption of the pump radiation. The activator ions are introduced into the core region of a fiber preform using either a vapor phase or aqueous solution technique, and then fibers are drawn using standard optical fiber fabrication methods.

5.2
### Optical Requirements for Solid-state Lasers

The key parameters of interest for intracavity optical components are optical loss, optical distortion, surface figure, surface quality, and birefringence (see OPTICAL FABRICATION).

Optical loss, due to either absorption or scattering, needs to be minimized for low-threshold and high-efficiency operation; this loss is typically ~0.1% per centimeter or less for high-quality optical components. Optical distortion and surface figure are measurements of the optical perfection of bulk optical elements and surfaces respectively. Optical distortion and imperfect surface figure cause optical wavefronts to become distorted, which leads to additional intracavity loss and laser beam distortion. Optical distortion in laser gain elements and on optical surfaces is typically specified to be less than $\lambda/4$ (at 633 nm) per optical element or surface. This requirement is easy to meet in glasses, but meeting this requirement in crystals can be more of a problem. Surface quality refers to residual scratches and digs in the surface of an optic after polishing. Poor surface quality leads to scatter loss, and scratches and digs are centers for optical damage under high-power operation. Optical elements that in principle are isotropic can act as birefringent elements owing to either residual stress or stress induced by mounting; unwanted birefringence can cause loss in laser cavities with intracavity polarizing elements. In crystals such as YAG, residual stress limits the fraction of the boule that is useable. Glass elements can have low residual birefringence, but small amounts of mounting stress can lead to significant birefringence that is easily observed under crossed polarizers.

Evaporated dielectric thin-film coatings are used for mirrors, polarizers, and antireflection coatings (see also OPTICAL COATINGS). These coatings are composed of single and multiple layers of materials such as $MgF_2$, $ThF_4$, $SiO_2$, $Ta_2O_5$, $ZrO_2$, ZnS, and $TiO_2$, which are a quarter wavelength thick at the wavelength of interest. Reflecting mirrors are made by alternating films of low refractive index, the first three materials, with films of high refractive index, the other materials. Simple antireflection coatings are made using a quarter-wave-thick coating of a film that has a refractive index equal to the square root of the substrate refractive index. Important characteristics for these thin-film coatings include good adhesion to the substrate, low optical absorption, good hardness, and high optical power damage threshold. A significant effort has gone into the development of high damage threshold optics for use with Q-switched and other short-pulse lasers, particularly around 1.06 μm; damage thresholds for these optics are in the range of a few GW $cm^{-2}$ peak power density for nanosecond-long pulses with energy density damage threshold on the order of 20 J $cm^{-2}$. Damage thresholds vary with pulse length, repetition rate, wavelength, coating technique, the coating's history, and beam size.

The support structure for the laser gain medium and for the optical resonator needs to be rigid and the temperature insensitive to varying degrees depending on the type of laser and the application. Many lasers for research applications have open structures with access to all the cavity components to allow modification of the laser. Military laser systems are enclosed in rugged structures that are compatible with large changes in environment and with the acceleration and vibration that are associated with moving platforms. Mode-locked lasers are often built with materials with low thermal expansion coefficient such as Invar and Zerodur because the cavity length needs to remain fixed within a few microns.

## 6 Applications of Solid-state Lasers

The number of applications of solid-state lasers is increasing as these devices become more efficient, more reliable, and more flexible. Many applications rely on the solid-state laser's unique combination of high peak and average power with good beam quality. This section discusses some of the more important solid-state laser applications.

### 6.1 Industrial Applications (see also LASERS, INDUSTRIAL USE OF)

Solid-state lasers have found industrial applications ranging from cutting, welding, drilling, and marking of metals to micromachining of integrated circuits. High-power Nd : YAG lasers, up to 2-kW average power, are used in metal working; in this application, the primary competition is the $CO_2$ laser. Nd : YAG lasers have the advantages of smaller spot sizes and somewhat higher absorption by metals, both of which are due to the shorter Nd : YAG laser wavelength. In addition, the 1-μm radiation can be delivered by optical fiber, whereas more cumbersome delivery systems are required for $CO_2$ lasers. On the other hand, $CO_2$ lasers currently offer an economic advantage of higher output power at a given price, which translates into greater workpiece speed. In practice, the choice of laser is determined by a complex trade of average power, energy per pulse, wavelength, spot size, metal-working application, and material.

Solid-state lasers are also used in the electronics industry in applications such as soldering, resistor trimming, and link blowing in integrated circuits. Both $CO_2$ and Nd : YAG lasers are used in laser soldering, with Nd : YAG lasers having the advantage of being able to make smaller solder joints. Q-switched Nd lasers are the lasers of choice in thick- and thin-film resistor trimming applications because there are fewer problems with cracking of the substrate; these thick- and thin-film hybrid circuits often use alumina substrates that weakly absorb 1-μm radiation but strongly absorb the 10-μm radiation from $CO_2$ lasers. Diode-pumped, frequency-doubled, Q-switched Nd : YAG and Nd : YLF lasers have found applications in integrated circuit manufacturing in link blowing (i.e., cutting conductive lines) for repair of high-density random access memory chips.

Ultrashort-pulse laser systems such as mode-locked Ti : $Al_2O_3$, and high pulse-repetition-frequency UV lasers such as frequency-tripled or quadrupled, Q-switched, Nd : YAG have been shown to be excellent systems for high-precision materials processing such as direct writing of fine surface features in materials and drilling of ceramics or thin films. Both these types of lasers minimize the peripheral thermal shock and damage, and the short wavelengths and use of multi-photon excitation processes enable small feature sizes.

### 6.2 Military Applications

The two primary military uses of solid-state lasers are in range finders and target designators. In the first application, Q-switched pulses from Nd : YAG or ruby lasers are aimed at a target, and the distance is measured by the round-trip time-of-flight. In the second application, a Nd : YAG laser illuminates a target, and a weapon equipped with a semiactive guidance system homes in on the illuminated spot. Solid-state lasers are particularly suited for these applications because of their combination of good beam quality, high energy per pulse, short-pulse length, and wavelength. The first two qualities are needed to allow operation over ranges of a few kilometers. Short pulses are needed for good range resolution in rangefinder applications, and the near-infrared wavelength of Nd : YAG lasers and the red wavelength of ruby lasers allow simple room temperature silicon photodetectors to be used for sensors.

Solid-state lasers are under active investigation for other military applications including electro-optic countermeasures, laser-radar systems, and directed-energy weapons. Electro-optic countermeasures are used to jam optical detection systems such as infrared sensors on heat-seeking missiles. Laser-radar systems can be used to track and identify targets and have higher resolution than microwave radar due to the shorter wavelength. Laser-radar systems can also be used to remotely identify atmospheric chemicals or sense wind velocity. In directed-energy weapons, high-power lasers (10s of kW to MW average power) are used to damage or destroy targets by delivery of optical energy to, for example, ignite or melt targets. Improvements in solid-state lasers including higher average power while maintaining near ideal beams are needed for directed-energy weapons to become a reality.

### 6.3
### **Medical Applications** (see also LASERS, MEDICAL USE OF)

Nd : YAG, along with $CO_2$ and argon ion, are the most common lasers in medicine. The 1.06-μm wavelength of Nd : YAG is not strongly absorbed by water, and thus it has found use in applications that require deep heat penetration such as photocoagulation of blood and tissue welding. In surgery, $CO_2$ lasers are more common and are preferred in applications that require precise cutting because 10.6 μm is strongly absorbed by water; Nd : YAG is preferred when tissue ablation and control of hemorrhage is desired because the requirements of these applications match the longer absorption depth for 1.06 μm. Nd : YAG has the advantage over $CO_2$ in the ease of delivery since 1.06 μm can be propagated down standard optical fiber. An example of one of the many Nd : YAG applications is in ophthalmology; here, Q-switched Nd : YAG lasers are used to remove opacified membranes that can develop after cataract surgery using optical breakdown.

There is increasing interest in $Ho^{3+}$ and $Tm^{3+}$ lasers in the 2.1-μm wavelength range and in $Er^{3+}$ lasers in the 2.9-μm wavelength range for ablating or cutting of tissues because these wavelengths are strongly absorbed by water, which helps in limiting the damage to the surrounding tissue. For 2.1-μm lasers, about 500 μm of surrounding tissue is affected, and for 2.9-μm lasers, this distance is only ~30 μm. One advantage of the 2.1-μm wavelength relative to 2.9-μm is in delivery of the laser radiation; 2.1-μm radiation can be propagated down currently available optical fiber. Fiber-delivery systems for 2.9-μm radiation are not currently available, but active research is ongoing in this area.

Tunable solid-state lasers are also beginning to find a role in medicine, particularly in research. The tunability allows investigation of laser–tissue interactions as a function of wavelength. Current applications of alexandrite lasers include tattoo removal and hair removal. The tunability of alexandrite lasers is not used in these applications; instead the ability to get high-energy pulses at the right wavelength from a relatively inexpensive laser determines the choice.

### 6.4
### Communications Applications (see also OPTICAL COMMUNICATIONS)

The primary solid-state laser used in communications is the EDFA, which provides a simple, all-optical technique for compensating losses in the 1.5-μm fiber-optic telecommunications window. Formerly, this loss was compensated by installing repeaters that contain a detector, signal-reconditioning electronics, and a diode-laser transmitter. The EDFA has enabled high-data-rate, long-haul optical communications systems, particularly those that utilize wavelength-division multiplexing (WDM) techniques to increase channel count. Tens to hundreds of wavelength channels with 50-GHz or more channel spacing, each with a separate information stream, can be fit under the EDFA gain curve. These EDFAs for telecommunications represent the single most economically important market for solid-state lasers.

Optical fiber amplifiers doped with neodymium and praseodymium were being investigated for the other primary optical communications band at 1.3 μm; however, the performance is not as good as that demonstrated in the EDFAs. Nd fiber amplifiers have been limited by ESA, and the Pr-doped fiber amplifiers are limited by the need to use specialty glass.

### 6.5
### Scientific Applications

Solid-state lasers have found wide use in the laboratory owing to their ability to run cw, Q-switched or mode-locked.

Q-switched Nd : YAG and Nd : YLF lasers are an important scientific tool. These Q-switched lasers are pumped either by pulsed flashlamps and run at relatively low repetition rates (10 Hz) and at high energy per pulse, up to ~1 J, or are pumped by cw lamps and Q-switched at high repetition rates, 100s of Hz, and at relatively low energy per pulse, 1 to 2 mJ. These lasers are used for nonlinear optics studies owing to their high peak power and as pump sources for other lasers. Frequency-doubled, Q-switched, and mode-locked $Nd^{3+}$ lasers are used to pump $Ti:Al_2O_3$ and dye lasers, and $Nd^{3+}$ lasers are also used to pump optical parametric oscillators that can provide tunable radiation. $Nd^{3+}$ lasers with high peak power and high energy per pulse are used to efficiently excite plasmas by focusing the radiation on metal targets. Soft X-ray lasers have been demonstrated using these laser-produced plasmas as the gain medium.

Tunable and short-pulse solid-state lasers are a powerful tool in spectroscopic research. $Ti:Al_2O_3$ lasers have become the tunable laser of choice in the near infrared, replacing dye lasers in this wavelength range. The $Ti:Al_2O_3$ laser offers broader band tunability than any single laser dye, higher power, and no problems with degradation over time. The large gain bandwidth of $Ti:Al_2O_3$ has been a boon for ultrashort-pulse generation. Pulses shorter than 5 fs have been demonstrated, and products for the scientific market with <100 fs pulses are readily available. These short-pulse systems are used in such areas as high-field physics and in physical chemistry to study reaction dynamics (see also ULTRASHORT-PULSE PHENOMENA).

## 7 Summary

The solid-state laser's unique ability to generate high peak and average power with good beam quality and moderate efficiency in the near and mid-infrared parts of the spectrum guarantees its place as an important technology in scientific, industrial, and medical applications. Current trends in solid-state laser technology include the development of efficient diode-pumped lasers, very high peak and average power systems, new wavelengths, optical fiber amplifiers, and improved solid-state laser materials. Progress in these areas promises to significantly expand the utility of solid-state lasers.

## Acknowledgment

This work was sponsored by the Department of the Air Force under Contract #F19628-00-C-0002. Opinions, interpretations, conclusions, and recommendations are those of the author and are not necessarily endorsed by the United States Air Force.

## Glossary

**Activator Ion:** A dopant ion that has the energy levels used to provide optical gain.

**Concentration Quenching:** A decrease in the metastable-level lifetime with increase in dopant concentration that is observed in many gain media.

**Host Material:** A crystalline or glassy solid into which activator ions are doped.

**Nonradiative Relaxation:** Transitions from a higher energy level to a lower energy level by emission of phonons (as opposed to photons).

**Rare-earth Ion:** Ions of elements with atomic number 57 to 71.

**Stress-fracture Limit:** The maximum level of heat that can be generated in a gain element limited by fracture due to stresses induced by temperature gradients.

## References

[1] Maiman, T. H. (**1960**), *Nature* **187**, 493, 494.
[2] Maiman, T. H., Hoskins, R. H., D'Haenens, J., Asawa, C. K., Evtuhov, V. (**1961**), *Phys. Rev.* **123**, 1151–1157.
[3] Sorokin, P. P., Stevenson, M. J. (**1960**), *Phys. Rev. Lett.* **5**, 557–559.
[4] Geusic, J. E., Marcos, H. M., VanUitert, L. G. (**1964**), *Appl. Phys. Lett.* **4**, 182, 183.
[5] Johnson, L. F., Guggenheim, H. J., Thomas, R. A. (**1966**), *Phys. Rev.* **149**, 179–185.
[6] Snitzer, E. (1964), in P. Grivet, N. Bloembergen (Eds.), *Quantum Electronics*, Paris: Dunod, pp. 999–1020.
[7] Walling, J. C., Peterson, O. G., Jenssen, H. P., Morris, R. C., O'Dell, E. W. (**1980**), *IEEE J. Quantum Electron.* **QE-16**, 1302–1315.
[8] Keyes, R. J., Quist, T. M. (**1964**), *Appl. Phys. Lett.* **4**, 50–52.
[9] Mears, R. J., Reekie, L., Jauncey, I. M., Payne, D. N. (**1987**), *Electron. Lett.* **23**, 1026–1028.
[10] Kaminskii, A. A. (1996), *Crystalline Lasers: Physical Processes and Operating Schemes.* Boca Raton: CRC Press.
[11] Henderson, B., Bartrum, R. H. (2000), *Crystal Field Engineering of Solid State Laser*

*Materials.* Cambridge: Cambridge University Press.
[12] Dieke, G. H., Crosswhite, H. M. (**1963**), *Appl. Opt.* **2**, 675–686.
[13] Becker, P. C., Olsson, N. A., Simpson, J. R. (1999), *Erbium-Doped Fiber Amplifiers: Fundamentals and Technology.* San Diego: Academic Press.
[14] Desurvire, E. (1994), *Erbium-Doped Fiber Amplifiers: Principles and Applications.* New York: Wiley.
[15] Moulton, P. F. (**1986**), *J. Opt. Soc. Am. B* **3**, 125–133.
[16] Endriz, J. G., Vakili, M., Browder, G. S., DeVito, M. A., Haden, J. M., Harnagel, G. L., Plano, W. E., Sakomoto, M., Welch, D. F., Willing, S., Worland, D. P., Yao, H. C. (**1992**), *IEEE J. Quantum Electron.* **28**, 952–965.
[17] Fan, T. Y., Byer, R. L. (**1988**), *IEEE J. Quantum Electron.* **24**, 895–912.
[18] Hughes, D. W., Barr, J. R. M. (**1992**), *J. Phys. D: Appl. Phys.* **25**, 563–586.
[19] Magni, V., Valentini, G., De Silvestri, S. (**1991**), *Opt. Quantum Electron.* **23**, 1105–1134.

## Further Reading

Brown, D. C. (1981), *High-Peak-Power Nd: Glass Lasers Systems.* Berlin: Springer-Verlag.
Koechner, W. (1999), *Solid-State Laser Engineering.* Berlin: Springer-Verlag.
Moulton, P. F. (**1992**), *Proc. IEEE* **80**, 348–364.
Pollock, C. R. (1991) in R. A. Meyers (Ed.), *Encyclopedia of Lasers and Optical Technology.* San Diego: Academic Press.
Weber, M. J. (Ed.) (1982), *Handbook of Laser Science and Technology,* Vol. 1. Boca Raton: CRC Press.
Weber, M. J. (Ed.) (1991), *Handbook of Laser Science and Technology,* Supplement 1. Boca Raton: CRC Press.

# Lidar (Laser Radar)

*P. S. Argall and R. J. Sica*
*Department of Physics and Astronomy, The University of Western Ontario, London, Ontario, Canada, N6A 3K7*
*Phone: (519) 661-2111 x86554; Fax: (519) 661-2111; e-mail: sargall@physics.uwo.ca*

**Abstract**
Lidar is an active remote sensing technique similar in principle to radar, except that it utilizes light rather than radio waves. Lidar finds applications in many areas, including atmospheric and oceanic research, digital mapping, chemical and biological agent detection, speed measurement, and targeting. This chapter describes the principles of lidar and provides examples of the application of lidar in measuring the properties of the Earth's atmosphere.

**Keywords**
lidar; laser–radar; laser; atmospheric optics; Rayleigh; aerosol; DIAL; Raman.

## 1 Introduction

Lidar (light detection and ranging) is an active optical technique in which a beam of light is used to make remote range-resolved measurements. A lidar emits a light beam that interacts with the medium or object under study; some of this light is scattered back and detected by the lidar. Analysis of the backscattered light allows some property of the medium or object to be determined.

The major scientific use of lidar is in the measurement of the properties of the Earth's atmosphere, while the major commercial uses of lidar are aerial surveying and bathymetry (water depth measurement). Lidar also finds extensive application in ocean research, forestry, and biomass research; it has numerous other applications including chemical and biological agent detection and vehicle speed measurement. Lidar-equipped binoculars are used by hunters and golfers for range finding, while imaging lidar is beginning to find applications in producing 3-D electronic representations of objects, such as buildings, for virtual reality applications.

## 2 Historical Overview

Synge [1] first proposed the method of determining atmospheric density by the detection of scattering from a beam of light projected into the atmosphere in 1930. Synge suggested a scheme in which an antiaircraft searchlight could be used as the source of the beam and a large telescope with photoelectric apparatus for detection of the scattered light. Ranging, determining the distance at which the scatter occurred, could be accomplished by operating in a bistatic configuration with the source and receiver separated by several kilometers, Fig. 1.

The first reported results obtained using the principles of this method are those of Duclaux [2] who made a photographic recording of the scattered light from a searchlight beam. The transmitted beam was visible on the photograph to an altitude of 3.4 km. Hulbert [3] extended these results in 1936 by photographing a beam to a 28-km altitude.

A monostatic lidar, the typical configuration of modern lidars, has the transmitter and receiver colocated, Fig. 1. Monostatic lidars use a pulsed light source and measure the round-trip time of the scattered light pulses to enable ranging, Fig. 2. The first reported use of a monostatic lidar was by Bureau [4] in 1938; this system was used for measuring cloud base heights.

In 1956, Friedland et al. [5] reported the first pulsed monostatic system for atmospheric density measurements. A pulsed monostatic lidar allows each light pulse fired to provide a complete

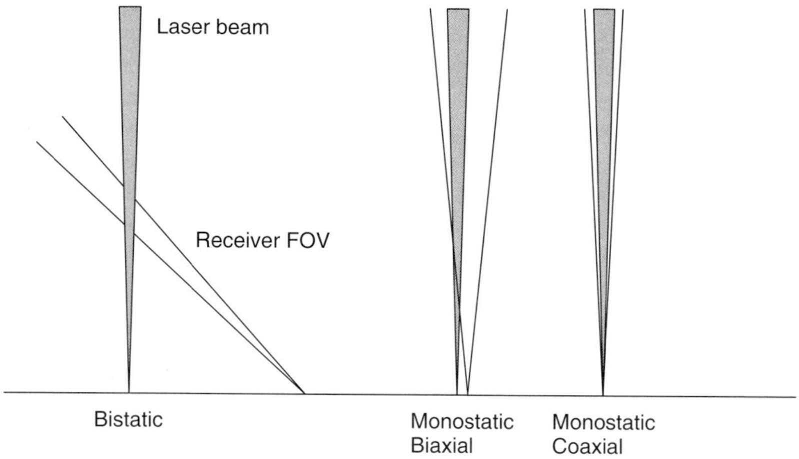


**Fig. 1** Geometric arrangements of lidar transmitter and receiver

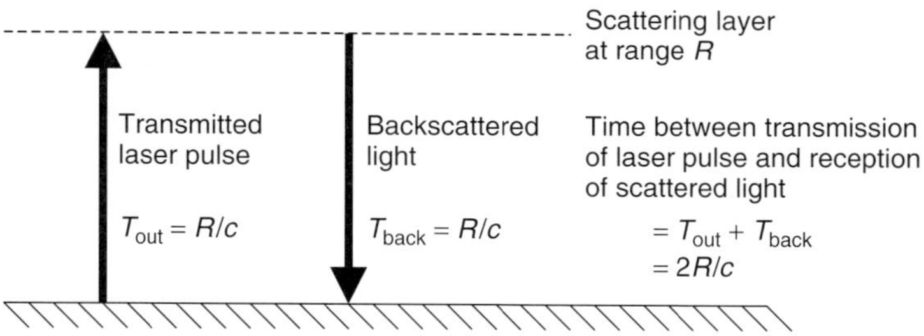


**Fig. 2** Relationship between backscatter interval and range

altitude-scattering profile; a bistatic system requires the alignment of the laser and the detector field-of-view (FOV) to be changed so that an altitude profile can be obtained. Although a monostatic lidar records a complete altitude profile for each pulse fired, commonly, many such profiles are required to obtain measurements with a useful signal-to-noise ratio (SNR).

The invention of the laser [6] in 1960 and the pulsed laser [7] in 1962 provided a powerful new light source for lidar systems. Since the invention of the laser, developments in lidar have been closely linked to advances in laser technology. The first use of a laser in a lidar system was reported in 1962 by Smullins and Fiocco [8], who detected laser light scattered from the lunar surface using a ruby laser that fired 0.5-J pulses at 694 nm.

## 3
## Lidar Basics

A lidar instrument can be conveniently divided into three subsystems: the transmitter, the receiver, and the detector. Each of these subsystems is discussed in the following sections in the context of an atmospheric lidar. Figure 3, a block diagram of a generic lidar system, shows how these subsystems fit together to form a complete lidar.

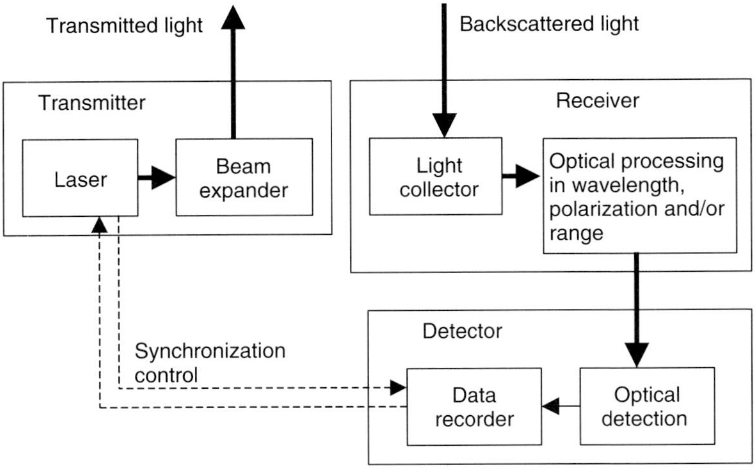


**Fig. 3** Schematic of a simple lidar

## 3.1 Transmitter

The transmitter of a lidar is the subsystem that generates light pulses and directs them into the atmosphere. Figure 4 shows the laser beam of the University of Western Ontario's Purple Crow Lidar against the night sky. Because of the special characteristics of the light they produce, pulsed lasers are ideal as the sources for lidar systems. Three properties of a pulsed laser – low beam divergence, extremely narrow spectral width, and short intense pulses – provide significant advantages over white light as the light source for a lidar.

**Fig. 4** The University of Western Ontario's Purple Crow Lidar operating at night. The sky near the horizon is illuminated by the sun

For most lidar systems, it is advantageous to have a transmitted beam with low divergence. The FOV of the detection system is one of the factors affecting the level of the background, measured scattered sunlight, moonlight, starlight, and so on in lidar measurements. A smaller FOV leads to a smaller measured background. Most lidars require that the transmitted laser beam be within the FOV of the detection system, thus a low FOV requires a low divergence laser beam. Many lidar systems incorporate a beam expander to reduce the divergence of the laser beam before transmission.

The inherent narrow spectral width of the laser has been used to advantage in many different ways in lidar. The simplest of these methods involves the spectral filtering of the backscattered light to reduce the background. Transmitting a beam with a narrow spectral width allows the detection optics of a lidar to spectrally filter incoming light and thus, selectively transmit photons at the laser wavelength. In practice, a narrowband interference filter is used; this transmits a relatively large fraction of the scattered laser light while transmitting only a very small fraction of background white light. This spectral selectivity allows the signal-to-background ratio of the measurement to be many orders of magnitude greater when a narrowband source and a detector system interference filter are used in a lidar.

The inherent pulsed properties of some types of laser systems make them ideal light sources for lidar; they allow ranging to be achieved by timing the scattered signal, Fig. 2. Other properties of the pulsed laser, namely, the pulse-repetition-frequency (PRF) and the pulse length, can have an influence on the operation of a lidar. The aliasing range of a lidar is the distance a laser pulse travels before the next pulse is transmitted. Usually, a lidar operates with a PRF such that at any time, no two pulses are within the range that causes scattering, that is, the aliasing range is much greater than the scattering range. The spatial length of the transmitted laser pulses ultimately limits the spatial resolution of the lidar measurements.

The type of laser used in a lidar system depends on the physical quantity the lidar is to measure. Some measurements require a very specific wavelength (i.e., resonance–fluorescence) or wavelengths (i.e., differential absorption lidar, DIAL) and can require complex laser systems to produce these wavelengths. Other lidars can operate over a wide wavelength range (i.e., Rayleigh, Raman, and aerosol lidars). The power, PRF, pulse length, and spectral width of a laser must also match the requirements of the measurements.

## 3.2
## Receiver

The receiver of a lidar collects and processes the scattered laser light and then directs it to a photodetector. The primary optic in a lidar receiver is the optical element that collects the light scattered back from the atmosphere. Along with the power of the laser, the size of the primary optic is an important factor in determining the effectiveness of a lidar system. A larger primary optic will collect a greater fraction of the scattered light and thus increase the signal measured by the lidar. The size of the primary optics used in lidars varies from about 10 cm, usually refractive systems, up to a few meters in diameter, usually reflective components. The smaller aperture optics are used in lidar systems that are designed to work at close range, for example, a few 100 m. Larger aperture primary optics are used in lidar systems that are designed to probe the middle and upper regions of the Earth's atmosphere in which the returned signal is a much smaller fraction of the transmitted power.

After collection by the primary optic, the backscattered light is usually processed in some way before being directed to the detector system. This processing can be based on wavelength, polarization, and/or range depending on the purpose for which the lidar has been designed.

The simplest form of spectral filtering is the use of a narrowband interference filter tuned to the laser wavelength. This is used to significantly reduce the background, as described in Sect. 3.1. More complex spectral filtering schemes are used for Doppler and high spectral resolution lidar [9–13]. These techniques require the measurement of spectral properties of the backscattered light.

Lidars used to study atmospheric aerosols and clouds often incorporate signal separation schemes based on the polarization of the backscattered light [14, 15] to obtain information about the scatterers, see Sect. 5.

Processing of the backscattered light based on range is usually performed in order to protect the detector from the intense near-field returns experienced by higher-power lidar systems. Exposing a photomultiplier (PMT) to a bright source such as near-field returns, even for a very short time, produces signal-induced noise, which affects the ability of the detection system to accurately record subsequent signals [16, 17]. Protecting the detector from these near-field returns can be achieved in a number of ways. One way is to use either a mechanical or electro-optical chopper that closes the optical path to the detector during and immediately after the laser fires. Another method involves removing the potential difference between two dynodes near the front of the PMT, effectively turning it off while the laser is fired; this technique is called *gating*.

## 3.3 Signal Detection and Recording

The signal detection and recording section of a lidar is that part of the instrument that takes the light from the receiver system and produces a permanent record of the backscattered intensity, and possibly wavelength and/or polarization, as a function of altitude. In the first lidar experiments, signal detection and recording was achieved using a photographic film [2, 3].

In modern lidars, detection and recording is achieved electronically. The detector converts the light into an electrical signal and the recorder is an electronic device or devices that process and store the electrical signal.

Photomultiplier tubes (PMTs) are generally used as detectors for incoherent lidar systems operating with visible or UV light. PMTs convert an incident photon into an electrical current pulse large enough to be detected by sensitive electronics (see SENSORS, OPTICAL). Other possibilities of detectors for lidar systems include multianode PMTs [18], avalanche photodiodes [19, 20], and charge-coupled devices (CCDs) [21, 22]. Coherent detection is discussed in Sect. 6.1.

The output of a PMT is in the form of current pulses that are produced both by photons entering the PMT as well as the thermal emissions of electrons inside the PMT, called *dark current*. The dark current is manifested as a background on the measured backscattered laser light.

There are two ways in which the output of a PMT can be recorded electronically. In one method, photon counting, the electrical pulses due to individual photons are counted; in the other technique, analog detection, the average current due to the PMT output pulses is measured and recorded. The most appropriate method to use depends on the average rate at which output pulses are produced.

### 3.3.1 Photon Counting

Photon counting is a two-step process. Firstly, the output of the PMT is filtered using a discriminator to remove a substantial number of the dark counts. Dark counts arise because of the thermal emission of electrons inside the PMT. These thermal emissions occur throughout the PMT and the location of the electron emission determines the gain it experiences before appearing as a pulse at the PMT output. This average dark-count gain will be less than the average photon-induced gain, that is, the average dark-count pulse amplitude is less than the average photon-induced pulse amplitude. This allows the use of a discriminator that selectively passes pulses that have amplitude greater than some preset level. A properly adjusted discriminator significantly increases the ratio of signal-to-dark counts in the pulse stream from the PMT.

The second step in the photon-counting process is counting of the electrical pulses; this is typically achieved using a Multichannel scaler (MCS). An MCS has a number of memory locations that become active sequentially, and for a fixed amount of time after the laser is fired. Each time a photon is detected, the number in the currently active memory location is incremented by 1. In this way, each memory location represents a measurement range bin, Fig. 5.

When the incident light is sufficiently intense that the electrical pulses output by the PMT start to significantly overlap, and as overlapping pulses are difficult to distinguish from individual pulses, photon counting becomes ineffective and analog detection becomes more appropriate.

### 3.3.2 **Analog Detection**

Analog detection is appropriate when the average count rate approaches the pulse-pair resolution of the detector system, usually on the order of 10 to 100 MHz depending on the PMT type and the speed of the discriminator and the MCS. Analog detection systems use a fast analog-to-digital converter to convert the average current from the PMT into digital form suitable for recording and manipulation on a computer. The digital-to-analog conversion must be done at a rate capable of giving the range resolution required in the measurements.

## 3.4 An Example of a Lidar Detection System

The Purple Crow Lidar [23, 24] at the University of Western Ontario detects scattering at four wavelengths, Fig. 6. An Nd:YAG laser operating at the second-harmonic frequency (532 nm) provides the light source for the Rayleigh (532 nm) and the two Raman shifted channels, $N_2$ (607 nm) and $H_2O$ (660 nm). The fourth channel is a sodium resonance–fluorescence channel that operates at 589 nm. Dichroic mirrors are used to separate light collected by the parabolic mirror into the four channels before being filtered by narrowband interference filters and directed onto the PMTs.

A rotating chopper is incorporated into the two high signal level channels (Rayleigh and sodium) to protect these PMTs from the intense near-field returns. The chopper comprises a rotating disk that has two teeth on the periphery; it rotates at 8400 rpm. The chopper is synchronized with the transmission of the laser pulses so

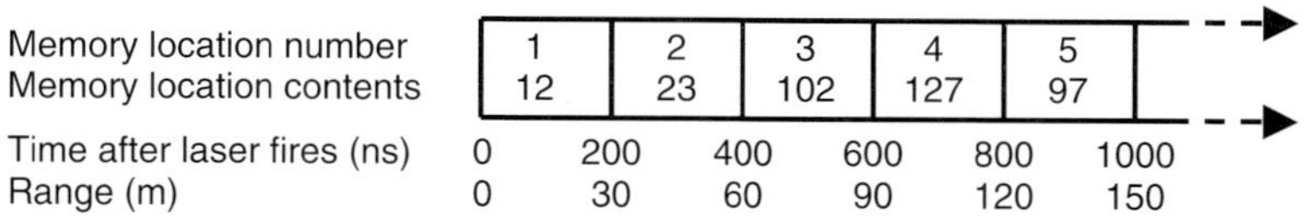


**Fig. 5** Schematic of MCS operation

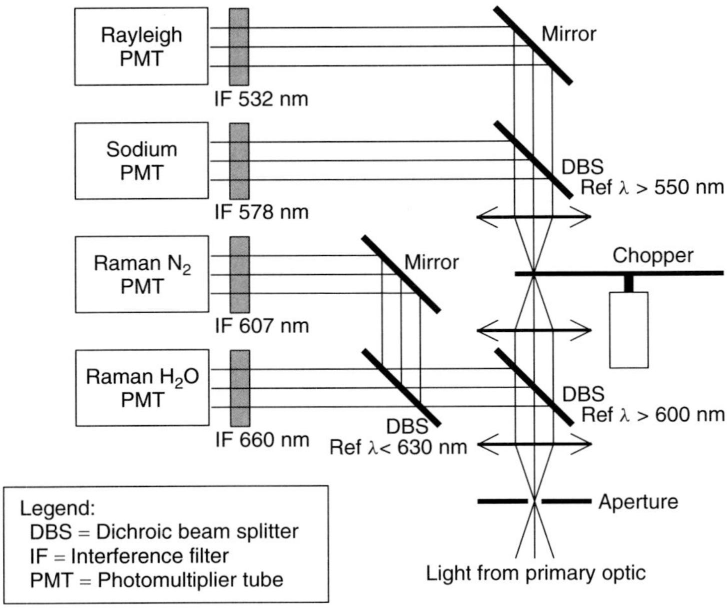


**Fig. 6** The University of Western Ontario's Purple Crow Lidar detection system

that scatter from below 20 km is blocked, however, the shutter is fully open by the time light scattered from 30 km returns to the instrument.

The two Raman channels are used for the detection of $H_2O$ and $N_2$ in the troposphere and stratosphere, allowing water vapor concentration and temperature profiles to be measured. Measurements from the Rayleigh and sodium channels are combined to provide temperature profiles from 30 to 110 km.

## 4
## Rayleigh Lidar

Rayleigh theory [25, 26] describes the scattering of light by particles that are small compared to the wavelength of the incident radiation and is applicable to the elastic scattering of light by atmospheric molecules (see also RADIATION INTERACTION WITH MOLECULES). The intensity of Rayleigh-scattered light is proportional to the density of the atmosphere and as the atmospheric temperature is related to density, the temperature can be determined from measurements of the intensity of the Rayleigh backscatter profile.

Rayleigh lidar is the name given to a class of lidar systems that measure the intensity of the Rayleigh backscatter from about 30 km up to around 80 or 100 km. The principle of operation of a Rayleigh lidar system is straightforward. A pulse of laser light is fired up into the atmosphere and any photons that are backscattered and collected by the receiving system are counted as a function of range. The measured Rayleigh backscatter intensity profile can be used to determine a relative density profile; absolute calibration of a Rayleigh lidar is difficult, due in part to the uncertainty in the transmission of the lower atmosphere. The relative density profile can be used to determine an absolute temperature profile by assuming that the atmosphere is in hydrostatic equilibrium and by applying the Ideal Gas Law. Details of the calculation and an error analysis for this technique can be found in both [27] and [28].

At altitudes below about 30 km, light is elastically scattered by aerosols in addition to the Rayleigh scattering from molecules; this scattering from aerosols contaminates the Rayleigh signal and sets the lower altitude limit for the application of the Rayleigh lidar. High spectral resolution techniques can be used to separate the scattering from these two sources [29].

The upper altitude limit for the Rayleigh lidar depends on the power of the lidar and the optical transmission of the lower atmosphere. Above about 90 km, the atmospheric density is extremely low resulting in low signal levels. Low signal levels give rise to large statistical uncertainties in the derived temperatures at the very top of the derived temperature profiles; at some point, the uncertainties become so large that the derived temperatures are not useful.

Rayleigh lidar systems are unable to operate when clouds obscure the middle atmosphere from their view. Additionally, most Rayleigh systems are able to operate only at nighttime due to the overwhelming presence of scattered solar photons during the day. However, the addition of an appropriately narrow band-pass filter in the receiver optics allows daytime measurements [30].

## 5
## Aerosol Lidar

The theory of scattering that was developed by Mie early this century is a general solution that covers the scattering of electromagnetic radiation by homogeneous spheres of all sizes and refractive indices.

The atmosphere contains particles with an infinite variety of shapes, sizes, and refractive indices, usually referred to as aerosols. The measurement of the properties of atmospheric aerosols is complicated by the small size and the fragility of these particles. The processes of evaporation, condensation, coagulation, absorption, desorption, and chemical reactions serve to change the atmospheric aerosol composition on short timescales. Care must be taken with direct sampling methods in order that the sampling process allows correct interpretation of the properties of the collected aerosols.

Aerosol concentrations in the atmosphere vary widely with altitude, time, and location; they also play an important role in the radiation budget of the earth by scattering and absorbing both incoming solar radiation and outgoing terrestrial radiation. Climate models require detailed information on the optical properties, as well as the spatial and temporal distribution of atmospheric aerosols. Typically, the aerosol concentration decreases with increasing altitude throughout the troposphere, however, there is a layer of aerosols in the atmosphere from about 15 to 23 km. This is known as the stratospheric aerosol layer or the Junge [31] layer. The aerosols in this layer are primarily volcanic in origin and lidar measurements have shown that the altitude range and density of the aerosols in this layer vary widely depending on recent volcanic activity [32].

Since the early 1960s, a large number of lidars have been used to study aerosols and clouds in the troposphere and lower stratosphere. Instruments using multiple wavelength transmitters and receivers as well as polarization techniques have been used to help quantify aerosol properties [33]. Lidars have been used to study polar stratospheric clouds (PSCs) [34, 35], to help understand the role they play in ozone depletion.

In September 1994, NASA flew a space shuttle mission, STS-64, which included the LITE experiment [36–38]. LITE was initially intended to serve as a technology development and validation exercise for future space lidar missions, however, it produced scientifically significant results [39, 40]. LITE incorporated an Nd:YAG laser operating simultaneously at three frequencies, the fundamental (1064 nm), the second harmonic (532 nm), and the third harmonic (355 nm). It also incorporated a system for automatic alignment of the laser beam into the FOV of the detector system.

## 6 Doppler Effects and Wind Lidar

Both random thermal motions and bulk mean flow (i.e., wind) contribute to the motion of the molecules and aerosols in the atmosphere. When a photon is scattered by a molecule or aerosol, it will suffer a change in frequency proportional to the component of its velocity in the direction of the laser beam, that is, line-of-sight (LOS) velocity, due to the Doppler effect.

Atmospheric aerosols have the same average velocity as atmospheric molecules as both types of scatterers are carried about by the wind, so that the average Doppler shift of the scattered light will be the same for scattering from both molecules and aerosols. The thermal velocity of molecules is much greater than that of aerosols due to their mass difference; this leads to the spectrum of the backscattered light having a distribution as shown in Fig. 7. The difference in the thermal (random) speeds of the molecules and aerosols leads to the two distinct regions of this distribution. The faster moving molecules result in a significantly wider spectral distribution than that due to aerosols [41, 42].

High spectral resolution measurements of backscattered light, in principle, allow the determination of LOS wind speed (from the frequency offset of the backscatter), the temperature (from the width of the molecular scattering function), and the separation of the signal due to molecules from that due to aerosols. In practice, making these high spectral resolution measurements with useful SNR is difficult, especially at large distances. However, several useful lidar techniques based on the measurement of Doppler shifts have been developed.

Separation of molecular and aerosol scattering is not necessary if only wind is to be measured since the average Doppler shift is the same for both molecular and aerosol scattering. The incoherent or direct detection wind lidar measures the LOS wind speed using optical techniques to

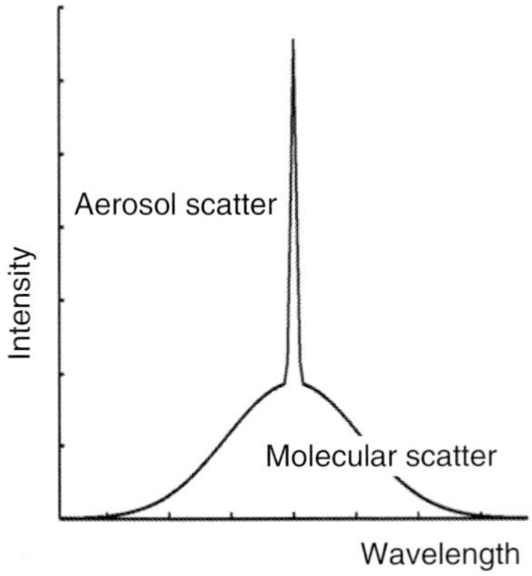


**Fig. 7** Modeled spectrum of monochromatic light scattered by the atmosphere

measure the spectrum of the backscattered light and the transmitted laser. The Doppler shift and LOS wind speed are calculated from the measured spectral information. Incoherent wind lidar has been applied to the measurement of wind fields in the stratosphere [43] and the troposphere [44], however, coherent detection techniques (Sect. 6.1) are more commonly used in the troposphere.

In principle, it is possible to determine atmospheric temperature by measuring the spectral shape of the light backscattered from air molecules in the middle atmosphere. However, using this Doppler technique, the SNR requirements for temperature measurement is much higher than that for measuring winds [45]. In practice, Rayleigh–Doppler temperature measurements are quite difficult. The advantage of this method of temperature determination is that the true kinetic temperature of the atmosphere is obtained without the need for the assumptions required by the Rayleigh lidar technique.

## 6.1
## Coherent Doppler Lidar

Coherent detection is used in a class of lidar designed for remote velocity measurement. This detection technique mixes the backscattered laser light with light from a local oscillator on a photomixer. The output of the photomixer is a radio frequency (RF) signal whose frequency is the difference between the frequencies of the two optical signals. Standard RF techniques are used to measure and record the photomixer output. The frequency of the measured RF signal can be used to determine the Doppler shift of the scattered laser light, which in turn allows the determination of the wind velocity [46].

Coherent lidar has more critical signal level requirements than the incoherent lidar. An incoherent lidar can make average measurements over an extended period of time to increase the SNR of the measurements. Although this decreases the temporal resolution of the measurements, this type of averaging cannot be applied to the coherent lidar in which a certain minimum signal level must be maintained to ensure valid measurements. With greater signal levels available in the lower atmosphere due to scattering from aerosols, the measurement of the Doppler shift via coherent detection becomes practical. Coherent Doppler lidar is used extensively in wind field mapping from the ground [47] and from the air [48]. There has been a great deal of interest in providing global tropospheric wind field maps based on space-based Doppler lidar systems; numerous studies [49] of both coherent and incoherent techniques have been undertaken in order to determine the best technique for this measurement.

# 7
# Differential Absorption Lidar (DIAL)

In 1964, Schotland [50] suggested a new lidar technique that has become known as the differential absorption lidar (DIAL). This technique finds application in measuring the concentration of trace species in the atmosphere, in particular, pollutants and other chemically important minor constituents. The DIAL technique takes advantage of the sharp variation in optical transmission of the atmosphere near the absorption line of an atomic or molecular species. A DIAL transmits two closely spaced wavelengths – one coincides with an absorption line of the constituent of interest ($\lambda_{on}$), while the other is in the wing

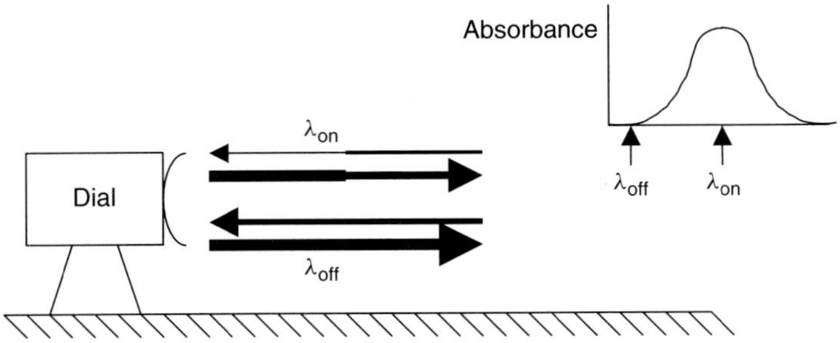


**Fig. 8** Operation of a differential absorption lidar (DIAL)

of this absorption line ($\lambda_{off}$), Fig. 8. During transmission of these two wavelengths through the atmosphere, the emission that is tuned to the absorption line will suffer greater attenuation than the emission in the wing of the line. The backscattered intensity at the two wavelengths can be used to determine the optical attenuation due to the species and thus the concentration of that species. A DIAL can transmit the two wavelengths simultaneously, or alternate pulses of the two frequencies. Whichever scheme is used, it is necessary to ensure that the system efficiency is the same at both wavelengths, or is well quantified.

The first use of a DIAL system was for the measurement of atmospheric water vapor concentration [51]. The DIAL technique has been extensively used for pollution monitoring in the lower regions of the atmosphere; it has been used for the detection of a number of species including NO, $O_3$, $SO_2$, and $CH_4$. An important application of DIAL has been the measurement of ozone profiles in the middle atmosphere [52, 53]. Traditional methods for measuring ozone give the amount of ozone integrated over the entire depth of the atmosphere; the DIAL technique gives height profiles of ozone concentration.

Temperature measurement with DIAL is possible when an absorption line has a temperature dependence, this technique has found limited applications in the atmosphere [54, 55].

## 8 Raman Lidar

When monochromatic light, or light of sufficiently narrow spectral width, is scattered by a molecular gas or liquid, the spectrum of the scattered light contains lines at wavelengths different from that of the incident radiation [56]. First observed by Raman [57], this effect is due to the interaction of the radiation with the quantized vibrational and rotational energy levels of the molecule (see also SPECTROSCOPY, RAMAN). Raman scattering involves a transfer of energy between the incident light and the molecule and is therefore an inelastic process. The cross sections due to Raman scattering are included by the Rayleigh scattering theory [58], although Raman spectroscopists use the term Rayleigh line to indicate only the elastically scattered central component of the scattered light. The intensity of the most intense vibrational shifted Raman lines is typically a factor of about 1000 lower than that for the elastic scattering by molecules; this varies somewhat from molecule to molecule. The spectral shift for vibrational Raman scattering is typically several hundred to a few thousand wave numbers. When scattering is accompanied by a change in the rotational state of the molecule only, pure rotational Raman lidar (PRRL), the scattered intensity is much greater, however, the spectral shift is much less, typically several to a few tens of wave numbers.

Each type of molecule has unique vibrational and rotational quantum energy levels and therefore, Raman scattering from each type of molecule has a unique spectral signature associated with these energy levels. This allows the presence of molecules to be determined by their scattered light spectra. Typically, the term *Raman lidar* refers to a lidar that makes use of a change in the vibrational state of the scattering molecule; this gives a much greater spectral shift in the scattered light than a change in the rotational state alone.

In the mid 1960s, Cooney [59] and Leonard [60] demonstrated the measurement of the Raman shifted light from nitrogen molecules in the troposphere by lidar. The Raman lidar has been most often applied to the measurement of atmospheric water vapor profiles [61, 62]. The determination of water vapor profiles is achieved by measuring the intensity of the light backscattered by both water vapor and nitrogen molecules, see Fig. 6. The measurement of the nitrogen Raman signal gives a reference that allows the absolute water vapor concentration to be determined. Raman lidar has also been applied to studying clouds [63]; there is a slight difference in the spectral signature of the scattering from the three phases of $H_2O$.

Raman lidar can be used to measure atmospheric temperature in two ways. As the vibrational Raman backscatter is spectrally distant from the scatter due to aerosols, the density profile of nitrogen can be measured. The mixing ratio of nitrogen in the atmosphere is a constant, except in the upper atmosphere; this allows the application of the Rayleigh lidar technique described above for the determination of temperature profiles. The Raman nitrogen signal is influenced by the presence of aerosols in that they affect the transmission of the atmosphere; this limits the application of this technique to the stratosphere and tropopause [64]. Aerosol concentrations in this region are typically relatively low and predictable, except after major volcanic events.

A second method of atmospheric temperature measurement utilizes the temperature dependence of the shape of the envelope of the pure rotational Raman spectrum [65]. High spectral resolution measurements allow the shape of the pure rotational Raman spectrum of air (effectively just nitrogen and oxygen) to be determined sufficiently well such that the temperature of the scatterers can be determined. As there is only a very small separation between the Raman pure rotational lines and the elastic scatter, careful attention must be paid to the effective separation of these signals in this technique [66].

## 9 Resonance Lidar

Resonant scattering occurs when the energy of an incident photon is equal to the energy of an allowed transition within an atom. This is an elastic process in which the atom absorbs the incident photon and instantly emits another photon at the same frequency. In general, the emitted photon will suffer a Doppler shift, however, the scatterer will not undergo a change in its internal energy level, that is, it is an elastic process. The fluorescence spectrum is unique to a particular atomic or molecular species. This can be used to identify and measure the concentration of a particular species [67, 68].

The constant ablation of meteors in the earth's upper atmosphere leads to the existence of extended layers of a variety of alkali metals in the 80 to 115 km region. While the concentration of these metals is low, their resonant-scattering cross sections are quite high. For instance, at 589 nm, the resonance–fluorescence cross section for sodium is about $10^{15}$ times larger than the cross-section for Rayleigh scattering from air. This large cross section allows these metals to be fairly easily exploited by ground-based lidars.

The atmospheric sodium layer is the most widely probed of the atmospheric alkali metal layers using lidar. The sodium abundance is relatively high and the transmitter frequency is relatively easy to generate.

The spectral shape of the sodium line at 589 nm, the $D_{2a}$ line, is temperature-dependent and the scattering cross section is proportional to the line shape. This allows the temperature of the sodium atoms and the atmosphere surrounding them to be determined from the spectral shape of the backscattered intensity. The measurement of the sodium $D_{2a}$ shape has been achieved by lidar in a number of ways [69, 70]. Usually, this measurement is achieved by transmitting narrow bandwidth laser pulses at two or three well-known frequencies within the sodium $D_{2a}$ line and recording the backscatter intensity at each of the transmitted frequencies separately. By knowing the frequency of the transmitted laser pulses and the intensity of the backscatter at each of the transmitted frequencies, the atmospheric temperature can be determined.

Resonance lidar temperature measurement is a direct spectral measurement and has associated uncertainties that are several orders of magnitude lower than those associated with Rayleigh temperature measurements in this altitude range.

Other alkali metals, including calcium [71], potassium [72], lithium [73], and iron [74] have also been used to study the mesopause region of the earth's atmosphere. Thomas [75] provides a review of the early work in this field. Resonance lidar requires laser transmissions at the precise frequency of the fluorescence line of the species being studied. Traditionally, dye lasers have been successfully used to probe many of these species, though working with these dyes is difficult in the field environment. Recently, solid-state lasers have been applied to resonance lidar systems [76].

Wind measurements are possible using a Na lidar. Transmission of laser pulses that are referenced to very narrow spectral features within the Na $D_{2a}$ line allow the Doppler shift of the backscattered light and hence, the LOS wind speed to be determined [77].

## 10
## Future

Lidar has established itself as one of the most important measurement techniques for atmospheric composition and dynamics from the surface to the upper atmosphere. It also has important uses in mapping, bathymetry, defense, oceanography, and natural resource management. Lidar solutions offer themselves to a wide range of environmental monitoring problems.

The technologies required for lidar are now mature enough that the past promises of global environmental monitoring by lidar are being fulfilled. Following two early space-based lidar systems – the French instrument ALISSA [78] on board the MIR platform and LITE [36] on the space shuttle, several space-based lidar systems are being planned, developed, or in operation. These include GLAS [79] for precision surface topography as well as cloud and aerosol studies, AMD-aelous [80] for global wind field measurement, the Vegetation Canopy Lidar [81] for measurements of the vegetation canopy and Earth surface topography, CALIPSO [82] to measure aerosols and clouds for climate studies, and MOLA [83], which has been used to measure the topography of Mars.

Another area where lidar has significant future potential is in digital 3-D imaging [84], particularly for including real objects, such as buildings, in virtual reality environments.

As technologies related to lidar such as laser and detector technologies develop further, lidar will continue to find application in a variety of remote sensing applications.

## Glossary

**Lidar:** Light detection and ranging.

**DIAL:** Differential absorption lidar.

**Active Remote Sensing:** A technique that transmits radiation that interacts with the object under study. A fraction of the radiation is returned to the instrument for analysis.

**Range:** Distance from lidar to object/region being monitored.

**Bistatic:** Lidar with the transmitter and receiver separated by a distance not insignificant compared to the operating range.

**Monostatic:** Having the transmitter and receiver colocated, that is, separated by a small distance compared to the operating range.

**Biaxial:** Having two axes; the optical axis of the transmitter is different from that of the receiver.

**Coaxial:** Having a single axis; the optical axis of the transmitter is coincident with that of the receiver.

**Bathymetry:** Water depth measurement, usually in coastal regions and in lakes.

**PRF:** Pulse-repetition frequency: the rate at which a laser produces output pulses.

**Middle Atmosphere:** The region of the atmosphere from about 10 km to about 100 km.

**Troposphere:** The region of the atmosphere from the ground to about 10 km varies with latitude and season. This region contains weather phenomena.

**Stratospheric Aerosol Layer:** A layer of aerosol (particulates) in the lower stratosphere (15 to 25 km) primarily volcanic in origin.

**Raman Lidar:** Lidar that uses the Raman component of the backscattered light.

**Doppler Lidar:** Lidar that measures wind and/or temperature based on the Doppler shift of the backscattered light.

**LOS:** Line of sight. Wind lidar typically measures the component of the wind velocity in the direction of the transmitted laser beam.

**PRRL:** Pure rotational Raman lidar: a lidar that utilizes the pure rotational part of the Raman scatter.

## References


[1] Synge, E. H. (**1930**), *Philos. Mag.* **52**, 1014–1020.
[2] Duclaux (**1936**), *J. Phys. Radiat.* **7**, 361.
[3] Hulbert, E. O. (**1937**), *J. Opt. Soc. Am.* **27**, 377–382.
[4] Bureau, R. (**1946**), *Meteorologie* **3**, 292.
[5] Friedland, S. S., Katzenstein, J., Zatzick, M. R. (**1956**), *J. Geophys. Res.* **61**, 415–434.
[6] Maiman, T. H. (**1960**), *Nature* **187**, 493.
[7] McClung, F. J., Hellworth, R. W. (**1962**), *J. Appl. Phys.* **33**, 828–829.
[8] Smullins, L. D., Fiocco, G. (**1962**), *Nature* **194**, 1267.
[9] Beneditti-Michelangeli, G., Congeduti, F., Fiocco, G. (**1972**), *JAS* **29**, 906–910.
[10] Abreu, V. J., Barnes, J. E., Hays, P. B. (**1992**), *Appl. Opt.* **31**, 4509–4514.
[11] Shipley, S. T., Tracey, D. H., Eloranta, E. W., Trauger, J. T., Sroga, J. T., Roesler, F. L., Weinman, J. A. (**1983**), *Appl. Opt.* **22**, 3716–3724.
[12] Fiocco, G., DeWolf, J. B. (**1968**), *JAS* **25**, 488–496.
[13] Chanin, M. L., Garnier, A., Hauchecorne, A., Porteneuve, J. (**1989**), *J. Geophys. Res.* **16**, 1273–1276.
[14] Carswell, A. I. (1983), in D. A. Killinger, A. Mooradian (Ed.), *Optical and Laser Remote Sensing*. Berlin: Springer-Verlag, pp. 318–326.
[15] Sassen, K., Benson, R. P., Spinhirne, J. D. (**2000**), *Geophys. Res. Lett.* **27**, 673–676.
[16] Cairo, F., Congeduti, F., Poli, M., Centurioni, S., DiDonfrancesco, G. (**1996**), *Rev. Sci. Instrum.* **67**, 3274–3280.
[17] Williamson, C. K., De Young, R. J. (**2000**), *Appl. Opt.* **39**, 1973–1979.
[18] McGill, M. J., Skinner, W. R. (**1997**), *Opt. Eng.* **36**, 139–145.
[19] Maruyama, T., Narusawa, F., Kudo, M., Tanaka, M., Saito, Y., Nomura, A. (**2002**), *Opt. Eng.* **41**, 395–402.
[20] Spinhirne, J. D. (**1993**), *IEEE Trans. Geosci. Remote Sens.* **31**, 48–55.
[21] Irgang, T. D., Hays, P. B., Skinner, W. R. (**2002**), *Appl. Opt.* **41**, 1145–1155.
[22] South, A. M., Povey, I. M., Jones, R. L. (**1998**), *J. Geophys. Res.* **103**, 31191–31202.
[23] Sica, R. J., Sargoytchev, S., Argall, P. S., Borra, E. F., Girard, L., Sparrow, C. T., Flatt, S. (**1995**), *Appl. Opt.* **34**, 6925–6936.
[24] Argall, P. S., Vassiliev, O. N., Sica, R. J., Mwangi, M. M. (**2000**), *Appl. Opt.* **39**, 2393–2400.
[25] Rayleigh, (J. W. Strutt) (**1871**), *Philos. Mag.* **41**, 274–279.
[26] Rayleigh, (J. W. Strutt) (**1871**), *Philos. Mag.* **41**, 447–454.
[27] Hauchecorne, A., Chanin, M. L. (**1980**), *Geophys. Res. Lett.* **7**, 565–568.
[28] Shibata, T., Kobuchi, M., Maeda, M. (**1986**), *Appl.Opt.* **25**, 685–688.
[29] Fiocco, G., Beneditti-Michelangeli, G., Maischberger, K., Madonna, E. (**1971**), *Nature* **229**, 79–80.
[30] Gille, S. T., Hauchecorne, A., Chanin, M. L. (**1991**), *J. Geophys. Res.* **96**, 7579–7587.
[31] Junge, C. E. (1963), *Air Chemistry and Radioactivity*. New York: Academic Press.
[32] McCormick, M. P., Swissler, T. J., Chu, W. P., Fuller Jr, W. H. (**1978**), *J. Atmos. Sci.* **35**, 1296–1303.
[33] Reagan, G. A., Spinhirne, J. D., McCormick, M. P. (**1989**), *Proc. IEEE* **77**, 433–448.
[34] Donovan, D. P., Fast, H., Makino, Y., Bird, J. C., Carswell, A. I., Davies, J., Duck, T. J., Kaminski, J. W., McElroy, C. T., Mittermeier, R. L., Pal, S. R., Savastiouk, V., Velkov, D., Whiteway, J. A. (**1997**), *Geophys. Res. Lett.* **24**, 2709–2712.
[35] Stefanutti, L., Morandi, M., Delguasta, M., Godin, S., Megie, G., Brechet, J., Piquard, J. (**1991**), *J. Geophys. Res.* **96**, 12975–12987.
[36] http://oea.larc.nasa.gov/PAIS/LITE.html.
[37] Winker, D. M., Couch, R. H., McCormick, M. P. (**1996**), *Proc. IEEE* **84**, 164–180.
[38] O'Connor, L. (**1995**), *Mech. Eng.* **117**, 77–79.
[39] Omah, A. H., Gardner, C. S. (**2001**), *J. Geophys. Res.* **106**, 1227–1236.
[40] Winker, D. M., Trepte, C. R. (**1998**), *Geophys. Res. Lett.* **25**, 3351–3354.

[41] Fiocco, G., Beneditti-Michelangeli, G., Maischberger, K., Madonna, E. (**1971**), *Nature* **229**, 79–80.

[42] Shimizu, H., Lee, S. A., She, C. Y. (**1983**), *Appl. Opt.* **22**, 1373–1382.

[43] Tepley, C. A., Sargoytchev, S. I., Hines, C. O. (**1991**), *Geophys. Res. Lett.* **18**, 167–170.

[44] Abreu, V. J., Barnes, J. E., Hays, P. B. (**1992**), *Appl. Opt.* **31**, 4509–4514.

[45] Tepley, C. A., Sargoytchev, S. I., Rojas, R. (**1993**), *IEEE Trans. Geosci. Remote Sens.* **31**, 36–47.

[46] Frehlich, R. (1976), in A. Consortini (Ed.), *Trends in Optics*. London: Academic Press.

[47] Hardesty, R. M. (1983), in D. K. Killinger, A. Mooradian (Eds.), *Optical and Laser Remote Sensing*. Berlin: Springer-Verlag.

[48] Bilbro, J., Fichtl, G., Fitzjarrald, D., Krause, M., Lee, R. (**1984**), *Bull. Am. Meteorol. Soc.* **65**, 348–359.

[49] Abreu, V. (**1979**), *Appl. Opt.* **18**, 2992–2997.

[50] Schotland, R. M. (1964), *Proc. of the Third Symposium on Remote Sensing of the Environment*, pp. 215–224.

[51] Schotland, R. M. (1966), *Proc. of the Fourth Symposium on Remote Sensing of the Environment*, pp. 273–283.

[52] Pelon, J., Godin, S., Megie, G. (**1986**), *J. Geophys. Res.* **91**, 8667–8671.

[53] McGee, T. J., Whiteman, D., Ferrare, R., Butler, J. J., Burris, J. F. (**1991**), *Opt. Eng.* **30**, 31–39.

[54] Browell, E. V., Ismail, S., Grossmann, B. E. (**1991**), *Appl. Opt.* **30**, 1517–1524.

[55] Theopold, F. A., Bosenberg, J. (**1993**), *J. Atmos. Ocean Tech.* **10**, 165–179.

[56] Herzberg, G. (1950), *Molecular Spectra and Molecular Structure I. Spectra of Diatomic Molecules*, (2nd ed.), New York: Van Nostrand Reinhold Company.

[57] Raman, C. V. (**1928**), *Indian J. Phys.* **2**, 387.

[58] Young, A. T. (**1981**), *Phys. Today* **35**, 42–48.

[59] Cooney, J. A. (**1968**), *Appl. Phys. Lett.* **12**, 40–42.

[60] Leonard, D. A. (**1967**), *Nature* **216**, 142–143.

[61] Cooney, J. A. (**1972**), *J. Geophys. Res.* **77**, 1078–1080.

[62] Melfi, S. H. (**1972**), *Appl.Opt.* **11**, 1605.

[63] Whiteman, D. N., Melfi, S. H. (**1999**), *J. Geophys. Res.* **104**, 31411–31419.

[64] Keckhut, P., Chanin, M. L., Hauchecorne, A. (**1990**), *Appl. Opt.* **29**, 5182–5186.

[65] Cooney, J. (**1971**), *J. Appl. Meteorol.* **11**, 108–112.

[66] Arshinov, Y., Bobrovnikov, S. (**1999**), *Appl. Opt.* **38**, 4635–4638.

[67] Chamberlain, J. W. (1961), *Physics of Aurora and Airglow*. New York: Academic Press.

[68] Measures, R. M. (1994), *Laser Remote Sensing: Fundamentals and Applications*. New York: John Wiley & Sons.

[69] Gibson, A., Thomas, L., Bhattacharyya, S. (**1979**), *Nature* **281**, 131–132.

[70] Frick, K. H., von Zahn, U. (**1985**), *J. Atmos. Terr. Phys.* **47**, 499–512.

[71] Granier, C., Jegou, J. P., Megie, G. (**1984**), *Geophys. Res. Lett.* **12**, 655–658.

[72] Megie, G., Bos, F., Blamont, J. E., Chanin, M. L. (**1978**), *Planet. Space Sci.* **26**, 27–35.

[73] Jegou, J. P., Chanin, M. L., Megie, G., Blamont, J. E. (**1980**), *Geophys. Res. Lett.* **7**, 995–998.

[74] Granier, C., Jegou, J. P., Megie, G. (**1989**), *Geophys. Res. Lett* **16**, 243–246.

[75] Thomas, L. (**1987**), *Philos. Trans. R.. Soc. London, Ser. A* **323**, 597–609.

[76] von Zahn, U., Hoffner, J. (**1996**), *Geophys. Res. Lett.* **23**, 141–144.

[77] Gardner, C. S., Tao, X., Papen, G. C. (**1995**), *Geophys. Res. Lett.* **22**, 2877–2880.

[78] Chanin, M. L., Hauchecorne, A., Malique, C., Nedeljkovic, D., Blamont, J. E., Desbois, M., Tulinov, G., Melnikov, V. (**1999**), *C R. Acad. Sci. II A* **328**, 359–366.

[79] http://glas.gsfc.nasa.gov/.

[80] http://www.esa.int/export/esaLP/aeolus.html.

[81] http://www.geog.umd.edu/vcl/.

[82] http://www-calipso.larc.nasa.gov/index.html.

[83] http://ltpwww.gsfc.nasa.gov/tharsis/mola.html.

[84] http://www.optech.on.ca/prodilris.htm.


## Further Reading


Frehlich, R. (1996), in A. Consortini (Ed.), *Trends in Optics: Research, Development and Applications*. London, UK: Academic Press, pp. 351–370.

Grant, W. B. (1995), in F. J. Duarte (Ed.), *Tunable Laser Applications*. New York: Marcel Dekker, pp. 213–305.

Killinger, D. K., Mooradian, A. (Ed.) (1983), *Optical and Laser Remote Sensing*. Berlin: Springer-Verlag.

Measures, R. M. (1984), *Laser Remote Sensing: Fundamentals and Applications*. New York: John Wiley & Sons.

Singh, U. H. (1997), in P. K. Rastogi (Ed.), *Optical Measurement Techniques and Application*. Norwood, MA: Artech House, pp. 369–396.

Thomas, L. (1995), in R. J. H. Clark, R. E. Hester (Eds.), *Spectroscopy in Environmental Science*. Chichester, UK: John Wiley & Sons, pp. 1–47.

Weitkamp, C. (1996), in E. Raschke (Ed.), *Radiation and Water in the Climate System*. Berlin, Germany: Springer-Verlag, pp. 217–247.

# Light and Lighting Systems

*John D. Bullough and Mark S. Rea*
*Lighting Research Center, Rensselaer Polytechnic Institute, 21 Union Street, Troy, New York 12180, USA*
*Phone: +1 518 687 7100; Fax: +1 518 687 7120; e-mail: bulloj@rpi.edu, ream@rpi.edu*

**Abstract**
Lighting is the application of visually effective optical radiation. This chapter provides the reader with some basic information about lighting terminology and technologies (lamps, ballasts, luminaires, and controls), and the results of recent research on the effective use of lighting. This article supplements the articles on COLOR VISION; COLORIMETRY; OPTICAL RADIATION SOURCES AND STANDARDS; and PHYSIOLOGICAL OPTICS.

**Keywords**
photometry; spectral sensitivity; lamps; luminaires; visual performance; efficiency; luminous efficacy.

## 1
## Definition of Light

### 1.1
### Luminous Efficiency Functions

Light is visually effective optical radiation [1]. Luminous efficiency functions are used to relate human spectral sensitivity to radiant power at different wavelengths.

The function that is nearly universally used in photometry is the *photopic* luminous efficiency function [2–5], denoted by $V(\lambda)$, as defined by the Commission Internationale de l'Éclairage (CIE). It is based on a series of studies mainly using the technique of flicker photometry to evaluate light at different wavelengths. In these studies, the stimulus was a small spot of light (subtending 2° at the observer's eye) alternating rapidly between a reference source and a test wavelength. Observers adjusted the intensity of the test wavelength until the flicker between the sources was least visible. At this point, the two sources were defined to have equivalent luminance. Several studies used this technique in the 1910s and 1920s to characterize light because it was a much more precise technique than brightness matching, where the intensity of a test field is adjusted until its brightness appears the same as a reference field of known fixed intensity.

Flicker photometry yields a peak spectral sensitivity at 555 nm, dropping off at shorter and longer wavelengths with a half-bandwidth of approximately 100 nm. The resulting function, shown in Fig. 1, closely matches the population-weighted average of the spectral sensitivities of the cone photoreceptors found in the human fovea, the central part of the retina used for fine spatial resolution (visual acuity).

Despite the esoteric nature of flicker photometry, $V(\lambda)$ has proven to be a remarkably robust characterization of light for visual performance [6, 7] in a broad range of applications such as those found in homes, schools, offices, and factories. Indeed, $V(\lambda)$ is the luminous efficiency function used by every standard-setting body in the world, including the CIE and the Illuminating Engineering Society of North America (IESNA) [1], and is used for all practical lighting measurements and calculations.

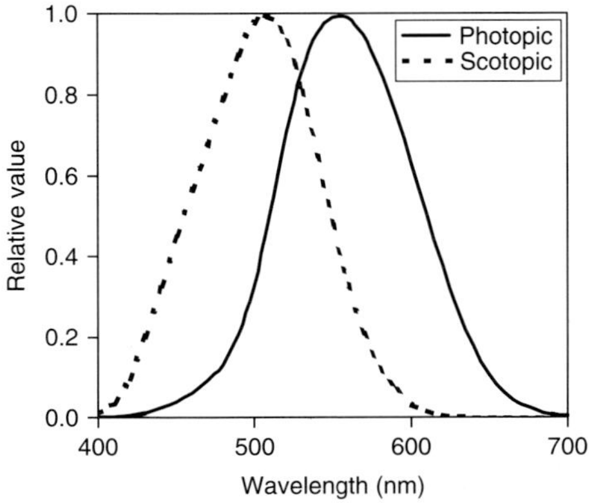


**Fig. 1** Photopic and scotopic luminous efficiency functions

A second luminous efficiency function is also sanctioned by the CIE to characterize light at very low levels. Rod photoreceptors have a spectral sensitivity that differs from that of the foveal cones, with a peak spectral sensitivity near 507 nm. This function was determined using brightness matching of different wavelengths (using a spatially divided field 30° in diameter), and is called the *scotopic luminous efficiency function*, $V'(\lambda)$ (Fig. 1) [2]. Since the dynamic operating range of rods corresponds to much lower light levels than that of cones, the scotopic function applies only at very low light levels. Consequently, it is virtually never used in practical photometry. At typical light levels used in outdoor lighting at night [1], both rods and cones simultaneously contribute to vision [8]. Section 3.3 describes issues pertaining to lighting at these mesopic levels between photopic and scotopic conditions.

### 1.2 Terminology

Light is characterized in photometric units. The lumen (lm) refers to the quantity of luminous flux without regard to a specific direction. The candela (cd) refers to the intensity of a light source in a particular direction. A point source having uniform luminous intensity of 1 cd will produce $4\pi$ lm; this corresponds to the solid angle (in steradians, sr) subtended by a sphere around the point source, so that the equivalent unit for cd is lm sr$^{-1}$. Illuminance is the density of luminous flux incident on a surface and is measured in lux (lx); 1 lx is equivalent to 1 lm m$^{-2}$. Luminance is a measure of the quantity of light coming from a surface toward an observer in a particular direction. It is typically measured in cd m$^{-2}$. Correlated color temperature (CCT) of a light source is measured in kelvins (K) and represents the temperature of an ideal blackbody that most closely matches the color of illumination emitted by a source.

## 2 Lighting Equipment

### 2.1 Lamps

Light-generating sources, or lamps, come in a variety of types and are based on different technologies. The types most commonly used in lighting systems include the following:

- Tungsten filament sources such as incandescent and halogen lamps.
- Fluorescent sources based on low-pressure mercury arcs.
- High-intensity discharge (HID) sources based on high-pressure arcs such as mercury vapor (MV), metal halide (MH), and high-pressure sodium (HPS) lamps.
- Low-pressure sodium (LPS) sources based on low-pressure sodium arcs.
- Light emitting diode (LED) sources based on semiconductor PN junctions.

Table 1 summarizes some important characteristics of common lamp types; Fig. 2 shows examples of several of these types of sources. Incandescent and halogen sources are common in residential interiors because they are relatively inexpensive products and retail display applications because of their color consistency and good optical control. Commercial interiors are illuminated primarily with fluorescent lamps because of their high luminous efficacy and good color-rendering properties. In exterior locations, HID sources are often used because of their high luminous

**Tab. 1** Important characteristics of common light sources [1, 9]

| *Property* | *Tungsten filament* | *Fluorescent* | *High-pressure sodium* | *Metal halide* | *Low-pressure sodium* | *White LED* | *Red LED* |
|---|---|---|---|---|---|---|---|
| Luminous efficacy (lm $W^{-1}$) | 14–20 | 70–104 | 90–125 | 81–100 | 150–183 | 15–20 | 25–40 |
| Median operating life (h) | 750–4000 | 10 000–20 000 | 24 000 | 10 000–20 000 | 18 000 | 10 000–50 000+[a] | 50 000–100 000+[a] |
| Lumen maintenance[b] | 85%–95% | 75%–95% | 90% | 80%–85% | 90% | 70%[a] | 70%[a] |
| Color rendering index (CRI) | 100 | 62–85 | 21 | 65–70 | Not applicable[c] | 50–90 | Not applicable[c] |
| Correlated color temperature (CCT) (K) | 2700–3200 | 2700–7500 | 2100 | 3200–4000 | Not applicable[c] | 4000–8500 | Not applicable[c] |
| Time from switching on to useful light output (min) | 0 | 0–1 | 1–5 | 1–10 | 7–15 | 0 | 0 |
| Minimum output when dimming | 0% | 1%–5% | 50% | 50% | Not usually dimmed | 0% | 0% |

[a] Operating life for most sources is defined as time to failure. LEDs will typically operate without failing for 100 000+ hours; the figures given reflect the time during which the source will maintain at least 70% light output under favorable operating conditions and are highly dependent upon LED packaging.
[b] Percentage of light output maintained by the lamp in an installation.
[c] CRI and CCT are not useful metrics for these sources, which produce saturated, nearly monochromatic light.

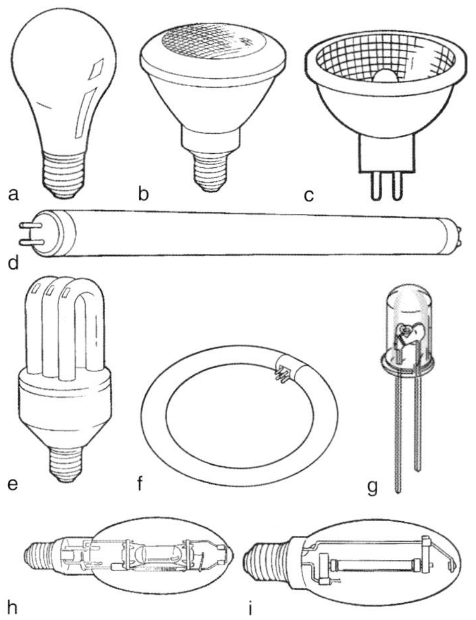

**Fig. 2** Common light sources (not to scale; a–c: tungsten filament lamps; d–f: fluorescent lamps; g: light emitting diode; h–i: high-intensity discharge lamps); (courtesy of the Lighting Research Center)

efficacy, good optical control, and high lumen packages that result in a need for fewer poles. These light sources are also described in greater detail in the articles on OPTICAL RADIATION SOURCES AND STANDARDS and DIODES, LIGHT-EMITTING.

The performance characteristics of LED sources for lighting applications have advanced significantly since the 1990s. Indicator-type colored LEDs have found their way into signage and signaling applications such as exit signs and traffic signals. The development of white LED products, based on blue LEDs in combination with phosphors that convert some of the LED's output to produce fluorescence in the phosphors, represents an opportunity for using these devices as illuminators. White LEDs at the present time are available that produce nearly 100 lm, with luminous efficacies exceeding those of incandescent lamps. These developments are beginning to make LEDs suitable for some low-light-level applications such as task lighting and outdoor path lighting.

The National Lighting Product Information Program (NLPIP) [9] provides performance information about several types of tungsten filament, fluorescent, HID and LED light sources.

## 2.2 Luminaires

A luminaire or a lighting fixture houses the lamp(s) within it and provides optical control through reflectors, lenses, refractors, and diffusers to distribute the light appropriately. Luminaires containing discharge lamps generally also house ballasts that provide the proper current and voltage required by those lamps. Luminaires are the building blocks of lighting systems, providing "layers of light." Some luminaires provide general ambient illumination enabling people to maneuver around a space and perform basic visual tasks; others provide location-specific lighting for demanding visual tasks such as food preparation, craft hobbies, and the like; still others are used for display lighting for such applications as appreciation of artwork.

Luminaires used indoors for general lighting are typically categorized by their overall distribution: *direct luminaires* direct light from the ceiling downward toward the surfaces being illuminated, *indirect luminaires* are suspended from the ceiling or are wall-mounted to reflect light from room surfaces to illuminate the entire room. *Direct–indirect* luminaires provide both types of lighting.

Interior luminaires that illuminate specific locations or areas include downlights, generally small, cylindrical units recessed into the ceiling; portable task lights, often found on desks and other work areas when additional lighting is desired; wall washers, which help to increase brightness of wall surfaces and can illuminate artwork on the walls. Track lighting can also be used to illuminate walls and artwork without the requirement for recessing the luminaire into the ceiling.

Luminaires used outdoors generally use HID or fluorescent lamps, and are categorized by the type of area they illuminate (e.g., Type I refers to long, narrow distributions for roadways; Type V refers to symmetrical patterns for lighting uniform areas such as parking lots [1]), and by their *cutoff distributions. Full cutoff* luminaires allow no upward light and have strict limits on the luminous intensity allowed at high angles from nadir (directly below the luminaire). These restrictions tend to make them attractive in locations where light pollution (increased sky glow that inhibits one's ability to see the stars at night), light trespass (light shining where it is not intended, such as onto adjacent properties or in residential windows), or glare are important design considerations. *Noncutoff* luminaires have no restrictions on upward light or light at high angles from nadir and tend to be used where the appearance of the luminaire (decorative or historic) is a more important consideration than illumination for visibility, such as in many downtown areas. The lighting industry is actively working to develop luminaires with attractive appearance and with cutoff characteristics to minimize light pollution, light trespass, and glare.

Luminaire selection depends upon the distribution of light provided by the luminaire as well as its efficiency for the intended task. Luminaires made of dark materials and with small apertures are less efficient than luminaires made of light-colored or specular materials with large openings. Section 3.4 provides additional information related to selecting the most efficient luminaires for various applications.

Figure 3 illustrates several types of luminaires for interior and exterior applications. NLPIP publications [9] have additional information on luminaires used in homes, offices, retail establishments, and outdoor areas, as well as ballasts for fluorescent systems.

## 2.3 Controls

In addition to simple switches, dimmers allow fine control of light levels from luminaires – motion sensors switch lighting off if no occupants are present, and

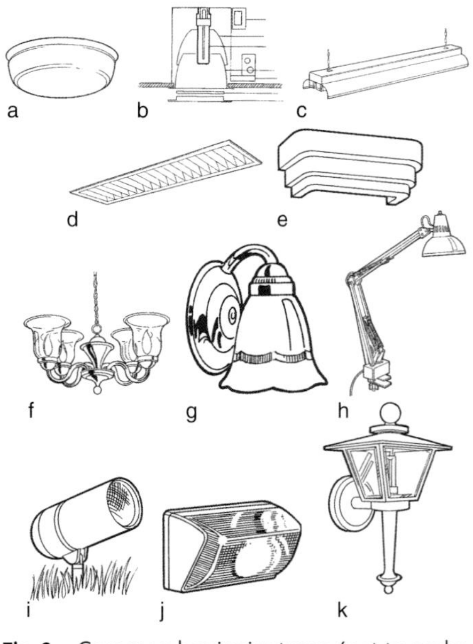


**Fig. 3** Common luminaire types (not to scale; a–h: interior luminaires; i–k: exterior luminaires); (courtesy of the Lighting Research Center)

photosensors dim lighting in response to the presence of daylight from windows or skylights. Lighting controls can be important elements in energy efficiency programs for lighting, also permitting greater flexibility in spaces; some government agencies require the use of motion sensors or photosensors as part of energy efficiency regulations. Contrary to common belief, fluorescent and HID lamps can be dimmed, although generally not to as low a light level as incandescent lamps.

One barrier to automated lighting controls in commercial spaces is the difficulty of setting the sensitivity and operational characteristics of the controls so that they operate without drawing attention to themselves. For example, motion sensors that do not detect small hand motions while typing and switch lights off unexpectedly, or photosensors that dim electric lighting too rapidly in response to daylight, can cause occupants to disable or deactivate them, thus preventing potential energy savings. Increasing the acceptability of lighting controls is still a major challenge for the lighting industry.

Information about motion sensors, photosensors, and more complex building automation systems that integrate lighting controls with controls for heating, ventilation, and other building systems is summarized by NLPIP [9].

## 3 Lighting Concepts

### 3.1 Visual Performance

Providing for adequate visual performance is one of the primary reasons for lighting. Different visual tasks require different amounts of light to ensure that performance of those tasks can occur well above the visual thresholds for seeing. This section describes the characteristics of visual tasks that affect visual performance and thus, the amount of light that will be needed to illuminate them.

Nearly every visual task can be characterized in terms of several factors that are related to its visibility [1]:

- time
- size
- contrast.

The presentation time of most visual tasks is long enough, therefore, this factor often need not be seriously considered in the design of lighting for visual performance. Still, for some visual detection tasks involving stimuli that are visible only briefly, temporal factors can be important. For very brief presentation times, up to 0.1 to 0.2 s, the product of a stimulus's luminance ($L$) and duration ($t$) are constant for equally visible stimuli. This is known as Bloch's law:

$$Lt = \text{constant}. \quad (1)$$

One application of Bloch's law in practical lighting situations occurs in automotive signaling, where sources such as LEDs have been shown to result in shorter response times than incandescent lamps [10].

As for size, visual performance improves with larger task sizes, but not in a linear manner. Doubling the size of printed text from 3 points to 6 points would have a significant effect on visibility, but further doubling to 12 points would not increase visibility by the same amount. Similarly, doubling the light level on the smaller text would result in a larger relative benefit than increasing the illuminance on large printed text.

Contrast refers to the difference in luminance or reflectance between a detail of interest and its surrounding background. Printed type such as that found in this book is designed to maximize contrast by using black ink on white pages. Some visual tasks are inherently low-contrast tasks, such as sewing white fabric with white thread. Just as with size, increasing the amount of light on a high-contrast task will have less benefit for visibility than the same increase in light level on a low-contrast task.

Another factor that will affect visual performance of any task is the age of the person. The light reaching the retina of an average 50-year old is only about half of that for an average 20-year old because of reduced pupil size and because of increased yellowing and thickening of the eye's crystalline lens, which reduces its transmittance and increases the amount of scattered light in the eye [1, 11].

A model of relative visual performance has been developed [12, 13] that allows the practitioner to precisely calculate the visual performance impact of changes in task illuminance, contrast, size, reflectance, and observer age. For most common visual tasks, though, light levels of just a few lux are sufficient for visual performance.

## 3.2 Brightness Perception

Light-level recommendations (see Table 2) are often much higher than those required for visual performance because it is also desirable to create perceptions of *brightness* in a space. People usually feel more comfortable and safer in brighter interior spaces. Generally, the brightness of a surface is correlated with its luminance, although this relationship is not precise. Saturated colors (e.g., red) appear brighter than unsaturated colors (e.g., pink) of equal luminance [1], indicating that brightness and color perception interact. In addition, the use of light sources having "cooler" ("blue–white") color appearance seems to increase the perception of brightness relative to "warmer" ("yellow–white") sources, even if matched for luminance [14], as seen in Fig. 4. Brightness is a complex phenomenon; for example, the presence of small, bright points of light, or "sparkle," in an environment will tend to increase the apparent brightness of the space [15].

**Tab. 2** Recommended illuminances for several common locations and tasks [1]

| *Task* | *Recommended illuminance (lx)* |
|---|---|
| Corridors | 50 |
| Graphic layout and artwork | 1000 |
| Kitchen sink (difficult seeing) | 500 |
| Postal mail processing | 500 |
| Reading photocopies | 300 |

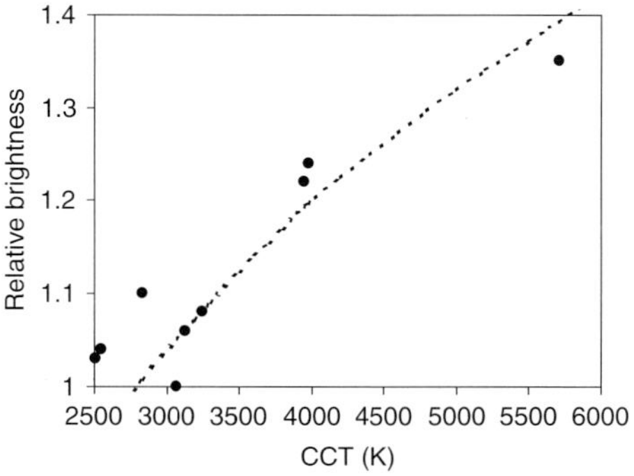


**Fig. 4** Relative brightness of light sources as a function of CCT [14]

### 3.3 Mesopic Vision

As mentioned in Sect. 1 of this chapter, many nighttime outdoor lighting applications result in light levels where both cone and rod photoreceptors operate to provide vision [1]. As with brightness, visual performance at these so-called mesopic light levels is not well correlated with luminance based on $V(\lambda)$ [8]. Interestingly, however, since the fovea does not contain rod photoreceptors, its spectral sensitivity is characterized by $V(\lambda)$ at all light levels, including very low levels [16].

At mesopic light levels, the spectral sensitivity of the peripheral retina can be characterized by various proportions of the photopic and scotopic luminous efficiency functions. A quantitative model of mesopic photometry based on reaction times has been developed [16, 17] that allows calculation of the relative proportion of photopic and scotopic luminous efficiency for a given adaptation luminance in the mesopic range. It has been shown that luminous efficiency at mesopic levels, $V_{\text{mes}}(\lambda)$ can be estimated in terms of a parameter $X$ as follows:

$$kV_{\text{mes}}(\lambda) = XV(\lambda) + (1 - X)V'(\lambda), \quad (2)$$

where $k$ is a constant such that the peak value of $V_{\text{mes}}(\lambda)$ is 1.

Consider the following example. HPS lamps are generally selected for roadway lighting because of their relatively high luminous efficacy. Table 3 lists, for several types of roadways, the (photopic) illuminance from HPS lamps required to meet IESNA [1] recommendations for roadway luminance, as well as the (photopic) illuminance from two other types of light sources that would result in equivalent mesopic luminances based on $V_{\text{mes}}(\lambda)$. The relative power requirement for each light source to provide that illuminance is also shown. For example, on major roads, HPS is the more efficient light source (results in less power used) than the MH source when providing equivalent mesopic light levels; but the opposite is true for local roads. Interestingly, the same analysis reveals that LPS is always the most efficient of the sources for the road types listed in Table 3, because of its inherently high light-conversion efficacy.

Work is underway to demonstrate the practical implications of the $V_{\text{mes}}(\lambda)$ formulation for applications such as visual searching [18] and driving [19–21]. These studies demonstrate that the performance of visual tasks depends upon many factors, and the applicability of $V_{\text{mes}}(\lambda)$ to task performance for targets of different sizes or contrast can vary widely. Nonetheless, $V_{\text{mes}}(\lambda)$ appears to be a useful tool for

**Tab. 3** Illuminances (photopic) to provide equivalent mesopic luminances as HPS on major and local roads

| *Light source* | *Major roads (luminance with HPS = 0.9 cd m$^{-2}$)* | | *Local roads (luminance with HPS = 0.3 cd m$^{-2}$)* | |
|---|---|---|---|---|
| | *Illuminance (lx)* | *Relative power (%)* | *Illuminance (lx)* | *Relative power (%)* |
| HPS | 40.4 | 100 | 13.5 | 100 |
| LPS | 41.2 | 71 | 15.8 | 82 |
| MH | 38.7 | 114 | 10.1 | 89 |

unifying photometry based on $V(\lambda)$ at high light levels with photometry based on $V'(\lambda)$ at very low levels.

### 3.4 Application Efficacy

Lighting is responsible for more than 30% of electricity use in commercial buildings and more than 10% in residences [22]. As such, significant attention has been paid to legislation and other regulatory measures to ensure that efficient lighting technologies and techniques are used. One example was the passage of the 1992 Energy Policy Act (EPACT), which effectively ended the manufacture of 25-mm diameter "cool white" and "warm white" halophosphor 40-W fluorescent lamps in the United States in favor of lamps with rare-earth phosphors and greater efficiency, or reduced-wattage (34 W) versions.

Indeed, luminous efficacy of light sources is generally seen as the primary measure of success in gauging energy efficiency. For the provision of general ambient illumination, luminous efficacy is a robust measure. One demonstration of the importance of luminous efficacy in lighting specification is the dominance of higher-efficacy fluorescent light sources in commercial and industrial facilities over incandescent lamps, which has occurred since the widespread development of fluorescent lamps in the 1940s.

Still, it is recognized among many lighting practitioners that luminous efficacy alone will not predict a light source's effectiveness for many lighting applications, especially when providing a second "layer of light" often desired to supplement general ambient illumination. One example in which sources with lower luminous efficacy are consistently employed rather than sources with higher luminous efficacy is retail display lighting, where tungsten filament-based reflector lamps are often recognized as more effective than more efficacious fluorescent lamps at highlighting merchandise for sale.

The concept of *application efficacy* [23] is an extension and generalization of luminous efficacy. It is based on the idea that the effectiveness of a lighting installation is dependent upon the area or location to be lighted, as well as a light source's ability to efficaciously generate luminous flux. Reflector lamps, as described above, are quite good at providing narrow light distributions such as those desired for retail display lighting, but luminous efficacy does not predict the effectiveness of these lamps because it is not defined with regard to where the luminous flux is to be delivered.

Application efficacy is defined as the average luminous flux from a lamp or lighting system, within a specific solid angle per unit power. It has units of lm $sr^{-1}$ $W^{-1}$; since lm $sr^{-1}$ is a measure of luminous intensity in cd, application efficacy can be defined in terms of cd $W^{-1}$. Application efficacy depends upon the location or object to be lighted. As an example [24], consider lighting a 0.6 m by 0.6 m kitchen sink top with a cylindrical downlight recessed into a ceiling located 2 m above the sink, as illustrated in Fig. 5. The sink would subtend about 0.09 sr, according to the formula for calculating solid angle ($\omega$, in sr),

$$\omega = \frac{A}{d^2}, \qquad (3)$$

where $A$ is the area of the sink ($0.6\ \text{m} \times 0.6\ \text{m} = 0.36\ \text{m}^2$) and $d$ is the distance from the sink to the luminaire (2 m).

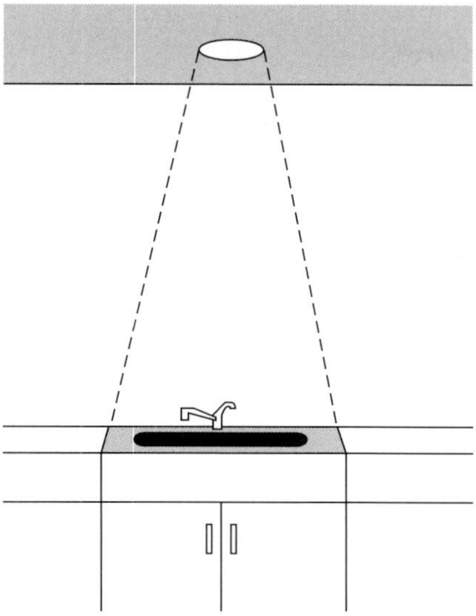

**Fig. 5** Illustration of application efficacy in lighting for a kitchen sink

The target illuminance for this application is 500 lx where difficult seeing is anticipated, according to Table 2 [1]. Downlight luminaires are available in versions that can contain incandescent lamps, halogen reflector lamps, or compact fluorescent lamps. If a downlight of each type was located above the sink, each would result in approximately 500 lx on the sink top:

- 175 W incandescent bulb downlight
- 35 W halogen reflector lamp downlight
- 52 W compact fluorescent lamp downlight.

It is clear that using the compact fluorescent lamp rather than the incandescent bulb would reduce the wattage needed. The reflector lamp, however, would achieve the target illuminance with the lowest wattage needed. This is despite the lower luminous efficacy of halogen reflector lamps in comparison to compact fluorescent lamps (see Table 1). The higher degree of optical control with reflector lamps than with compact fluorescent lamps allows the luminaire to have greater effectiveness at putting light where it is needed.

## 4 Discussion

Lighting is a practical application of optical science and engineering in the modern world. Selection of lighting systems goes beyond light source and luminaire selection, and even aesthetic appearance. An understanding of human vision, combined with knowledge about the most efficient uses of technologies in different locations are also key elements in the successful deployment of lighting systems.

Interestingly, even this understanding will likely be insufficient for the most effective application of lighting in the future. A rapidly growing area of research involves the use of light not for vision but for the regulation of physiological functions such as hormone secretion and body temperature that exhibit circadian (daily) rhythms. The characteristics of light for effective maintenance of circadian rhythms are very different from that for vision [25]; for example, the spectral sensitivity for circadian regulation is maximal at short visible wavelengths ("blue light"), and the circadian system is relatively insensitive to spatial variations in the field of view. As this research emerges, lighting engineering will certainly be impacted.

This overview is limited primarily to the application of lighting for buildings and exteriors in fixed systems, but the principles described here apply just as surely to the use of lighting in transportation. The increasing use of HID headlamps in vehicles is an example where the optimization between increased visibility with these more efficient, higher-output sources and the potential for increased glare has captured the attention of consumers, industry, and government [26]. LEDs are also finding their way onto vehicles in signal lighting and interior display lighting where their relatively high efficiency at generating light of saturated color is an asset.

The design of systems for effectively and efficiently delivering light thus has far-reaching implications for energy use, productivity, and aesthetic enjoyment of the built environment and of highways and vehicles, all of which can have significant impacts on the international economy.

## Acknowledgment

The authors gratefully acknowledge Dennis Guyon of the Lighting Research Center, Rensselaer Polytechnic Institute for his editorial review of this chapter.

## Glossary

Where noted (†), glossary terms are from the National Lighting Product Information Program [9].

**Ballast†:** A device required by electric-discharge light sources such as fluorescent or HID lamps to regulate voltage and current supplied to the lamp during start and throughout operation.

**Color Rendering Index (CRI)†:** A measure of the degree of color shift that objects undergo when illuminated by a lamp, compared with those same objects when illuminated by a reference source of comparable correlated color temperature. A CRI of 100 represents the maximum value. A lower CRI value indicates that some colors may appear unnatural when illuminated by the lamp.

**Contrast†:** Also known as luminance contrast, it is the relationship between the luminances of an object and its immediate background.

**Correlated Color Temperature (CCT)†:** A specification of the apparent color of a light source relative to the color appearance of an ideal incandescent source held at a particular temperature and measured on the kelvin (K) scale. As CCT increases, the appearance of the source shifts from reddish white toward bluish white. Lamps with a CCT rating below 3200 K are usually considered "warm" in appearance, whereas those with a CCT above 4000 K are usually considered "cool" in appearance.

**Cosine Distribution†:** A property of a light source such that its luminous intensity in a particular direction is proportional to the cosine of the angle from the normal to the source.

**Efficacy†:** The ratio of the light output of a lamp (lumens) to its active power (watts), expressed as lumens per watt.

**Glare†:** The sensation produced by luminances within the visual field that are sufficiently greater than the luminance to which the eyes are adapted, which causes annoyance, discomfort, or loss in visual performance and visibility.

**Illuminance†:** The amount of light (luminous flux) incident on a surface area. Illuminance is measured in lux (lumens/square meter) or footcandles (lumens/square foot). One footcandle equals 10.76 lux.

**Lamp†:** A radiant light source.

**Lamp Life†:** The median life span of a very large number of lamps (also known as the average rated life). Half of the lamps in a sample are likely to fail before the rated lamp life, and half are likely to survive beyond the rated lamp life.

**Light Pollution†:** Brightening of the sky caused by outdoor lighting and natural atmospheric and celestial factors.

**Light Trespass†:** An undesirable condition in which spill light is cast where it is not wanted.

**Lumen Maintenance†:** The lumens produced by a light source at any given time during its operating life as a percentage of its lumens at the beginning of life.

**Luminaire†:** A complete lighting unit consisting of a lamp or lamps and the parts designed to distribute the light, to position and protect the lamp(s), and to connect the lamp(s) to the power supply.

**Luminance†:** The photometric quantity most closely associated with the perception of brightness, measured in units of luminous intensity (candelas) per unit area (square feet or square meter).

**Luminous Flux†:** The rate of flow of light, measured in lumens.

**Luminous Intensity†:** The luminous flux on a small surface centered on and normal to the direction divided by the solid angle (in steradians) so that the surface subtends at the source. Luminous intensity can be expressed in candelas or in lumens per steradian.

**Mesopic:** The adaptation level in which both rods and cones contribute to vision, generally associated with luminances between 0.001 and 3 candelas per square meter (cd $m^{-2}$).

**Nadir†:** The angle pointing directly downward from the luminaire, or $0^\circ$.

**Phosphor†:** The white, powdered material coating the inside of the glass tube of a lamp. The phosphors fluoresce (emit visible light) when excited by the ultraviolet radiation produced by the mercury vapor that is energized by the electric arc sustained inside the lamp.

**Photopic:** The adaptation level in which only cones contribute to vision, generally associated with luminances higher than 3 candelas per square meter (cd $m^{-2}$).

**Reflectance:** The ratio of the luminous flux reflected from a surface to the luminous flux incident upon it.

**Scotopic:** The adaptation level in which only rods contribute to vision, generally associated with luminances lower than 0.001 candelas per square meter (cd $m^{-2}$).

**Spectral Power Distribution†:** A representation of the radiant power emitted by a light source as a function of wavelength.

**Transmittance:** The ratio of the luminous flux transmitted through a material to the luminous flux incident upon it.

**Visual Performance†:** The quantitative assessment of the performance of a visual task, taking into consideration speed and accuracy.

## References


[1] Rea, M. S. (Ed.) (2000), *IESNA Lighting Handbook: Reference and Application*, (9th ed.), New York: Illuminating Engineering Society of North America.

[2] Commission Internationale de l'Éclairage (1978), *Light as a True Visual Quantity: Principles of Measurement.* Paris: CIE.

[3] Judd, D. B. (1951), *CIE Proceedings*, Vol. 1, Part 7, Stockholm: CIE, p. 11.

[4] Commission Internationale de l'Éclairage (1990), *CIE 1988 2° Spectral Luminous Efficiency Function.* Vienna: CIE.

[5] Schanda, J., Morren, L., Rea, M., Rositani-Ronchi, L., Walraven, P. (**2002**), *Lighting Res. Technol.* **34**(1), 69–78.

[6] Smith, S. W., Rea, M. S. (1980), *CIE Proceedings.* Kyoto: CIE, pp. 207–211.

[7] Boyce, P. R., Akashi, Y., Hunter, C. M., Bullough, J. D. (**2003**), *Lighting Res. Technol.* **35**(2), 141–161.

[8] Commission Internationale de l'Éclairage (1989), *Mesopic Photometry: Special Problems and Practical Solutions.* Vienna: CIE.

[9] National Lighting Product Information Program. Accessed online March 31, 2003 at http://www.lrc.rpi.edu/nlpip.

[10] Sivak, M., Flannagan, M. J., Sato, T., Traube, E. C., Aoki, M. (**1994**), *Ergonomics* **37**(6), 989–994.

[11] Weale, R. A. (**1961**), *Trans. Illum. Eng. (London)* **26**(2), 95–100.

[12] Rea, M. S., Ouellette, M. J. (**1988**), *Lighting Res. Technol* **20**(4), 139–153.

[13] Rea, M. S., Ouellette, M. J. (**1991**), *Lighting Res. Technol* **23**(3), 135–144.

[14] Alman, D. H. (**1977**), *J. Illum. Eng. Soc* **7**(1), 55–62.

[15] Akashi, Y., Tanabe, Y., Akashi, I., Mukai, K. (**2001**), *Lighting Res. Technol.* **32**(1), 19–26.

[16] He, Y., Rea, M. S., Bierman, A., Bullough, J. (**1997**), *J. Illum. Eng. Soc* **26**(1), 125–138.

[17] He, Y., Bierman, A., Rea, M. S. (**1998**), *Lighting Res. Technol.* **30**(4), 175–181.

[18] Mehra, M., Rea, M., Maniccia, D. (1998), *IESNA Conference Proceedings.* San Antonio: IESNA, pp. 171–198.

[19] Bullough, J. D., Rea, M. S. (**2000**), *Lighting Res. Technol.* **32**(4), 194–198.

[20] Lingard, R., Rea, M. (**2002**), *J. Illum. Eng. Soc.* **31**(1), 33–39.

[21] Akashi, Y., Rea, M. (**2002**), *J. Illum. Eng. Soc.* **31**(1), 85–94.

[22] Goeke, K. (1998), *1998 Baseline Energy Outlook.* Sacramento: California Energy Commission.

[23] Rea, M. S., Bullough, J. D. (**2001**), *J. Illum. Eng. Soc.* **30**(2), 73–96.

[24] Banwell, P., Figueiro, M. G. (2001), *IESNA Conference Proceedings.* Ottawa: IESNA, pp. 149–166.

[25] Rea, M. S., Figueiro, M. G., Bullough, J. D. (**2002**), *Lighting Res. Technol.* **34**(3), 177–190.

[26] Bullough, J. D., Van Derlofske, J. (**2003**), *OE Mag.* **3**(6), 17–18.

[27] Ikeda, M., Yaguchi, H., Sagawa, K. (**1982**), *J. Opt. Soc. Am* **72**, 1660–1665.


## Further Reading


Boyce, P. R. (2003), *Human Factors in Lighting*, (2nd ed.), New York: Taylor & Francis.

Bullough, J. (1997), *Time-Saver Standards for Architectural Design Data*, (7th ed.), New York: McGraw-Hill, pp. D231–D254.

Leslie, R. P., Conway, K. M. (1996), *The Lighting Pattern Book for Homes.* New York: McGraw-Hill.

Leslie, R. P., Rodgers, P. A. (1996), *The Outdoor Lighting Pattern Book.* New York: McGraw-Hill.

Moyer, J. L. (1992), *The Landscape Lighting Book.* New York: John Wiley & Sons.

Rea, M. S. (Ed.) (1992), *Selected Papers on Architectural Lighting, SPIE Milestone Series Volume MS 58*, Bellingham, Washington: SPIE Optical Engineering Press.

Rea, M. S. (Ed.) (2000), *IESNA Lighting Handbook: Reference and Application*, (9th ed.), New York: Illuminating Engineering Society of North America.

# Linear Optics

*Andreas Dorsel*
*Carl Zeiss SMT AG, Oberkochen, Germany*
*Phone: +49-7364-20-7915; Fax: +49-7364-20-8800; e-mail: a.dorsel@zeiss.de*

**Abstract**
This article gives a condensed overview of the key areas of linear optics, addressing fundamental aspects of light propagation and their application. After first principles like the speed of light and the laws of reflection and refraction, as well as the Fresnel formulas for transmission and reflection, total internal reflection, and linear absorption, geometric optics is briefly covered with the basic descriptions of imaging using simple mirrors and lenses. As examples of simple optical instruments, the single-lens eye, the magnifier and microscope, as well as astronomical and terrestrial telescopes are included. Subsequently, phase-dependent phenomena like interference and diffraction are described, including examples of a few basic cases that can be used to address more complex cases. Anisotropic media and polarization effects are covered next, including brief notes on electro-optic, magneto-optic, and acousto-optic effects. Finally, examples of recent developments are described as an outlook to contemporary optics work.

**Keywords**
optics; refraction; diffraction; reflection; geometric optics; interference; polarization; light.

## 1 Introduction

When optical pumping and especially the invention of the laser opened the vast and fascinating field of *non*linear optics in 1960, it became meaningful if not necessary to introduce the distinction of linear optics for all optics that had been dealt with until then. Linear optics can be defined as that field of science that covers all phenomena of light propagation in media (including vacuum) in which the response of the medium (i.e., the polarization of that medium) is directly proportional to the field strength of the electromagnetic wave used to describe the light. What then *is* light? Originally restricted to the visible part of the spectrum ranging from 380 to 780 nm, the notion of light is commonly used to describe an extremely wide range of wavelengths of electromagnetic waves reaching from tens and hundreds of micrometers in the far-infrared (FIR) region down to the extreme ultraviolet (XUV) or even to x rays with nanometer wavelengths. The theoretical descriptions of light follow a dual approach in trying to describe its nature: the particle model and the wave model. In addition, geometrical optics may be regarded as a third less fundamental but most successful way of solving optical problems.

A most important feature of linear optics is the principle of linear superposition, which states that the electric field at a given point in space and time is the vector sum of the fields from all sources contributing:

$$E = E_1 + E_1 + E_3 + \cdots. \quad (1)$$

Historically, optics may be traced back to ancient Greece where Euclid noticed that "light travels along straight lines in homogeneous media." Hero of Alexandria was able to derive the law of reflection in the first century B.C. starting from the additional assumption that light travels the shortest possible path between two points, but it took more than 15 centuries until Willebrord Snell (1580–1626) discovered the correct law of refraction as the third principle of geometrical optics.

Christiaan Huygens (1629–1695) introduced a first description of light using waves that was able to explain phenomena like birefringence. Thomas Young (1773–1829) and Augustin Fresnel (1788–1827) made most important contributions to the development of the wave model, introducing the concepts of interference and diffraction. The model of geometrical or ray optics was included in the wave theories as a special case with wavelengths approaching zero. James Clerk Maxwell (1831–1879) set the foundations for the description of light as an electromagnetic wave, which is still valid today.

In the 20th century, the emission and absorption or detection of radiation were included in the theoretical treatment of quantum optics. Here the particle model employing the photon to describe light and dating back to René du Perron Descartes (1596–1650), Pierre de Fermat (1601–1665), and Isaac Newton (1642–1727) has found renewed interest among physicists. Linear optics can be generalized to include effects of special and general relativity, although these are often not observable with present equipment.

## 2 Fundamental Properties of Light

### 2.1 The Velocity of Light

The determination of the velocity of light has been the object of many ingenious experiments. Interest in measurements of the velocity of light was caused by Einstein's theory of relativity, which declared it to be a fundamental constant, an unattainable limiting speed for all objects with finite rest mass, and, in its general form, defined the energy of any object as the product of its mass and the square of the velocity of light in vacuum.

As early as 1675, the Danish astronomer Ole Rømer found the velocity of light to be finite as he determined it from the apparent times of revolution of the moons of Jupiter; these are apparently changing as Jupiter approaches or flees a terrestrial observer, because of the relative orbital motion of the two planets. Many experiments were performed to get a more and more accurate value of the velocity of light. Only about 50 years later, Bradley used stellar aberration and obtained a much more precise result. Fizeau, Foucault, and Michelson each made valuable contributions in this field until this noble competition was ended in 1983 by the *Conférence Générale des Poids et Mésures*, which defined the velocity of light in vacuum at its then-known value of $c = 299\,792\,458 \text{ m s}^{-1}$.

This permits one to measure distance (as a multiple of the wavelength of an optical radiation source) by measuring the frequency of that source; this, however, can be traced back to radiofrequency counting and thus to one of the most advanced precision measurement techniques available today (see also [1] and OPTICAL METROLOGY). These measurements yielded many other interesting insights. When Foucault extended the measurement of the velocity of light to other media than air and vacuum – if we allow ourselves for simplicity to call vacuum a medium – he found that the velocity in water, an optically denser medium than air, was lower than in air. This was in agreement with the predictions of the wave model and contradicted the particle model of Newton, which explained refraction by an *increased* velocity of light in denser media.

These remarks refer to the phase velocity of a signal only; it is the velocity of a point of constant phase. For a modulated (i.e., information-carrying) wave, the group velocity describes the speed at which a point of constant amplitude is advancing; only in nondispersive media (i.e., strictly speaking only in vacuum) are these two velocities equal.

The ratio of the velocity of light in vacuum, $c_0$, as compared to its velocity in a medium, $c$, is defined as the refractive index $n$ of that medium:

$$n = \frac{c_0}{c}. \qquad (2)$$

## 2.2
## Dispersion

It was soon found out that this index varied with wavelength and thus (angular) frequency for all media except vacuum: $n = n(\omega)$. This effect is called dispersion, an increase of refractive index with frequency corresponding to normal dispersion and a decrease to anomalous dispersion. Today, quantum mechanics can explain quantitatively this characteristic behavior of the index of refraction from the behavior of atoms and molecules acting as forced oscillators under the influence of the radiation field.

In addition, a distinction had to be made between phase velocity and signal velocity as these are only identical in the absence of dispersion, i.e., in vacuum. Also, these effects should not be confused with modal dispersion in optical wave guides, which is caused by bulk effects (i.e., material dispersion as just described) *and* spatial boundary conditions.

## 2.3
## The Doppler Effect

When a source of light moves relative to an observer, its spectrum is shifted for that observer: When it approaches the observer, all frequencies are increased in his frame of reference; when it moves away from him, they are decreased. This is due to the Doppler effect, the latter describing similar phenomena for all wave processes. For a monochromatic source with frequency $f$ and vacuum wavelength $\lambda_0 = c_0/f$ decreasing its distance from the observer with a velocity $v$, the observed wavelength is

$$\lambda_0' = \lambda_0 \sqrt{\frac{c_0 - v}{c_0 + v}} \tag{3}$$

and the corresponding frequency is

$$f' = f \sqrt{\frac{c_0 + v}{c_0 - v}}. \tag{4}$$

The physical explanation, e.g., for a receding source is that consecutive crests or hollows of the wave have to travel an increasing distance to the observer, so that the time between the arrival of two crests is greater than for a source at rest; thus, the measured oscillation period is longer for that observer, the frequency lower, and the wavelength longer. For an approaching source, all these effects are simply reversed.

For values of $v$ small compared to $c_0$, these expressions may be approximated by the simpler forms

$$\lambda_0' = \lambda_0 \left(1 - \frac{v}{c_0}\right) \tag{5}$$

and

$$f' = f \left(1 + \frac{v}{c_0}\right), \tag{6}$$

whence this effect is sometimes also called the linear or first-order Doppler effect. This attribute became necessary as the theory of special relativity yields an additional effect, the quadratic or second-order Doppler effect, sometimes also called the transverse Doppler effect as it is more easily observed when the source moves transverse to the observer's line of sight, thus suppressing any first-order effect. In this case, the formulas are

$$f' = f \sqrt{1 - \frac{v^2}{c_0^2}} \tag{7}$$

and

$$\lambda_0' = \frac{\lambda_0}{\sqrt{1 - v^2/c_0^2}}, \tag{8}$$

showing that only a redshift is possible.

## 2.4 Polarization

As light waves are transverse, they may have specific states of polarization, i.e., of the direction of oscillation of the electric field vector:

If the field vector oscillates (almost) exclusively in one linear direction, thus defining a plane together with the propagation direction, the light wave is called linearly or plane polarized. For a given direction of propagation, any state of linear polarization can be described as the linear superposition of two orthogonal components of plane-polarized light that are in phase.

If two orthogonal components are superposed at a phase difference different from zero, elliptically polarized light will result. In case the phase difference is $\pm 90°$ and the magnitudes of the two components are identical, we have circularly polarized light: The tip of the electric field vector moves around the origin on a circle either clockwise or counterclockwise.

When the polarization characteristics of light were first described, the electromagnetic wave model was yet to come, and unfortunately, the plane of polarization was at that time defined in such a way that it is normal to the electric field vector. Naturally, ignoring the distinction between these two definitions quite often has led to confusion. In this article (see, e.g., Sect. 3.3), this historical convention is followed simply because it seems to be most commonly used today.

Originally, reflection phenomena or crystals were employed to polarize light. Today, light is often polarized using polarizing films (Polaroid) that absorb light of one linear polarization while transmitting the orthogonal polarization. Thin-film beam splitters can be designed to show polarizing behavior, too (see POLARIZATION OPTICS, POLARIMETERS AND POLARIZATION SPECTROMETERS).

## 2.5 Linear and Angular Momentum of Light Waves

### 2.5.1 Linear Momentum of Electromagnetic Waves

Electromagnetic waves carry linear momentum, which is changed when a light wave is changed in its direction – e.g., when reflected by a plane mirror. As conservation of momentum must be satisfied, an ideally reflecting mirror will experience a resulting force directed along the normal of the mirror. This effect, often referred to as radiation pressure or photon pressure, can be explained by classical electrodynamics, but is more easily explained in the photon model: A light beam with power $P$ and of frequency $\nu$ has a photon rate $n$ of

$$n = \frac{P}{h\nu}, \tag{9}$$

where $h$ is Planck's constant, $h = 6.626 \times 10^{-34}$ J·s. Each of these photons has a linear momentum of $p = E/c_0 = h\nu/c_0$. Thus, for normal incidence on a mirror (where $\mathbf{p}$ changes to $-\mathbf{p}$) we have a momentum transfer rate or force $F$ of

$$F = 2np = \frac{2P}{c_0}. \tag{10}$$

For a perfect absorber, the force would be half this value. Although these forces are usually very small, they were detected near the beginning of the 20th century, and the invention of the laser was followed by particle levitation and handling using laser beams; applications in spacecraft propulsion based on photon recoil are still in the research stage.

### 2.5.2 Angular Momentum of Electromagnetic Waves

The photons in a circularly polarized light beam each carry an angular momentum $l = h/2\pi$. Transition through a half-wave plate (see section "Half-wave Plate") changes light of left circular polarization to right circular polarization and vice versa. The conservation of angular momentum then requires a torque to be exerted on the half-wave plate, which has been detected experimentally [2].

# 3 Optics of Plane Interfaces and Isotropic Media

As wave properties of light are more evident in phenomena like interference (see Sect. 5.1) and diffraction (see Sect. 5.2), we will restrict ourselves to ray optics in this section. In general, it is quite justified to identify a ray with the wave normal of an (almost) infinite plane wave.

## 3.1 Reflection

When a light ray is incident on an interface, i.e., on the (locally) plane boundary surface

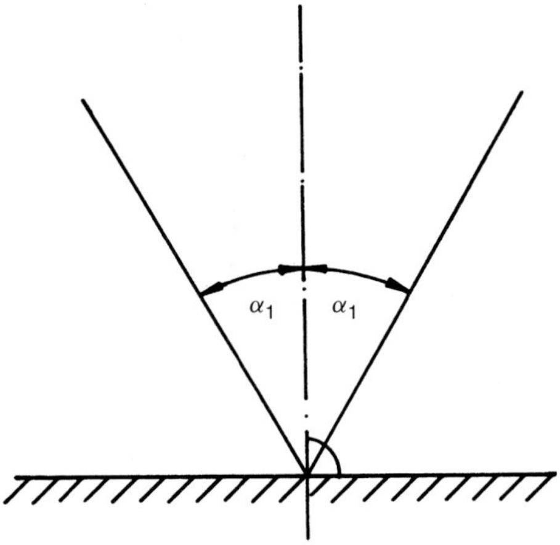


**Fig. 1** The law of reflection

between two adjacent media, at least some portion of it may be reflected. The plane containing the normal to the interface and the incident ray is called the plane of incidence. The fundamental law of reflection in geometrical optics states that the reflected beam will lie in the plane of incidence and will form the same angle with the normal to the interface as the incident ray (Fig. 1).

Let us assume a vector representation that uses the normal of the interface as one of its basis vectors. All components of the reflected wave's unit vector are then identical to those of the incident wave's unit vector, except that the normal component changes sign.

### 3.2 Refraction

The law of refraction or Snell's law is somewhat more complicated. At a plane interface separating regions with different indices of refraction $n_1$ and $n_2$, a transmitted ray will be changed in its direction (see Fig. 2). It will still lie in the plane of incidence, and for the values of the angles between the incident (in $n_1$) and

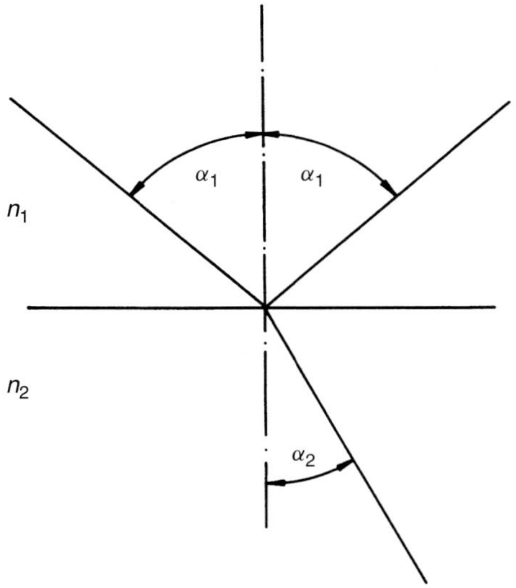


**Fig. 2** The law of refraction

transmitted (in $n_2$) beams and the interface normal the relation

$$n_1 \sin\alpha_1 = n_2 \sin\alpha_2 \qquad (11)$$

is valid. If no solution of this equation for real values of $\alpha_1$ or $\alpha_2$ exists, total internal reflection occurs (see Sect. 3.4).

### 3.3 Transmission

The transmission and reflection of light waves at interfaces are dependent on the state of polarization. If the plane of incidence is coincident with the plane of oscillation of the electric field vector, the light is called $\pi$ polarized. If, instead, the plane of polarization coincides with the plane of incidence, the light is called $\sigma$ polarized (from the German *senkrecht*). The reflection and transmission of energy at an interface separating regions of refractive indices $n_1$ and $n_2$ are described by the Fresnel formulas; for light incident from the side with refractive index $n_1$ with

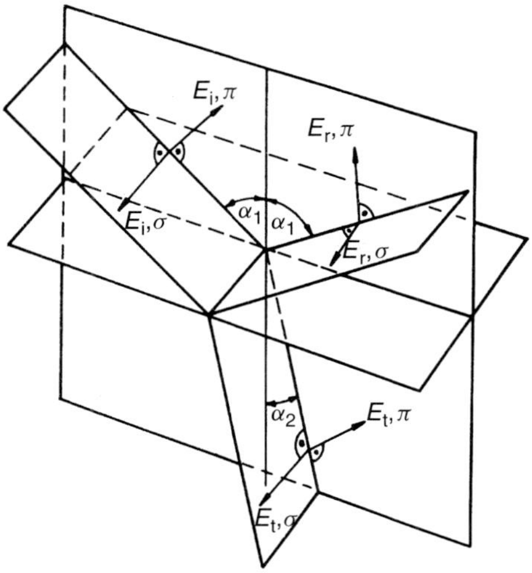


**Fig. 3** Transmission and reflection (after Alonso, M., Finn, E. J. (1976), *Fundamental University Physics*, Vol. 2. Reading, MA: Addison-Wesley, p. 789ff [3])

an angle of incidence $\alpha_1$ and an angle $\alpha_2$ between the refracted ray and the interface normal, these are (compare Fig. 3)

$$R_\pi = \left(\frac{n_1 \cos\alpha_2 - n_2 \cos\alpha_1}{n_1 \cos\alpha_2 + n_2 \cos\alpha_1}\right)^2, \tag{12}$$

$$R_\sigma = \left(\frac{n_1 \cos\alpha_1 - n_2 \cos\alpha_2}{n_1 \cos\alpha_1 + n_2 \cos\alpha_2}\right)^2, \tag{13}$$

$$T_\pi = \frac{n_2 \cos\alpha_2}{n_1 \cos\alpha_1}\left(\frac{2n_1 \cos\alpha_1}{n_1 \cos\alpha_2 + n_2 \cos\alpha_1}\right)^2, \tag{14}$$

$$T_\sigma = \frac{n_2 \cos\alpha_2}{n_1 \cos\alpha_1}\left(\frac{2n_1 \cos\alpha_1}{n_1 \cos\alpha_1 + n_2 \cos\alpha_2}\right)^2. \tag{15}$$

One should note that these formulas are describing *energy* transfer. When light with mixed polarization is incident, the fields have to be described as superpositions of $\sigma$-polarized and $\pi$-polarized waves.

For nonabsorbing media, $n_1$ and $n_2$ are real and $R_\pi$ vanishes for $\tan\alpha_1 = n_2/n_1$. This angle is called Brewster's angle or polarizing angle. The physical model for this situation is that the dipoles in the refracted ray cannot radiate in their direction of oscillation, i.e., orthogonally to the direction of propagation of that ray; thus, Brewster's angle is equivalent to the refracted and reflected rays being orthogonal to each other. If a plane wave is incident on an interface at Brewster's angle, the reflected wave is completely polarized. Stacks of plates or thin films at or near Brewster's angle can be used as polarizers.

### 3.4
### Total Internal Reflection

The relation between angles in the law of refraction (Snell's law) shows that for $n_1 > n_2$ a choice of $\sin\alpha_1 > n_2/n_1$ will lead to $\sin\alpha_2 > 1$, which is not defined for a real value of $\alpha_2$. For these values, total internal reflection occurs: Only a sharply decaying evanescent field probes into the region with index $n_2$, and the entire energy of the wave is reflected at the interface. The reflected beam may be attenuated through absorption or scattering, however. In the wave picture, the evanescent wave advances so much faster than the incident wave that the wave path (ray) is bent back into the medium from which it is incident.

Frustrated total internal reflection occurs when a second interface to a medium with sufficiently high index of refraction (e.g., $n_1$) is brought within the region where the evanescent field is still significant. In this case, the Fresnel equations have to be used with the complex angle that satisfies the law of refraction, and interference from the two closely spaced interfaces has to be taken into account.

### 3.5
### Absorption

Materials absorb light waves in different regions of the electromagnetic spectrum. The absorbed light may be reemitted at the same frequency or at one or more different frequencies, or its energy may be used for completely different processes – e.g., simply heating the medium. When these processes are due to electronic, vibrational, or rotational excitation of crystals, molecules, or atoms, they are called absorption.

In a homogeneous absorbing medium with absorption coefficient $\alpha$, the irradiance $I$, i.e., the energy per unit area per unit time, of a plane wave traveling in the positive $x$ direction is described by Beer's law,

$$I = I_0 \exp(-\alpha x), \qquad (16)$$

where $I_0$ is the irradiance at $x = 0$. It is derived from the assumption that the decrease $\mathrm{d}I$ caused by passing through an infinitesimal length $\mathrm{d}x$ of absorbing material is the product of the local irradiance $I$ and of the absorption coefficient: $\mathrm{d}I = -\alpha I \,\mathrm{d}x$.

### 3.6
### Extinction

The attenuation of light by absorption *and* scattering is called extinction. The processes most relevant in scattering normally are Rayleigh and Mie scattering. These two are generally characterized by the relative sizes of the scatterers and the wavelength of the light involved. For Rayleigh scattering to occur, the scattering particles must be small compared to the wavelength of the light scattered. In this case, the strength of the scattering is proportional to the fourth power of the light frequency. Although originally developed to cover particles of arbitrary size, the expression Mie scattering is mostly reserved for scatterers of a size similar to the wavelength of the light under consideration.

The scattered light is normally changed in frequency by quite small amounts by the Doppler effect from the moving scatterers. As the angular distribution of the scattered light is normally not isotropic and may have a pronounced peak in the forward direction, care has to be taken in experiments to discern the unscattered light from light scattered through small angles.

Other scattering processes such as Brillouin scattering and Raman scattering may be accompanied by significant amounts of frequency shifts; these imply energy transfer from solids or molecules.

## 4
## Geometrical Optics

Optics is closely linked to vision and thus to images. Accordingly, imaging optics play a major role in the applications of optics. Later on, we shall see how diffraction limits the performance of different optical devices. In the limit of the wavelength $\lambda$ approaching zero, the wave-optical description converges toward the ray-optical image, i.e., geometrical optics, as diffraction effects become negligible. For the time being, we will make the approximation that the wavelength of light is negligibly small compared to system dimensions so that a ray-optical treatment characteristic of geometrical optics is sufficient (see GEOMETRIC OPTICS).

Most of the treatment presented here is limited to the so-called paraxial case, where the light rays considered deviate so little from the axis of the optical system that the sine of the angle they form with the optical axis is sufficiently approximated by its argument: $\sin(x) \approx x$. For a more complete approach, the article GEOMETRIC OPTICS and the literature, e.g., Born and Wolf [4], should be consulted.

### 4.1 Plane Mirrors

The most simple kind of imaging is performed by a plane mirror: The image position can be derived by geometrically mirroring the object at the plane of the mirror, which is then the plane of symmetry between the object and its image. The image thus created is a virtual image: If a screen is brought to its apparent position, the image will not be apparent on it.

### 4.2 Spherical Mirrors

Spherical surfaces for mirrors and lenses are most common as they are comparatively easy to produce. Spherical mirrors can be convex or concave (this distinction naturally is only justified for a nontransmitting mirror): A convex mirror has a protruding center, while a concave mirror has a receding center.

#### 4.2.1 Concave Mirrors

When a bundle of parallel rays of light is incident on a concave mirror parallel to its axis, these will be deflected so that they intersect at a common point $F$. This point is called the focus of the mirror. Its distance from the center of the mirror is the focal length (cf. Fig. 4). A concave mirror of radius $R$ has a focal length $f = R/2$.

Vice versa, rays impinging on the mirror surface after going through the focus will afterward be parallel to the axis. In

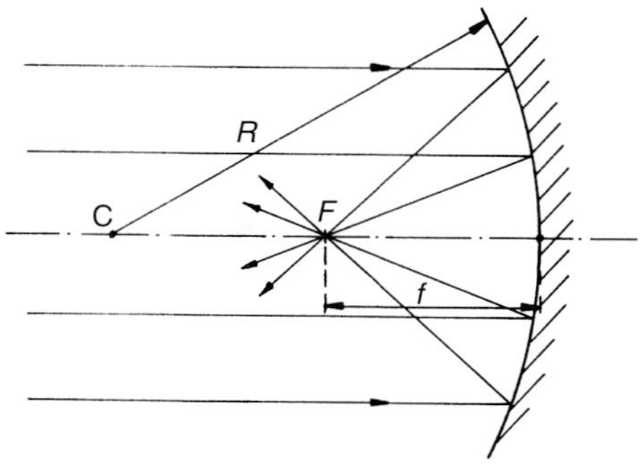


**Fig. 4** Focusing of parallel light rays by a concave mirror (after Hecht, E., Zajac, A. (1974), *Optics*. Reading, MA: Addison-Wesley [5])

addition, a ray going through the center of curvature of the mirror will always be reflected into itself, as it is always perpendicular to the mirror surface. With these rules, it is easy to construct the image point for any given object point, as follows.

Rays are drawn toward the mirror from the object point in directions parallel to the mirror axis, toward the focus, and toward the center of curvature. At their point of incidence on the mirror, they are then prolonged according to the above rules. The intersection of the reflected rays is the image point. If these do not intersect, either they are parallel, which means that the image is at infinity, or their prolongations on the opposite side of the mirror will intersect to form the virtual image point.

The mathematical treatment of this simple imaging situation yields the relation between the object and image distances (measured along the mirror axis) $s_1$ and $s_2$ and the focal length:

$$\frac{1}{f} = \frac{1}{s_1} + \frac{1}{s_2}. \quad (17)$$

Positive values for $s_1$ and $s_2$ correspond to an image or object position in front of the mirror; negative values indicate

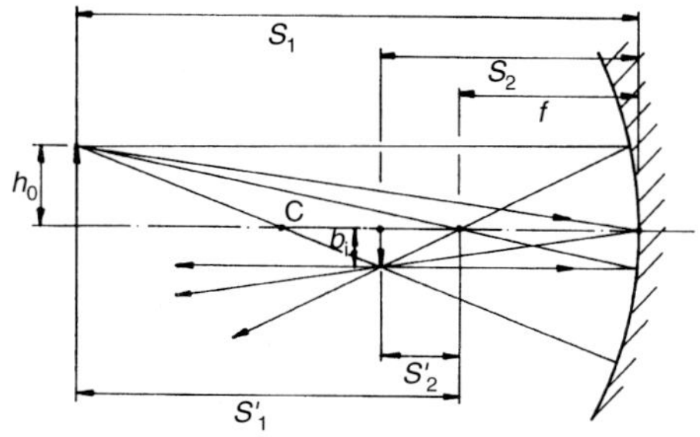


**Fig. 5** Image formation for a concave mirror (after Hecht, E., Zajac, A. (1974), *Optics*. Reading, MA: Addison-Wesley [5])

that the (virtual) object or image seems to lie behind the mirror. With $s_i' = s_i - f$ for $i = 1, 2$, this imaging equation can be written in a simpler form as

$$s_1' s_2' = f^2. \tag{18}$$

The magnification (or demagnification) can be derived from Fig. 5 directly:

$$M = \frac{h_i}{h_o} = -\frac{s_2}{s_1}. \tag{19}$$

If $s_1 < f$, we get a virtual upright magnified image.

#### 4.2.2 Convex Mirrors

If a bundle of parallel rays is incident on a convex mirror parallel to the mirror axis, it will be reflected in such a way that the emerging rays seem to have a common point of origin; this point is called the focus of that mirror. For a mirror with a radius of curvature $R$, this point is situated at a distance $-f = R/2$ behind the mirror measured along the ray that defines the optical axis. For reasons of symmetry, the optical axis of a spherical mirror alone is not defined but has to be derived from other components in a system or from the illumination conditions under which the mirror is used. For a mirror alone and an incident bundle of parallel rays, it is given by the ray that impinges on the mirror at normal incidence.

Figure 6 shows the situation just described. As all light paths can be reversed, it is immediately obvious that a bundle of rays converging toward the focus of such a mirror will be collimated to a parallel bundle after reflection.

Imaging of a real object is represented in Fig. 7. A convex mirror will always yield a virtual image for a real object. Geometrical construction of the image point for a given object point is simple when referring to the situation in Fig. 6: A ray parallel to the mirror axis and a ray toward the focus are drawn from those points to the mirror surface, and from there they are drawn further according to the rules given above. They then seem to emerge from a common origin behind the mirror, the image point.

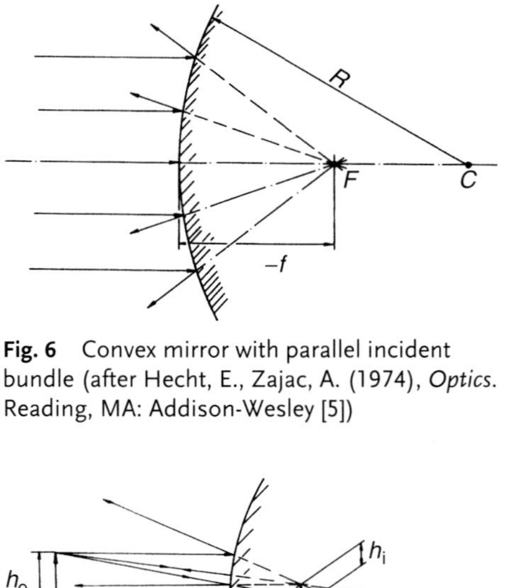


**Fig. 6** Convex mirror with parallel incident bundle (after Hecht, E., Zajac, A. (1974), *Optics*. Reading, MA: Addison-Wesley [5])

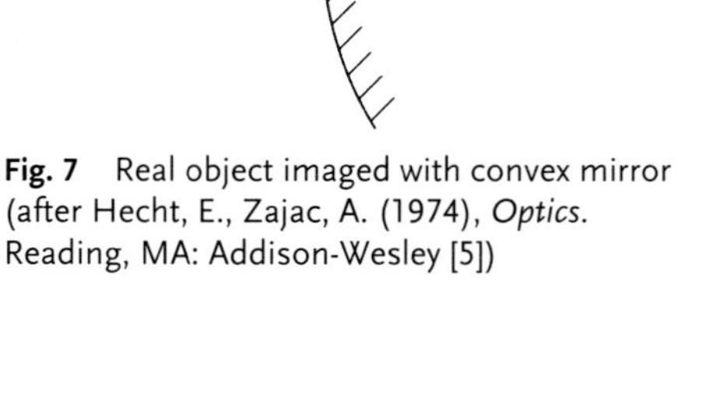


**Fig. 7** Real object imaged with convex mirror (after Hecht, E., Zajac, A. (1974), *Optics*. Reading, MA: Addison-Wesley [5])

This procedure can be performed for all object points so that a complete image can be constructed.

Sometimes it may be helpful to use a third kind of ray in constructing the image of a given object: A ray incident toward the mirror's center of curvature $C$ will always simply be reversed in direction upon (normal incidence) reflection. The mathematical description of the image position is quite simple:

$$\frac{1}{f} = \frac{1}{s_1} + \frac{1}{s_2}. \tag{20}$$

Here $s_1$ and $s_2$ are the object and image separations from the mirror (measured along the mirror axis in paraxial optics). The term $f$ is negative for a convex mirror. A negative value for $s$ means that an image (or a virtual object) is situated behind the mirror. Note that Eqs. (17) and (20) for concave and convex mirrors are identical and that only the focal length changes its sign.

A Virtual object (VO) may be created using a concave mirror. If the position of such a virtual object lies between the focus and the mirror surface, even a convex mirror will yield a real image (Fig. 8).

## 4.3 Spherical Lenses

When refraction occurs at a spherical interface, (de)focusing effects will occur.

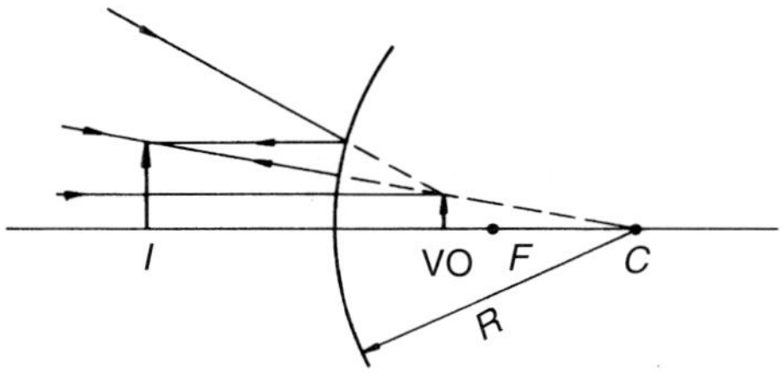


**Fig. 8** Virtual object imaged with convex mirror. VO denotes the virtual object and $I$ the image obtained therefrom

Here we deal with spherical surfaces only, but advanced optical systems often employ aspherical surfaces also. However, the deviations from a sphere are often small for imaging systems with $f$ numbers corresponding to paraxial optics. (The $f$ number is the ratio of the focal length to the diameter of the entrance pupil. In paraxial optics, we typically encounter $f$ numbers of about 10 or more.)

### 4.3.1 Single-interface Lens

Although most technical lenses have two surfaces, there are two reasons why the single-interface lens is a worthwhile object for study: First, any given lens can be treated mathematically as an arrangement of consecutive single-surface lenses, and second, the single-interface lens is a first approximation to the imaging system of the human eye, where object space and image space are not filled with the same medium.

Figure 9 shows the imaging situation for a lens consisting of a spherical surface with radius $R$ separating the object space with refractive index $n_1$ from the image space with refractive index $n_2$ where $n_1 > n_2$ is assumed (the situation may, e.g., be given for a spherical fish bowl). In this case, two focal lengths appear, one for each medium:

$$f_1 = \frac{Rn_1}{(n_1 - n_2)} \tag{21}$$

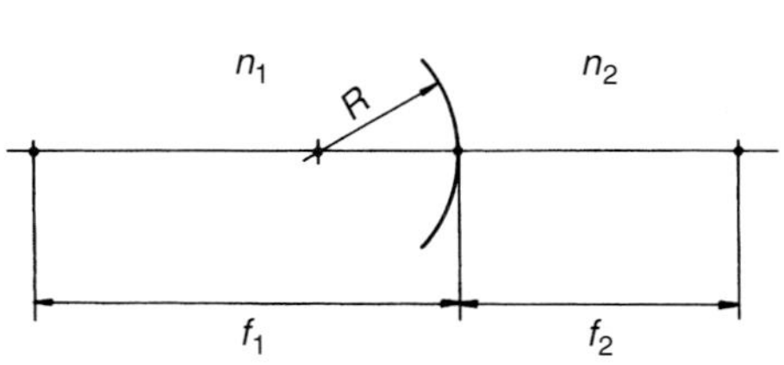


**Fig. 9** Single-interface lens

and

$$f_2 = \frac{Rn_2}{(n_1 - n_2)}. \quad (22)$$

The imaging equation still has a form similar to that of mirrors though some changes become necessary:

$$\frac{n_1}{s_1} + \frac{n_2}{s_2} = \frac{n_1 - n_2}{R}. \quad (23)$$

If one medium is air, its refractive index can be replaced by 1 for most practical purposes, and we can omit the suffix for the refractive index of the other medium; Eq. (23) then simplifies to

$$\frac{n}{s_1} + \frac{1}{s_2} = \frac{n - 1}{R}. \quad (24)$$

The lateral magnification accordingly has to be changed to

$$M = -\frac{n_1}{n_2}\frac{s_2}{s_1}. \quad (25)$$

This quantity gives the ratio of image size to object size; it is positive for an erect image and negative for an inverted image.

### 4.3.2 Double-interface Lens; Thin Lens

A double-interface lens is usually called a thin lens if its center thickness $D$ is small compared to its focal length and the object and image distances. For such a thin lens of index $n$ in air as shown in Fig. 10, the focal length is given by

$$\frac{1}{f} = (n - 1)\left(\frac{1}{R_1} + \frac{1}{R_2}\right), \quad (26)$$

where the ambient medium is assumed to be vacuum. Comparing this to Eq. (24) shows that this is simply the sum of the refractive powers of the two interfaces. The quantity $1/f$ is often called the power of a lens. It should be noted that $R_1$ and $R_2$ are positive for a biconvex lens.

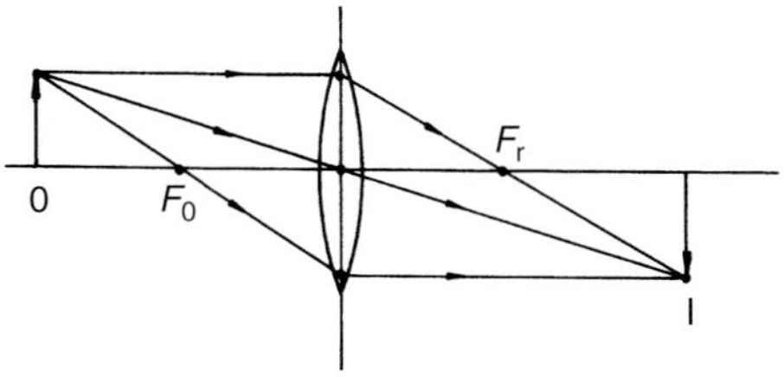


**Fig. 10** Double-interface lens. $F_0$ and $F_1$ denote the focal points on the object side and on the image side. (after Klein, M. V., Furtak, T. E. (1986), *Optics*. New York: Wiley [6])

The magnification is then simply given by $M = -s_2/s_1$.

## 4.4 Thick Lenses

Whereas for thin lenses in paraxial treatment it is normally sufficient to assume that a ray is deflected at the central plane of the lens, for thick lenses refraction at the entrance and exit faces of the lens must be treated separately as well as the propagation within the lens. The situation is depicted in Fig. 11 where the (paraxial) tracing is shown for the three rays normally used in geometrical constructions: the ray through the focus on the object side, the ray through the focus on the opposite side, and the ray through the lens center. The central plane used for thin lenses is split into two principal planes; between these, all rays are treated as if they traveled parallel to the optical axis (which they usually do not), and the figure indicates the directional changes for the different rays at the principal planes.

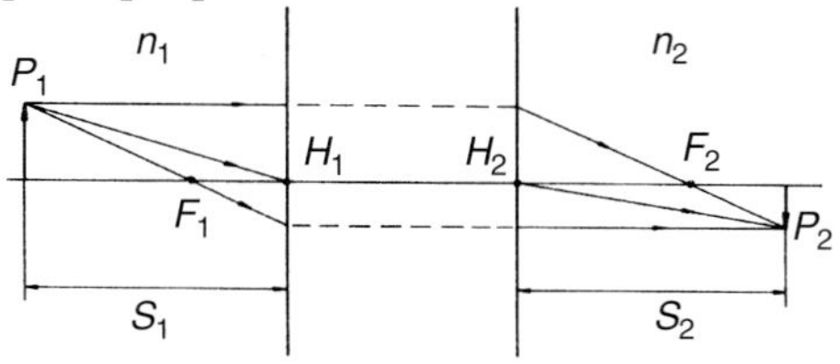


**Fig. 11** Fundamental rays through a thick lens

For a lens with center thickness $t$ and refractive index $n$ and surfaces with radii of curvature $R_1$ and $R_2$ of the surfaces facing the adjacent media with refractive indices $n_1$ and $n_2$, the lens power is then

$$P = P_1 + P_2 - P_1 P_2 \frac{t}{n}. \tag{27}$$

Here $P_1$ is the refractive power of the surface with radius $R_1$ and $P_2$ defined analogously. Here we have used $P_1 = (n_1 - n)/R_1$ and $P_2 = (n - n_2)/R_2$ with $R_i$ negative for a surface convex to the left. Then $P$ is used in the relation $P = n_1/f_1 = n_2/f_2$.

The separations between the vertices and the principal planes are given by $D_1 = -n_1 P_2 t/P_n$ and $D_2 = -n_2 P_1 t/P_n$ where these values are positive when the separation points out of the lens as seen from the vertex.

For the common simple case of $n_1 = n_2 = 1$, the magnification is the same as that for the thin lens described in Sect. 4.3.2, but the focal length changes to

$$\frac{1}{f} = (n-1)\left(\frac{1}{R_1} + \frac{1}{R_2} - \frac{(n-1)D}{nR_1R_2}\right). \tag{28}$$

For a more detailed treatment, the reader should refer to GEOMETRIC OPTICS.

## 4.5 Matrix Optics

High-performance optics are usually characterized by diffraction-limited performance for large numerical apertures (NAs) (see Sect. 5.2.3 for a definition of numerical aperture). For the design of these, it is always necessary to trace a ray through the system without approximations, i.e., by determining the ray's intersections with different interfaces and applying the law of refraction (or reflection) for each change of ray direction caused at such an interface. For many applications, however, it is sufficient (at least for a first sketch) to use paraxial optics with principal planes and a linear approximation of the law of refraction.

The resulting approximate equations describe the change of two variables characterizing a ray passing through an element: the distance of that ray from the optical axis and the angle between that ray and the optical axis times the refractive index. The equations are linear in these two variables, and, therefore, 2 × 2 matrices may be used to calculate free-space propagation and directional changes due to optical elements. This greatly simplifies paraxial ray tracing, as it permits a modular treatment of optical systems: The matrix representing the entire system is nothing but the (matrix) product of the matrices representing the optical elements constituting that system. Thus, after determining that system matrix, a single matrix multiplication suffices for tracing a ray through the entire system (see GEOMETRIC OPTICS).

## 4.6 Simple Systems

Vision is man's "optical" sense. Thus, we have chosen as the most prominent examples of (simple) optical systems, the eye, and the instruments devised to extend its reach into the microscopic and macroscopic world. For a much more detailed presentation, the reader can consult OPTICAL COMPONENTS AND SYSTEMS.

### 4.6.1 The Eye

The eye consists of a single air-solid interface lens, namely the cornea, and an embedded variable-power lens. The lens power is changed in a unique way: The elastic lens is deformed by muscles and thus changes its radii of curvature. Image space is filled by the aqueous body, and in the image plane – which actually is a spherical surface in this case; cf. Fig. 12 – we find the retina, which acts as a detector. A variable stop contributes a little to light regulation, whereas the main task here is performed by adaption (see PHYSIOLOGICAL OPTICS). Some typical data of the eye are given in Table 1.

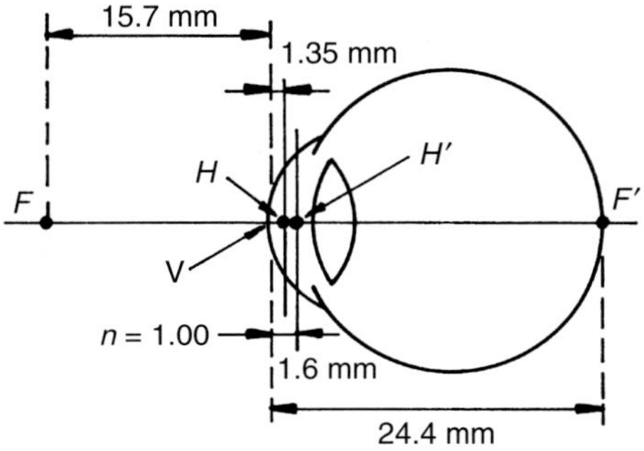


**Fig. 12** The human eye. $H$ and $H'$ are the two principal planes and $F$ and $F'$ are the focal points. (after Klein, M. V., Furtak, T. E. (1986), *Optics*. New York: Wiley [6])

**Tab. 1** Some typical data of the human eye

| ***Property*** | ***Value*** | ***Unit*** |
|---|---|---|
| Back focal length | 20 | mm |
| Pupil diameter | 1–7 | mm |
| Refractive index of cornea | 1.376 | |
| lens | 1.386 | |
| aqueous body | 1.336 | |
| Length | 25 | mm |
| Lens radii of curvature, far accommodation | | |
| front | 10 | mm |
| rear | 6 | mm |
| center thickness | 3.6 | mm |
| Near accommodation | | |
| front | 5.33 | mm |
| rear | 5.33 | mm |
| center thickness | 4 | mm |
| Near object distance | 250 | mm |

The properties of the eye vary considerably with age and from one individual to another. Thus, small children may be able to focus at distances smaller than 100 mm. (See also PHYSIOLOGICAL OPTICS.)

An optical camera is quite similar to the eye: The main differences are the absence of an aqueous body (which would lead to unacceptable weight), focusing by changing the separation between lens and film or detector plane, and the replacement of the curved retina by other, usually planar, detection or registration means (film, CCD array, etc.)

### 4.6.2 The Magnifying Glass or Simple Microscope

The magnifying glass is one of the simplest optical systems: It consists of a single (possibly compound) lens of positive power with a focal length that is smaller than the nearest object distance for the eye. If the object is in the focal plane of the lens, the image lies at infinity and a magnification factor

$$M = \frac{s_0}{f} \tag{29}$$

results. Alternatively, the object distance is chosen so that a virtual image at about 250 mm behind the lens results. Thus, the image can just be focused when the eye is brought close to the lens (cf. Fig. 13). In

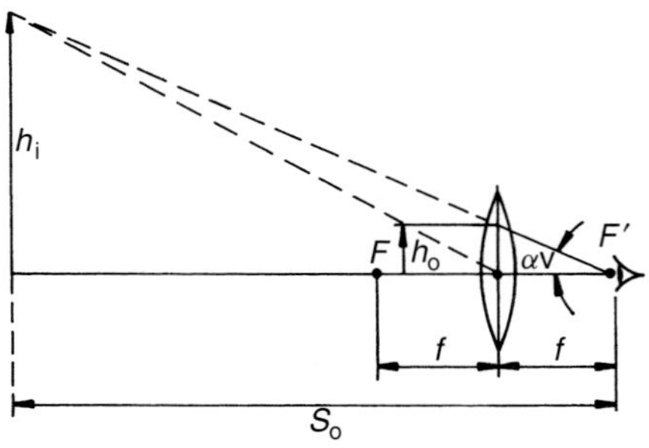


**Fig. 13** The magnifying glass (after Klein, M. V., Furtak, T. E. (1986), *Optics*. New York: Wiley [6])

this case, we have the magnification

$$M = 1 + \frac{s_0}{f}. \tag{30}$$

For weak magnifiers, this contribution can be useful, whereas it is normally neglected for strong magnifiers with small focal length, e.g., microscope or telescope eyepieces.

### 4.6.3 The Compound Microscope

The microscope (see also MICROSCOPY) normally has two lenses or rather lens systems (cf. Fig. 14): the objective and the eyepiece. The objective is used to create a real magnified image of the object. This intermediate image is then viewed with the eyepiece serving as a magnifying glass. It is common usage to call the distance $t$ from the back focal plane of the objective to the intermediate-image plane the tube length. This leads to a simple expression for the magnification factor of the objective:

$$M_{\text{obj}} = \frac{t}{f_{\text{obj}}}. \tag{31}$$

The total magnification of an optical system is the product of the magnifications of its subsystems. Thus, the total magnification of the microscope is

$$M_{\text{mic}} = \frac{t}{f_{\text{obj}}}\frac{s_0}{f_{\text{ep}}}, \tag{32}$$

where $f_{\text{ep}}$ is the focal length of the eyepiece.

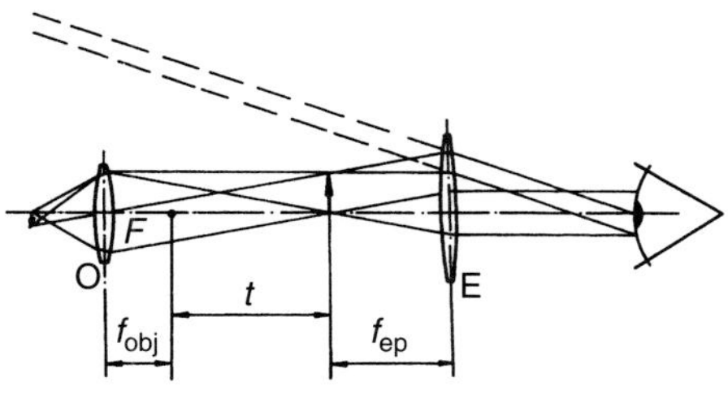


**Fig. 14** The compound microscope. *O* is the objective, *E* is the eyepiece, and *F* denotes the back focal point of the objective. (after Hecht, E., Zajac, A. (1974), *Optics*. Reading, MA: Addison-Wesley [5])

### 4.6.4 The Telescope

A telescope usually consists of two lenses with a common focus. Thus, a parallel bundle of rays will remain a parallel bundle when passing through a telescope and will not be focused outside of it; therefore, telescopes and telescopical subsystems are sometimes called afocal systems.

**The Astronomical Telescope** The astronomical telescope consists of an objective lens of large focal length and diameter and an eyepiece of small focal length and diameter (Fig. 15). Both the object and the image then lie at infinity, and the angular magnification is given by

$$M = -\frac{f_{\text{obj}}}{f_{\text{ep}}}. \tag{33}$$

For astronomical telescopes, both lenses often have positive power so that a negative magnification (i.e., an inverted image) results. As the amount of light that enters the telescope from a given star depends only on the apparent luminance of that star and the objective diameter, the

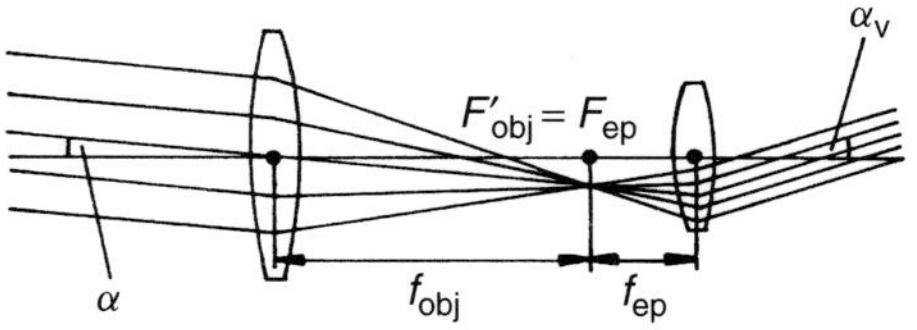


**Fig. 15** The astronomical telescope. $\alpha$ denotes the angular size of the object without and $\alpha_v$ with the use of the telescope. (after Klein, M. V., Furtak, T. E. (1986), *Optics*. New York: Wiley [6])

number of stars visible will not depend on magnification but on that diameter (and optical transmission) only.

Larger telescopes are normally built using mirrors that can be supported from the back; larger lenses would experience sag under their own weight, accompanied by severe image-quality deterioration. (See also ASTRONOMICAL TELESCOPES AND INSTRUMENTATION.)

**The Terrestrial Telescope** The terrestrial telescope uses an eyepiece with negative focal length and thus positive magnification, i.e., upright images (Fig. 16). The magnification is given by the same formula as for the astronomical telescope. However, this construction yields an upright image and a shorter system, which besides is often folded using prisms for further compactness. These prisms may also assure an upright image for an eyepiece of positive power. For more comfortable observation, two telescopes are often aligned in a parallel pair, called binoculars. Two figures are normally used to characterize terrestrial telescopes. The first gives the magnification and the second the objective diameter in millimeters.

The useful diameter of the input pupil is the maximum diameter of the eye's pupil times the magnification. Light entering larger input pupils will be blocked by the iris. However, larger exit pupils may be useful for a telescope used in a vibrating environment, as they permit relative motion of the eye and the telescope.

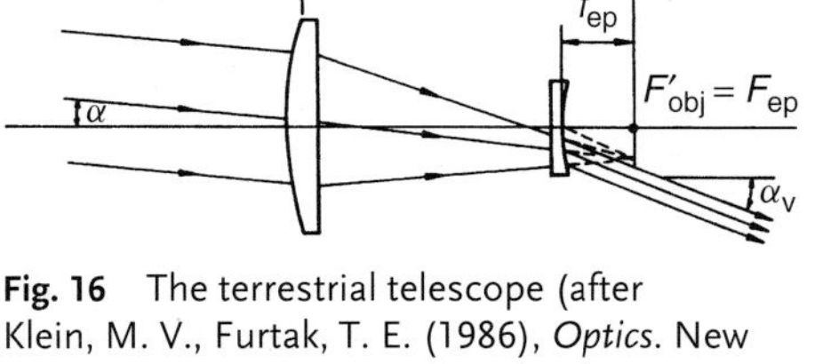


**Fig. 16** The terrestrial telescope (after Klein, M. V., Furtak, T. E. (1986), *Optics*. New York: Wiley [6])

It is not even possible to give an appropriate overview of the many designs that have been invented for optical systems. For further information, we have to refer the reader to the articles on OPTICAL INSTRUMENTATION; GEOMETRIC OPTICS; OPTICAL COMPONENTS AND SYSTEMS and OPTICAL DESIGN of this Encyclopedia and to the literature, e.g., Kingslake [7, 8].

## 5 Interference Phenomena

Interference phenomena are very widespread in their applications: From reflective (or antireflective) coatings to gratings and zone plates, they demonstrate the wave properties of light. One of the most important consequences of these is the limits set to the resolution of optical instruments like microscopes, telescopes, or spectrographs as well as to the propagation of diffraction-limited beams.

One important characteristic of interference phenomena is that the wavelength dependence here is often much stronger than for refractive systems. For example, whereas the focal length of a refractive lens may change by a few percent over the visible spectrum range, the focal length of a diffractive lens will vary almost by a factor of 2 for this case.

### 5.1 Interference

It is extremely difficult to observe interference of the waves from two different sources, as explained in Sect. 5.1.1. Therefore, in most practical cases the different sources are derived from the same source using beam splitters, etc. Unless otherwise stated, the treatment in this section restricts itself to monochromatic plane waves.

In order for interference to occur, several conditions must be fulfilled, as follows.

The interfering waves must have nonorthogonal polarizations. For identical polarization, fringe contrast is maximum, and for orthogonal states of polarization the interference pattern vanishes.

For static interference patterns to occur for finite path differences, the interfering waves usually must have the same constant frequency, i.e., the light should be monochromatic. (Note, however, that for zero path difference interference may be even observed using white light.) Otherwise, the interference pattern will change with time. If, in this case, the oscillation period associated with the difference frequency becomes shorter than the integration time of the device used for the detection of the interference pattern (eye, CCD camera, film, etc.), the interference pattern will be washed out. This is also the reason why it is quite difficult to observe interference between waves from different sources. Normally, their frequencies will differ considerably so that the interference pattern changes so rapidly that the interference fringes are washed out for time-integrating detectors like the eye.

Finally, the interfering waves have to be coherent. As ideal monochromaticity is never really obtained, the phase of a light source (LS) varies constantly. As the different waves for an interference setup are normally derived from the same source using beam splitters, the phase is controlled by choosing different optical path lengths to the point of observation for different beams. If the phase of the light source changes considerably during the time it takes the light to pass through the difference of optical path lengths, the interference pattern will also change in phase and will again be washed out for a time-integrating detector. As a quantitative measure for this property, the coherence length defines that path difference at which the fringe contrast drops to half of its maximum value. Coherence lengths vary from a couple of microns for white light to several thousand kilometers for highly stabilized lasers.

It should be noted that these last two conditions – monochromaticity and coherence – are closely linked as the frequency (difference) is directly proportional to the time derivative of the phase (difference) of the waves.

### 5.1.1 Two-beam Interference

In many cases, just two beams interfere or it is sufficient to take two of the interfering beams into account. The irradiance for a given point is proportional to the time-averaged square of the electric field at that point

$$I = \varepsilon\langle \mathbf{E}\cdot\mathbf{E}\rangle. \tag{34}$$

For $\mathbf{E} = \mathbf{E}_1 + \mathbf{E}_2$, this yields

$$\begin{aligned} I &= \varepsilon\langle(\mathbf{E}_1^2 + \mathbf{E}_2^2 + 2\mathbf{E}_1\cdot\mathbf{E}_2)\rangle \\ &= I_1 + I_2 + 2\varepsilon\langle\mathbf{E}_1\cdot\mathbf{E}_2\rangle, \end{aligned} \tag{35}$$

where we have used the abbreviation $\mathbf{E}_i^2 = \mathbf{E}_i\cdot\mathbf{E}_i$ for the scalar product of a vector with itself.

Often complex numbers are used instead of vectors to represent the fields. The physical field is then regarded as the real part of these quantities. For two waves with amplitudes $A_1$ and $A_2$, the fields may be represented as

$$E_1 = A_1\exp(i\phi_1), \quad E_2 = A_2\exp(i\phi_2) \tag{36}$$

with the phases $\phi_1 = \omega t$ and $\phi_2 = \omega t + \delta$. In this case, the resultant field is given by

$$E = E_1 + E_2 = A \exp(i\Phi) \qquad (37)$$

with

$$A = [A_1^2 + A_2^2 + 2A_1A_2 \cos\delta]^{1/2} \qquad (38)$$

and

$$\Phi = \omega t + \Phi'. \qquad (39)$$

The value of $\Phi'$ can be determined from the relations

$$\cos\Phi' = \frac{A_1 + A_2\cos\delta}{A} \qquad (40)$$

and

$$\sin\Phi' = \frac{A_2 \sin\delta}{A}. \qquad (41)$$

For $A_1 = A_2$, we obtain for the irradiance

$$I = |E^2| = 2A_1^2(1 + \cos\delta) = 4A_1^2 \cos^2\left(\frac{\delta}{2}\right), \qquad (42)$$

which is represented in Fig. 17.

Figure 18 shows two setups for two-source interference. In both cases, the phase difference is given by

$$\delta \approx \frac{2\pi}{\lambda}\frac{ax}{D}, \qquad (43)$$

and a pattern of $\cos^2$ shape on the screen can be detected. Here we have neglected possible phase changes upon reflection at the mirror. The term $a$ is the separation of the two sources, $D$ is the distance between the planes of the sources and that of the screen, and $x$ is the position in the screen plane measured from the axis of symmetry.

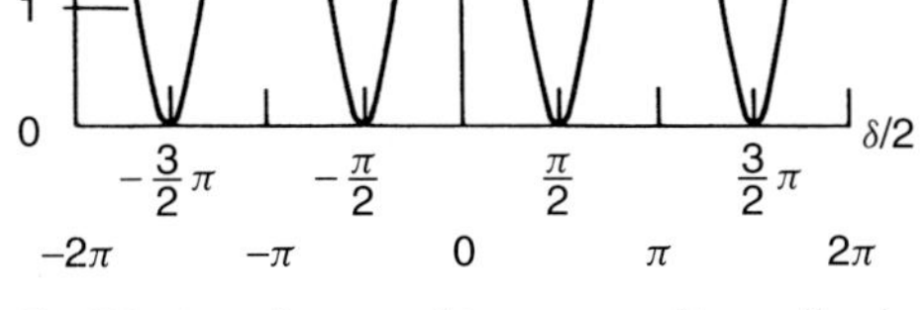


**Fig. 17** Interference of two waves. Normalized irradiance as a function of phase difference. (after Klein, M. V., Furtak, T. E. (1986), *Optics*. New York: Wiley [6])

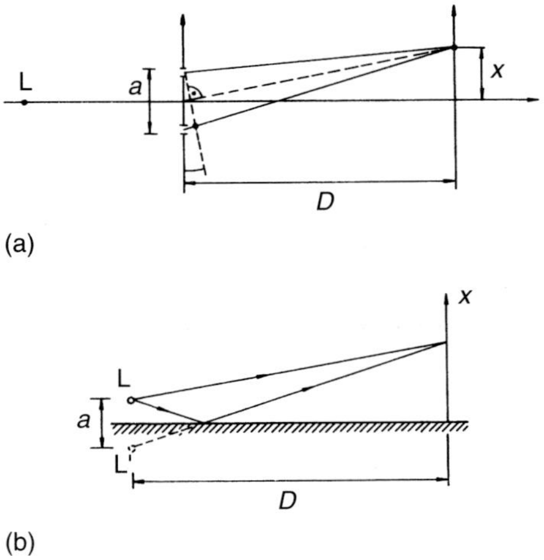


**Fig. 18** (a) Young's double-slit experiment and (b) Lloyd's mirror. For details see text. (after Klein, M. V., Furtak, T. E. (1986), *Optics*. New York: Wiley [6])

**Newton's Rings and Fizeau Fringes** When a (plano-)convex lens is placed on a plane glass surface, rings around the point of contact will occur, which are especially well pronounced in reflection and for monochromatic light (Fig. 19). The radius of the $m$th dark ring can be shown to be approximately given by $x_m = \sqrt{m\lambda R}$ where $R$ is the radius of curvature of the lens side facing the plane. This is valid for near-normal incidence.

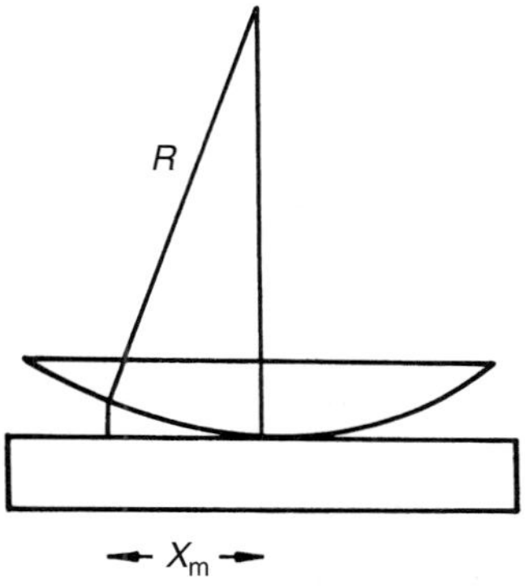


**Fig. 19** Setup for the observation of Newton's rings (after Klein, M. V., Furtak, T. E. (1986), *Optics*. New York: Wiley [6])

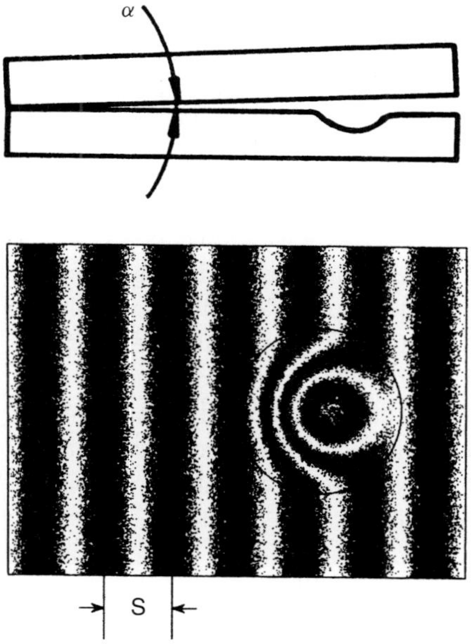


**Fig. 20** Calculated Fizeau fringes caused by an air wedge with a circular depression. The depth of the depression is strongly exaggerated for clarity. The actual depth used for computing the fringe pattern is approximately one wavelength. (after Klein, M. V., Furtak, T. E. (1986), *Optics*. New York: Wiley [6])

Figure 20 shows another situation that is normally described with sufficient precision using two-beam interference. Again near-normal incidence lighting and viewing with monochromatic light are supposed. In this case, the relation between wedge angle $\alpha$ and fringe separation $s$ is given by

$$\alpha s = \frac{\lambda}{2}. \tag{44}$$

In both cases, the waves reflected from the two interfaces will enhance or (partially) compensate each other according as the phase difference due to one of them passing the gap twice is an odd or even multiple of $\lambda/2$.

**Antireflection Coating** Figure 21 shows the most simple antireflection coating. To reduce Fresnel reflection losses at (air-glass) interfaces, a thin film of transparent material can be deposited on the surface using, e.g., vacuum deposition. For normal incidence and a given wavelength, the reflection vanishes if the film thickness $d_f$ satisfies the condition

$$d_f = \frac{\lambda}{4n_f}, \tag{45}$$

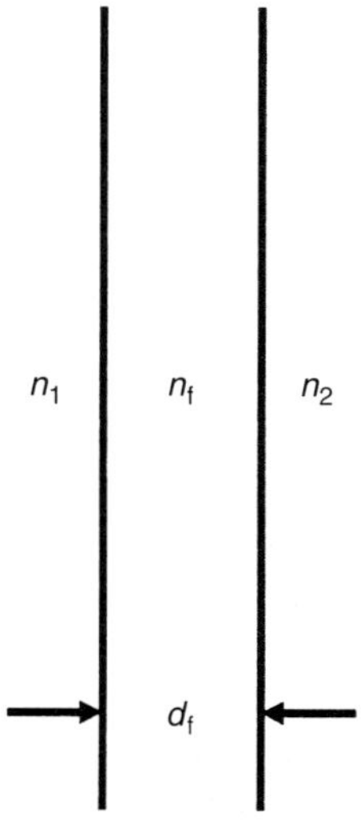


**Fig. 21** Single-layer antireflection coating

where $n_f$ is the refractive index of the film material and is chosen to be

$$n_f = \sqrt{n_1 n_2}, \qquad (46)$$

where $n_1$ and $n_2$ are the refractive indices of the adjacent media. The reflections from the two interfaces are then approximately equal in amplitude but out of phase by half a wavelength and thus interfere destructively, canceling almost completely. In practical systems, more elaborate coatings consisting of a stack of several thin-film layers are used to achieve reflex suppression over an extended wavelength range. At the same time, the permissible range of angles of incidence normally increases.

For other thin-film coatings such as high-reflection coatings or dichroic mirrors, the reader is referred to the article on OPTICAL COATINGS of this Encyclopedia.

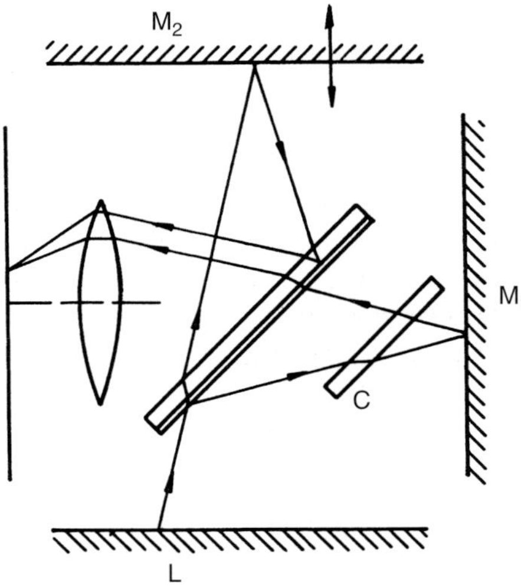


**Fig. 22** The Michelson interferometer. $M_1$ is the fixed mirror and $M_2$ is the moving mirror. $C$ denotes the compensator plate. (after Klein, M. V., Furtak, T. E. (1986), *Optics*. New York: Wiley [6])

**The Michelson Interferometer** Interferometers have found widespread applications in precision measurements of lengths, positions, and shapes. Microscope interferometers are used for surface analysis or process quality control. As these are handled in detail in a dedicated entry (see INTERFEROMETRY), only one type of interferometer is presented here.

The Michelson interferometer is depicted in Fig. 22. Light waves coming from the source are split by a semireflecting beam splitter and propagate through the two arms of the interferometer; here the reflected wave passes through a compensator plate, is reflected from a fixed mirror $M_1$, and traverses the compensator plate once again, while the transmitted wave is reflected from a movable mirror $M_2$ forming the second arm of the interferometer.

The compensator plate, made of the same material as the beam splitter substrate, assures symmetry of optical path lengths. Otherwise, dispersion effects would complicate considerably the theoretical description of this device under broadband illumination. Besides, this plate symmetrizes the interferometer paths. Although this can be done by simply moving one of the mirrors for plane waves, this is not possible for divergent (or convergent) waves without such a plate.

After passing through the arms of the interferometer, the two waves are recombined by the same partially transmitting mirror, thus creating interference. Tiny differences in the surface profiles of the two mirrors appear as interference fringes if the system is adjusted at a mean path difference of zero between the two arms. Minute tilts of the mirrors will result in parallel fringes in the output.

Alternatively, with a collimated source, length measurements are possible. For simplicity, normal incidence on the plane mirrors is assumed. Moving one mirror will then result in a modulation of output light intensity following a $\cos^2$ law. The zeros of this signal being separated by mirror movements of $\lambda/2$ can be counted or even interpolated for length measurements. For extreme precision, it is necessary to account for the refractive index of air, if the measurements are not performed in vacuum.

For a plane wave incident normally on the mirrors, the irradiance on the output screen will be given by

$$I = I_0 \cos^2\left(\frac{\delta}{2}\right) \quad (47)$$

with $\delta = 2\pi\Delta x/\lambda$ and $\Delta x$ the difference in path length in the two arms.

Michelson interferometers are often used to measure the wavelengths of lasers by comparing them to known lengths (e.g., gauges) or to the known wavelength of another laser. These devices are often called wave meters.

### 5.1.2 Multiple-beam Interference

The case of multiple-beam interference is treated similarly. With the extension of the above notation to $N$ beams, the total field is

$$E = A_1 \exp(i\phi_1) + A_2 \exp(i\phi_2) + \cdots + A_N \exp(i\phi_N). \quad (48)$$

For equal amplitudes and with $\phi_n = \omega t + (n-1)\delta$, this yields

$$E = A_1 \exp\left[i\left(\omega t + \frac{(N-1)\delta}{2}\right)\right] \times \frac{\sin(N\delta/2)}{\sin(\delta/2)}. \quad (49)$$

For the irradiance for a given value of $\delta$ there follows hence

$$I(\delta) = I(0)\left[\frac{\sin(N\delta/2)}{N\sin(\delta/2)}\right]^2. \quad (50)$$

This distribution has maxima for $\delta = 2m\pi$ where $m$ is an integer. Between these principal maxima, $N-1$ minima occur, separated by $N-2$ minor maxima. For large values of $N$, the distribution $I(\delta)$ approaches limiting values in the (minor and major) maxima that depend only on $\delta$.

**Diffraction Gratings** A diffraction grating is a periodic array of transmitting or reflecting structures. With these spaced at equal distances [Fig. 23(a)], the light

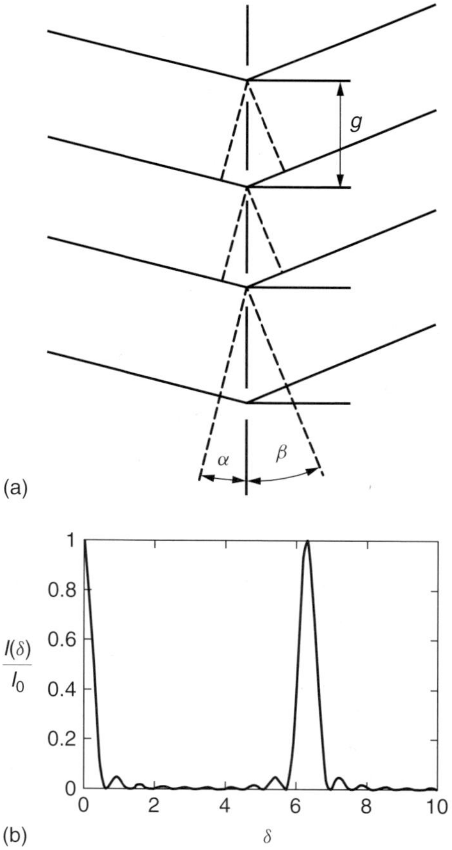


**Fig. 23** (a) Schematic view of a diffraction grating (after Klein, M. V., Furtak, T. E. (1986), *Optics*. New York: Wiley [6]). (b) Irradiance distribution for a grating with $N = 10$ grooves. $N = 10$, $I_0 = 1$, $I(\delta) = I_0[\sin(N\delta/2)/N\sin(\delta/2)]^2$, $\delta = 0.00001, 0.00501 \ldots 10.000$

that is detected at an angle $\beta$ for a plane wave incident at an angle $\alpha$ is given by the equations presented in Sect. 5.1.2 with $\delta = g(\sin\alpha + \sin\beta)$. The separation of the repetitive structures $g$ is called the grating period. The irradiance distribution according to Eq. (49) for $N = 10$ is shown in Fig. 23(b). Usually, $N$ is much larger, however. For higher values of $\delta$, additional maxima in this curve occur; the diffraction order is the corresponding value of $\delta$ divided by $2\pi$ (e.g., first order). For negative $\delta$, negative diffraction orders occur (e.g., second order).

Gratings can be amplitude or (for higher efficiency) phase gratings. An amplitude-transmission grating has transmitting and opaque stripes, whereas for a phase-transmission grating grooves are produced as the surface profile of a transparent carrier. An amplitude-reflection grating has reflecting and nonreflecting stripes, whereas a reflection phase grating is a mirror with a periodic surface profile. The spectral resolution of gratings is discussed in Sect. 5.2.3.

Phase gratings with planar groove surfaces can be *blazed* for a given wavelength by making the angle of the triangular groove profile satisfy the law of refraction for a transmission grating or the law of reflection for a reflection grating. In this case, a reflection grating, e.g., consists of many microscopic strip mirrors reflecting the incident light in the direction of the desired order. Blazing implies that the angle of incidence also is known at least approximately. Blazing concentrates the diffracted light into one desired diffraction order.

**The Fabry-Pérot Interferometer** A Fabry-Pérot interferometer consists, e.g., of two plane high-quality mirrors adjusted parallel to each other with a separation $d$ (Fig. 24). Again, multiple-beam

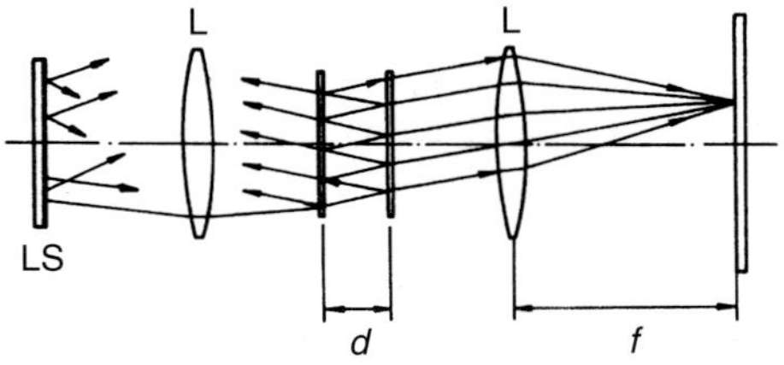


**Fig. 24** A plane Fabry-Pérot interferometer. *LS* is the light source, *L* is used to denote lenses, and *f* is the focal length of the second lens, which is used for imaging the interference fringes onto a screen. (after Hecht, E., Zajac, A. (1974), *Optics*. Reading, MA: Addison-Wesley [5])

interference occurs, but in this case, each consecutive beam is weaker as a result of the reflection coefficient, which will always be smaller than unity. The mathematical treatment is a bit different as an infinite number of beams has to be taken into account, and for (identical) transmission $T_m$ and reflection $R_m$ of the two mirrors the transmission $T$ of the arrangement is

$$T = \frac{T_m^2}{1 + R_m^2 - 2R_m \cos\delta}. \qquad (51)$$

The total phase difference $\delta$ between successive beams is the sum of the propagation difference $\delta_1 = (2\pi/\lambda)nd\cos\theta$ and twice the phase change $\delta_2$ experienced upon reflection. The maximum transmission is then given by

$$T_{\max} = \frac{T_m^2}{(1 - R_m)^2}. \qquad (52)$$

Using $T_{\max}$ and $F = 4R_m/(1 - R_m)^2$, the expression for $T$ can be rewritten as

$$T = \frac{T_{\max}}{1 + F\sin^2(\delta/2)}. \qquad (53)$$

The ratio of the separation of two adjacent transmission maxima to the width of these maxima for monochromatic light (both measured in units of delta) is called the finesse $\mathcal{F}$:

$$\mathcal{F} = \frac{\pi\sqrt{F}}{2} \approx \frac{\pi}{1 - R_m}. \quad (54)$$

## 5.2 Diffraction

Diffraction can be regarded as the interference of infinitely many infinitely weak sources. So most of the sums in the previous section turn into integrals. However, diffraction is more than that. The diffraction integral is one of the cornerstones of linear optics. It should be noted that diffraction also limits the validity of the fundamental statement of geometrical optics that light travels in straight lines. Light may actually "bend" around obstacles and thus illuminate the geometrical shadow area.

A rigorous theory based on Maxwell's equations and appropriate boundary conditions is available for a vectorial treatment. However, this rigorous treatment is often dispensable and various simplified scalar treatments have been developed, e.g., by Young, Fresnel, and Kirchhoff. The necessity of using a vectorial (i.e., polarization-sensitive) treatment is self evident in the case of deflection angles where the electric field vectors from different elementary waves are not sufficiently parallel anymore. Other examples are pinholes with openings comparable to the wavelength of the diffracted light and gratings with a transmission function varying considerably over dimensions of about one wavelength. For these demanding applications of diffraction theory, the reader is directed to the Further Reading.

### 5.2.1 Fresnel Diffraction

For most practical purposes, the Fresnel–Kirchhoff integral of diffraction is sufficient. It states the scalar field relation

$$E(r) = \frac{i}{\lambda}\int E_{\text{in}}(\tilde{r})\frac{\mathrm{e}^{-ikR'}}{R'} \times \frac{\cos\theta_{\text{in}} + \cos\theta_{\text{out}}}{2}\,\mathrm{d}\sigma. \quad (55)$$

For monochromatic light of wavelength $\lambda$, it yields the field at a given point $r$ in space, if the field of the incident wave $E_{\text{in}}$ is known for all points $\tilde{r}$ on the surface $\sigma$ of the diffracting aperture(s), these points being separated by distances $R'$ from the point $r$. The wave number $k$ is defined as $k = 2\pi/\lambda$. The terms $\theta_{\text{in}}$ and $\theta_{\text{out}}$ are the angles between the directions of propagation of the incident or diffracted wave relative to the normal to $\sigma$ at $\tilde{r}$. The only deviation from a straightforward extension of interference (summation) to diffraction (integration) seems to lie in the directional cosines that describe the component of the incoming and outgoing waves and the normal to the surface of integration.

The factor $\mathrm{e}^{-ikR'}/R'$ in the diffraction integral represents a spherical wave emerging from the aperture plane toward the detection plane. Several approximations to the actual phase differences for different coordinate pairs in these two planes can be made depending on the particular geometric limits under consideration.

A first approximation neglects all terms contributing to $R'$, and thus to the phase, that are of higher than quadratic order in the aperture coordinates. This parabolic or Fresnel approximation for the phase is often used when the exact solution is too complicated and a linear or Fraunhofer approximation would result in insufficient accuracy. For the arrangement depicted in Fig. 25, the Fresnel integral then takes the

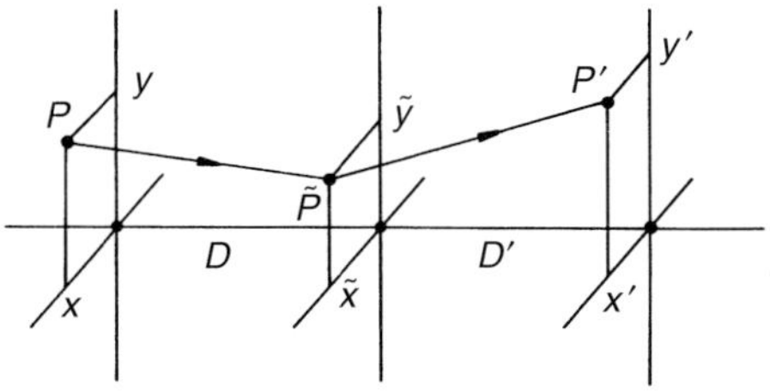


**Fig. 25** Configuration for Fresnel diffraction for a point source at $P$ (after Klein, M. V., Furtak, T. E. (1986), *Optics*. New York: Wiley [6])

simplified form

$$E(P') = \frac{i}{2} E_0(P') \iint \tau(\eta_x, \eta_y) \times \exp\left[-\frac{i\pi}{2}(\eta_x^2 + \eta_y^2)\right] \mathrm{d}\eta_x \, \mathrm{d}\eta_y \quad (56)$$

with the normalized aperture-plane coordinates

$$\eta_x = \sqrt{\frac{2}{\lambda d}}(\tilde{x} - \tilde{x}_u), \quad (57)$$

$$\eta_y = \sqrt{\frac{2}{\lambda d}}(\tilde{y} - \tilde{y}_u), \quad (58)$$

and

$$\frac{1}{d} = \frac{1}{D} + \frac{1}{D'}. \quad (59)$$

Here, $\tau$ is the transmission function of the aperture, whose value determines the amplitude-transmission coefficient, which may be complex to include phase shifts. The tilde denotes aperture-plane coordinates, and the suffix $u$ corresponds to that point in the aperture plane that lies on the straight line connecting the source point and the observation point. The terms $D$ and $D'$ are the distances of the source plane and of the plane of observation from the aperture plane.

Here we limit our presentation to three well-known examples of Fresnel diffraction.

**Rectangular Aperture** In this case, the transmission function can be factored:

$$\tau(\eta_x, \eta_y) = \tau_x(\eta_x)\tau_y(\eta_y), \quad (60)$$

and, thus, the Fresnel integral can be solved separately for the two coordinates $x$ and $y$:

$$E(P) = \frac{i}{2} E_0(P) I_x I_y, \quad (61)$$

with

$$I_x = \int_{-\infty}^{+\infty} \tau_x(\eta_x) \exp\left(-\frac{i\pi}{2}\eta_x^2\right) \mathrm{d}\eta_x, \quad (62)$$

$$I_y = \int_{-\infty}^{+\infty} \tau_y(\eta_y) \exp\left(-\frac{i\pi}{2}\eta_y^2\right) \mathrm{d}\eta_y. \quad (63)$$

The complex Fresnel integral

$$I(\eta) = \int_0^{\eta} \mathrm{e}^{-(i\pi/2)u^2} \, \mathrm{d}u = C(\eta) + iS(\eta) \quad (64)$$

with

$$C(\eta) = \int_0^{\eta} \cos\left(\frac{\pi}{2}u^2\right) \mathrm{d}u \quad (65)$$

and

$$S(\eta) = \int_0^{\eta} \sin\left(\frac{\pi}{2}u^2\right) \mathrm{d}u \quad (66)$$

may be used to express these integrals as

$$I_x = I(\eta_{x2}) - I(\eta_{x1}), \quad (67)$$

$$I_y = I(\eta_{y2}) - I(\eta_{y1}), \quad (68)$$

where $x1, x2, y1, y2$ denote the upper and lower limits of the interval wherein the transmission function is unity, whereas its value is zero outside. The plot of the Fresnel integral in the complex plane yields the well-known Cornu spiral. The values of the integral can be found in tabulated form. The Fresnel integral simply gives the relative amplitudes of the diffracted wave's field as compared to the amplitude that would result with the incident wave reaching the point of observation without a diffracting obstacle.

**Half-plane Obstruction** An opaque screen ranging from $\tilde{x} = -\infty$ to $\tilde{x} = \tilde{x}_1$ and from $\tilde{y} = -\infty$ to $\tilde{y} = +\infty$ can be treated as a special case of a rectangular aperture by using Eqs. (56), (57), (66), and (67) to calculate $I_x$ and $I_y$. We then have $I_y = 1 - i$ and $\eta_{x2} = \infty$. Changing to the representation in observation-plane coordinates using

$$\eta_{x1} = \frac{(x_1' - x')}{F} \tag{69}$$

with

$$F = \left[\frac{\lambda D'(D + D')}{2D}\right]^{1/2}, \tag{70}$$

we can then calculate $I_x$ for any point in the observation plane. The resulting field is obtained simply as the vector joining the point corresponding to $\eta_{x1}$ with the center of the $+\infty$ lobe of the Cornu spiral (Fig. 26). When $x'$ has large negative values, $\eta_{x1}$ is quite large and the resulting points on the curve lie close to those for $\eta_{x1} = +\infty$; the vector representing $I_x$ is small and increases monotonically with increasing $x'$. For $x_1' = x'$, the value of the integral is $\frac{1}{2}(1 - i)$ with an absolute value half as big as that obtained in the absence of the obstacle (as is to be expected from symmetry arguments); for further increase of $x'$, the absolute value of the integral converges with decreasing oscillations toward the limiting value for $x' = +\infty$. At the point of the geometrical shadow edge, the field has half its maximum value resulting in an irradiance of a quarter of the maximum value. This result is important when determining the position of an edge from its measured shadow.

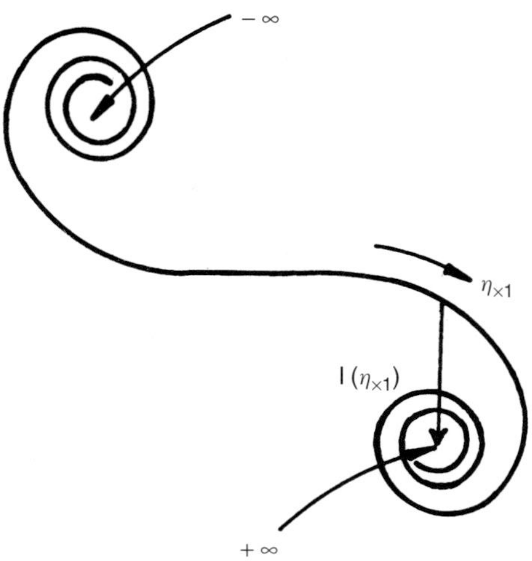


**Fig. 26** Discussion of half-plane diffraction using the Cornu spiral (after Klein, M. V., Furtak, T. E. (1986), *Optics*. New York: Wiley [6]). Actually, $I(\eta_{x1})$ stands for $I_x$ here

**Arago's Bright Spot** An interesting phenomenon occurs for circular obstacles: At a distance behind the obstacle large enough to assure that the *inclination factor*, i.e., the term $\frac{1}{2}(\cos\theta_{in} + \cos\theta_{out})$ in Eq. (54), is still approximately unity, Fresnel diffraction theory correctly predicts that the intensity on the axis be practically equal to the value without an obstacle, a phenomenon called Arago's bright spot, a reference to the French researcher who first demonstrated this somewhat surprising phenomenon.

### 5.2.2 Fraunhofer Diffraction

When it is sufficient to take into account the phase changes that are linear in the aperture coordinates, the quadratic terms are discarded and Fraunhofer diffraction is obtained. This is true, for example, in a situation where both the source and the detection planes are at very large distances from the diffracting aperture so that the waves impinging on the aperture are virtually plane waves (these correspond to

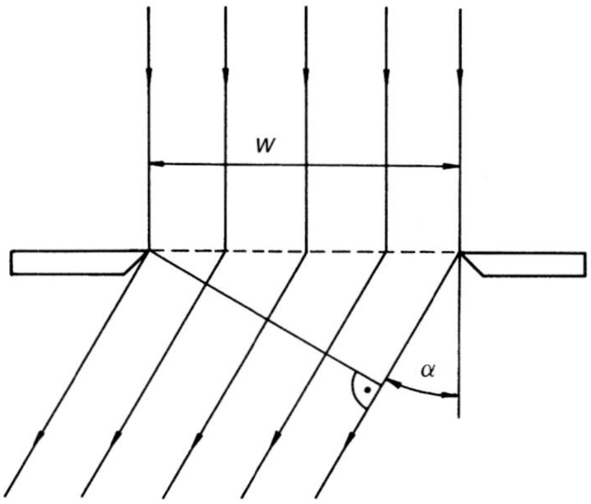


**Fig. 27** Fraunhofer diffraction. The incident wave is normal to the aperture plane for reasons of simplicity only. (after Alonso, M., Finn, E. J. (1976), *Fundamental University Physics*, Vol. 2. Reading, MA: Addison-Wesley, p. 789ff [3])

bundles of parallel rays; see section "Linear Slit, Rectangular Aperture" and especially Fig. 27) and so that waves converging toward a given point in the detection plane are essentially plane over the area of the aperture. This situation is quite similar to the one treated above in discussing the diffraction grating.

For simplicity, we assume a plane wave impinging normally on the aperture plane with coordinates $\xi$ and $\eta$. In this case, the amplitude of a plane wave diffracted with directional cosines $p$ and $q$ in the directions defined by the aperture-plane coordinates is given by the integral

$$U(p,q) = C \int\int G(\xi,\eta)\mathrm{e}^{-(2\pi i/\lambda)(p\xi+q\eta)} \times \mathrm{d}\xi\,\mathrm{d}\eta, \tag{71}$$

where $C$ is a constant and $G$ is the pupil function describing the phase and amplitude transmission of the aperture plane. The physical meaning of this equation is very important: The far-field diffraction pattern is the Fourier transform of the diffracting aperture. This has led to important applications of Fraunhofer diffraction in Fourier optics and other fields.

Again, examples are given for the simplest cases only, i.e., for a linear slit and for a circular and an annular aperture.

**Linear Slit, Rectangular Aperture** The diffraction integral is a function deriving the field in an arbitrary point from the known field distribution on a given surface, usually a plane or sphere. It is useful for calculating the resulting field distribution on another surface. For an infinitely long straight slit (see Fig. 27), the diffraction integral in linear approximation is

$$I = I_0\left(\frac{\sin u}{u}\right)^2 \tag{72}$$

with $u = \pi w \sin\alpha/\lambda$ where $w$ is the slit width, $\alpha$ is the diffraction angle shown in Fig. 27, and $\lambda$ is the wavelength. This is evidently proportional to the square of the Fourier transform of the transmission function, so that a rectangular transmission profile yields the well-known sinc function.

A rectangular aperture can be factorized to be described by the product of two orthogonal slits. The resultant field distribution is then given by the product of two functions of the above form, each with a value of $w_1$ or $w_2$ that corresponds to the width of the opening in that direction and with its value of $\alpha_1$ or $\alpha_2$ representing the deviation in that coordinate.

**Circular Aperture** For a circular aperture such as the common stops in optical instruments like telescopes and microscopes, the far-field diffraction pattern is given by

$$I = I_0 \left[ \frac{2J_1(ka \sin \alpha)}{ka \sin \alpha} \right]^2, \qquad (73)$$

where the common abbreviation $k = 2\pi/\lambda$ has been employed. Here, $a$ is the radius of the diffracting aperture supposedly illuminated by a plane wave at normal incidence, $J_1$ is the first-order Bessel function, and $\alpha$ is the diffraction angle. This formula describes diffraction for a plane wave incident normally on the aperture plane. It is valid if the angular deviation is not too large, so that the inclination factor $Q = \frac{1}{2}(\cos \theta_i + \cos \theta_d)$ is almost equal to 1. Here $\theta_i$ and $\theta_d$ are the angles that the wave normals of the incident and diffracted wave form with the surface normal of the aperture plane. This well-known diffraction pattern is characterized by a central peak encircled by a dark ring and containing some 84% of the total energy. It is usually called the Airy disk when observed in the image plane of the source.

**Annular Aperture** An annular aperture can be handled with this formula quite simply as the weighted difference of the fields for two circular apertures. The irradiance is then obtained from the square of that difference:

$$I = I_0 \left[ \frac{a^2}{a^2 - b^2} \frac{2J_1(ka \sin \alpha)}{ka \sin \alpha} - \frac{b^2}{a^2 - b^2} \frac{2J_1(kb \sin \alpha)}{kb \sin \alpha} \right]^2, \qquad (74)$$

where $a$ and $b$ are the outer and inner radii of the aperture.

### 5.2.3 Fundamental Consequences of Diffraction

When the lens edge is the diffracting aperture, as for the objective lens of an astronomical telescope with focal length $f$ and diameter $d$, the diameter of the Airy disk is given by $D = 2f\alpha$ where $\alpha = 1.22\lambda/d$ is the angle at which the first minimum of the diffraction pattern occurs. According to a widespread resolution criterion (or convention), two equally strong point sources are resolved if their mutual separation is equal to this angle. This is the angular resolution of a telescope with a circular aperture.

Similarly, for a microscope, the spatial resolution $D$ is given by the same formula as above, where the object distance is assumed to be about equal to the focal length of the objective lens. Then $D$ is given by $D = 2.44\lambda f/d$. The product of the refractive index $n$ of the medium filling the object space and of the sine of the angle $\alpha$ subtended by the system axis and the most oblique rays passing the objective to contribute to the image is called the numerical aperture NA of the system. For small values of $\alpha$, we have $\sin\alpha \approx d/2f$ so that NA is roughly given by NA $= d/2f = \mathbf{r}/f$ and the resolution is $D = 1.22\lambda/\mathrm{NA}$.

The resolution $R = \lambda/\Delta\lambda$ of spectrographs is also fundamentally based on the limiting effect of grating or prism boundaries; the results are

$$R = N \times m \qquad (75)$$

for the resolution of a grating spectrometer where $N$ is the number of grooves illuminated and $m$ is the diffraction order used. For a prism spectrometer, we obtain

$$R \approx b \frac{\mathrm{d}n}{\mathrm{d}\lambda} \qquad (76)$$

for the resolution, where $b$ is the maximum geometrical path differences for the limiting rays in the prism.

Obviously, these theoretical limits are only attained for high-quality equipment with wave fronts deviating by $\lambda/10$ or even less from their theoretical shape, which is therefore often characterized as "diffraction limited" in its performance.

For completeness, we include the resolution of the Fabry-Pérot interferometer also:

$$R = \frac{\pi\sqrt{F}}{2}m \qquad (77)$$

with $\sqrt{F}/2 \approx \pi/(1-r)$ for mirror reflectivities $r$ close to unity and $m$ the interference order.

## 5.3 Thin-film Optics

Dielectric antireflection coatings of a single layer are a very simple example of thin-film optics. With stacks of thin layers, almost arbitrary spectral reflection/transmission curves can be designed. With a matrix formalism, the calculation of these curves for thin-film systems with known parameters (film thicknesses and refractive indices including dispersion) is possible. For any interface, the relation between the (complex) amplitudes of waves impinging from and leaving toward both sides is described by a $2 \times 2$ matrix. In the same way, the passage of waves through a homogeneous (film) slab of material can be described. Simple matrix multiplication then permits calculation of the performance of more complex coatings; for a more detailed treatment, the reader should refer to the literature.

In practice, however, thin-film properties may differ from bulk parameters because of voids, film tension etc. For coatings used at nonnormal incidence, polarization-dependent designs are also possible. Computer programs for coating design are commercially available.

High-pass, low-pass, and band-pass filters, dichroic mirrors, polarizing beam splitters, and dielectric high reflectors for lasers are only some examples of useful specialized coating designs. However, until now, no simple algorithm allows even the approximate inverse determination of the coating design from a desired or measured spectral reflectivity/transmission.

## 5.4 Holography

As a result of interference, it is possible to record not only the amplitude or intensity information of an electric field but also its phase distribution so that the information stored, e.g., in a photographic emulsion, may be used to restore the wave front completely. Such wavefront recordings are called holograms. Figure 28(a) shows a setup that may be used to record such a hologram. A monochromatic source is split into two waves. One serves as a reference wave, and the other is used to illuminate the object of which a hologram is to be taken. The scattered wave from the object is the wave to be recorded, with the other wave serving as a phase reference.

In Fig. 28(b), the inverse process is shown. The processed emulsion is illuminated by the reference wave, and the interference pattern recorded on the film diffracts the incident wave in such a way that a wave with the recorded phase front (and approximately the recorded amplitude) is generated. Thus, an apparently three-dimensional image of the object used in recording the hologram is obtained. These reconstructed wave fronts

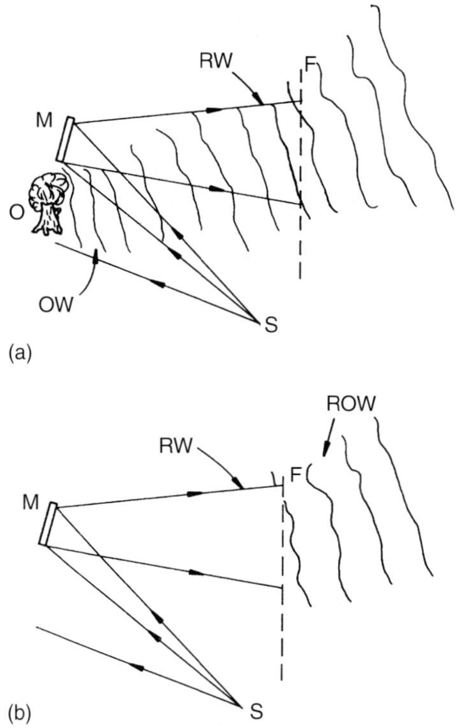


**Fig. 28** Holography: (a) recording setup and (b) reconstruction setup. *S* denotes the source, *O* the object, *M* a mirror, *F* the recording film, *OW* the object wave, *RW* the reference wave, and *ROW* the reconstructed object wave

permit depth perception in the image, unlike normal photographs. For a photograph, the source points for (spherical) elementary waves will lie in the plane of the emulsion, whereas for a hologram wavefront reconstruction permits them to lie behind or in front of that plane as well.

As a film permits sequential recording of several wave fronts, these may be superimposed using holography. This effect is used in interferometric holography where an object wave is recorded before and after deformation. Thus, deformations become directly visible as interference fringes in the reconstructed holographic recording.

### 5.5 Fourier Optics

As mentioned above, the far field of a given structure is proportional to its Fourier transform assuming plane-wave illumination. (The diffraction pattern is therefore proportional to the square of this transform.) If a lens is inserted, the far field is imaged from infinity to the focal plane of that lens. This plane is therefore sometimes called the Fourier plane. Insertion of stops in the Fourier plane and retransformation with a second lens thus permit changing the distribution of spatial frequencies for a given image. Two applications shall be considered here, as follows.

If a pinhole is placed in the Fourier plane, all spatial noise, i.e., high spatial frequency deviations, e.g., from a Gaussian profile, is suppressed, whereas the $TEM_{00}$ mode of a laser (see LASER PHYSICS), having an essentially Gaussian field distribution, would pass without attenuation. Such a device is therefore called a *spatial filter* or *mode filter*.

If a picture has a periodic structure, e.g., from rastering or from successive recording (like many space photographs), these periodic disturbances will be manifest in the Fourier plane as pronounced peaks spaced regularly in the corresponding direction(s). If these are suppressed by an appropriate stop, the periodic disturbance in the picture will vanish almost without loss of information but with a tremendous increase in subjective image quality.

## 6 Anisotropic Media

Anisotropic media have opened up many new fields of applications for optics, e.g., in polarization optics and electro-optic modulation and switching. The most important anisotropic media are crystals of various symmetry classes. However, fluids may become anisotropic under external fields, too. Recently, liquid crystals have found an increasing number of applications for which modulating speed requirements are less stringent.

## 6.1
## Linear Birefringence

Linear birefringence or double refraction, as it is sometimes called, is observed in media that are anisotropic by nature (crystals) or by external effects (e.g., electric fields). Birefringent crystals have an optical axis; light traveling along the optical axis has a refractive index, and thus a velocity, independent of its polarization. Any plane containing the optical axis is called a principal plane. For light waves traveling in such a plane, two states of polarization can be identified. The ordinary ray is characterized by its electric field vector being normal to the optical axis and thus, also perpendicular to the principal plane. The electric field vector of the extraordinary ray, however, lies in the principal plane. For the ordinary ray, the propagation velocity is independent of direction, the elementary waves and the index surfaces being spheres. For the extraordinary beam, the elementary waves and the index surfaces are ellipsoids tangent to this sphere at the points determined by the optical axis through the sphere's center (Fig. 29). If the extraordinary ray has a larger wave velocity than the ordinary ray, the material is said to exhibit positive birefringence (e.g., crystalline quartz), whereas the opposite situation is called negative birefringence (e.g., calcite).

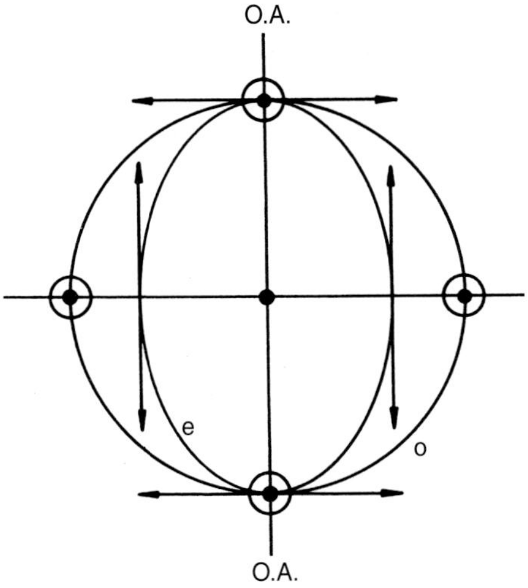


**Fig. 29** Elementary wave surfaces for ordinary (*o*) and extraordinary (*e*) rays in a uniaxial birefringent crystal. *O.A.* denotes the optical axis. (after Klein, M. V., Furtak, T. E. (1986), *Optics*. New York: Wiley [6])

When a ray of unpolarized light is incident normally on a birefringent plate, the ordinary beam is not deflected upon transmission through each surface, just as one would expect from Snell's law. The extraordinary ray behaves quite differently: The ray is deflected inside and displaced outside the crystal. Only when the optical axis lies in the face plane of the crystal or is orthogonal to it will the transmitted ray be normal to the surface inside the crystal.

This behavior of the extraordinary ray is illustrated in Fig. 30. While the light travels from A to B outside the crystal (wave fronts normal to ray vector), its ordinary and extraordinary components travel from B to C and from B to D inside the crystal (wave fronts not normal to ray vector in the latter case). So for this situation, the behavior of the extraordinary ray is really most extraordinary. Much more complicated phenomena are observed for biaxial crystals, which we leave to be studied in the literature.

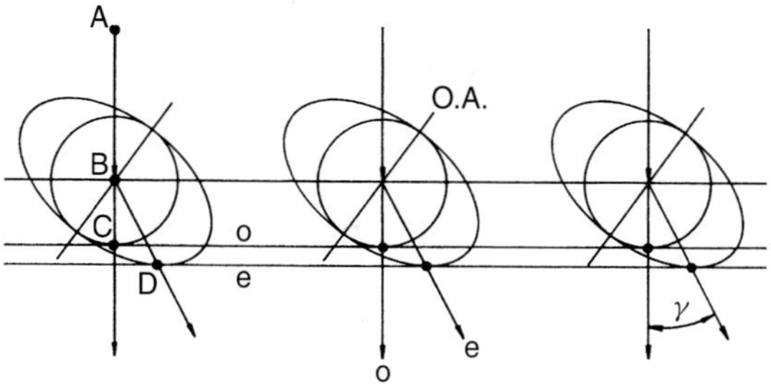


**Fig. 30** Refraction of an extraordinary ray entering a birefringent crystal. $\gamma$ is the deviation from straight transition. The crystal surface is parallel to the drawn wave fronts. (after Klein, M. V., Furtak, T. E. (1986), *Optics*. New York: Wiley [6])

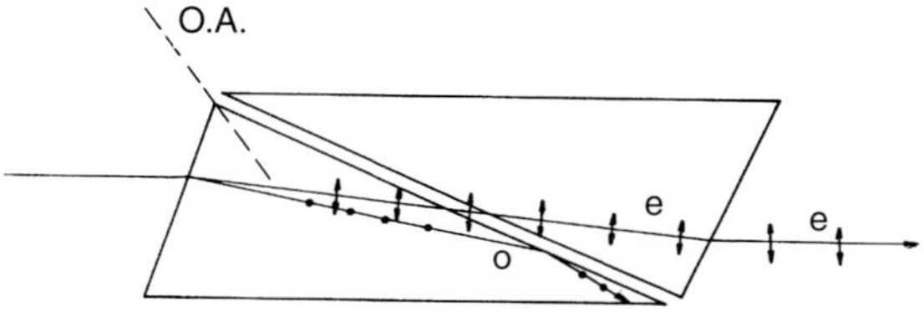


**Fig. 31** The Nicol prism (after Klein, M. V., Furtak, T. E. (1986), *Optics*. New York: Wiley [6])

### 6.1.1 Crystal Polarizers

As stated above (see Sect. 3.3), Fresnel reflection can be used to polarize a light wave. For more demanding applications, crystal polarizers are used. These can be divided into two groups: those that use total internal reflection and those that use birefringence to transmit both polarization states at different angles through an internal interface.

For brevity, only one example of each group will be shown. The Nicol prism is one of the oldest polarizers, though less common nowadays. Two prisms are cut from a birefringent calcite crystal as shown in Fig. 31. A beam entering the prism is split into the ordinary and extraordinary components as shown. At the interface between the first crystal and the optical-cement layer, the ordinary beam is totally internally reflected, whereas the extraordinary beam passes on through the second prism and exits parallel to the input beam though with a lateral offset.

The Rochon prism depicted in Fig. 32 passes the ordinary beam undeflected and without an offset whereas the

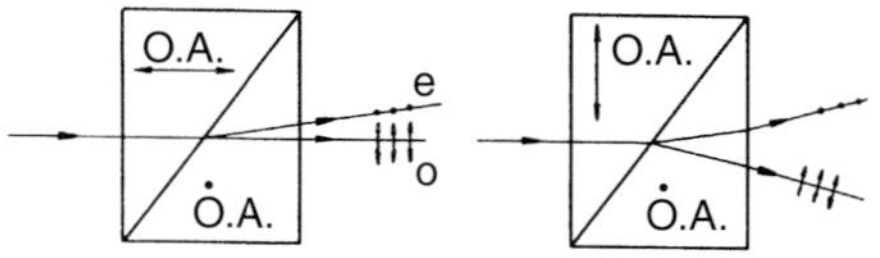


**Fig. 32** The Rochon prism. A bold dot indicates that the optical axis *O.A.* is normal to the plane of the drawing. (after Klein, M. V., Furtak, T. E. (1986), *Optics*. New York: Wiley [6])

extraordinary beam is deviated in this case. Most polarization optical elements are made from quartz or calcite. For polarization prisms, calcite is usually advantageous because of its larger difference of refractive indices for the ordinary and extraordinary beams.

### 6.1.2 Retardation Plates

Retardation plates are plates cut from birefringent crystals with the optical axis usually lying in the plane of the parallel faces. The phases of the ordinary and extraordinary rays change by different amounts when passing through such a plate:

$$\Delta\phi_o = \left(\frac{2\pi}{\lambda}\right) n_o d \tag{78}$$

and

$$\Delta\phi_e = \left(\frac{2\pi}{\lambda}\right) n_e d. \quad (79)$$

Two cases for the value of the phase difference $\Delta\phi = \Delta\phi_e - \Delta\phi_o$ are of special interest, namely, $\Delta\phi = \pm\pi/2$ and $\Delta\phi = \pm\pi$. A retardation plate satisfying the first condition is called a quarter-wave plate: a plate satisfying the second is called a half-wave plate.

**Quarter-wave Plate** For a quarter-wave plate, the equation

$$|n_e - n_o|d = \frac{(4n+1)\lambda}{4} \quad (80)$$

holds. If the integer $n$ is zero, the quarter-wave plate is said to be a "zero-order" plate; this indicates a relatively small chromatic dependency of the phase difference. [It should be noted, however, that the chromaticity is still much more pronounced than for a Fresnel rhomb (Fig. 33). There, total internal reflection is used to create the phase difference between the polarization component with the electric field vector in the plane of incidence and the component with the electric field vector orthogonal to that plane. Such a rhomb, however, has the disadvantage of introducing a lateral offset between input and output rays.] In practice, the thickness of retardation plates may be much larger than the value corresponding to the above equation. Two plates for which the phase differences differ by $\pi/2$ are cemented together with their optical axes orthogonal to each other, thus yielding a net phase difference of $\pi/2$ and greater mechanical stability.

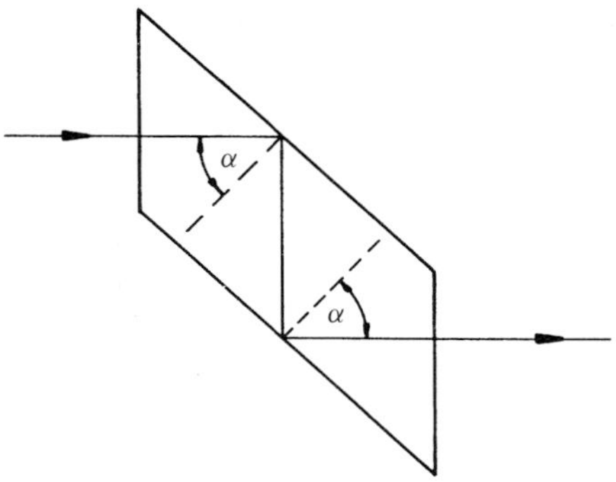


**Fig. 33** A Fresnel rhomb. The phase shift upon reflection is obtained by inserting the complex angles that correspond to total internal reflection into the equations for the coefficients of reflection. (after Klein, M. V., Furtak, T. E. (1986), *Optics*. New York: Wiley [6])

Linearly polarized light incident on such a plate will maintain its state of polarization when the electric field is parallel or orthogonal to the optical axis. For all other orientations of the plate, the light is split into two components, which will be retarded by $\pm\pi/2$ relative to each other after passing through the plate; thus, generally elliptically polarized light will result. When the angle between the electric field vector and the optical axis is 45°, the two components will be equal (neglecting slightly different Fresnel losses), and circularly polarized light results. The transformation of linear into circular polarization states and vice versa is one of the most important applications of quarter-wave plates.

**Half-wave Plate** A half-wave plate has a thickness satisfying the equation

$$|n_e - n_o|d = \frac{(2n+1)\lambda}{2}. \quad (81)$$

Naturally, two Fresnel rhombs in series will serve the same purpose and, in this case, may compensate each other's offset. For linearly polarized light passing through such a plate, the plane of polarization will be rotated by $2\alpha$ when the plate is rotated by $\alpha$. Thus, when the plate's axis is oriented at 45° to the electric field vector, the outgoing ray will be polarized orthogonal to the incident ray.

For circularly polarized light, the same exchange between the two orthogonal polarization states is obtained. Left circular light will be transformed into right circular light and vice versa.

### 6.1.3 Compensators

Compensators are retardation plates with adjustable phase difference. Their main application is the measurement of unknown birefringences. A sample of the material under inspection and the compensator are placed between crossed polarizers, and the compensator is adjusted for minimum transmission of the analyzer. For this adjustment, the (known) retardation introduced by the compensator compensates the retardation due to, e.g., the crystal under study modulo an integer multiple of $2\pi$; a measurement with a second plate of suitable thickness is then used to exclude this modulo uncertainty.

A widespread example of a compensator is the Soleil–Babinet compensator (Fig. 34). Two wedges with equal wedge angles and parallel optical axes form a plate of variable thickness. A plane-parallel plate with its optical axis orthogonal to the wedges allows adjustment to both positive and negative retardation values.

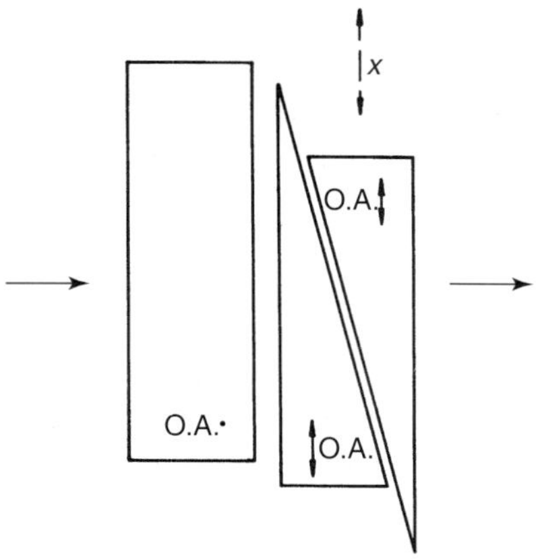


**Fig. 34** A Soleil–Babinet compensator. $x$ is the lateral displacement between the two wedges

## 6.2 Circular Birefringence

For chiral systems (i.e., systems that are not invariant under inversion), circular birefringence occurs. The refractive indices for left and right circular polarized light are different. This effect is isotropic. Instead of circular birefringence, the expression optical activity is often used for this phenomenon.

For linearly polarized light passing through a medium exhibiting optical activity, the plane of polarization is rotated by an amount proportional to the path traveled. According to the sign of this rotation, right- and left-rotating substances are distinguished. Optical activity is usually caused by a substance in a solution. In this case, the rotation angle is also proportional to the concentration of that substance. Thus, polarimetry, i.e., the measurement of the rotation angle, is an important means of determining the concentration of such a substance in a solution (see POLARIZATION OPTICS, POLARIMETERS AND POLARIZATION SPECTROMETERS).

Some materials have different modifications that have the same chemical constitution but differ in molecular structure and thus may be left, right, or even nonrotating. In this case, the different effects add up to form a net rotation. Other substances, e.g., quartz, exhibit both linear birefringence and optical activity. In this case, an exact treatment is more complicated as the energy of a light beam is continuously redistributed among different polarization states.

## 6.3
## Electro-optical Effects

Crystals and liquids can be polarized or oriented by electric fields that are constant or change with frequencies much lower than those of light oscillations. These phenomena have many important applications for modulation of the phase and the amplitude of light waves.

### 6.3.1 The Kerr Effect

The Kerr effect was the first electro-optic effect to be discovered. When isotropic substances are placed in a homogeneous electric field, they behave like a uniaxial crystal with its optical axis parallel to the applied field. The difference between the two refractive indices – which can be seen in analogy to $n_o$ and $n_e$ – is given by

$$\Delta_n = n_e - n_o = \lambda K E^2, \qquad (82)$$

where λ is the wavelength, $E$ is the electric field, and $K$ is the Kerr constant, which is usually, but not necessarily, positive. Because of the quadratic field dependence, the Kerr effect is sometimes called the quadratic electro-optic effect.

Today, crystals are commonly employed in Kerr modulators. The medium exhibiting the Kerr effect is placed in a parallel-plate capacitor so that an applied voltage $U$ results in a phase shift

$$\Delta\phi = 2\pi KLE^2 = 2\pi KL\left(\frac{U}{d}\right)^2 \qquad (83)$$

between the two polarization components (cf. Fig. 35). Here, $L$ is the length of the Kerr cell and $d$ is the electrode separation. If such a cell is placed between crossed polarizers oriented at 45° relative to the electric field, the transmitted light power $P$

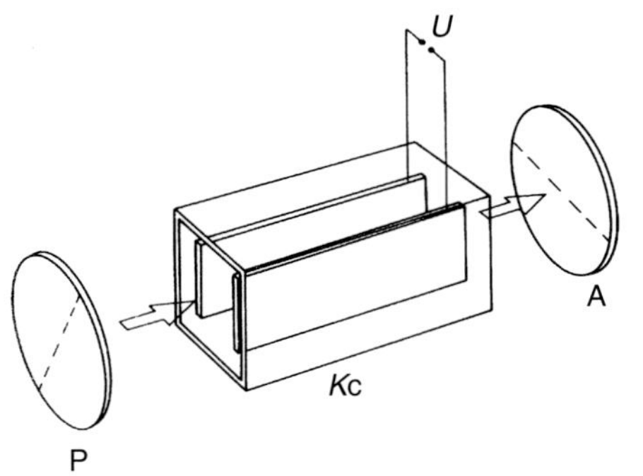


**Fig. 35** A Kerr cell *Kc* between polarizer *P* and analyzer *A* (after Hecht, E., Zajac, A. (1974), *Optics*. Reading, MA: Addison-Wesley [5])

as a function of input power $P_0$ is given by

$$P = P_0 \sin^2\left(\frac{\Delta\phi}{2}\right). \qquad (84)$$

Unfortunately, typical values for a change from complete blocking to complete transmission, called half-wave voltage as the cell then behaves like a half-wave plate, are often in the high voltage range. Possibly the diminishing of the required modulation voltages is an important advantage of integrated optics (see below).

### 6.3.2 The Pockels Effect

The Pockels effect or linear electro-optic effect is restricted to crystals lacking a center of symmetry. For these, the change in birefringence is linearly dependent on the applied fields and typically requires much lower voltages for device operation than the Kerr effect.

Figure 36 gives a sketch of Pockels cells. In (a) a longitudinal device is shown for which the electric field causing the retardation change is parallel to the propagation direction of the modulated light. In (b) the light travels orthogonal to the applied modulation field, thus forming a transverse configuration. These devices are advantageous as the voltage required for a phase

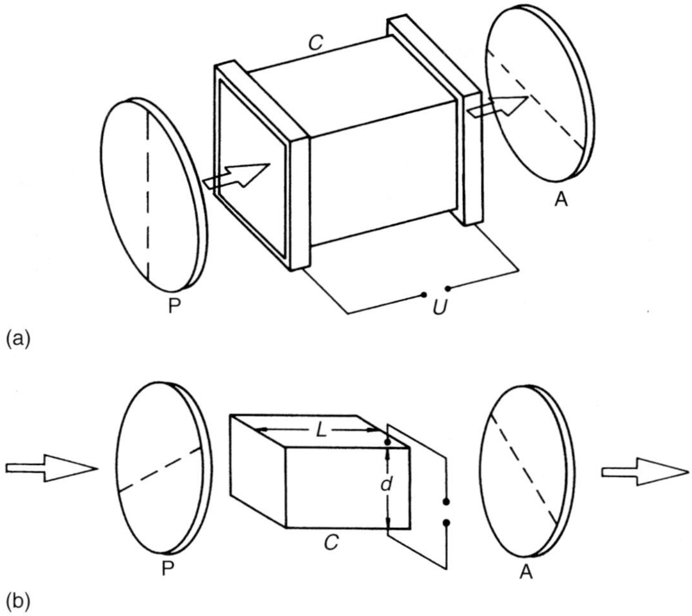


**Fig. 36** Pockels cells: (a) longitudinal cell (after Hecht, E., Zajac, A. (1974), *Optics*. Reading, MA: Addison-Wesley [5]), (b) transverse cell (after Yariv, A. (1991), *Optical Electronics*. Fort Worth: Saunders College Publishing)

change of $\pi$ decreases with increasing cell length. Transit-time limitations to the modulation frequency can be overcome by using a traveling-wave electrode arrangement. Enormous progress has been achieved in integrated optical modulators requiring only modulation voltages of a few volts, which can be provided easily (see below).

### 6.4 Magneto-optical Effect, Faraday Effect

The magneto-optical effect having presumably the largest number of applications is the Faraday effect. Light traveling (anti-)parallel to a magnetic field experiences polarization rotation in media that exhibit this effect. For a homogeneous magnetic field, the rotation angle is given by

$$\beta = VBL, \tag{85}$$

where $V$ is the Verdet constant characteristic for a given material, $B$ is the value of the magnetic field, and $L$ is the path length in the material.

A positive Verdet constant corresponds to a substance for which a magnetic field parallel to the direction of propagation seemingly turns the direction of the electric field counterclockwise when one looks toward the source.

A very important feature of the Faraday effect is its nonreciprocity. When linearly polarized light passing through a half-wave plate is reflected off a mirror to pass back through the same retardation plate, the net rotation of the plane of polarization is zero. When the retardation plate is replaced by a

Faraday rotator, i.e., a medium exhibiting the Faraday effect placed in a magnetic field, however, the rotation is not zero but twice the value for a single pass.

This effect is used in optical isolators. Light with linear polarization passing first through a polarizer adjusted for maximum transmission, then through a Faraday rotator, and finally through a half-wave plate compensating the rotation caused by the Faraday rotator will pass through an analyzer oriented parallel to the first polarizer. If the light passes through the arrangement in the opposite direction, however, the effects of half-wave plate and Faraday rotator add up; if both elements rotate the plane of polarization by 45°, the light is then blocked by the polarizer. Therefore, these devices are also called optical diodes.

### 6.5 Acousto-optical Effects

Acoustic waves in gases, liquids, and (preferably) solids are accompanied by density variations, and usually a monochromatic acoustic plane wave will create a phase grating due to the pressure (or density) dependence of the refractive index. Such a grating can be used for the diffraction of light waves when the grating constant and angle of incidence are chosen appropriately. Acousto-optical modulators based on this effect are used to modulate the amplitude of (laser) light beams or for shifting the frequency of a monochromatic light ray (see MODULATORS AND DEMODULATORS, OPTICAL). They may also be employed as variable deflectors as the diffraction angle can be controlled by changing the frequency of the acoustic frequency and thus, the grating constant.

## 7 Recent Developments

In this section, we cannot even attempt to be exhaustive in our treatment of the subject; thus, it is rather intended as an arbitrary choice of some examples of developments that played a major role in recent years. The foundations of many of the fields treated date back much further though. The reader is encouraged to take this section as a kind of appetizer to interest him in further studies in this field of physics which – not only because of its close link to one of man's most important senses, namely vision – is still full of new possibilities for applications.

### 7.1 Fiber Optics and Light Guides

The principle of total internal reflection can be used to guide light in conduits made of glass, plastics, and other materials transparent for the radiation under consideration. Development of new fibers has to a large extent been promoted by the aim to use optical transmission through fibers for telecommunications. Today, optical fibers can be divided into three basic types: step-index multimode fibers and gradient-index multimode and monomode fibers (see Fig. 37), the latter permitting polarization-maintaining designs (see FIBER OPTICS).

Optical fibers already play a major role in communication and data transfer. They are employed in many new sensor types. Endoscopes are another important field of application for fiber optics. They may be used as mode scramblers or simply as light guides. Monomode fibers are implemented in interferometers, e.g., of Sagnac or Mach–Zehnder type (see Fig. 24 in INTERFEROMETRY). Directional couplers

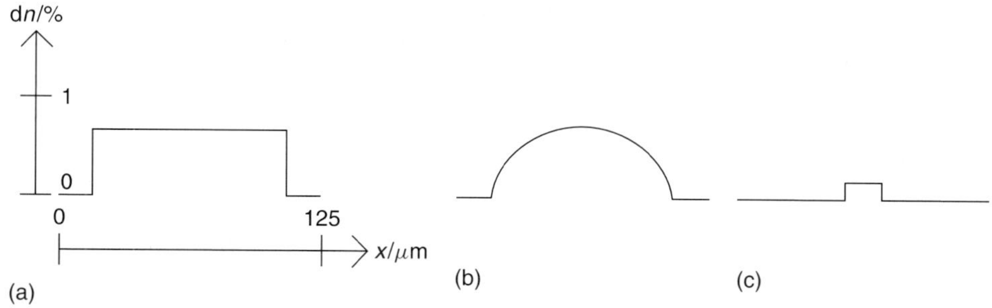


**Fig. 37** Optical fibers: (a) step-index multimode; (b) gradient-index multimode; and (c) monomode fiber (refractive index profiles)

can be produced for optical fibers in close analogy to microwave techniques.

## 7.2 Integrated Optics

Integrated optics has the aim to transfer the success of planar mass production processes known from semiconductor electronics to the field of optics (see OPTOELECTRONICS). Using photolithography, planar or linear waveguides can be created, e.g., on a glass substrate using ion-exchange processes (Fig. 38).

Components produced in this way usually look and act either like projections of optical systems into the substrate plane (e.g., grating spectrometers) or like analogies to microwave components (e.g., directional couplers). Combining such elements can lead to novel devices for communication systems or sensors (see, e.g., Fig. 39).

One of the fundamental advantages of integrated optics is the ability to confine optical waves to a narrow cross section (i.e., a single-mode waveguide) over an extended length. This makes possible the construction of devices unequaled in classical optics.

A single-mode waveguide in a suitable material should yield much higher conversion efficiencies for frequency doubling (see NONLINEAR OPTICS) than bulk systems. It may be necessary to change substrate properties periodically to achieve phase

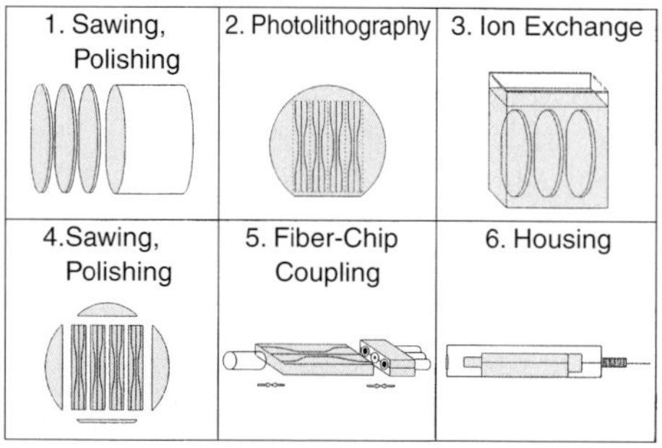


**Fig. 38** Integrated-optics production process (picture courtesy of IOT Integrierte Optik GmbH, Jena)

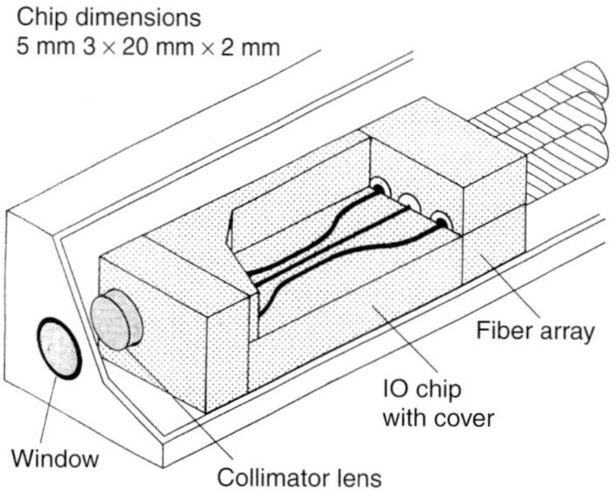


**Fig. 39** An integrated optical interferometer system (picture courtesy of IOT Integrierte Optik GmbH, Jena)

matching here. The effect has been demonstrated though so far technical problems seem to have blocked superior results.

In a material exhibiting the electro-optic effect, single-mode waveguides permit a very close electrode spacing for an electro-optic modulator (see MODULATORS AND DEMODULATORS, OPTICAL). Using this advantage has led to modulators using voltages of a few volts for a phase change of $\pi$ and multimegahertz modulation frequencies.

Micro-Optics is often combined with integrated and/or fiber optics, for example, for coupling light into waveguides or collimating light exiting from these. Micro-optics has also found different applications like endoscopic imaging or beam homogenization with lens arrays.

## 7.3 Diffractive Optics

Diffractive optics extends the applications of diffraction from gratings and holography to two different areas: the correction of optical imaging systems using diffractive structures and the use of diffractive structures to reconfigure light distributions in many ways.

### 7.3.1 Optical-imaging System Correction

Refractive lenses and diffractive elements (e.g., blazed Fresnel-zone plates) both exhibit chromatic aberration. However, normally the two aberrations differ considerably in their value and are of opposite sign. It is therefore obvious to try to combine a strong refractive lens with a weak diffractive lens in order to obtain a system in which these chromatic aberrations compensate each other.

This approach has been proven useful, and diffractive structures are not only used to compensate chromatic aberration, but play an important role as an additional (set of) free parameter(s) for minimizing the various errors present in optical systems. However, practical applications are predominantly in the infrared region of the spectrum, as production of diffractive structures with the required precision using, e.g., diamond turning is much easier at the longer wavelengths. As diamond turning is expensive and lithographic production of diffracting elements is in practice mostly limited to plane surfaces, applications are presently mainly in areas where high cost is acceptable because the required performance cannot be achieved otherwise.

### 7.3.2 Unconventional Applications

Possibly, purely diffractive elements are even more promising. Using photolithography from computer-generated mask patterns, they permit virtually any change of wave front one can think of – provided that the light used is monochromatic. The borderline between diffractive optics and (computer-generated) holography is somewhat arbitrary but will not be considered here.

Laser-beam shaping is a prominent example of such application. In material processing, the beam profile of the laser employed is often inappropriate for a specific application. Laser beams are mostly of three different types: $TEM_{00}$ (i.e., with a transverse field distribution described by a Gaussian), doughnut modes, and multimode beams. For beams with a well-defined mode (and to a lesser extent for multimode beams), diffractive optical elements (DOEs) permit the generation of virtually any desired intensity distribution. Multiple-spot illumination for simultaneous welding of all leads of an integrated circuit, specific intensity profiles advantageous for cutting and hardening, or even character generation for marking can be obtained. For a multimode beam, these possibilities may be somewhat limited, but still beam homogenization is possible.

### 7.3.3 Binary Optics

Binary optics is a special type of diffractive optics. It uses lithography to create blazed diffractive optical elements. Binary masks, that is, either transmitting or opaque locally, are used to create binary, that is, two-level, profile structures in resist layers, and from these the profile may be transferred to the substrate using etching techniques. By stacking consecutive two-level profiles, while reducing the profile step by a factor of 2 for consecutive masks, multilevel profiles with higher efficiency can be obtained in a way very similar to the generation of binary numbers. When a light wave passes through such a structure, the surface profile results in a phase profile of the wave front.

Figure 40 shows different gratings having the same grating constant. In (a) an amplitude grating is shown, and about 10% of the power incident on it is diffracted into the first order. With a blazed phase grating as shown in (b), almost 100% can be concentrated into this diffraction order. Binary optics uses consecutively refined mask structures to approximate this blaze by a staircaselike structure. In (c) with two phase levels an efficiency of about 40% is achieved; this value is roughly doubled when two masks are used to generate four different depth levels. With four levels as in (d), a value of about 80% is reached, but the finest features on the required mask are also two times smaller than on the mask used for the two-level structure.

Binary optical structures are frequently used as phase masks in high end photolithography used in microelectronics chip fabrication.

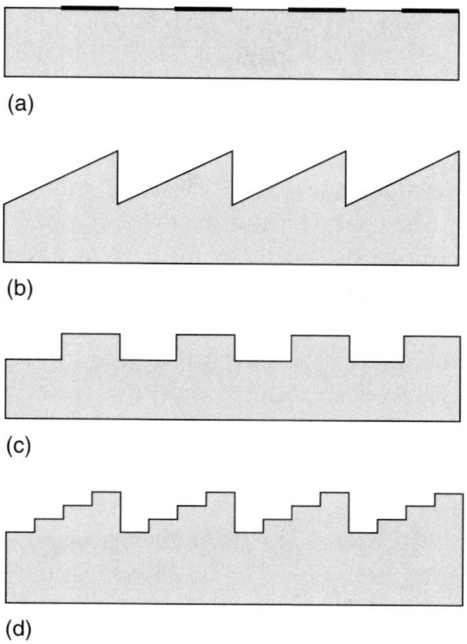


**Fig. 40** Binary optics. For details see text

## 7.4 Point Spread Function Engineering

Various approaches have resulted in an improvement of resolution, seemingly, in contradiction to traditional expectations. However, there is no true paradox here: What is often forgotten is the assumptions or boundary conditions for those resolution criteria. In microscopy, measured image data is used to calculate the ideal image from real data using the available knowledge of the point spread function and aberrations of the system under consideration. The aspects of coherence and polarization for resolution are also being taken into account. Phase information can also be combined with data processing to yield novel results.

## 7.5 EUV- and X-Ray Optics

Simple telescopes for EUV and X-rays using grazing incidence have been around for a while (see X-RAY OPTICS). In recent years, significant progress has been made for multilayer coatings here, and relevant reflectivities now allow for multi mirror designs with near-normal incidence at wavelengths of around 10 to 20 nm. In the same context, improvements in optics manufacturing technologies have resulted in the availability of fully aspheric off-axis surfaces that add many new design options for such mirror systems useful for example in advanced microlithography.

## 7.6
## Adaptive Optics

Adaptive optics are used to change the imaging behavior, e.g., of an optical telescope in order to compensate for disturbances whose origins lie in the density variations and turbulence of the atmosphere (seeing). Because of these disturbances, earthbound telescopes normally achieve a much lower effective resolution than the diffraction limit due to finite aperture diameter.

Two subsystems are necessary for the implementation of an adaptive optical system: a wavefront sensor for the detection of deviations of the incoming wave front from its ideal shape, and a servo element such as a deformable mirror, for the correction of these.

As this technique is easier to implement for longer wavelengths, it was first successfully tested in near-infrared astronomy, but it is being transferred to the visual region as well with impressive results. However, this does not mean that space telescopes will be unnecessary in the future. Absorption by the earth's atmosphere will always limit the wavelength range accessible to earthbound telescopes.

A different kind of "adaptive" optics is used in modern microlithography: here, actuators in high-performance lens systems allow optimization of performance in the presence of disturbances such as lens heating, air pressure changes, etc. or for imaging optimization for various criteria.

## 8
## Conclusions

It is not possible to give a complete impression of the wide variety of optical systems and applications. Interesting fields like stress birefringence or intensity correlation experiments have not even been sketched in this article. However, this outlook may motivate readers to engage themselves more deeply in the study of this fascinating field of physics and the study of light.

## Glossary

**Aperture:** A stop that limits the object area that is imaged by an optical system.

**Chromatic Aberration:** The dependence of the focal length on wavelength due to the dispersion of refractive materials.

**Coherence:** The ability of light (waves) to interfere, i.e., their phase relation and its spatial and temporal behavior.

**Diffraction:** The deviation of light waves from straight propagation that is caused by obstacles.

**$f$ Number:** The ratio $r$ of focal length and diameter of an objective. Often, this value is written as $f/r$, e.g., $f/4$ for a 200-mm lens with 50-mm diameter.

**Interference:** The superposition of coherent fields yielding phase-sensitive phenomena.

**Kerr Cell:** An electro-optic modulator based on the Kerr or quadratic electro-optic effect.

**Magnification:** Two similar quantities are called magnification: The ratio of the sizes of image and object is usually called magnification. The same expression is used for the ratio between the apparent sizes (corresponding to the viewing angles) of an object with and without an optical system like a microscope or a telescope. For clarity, it is useful to speak of linear magnification in the first case and of angular magnification in the second.

**Numerical Aperture:** This expression can have two different significations depending on the field of optics in which it is employed:

1. The sine of the half-angle of the cone of light entering a microscope objective from an object point. Its value sets a limit to the resolution that can be obtained.
2. The maximum value for the sine of the angle of incidence for light rays incident upon the front fact of an optical fiber at which these rays will still be guided by total internal reflection.

**Optical Activity:** Circular birefringence.

**Optical Axis:** Axis of a uniaxial crystal along which the refractive index for the extraordinary ray has the same value as for the ordinary ray.

**Phase Velocity:** The velocity with which a point of constant phase travels within a wave.

**Plane of Incidence:** Plane containing the propagation vector of a plane wave and the normal to an interface upon which this wave is incident.

**Pockels Cell:** An electro-optic modulator based on the pockels or linear electro-optic effect.

**Polarizer:** A device selectively transmitting one linearly polarized component of light, or both (orthogonally polarized) components into different directions thus separating them.

**Pupil:** A stop limiting the solid angle of rays contributing to the image of a point for a given imaging situation.

**Refractive Index:** The ratio of the velocity of light in vacuum and in a medium. A complex value indicates extinction (or amplification).

**Resolution:** The ability of an optical system to distinguish two neighboring points in the image of an object or of a spectrum. According to a criterion given by Lord Rayleigh, these are considered to be resolved if the intensity decreases to about 80% of the maximum between two maxima of equal strength.

**Scattering:** The generation of secondary waves by light traveling in a medium and the propagation of these. The secondary waves do not necessarily have the same frequency as the exciting wave.

**Signal Velocity:** The velocity with which a point of constant relative amplitude travels in a modulated wave.

**Spectrograph:** A device (usually employing prisms or gratings) for the spectral analysis of light, and a film or Opto-electronics detector for the recording of the spectral distribution.

**Stellar Aberration:** The apparent movement of the stars due to observation from the Earth moving in its orbit. A relativistic vector addition of the Earth's velocity and the velocity vector of the impinging photons yields the actual telescope pointing direction.

**Verdet Constant:** A coefficient characterizing the strength of the Faraday effect for a given material and for a given wavelength.

### References


[1] Quinn, T. J. **(1992)**, *Metrologia 1993/1994* **30**, 523–541.
[2] Beth, R. A. **(1936)**, *Phys. Rev.* **50**, 115–125.
[3] Alonso, M., Finn, E. J. (1976), *Fundamental University Physics*, Vol. 2. Reading, MA: Addison-Wesley, p. 789ff.
[4] Born, M., Wolf, E. (1980), *Principles of Optics*, (6th ed.), Oxford: Pergamon Press.
[5] Hecht, E., Zajac, A. (1974), *Optics*. Reading, MA: Addison-Wesley.
[6] Klein, M. V., Furtak, T. E. (1986), *Optics*. New York: Wiley.
[7] Kingslake, R. (1978), *Lens Design Fundamentals*. New York: Academic Press.
[8] Kingslake, R. (1983), *Optical System Design*. Orlando, FL: Academic Press.


### Further Reading

### General


Born, M., Wolf, E. (1980), *Principles of Optics*, (6th ed.), Oxford: Pergamon Press.
Hecht, E. (1975), *Optics*. New York: McGraw-Hill.
Jenkins, F. A., White, H. E. (1981), *Fundamentals of Optics*. Singapore: McGraw-Hill.
Klein, M. V., Furtak, T. E. (1986), *Optics*. New York: Wiley.


### Selected Topics


Anan'ev, Y. A. (1991), *Laser Resonators and the Beam Divergence Problem*. Bristol: Adam Hilger.
Arnaud, J. A. (1976), *Beam and Fiber Optics*. New York: Academic Press.
Beckmann, P., Spizzichino, A. (1987), *The Scattering of Electromagnetic Waves from Rough Surfaces*. Norwood, MA: Artech House.
Caulfield, H. J., Lu, S. (1970), *The Applications of Holography*. New York: Wiley.
Collett, E. (1993), *Polarized Light*. New York: Marcel Dekker.
Cox, A. (1964), *A System of Optical Design*. London: Focal Press.
Duffieux, P. M. (1983), *The Fourier Transform and its Applications to Optics*. New York: Wiley.
Felsen, L. B. (Ed.) (1984), *Hybrid Formulation of Wave Propagation and Scattering*. Dordrecht, The Netherlands: Martinus Nijhoff.
Flügge, S. (Ed.) (1967), *Encyclopedia of Physics, Optical Instruments*, Vol. XXIX. Berlin: Springer-Verlag.
Frieden, B. R. (Ed.) (1980), *The Computer in Optical Research*. Berlin: Springer-Verlag.
Gaskill, J. D. (1978), *Linear Systems, Fourier Transforms, and Optics*. New York: Wiley.
Gerrard, A., Burch, J. M. (1975), *Introduction to Matrix Methods in Optics*. London: Wiley.
Ghatak, A. K., Thyagarajan, K. (1989), *Optical Electronics*. Cambridge, UK: Cambridge University Press.
Goodman, J. W. (1984), *Statistical Optics*. New York: Wiley.
Guenther, R. D. (1990), *Modern Optics*. New York: Wiley.
Haferkorn, H. (1986), *Bewertung Optischer Systeme*. Berlin: VEB Deutscher Verlag der Wissen-schaften.
Hall, D. R., Jackson, P. E. (1989), *The Physics and Technology of Laser Resonators*. Bristol: Adam Hilger.
Haus, H. A. (1984), *Waves and Fields in Optoelectronics*. Englewood Cliffs, NJ: Prentice Hall.
Hecht, E., Zajac, A. (1974), *Optics*. Reading, MA: Addison-Wesley.
Hecht, E. (1975), *Optics*. New York: McGraw-Hill.
Horne, D. F. (1975), *Lens Mechanism Technology*. London: Adam Hilger.
Hutley, M. C. (1982), *Diffraction Gratings*. London: Academic Press.
Iizuka, K. (1983), *Engineering Optics*. New York: Springer-Verlag.
Kersten, R. Th. (1983), *Einführung in die optische Nachrichtentechnik*. Berlin: Springer-Verlag.

Kingslake, R. (Ed.) (1965), *Applied Optics and Optical Engineering*, Vol. 1. New York: Academic Press.

Kingslake, R. (1978), *Lens Design Fundamentals*. New York: Academic Press.

Kingslake, R. (1983), *Optical System Design*. Orlando, FL: Academic Press.

Kliger, D. S., Lewis, J. W., Randall, C. E. (1990), *Polarized Light in Optics and Spectroscopy*. Boston: Academic Press.

Korsch, D. (1991), *Reflective Optics*. Boston: Academic Press.

Laikin, M. (1990), *Lens Design*. New York: Marcel Dekker.

Lauterborn, W., Kurz, T., Wiesenfeldt, M. (1993), *Kohärente Optik*. Berlin: Springer-Verlag.

Mahajan, V. N. (1991), *Aberration Theory Made Simple*. Bellingham, WA: SPIE.

Malacara, D. (1978), *Optical Shop Testing*. New York: Wiley.

Marathay, A. S. (1982), *Elements of Optical Coherence Theory*. New York: Wiley.

Meyers, R. A. (Ed.) (1991), *Encyclopedia of Lasers and Optical Technology*. San Diego, CA: Academic Press.

Neumann, E.-G. (1988), *Single-Mode Fibers*. Berlin: Springer-Verlag.

Nieto-Vesperinas, M. (1991), *Scattering and Diffraction in Physical Optics*. New York: Wiley.

Ogilvy, J. A. (1991), *Theory of Wave Scattering from Random Rough Surfaces*. Bristol, UK: Adam Hilger.

O'Neill, E. L. (1963), *Introduction to Statistical Optics*. Reading, MA: Addison-Wesley.

Parker, S. P. (Ed.) (1988), *Optics Source Book*. New York: McGraw-Hill.

Petit, R. (Ed.) (1980), *Electromagnetic Theory of Gratings*. Berlin: Springer-Verlag.

Petykiewicz, J. (1990), *Wave Optics*. Dordrecht: Kluwer Academic Publishers.

Reynolds, G. O., DeVelis, J. B., Parrent, G. B., Thompson, B. J. (1989), *The New Physical Optics Notebook*. Bellingham, WA: SPIE.

Saleh, B. E. A., Teich, M. C. (1991), *Fundamentals of Photonics*. New York: Wiley.

Schroeder, D. J. (1987), *Astronomical Optics*. San Diego: Academic Press.

Slyusarev, G. G. (1984), *Aberration and Optical Design Theory*. Bristol: Adam Hilger.

Snyder, A. W., Love, J. D. (1983), *Optical Waveguide Theory*. London: Chapman and Hall.

Solimeno, S., Crosignani, B., Di Porto, P. (1986), *Guiding, Diffraction, and Confinement of Optical Radiation*. Orlando, FL: Academic Press.

Stamnes, J. J. (1986), *Waves in Focal Regions*. Bristol: Adam Hilger.

Stover, J. C. (1990), *Optical Scattering*. New York: McGraw-Hill.

Tamir, T. (Ed.) (1979), *Integrated Optics*. Berlin: Springer-Verlag.

Tamir, Th. (Ed.) (1988), *Guided-Wave Optoelectronics*. Berlin: Springer-Verlag.

Tyson, R. K. (1991), *Principles of Adaptive Optics*. Boston: Academic Press.

van Etten, W., van der Plaats, J. (1990), *Fundamentals of Optical Fiber Communications*. New York: Prentice Hall.

Voronovich, A. G. (1994), *Wave Scattering from Rough Surfaces*. Berlin: Springer-Verlag.

Weichel, H. (1990), *Laser Beam Propagation in the Atmosphere*. Bellingham, WA: SPIE.

Welford, W. T. (1986), *Aberrations of Optical Systems*. Bristol: Adam Hilger.

Welford, W. T., Winston, R. (1989), *High Collection Nonimaging Optics*. San Diego: Academic Press.

Williams, C. S. (1989), *Introduction to the Optical Transfer Function*. New York: Wiley.

Yariv, A., Yeh, P. (1983), *Optical Waves in Crystals*. New York: Wiley.

Yariv, A. (1991), *Optical Electronics*. Fort Worth: Saunders College Publishing.

Yu, F. T. S. (1985), *White-Light Optical Signal Processing*. New York: Wiley.

## Appetizers

Falk, D., Brill, D., Stork, D. (1986), *Seeing the Light*. New York: Harper & Row.

Léna, P., Blanchard, A. (1990), *Lumières*. Paris: InterEditions.

Minnaert, M. (1954), *Light & Colour*. Dover Publications.

Schulz, G. (1974), *Paradoxa aus der Optik*. Leipzig: Johann Ambrosius Barth.

# Luminescence

*Detlev M. Hofmann and Manfred Böhm*
*Justus-Liebig-Universität Giessen, I. Physikalisches Institut, Heinrich-Buff-Ring 16, D-35692 Giessen, Germany*
*Phone: +49-(0)641-9933105; Fax: +49-(0)641-9933119;*
*e-mail: detlev.m.hofmann@exp1.physik.uni-giessen.de*

**Abstract**
The article gives an introduction to the widespread phenomenon of light emission from matter under the aspect of the most common stimulation sources, that is, the use of photons for photoluminescence, or electron beams for cathodoluminescence, electric fields in the case of electroluminescence, chemical energy for chemiluminescence, sound waves for sonoluminescence, or thermal energy for the stimulation of the thermoluminescence phenomena. It focuses mostly on the respective principles of the luminescence excitation, rather than an overview of the uncountable applications of the methods in all areas of science, technology, and engineering.

**Keywords**
luminescence; photoluminescence; cathodoluminescence; thermoluminescence; chemiluminescence; sonoluminescence; electroluminescence.

## 1 Introduction

Luminescence is the emission of optical radiation (infrared, visible, or ultraviolet light) by matter. This phenomenon is to be distinguished from incandescence, which is the emission of radiation by a substance by virtue of being at a high temperature. Luminescence can occur in a wide variety of matter and under many circumstances. Thus, atoms, various kinds of molecules, polymers, organic- or inorganic crystals, amorphous substances or even biological units can luminesce under appropriate conditions. The common aspect to observe luminescence is when an atom or molecule emits a photon (a quantum of light), there is a change in its electronic structure. When electrons are confined to a region of space that is defined by the position of the atomic centers of a molecule (or just the nucleus, in the case of an atom), a set of discrete energy levels is imposed on the system. In the case that some process delivers an appropriate amount of energy to the system, the electronic structure can be changed to an excited state. The release of this extra energy, and thus the way back to the initial state (ground state) can occur via the emission of a photon. Dependent on whether the spin quantum number of the electrons is changed in the emission process or not, one may distinguish two different types of luminescence: fluorescence, in which emission occurs with no net change of the spin quantum number, and phosphorescence, in which emission is accompanied by a change in the spin quantum number. However, to apply this rule correctly, the spin state of the system under investigation has to be known in detail, which is, especially for complex systems, rarely the case.

Another way to get some sort of order in the widespread phenomenon of light emission of matter is to distinguish the modes of luminescence production, that is, the way in which the electronic system is excited. We follow this approach, and give in the following chapters a brief description of the methods of photoluminescence (PL), electroluminescence (EL), thermoluminescence (TL), chemiluminescence (ChL), cathodoluminescence (CL), and sonoluminescence (SL). They represent the most common techniques to achieve luminescence from matter, independent of the final goal of the interest, whether it is to understand the electronic structure of the substances or the luminescence excitation/deexcitation mechanisms, or to develop knowledge and technologies for novel light-emitting devices.

## 2 Photoluminescence (PL)

In PL, the excitation results from light absorption. Because of the quantum nature of light and considering only low excitation conditions, one absorbed photon can induce at best one luminescence photon whose energy is necessarily smaller than (or equal to) that of the exciting photon (Stokes' law). Historically, luminescence was restricted to the visible part of the spectrum, but nothing prevents an extension toward the ultraviolet or infrared regions. By comparison with other excitation processes, PL allows the best control over the excitation conditions and for that reason, is the most useful process for studying the fundamental properties of luminescent materials. Development of luminescent materials for fluorescent tubes and cathode-ray tubes took place in the 1940s. Study of PL in organic and semiconductor materials started at the end of the 1950s and developed very fast with the advent of lasers as excitation sources. To date, PL has become a very common nondestructive test for material properties in physics, chemistry, and biology as well as in the applied sciences.

In general, PL can be described as a three-step process:

1. absorption of one photon of the excitation source creating in the material a primary excitation;
2. evolution of this primary excitation by energy relaxation or energy transfer toward secondary excitation at lower energy; and
3. coupling of this secondary excitation to light, giving rise to the spontaneous emission of one luminescence photon.

Steps 1 and 3 require the knowledge (1) of the electronic energy spectrum of the material and (2) of the strength of optical transition between any pair of energy levels. Traditionally, the energy spectrum of condensed matter is described in the framework of the Born–Oppenheimer approximation, which relies upon the fact that the movements of the nuclei are much slower than those of the electrons, because of the large mass ratio. Actually, electronic excitations and lattice-

or molecular-vibrations are coupled with a strength depending on the material under consideration. Luminescence is linked mainly with electronic excitations; coupling between electronic and lattice excitations enters into play mainly in Step 2 to allow for some energy relaxation. Let us note that Stokes law is strictly valid only at temperature $T = 0$; at finite $T$, there is a finite probability to see the excitation jumping to a higher energy by phonon absorption: This leads to anti-Stokes emission.

The PL quantum efficiency $\eta$ is defined as the ratio of the number of photons emitted $n_l$ to the number of excitation photons absorbed $n_a$ : $\eta = n_l/n_a$. The fact that $\eta < 1$ is due to nonradiative processes that allow the complete energy dissipation of the exciting photon. Among the nonradiative processes are the multiphonon processes in which an electronic excitation $\Delta E$ is dissipated into $n$ lattice vibrations. Since the probability of such process decreases very fast with $n$, an energy gap large enough is required between the emitting state and the ground state in order to have a sizable quantum efficiency. This explains why metals, which exhibit a continuum of electronic excited states, are not luminescent materials.

More specifically, one can consider three main types of luminescent materials:

1. *Molecular organic materials.* In that case, the building block is the isolated molecule, which is only weakly perturbed by its environment, for example, a solvent or its inclusion in a molecular solid, tied by van der Waals interactions.

2. *Luminescent centers in large-gap (hence transparent) insulators,* like oxides, tungstates, silicates, or glasses. The most studied are transition-metal or rare-earth ions; the isolated ions have usually a rich spectrum of excitations linked to their partially filled $d$ or $f$ shell, providing opportunities to obtain luminescence in the visible. Perturbation of these ions by the crystalline matrix is usually treated in the crystal-field approximation. Another example is the $F$ center in alkali halides, where an electron is trapped in an anion vacancy.

3. *Semiconductors.* They present an energy gap $E_g$ between a filled valence band and an empty conduction band. In that case, luminescence takes place at an energy close to the band gap, which spans a range of values from near ultraviolet to infrared, depending on the material. The building blocks are now free electrons in the conduction band and free holes in the valence band, characterized by their effective masses. When Coulomb interaction between these particles with impurities is taken into account, weakly bound states are built, which can be described in the effective-mass approximation: free excitons or bound excitons or related quasi-particles. They dominate the luminescence spectrum in semiconductors at low temperature.

In comparison to the other sources for the excitation of the luminescence, that is, the cathodo-, electroluminescence and so on, the use of monochromatic excitation for the PL has the advantage that very specific energy levels can be excited; this can be used either to optimize the luminescence quantum efficiency or, for fundamental research, in order to get a better understanding of Step 2. In addition, one can take advantage of light polarization, which allows additional selectivity in anisotropic materials.

## 2.1 Instrumentations

A typical experimental PL setup is shown in Fig. 1. It is composed of an excitation source, optical elements to focus excitation on the sample and collect the luminescence light, a spectrometer (or filter) to analyze the luminescence, and a detector followed by a data acquisition system.

Before the advent of lasers, the most useful sources were high-pressure mercury or xenon lamps. Both have a broad continuous spectrum extending from 300 nm to near infrared. For short wavelength excitations, deuterium lamps could be used. Appropriate filters have to be inserted in order to cut source emission in the spectral region of sample luminescence. Lasers are excitation sources much more convenient for several reasons:

1. They are much more powerful.
2. Since they are well collimated, it is easy to use 100% of their power, as compared to 15% at best for conventional sources.
3. Their emission is essentially monochromatic (Although filters may be required to suppress plasma emission lines in gas lasers).

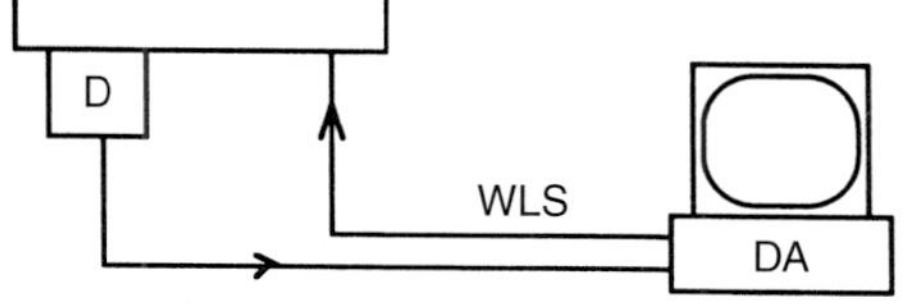


**Fig. 1** Typical photoluminescence setup. L, laser; F, filter; M, mirror; S, sample; C, condenser; SP, spectrometer; D, detector; DA, data acquisition system; WLS, wavelength scanning

A wide range of wavelengths is now available from the infrared ($CO_2$ near 10 μm, CO near 5 μm, and Nd solid-state laser at 1.06 μm, through gas lasers in the visible (He–Ne, Ar, Kr, He–Cd) and gas lasers with intracavity second-harmonic generation for the 250-nm region as well as excimer lasers down to about 190 nm, which are, however, pulsed excitation sources.

If in some cases a filter could be enough to select a luminescence line, in most cases a spectrometer is needed to get the luminescence spectrum. The most widely used are grating spectrometers, either single or double. The light emitted by the sample should be focused on the entrance slit of the spectrometer by an objective or a combination of mirrors in the infrared to avoid absorption and dispersion.

In the ultraviolet, visible, and near infrared down to 1 μm, the most useful detectors are photomultipliers. Semiconductor detectors (photovoltaic, photoconductor, avalanche $p-n$ junction) cover the visible and infrared range down to 10 μm. Chopping of the excitation at a frequency $f$ followed by synchronous detection is common practice. Let us point out that this is mandatory only when using detectors exhibiting $1/f$ noise, since the detector contributes to the noise in a window $\delta f$ around the frequency $f$. With photomultipliers, the best is to use a discriminator followed by a counter. Semiconductor-Detector arrays, featuring up to 2048 (or even more) elements, provide a very efficient detection method, since wavelength scanning is suppressed. Fourier transform spectrometers have also been used in the infrared; for a given spectral resolution, they can accommodate a luminescence beam of larger extension.

Sample temperature is an important parameter in PL experiments, since, in general, line narrowing is expected on cooling. Moreover, new ground or excited states can appear at low temperature. Optical cryostats allow easy operation down to 1.5 K, the best cooling conditions being achieved by immersion of the sample in superfluid liquid helium. In peculiar cases, operation at temperatures below can be obtained with dilution refrigerators. Experiments between helium and room temperature are also easily performed – for example, using commercial closed-cycle refrigerators.

## 2.2
## Photoluminescence Kinetics and Excitation Spectra

It is often useful to study PL *kinetic*, that is, how the luminescence intensity depends upon the excitation intensity, taking care of the fact that different luminescence lines may exhibit different kinetics. Another powerful way to obtain information about the PL process is the study of the dependence upon excitation wavelength of the luminescence spectrum. This is performed usually by recording the *PL excitation spectrum*, the excitation wavelength being scanned at fixed luminescence wavelength. A tunable source is required; this can be done either with a broadband source and a monochromator or, more conveniently, with a tunable laser, like a dye laser or a solid-state tunable laser (titanium–sapphire). Often, the excitation spectrum is close to the absorption spectrum of the sample. Light *polarization* is an additional useful feature: by inserting a polarizer (either linear or circular) in the excitation and/or the luminescence beam, one can gain information about the symmetries of the primary excited state and the final emitting state, when the sample presents some kind of anisotropy, either natural or induced by external perturbations. Selection of the luminescence photon wave vector, realized by inserting a slit out of focus to limit the solid angle of the luminescence beam, can also be useful with anisotropic samples. Even with isotropic samples, a polarized excitation can induce some degree of polarization of the luminescence, a technique known as *optical pumping*.

## 2.3
## Time Resolution

Time resolution adds new capabilities to directly measure characteristic times in the sample, like lifetime of states or transfer time between states. It requires first a pulsed excitation. For decay time longer than about 10 ns, modulation by electro-optical or acousto-optical devices can be used. For shorter time scales down to 0.1 ns, mode-locked lasers can be used. Synchronously pumped dye (or titanium – sapphire) lasers can provide pulses reaching the subpicosecond range, tunable in a substantial wavelength range.

Detection was done first using fast oscilloscopes or boxcar techniques, which require a scanning of the delay between excitation and gate defining the measurement window. A better method, in conjunction with photomultiplier detectors, is the so-called *single-photon counting* technique: the time delay between the excitation pulse and the first luminescence photon detected is measured with a time-to-amplitude converter, and the data are stored in the multichannel mode. This method is very efficient at low levels of excitation since essentially all luminescence photons are detected. With fast discriminators, time resolution below 50 ps is achievable. A second method is based on *up-conversion* in a nonlinear medium: part of the excitation pulse is combined with the luminescence beam in a nonlinear crystal under phase-matching conditions; photons at an energy that is the sum of excitation and luminescence photon energies are detected. This method, less sensitive than the previous one, allows a time resolution down to the subpicosecond range; moreover, it can be used for infrared luminescence. A third method uses a *streak camera*: The luminescence light focused on a transparent photocathode creates electrons focused on a fluorescent screen after deflection by a sawtooth-like electric field synchronized with the excitation pulse. The bright trace on the screen, which is an image of the time-dependent luminescence, is read by an array of detectors. This method is fairly sensitive with a typical time resolution of about 10 ps and a spectral domain of wavelengths below 850 nm. In all cases in which the expected time resolution is below 10 ps, time broadening due to the passage on one grating has to be suppressed with the use of a double-grating spectrometer in the substractive mode.

Detection of *quantum beats* occurs when two nearby energy states separated by $\Delta E$ have been both excited by a short pulse; during their coherent decay, beats with a period $\Delta t = h/\Delta E$ can be detected on the luminescence signal, under appropriate polarization conditions, thus allowing measurement in the time domain of small energy splitting.

## 2.4 Spatial Resolution

Using well-focused excitation in a spot of a few microns, any extension of the luminescence spot gives information about the diffusion of the excitations. This can be done by inserting a slit in the plane of an enlarged image of the emitting region. Such measurements can be performed in the time-resolved mode, allowing the measurement of a signal $S(\Delta x, \Delta\tau)$, where $\Delta x$ is the distance to the exciting spot and $\Delta\tau$ is the delay after the exciting pulse. Under typical conditions, the analysis of $S(\Delta x, \Delta\tau)$ provides access to the lifetime and the diffusion coefficient of the excitations.

With the rapid development of technologies for the fabrication of as well as application of, low dimensional semiconductors and nano-structured materials, the demand on local probe techniques has widely increased. This has led to the development of the so-called NSOM (near-field scanning optical microscopy) technique. The diffraction limit of spatial resolution, is given by the well-known Rayleigh criterion for the minimum spot size $d$ of a laser beam:

$$d = \frac{1.22\lambda}{2n \sin\theta} \tag{1}$$

where $\lambda$ is the wavelength of light, $n$ is the refractive index of the medium, $\theta$ is the half angle of the solid angle passing through the focussing lens, and $n \sin\theta$ is referred to as the numerical aperture. In practice, the lower limit of d is usually considered to be $\lambda/2$, that is, using light of 500 nm wavelength objects in size below 250 nm cannot be resolved. To overcome this limitation of the spatial resolution, one has to work very close to the sample surface, where the diffraction limit is not yet established. This region is generally known as the near field, as opposed to the far field where the diffraction limit has been established. The basis for the NSOM technique is to utilize a small subwavelength-sized aperture placed very close to the sample. If both the aperture diameter and the aperture-to-sample distance are made smaller than the wavelength of light, the resolution is primarily given by the size of the aperture, and it is possible to obtain subwavelength spatial resolution. To obtain the small aperture, typically single-mode optical fibers with a tapered end are used. To regulate the aperture to sample distance, several techniques were developed, which are based on regulating the distance by electric tuning of piezo-elements, comparable to atomic force (AFM) or scanning tunneling microscopy (STM, see also MICROSCOPY). The tapered-end optical fiber can be used either to collect the luminescence from a local spot on the sample (with conventional excitation) or the excitation laser light is fed in the fiber, for a local excitation and a conventional detection via a luminescence collecting mirror.

Another technique that is more powerful is provided by *four-wave mixing* experiments, but this is outside the scope of this article.

## 2.5
## External Perturbations

The purpose of external perturbations applied to the sample is to induce shift or splitting of the energy levels in order to help their theoretical identification. On solid samples, hydrostatic pressure up to 10 GPa can be applied by the use of a diamond anvil cell: this induces shifts and in some cases, phase transitions. Uniaxial stress of much lower intensity can be applied; it can induce splitting associated with polarization effects. It can also be applied as a modulation, using piezoelectric elements.

Magnetic fields up to about 15 T can be applied by the use of split-coil superconducting magnets to allow light propagation perpendicular or parallel to the field. Different effects are induced, like Zeeman splitting, diamagnetic shift, or Landau level splitting; they can be detected on the PL spectrum, mainly at low temperature. Photoluminescence and electron paramagnetic resonance can be combined into a technique known as *optical detection of magnetic resonance*: The change in spin populations at resonance is detected through modification of the intensity of some polarized components of PL lines. This method is quite sensitive and provides a correlation between different magnetic resonance components (i.e., different $g$ factors) and different luminescence components. A related method, level-crossing spectroscopy, does not require microwaves: When two magnetic sublevels happen to cross, transfer between these two levels is enhanced and changes in polarized PL intensity can be recorded as bump or dip as a function of the field strength.

## 3
## Cathodoluminescence (CL)

By the use of an incident electron beam (*cathode rays*) as the source of excitation for the luminescence a very wide range of materials can be studied, including semiconductors, ceramics, geological minerals, polymers, and biological and medical samples. Because the beam is easily focused to a small exploratory probe, the technique is not only very versatile but also capable of providing information with high spatial resolution. Since the use of an electron beam effectively avoids the resolution limit applying to conventional far-field optical microscopy techniques (see Sect. 2.4). Furthermore, an electron beam is extremely flexible because of the ease of scanning, focusing, and the large depth of field. This depth of field simplifies imaging specimens that are not flat, especially compared to near-field techniques, in which even small surface topography can limit the techniques. The high vacuum of conventional electron microscopes can also provide a good environment for cryogenic cooling of the specimens. In addition, modern environmental or variable pressure scanning electron microscopes (SEM) open possibilities to study specimens in gaseous environments, or specimens traditionally avoided with high vacuum conditions, for example, food or biological samples.

### 3.1 CL Generation Mechanism

One of the most important factors for the luminescence generation mechanism in CL is the effective ionization energy ($E_i$), which is required to generate one electron hole pair, for semiconductors it requires roughly three times the band gap energy [1]. For a given acceleration voltage $E_{acc}$, the number of e–h pairs $G$ generated per primary electron (PE) is given by [2]:

$$G = \frac{(1-\gamma)E_{acc}}{E_i} \tag{2}$$

where $\gamma$ is the fraction of energy transmitted through or scattered back out of the material. The interaction between the PEs and the material can be either elastic or inelastic. The PE can be elastically scattered by the nuclei [3]. The main result is a change of direction of the PEs of up to 180°. PEs that are scattered back from the material are denoted as backscattered primary electrons. In the SEM mode, they are used to form the images that exhibit atomic number contrast caused by the monotone increase of the backscattering coefficient with atomic number. The PE can also interact with the electrons of the material. This interaction can take place with electrons of the inner electron shell of the atoms as well as with electrons in the valence states of the material. This interaction generates the secondary electrons (SEs). The interaction with loosely bound electrons has the largest cross section, in the case of semiconductors, electrons in the valence- and conduction-bands. This interaction leads to SEs with low energies (<50 eV). These SEs travel a short distance before they are recaptured and are referred to as *slow SEs*. Fast SEs can also be created as a result of interaction with more tightly bound electrons in the atomic levels. These SEs are not as numerous as the slow SEs, but they have larger kinetic energy and can travel much longer distances before they lose their kinetic energy and are recaptured. The energy of the fast SEs can reach almost that of the PEs they interact with. Furthermore, they can be scattered at almost right angles from the trajectory of the PE. This, together with the larger traveling distance, contributes to the spreading of the excitation within the material, which detoriates the resolution of the measurement. The PEs can also interact with the tightly bound electrons of the inner shells, for example, K and L. Here a large portion of energy is transferred to the atoms. This process also leads to the generation of SEs, but a more important consequence is the emission of X rays.

The SEs that are generated near the surface of the samples have enough energy to overcome the work function and thereby the possibility to escape from the surface, which results in SE emission. In the SEM mode, the detection of SEs is the most common detection mode. In this mode, the topography of the surface is recorded. In the case where the SE remains in the semiconductor, it can combine with a hole to form an e–h pair, where the number of e–h pairs per PE is given by Eq. (2). One of the ways this e–h pair can recombine is via emission of a photon: cathodoluminescence. Other recombination ways are nonradiative and include emission of cascades of phonons and various Auger processes.

The inelastic scattering includes two more processes, both relying on Coulomb interaction. When the PE passes through the coulomb field of the atoms in the material, the PE velocity can be reduced and its direction can be changed. In this process, the excess energy is emitted in the form of a photon. This phenomenon is known as *bremsstrahlung* (braking radiation). The energy of the emitted photon depends on the energy of the PE and the deviation from its original trajectory. The possible energy of the emitted photon ranges from virtually zero to the total energy of the PE. This radiation is mainly observed in the X-ray spectrum of semiconductors in which it appears as a continuous background, but it can also be observed in the CL spectrum, especially when the CL yield is low. The second kind of Coulomb interaction is between the Coulomb field of the PE and free electron gases in the conduction bands that induce oscillations in the free-electron gas (plasmons). The cross section of this type of scattering is directly proportional to the density of free electrons and therefore mainly significant in metals. For details of the different scattering processes and the discussion of the average energy loss, Ref. [3] is recommended.

An interesting parameter is the penetration depth of the PEs into the material; it can be estimated as a function of the acceleration voltage by a modified expression originally developed by Grün [4, 5]:

$$R_G = \left(\frac{0.0398}{\rho}\right)(E_{acc})^{1.75}(\mu m) \qquad (3)$$

with $\rho$ as the density in g/cm$^3$. The expression was derived for the range of atomic numbers between 10 and 15 and $E_{acc}$ between 5 and 25 keV. It was also concluded that 98% of the energy is dissipating within $R_G$. Another factor that is important for the ultimate resolution of the CL technique is the spreading of the PE parallel to the surface of the semiconductor, the lateral spreading. This, together with the penetration depth, determines the volume that is excited by the PEs. This volume is generally referred to as the generation volume. A good first approximation of the lateral extension of the excitation volume is a circle with the same diameter as the penetration depth. In fact, the first approximation of the excitation volume is a cylinder with a diameter of, as well as length, $R_G$; it gives a first estimation of the resolution that can be obtained in CL. In reality, the excitation density varies dramatically within this cylinder. Several attempts have been made to obtain more detailed expressions analytically [6] as well as from the side of Monte Carlo calculations [7]. Experimentally, the depth-dose functions as well as the shape of the excitation volume have been verified by several methods [8].

### 3.2
### Instrumentations

The excitation source in a CL system is an electron gun, in which the electrons have been accelerated by an electric field to several kilovolts. The high excitation energy of conventional CL makes it possible to study radiative processes in most semiconductors and insulators. One of the strong points with CL is the ability to excite materials with large band gaps, which require lasers in the UV, or even in the deep UV, as an excitation source in PL. The excitation source used normally is the electron beam of an SEM or a scanning transmission electron microscope (STEM). This means that the beam is well-defined in energy and it is also focused to a small spot, with the possibility to scan the beam over the sample to obtain a map of the intensity variations of the emission. The use of a SEM or a STEM means that the wealth of information obtained from the ordinary detection system can be used as a complement to the CL signal, for example, the topological information. As the electron beam is easily guided and focused by electrical and magnetic fields, it is possible to scan an electron probe with a diameter of typically 1–100 nm over large areas of the sample with high precision. With an electric field near the electron source, it is also possible to deflect the beam and to turn the electron source on and off. Rise and fall times, as well as pulse, with a duration in the sub-ns range can be obtained and can be used to study the transient behavior of the luminescence.

The nature of the very local excitation means that, like in all local probe techniques, the total intensity of the generated luminescence can be very limited. This puts high demands on the efficiency of the light collection and detection systems. In its simplest version, the light collection system consists of a lens with the sample at one focal point, with a detector after the lens. This type of system will only provide panchromatic images. In a slightly more advanced system, a filter can be placed between the lens and the detector for monochromatic imaging. A simple alternative is to place a detector directly under the sample in a mode that is known as *transmission CL*. Another alternative is to place a large-area detector, with a hole for the incoming electron beam above the sample. For more efficient light collection, a mirror is usually employed. The most commonly used systems were based on an ellipsoidal mirror with the specimen in one focal point and a fiber bundle in the other, which guides the luminescence to the spectrometer/detector. This configuration allows a very large collection angle for the emitted light, however, has some drawbacks related to the losses in the fiber, and highly efficient systems have been developed using parabolic mirrors. This type of mirror features parallel light from the sample, that is, the following adjustment of the detector or other parts requires more efforts than for the fiber systems.

Many CL systems used with semiconductors have a cold stage, operated either with liquid nitrogen or liquid helium, where the temperature range between 300 K and 4 K can be covered. The most common systems have a flow cryostat. Care has to be taken to prevent vibrations caused by the boiling liquid from ruining the resolution of the measurement. A solution to this problem is to connect the cryostat via a flexible braid to the specimen holder.

The detector systems used for CL are basically the same as those that are in use for the PL, and its choice depends on the application. If the desired spectral range allows the use of parallel detectors, like charge-coupled devices (CCDs) or diode arrays it enables to measure a whole spectrum, or at least a part, at one time. This considerably reduces the total measurement time, and maps can be obtained where each point contains the information of the whole spectrum.

In CL, this is known as *CL wavelength imaging* (CLWI). The data can be presented either as a normal intensity map for a certain energy or as a peak energy position map, where the energy position of the most intense energy in the spectrum is plotted. Further, the use of a streak camera could be added the time dependence to the data. Using these types of modes, one should realize the amount of data which is produced, as an example: A 1024-element CCD array with 16-bit resolution (65 536 intensity levels) generates one spectrum containing 2 kb of information per pixel. A 128 × 128 pixel image then results in 32 Mb.

3.3
## Comparing PL and CL

Although CL and PL are superficially similar, there are some important differences, one is the electron energies normally used in CL in comparison with the photon energies used in photoluminescence. The high electron energy means that CL is very versatile and can be applied both to insulators with large band gaps and to small band-gap semiconductors. Another attractive feature of the high voltage is the ease of control of electron range by altering the operating voltage. However, there is no equivalent to the technique of tuning the photon energy to near the band gap of the semiconductor being studied or to an absorption peak of an insulator. This technique, photoluminescence excitation spectroscopy (PLE), provides some of the richest spectral detail and permits the measurement of Stokes shifts. The range of excitation can also be controlled in PL by this technique since the absorption coefficient varies very rapidly with change of wavelength near an absorption edge.

An additional aspect in performing CL as compared with PL is the much greater likelihood of beam-induced changes to the specimen. This can occasionally be advantageous when the electron beam is used to create luminescence centers that did not exist before irradiation.

Finally, economical considerations have to taken into account: to use a complete SEM or STEM system as an excitation source for luminescence may be justified for scientific purposes, but may be out of range for more commercial purposes.

4
## Thermoluminescence (TL)

During the occurrence of luminescence, the emission of electromagnetic radiation results from a change in electronic states due to an external energy supply. The expression *thermo* in TL is no hint to the kind of excitation as in the case of PL or EL but is rather referred to the kind of stimulating procedure. Thermal stimulation of a solid previously excited by energetic radiation possibly leads to the appearance of luminescence. The intensity in dependence on heating temperature gives rise to a glow curve of TL, *thermally stimulated luminescence* would thus be a more correct label for this type of luminescence spectroscopy.

Observations of this effect have been reported as early as the seventeenth century. Sir Robert Boyle [9] described in 1663 a "glimmering light" when a diamond is warmed up in the dark of the bedroom. TL measurements in the modern sense were carried out for the first time by Urbach and Frisch in 1930 [10]. In addition, first results of a theory are introduced in the reports.

## 4.1 Excitations and Defects

The phenomenon of TL arises from effects induced by the interaction of solid matter with exciting radiation. In principle, two general classes of radiation damage processes can be distinguished: the electronic processes and the elastic collisions.

The electronic class includes all processes in which an electronic state is changed or a charge is moved by the absorption of radiant energy. The absorption occurs somewhat differently for various types of radiation.

Heavy particles passing through matter are usually stripped off some or all of its electrons. The interaction with crystal electrons is controlled by the Coulomb potential barrier. Fast neutrons, since they are not charged, do not excite the crystal electronically as do charged particles. However, when a fast neutron displaces a crystal ion, the ion gives up some of its kinetic energy to the electronic structure of the crystal. Finally, photons can transfer their energy to the electronic system of a crystal by three different processes: the photoelectric effect, the Compton effect and – at energies above 1.02 MeV – the production of electron–positron pairs.

In order to create defects through elastic collisions, it is necessary for the incident particle to transfer sufficient energy to a lattice particle to displace it from its regular site. Thus, the effectiveness of an incident particle in creating damage depends on its energy and mass as well as on the mass of the lattice ion.

The interaction of radiation with matter is found to be most effective in creating lattice defects, which in turn give rise to some change in the fundamental properties of the crystal. Consequently, extensive investigations of these changes prove to be efficient in obtaining some information on defect structures. The end products may be classified in terms of two categories of defects.

On the one hand, there are electronic defects involving both the impurities (extrinsic defects) and the lattice ions (intrinsic defects). They arise from a change of the valence states by trapping electrons and holes elsewhere in the lattice because of radiation. On the other hand, we have ionic defects that consist of displaced lattice ions. The most important role is placed by vacancies (e.g., $F$-, $F'$- and $F_A$-centers). Other ionic defects are formed by interstitials. All defects may be considered as traps and activators respectively, and consequently are involved in TL processes.

## 4.2 Thermal Excitations and Recombinations

Supposing the solid previously excited is heated, then the energy supply by means of thermal energy gives rise to thermal relaxation. In the case of TL, electrons and holes, respectively, are liberated from metastable states during thermal stimulation. The probability of thermal excitation of a carrier, the so-called escape probability $\alpha$, is assumed to be given by a Boltzmann factor

$$\alpha = \alpha_0 \exp\left(-\frac{E}{kT}\right) \tag{4}$$

($\alpha_0$: preexponential factor [11], $E$: thermal activation energy or "trap depth").

Thus, the free charge carrier density is expected to increase as the temperature rises and the phenomenon of *thermally stimulated (electrical) conductivity* (TSC) may be consequently observed, too. Moreover, if some of the carriers close to the surface have a chance of leaving the solid, the additional phenomenon of *thermally stimulated exoelectron emission* (TSEE) associated with TL and TSC is expected to appear.

Although expression (4) is well supported by experiment, a detailed theoretical treatment is still lacking. In a semiclassical approach, a trapping model with levels equidistant in energy and a successive absorption of phonons is assumed [12]. Another model is based on thermodynamic concepts as they are assumed to be valid in chemical reactions [13]. Analyzing the phenomenological processes and presuming the law of detailed balancing concerning the transitions into the conduction band and into the traps, the density of free carriers may be evaluated from the condition that the free energy reaches a minimum [14]. Finally, Eq. (4) can be approximately deduced from a quantum-mechanical treatment, involving the temperature dependence of nonradiative transitions [15].

When charge carriers have been thermally excited into the allowed bands, they will be captured by traps again or will recombine with opposite carriers. In nonradiative processes, the energy may be transferred to another electron (Auger recombination) or to the lattice in multiphonon processes. If recombination occurs with the emission of light, a TL glow curve can be observed.

As it is known from optical absorption, energy and momentum must be conserved in each recombination process. Assuming the momentum of the photon ($h/\lambda$) to be negligible, these requirements provide the selection rule for optical transitions

$$\hbar k - \hbar k' = 0 \qquad (5)$$

($k$, $k'$: wave number of the carrier in the initial and final state, respectively).

Therefore, direct recombination occurs only between states with equal wave numbers. In the energy band model, the carriers may only undergo vertical transitions. The change in momentum and energy due to electron–phonon interaction occurs in a time that is much smaller than the carriers' mean lifetime. Thus, a direct band–band recombination is very unlikely.

A higher probability of radiative transitions is obtained in the presence of defects such as vacancies, interstitials, or impurities. When the optical transition involves an imperfection, the selection rule (5) is satisfied even for different wave numbers of the carriers.

### 4.3 Analysis of TL Spectra

The phenomenological analyses are not interested in considering the physical processes and interactions within the atomistic regions. This means that one does not investigate the kinds of transition involved and the contribution of phonons during thermal stimulation. One is also not interested in the structure of centers. However, the possible transitions of carriers are discussed by means of kinetic rate equations as time goes on during thermal stimulation. The possible transitions are represented by arrows (Fig. 2). The simplest model involves only a single trap level and a single recombination level. It can be extended by assuming an additional kind of trap level ($M$) temporarily not involved in the thermal activation in question [16]. The symbols used here are

$h$: density of trapped charge carriers
$H$: density of trap levels
$M$: density of additional trap levels

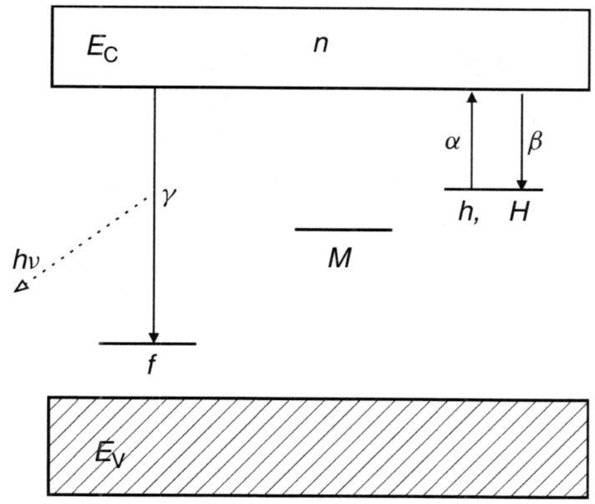


**Fig. 2** Schematic energy-band diagram with one type of trap level and recombination level as is used for phenomenological analysis of TL and TSC ($E_C$: conduction band, $E_V$: valence band)

$n$: density of free charge carriers in the conduction band
$f$: density of recombination levels

The transition probabilities are replaced by transition coefficients:

$\alpha$: the escape coefficient for trapped carriers
$\beta$: retrapping coefficient
$\gamma$: recombination coefficient

It is not probable that such a model accurately represents an actual situation occurring in solids since essential simplifications have been introduced. For instance,

1. the free carriers are electrons and there is no thermal quenching by the presence of holes;
2. no interaction exists between centers;
3. there is no trap distribution with respect to energy;
4. only one kind of center is involved in the recombination process;
5. recombination of trapped electrons via excited states of defect centers does not occur.

The phenomenological analysis now deals with the kinetic balance and the following transition rates:

1. Rate of liberation $= \alpha h$
2. Rate of retrapping $= \beta n(H - h)$
3. Rate of recombination $= \gamma n f$.

In this simple model, the kinetic processes involving the change in the density of trapped and free carriers during thermal stimulation are described by the following system of differential equations within the meaning of Master equations

$$\frac{dh}{dt} = -\alpha h + \beta n(H - h), \qquad (6)$$

$$\frac{dn}{dt} = \alpha h - \beta n(H - h) - \gamma n f. \qquad (7)$$

Since the solid is electrical neutral, the density of carriers in the conduction band and in the traps must be equal to the density of empty recombination centers. This condition of charge neutrality then yields

$$f = n + h + M. \qquad (8)$$

The TL intensity $I$ is given by the number of radiative transitions per unit time and volume and is therefore proportional to the recombination rate

$$I \propto \gamma n f. \qquad (9)$$

The electrical conductivity $\sigma$ is determined by the free charge carriers in the conduction band when the mobility $\mu$ is assumed to be constant

$$\sigma = \mu e n \qquad e: \text{electron charge}. \qquad (10)$$

The purpose of phenomenological analysis is to discuss the temperature dependence of luminescence and current intensities, respectively, assuming a constant heating rate $q$ $(= dT/dt)$. This can be done either by means of explicit functions or with the results of numerical solutions (Fig. 3). The experimental glow curves are then fitted to the calculated ones by variation of some trapping parameters. This procedure possibly supplies some physical quantities such as the activation energy $E$ and the preexponential factor $\alpha_0$. However, all attempts to find explicit functions are based on the introduction of some approximations or simplifying models.

The thermal activation energy $E$ is of particular interest because of its importance concerning the lifetime of the excited state of a solid. According to the approximative kinetic models that allow the discussion of explicit temperature-dependent functions of TL and TSC, one may derive many methods for the determination of activation energies and preexponential factors.

In a first approximation, it is possible to calculate a formula that provides a linear dependence of the trap depth $E$ on the maximum temperature $T_m$ [17, 18]

$$E \approx 25kT_m. \tag{11}$$

The derivation uses a preexponential factor of about $\alpha_0 = 2.9 \cdot 10^9\ s^{-1}$. The rough formula is quite useful for a quick first estimation. However, the numerical factor in Eq. (11) is dependent on the value $\alpha_0$, and hence the values of $E$ thus obtained only have the right order of magnitude. The preexponential factor, in fact, may be different for each kind of trap in the same substance.

A rather more accurate relation is given by

$$\frac{E}{kT_m^2} = \frac{\alpha_0}{q}\exp\left(-\frac{E}{kT_m}\right). \tag{12}$$

From this equation, the peak position is influenced by the thermal activation energy as well as by the preexponential factor. A decrease of $\alpha_0$ or an increase of $E$, respectively, gives rise to a shift of the maximum to higher temperatures. If one of the parameters $E$ or $\alpha_0$ are known, Eq. (12) can be used to determine the other unknown parameter.

The various methods can be divided into several groups according to the experimental procedure

1. Methods of heating rate variation.
2. Methods of geometrical approximation.
3. Other methods.

Frequently the so-called *initial rise method* is used to determine $E$. This technique is based on the fact that the initial part of a glow curve increases exponentially

$$I \propto \exp\left(-\frac{E}{kT}\right) \quad \text{or} \quad \ln I = -\frac{E}{kT} + \text{const.} \tag{13}$$

for either type of kinetics. Thus, the plots of $\ln I$ against $1/T$ are linear, having slopes equal to $(-E/k)$. This provides a quick analysis of the initial rising part of the glow curve that yields the value of $E$ without any knowledge of $\alpha_0$ [19–21]. However, the peaks will become wider and shallower with increasing retrapping. The temperature limiting the validity of Eq. (13) is therefore decreasing [22]. In addition, the area under the glow curve can be used for parameter determination [23–25]. Finally, from numerical methods of analysis, some initial approximate values of $E$ and $\alpha_0$ are selected and they are suitably varied to determine the values giving the best fit to the experimental results [26–28]. A survey of methods from literature is given in Refs. [29–31].

A rigorous solution of the differential Eqs. (6) and (7) succeeds by means of an analog computer without any approximations and without regard to the temperature dependence of the coefficients [16].

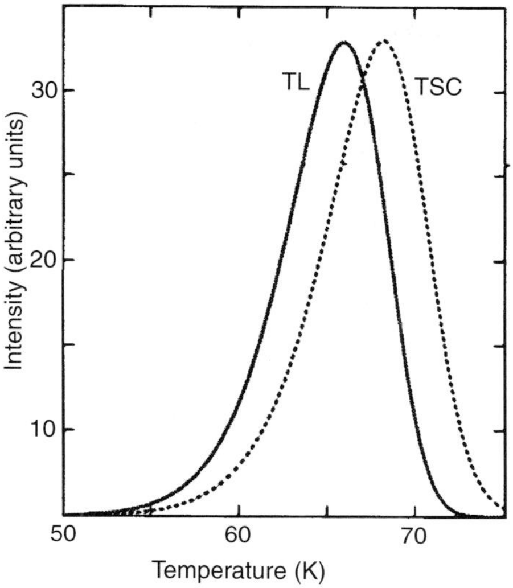


**Fig. 3** Computer simulation of TL (———) and TSC (· · · ·) with the parameters $E = 0.13$ meV, $\alpha_0 = 2 \cdot 10^9\ \text{s}^{-1}$, $\gamma H = 2\ \text{cm}^3\text{s}^{-1}$, $M/H = 0.5$, $\beta/\gamma = 0$, and the heating rate $q = 1\ \text{Ks}^{-1}$

A numerical calculation with single-step techniques provides similar results, too. From these calculations, the densities of carriers may be evaluated as a function of temperature and a simulation of both the TL and the TSC may be performed (Fig. 3).

In the TL glow curve, the maximum temperature differs to some extent from that curve showing the change in the trapped carrier density ($-dh/dT$) and is often mistaken for the glowcurve. Moreover, the exact calculation gives rise to a shift of the TSC against the TL of about 2 to 5 K to higher temperatures as it is expected especially for alkali halides, earthalkali halides, and oxygen-dominated phosphors.

As an additional important result, the exact simulation gives evidence to the fact that all methods used for determining trapping parameters are only approximately valid. However, parameters obtained from the initial rise method are relatively independent of the kinetic reaction type and mechanism, as well.

## 4.4 Instrumentations

The method of observing the TL intensity is simple and straightforward. Thus, the experimental set up consists essentially of two basic parts

(1) a heating device
(2) a light detecting system.

Three different types of heating method can be found in widespread use. The objectives are reproducible and controllable heating profiles. Oven and sample chambers are most often filled with inert gas during heating to enhance thermal contact between the heater and sample, and to reduce spurious emissions of light from chemical effects.

At the resistance heating, one passes high current through a thin, low-impedance load. The sample is placed in direct contact with the planchet, the temperature of which is monitored via thermocouples. Alternative designs involve the use of wire coils in contact with more massive metal blocks, or in direct contact with the sample itself. Rapid heating of solid samples, with precision temperature control, can be achieved via the use of hot gas as a heat-exchange medium. In addition, the samples are heated quite uniformly. By the use of infrared cw lasers, it is possible to heat either the sample or a substrate to which the sample has been intimately bonded. Extremely high heating rates can be achieved with short laser pulses.

The low intensities of luminescence emitted during thermal stimulation are normally detected with photomultipliers. Either DC current or photon-counting modes may be used. Dark currents may be reduced by the use of Peltier-cooled tube housings to limit the thermionic emission from the photocathode.

The schematic energy-band diagram (Fig. 2) suggests that only one emission wavelength will be observed during TL readout since only one type of recombination center is present. However, real materials often contain several types of recombination sites, each of which may be represented by its own characteristic emission wavelength. Detection through broadband filters, centered in the visible, is a widespread technique. Some other attempts are made using narrow-band interference filters. More detailed information can be obtained by measuring the TL intensity as a function of temperature and wavelength simultaneously, which provides 3-D plots. The spectrometers to achieve these results include rapid-scanning monochromators. However, most information is obtained by multiplexing the data and dispersing the signal over a defined wavelength range, and measuring the intensity over each wavelength interval separately but simultaneously. Modern techniques include Fourier-transform spectroscopy or the use of position-sensitive photodetectors. In the latter, the light is dispersed over the wavelength range of interest and the dispersed light is detected over the active area of a device such as a CCD camera, diode array, or position-sensitive multiplier.

### 4.5
### Applications

The application of TL for dosimetric purposes has become of increasing interest. There are various materials corresponding to the practical problem. The reasonable dose ranges from $10^{-3}$ rd to $10^4$ rd. The energies of $\gamma$-radiation are found to be detectable between 10 keV and 30 MeV. Low fading of only 1% per month provides reliable results. Some more advantages are the low energy dependence and the unlimited storage capacity. On the other hand, it is to be noted that the process of TL is found to be nonreproducible and consequently does not allow a repeated interpretation after first studies of the information.

A second important application is done at the age determination of ancient pottery. The zero point of the "TL-clock" is equated with the date of firing. Subsequent exposure to exciting radiation from especially radioactive isotopes ($^{238}U$, $^{232}Th$, $^{40}K$) gives rise to a natural TL. After calibration with respect to the sensitivity – an artificial TL is induced by a well-known dose – and after evaluation of the dose rate, the age may be roughly estimated from:

$$\text{age} = \frac{TL_{\text{natural}}}{TL_{\text{artificial}}} \frac{\text{dose}_{\text{well-known}}}{\text{dose/time}}. \qquad (14)$$

## 5
## Chemiluminescence (ChL)

Chemiluminescence (ChL) refers to emission from excited molecules generated as products of a chemical reaction. If this reaction is derived from biological organisms such as bacteria, fungi, and insects, the luminescence is called *bioluminescence*. Although ChL phenomena were reported more than a half century ago [32], only in the last ten years have there been extensive basic studies as well as practical applications. The dynamic field of ChL is experiencing considerable growth due to the advent of new instrumentation, sensitive detection methods, and efficient automation technology. The use of ChL is also receiving increased interest due to its simplicity, sensitivity, and dynamic range.

## 5.1 Excitations and Emissions

Most organic molecules contain an even number of electrons. In the ground state, the electrons fill the various atomic orbitals with the lowest energies in pairs. According to the Pauli exclusion principle, which requires a totally antisymmetric state function with respect to particle exchange, two electrons in one given orbital must have spins in opposite directions and as a consequence, the total spin disappears. For this reason, the singlet ground state $S_0$ has no net electron spin. Excitation of a molecule, for example, by the absorption of light, usually results in the promotion of one electron from the highest occupied orbital to a previously unoccupied orbital $S_n$ $(n > 0)$ having a higher energy (Fig. 4).

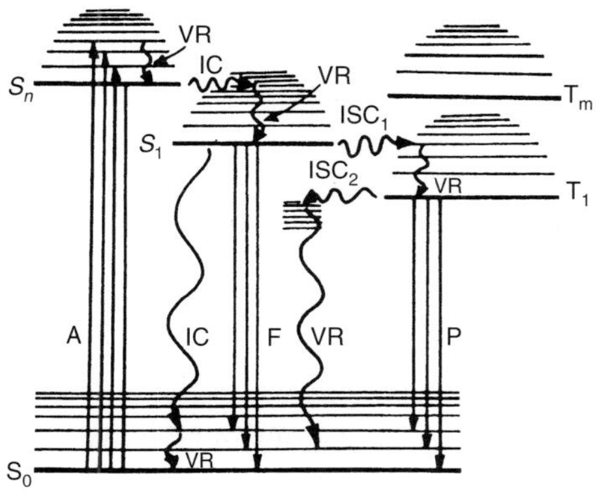


**Fig. 4** Jablonskii diagram for molecular photophysical processes. $S_0$, singlet ground state; $S_1$, first-excited singlet state; $S_n$, *n*th-excited singlet state; $T_1$, first-excited triplet state; $T_m$, *m*th-excited triplet state; A, absorption; IC, internal conversion; F, fluorescence; VR, vibrational relaxation; P, phosphorescence; ISC, intersystem crossing

At thermal equilibrium, molecules have the greatest probability of having the thermal energy of the lowest vibrational and rotational levels of an electronic sate in accordance with the Boltzmann distribution. In condensed media, a molecule in the $S_n$ state deactivates rapidly to lower vibronic levels of $S_n$ within $10^{-10}$ to $10^{-13}$ s via vibrational relaxation (VR) processes and acquires the thermal energy distribution of the lowest vibrational levels of $S_n$. This VR process is faster than electronic transitions ($10^{-8}$ s). The VR excess energy is released as thermal energy to the lattice. In another process known as *internal conversion* (IC) the molecule deactivates rapidly to the isoenergetic vibrational level of a lower electronic state such as $S_{n-1}$. By a succession of VR and IC processes, the molecule deactivates rapidly to the lowest excited singlet state $S_1$.

From $S_1$ the molecule can arrive at the ground state $S_0$ in three different ways. Firstly, the energy is released to the lattice and the transition to $S_0$ takes place in a nonradiative manner. Secondly, the transition probability for dipole radiation is high enough. Consequently, the transition to the ground state $S_0$ gives rise to the emission of a photon that is known as *fluorescence* [33, 34]. Finally, the molecule may also undergo a transition to some vibrational level of one of the lower triplet states – states with total spin equal to 1 – via a mechanism known as *intersystem crossing* (ISC). The molecule undergoes deactivation to the lowest excited triplet state $T_1$ via VR and IC processes. From $T_1$, the molecule may return to the ground state $S_0$ in a radiationless fashion via an ISC process or by emitting a photon. This latter process is known as *phosphorescence* [33, 34].

The nature of the orbitals involved in an electronic transition is an important factor in determining the luminescent properties of the molecule. Organic compounds that are known to be strongly luminescing are molecules with extended $\pi$-electron systems. It is noteworthy that, in the gas phase, fluorescence may be observed from higher vibrational levels of the excited state. These emission mechanisms, known as *single-vibronic-level* (SVL) luminescence processes, do not occur in solution.

## 5.2 Chemical Reactions

Chemiluminescence emission is usually a fluorescence process rather than a phosphorescence process. This is because the triplet state has a relatively long lifetime and is subject to radiationless deactivation by other molecules present such as oxygen. Chemiluminescence in solution generally involves an oxidative reaction. Molecular oxygen and hydrogen peroxide are most often involved.

The production of chemiluminescence usually requires three conditions:

1. adequate chemical energy to populate an excited state of a molecule;
2. an efficient chemical reaction with high yield;
3. deactivation of the excited state via radiative ChL emission.

In some cases, the excited molecule does not emit radiation but transfers the energy to another molecule that subsequently emits ChL. The efficiency of ChL emission may be determined by the number of photons produced per molecule undergoing a chemical reaction, defined as follows:

$$M \xrightarrow{\Omega_E \Omega_C} M^* \quad (15a)$$

$$M^* \xrightarrow{\Omega_L} M + h\nu_{ChL} \quad (15b)$$

$$\Omega_{ChL} = \Omega_E \times \Omega_C \times \Omega_L, \quad (15c)$$

where $M$ denotes a molecule in the ground state, and $M^*$, a molecule in the excited state. The excitation yield $\Omega_E$, which denotes the proportion of molecules promoted to the excited state, depends on the chemical reaction mechanism. A high ChL yield $\Omega_{ChL}$ depends on a high chemical yield $\Omega_C$ and an efficient luminescence yield $\Omega_L$. The values of $\Omega_{ChL}$ vary from $10^{-8}$ up to 0.88, the latter for the *in vitro* firefly reaction, which has one of the most efficient ChL yields [35].

## 5.3 Gas Phases

When a reagent gas is mixed with another gas in a reaction cell, and if the reaction is sufficiently exothermic, one or more of the products may be promoted to an excited electronic state, which can subsequently decay to the ground state with emission of ChL. By use of different reagent gases, a wide variety of ChL reactions may be induced. Chemiluminescence reagents are not limited to stable molecules. Highly exothermic reactions of atoms, radicals, ions, and electronically excited species can be produced in gas discharges and detected by photometric detectors.

An active-nitrogen detector has been developed for monitoring hydrocarbons [36]. High-energy electrical discharges – for example, electrical or microwave discharges – can be used to produce a mixture of nitrogen atoms and highly excited nitrogen molecules $N_2$ ($A^3\Sigma_u^+$). These excited-state nitrogen molecules (active nitrogen) can induce collision energy transfer to analyze molecules present, providing a pathway to electronic and vibrational excitation of the analytes. For example, active nitrogen reacts with hydrocarbons to produce an excited state of the cyanogen radical ($CN^*$), which emits intense ChL emission.

Since carbon compounds are found to react with active nitrogen to produce excited states of cyanogen radicals, the active-nitrogen detector can be used as a universal monitor for hydrocarbons.

Ozone ($O_3$) is one of the most widely used ChL reagents since most gas-phase reactions with ozone have been found to be chemiluminescent. The basis of ozone ChL reactions is attributed to the weak O−O bond (24.9 kcal/mol), which is broken to form the strong O=O bond (119 kcal/mol) in molecular oxygen and strong C=O bonds (172 kcal/mol) in reactions with many organic compounds [37]. Ozone reacts rapidly with olefins, producing ChL emission from various species.

Various nitrogen-containing compounds, oxygenated compounds, and reduced sulfur compounds can be detected by use of ozone ChL reactions. Sulfur compounds and their interactions with ozone have also been extensively investigated. Reduced sulfur compounds usually become oxidized to produce electronically excited $SO_2^*(^3B_1)$, which exhibits ChL emission between 300 and 400 nm [37].

Fluorine molecules, $F_2$, react with reduced sulfur compounds to produce several chemiluminescent species, providing an important detection method highly selective to organosulfur compounds. When fluorine atoms are used in place of fluorine molecules, the reactions occur at a much faster rate, and the chemical processes are similar to those occurring with active nitrogen. Fluorine atoms can be produced via various methods such as microwave discharges of $F_2$, $CF_4$, and $SF_6$ [37]. Other compounds that have been used as gas-phase ChL reagents include sodium atoms [38] and chlorine dioxide [39].

## 5.4 Liquid Solutions

Luminol or 5-amino-2,3-dihydrophthalazine-1,4-dione is one of the most investigated ChL systems. The oxidation process of luminol has been extensively investigated although the mechanisms have not yet been completely elucidated. In aqueous alkaline solution, luminol is oxidized to produce an excited-state species, viz. 3-aminophthalate, which exhibits ChL emission centered at approximately 425 nm [40]. This reaction is depicted in Fig. 5. Various oxidants (e.g., permanganate, hypochlorite, iodine, and perhaps the most common oxidant, hydrogen peroxide) can be used. When peroxide is used as the oxidant, a catalyst (e.g., peroxidase, hemin, ferricyanide, and transition-metal ions, such as $Co^{2+}$, $Cu^{2+}$, $Fe^{3+}$, etc.) is required. The ChL luminol reaction may be summarized as follows:

$$\text{Luminol} + H_2O_2 + \text{Catalyst} + OH^- \longrightarrow \text{Product} + \text{Light}. \qquad (16)$$

Equation (16) indicates that the ChL emission depends upon the concentrations of luminol, $H_2O_2$, and the catalyst. Since the measurement of the ChL signal intensity can be used to quantitate any of these species, the luminol ChL method can be used to determine the following species:

1. Luminol or species labeled with luminol.
2. Peroxide or species that can be converted to peroxide.

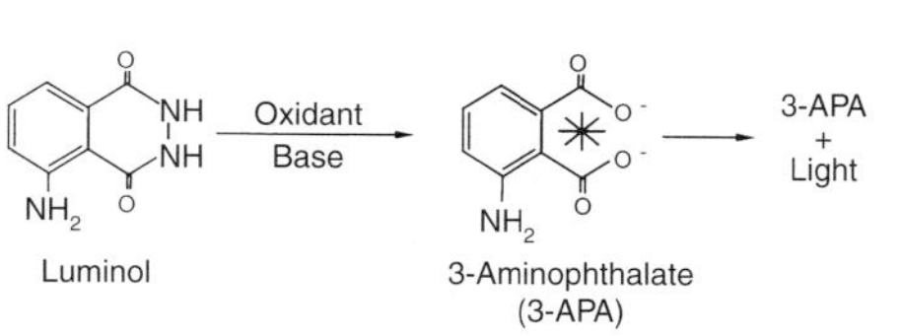


**Fig. 5** Chemiluminescence reaction for luminol (asterisk indicates luminescent excited state)

3. Catalysts (e.g., $Cu^{2+}$, $Co^{2+}$, $Ni^{2+}$, $Fe^{3+}$, etc.) or species labeled with catalysts.

Some commonly used ChL reactions are peroxyoxalate reactions that involve the hydrogen peroxide oxidations of aryl oxalate esters. The quantum efficiencies of these reactions are very high (usually 20%–30%), with the production of a high-energy intermediate. This intermediate and a fluorophore (fluorescent product) then form a charge-transfer complex that dissociates to produce an excited-state fluorophore. The fluorophore subsequently emits ChL. Peroxyoxalate ChL has useful applications in the quantification of hydrogen peroxide or fluorophores such as polycyclic aromatic hydrocarbons (PAHs).

Lucigenin (N,N′-dimethyl-9,9′-diacridinium nitrate) is another commonly used system that undergoes ChL reaction in alkaline solution to produce N-methylacridone with emission of ChL light. The ChL emission originates from an excited singlet state of N-methylacridone, which exhibits a ChL spectrum from 420 to 500 nm with a maximum at about 440 nm [41]. Acridinium esters and acridans react similarly to lucigenin to also yield N-methylacridone- and ChL light.

Various ChL reactions involve singlet oxygen that can be formed by the following reactions:

$$S + h\nu \longrightarrow {}^1S^* \longrightarrow {}^3S^* \quad \text{(17a)}$$

$$^3S^* + {}^3O_2 \longrightarrow S + {}^1O_2, \quad \text{(17b)}$$

where $S$ denotes the ground state of a sensitizer for singlet oxygen; $^1S^*$, an excited singlet state of the sensitizer; $^3S^*$, an excited triplet state of the sensitizer; $^1O_2$, the singlet state of molecular oxygen; and $^3O_2$, the triplet state of molecular oxygen.

In the presence of a singlet-oxygen acceptor, $A$, the following reaction, which involves direct energy transfer from singlet-oxygen dimers to a dye capable of fluorescence, may occur:

$$2\,{}^1O_2 + A \longrightarrow 2\,{}^3O_2 + {}^1A^* \longrightarrow A + h\nu, \quad \text{(18)}$$

where $A$ denotes the state of acceptor $A$, and $^1A^*$, an excited singlet state of $A$. The formation of singlet-oxygen dimers in Eq. (18) is well-studied and many investigators have reported ChL by this process. Acceptors have included violanthrene, methylene blue, and chlorophyll [37]. Another reaction that can produce ChL involves formation of a fluorescent product:

$$^1O_2^*A \longrightarrow (AO_2) \longrightarrow \text{Product}^* \longrightarrow \text{Product} + h\nu. \quad \text{(19)}$$

In this reaction, the acceptors may form dioxetane intermediates that can decompose to yield ChL. Although most dioxetanes decompose in solution to yield predominantly excited triplet states producing little ChL, there are acceptors that form dioxetanes exhibiting good excited singlet yields. These include lophine (2,4,5-triphenylimidazole) and some of its derivatives [42].

## 5.5 Instrumentations

The instrumentation commonly used to detect ChL is relatively simple and is essentially similar to that used for luminescence. Most of the discussion on detection instrumentation in this section also applies to conventional luminescence. A typical ChL detection system may have the following basic components:

1. a dispersive element, such as an optical filter, a monochromator, or a polychromator;
2. an appropriate detector.

No excitation light source is required since the chemical reactions provide the necessary excitation energy.

There are two basic classes of luminescence spectrometers: single-channel instruments and multichannel instruments. The first type of instrument generally uses a monochromator coupled to a photomultiplier, whereas the latter uses a polychromator interfaced to a multichannel detector, such as a vidicon, a photodiode array, or a charge-coupled device. The single-photon counting method can be used for detection, too. Unlike the more conventional analog detection method, the single-photon counting signal output is digital in nature, producing discrete pulses of charge.

Photomultipliers are widely used as ChL detectors for their high spectral sensitivity, wide operating range, low cost, and relatively simple electronics. Whenever low-intensity signals are observed in ChL measurements, it is desirable to improve the sensitivity of detection by recording the entire spectrum simultaneously. Interferometric methods, for example, Fourier-transform techniques, can enhance the detection sensitivity only if the noise of the measured signal is detector-limited.

Chemiluminescent reactions in the gas phase and in solution have been successfully used to develop an extremely wide variety of detectors in gas and liquid chromatographs. In general, ChL-based detectors are very sensitive and selective to the specific species that undergo ChL reactions.

A variety of ChL reaction schemes in solution have been used to develop detectors for liquid chromatography (LC). The ChL systems include luminol, lucigenin, peroxyoxalates, Ru(bpy), electrochemiluminescence, and the luciferin/luciferase (bioluminescence) reactions. Another example described in a previous section involves the use of ChL reactions of hydrogen peroxide for LC detection.

A variety of ChL-based detection systems can also be operated as post-column reactions for flow injection as well as chromatographic systems [37]. The instrumental device involves a method to deliver and mix the ChL reagents with the column effluent and an appropriate detection flow cell that permits monitoring the ChL signal.

Because of the reaction kinetics involved in the ChL process, the ChL emission intensity is usually a transient signal rather than a constant signal. The ChL emission intensity $I_{\mathrm{ChL}}$ (photons/s) is directly proportional to the rate of chemical reaction $\mathrm{d}M/\mathrm{d}t$ [(molecules reacting)/s]:

$$I_{\mathrm{ChL}} = \Omega_{\mathrm{ChL}} \left(\frac{\mathrm{d}M}{\mathrm{d}t}\right). \qquad (20)$$

Since the rate of the ChL reaction varies with time owing to consumption of reagents, the ChL intensity also varies with time. In order to optimize the sensitivity of detection, one needs to measure the ChL signal at its maximum level, which occurs after a specific time delay following reagent mixing. In general, the ChL signal reaches the maximum level in a second or less after reagent mixing and decays to the background level in a few seconds. The nature of the solvents used in flow-injection analysis (FIA) or LC is an important parameter affecting the ChL intensity.

## 5.6 Applications

There is an extremely wide variety of analytical instruments for monitoring gaseous species using ChL reactions. These devices are based on various schemes (e.g., $NO + O_3$, $CH_2 + N$, $CH + N$, etc.). A solid-state ChL detector has been developed for hydrogen peroxide for potential application to acid rain analysis [43]. The acidity of rain and cloudwater is affected by the levels of sulfur oxides and nitrogen oxides released into the atmosphere and subsequently dissolved in the water. One of the main reactants in the decomposition cycle of sulfur oxides is believed to be hydrogen peroxide. An important problem related to monitoring hydrogen peroxide is the low concentrations of these species, which are unstable in aqueous samples. The ChL monitor used peroxylate ChL and consisted of a two-bed reactor packed with solid trichlorophenyloxalate (TCPO) and 3-aminofluoranthene immobilized on controlled pore glass beads.

The ChL detection method in LC has used the luminol reaction for trace metals separated by ion-exchange chromatography [37]. Mixtures of $Cu^{2+}$ and $Co^{2+}$ and of $Cu^{2+}$, $Co^{2+}$, and $Ni^{2+}$ were separated on an anion-exchange column as their chloride complexes. Manganese(II) has been determined at part-per-billion (ppb) levels by detection of the ChL emission produced by oxidation of luminol with hydrogen peroxide [43].

The approach involving peroxide has been used to detect a variety of aliphatic oxygen-containing compounds, such as alcohols, ethers, aldehydes, and saccharides. This general method is based on the principle that if an analyte of interest can be converted into a specific amount of hydrogen peroxide, then addition of a luminol-catalyst reagent mixture will permit quantitation of the peroxide produced and thus the original analyte.

Chemiluminescence of PAH compounds, emitted at a platinum electrode, can be utilized to determine these compounds in aprotic (i.e., containing no protons) solvents [44]. This ChL method has been applied for detection in liquid chromatography [45]. Electrogenerated ChL of PAHs was observed in micellar solution [46].

Species that do not participate directly in ChL reactions can still be detected by their interactions with other species that are catalysts for the reaction. This indirect approach has been used in several variations to detect a wide range of analytes.

Clinical methods for protein analysis have been developed using changes in either absorption or fluorescence characteristics induced by the formation of complexes with the probe molecule. One of the major advantages of such methods is that chemical modification of the analyte is not required.

Aminoglycoside antibiotics are good $Cu^{2+}$ ligands, and the three components of gentamicin-C in commercial gentamicin sulfate have been determined via ion-pair chromatography using ChL detection [46]. Many organic compounds such as drugs and food components are slowly oxidized in air and produce an extremely weak ChL emission. This phenomenon has been used to investigate the stability of organic compounds [47]. The study indicates that ChL may be useful in evaluating the stability of crude drugs extracted from natural resources that contain many amino acid derivatives including proteins and sugars.

Several oxidizing systems have been used to determine sulfides by monitoring their ChL emission [48]. These ChL systems involved sodium hypochlorite with and without a sensitizer, and hydrogen peroxide with osmium tetroxide, peroxidase, or catalase as catalyst.

Chemiluminescence immunoassays can be performed by labeling either antibodies or antigens with a ChL compound of interest. These systems offer several advantages, including high sensitivity, excellent stability of the label, simplicity of the instrumentation, and flexibility of the measurement. Several types of ChL immunoassays can be performed. The simplest method involves direct replacement of the radioactive label, for example, $^{125}I$, by a ChL label.

One of the most widely used techniques in molecular biology deals with nucleic acid hybridization to detect specific DNA sequences immobilized on membranes. In addition, the increasing interest in human genome mapping and sequencing has induced the rapid development of many DNA cloning and sequencing techniques. The usual procedure for detection and quantification of specific DNA or RNA sequences is mainly based on a procedure, often referred to as *Southern blotting* [49]. This method involves transferring electrophoretically separated DNA fragments from a gel to a membrane support followed by hybridization with a labeled probe. The blot is washed and the DNA of interest is detected by autoradiography or other detection techniques. Conventional methods use radioactive labeled probes that give high sensitivity for detection. Separated DNA fragments are most often detected with radioisotopic labels such as $^{125}I$, $^{32}P$, and $^{35}S$. Radiometric techniques generally suffer from several disadvantages: radioisotopes have limited shelf life (e.g., about 60 days for $^{125}I$); instruments for detection and reagents are relatively expensive; and special precautions are required for the shipping, handling, and waste disposal of radioactive materials. Consequently, a number of techniques have been developed for nonradiative probe labeling and detection. Various nonradioactive labels include fluorophors (e.g., fluorescein and tetramethylrhodamine), metal atoms (e.g., colloidal gold), enzymes (e.g., alkaline phosphatase (AP), acid phosphatase, microperoxidase, and horseradish peroxidase), and antigen: antibody or biotin: avidin enzyme-conjugated systems [50]. Because the luminol moiety can be quantitated conveniently at trace levels, there is current interest in the use of luminol or related derivatives as labels in immunoassays [51].

Recently, enzyme-based chemiluminescence and bioluminescence assays have been developed for detecting DNA [52]. Enzyme-based labels offer a nonradioactive technique that is very sensitive because of their catalytic amplification process. Alkaline phosphatase is an enzyme system that has been used recently for ChL assays. Another nonradiative detection technique for DNA sequencing by means of ChL uses biotinylated alkaline phosphatase [53]. The ChL detection methodology is similar to colorimetric staining and involved a biotin/streptavidin interaction.

## 6 Sonoluminescence (SL)

*Sonoluminescene* is the phenomenon of light emission that occurs when high-intensity ultrasound is transmitted in liquids containing dissolved gases. In the presence of ultrasound, the minute inhomogeneities in the bulk of the material grow into microscopic bubbles. These bubbles oscillate radially and nonlinearly. In some cases, the wall motion becomes unstable and the bubbles collapse. It takes bubbles longer to expand than to collapse. The bubbles acquire energy over a relatively long period and release it in a very short interval. Through this process, the bubbles are able to concentrate acoustic energy in small local volume elements by as much as 12 orders of magnitude. The unleashing of the intense energy by the bubbles generates short flashes of light,–SL.

The phenomenon of SL was first observed in the 1930s [54, 55]. Early work has been covered extensively in several review articles. The periodic nature of SL was first noted in 1950–1960 when it was demonstrated that the light emission occurs during the last phases of bubble collapse [56, 57]. Further, the periodicity of SL was studied under better-controlled experimental conditions [58]. The SL was noted to be from stable cavitation with flashes repeating at the frequency of the acoustic wave. Prior to this study, the emission was believed to occur only during the implosion of transient cavities. In the 1970s, the interest in SL and sonochemistry was revived in North America [59, 60] and in Russia [61, 62]. These studies elucidated the free-radical chemistry underlying the phenomenon of cavitation and SL in aqueous media. This generated a widespread interest in the field of sonochemistry. Detailed spectra of nonaqueous liquids with bands characteristic of the chemical species formed by the breaking of chemical bonds during cavitation were reported [63]. One problem in studying SL has been that the observed light emission is generally from multiple bubbles in a cavitation field. The variable emission from multiple sources makes the analysis of SL difficult. This difficulty was overcome by Gaitan et al. [64] by trapping a single bubble in a resonance field. After Gaitan's discovery, a search for the correct SL mechanism began. For several years, experimental information accumulated about the properties of sonoluminescing bubbles. In sharp contrast to the spectrum of multi-bubble SL, single-bubble SL (SBSL) shows a smooth continuum, without spectral lines. The presence of spectral lines points to lower temperatures, since the atomic transitions leading to lines tend to be overwhelmed by the thermal (Black body) emission processes at high temperatures. By fitting the observed spectra to that of a blackbody emitter, Hiller et al. [65] concluded that the temperature of the gas was at least 25 000 K. Further, they found a sensitive dependence on the type of gas within the bubble: when the air dissolved in the liquid was replaced with pure nitrogen, the characteristically stable SBSL disappeared [66]. Only when the inert gas argon was added did the SBSL light emission return. SBSL can be achieved with a pure noble gas as well, but in a vastly different range of gas concentrations. During this time the predominant belief in the field was that shocks were somehow important for the energy focusing and light emission of sonoluminescence [67]. However, there was little agreement as to the details of how this worked, and many other physical mechanisms were suggested, including dielectric breakdown of the gas [68], bremsstrahlung [69], collision-induced emission [70], and even the quantum-electrodynamical Casimir effect [71, 72]. However, it became clear that at least some of the experimental facts of SL were direct consequences of the classical theory of bubble dynamics.

## 6.1 Instrumentations

A sketch of a typical setup for generating SL is shown in Fig. 6. A flask of liquid lined with transducers (piezo's) that are tuned to set up an acoustic standing wave at resonance frequency with the flask. Photon counter, spectrometer or other analytical tools are used to investigate the emitted light. Typical oscillation frequencies $f$ are 20 to 40 kHz. When the pressure amplitude of the sound waves ($P_a$) is larger than the ambient pressure, ($P_0 =$ 1 bar) the pressure in the flask becomes negative, putting the liquid under tension. At large enough tension, the liquid breaks apart (cavitation), creating unstable bubble clouds in which the bubbles often self-organize into dentritic structures. These cavitation clouds collapse with enormous force, powerful enough to do serious damage to the surface of solid bodies in their vicinity. At moderate forcing pressures ($P_a/P_0 \sim 1.2 - 1.4$) and with water degassed to around 20% of its saturated concentration of air, it was observed that the degassing action of the sound field was reducing the number of bubbles until only a single bubble remained; it was remarkably stable in size and position. This observation was the birth of the field of single-bubble SL, from which all areas of the field of SL benefited because of the well-defined experimental conditions for its observation. Once during each oscillation period, the bubble, whose undriven radius is typically around 5 μm, collapses very rapidly from its maximum radius of about 50 μm to a minimum radius of ~0.5 μm, changing its volume by a factor of one million. Figure 7 shows the radius, forcing pressure, and light intensity/top to bottom) during this process. The bubble expansion is followed by a violent collapse, during which light is emitted. The process repeats itself with extraordinary precision, as demonstrated by measuring of the phase of the light emission relative to the driving.

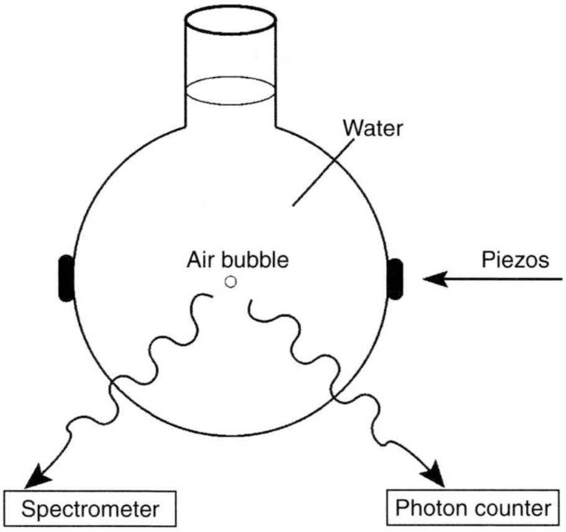


**Fig. 6** Sketch of a typical setup for generating sonoluminescing bubbles. (Adapted from Brenner, M. P. et al. [73].)

## 6.2 Bubble Dynamics and Light Emission

The bubble dynamics is an essential point for the understanding of the SL phenomenon and theories based on the Rayleigh-Plesset equation are meanwhile able to describe the expansion/collapse cycle (for a detailed description see [73] and reference therein). A representative example is given in Fig. 8. The various parts of the oscillation cycle of a typical single bubble containing argon in water are labeled 1 to 10 (for definiteness, we use an example from [74], with $R_0 = 4.5$ μm, driven at $f = 26.5$ kHz and $P_a = 1.2$ atm). The cycle begins, by definition, when the driving pressure begins to dip into the negative half of its cycle, and the bubble is therefore allowed to expand. The surrounding liquid far from the bubble

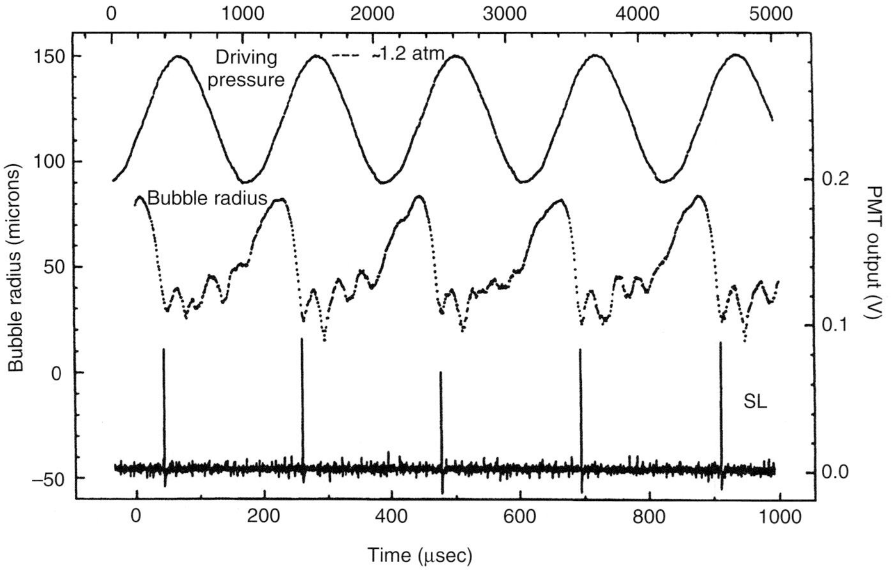


**Fig. 7** Correlation between sound wave (top trace), bubble radius (middle trace) and measured sonoluminescence (bottom trace) from a 21% (by weight) glycerin-water mixture. Ambient pressure 1.2 atm insonification frequency 22.3 kHz [64]

is degassed to some level and maintained at room temperature. Consequently, the bubble contains about $10^{10}$ argon atoms and about $2 \times 10^8$ water molecules at the outset.

1. *Expansion*: The bubble expansion is comparatively slow and the growth is sustained for almost half a cycle (~15 μs). In this phase, the bubble is in both thermal and mass transfer equilibrium with the liquid. Because of the falling pressure inside the bubble, it gains large numbers of water-vapor molecules (evaporating from the wall) and also some gas molecules from the liquid.

2. *Turnaround at maximum radius*: The driving pressure begins to increase again, and the expansion comes to a halt. At maximum radius ($R_{\max} \approx 7R_0$), the bubble contains little more than the initial $10^{10}$ argon atoms, but has collected up to $10^{11}$ water molecules.

3. *Rayleigh collapse*: As the external pressure increases, the inertial collapse of the liquid layers around the bubble begins. Even with the increased number of molecules, the internal pressure is still very low, and the collapse proceeds almost exactly like the classical collapse of an empty cavity treated by Lord Rayleigh [75]. The radius decreases quickly (over about 4 μs) from $R_{\max}$ to a value comparable to $R_0$. During this collapse, water vapor recondenses at the wall and the argon atoms again become the dominant species inside the bubble.

4. *Decoupling of water vapor*: About 50 ns before the minimum radius is reached, the time scale of the bubble collapse becomes smaller than the time scale for the diffusion of water vapor. The water vapor still left inside the bubble is now trapped until the reexpansion. Calculations show that about 20% water vapor should be mixed

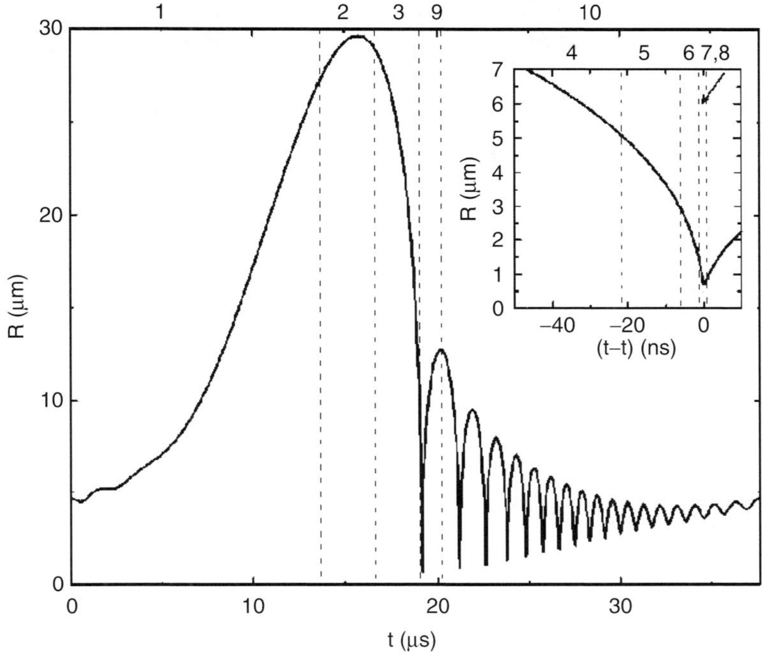


**Fig. 8** Classical bubble dynamics calculation for a driving pressure amplitude of 1.2 atm, a frequency of 26.5 kHz, and an ambient bubble radius of 4.5 μm. One oscillation cycle R is shown as a function of time t. The bubble expands to nearly ten times its ambient radius, and then collapses extremely quickly, followed by afterbounces with roughly the eigenfrequencies of the bubble. The vertical dashed lines and the printed numbers indicate the intervals at which different physical processes are important, which are described in the text. The inset shows the innermost 60 ns around the time of maximum compression where the light is emitted. (Adapted from Brenner, M. P. et al. [73].)

with the argon. Up to this moment, the polytronic exponent of the gas mixture has not increased significantly above 1, and the temperature has only risen to about 500 K.

5. *Thermal decoupling*: Only ≈30 ns later, the accelerating bubble wall becomes fast enough so that heat can no longer escape the bubble. Until reexpansion, the bubble is now thermally isolated from the liquid as well, and the polytropic exponent $\gamma$ rises quickly to its adiabatic value. The latter is determined by the mixture of 80% Ar and 20% $H_2O$ to $\gamma \approx 1.6$. From now on, the temperature increases rapidly.

6. *Onset of dissociation reactions*: Once the temperature exceeds roughly 4000 K, water-vapor molecules start dissociating into OH and H radicals. While the dissociation products have lower adiabatic exponents than $H_2O$, this endothermic reaction also consumes much energy and curbs the temperature rise. At this stage, molecular light emission is a possibility.

7. *Onset of light emission*: Despite the temperature-limiting influence of water vapor, about 10 000 K temperature is finally reached in the bubble about 100 ps before maximum compression. At this

temperature, a small fraction of the argon as well as of the oxygen and hydrogen atoms now present undergo ionization and release free electrons. The electrons interact with the ions and neutral atoms, and emission of electromagnetic radiation (thermal bremsstrahlung and radiative recombination) begins, which spans the spectrum of visible and ultraviolet wavelengths. SL is observed.

8. *Maximum compression*: At this point, the gas density reaches (almost) solid-state values. The deceleration of the bubble wall down to zero speed has begun to enhance random shape perturbations (Rayleigh-Taylor instability) and leads to massive energy loss through acoustic wave emission. The temperature and light emission peak, helped by the high densities that prevent further endothermic dissociation reactions.

9. *Reexpansion*: The bubble loses about 90% of its energy in the collapse, mostly due to acoustic emission. The reexpansion is much slower than the collapse. The Rayleigh–Taylor instability grows and may overwhelm a strongly driven bubble during this stage. Only a small increase in radius and decrease in temperature are sufficient to dramatically reduce the photon absorption coefficient and quench the light emission uniformly for all wavelengths, about 100 to 200 ps after it has begun. Subsequently, reaction chemistry stops and thermal as well as diffusive equilibria are reestablished.

10. *Afterbounces*: The bubble rebounds to a much smaller size than the maximum radius before the main collapse, and for no parameter combination realized so far is there enough energy left to induce SL during the afterbounces. The afterbounces provide a parametric excitation that can accumulate and render the bubble shape unstable. The radial motion, however, is damped rapidly until the driving pressure dips into its negative cycle once again, and the oscillation starts anew. Over the whole cycle, shape perturbations may have been enhanced (then the bubble is parametrically unstable), or a net gain or loss of gas may have resulted (diffusive instability). In the correct parameter range, the bubble is stable with respect to both processes and continues to oscillate and emit light in exactly the same fashion.

The above 10 steps hold for SBSL bubbles in water with only argon (or any other noble gas) dissolved. If in addition, molecular gases such as nitrogen and oxygen are dissolved, not only water dissociates at Step 6 but also these gases (at around 7000 K for $N_2$ and $O_2$). The reaction products subsequently dissolve in water. "Cleaning" the bubble from molecular gases in this fashion can take thousands of cycles [76]. Thermal emission due to the high temperatures in the collapsing bubble can be expected. For the estimated 10 000 K, it will cover the complete visible and near-ultraviolet spectral range. This thermal emission is indeed observed in most of the single-bubble experiments and was in contrast to experiments using multibubbles (Fig. 9).

In the multibubble experiments, in addition to the thermal emission, molecule-specific narrow-emission lines could be detected. These observations led in the past to a widespread discussion whether the physical mechanisms underlying the single-bubble phenomenon are the same as in the multibubble case. However, the years 2000 and 2001 brought the clarifying experiments [77, 78]. Reducing

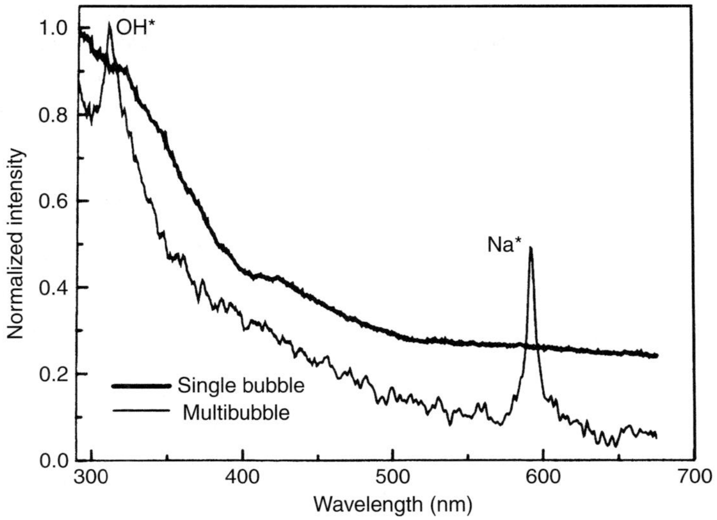


**Fig. 9** Comparison of spectra of multibubble and single-bubble sonoluminescence in 0.1 M sodium chloride solution [73]

the driving pressure in SBSL experiments revealed that molecular emission can also be observed, however, with orders of magnitude lower intensity with respect to the thermal emission intensities, that required detection times of days, earlier experiments have thus overseen this small fraction of the total emitted light.

The future of SL can clearly be seen in the fields of human activities where ultrasound is used: (1) Sonochemistry with its origin in chemistry and physics; this includes synthesis, and catalysis under ultrasound conditions; (2) Power ultrasound, with its origins in engineering and processing; and (3) Diagnostic ultrasound involving nondestructive testing and medical scanning. In these fields, SL may find its applications as an analytical tool, keeping in mind that a large industry is concerned with avoiding violent bubble collapses that can do damage by shear as well as with the shock waves they emit.

## 7
## Electroluminescence (EL)

Electroluminescence is a nonthermal generation of light resulting from the application of an electric field to a substance. Although in general this definition includes light-emitting diodes (LEDs), and most of the LASERs, they are not discussed here; they are considered elsewhere in the Encyclopedia (see DIODES, LIGHT-EMITTING and LASER PHYSICS). EL solid-state devices can be divided into various types according to the material that is the emission-active layer, typically called the *phosphor*, or the drive-voltage wave forms: AC or DC powder EL devices. The materials classically used for EL devices are doped wide-bandgap semiconductors of II–VI compounds, as well as group IIa–VIb and IIb–VIb type. Besides these, EL form organic materials like conjugated polymers that becomes increasingly important [79].

The great impact to the field in the last decades is mostly due to two reasons: first, the economical demand: electricity is the most widely reliable, available, and safe and cheap energy form to create visible light, and second, the rapid development in semiconductor and thin-film technologies. The latter enabled to miniaturize and to integrate the EL devices to flat-panel-displays with year-by-year increasing size and resolution.

The first observation of the EL phenomenon was reported in 1929 by Gudden and Pohl [80]. It was followed by the discoveries of Destriau in 1936 [81]. Light emission from a ZnS phosphor powder dispersed in an insulator and sandwiched between two electrodes was observed when a high AC voltage was applied. Until 1950, when transparent, electrically conductive films made of $SnO_2$ have been developed, no effort was made to develop practical devices. With the use of these $SnO_2$ transparent conductive films, AC powder EL devices have been developed, which triggered worldwide research and development of EL devices.

In 1968, a second period of EL development began with DC powder EL devices [82] and AC thin-film EL (TFEL) devices using Lumocen (luminescence from molecular centers) [83, 84]. The former were made possible by coating powder EL phosphor surfaces with copper ions. In the latter, rare-earth fluorides as molecular centers were doped into ZnS thin films. Thin-film EL devices benefited, in particular, from development of the double-insulating-layer structure [85], which overcame the problems of short lifetime and low luminance. With the use of these technologies, the mass production of 6-in.-diagonal matrix EL panels started in 1983 [86].

To show more information on the screen, multicoloring of EL displays is necessary. For this purpose, phosphor materials that emit three primary colors, that is, red, green, and blue, are necessary. Beginning with the pioneering work by Kahng and coworkers [83, 84], extensive studies have been conducted on materials and processing technologies. In order to obtain efficient color EL phosphors, many studies have been directed to new host materials and luminescent centers. It was found that alkaline-earth sulfides, such as SrS and CaS, doped with rare-earth luminescent centers, good phosphor, and multicolor display panels were fabricated with them [87], and by 1994 Planar Systems introduced a full color AC-TFEL display [88].

In the following text, we will describe the basic device structure of TFEL devices and briefly discuss the underlying device physics. For more advanced aspects, readers are referred to the review articles [89–92].

## 7.1
## EL Device Structures and EL Emission Mechanisms

The standard structure is illustrated in Fig. 10. The metal-insulator-semiconductor-insulator metal (MISIM) structure is usually deposited on a transparent glass substrate. Thus, the EL

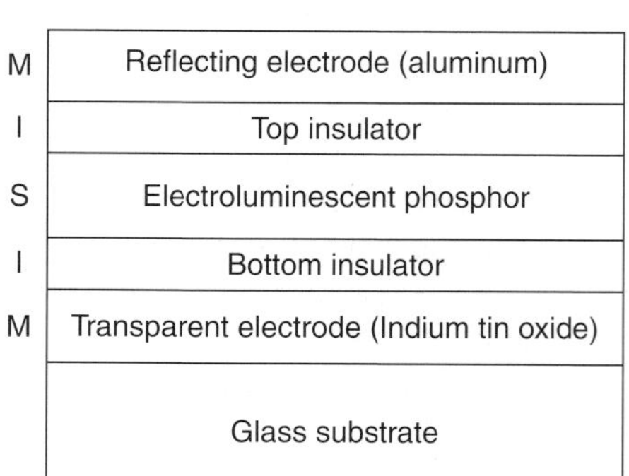


**Fig. 10** Standard MISIM (metal-insulator-semiconductor-insulator-metal) device structure

light is visible through the glass substrate that in addition protects the active layers. Another, the so-called *inverted* structure, is also in use. In comparison to the standard structure, the top-reflecting electrode is then interchanged with the bottom transparent electrode. It allows using non-transparent substrates such as silicon. Regardless of whether the standard or the inverted structure is used, the basic function of each layer is the same. First, the middle semiconducting layer (commonly called the *phosphor layer*) is the light-emitting material. This material consists of two constituents: (1) the host material that dominates the electrical properties of the phosphor layer, and (2) the luminescent centers that dominate the optical properties of the phosphor layer. When a large electric field is applied across the electrodes, electrons are injected from the phosphor/insulator interface and accelerate through the phosphor layer. These electrons then excite the luminescent centers that subsequently generate light. The phosphor layer is sandwiched between two insulators that limit the current transport across the device. These insulating layers have been found to effectively prolong the lifetime of the device. Finally, the top and bottom electrodes simply complete the device structure and make the electrical contact to the drive electronics.

The basic device structure can be modeled as a simple circuit in which the insulators are represented by perfect capacitors and the phosphor layer is represented by a capacitor below threshold and a pair of back-to-back diodes above threshold. The threshold is the voltage or electric field at which the phosphor layer experiences a Zener breakdown. Schematically, the equivalent circuit is shown in Fig. 11. Below threshold, the device is simply a three-series capacitor circuit in which the voltage drop across each layer is proportional to the dielectric constant and the layer thickness. At threshold, the phosphor layer experiences a Zener breakdown, that is, electrons are injected by tunneling into the phosphor layer. This produces a real current that is transported across the phosphor layer and is represented by the back-to-back Zener diodes in the diagram. The devices are driven at high voltages (200–250 V) that induce large electric fields (1–2 MeV $cm^{-1}$) across the insulators and the phosphor layer. While this simple model describes most of the device characteristics, more sophisticated models have been suggested that account for the time [93] and waveform dependence [94].

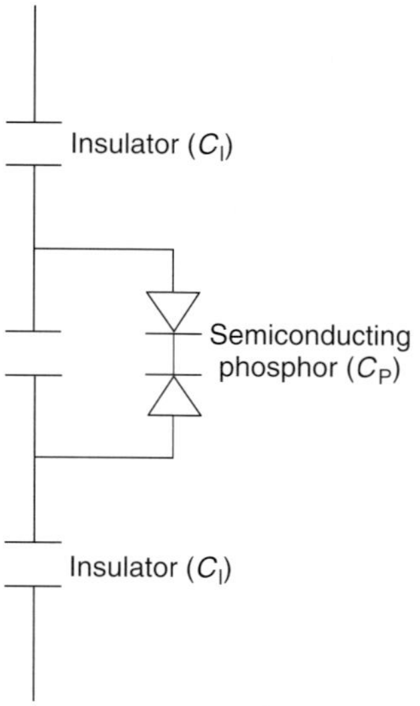


**Fig. 11** Equivalent circuit diagram for an AC thin-film electroluminescence device. $C_I$ and $C_P$ are the insulator and phosphor capacitance, respectively

The optical response to the electrical excitation is the most critical point for the device performance. Probably the most important measurement used to evaluate an EL device is the luminance versus

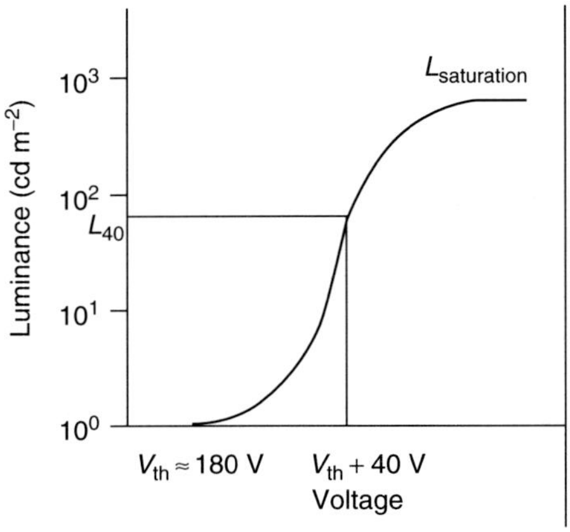


**Fig. 12** Schematic diagram of a luminance-voltage (L–V) plot

voltage (L–V) plot (Fig. 12). To make these measurements, all one needs is a variable AC-voltage supply to ramp the external voltage (typically 0–300 V) and a spectrometer/calorimeter to measure the luminance as a function of the applied voltage. From this simple measurement, several important parameters can be extracted, namely, the threshold voltage, the luminescence color, and its efficiency.

The basic physics of an EL device can be divided into four main topics, namely:

1. *Tunnel emission of the electrons from the interface states*: In general, there are two mechanisms responsible for carrier injection, namely thermoionic emission, and field emission or tunneling. In the devices, one must also consider the origin of these carriers, which either come from the bulk or from the phosphor/insulator interface. The accepted theory for carrier injection in most EL phosphors is thermally assisted tunneling from interface states. The description of the tunneling current follows that developed for Shottky barriers [95]. For good device performance, one has to find the optimum conditions between the interface-trap-depth and its density and the semiconductor/metal properties like the electron affinity, the Fermi level, the band-gap energy and the interface density of states as well as the metal work function. These properties basically define the tunneling barrier height and its width, and the conditions under which the carriers are injected into the EL phosphor.

2. *Acceleration of electrons to high energies*: Once the electrons are injected into the phosphor-layer conduction band, they are accelerated under the influence of the electric field. Using Maxwell's equations rearranged for series capacitors, one can estimate that for typical device geometries (layer widths of $\sim$0.5 μm), an applied voltage of about 200 V gives electric fields in the range of 2.0 to 3.0 MV $cm^{-1}$ in the phosphor layer. Under these conditions, an electron would gain sufficient energy to impact a blue ($\sim$450 nm) EL center in less than 11 nm. However, because the electrons scatter and lose energy by several mechanisms, the electron energy distribution does not obey this simple relationship. The real electron energy distribution is a complicated function of high-field scattering mechanisms such as intervalley scattering, polar optical scattering, acoustic phonon scattering, ionized impurity scattering, band-to-band impact ionization, and impact excitation [96]. To account for this situation, simplifying analytical models that use the mean free path of the electrons to describe the high-field transport properties have been developed [97], as well as extensive numerical Monte Carlo computer simulations are nowadays used and have made significant contributions to the understanding of the high field electron transport phenomenon in EL devices.

3. *Impact excitation or impact ionization of the luminescent centers*: After the electrons accelerate to sufficiently high energies,

they can interact with a luminescent center and promote one of the luminescent centers ground-state electrons. The electron can either be promoted into an excited state of the luminescent center, or into the conduction band of the host material, which is called *impact excitation* or *impact ionization*, respectively [97, 98]. The electron-luminescent center interaction probability is related to the luminescent center radiative cross section. When the electron is excited into a higher atomic state (impact excitation), it can relax back to its ground state either radiatively or nonradiatively.

When an electron is promoted into the conduction band of the solid (impact ionization), the excited electron is immediately accelerated toward the anode by the electric field. This type of impact causes electron multiplication (avalanche breakdown). Subsequent light emission from the ionized luminescent centers requires that an electron be recaptured from the conduction band, which is a low-probability process for highly accelerated electrons, thus occuring predominately in the low-field regions. This type of EL is a two-step process that changes the kinetics as compared to the impact excited centers.

4. *De-excitation of the excited electrons by radiative (photon generation) or nonradiative recombination*: As mentioned, deexcitation can occur either radiatively (which produces a photon) or nonradiatively. Typically, nonradiative relaxation occurs by phonon (lattice vibrations) generation, which produces heat. Local lattice imperfections such as point defects (interstitials, vacancies...), line defects (dislocations), and volume defects (grain boundaries) also act as nonradiative recombination centers that reduce the radiative efficiency $\eta$. It is given by

$$\eta = \frac{W_{\text{rad}}}{W_{\text{rad}} + W_{\text{nonrad}}} \tag{21}$$

with $W_{\text{rad}}$ and $W_{\text{nonrad}}$ being the radiate and nonradiative recombination rates, respectively. $W_{\text{rad}}$ is inversely proportional to the lifetime of the luminescence, which is directly obtainable from time-resolved measurements. However, a fast radiative decay time of a luminescent center does not necessarily mean that it will be a good EL material. Similarly, a slow decay does not mean that a luminescent center will necessarily be a poor material. *A priori*, it is difficult to know whether a material will be efficient or not because one must consider both the electrical and optical characteristics of the material. Thus, there is ongoing research and development to find and improve all the materials and technologies required for the EL devices production.

## 8 Summary

In the present article, we have tried to describe some aspects of the widespread phenomenon of light emission from matter under the aspect of the most common stimulation sources. We are aware of the fact that such an attempt has to be incomplete because of the limited knowledge of the authors, and other circumstances, as well as the ongoing progress in the subject. We apologize if we have missed some aspects and are aware of the fact that current research, especially in the life science and nanotechnologies have already lead to further developments. The detection of luminescence from single molecules and the excitation of luminescence via a tunneling current from a scanning tunneling microscope are only two examples. The enclosed reference list and the "Further Reading" section may be helpful in overcoming such shortcomings.

## Glossary

**Activator:** An element added to a phosphor to enhance its light-emission properties.

**Auger Transition:** A transition in which an excited electronic state returns to its ground state by imparting its energy to a second electron that is emitted.

**Band-edge Luminescence (Edge Emission):** The light emission from an excited semiconductor of photons with energy close to that of the band-gap energy.

**Boxcar:** Detection mode using a gate to select an electronic signal in a time window.

**Carriers:** A collective name for mobile charged particles that may be electrons or holes or both.

**Cavitation:** A process involving nucleation, growth, and subsequent oscillation or collapse of bubbles in a liquid. Cavitation can be induced by acoustic waves, hydrodynamics, lasers, or electrical sparks. The emphasis of this article is on SL from acoustic cavitation.

**Cavitation Field:** A cloud of bubbles with the individual members distributed over the entire acoustic field.

**Deep-level (Deep Center) Emission:** Emission from crystal defects that create energy levels within the band gap, well separated from either the conduction or valence-band edges. Some deep levels emit light when excited, but many do not. They cannot be treated by hydrogenic theory.

**Dose (Unit – Gray, Gy):** The amount of energy absorbed by the target per unit mass of target. 1 Gy = 1 J/kg.

**Edge Emission:** Same as band-edge luminescence.

**Excitons:** Bound electron-hole pairs. They may vary in size from atomic dimensions to rather large, weakly bound pairs with radii of some tens of atomic spacings that can be treated, to a reasonable approximation, by Bohr theory.

**Fluorophore:** A species that emits fluorescence.

**Impact Excitation:** Excitation of ions in a crystal field via a strong interaction with high-energy electrons or hot electrons.

**Interstitial Sites:** Positions in a crystal structure that lie between the sites occupied by the atoms that comprise the ideal structure.

**Luminance:** Photometric term indicating the radiated power in terms of human eye sensitivity, that is, a physical measure of luminous intensity per unit area, in units of $cd/m^2$. Other units, ft L (foot-lambert), are also used: 1 ft L = 3.43 cd $m^{-2}$.

**Luminescence:** Physical phenomenon indicating emission of light.

**Luminous Efficiency:** Number of lumens emitted per watt electrical power.

**Multiple-bubble Sonoluminescence:** SL from many bubbles distributed in space.

**Multiple-bound-exciton Complexes:** Just as atoms can sometimes bind together to form diatomic, triatomic, or more molecules, so can excitons bind together to form complexes that contain two, three, four, or more excitons.

**Passivation:** Surface treatments that reduce the nonradiative surface recombination of carriers.

**Phonon:** A mode of distortion of the lattice during thermal vibration.

**Photoluminescence:** Luminescence excited by photons (light) that are sufficiently energetic to excite optical centers in a crystal or to lift electrons out of the valence band into the conduction band.

**Pixel (Picture Element):** Basic building blocks for displays. One monochromatic pixel is composed of a single dot and one color pixel is composed of three primary color dots, that is, in red, green, and blue triads.

**Plasmon:** A quantized oscillation of the conduction-electron gas relative to the positive ion lattice.

**Powder EL Device:** ZnS phosphor powder dispersed in an insulator with a binder, sandwiched between two electrodes.

**Radiation Dosimetry:** The measurement of absorbed radiation *dose*.

**Stable Cavitation:** Bubbles oscillating about a mean size.

**Sensitizer:** An element added to a phosphor to enhance its ability to absorb incident radiation and transfer it to an emitting center.

**Single-bubble Sonoluminescence:** Emission from an isolated bubble entrapped in a specific region of the acoustic field.

**Singlet State:** Electronic state having a spin equal to 0.

**Substitutional Sites:** Sites in a crystal structure normally occupied by atoms of the host material but capable of replacement by impurities or dopant atoms.

**Threshold Voltage:** The voltage at which luminance becomes 1 cd $m^{-2}$.

**Transient Cavitation:** The violent collapse of bubbles.

**Triplet State:** Electronic state having a spin equal to 1.

## References


[1] Klein, C. A. (**1968**), *J. Appl. Phys.* **39**, 2029.
[2] Yacobi, B. G., Holt, D. B. (1990), *Cathodoluminescence Microscopy of Inorganic Solids.* New York: Plenum Press.
[3] Newbury, D. E., Joy, D. C., Echlin, P., Fiori, C. E., Goldstein, J. I. (1986), *Advanced Scanning Electron Microscopy and X-ray Microanalysis.* New York: Plenum Press.
[4] Grün, A. (**1957**), *Z. Naturforsch., A* **12a**, 89.
[5] Everhart, T. E., Hoff, P. H. (**1971**), *J. Appl. Phys.* **42**, 5837.
[6] Donolato, C. (**1981**), *Phys. Status Solidi A* **65**, 649.
[7] Holt, D. B., Napchan, E. (**1994**), *Scanning* **16**, 78.
[8] Gustafsson, A., Pistol, M.-E., Montelius, L., Samuelson, L. (**1998**), *J. Appl. Phys.* **84**, 1715.
[9] Boyle, R. (1664), *Light and Colors.* London: Herringman.
[10] Urbach, F. (**1930**), *Wien. Ber. Akad. IIa* **139**, 353–363.
[11] Böhm, M., Scharmann, A. (**1987**), *Appl. Phys.* **A43**, 29.
[12] Curie, C. (1963), *Luminescence in Crystals.* London: Methuen.
[13] Gibbs, J. H. (**1972**), *J. Chem. Phys.* **57**, 4473.
[14] Mott, N. F., Gurney, R. W. (1964), *Electronic Processes in Ionic Solids.* New York: Dover Publications.
[15] Böhm, M., Erb, O., Scharmann, A. (**1985**), *Appl. Phys.* **A37**, 165.
[16] Böhm, M., Scharmann, A. (**1971**), *Phys. Status Solidi* **5**, 563.
[17] Randall, J. T., Wilkins, M. H. F. (**1945**), *Proc. R. Soc.* **A184**, 366.

[18] Hearing, R. R., Adams, E. N. (**1960**), *Phys. Rev.* **117**, 451.
[19] Halperin, A., Braner, A. A., Ben-Zvi, A., Kristianpoller, N. (**1960**), *Phys. Rev.* **117**, 416.
[20] Townsend, P. D., Clark, C. P., Levy, P. W. (**1967**), *Phys. Rev.* **155**, 908.
[21] Gobrecht, H., Hofmann, D. (**1966**), *J. Phys. Chem. Solids* **27**, 509.
[22] Bräunlich, P. (**1967**), *J. Appl. Phys.* **38**, 2516.
[23] Mintoni, C., Rucci, A., Sepri, A. (**1968**), *Ricerca Sci.* **38**, 762.
[24] de Muer, D. (**1970**), *Physica* **48**, 1.
[25] Maxia, V., Onnis, S., Rucci, A. (**1971**), *J. Lumin.* **3**, 378.
[26] Chemmy, P. J., Townsend, P. D., Levy, P. W. (**1967**), *Phys. Rev.* **155**, 917.
[27] Mohan, N. S., Chen, R. (**1970**), *J. Phys. D: Appl. Phys.* **3**, 243.
[28] Satzhyamoorthy, A., Bhalla, K. C., Luthra, J. M. (**1976**), *J. Lumin.* **11**, 357.
[29] Chalgaonkar, C. S., Narlikar, A. V. (**1972**), *J. Mater. Sci.* **7**, 1465.
[30] Chen, R. (**1976**), *J. Mater. Sci.* **11**, 1521.
[31] Kivits, P., Hagebenk, H. J. L. (**1977**), *J. Luminescence* **15**, 1.
[32] Albrecht, H. O. (**1928**), *Z. Phys. Chem.* **136**, 321.
[33] Harmony, M. D. (1972), *Introduction to Molecular Energies and Spectra*. New York: Holt, Rinehart & Winston.
[34] Vo-Dinh, T. (1984), *Room Temperature Phosphorimetry for Chemical Analysis*. New York: Wiley.
[35] Seliger H. H., McElroy, W. D. (**1960**), *Radiat. Res., Suppl.* **2**, 528.
[36] Sutton, D. G., Westberg, K. R., Melzer, J. R. (**1979**), *Anal. Chem.* **51**, 1399.
[37] Birks, J. R. (Ed.) (1989), *Chemiluminescence and Photochemical Reaction Detection in Chromatography*. New York: VCH Publishers.
[38] Yamada, M., Ishiwada, A., Hobo, T., Suzuki, S., Araki, S. (**1982**), *J. Chromatogr.* **238**, 347.
[39] Spurlin S. R., Yeung, E. S. (**1982**), *Anal. Chem.* **54**, 318.
[40] Seliger, H. H., Lee, J. (**1972**), *Photochem. Photobiol.* **15**, 227.
[41] Totter, J. R. (**1975**), *Photochem. Photobiol.* **22**, 203.
[42] White, E. H., Harding, J. J. C. (**1964**), *J. Am. Chem. Soc.* **86**, 5686.
[43] Van Zoonen, P., Kamminga, D. A., Gooijer, C., Velthorst, N. H., Frei, R. W., Gubitz, G. (**1985**), *Anal. Chim. Acta* **174**, 151.
[44] Cruser, S., Bard, A. (**1969**), *J. Am. Chem. Soc.* **91**, 267.
[45] Hill, E., Humphreys, E., Malcolm-Lawes, J. (**1986**), *J. Chromatogr.* **370**, 427.
[46] Haapakka, K., Kankare, J., Lipiainen, K. (**1988**), *Anal. Chim. Acta* **215**, 341.
[47] Koerner, Jr, P. J., Nieman, T. A. (**1987**), *Mikrochim. Acta* **2**, 79–90.
[48] Edo, K., Sato, H., Kato, M., Mizugzki, M., Uchiyama, M. (**1985**), *Chem. Pharm. Bull.* **33**, 3042.
[49] Klockow, D., Teckentrup, J. (**1976**), *Talanta* **23**, 889.
[50] Southern, E. M. (**1975**), *J. Mol. Biol.* **98**, 503.
[51] Wilchek, M., Bayer, E. A. (**1988**), *Anal. Biochem.* **171**, 1.
[52] Kricka, L. J., Carter, T. J. N. (Eds.) (1982), *Chemical and Biological Luminescence*. New York: Marcel Dekker.
[53] Bronstein, I., Voyta, J. C., Lazzari, K. G., Murphy, O., Edwards, B., Kricka, L. J. (**1990**), *Biotechniques* **8**, 310.
[54] Frenzel, H., Schultes, H. (**1935**), *Z. Phys. Chem.* **27B**, 421–424.
[55] Zimakov, P. V. (**1934**), *Dokl. Akad. Nauk SSSR* **3**, 450–452.
[56] Jarman, P. (**1960**), *J. Acoust. Soc. Am.* **32**, 1459–1462.
[57] Negishi, K. (**1960**), *Acoustica* **10**, 124.
[58] Saksena, T. K., Nyborg, W. L. (**1970**), *J. Chem. Phys.* **53**, 1722–1734.
[59] Mead, E. L., Verrall, R. E., Sutherland, R. G. (**1975**), *Can. J. Chem.* **53**, 2394–2399.
[60] Sehgal, C. M., Steer, R. P., Sutherland, R. G., Verrall, R. E. (**1977**), *J. Phys. Chem.* **81**, 2618–2620.
[61] Margulis, M. A. (**1969**), *Akust. Zh.* **15**, 153–173; *Sov. Phys. Acoust.* **15**, 135–151.
[62] Margulis, M. A. (**1978**), *Zh. Fiz. Khim.* **52**, 605–609; *Russ. J. Phys. Chem.* **52**, 342–344.
[63] Suslick, K. S., Flint, E. B. (**1987**), *Nature* **330**, 553–555.
[64] Gaitan, D. F., Crum, L. A., Church, C. C., Roy, R. A. (**1992**), *J. Acoust. Soc. Am.* **91**, 3166–3183.
[65] Hiller, R., Puttermann, S. J., Barber, B. P. (**1992**), *Phys. Rev. Lett.* **69**, 1182.
[66] Hiller, R., Weninger, K., Puttermann, S. J., Barber, B. P. (**1994**), *Science* **266**, 248.

[67] Barber, B. P., Hiller, R. A., Arisaka, K., Fettermann, H., Puttermann, S. (**1997**), *Phys. Rep.* **281**, 65.
[68] Garcia, N., Levanyuk, A. P. (**1996**), *Pis'ma Zh. Eksp. Teor. Fiz.* **64**, 849; (**1996**), *JETP Lett.* **64**, 907.
[69] Moss, W. C. (**1997**), *J. Acoust. Soc. Am.* **101**, 1187.
[70] Frommhold, L., Atchley, A. A. (**1994**), *Phys. Rev. E* **73**, 2883.
[71] Schwinger, J. (**1992**), *Proc. Natl. Acad. Sci. U. S. A.* **89**, 11118.
[72] Eberlein, C. (**1996**), *Phys. Rev. Lett.* **76**, 3842.
[73] Brenner, M. P., Hilgenfeldt, S., Lohse, D. (**2002**), *Rev. Mod. Phys.* **74**, 425.
[74] Storey, B. D., Szerri, A. J. (**2001**), *Proc. R. Soc. London, Ser. A* **457**, 1685.
[75] Rayleigh, L. (**1917**), *Philos. Mag.* **34**, 94.
[76] Matula, T. J., Crum, L. A. (**1998**), *Phys. Rev. Lett.* **80**, 865.
[77] Didenko, Y. T., McNamara III, W. B., Suslick, K. S. (**2000**), *Nature (London)* **407**, 877.
[78] Young, J. B., Nelson, J. A., Kang, W. (**2001**), *Phys. Rev. Lett.* **86**, 2673.
[79] Friend, R. H., Gymer, R. W., Holmes, A. B., Burroughes, J. H., Marks, R. N., Taliani, C. Bradley, D. D. C., Dos Santos, D. A., Bredas, J. L., Lögdlund, M., Salaneck, W. R. (**1999**), *Nature* **397**, 121.
[80] Guden, H., Pohl, R. W. (**1929**), *Z. Phys.* **2**, 192.
[81] Destriau, G. (**1936**), *J. Chim. Phys.* **33**, 587.
[82] Vecht, A., Werring, N. J., Smith, P. J. F. (**1968**), *Br. J. Appl. Phys.* **1**, 134.
[83] Kahng, D. (**1968**), *Appl. Phys. Lett.* **13**, 210.
[84] Chase, E. W., Heppelwhite, R. T., Krupka, D. C., Kahng, D. (**1969**), *J. Appl. Phys.* **40**, 2512.
[85] Russ, M. J., Kennedy, D. I. (**1967**), *J. Electrochem. Soc.* **114**, 1066–1072.
[86] Takeda, M., Kanatani, Y., Kishishita, H., Uede, H. (1983), *Advances in Display Technology III*, SPIE Proceedings Vol. 386, Bellingham, WA: SPIE, pp. 34–38.
[87] Barrow, W. A., Tuenge, R. T., Ziuchkovski, M. J. (1986), *Digest of the 1986 SID International Symposium.* Playa del Rey, CA: Society for Information Display, pp. 25–28.
[88] Barrow, W. A., Coovert, R. C., Dickey, E., King, C. N., Laakso, C., Sun, S.-S., Tuenge, R. T. (**1993**), *SID Dig.* 761.
[89] Kobayashi, H. (1983), in M. A. Herman (Ed.), *Optoelectronic Materials and Devices.* Warszawa: PWN-Polish Scientific Publishers, Chap. 13.
[90] Alt, P. M. (**1984**), *Proc. Soc. Inf. Disp.* **25**, 123–146.
[91] Ohnishi, H. (**1989**), *Annu. Rev. Mater. Sci.* **19**, 83–101.
[92] Rack, P. D., Holloway, P. H. (**1998**), *Mater. Sci. Eng.* **R21**, 171.
[93] Davidson, J. D., Wager, J. F., Khormaei, I., King, C. N., Williams, R. (**1992**), *IEEE Trans. Electron Devices* **ED-39**, 1122.
[94] Douglas, A. A., Wager, J. F. (**1992**), *SID Dig.* 356.
[95] Sze, S. (1981), *Physics of Semiconductor Devices.* New York: Wiley, p. 403.
[96] Bhattacharyya, K., Goodnick, S. M., Wager, J. F. (**1993**), *J. Appl. Phys.* **73**, 3390.
[97] Bringuier, E. (**1991**) *J. Appl. Phys.* **70**, 4505.
[98] Bringuier, E. (**1990**) *J. Appl. Phys.* **67**, 7040.

## Further Reading

Aitken, M. J. (1998), *An Introduction to Optical Dating.* London: Oxford University Press.
Birks, J. W. (Ed.) (1989), *Chemiluminescence and Photochemical Reaction.* Weinheim: Wiley-VCH.
Blasse, B., Grabmaier, B. C. (1994), *Luminescent Materials.* Heidelberg: Springer-Verlag.
Burr, J. G. (Ed.) (1985), *Chemi- and Bioluminescence.* New York: Marcel Dekker.
Cantow, H. J. (1981), *Advances in Polymer Science: Luminescence.* Berlin: Springer-Verlag.
Case, J. F. (Ed.) (2000), *Bioluminescence & Chemiluminescence 2000.* Singapore: World Scientific Publishing Company.
Chen, R., Kirsh, Y. (1981), *Analysis of Thermally Stimulated Processes.* Oxford, UK: Pergamon Press.
Chirac, J. D. (Ed.) (2003), *Handbook of Thermoluminescence.* Singapore: World Scientific Publishing Company.
Garcia-Campana, A. M., Baeyens, W. R. G. (2001), *Chemiluminescence in Analytical Chemistry.* New York: Marcel Dekker.
Horowitz, Y. S. (Ed.) (1983), *Thermoluminescence and Thermoluminescent Dosimetry*, Vols. 1–3. Boca Raton, FL: CRC Press.
Kitai, A. H. (1993), *Solid State Luminescence: Theory, Materials and Devices.* Dordrecht: Kluwer Academic Publishers.

Klingshirn, C. F. (1995), *Semiconductor Optics.* Berlin: Springer-Verlag.

Krasovitskii, B. M., Bolotin, B. M. (1998), *Organic Luminescent Materials.* Weinheim: Wiley-VCH.

Kricka, L. J., Stanley, P. E., Thorpe, G. H. G., Whitehead, T. P. (Eds.) (1984), *Analytical Applications of Bioluminescence and Chemiluminescence.* London: Academic Press, pp. 149–158.

Lumb, M. D. (Ed.) (1978), *Luminescence Spectroscopy.* New York: Academic Press.

Marfunin, A. S. (1979), *Spectroscopy, Luminescence and Radiation Centers in Minerals.* Heidelberg: Springer-Verlag.

Marshall, D. J. (1988), *Cathodoluminescence of Geological Materials.* Boston (MA): Unwin Hyman.

McKeever, S. W. S. (1985), *Thermoluminescence of Solids.* Cambridge, UK: Cambridge University Press.

Meier, F., Zakharchenya, B. P. (Eds.) (1984), *Optical Orientation.* Amsterdam: North Holland.

Mueller, G. (1999), *Electroluminescence (Semiconductors & Semimetals).* San Diego: Academic Press.

Oberhofer, M., Scharmann, A. (Eds.) (1981), *Applied Thermoluminescence Dosimetry.* Bristol, UK: Adam Hilger.

Ozawa, L. (1998), *Cathodoluminescence. Theory and Application.* Weinheim: Wiley-VCH.

Pagel, M. (Ed.) (2000), *Cathodoluminescence in Geoscience.* Berlin: Springer-Verlag.

Pankove, J. I. (Ed.) (1971), *Electroluminescence.* Heidelberg: Springer-Verlag.

Pankove, J. I. (1971), *Optical Processes in Semiconductors.* Englewood Cliffs, NJ: Prentice Hall.

Perkowitz, S. (1993), *Optical Characterisation of Semiconductors (Techniques of Physics 15).* San Diego: Academic Press.

Pope, M., Swenberg, C. E. (1982), *Electronic Processes in Organic Crystals.* New York: Oxford University Press.

Rashba, E. I., Sturge, M. D. (Eds.) (1982), *Excitons.* Amsterdam: North Holland.

Shionoya, S., Kobayashi, H. (Ed.) (1989), *Electroluminescence.* Heidelberg: Springer-Verlag.

Simon, H. (1971), *Living Lanterns.* USA: Penguin.

Singh-Nalwa, H., Rohwer, L. S. (2003), *Handbook of Luminescence, Display Materials and Devices.* Stevenson Ranch, USA: American Scientific Publ.

Srivastava, R. C. (Ed.) (2002), *Shock Focussing effect in Medical Science and Sonoluminescence.* Berlin: Springer-Verlag.

VanDyke, K. (Ed.) (1982), *Analytical Applications of Bioluminescence and Chemiluminescence.* New York: AcademicVo-Dinh T; (1984), *Room Temperature Phosphorimetry for Chemical Analysis.* New York: Wiley.

Vij, D. R. (Ed.) (1998), *Luminescence of Solids.* New York: Plenum Publishing Company.